APPLIED
FLOW
and SOLUTE
TRANSPORT
MODELING
in AQUIFERS

Fundamental Principles and Analytical and Numerical Methods

APPLIED FLOW and SOLUTE TRANSPORT MODELING in AQUIFERS

Fundamental Principles and Analytical and Numerical Methods

Vedat Batu

CRC Press
Taylor & Francis Group
Boca Raton   London   New York

CRC Press is an imprint of the
Taylor & Francis Group, an **informa** business

CRC Press
Taylor & Francis Group
6000 Broken Sound Parkway NW, Suite 300
Boca Raton, FL 33487-2742

First issued in paperback 2019

© 2005 by Taylor & Francis Group, LLC
CRC Press is an imprint of Taylor & Francis Group, an Informa business

No claim to original U.S. Government works

ISBN-13: 978-0-8493-3574-7 (hbk)
ISBN-13: 978-0-367-39250-5 (pbk)

Library of Congress Cataloging-in-Publication Data

Batu, Vedat
    Applied flow and solute transport modeling in aquifers : fundamental principles and analytical and numerical methods / Vedat Batu.
        p. cm.
    Includes bibliographic references and index.
    ISBN 0-8493-3574-4 (alk. paper)
    1. Groundwater flow—Mathematical models. 2. Aquifers. I. Title.

GB1197.7.B375 2005
551.49—dc22   2005041787

**Visit the Taylor & Francis Web site at
http://www.taylorandfrancis.com**

**and the CRC Press Web site at
http://www.crcpress.com**

To Nevin

Preface

The ground water resources of the Earth have long been subject to degradation as a result of man's increasing utilization of natural resources and worldwide industrialization. Potential anthropogenic sources of ground water contamination include accidental spills, landfills, industrial facilities, septic tanks, agricultural practices, and spent nuclear fuel repositories. The identification and delineation of contaminants and their respective concentrations in an aquifer, as well as the cleanup of the aquifer once the contamination is delineated, is a much more expensive and time-consuming process when compared with dealing with similar surface water contamination.

Beginning in the 1960s, contaminated aquifers were cleaned up and protected from further degradation in various countries around the world because governmental agencies identified ground water as a valuable and increasingly important potable water resource. During this time, it was found that mathematical ground water flow and solute transport modeling could be used as a powerful and cost-effective tool in the investigation and remedial design decision-making process. During the past four decades, improvements in our knowledge and understanding of ground water flow and contaminant transport mechanisms, scientific developments in solving partial differential equations numerically, and the exponential growth in the computational speeds of the personal computer have combined to make solving complex aquifer flow and contamination problems both cost-effective and efficient.

Fundamental principles of flow and solute transport in aquifers, and the analytical and numerical methods used to model these mechanisms, are the focus of this book. This book is an outgrowth of my research and practical experiences in the areas of flow and solute transport in aquifers conducted during the past quarter of this century. *Applied Flow and Solute Transport Modeling in Aquifers: Fundamental Principles and Analytical and Numerical Methods* is designed to be used by practicing hydrologists, hydrogeologists, environmental engineers, and others. Portions of this book may be suitable for use in upper level undergraduate courses; whole chapters of the book may be used in graduate level courses for students in the disciplines of civil and environmental engineering, chemical engineering, geology, hydrogeology, soil physics, agricultural engineering, geography, forestry, and others.

While writing this book, I tried to simplify the necessary theoretical background as much as possible with enriched practical applications. In almost every chapter, all derivational details of the theoretical backgrounds are included as worked examples to meet the needs of those readers who want to learn in detail how the derivations were generated. If the reader is not interested in the derivational details, he or she can easily skip them and still benefit and learn from the practical applications of the mathematical approaches. It is my belief that worked examples and their applications to real-world project cases are very important components of learning new subjects. For this reason, many worked examples are included in this book based on my academic and industrial experience. I tried to design this book in a format that would allow the reader to follow and understand the subject matter presented without the need for secondary references. In other words, I tried to pay special attention to making this book a self-contained document.

I express my thanks to Martin N. Sara and Patrick W. Dunne, who provided reviews on the manuscript; Constantinos V. Chrysikopoulos of the University of California-Irvine; M. Yavuz Corapçıoğlu of the Texas A&M University; and Martinus Th. Van Genucten of the U.S. Salinity Laboratory for various discussions on the subjects of this book during the past decade. My special thanks also go to David Fausel of CRC Press who provided timely effort in achieving this project. Finally, I thank my wife, Nevin, and my sons, Özer and Eren, for their patience during the long hours I spent daily writing this book.

The computer programs mentioned in Chapter 3 (ST1A, ST2A, ST3A, ST1B, ST2B, ST3B, ST1C, ST2C, and ST3C) and Chapter 5 (HDSF2) are parts of the ST123ABC software package that I developed. Purchasing information and other pertinent information about this software package can be found on the CRC Press website at: http://www.crcpress.com/e_products/downloads/download. asp?cat_no=3574 or by contacting me directly at vbatu@msn.com.

Vedat Batu
Naperville, Illinois
August 2004

About the Author

Dr. Vedat Batu received his Dipl. Ing. in civil engineering and his Ph.D. in hydraulic engineering from the Civil Engineering Faculty of Istanbul Technical University, Istanbul, Turkey. He also received his Doçent (University Associate Professor) degree (gained with an additional approved thesis and by passing some examinations) in the hydraulic engineering area by the Interuniversities Institution of Turkey; his thesis was based on his research work carried out at the University of Wisconsin-Madison. All of the theses of Dr. Batu are on the subject of ground water hydrology, and he is intensely involved in different ground water-related problems, including research and practice on saturated and unsaturated flows, aquifer hydraulics, contaminant transport, and analytical and numerical modeling techniques. He is also involved in project management, surface water hydraulics and  hydrology, storm water management, design of storm water management facilities, and other site development activities. Dr. Batu served as a faculty member and head of the Department of Civil Engineering, Karadeniz Technical University, Trabzon, Turkey. He has had visiting positions at several universities in the United States and United Kingdom, and for more than two decades he has been working with private corporations in the United States, mainly on aquifers, ground water-, and surface water–related projects.

Dr. Batu's own research and publications have continued throughout his professional life in the academy and industry. He is the author of the book entitled *Aquifer Hydraulics: A Comprehensive Guide to Hydrogeologic Data Analysis* published in 1998 by John Wiley & Sons, Inc., New York (727 pages). This book has been well received nationally and internationally; some reviews about it can be viewed at www.amazon.com by searching for the author's name. Dr. Batu has also published many papers in internationally recognized journals such as *Ground Water*, *Journal of Hydraulic Engineering*, *Journal of Irrigation and Drainage Engineering*, *Journal of Hydrology*, *Soil Science Society of America Journal*, and *Water Resources Research* and in numerous symposia proceedings. He served as an associate editor of the *Journal of Hydrologic Engineering* of the American Society of Civil Engineers.

Dr. Batu is a registered Professional Engineer in the states of Illinois and Pennsylvania, and he is currently with URS corporation in Chicago, Illinois, United States of America.

Table of Contents

Chapter 3

Analytical Solute Transport Modeling in Aquifers

Chapter 4
Numerical Flow and Solute Transport Modeling in Aquifers

Chapter 6
Stochastic Analysis of Flow and Solute Transport in Heterogeneous Aquifers and Determination
of Expressions for Dispersivities Using Stochastic Approaches

1 Introduction

1.1 IMPORTANCE OF FLOW AND SOLUTE TRANSPORT MODELING IN AQUIFERS

The water resources of the Earth, which are composed mainly of ground water and surface water, are subject to pollution from a variety of sources. Anthropogenic activities are the primary causes of pollution. Accidental spills, old landfills and underground storage tanks, industrial facilities, septic tanks, agricultural practices, spent nuclear fuel repositories, and poor waste management and disposal practices are well-known sources of contamination caused by man. The quality of the Earth's water resources has been declining since the middle of the 19th century, i.e., approximately since the beginning of the industrial revolution. However, the importance of this problem was recognized only after the mid-20th century, particularly in the case of the world's increasingly utilized ground water resources. The delineation and subsequent cleanup of contaminated ground water in the world's aquifers is a far more complex and expensive process when compared to a similar restoration of contaminated surface water. Therefore, it is very important to clean up currently degraded aquifers and protect the existing aquifers from further contamination to mitigate future cleanup costs and improve ground water quality for future generations.

The growing economic importance of aquifers as a potable water supply and the potential adverse effects that contamination can pose to users of aquifers necessitate the development of improved methods of predicting the transport of contaminants in aquifers both spatially and temporally. The scientific developments in solving partial differential equations numerically, and the technological developments in the computer industry during the past four decades, have collectively made it possible to solve complex aquifer flow and aquifer contamination problems in a cost-effective and efficient way. Owing to these techniques and developments, variable velocity fields in an aquifer can be handled effectively. For some aquifer problems, such as wellhead protection and water supply design, ground water flow modeling is sufficient. However, for a solute transport problem both ground water flow and solute transport modeling are required. Ground water flow and solute transport models are mathematical tools that can range from simple analytical solutions, which can be solved using a calculator, to far more complex numerical models that can only be solved using a digital computer.

This book is an outgrowth of the research and practical experiences of the author in the areas of flow and solute transport in aquifers. It should be a useful reference work for practicing hydrologists, hydrogeologists, environmental engineers, and others. Some portions of this book may be suitable for use at undergraduate-level courses, and all the chapters of the book can also be used in graduate-level courses in the disciplines of civil and environmental engineering, chemical engineering, geology, hydrogeology, soil physics, agricultural engineering, geography, forestry, and others.

The use of flow and solute transport models for predicting the fate and transport of contaminant constituents in ground water and surface water flow systems has intensified during the past three decades because of environmental and human health concerns arising from accidental spills and municipal, industrial, and hazardous waste disposal activities. Both analytical and numerical flow and solute transport models have proved to be cost-effective tools in predicting the extent of groundwater contamination as well as in developing remedial solutions to address the contamination. For some relatively simple cases, analytical models may be sufficient. Numerical models are capable of making predictions about the transport of contaminants in nonhomogeneous and anisotropic media under different stress conditions, including extraction wells, injection wells, and others, but certain requirements need to be satisfied to produce converged and numerically stable results. A number of sophisticated numerical models (finite difference, finite element, and others) are available for solving both the flow and the convective–dispersive solute transport differential equations for predicting both ground water flow patterns and spatial and temporal variations in contaminant concentrations in aquifers. For complicated hydrogeologic or contaminant conditions, numerical models have a significant advantage over analytical models because they have the capacity to simulate solute transport under nonuniform aquifer flow conditions with spatially variable solute transport parameters, arbitrary initial conditions, and complicated boundary conditions.

However, numerical models are relatively expensive, and in many cases, at least initially, insufficient data limit the effective application of a numerical model. For many projects, numerical models may not be feasible because of budgetary constraints. In these cases, simpler and cheaper modeling efforts are preferred. In these situations, analytical flow and solute transport models may be more useful in solving aquifer problems. Potentially, aquifer contamination problems can be solved using sophisticated numerical models or simple analytical techniques, depending on the complexity of the problem and the availability of data. However, many of the solute transport problems faced by practitioners can usually be handled using analytical modeling techniques, depending on the complexities and purpose of study. Analytical solute transport models can be useful tools for estimating the fate and transport of various constituents (including radionuclides) in aquifers. Their application is generally limited to situations with steady and uniform flow fields and to problems of relatively simple initial and boundary conditions. Compared to numerical models, analytical models often provide more insight into the conceptual–mathematical behavior of the system. Their computer solutions are easier to operate, cheaper to run, and contain significantly fewer parameters than the numerical solutions.

1.2 PURPOSE OF THIS BOOK

The main purpose of this book is to present the fundamental principles of flow and solute transport mechanisms in aquifers, and to present analytical and numerical modeling methods for predicting the migration of contaminants in space and time. The title of this book, *Applied Flow and Solute Transport Modeling In Aquifers: Fundamental Principles and Analytical and Numerical Methods* concisely summarizes the contents of this book.

It is important to emphasize that every chapter of this book has a common goal, which is to present the necessary theoretical background in as simplified a manner as possible, along with enriched practical applications. In almost every chapter, all the derivational details of the theoretical backgrounds are included as worked examples for readers who require the details as how the derivations were generated. If a reader is not interested in the details of the derivation, he or she can easily skip them without affecting the learning process regarding the practical usage of the approaches. The author of this book believes that the study of worked examples and their applications to real-world project cases are very important components of learning new subjects. For this reason, many worked examples, based on the author's academic and industrial experience, are included in this book. The book is designed such that the readers in the previously mentioned disciplines will be able to follow them without the need for additional reference. In other words, the book is designed to be a self-contained document.

The numerical flow and solute transport modeling approaches that are presented are based on practical applications in accordance with the publicly available MODFLOW family of codes. However, some limited information about finite-difference numerical solution techniques for flow and solute transport partial differential equations are also presented. The main purpose of the corresponding chapters of the book is to present the general principles of model code usage from a practical point of view.

The subjects regarding flow and solute transport modeling in aquifers in this book are all based on the *porous media flow theory*. Topics and applications pertaining to the *fractured media flow theory* are not included. There are several reasons for this: flow and solute transport modeling methods for fractured media are still in the development process in the literature (although a limited number of methods for some relatively simple cases are available); the available methods do not have widespread practical application; and models based on porous media flow theory may also be used for most fractured aquifers depending on the size and frequency of fractures (i.e., the equivalent porous media approach).

During the past two decades a number of software packages on flow and solute transport modeling in aquifers have been released by various public and private organizations. Almost all of these software packages assume that the user is experienced enough to select the appropriate model or models, to construct the modeling study based on the project goals, to select the critical flow and solute transport parameters, to calibrate the flow and solute transport models, and to interpret the modeling results. A certain level of familiarity with the fundamentals of flow and solute transport modeling techniques in aquifers is required to handle a flow or solute transport problem for an aquifer either by hand calculations or by using a software package. Therefore, the main purpose of this book is to provide sufficient knowledge to the readers in order for them to be able to solve aquifer flow and contamination problems by hand calculations or by software packages.

1.3 SCOPE AND ORGANIZATION

This book is composed of eight chapters. Each chapter is briefly described below.

Chapter 1 is the introduction to this book and it presents the importance of flow and solute transport modeling in aquifers, the purpose of the book, scope and organization of the book, and how to use the book.

Chapter 2 presents the fundamental principles of solute transport (or contaminant transport) in aquifers. This is a stand-alone chapter and presents the general laws of solute transport in aquifers, the physical meaning of various parameters, derivation of governing equations, and initial and boundary conditions commonly used in solving aquifer solute transport problems.

Chapter 3 presents a few one-, two-, and three-dimensional analytical solute transport models for aquifer contamination. These models are based on first- (or constant source), third- (or flux-type source), and instantaneous-type solute source cases. For the two- and three-dimensional analytical solute transport models, the medium is assumed to be bound by impermeable boundaries and capable of handling strip and rectangular sources. For the three-dimensional solute transport models, the thickness of the aquifer can also be taken into account. For all of these models, the ground water flow field is assumed to be uniform. In presenting these models, the following format is used: (1) problem statement and assumptions; (2) governing differential equations; (3) initial and boundary conditions; (3) unsteady-state solution for concentration distribution; (4) steady-state solution for concentration distribution; (5) convective–dispersive flux components under unsteady- and steady-state cases; (6) model evaluation; and (7) application to practical cases. The majority of these analytical solute transport models were developed by the author of this book during the past two decades, and most of them have been published in journals such as *Ground Water, Journal of Hydrology*, and *Water Resources Research*. These models were extensively used with high client satisfaction by the author and his co-workers during his consulting practice on real-world projects in the past two decades. For the interested readers, some details of the solution methods for the analytical models are presented as worked examples in this chapter. If a reader is not interested in how the solution for an analytical model is derived, he or she can easily skip them without affecting the learning process. The computer programs, written by the author, for these analytical solute transport models can be purchased by the user.

Chapter 4 presents numerical flow and solute transport modeling approaches in aquifers. Before the presentation of numerical flow modeling techniques, the fundamental principles of flow in aquifers are presented. This presentation includes the following topics: (1) definition of various aquifer flow quantities; (2) classification of aquifers with respect to hydraulic conductivity; (3) equations of motion in aquifers; (4) principal hydraulic conductivities; (5) directional hydraulic conductivities; (6) derivation of the governing flow differential equation in the rectangular coordinates; and (7) derivation of the governing flow differential equation in the cylindrical coordinates. After the fundamental principles of ground water flow are discussed, the finite-difference modeling approach for flow in aquifers is presented in accordance with the MODFLOW family of codes, and the following topics are covered: (1) governing differential equation for the finite-difference approach: (2) finite-difference grid and discretization convention; (3) derivation of the finite-difference equation; (4) boundary and initial conditions; (5) finite-difference discretization of space and time; (6) implementation of the boundary conditions; and (7) solution techniques for the finite-difference equations for flow. Examples also are included to clarify the associated methods and concepts. The finite-difference modeling approach for solute transport is then presented with the same format for flow in accordance with the MODFLOWT code, which is in the family of the MODFLOW code. Based on the above-mentioned details, the application of numerical flow and solute transport models is presented with the following topics: (1) model conceptualization; (2) model calibration; (3) application of a calibrated model; and (4) sensitivity analysis. Finally, model testing, model misuse, modeling limitations and sources of error, and a case study for a real-world example is presented to show the application of flow and solute transport modeling approaches.

Chapter 5 presents analytical and numerical modeling approaches for solute travel times and path lines in aquifers. First, path lines of solute under convective (or advective) transport conditions are presented under the following headings: (1) kinematics of a solute particle; (2) analytical methods for the determination of solute travel times; and (3) numerical methods for solute path lines and travel times determination. The finite-difference modeling approach for solute path lines and travel times is presented in accordance with the MODPATH code, which is in the family of the MODFLOW code. In the second part of the chapter, path lines of solute particles under convective–dispersive transport conditions are presented. This approach was introduced into the literature by the author about two decades ago and its foundations and applications are discussed in several papers that have been published in *Ground Water* and *Water Resources Research*. The concepts of *hydrodynamic dispersion stream function* and *hydrodynamic dispersion streamlines* are discussed in detail with some analytical solutions for strip solute sources.

The determination of dispersivities for solute transport in aquifers is one of the most difficult subjects in this area of study. During the past three decades, many advancements in understanding of dispersivities were made using stochastic approaches, despite the fact that some controversies still exist regarding the quantification of dispersivities. Chapter 6 presents various methods developed during the past three decades for the determination of dispersivities for solute transport in aquifers using stochastic approaches. Dispersivity expressions based on these methods are functions of particular statistical parameters of aquifers. Therefore, a certain level of knowledge regarding the statistical determination of these parameters as well as the foundations of stochastic approaches are required in order to effectively apply these methods. Under this framework, the physical mechanism of convective–dispersive solute transport in heterogeneous porous media is discussed. Determination of dispersivities from solute plume spatial moments are then included under the following topics: (1) one-dimensional convective–diffusive solute transport and plume spatial moments; (2) three-dimensional plume spatial moments; (3) mechanical dispersion coefficient and dispersivity tensors for general case; (4) dispersivity and mechanical dispersion coefficient tensors for uniform ground water flow case; (5) important points for the application of spatial moments; (6) stochastic description of hydraulic conductivity variation; and (7) illustrative examples for dispersivity determination from natural gradient experiments using the spatial moments method. For the last topic, some worked examples based on the three well-known experiments in the literature, the Borden, Twin Lake, and Cape Cod experiments, are also included. In the remaining part of Chapter 6, methods of dispersivities determination by applying perturbation analysis using Fourier–Stieltjes integral representation are included. For this purpose, the following main topics are included: (1) definition of key concepts and quantities; (2) one-dimensional stationary random process representation; (3) three-dimensional stationary random process representation; (4) stochastic analysis of one-dimensional steady-state flow; (5) stochastic analysis of three-dimensional steady-state flow in isotropic porous media; (6) analysis of macrodispersion in three-dimensionally heterogeneous aquifers under steady flow conditions using stochastic solute transport theory: the Gelhar and Axness first-order analysis approach; and (7) analysis of macrodispersion in three-dimensional, heterogeneous aquifers under unsteady flow conditions by applying stochastic solute transport theory using the Rehfeldt and Gelhar first-order analysis approach. Some worked examples are also included.

Chapter 7 presents estimation methods of dispersivities for solute transport in aquifers, based on the measured values of dispersivities determined by various investigators and experience. For this purpose, the following topics are covered: (1) laboratory vs. field scale dispersivities; (2) approaches for the determination of dispersivities; and (3) estimation of dispersivities using relationships between dispersivities and field scale for saturated flow. Diagrams and some empirical equations are included in quantifying dispersivity values for a given aquifer under certain scale conditions.

Finally, Chapter 8 presents determination methods of solute sorption parameters in aquifers. For this purpose, the following topics are included: (1) the K_d/K_{oc} approach for the determination of retardation factor; (2) compiled data for the organic carbon partition coefficient (K_{oc}); and (3) estimation of the organic carbon partition coefficient (K_{ow}) from the regression equations.

This book includes more than 300 figures and nearly 70 tables. In figures and tables, regular and scientific notation (E format) are used simultaneously, as appropriate. For example, 2.83E3 is equivalent to 2.83×10^3, and 1.16E$-$5 is equivalent to 1.16×10^{-5}. Besides these, this book includes nearly 70 worked example problems, which were carefully selected to simplify the process of learning many principles and methods. Special care has been taken to present the solutions of the example problems as clearly as possible. This book also includes nearly 70 problems to be solved by the reader. These problems are added at the end of the chapters and their solutions are included in the solutions' manual for instructors that accompanies this book.

The world is in the process of adopting a single *International System of Units*. This measurement system is commonly referred to as the metric system or "SI", based on the first two initials of its French name *Système International d'Unités*. This system of units has already been adopted by almost all countries around the world. However, in the United States, the English system (or U.S. customary system) is still used in practice; the SI system has been partially adopted by some research organizations. In this book, both the metric SI system and the English system of units are used.

1.4 HOW TO USE THIS BOOK

As previously mentioned, this book has been written to present the methods for applied flow and solute transport modeling in aquifers using analytical and numerical methods. For an effective presentation, the fundamental principles of flow and solute transport in aquifers are also included. This book is designed to be a self-contained document. If a reader does not have enough background for the fundamental principles for flow and solute transport, he or she is advised to study Chapter 2 first, for the fundamental principles of solute transport in aquifers, and Sections 4.3.1–4.3.7 of Chapter 4 for the fundamental principles of flow in aquifers. The reader should then be able to follow the rest of the chapters.

There are some advanced sections, particularly in Chapters 3 and 6. Chapter 3 has been designed for practitioners as well as for advanced-level readers. If a reader is not interested in how the analytical solutions are derived but is interested in their potential application areas, he or she can skip those solution details and focus on their companion application areas. This can be done because almost all solution details are included as example problems. The solution details are included for those readers interested in handling similar problems for aquifer solute transport problems. A similar approach is also adopted for Chapter 6 by presenting additional details for stochastic approaches to solute transport in aquifers. This has been done intentionally because the stochastic approaches are relatively new for aquifer problems and published documents and books in this area are very limited.

A reader may already have sufficient background in the areas of flow and solute transport principles in aquifers and may be willing to learn the principles of flow and solute transport modeling using numerical methods. For such readers, it is advised to read Chapter 4 for numerical flow and solute transport modeling in aquifers, and Chapter 5 for solute path lines and travel times in aquifers.

Estimation of flow and solute transport input parameters is one of the most important stages of any aquifer modeling project. Estimation procedures for required flow modeling input parameters can be found in Chapter 4 (e.g., porosity, hydraulic conductivity, specific yield, recharge rate, water loss, etc.). Estimation procedures of solute transport parameters can be found in Chapter 2 (e.g., molecular diffusion coefficient, distribution coefficient, half-life, degradation rate coefficient, etc.), Chapter 6 (e.g., statistical parameters for log normal frequency distribution of hydraulic conductivity, aquifer covariance data, etc.), Chapter 7 (e.g., longitudinal dispersivity, transverse horizontal dispersivity, transverse vertical dispersivity, etc.), and Chapter 8 (e.g., solubility, K_{ow}, K_{oc}, etc.).

2 Fundamental Principles of Solute Transport in Aquifers

2.1 INTRODUCTION

Dissolved substances are called *solutes*, which may be natural constituents, artificial tracers, or contaminant constituents. Quantification of solute transport for aquifers or ground water systems is based on the convective–dispersive (or advective–dispersive) solute transport theory, which is also applicable in other transport media, such as surface water. Basically, the convective–dispersive solute transport theory is based on Fick's first law (see Eq. [2-20]), which was established by the mid-19th century. The original Fick's first law was established for molecular diffusion in surface water. Later, by the mid-20th century, Fick's first law was extended to solute transport in ground water by including the dispersion effects. There is no doubt that the migration mechanism of solutes in ground water systems is more complex than the one in surface water flow systems. Different solute constituents have different migration rates, depending on their physical and chemical characteristics as well as the characteristics of the ground or aquifer materials. Also, some solute constituents may be subjected to decay or degradation. These phenomena make the migration of solutes in ground water relatively complex as compared with the solutes for surface water.

The selection of predictive solute transport models as well as solute transport parameters requires a thorough knowledge of the physical mechanism of solute transport theory along with the limitations of the models. In this chapter, the fundamental principles of solute transport in ground water are presented. First, the definitions of the prime quantity, concentration, are included with examples. Then the physical mechanisms of solute transport in porous media and their governing equations are presented.

2.2 CONCEPTS FOR CONCENTRATION AND ITS UNITS

2.2.1 VOLUME-AVERAGED (RESIDENT) CONCENTRATION

2.2.1.1 Solute and Solvent

Let us consider two substances: substance 1 and substance 2. In a solution of Substance 1 in Substance 2, Substance 1 is the dissolved substance and is called the *solute*. Substance 2 is the one in which the solute (Substance 1) is dissolved and is called the *solvent*. When the relative amount of one substance in a solution is much greater than the amount of the other substance, the substance present in greater amounts is generally regarded as the solvent. If the relative amounts of the two substances are about the same order, it becomes difficult to specify which one is the solvent (Rosenberg, 1980).

In the case of ground water, the solvent is almost always water; the dissolved inorganic and organic constituents are the solutes. To analyze ground water quality problems efficiently, the relative amounts of solute (the dissolved parts) and solvent (water) must be specified, which can be achieved through the use of concentration concept and its units. Various units of concentration are given in the following section.

Concentrations in the subsequent sections are all volume-averaged concentrations unless otherwise stated. However, for simplicity, the subscript "r" is dropped and concentration is denoted simply by C instead of C_r.

2.2.1.2 Definitions for Concentration

Mass Concentration

Mass concentration is the mass of solute dissolved in a specified unit volume of solution. This concentration is also called *volume-averaged concentration* or *resident concentration* and is denoted by C_r. The definition of the resident concentration is shown in Figure 2-1. The resident concentration must be defined in the context of representative elementary volume (REV) or, equivalently, a representative elementary area since they are macroscopic quantities. Resident concentrations, C_r, may be viewed by their local or microscopic volume (Parker and van Genuchten, 1986). If in a unidirectional ground water flow, Δm_A is the mass of solute for a constituent A in a volume of fluid ΔV, the definition of the volume-averaged or resident concentration is

$$C_r = \frac{\Delta m_A}{\Delta V} \qquad (2\text{-}1)$$

where

$$\Delta V = \varphi S \Delta x \qquad (2\text{-}2)$$

in which φ is the total porosity, S the cross-sectional area of the tube (see Figure 2-1), and Δx the length of the fluid volume.

The SI unit for this quantity is kilograms per cubic meter (kg/m^3); also grams per liter (g/L) is permitted as an SI unit. The most common mass concentration unit reported in the ground water literature is milligrams per liter (mg/L). It is clear that 1 mg/L equals 1 g/m^3; therefore, there is no difference in the magnitude of mg/L and g/m^3, both of which are the permitted SI concentration units.

Parts per million (ppm) is the number of grams of solute per million grams of solution:

$$1 \text{ ppm} = \frac{\text{g solute}}{10^6 \text{ g of solution}} \qquad (2\text{-}3)$$

For nonsaline waters,

$$1 \text{ ppm} = 1 \text{ g/m}^3 = 1 \text{ mg/L} \qquad (2\text{-}4)$$

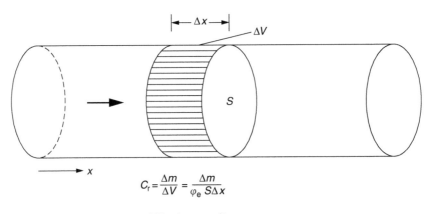

$$C_r = \frac{\Delta m}{\Delta V} = \frac{\Delta m}{\varphi_e \, S \Delta x}$$

φ_e = Effective porosity
Δm = Solute mass in ΔV of fluid
$\Delta V = \varphi_e \, S \Delta x$ = Volume of fluid
S = Cross-sectional area of the tube

FIGURE 2-1 Definition of volume-averaged (resident) concentration for unidirectional flow. (Adapted from Kreft, A. and Zuber, A., *Water Resources Res.*, 22(7), 1157–1158, 1986.)

Parts per billion (ppb) is the number of grams of solute per billion grams of solution:

$$1 \text{ ppb} = \frac{\text{g solute}}{10^9 \text{ g of solution}} \tag{2-5}$$

For nonsaline waters,

$$1 \text{ ppb} = 1 \text{ mg/m}^3 = 1 \text{ } \mu\text{g/L} \tag{2-6}$$

Parts per trillion (ppt) is the number of grams of solute per million grams of solution:

$$1 \text{ ppt} = \frac{\text{g solute}}{10^{12} \text{ g of solution}} \tag{2-7}$$

For nonsaline waters,

$$1 \text{ ppt} = 1 \text{ } \mu\text{g/m}^3 = 10^{-6} \text{ } \mu\text{g/L} \tag{2-8}$$

Mass Fraction
If Δm_A is the mass of solute for a constituent A in a mass of fluid Δm, the definition of the mass fraction is

$$C_r^* = \frac{\Delta m_A}{\Delta m} \tag{2-9}$$

where C_r^* is an intrinsic quantity that does not depend upon the physicochemical ways of mixing. Dividing Eq. (2-1) by (2-9),

$$\frac{C_r}{C_r^*} = \frac{\Delta m_A / \Delta V}{\Delta m_A / \Delta m} = \frac{\Delta m}{\Delta V} = \rho \tag{2-10}$$

or

$$C_r = \rho C_r^* \tag{2-11}$$

where ρ is the density of the mixture.

The *molecular concentration* C_M is the number of molecules of constituent A (n_A) divided by the volume of mixture V that contains the molecules:

$$C_M = \frac{n_A}{V} \tag{2-12}$$

The *molecular fraction* C_M^* is the number of molecules of constituent A (n_A) in a volume of mixture divided by the total number of mixture molecules A (n_M) in the same volume, i.e.,

$$C_M^* = \frac{n_A}{n_M} \tag{2-13}$$

Dilution
As seen from the above definition of concentration, the concentration is expressed as the amount of solute per fixed volume of solution. Therefore, the amount of solute mass contained in a given volume V of solution is equal to the product of the volume and the concentration, C_r:

$$\text{mass of solute} = (\text{volume}) (\text{concentration}) = VC_r \tag{2-14}$$

If the volume is increased and the concentration is decreased, but the total amount of solute mass remains constant, the solution is diluted, and this phenomenon is called *dilution*. Let V_1 and V_2 be two different volumes of solution and C_{r_1} and C_{r_2} be their concentrations, respectively. The two solutions have different concentrations, but the amounts of solute mass are the same. Therefore,

$$V_1 C_{r_1} = V_2 C_{r_2} \qquad (2\text{-}15)$$

If any three terms in the above equation are known, the fourth one can be calculated. The quantities on both sides of Eq. (2-15) must be expressed in the same units.

EXAMPLE 2-1

This example is adapted from Rosenberg (1980). The solution of concentration of chloride in a container is 60 mg/m³. To what extent must this solution be diluted to yield a concentration of 15 mg/m³?

SOLUTION

Let V_1 be the volume to which 1 m³ of the original solution must be diluted to yield a solution of the concentration. Substitution of the given quantities into Eq. (2-15) gives

$$(1 \text{ m}^3)\ (60 \text{ mg/m}^3) = V_2(15 \text{ mg/m}^3)$$

which results in $V_2 = 4$ m³. This implies that the original solution ($V_1 = 1$ m³) must be diluted to a volume of 4 m³. In other words, 4 m³ of the diluted solution will contain as much solute mass as 1 m³ of the original solution. It must be emphasized that 4 m³ is not the amount of water that is to be added, but the final volume of the solution obtained after water has been added to 1 m³ of the original solution.

2.2.2 FLUX-AVERAGED CONCENTRATION

The definition of flux-averaged concentration is given in Figure 2-2. Let F_x be the mass flux (mass per unit area per unit time). Then the definition of flux-averaged concentration C_f is (Kreft and Zuber, 1986)

$$C_f = \frac{F_x \Delta t S}{Q \Delta t} = \frac{F_x S}{Q} \qquad (2\text{-}16)$$

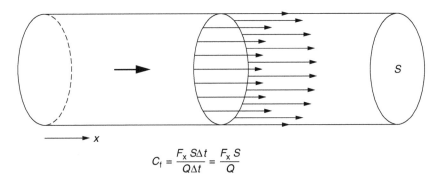

$$C_f = \frac{F_x S \Delta t}{Q \Delta t} = \frac{F_x S}{Q}$$

F_x = Mass flux of solute (mass per unit area per unit time)

Q = Volumetric flow rate

S = Cross-sectional area of the tube

FIGURE 2-2 Definition of flux-averaged concentration for unidirectional flow. (Adapted from Kreft, A. and Zuber, A., *Water Resources Res.*, 22(7), 1157–1158, 1986.)

where Q is the volumetric flow rate in the tube, S the cross-sectional area of the tube, and Δt the time interval. Since $Q = qS$, in Eq. (2-16), we have (Parker, 1984)

$$C_f = \frac{F_x S}{Q} = \frac{F_x}{q} = \frac{F_x}{\varphi_e U} \tag{2-17}$$

where q is the Darcy velocity, φ_e the effective porosity, and U the ground water velocity (or seepage velocity). In Eq. (2-17), since the mass flux F_x is divided by the Darcy velocity q, C_f is called the *flux-averaged concentration*.

2.2.3 DISTINCTIONS BETWEEN VOLUME- AND FLUX-AVERAGED CONCENTRATIONS

Distinctions between volume- and flux-averaged concentrations have been made by scientists in the field of petroleum and chemical engineering (Brigham, 1974; Kreft and Zuber, 1978, 1979). Parker and van Genuchten (1984) showed that it is of fundamental importance to make such a distinction in order to stipulate boundary conditions appropriate for specific experimental solute detection modes. The authors also noted that when the assumptions implicitly invoked by various boundary conditions are considered and mass balance constraints are carefully adhered to, one will find that certain perceived limitations of the convection–dispersion model appear to be mitigated. Parker and van Genuchten (1986, p. 1159) also noted that both types of concentrations must be defined in the context of an REV (or equivalently a representative elementary area [Bear and Bachmat, 1983]), since they are macroscopic quantities. They also added that volume-averaged concentrations C_r may be viewed as averages of local concentrations within the REV weighted by their local or macroscopic volume, and flux-averaged concentrations C_f represent averages weighted by the local pore water velocity.

2.3 PHYSICAL MECHANISM OF SOLUTE TRANSPORT IN SATURATED POROUS MEDIA WITHOUT SORPTION AND DECAY: ONE-DIMENSIONAL CASE

2.3.1 DIFFUSIVE TRANSPORT

The effects of diffusion can easily be observed by a simple experiment. Take a glass of water and drop a small amount of dye into it. During the early stages, a sharp interface can be observed between the colored water and the rest of the water body in the glass. Gradually, the sharp interface disappears and the whole glass of water eventually takes on the same color. This phenomenon is called *molecular diffusion* and the transport mechanism is called *diffusive transport*. Diffusion occurs even in a fluid at rest (Fried and Combarnous, 1971).

2.3.1.1 Physical Explanation of Molecular Diffusion

The molecules of a fluid undergo random displacements and collisions, and this phenomenon is called *Brownian motion*. Consider a mixture of two liquids, A and B. Since the liquids are different, one may expect that the mechanical conditions of collisions between couples of molecules A–A, A–B, and B–B are not the same. If the concentrations of liquids A and B are uniform, each molecule of A will have the same mean number of collisions per unit time as other A molecules would have with other A and B molecules, respectively. If concentrations are not uniform such that in some zones with high concentrations of liquid A molecules, the mechanical conditions of liquid A molecules will be different from the mechanical conditions of the liquid A molecules in zones having lower concentrations. As a result of this unbalanced situation, fluid particles will be able to migrate from zones of high concentration to zones of lower concentration. This phenomenon is called molecular diffusion (Fried and Combarnous, 1971).

Diffusive transport in soils and aquifers tends to decrease the existing concentration gradients. As explained in the above paragraph, diffusive transport of solutes can occur in fluids at rest or in the absence of flow of the solution. If the solution is flowing, diffusive transport can still occur and become part of hydrodynamic dispersion along with mechanical dispersion (see Section 2.3.4). The process of diffusion is referred to as molecular diffusion.

2.3.1.2 Quantification of Molecular Diffusion: Fick's First Law

Fick's First Law for Pure Water
In order to describe the diffusion of a solute constituent quantitatively, a relationship between the mass flux per unit time per unit area F_s and concentration gradient $\Delta C/\Delta s$ (where s is the distance) must be established. The first quantitative study of diffusion was made by a German physician Adolf E. Fick (1829 to 1901) in 1855, who noticed an analogy between molecular diffusion and heat transfer by conduction. Fick adopted the heat flux equation to describe the diffusion process by stating that the rate of transfer of diffusing substance ($F_{s_{dif}}$) through the unit area of a section is proportional to the concentration gradient $\Delta C/\Delta s$ normal to the section (see Figure 2-3):

$$F_{s_{dif}} \propto \frac{\Delta C}{\Delta s} \qquad (2\text{-}18)$$

Introducing a proportionality constant, D^*, leads to the equation

$$F_{s_{dif}} = -D^* \frac{\Delta C}{\Delta s} = -D^* \frac{C_2 - C_1}{\Delta s} \qquad (2\text{-}19)$$

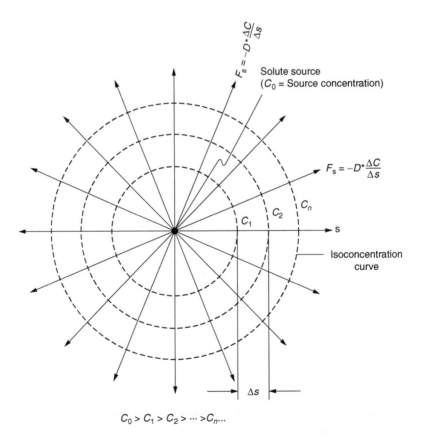

$$C_0 > C_1 > C_2 > \cdots > C_n \ldots$$

FIGURE 2-3 Solute source in a stagnant liquid and schematic representation of Fick's first law.

or, in the limiting case, $\Delta C/\Delta s$ becomes dC/ds, and hence

$$F_{s_{dif}} = -D^* \frac{dC}{ds}$$ (2-20)

which is called Fick's first law.

Fick's Modified First Law for Porous Media

By modifying Fick's first law (Eq. [2-20]), in the presence of the solid phase, the diffusion flux in a saturated porous medium becomes (see Figure 2-3)

$$F_{s_{dif}} = -\varphi_e D^* \frac{dC}{ds}$$ (2-21)

where φ_e is the effective porosity of the medium. For unsaturated porous media flow case, $\varphi_e = \theta$, where θ is the water content. Here, $F_{s_{dif}}$ is the mass flux (the mass of solute per unit area per unit time and its dimensions are $M/[L^2T]$), C is the solute concentration (M/L^3), and dC/ds is the concentration gradient. The negative signs in the above equations indicate that solute migration is in the direction of decreasing concentration. D^* is the ionic or effective molecular diffusion coefficient of the porous medium and s the distance. From dimensional analysis, it can be shown that the dimension of D^* is L^2/T. Due to a tortuous diffusion pathway and the presence of the solution–solid interface, D^* in granular materials is comparatively lower than the molecular diffusion coefficient in pure water D_0 or diffusion coefficient in aqueous solution:

$$D^* = D_0\tau, \quad 0 < \tau < 1.0$$ (2-22)

where τ is a dimensionless tortuosity factor, approximately ranging from 0.3 to ~0.7 for most soils (van Genuchten and Wierenga, 1986). From a review of data on unconsolidated granular media obtained by several investigators, Perkins and Johnston (1963) suggested that the value of τ is approximately 0.707 as presented in Gillham and Cherry (1982a). Bear (1972) suggests that for an actual porous medium, $\tau = 2/3 = 0.67$ is a better value for granular materials (e.g., for loose sand).

The differential equation for the one-dimensional diffusion is called *Fick's second law* and is expressed by Eq. (6-24).

2.3.1.3 The Effective Molecular Diffusion Coefficient D^*

Effective Molecular Diffusion Coefficients in Pure Liquids

For liquids, $\tau = 1.0$, and according to Eq. (2-22), $D^* = D_0$. The values of diffusion coefficient D_0 for electrolytes are well known. The major ions in ground water (Na^+, K^+, Mg^{2+}, Cl^-, HCO_3^-, and SO_4^{2-}) have diffusion coefficients in the range 1.0×10^{-9} to 2.0×10^{-9} m^2/sec at 25°C (Robinson and Stokes, 1965). The coefficients depend on temperature. At 5°C, for example, the coefficients are about 50% smaller. The effect of ionic strength is very small (Freeze and Cherry, 1979, p. 103). Values of D_0, in general, vary with concentration. In Table 2-1, values of D_0 for different solute constituents in water and other solvents are given as functions of both solute concentration and temperature (Treybal, 1980, p. 36).

Effective Molecular Diffusion Coefficients in Porous Media

Owing to the tortuosity factor, the effective molecular diffusion coefficient D^* is the dominant parameter for diffusive transport in porous media. In other words, D^* is the effective diffusion coefficient and is smaller than in water because the ions follow longer paths of diffusion by the granular structure of porous media. With an average tortuosity factor, for example, $\tau = 0.5$, the effective molecular diffusion coefficient D^* can be determined from Eq. (2-22) using the D_0 values given in Table 2-1. For example, from Table 2-1, for chloride (Cl_2) in ground water at 16°C, $\tau = 0.5$ and $D_0 = 1.26\times10^{-9}$ m^2/sec; substituting these values into Eq. (2-22) gives $D^* = 6.3\times10^{-10}$ m^2/sec. According to Freeze and Cherry (1979), $D^* = 1.0\times10^{-11}$ to 1.0×10^{-10} m^2/sec, which are representative of a range typical

TABLE 2-1
Molecular Diffusion Coefficients (D_0) of Some Liquids

Solute	Solvent	Temperature($^\circ$C)	Solute Concentration (kmol/m^3)	Diffusion Coefficient $D_0 \times 10^9$ (m^2/sec)[a]
Cl$_2$	Water	16	0.12	1.26
HCl	Water	0	9	2.7
			2	1.8
		10	9	3.3
			2.5	2.5
		16	0.5	2.44
NH$_3$	Water	5	3.5	1.24
		15	1.0	1.77
CO$_2$	Water	10	0	1.46
		20	0	1.77
NaCl	Water	18	0.05	1.26
			0.2	1.21
			1.0	1.24
			3.0	1.36
			5.4	1.54
Methanol	Water	15	0	1.28
Acetic acid	Water	12.5	1.0	0.82
			0.01	0.91
		18.0	1.0	0.96
Ethanol	Water	10	3.75	0.50
			0.05	0.83
		16	2.0	0.90
n-Butanol	Water	15	0	0.77
CO$_2$	Ethanol	17	0	3.2
Chloroform	Ethanol	20	2.0	1.25

[a] For example , D_0 for Cl$_2$ in water is 1.26×10^{-9} m^2/sec.

Adapted from Treybal, R.E., *Mass-Transfer Operations*, 3rd ed., McGraw-Hill, New York, 1980.

of nonreactive chemical species in clayey geological deposits. Values of effective molecular diffusion coefficients for coarse-grained unconsolidated materials can be higher than 1.0×10^{-10} m^2/sec, but are less than D_0 values in water (i.e., $< 2.0 \times 10^{-9}$ m^2/sec) (Freeze and Cherry, 1979, p. 303). For some porous materials, the tortuosity factor τ may have relatively smaller values. For example, Baron et al. (1990) conducted an experimental investigation on the diffusive transport of nonreactive solute (chloride) in saturated, intact Queenston Shale (in Canada) in laboratory and found that the experimental effective diffusion coefficient D^* for chloride at a temperature of $22 \pm 1^\circ$C ranged from 1.4×10^{-10} to 1.6×10^{-10} m^2/sec, which correspond to a tortuosity factor τ ranging from 0.095 to 0.108.

Importance of Diffusive Solute Transport
As will be discussed later in detail in Section 2.3.4, the recognized mechanisms that affect the transport of solutes through saturated geologic materials include transport as a result of the bulk motion of the fluid phase (convection or advection), dispersive transport caused by velocity variations in the mean velocity, and molecular diffusion. In fine-grained geologic materials, the principal mechanism of solute transport is the molecular diffusion.

Gillham and Cherry (1982a, 1982b) showed that if the average ground water velocity is less than about 1.6×10^{-10} m/sec, then molecular diffusion would be the dominant transport mechanism. Assuming a hydraulic gradient of 0.01 m/m and a porosity of 0.4, this would correspond to a hydraulic conductivity of the order of 5.0×10^{-8} m/sec. Since natural clay materials have hydraulic conductivity values in the range of $\sim 5.0 \times 10^{-9}$ to 5.0×10^{-10} m/sec, and the hydraulic conductivity of bentonite is generally less than $\sim 1.0 \times 10^{-9}$ m/sec, it is reasonable to expect that the principal mechanism of solute transport through many liner materials will be molecular diffusion (Gillham et al., 1984).

2.3.2 CONVECTIVE (ADVECTIVE) TRANSPORT

Convective transport (or *advective transport*) refers to the process by which solutes are transported by the motion of the flowing water in ground water. Under pure convection conditions, ground water and dissolved solute move at the same average velocity, and the flux for convection $F_{s_{con}}$ is

$$F_{s_{con}} = v_s \varphi_e C = q_s C \qquad (2\text{-}23)$$

where $F_{s_{con}}$ represents the mass of solute per unit cross-sectional area transported in the s direction per unit time, q_s is the Darcy velocity in the same direction, φ_e the effective porosity, and v_s the ground-water velocity or pore water velocity in the same direction. The effective porosity is the portion of pore space in a saturated porous material in which water flow occurs. In other words, it is the volume of the interconnected voids of a sample. This definition is based on the fact that the entire pore space of a porous medium filled with water is not open for water flow. Mathematically, the effective porosity is the ratio between "the volume of the interconnected voids of a sample" and "the total volume of the sample" (see, e.g., Batu, 1998). Effective porosity values for some porous materials are given in Table 5-2. Unless otherwise stated, in this book, φ will denote the effective porosity.

2.3.3 DISPERSIVE TRANSPORT

For simplicity, let us consider an average one-dimensional ground water flow as shown in Figure 2-4. This can also be visualized as a Darcy experiment device. In Figure 2-4a, the mean flow directions are shown in the sense of Darcy velocity. In Figure 2-4b, the path lines of individual fluid particles around the grains of a porous medium, which is filled by water, are shown. Due to the irregular shapes of grains and pores, local fluid velocities inside individual pores deviate from the average pore velocity. Such velocity variations cause the solute to be transported down-gradient at different rates, thus leading to a mixing process that is macroscopically similar to mixing caused by the diffusive transport. Because of the passive nature of the dispersion process, the term mechanical dispersion is often used to describe mixing caused by local velocity variations (Fried and Combarnous, 1971; Bear, 1972; Freeze and Cherry, 1979; van Genuchten and Wierenga, 1986).

Laboratory and field experiments have shown that mechanical dispersive flux (or dispersive flux) can be described by an equation similar to the equation for diffusive flux, given by Eq. (2-21),

$$F_{s_{disp}} = -\varphi_e D_m \frac{dC}{ds} \qquad (2\text{-}24)$$

where D_m is the mechanical dispersion coefficient (Bear, 1972). The mechanical dispersion coefficient is the most complex and controversial component for transport of solutes in porous media and will be discussed later in Section 2.11.2 in detail. Equation 2-24 is based on the assumption that the dispersive transport can be expressed similar to that of Fick's first law as given by Eq. (2-21). In other words, dispersive transport is a Fickian phenomenon or simply Fickian. The negative sign in Eq. (2-24) indicates that the solute moves toward the zone of lower concentration.

2.3.4 HYDRODYNAMIC DISPERSION: ONE-DIMENSIONAL CASE

As mentioned in Sections 2.3.1–2.3.3, the solute flux in porous media comprises three components: $F_{s_{dif}}$ (diffusive flux), $F_{s_{con}}$ (convective flux), and $F_{s_{dif}}$ (dispersive flux). Therefore, the total flux F_s will be the sum of the above three:

$$F_s = F_{s_{dif}} + F_{s_{con}} + F_{s_{disp}} \qquad (2\text{-}25)$$

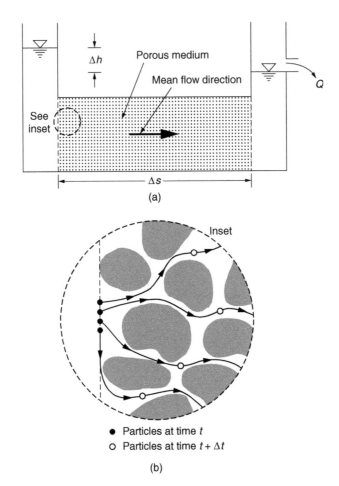

(a)

(b)

FIGURE 2-4 Physical mechanism of mechanical dispersion in a one-dimensional ground water flow: (a) one-dimensional Darcy experiment-type ground water flow and (b) fluctuating streamlines around the porous medium grains on a pore scale.

Substitution of Eqs. (2-21), (2-23), and (2-24) into Eq. (2-25) gives

$$F_s = \varphi_e v_s C - \varphi_e D_s \frac{dC}{ds} \tag{2-26}$$

where

$$D_s = D^* + D_m \tag{2-27}$$

with D_s being the longitudinal dispersion coefficient (Bear, 1972), further referred to simply as the dispersion coefficient. Other terms frequently used for D_s are the apparent diffusion coefficient (Nielsen et al., 1972; Boast, 1973) and the diffusion–dispersion coefficient (Hillel, 1980). Sometimes, the term hydrodynamic dispersion coefficient is reserved for only D_m (Shamir and Harleman, 1966; Nielsen et al., 1972; Boast, 1973).

2.3.4.1 The Effects of Convection and Dispersion

As can be seen from Eq. (2-25), solute transport is generally viewed as the net effect of two processes, convection and dispersion. Dispersion represents the effects of molecular diffusion and

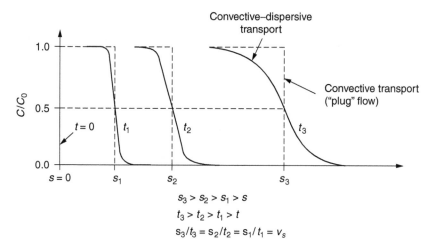

FIGURE 2-5   Schematic representation of the convective and dispersive solute transport processes. (Adapted from Gillham, R.W. and Cherry, J.A., *Proceedings of the Symposium on Low-Level Waste Disposal: Facility Design, Construction and Operating Practices*,1982a.)

mechanical dispersion. Convective transport is attributed to the average motion of the fluid. For miscible fluids that involve one-dimensional displacement processes, with convection being the only transporting mechanism, a sharp front will be maintained between the initial and displacing fluids, and this front will move at a velocity equal to the average linear ground water velocity. The effects of convection and dispersion are shown schematically in Figure 2-5 for one-dimensional displacement of miscible fluids (Gillham and Cherry, 1982a).

2.4   GENERALIZATION OF THE CONVECTIVE–DISPERSIVE SOLUTE FLUX EQUATIONS WITHOUT SORPTION AND DECAY FOR SATURATED POROUS MEDIA

Using the concepts and equations for the one-dimensional convective–dispersive solute flux in porous media, the flux equations for two- and three-dimensional solute transport cases can be determined. These are presented in the following sections.

2.4.1   THREE-DIMENSIONAL CONVECTIVE–DISPERSIVE FLUX COMPONENTS FOR GENERAL CASE: ANISOTROPIC DISPERSION

2.4.1.1   Tensorial Forms of the Convective–Dispersive Flux Components

Experiments in two dimensions indicate that the magnitude of the dispersion depends on the reference direction, with the larger value oriented in the direction parallel to the flow. In order to include this directional dependency in the equations, the longitudinal dispersion coefficient D_s in Eq. (2-26) should be represented by a tensorial quantity. Based on the studies of Bear (1961) and Scheidegger (1961), for simplicity, it is assumed that the dispersion coefficient is approximated by a tensor composed of, at most, nine components. Under the framework of the above points, using the standard notation for second-order tensors, the three-dimensional form of Eq. (2-26) is (Ogata, 1970)

$$F_i = \varphi_e v_i C - \varphi_e D_{ij} \frac{\partial C}{\partial x_j}, \quad i,j = 1, 2, 3 \tag{2-28}$$

In other words, the three convective–dispersive flux components are as follows:

$$F_1 = \varphi_e v_1 C - \varphi_e \left(D_{11} \frac{\partial C}{\partial x_1} + D_{12} \frac{\partial C}{\partial x_2} + D_{13} \frac{\partial C}{\partial x_3} \right) \tag{2-29a}$$

$$F_2 = \varphi_e v_2 C - \varphi_e \left(D_{21} \frac{\partial C}{\partial x_1} + D_{22} \frac{\partial C}{\partial x_2} + D_{23} \frac{\partial C}{\partial x_3} \right) \tag{2-29b}$$

$$F_3 = \varphi_e v_3 C - \varphi_e \left(D_{31} \frac{\partial C}{\partial x_1} + D_{32} \frac{\partial C}{\partial x_2} + D_{33} \frac{\partial C}{\partial x_3} \right) \tag{2-29c}$$

and the dispersion tensor D_{ij} can be represented by the matrix

$$[D_{ij}] = \begin{bmatrix} D_{11} & D_{12} & D_{13} \\ D_{21} & D_{22} & D_{23} \\ D_{31} & D_{32} & D_{33} \end{bmatrix} \tag{2-30}$$

2.4.1.2 Convective–Dispersive Flux Components in Cartesian Coordinates

In x–y–z Cartesian coordinates, Eq. (2-29) takes the form

$$F_x = \varphi_e v_x C - \varphi_e \left(D_{xx} \frac{\partial C}{\partial x} + D_{xy} \frac{\partial C}{\partial y} + D_{xz} \frac{\partial C}{\partial z} \right) \tag{2-31a}$$

$$F_y = \varphi_e v_y C - \varphi_e \left(D_{yx} \frac{\partial C}{\partial x} + D_{yy} \frac{\partial C}{\partial y} + D_{yz} \frac{\partial C}{\partial z} \right) \tag{2-31b}$$

and

$$F_z = \varphi_e v_z C - \varphi_e \left(D_{zx} \frac{\partial C}{\partial x} + D_{zy} \frac{\partial C}{\partial y} + D_{zz} \frac{\partial C}{\partial z} \right) \tag{2-31c}$$

and the dispersion tensor D_{ij}, given by Eq. (2-30), can be written as

$$[D_{ij}] = \begin{bmatrix} D_{xx} & D_{xy} & D_{xz} \\ D_{yx} & D_{yy} & D_{yz} \\ D_{zx} & D_{zy} & D_{zz} \end{bmatrix} \tag{2-32}$$

Equation (2-29) or (2-31) describes the state-of-the-art forms of the convective–dispersive flux components, which constitute the mechanical and mathematical foundations of almost all analytical and numerical solute transport models existing in the literature during the past half century.

2.4.2 THREE-DIMENSIONAL CONVECTIVE–DISPERSIVE FLUX COMPONENTS FOR SPECIAL CASES

There are some important special cases of the convective–dispersive flux components that have practical importance. These are mentioned below.

2.4.2.1 The Principal Axes of the Dispersion Tensor can be Defined: the Porous Medium is Homogeneous and Anisotropic

A homogeneous and anisotropic aquifer is schematically defined in Figure 4-8. Consider points 1 and 2 in the aquifer whose coordinates are (x_1, y_1, z_1) and (x_2, y_2, z_2), respectively. If the hydraulic conductivities in the x-, y-, and z-coordinate directions satisfy that $K_x(x_1, y_1, z_1) = K_x(x_2, y_2, z_2) = K_x$, $K_y(x_1, y_1, z_1) = K_y(x_2, y_2, z_2) = K_y$, and $K_z(x_1, y_1, z_1) = K_z(x_2, y_2, z_2) = K_z$, for points 1 and 2, respectively, the aquifer is said to be a homogeneous and anisotropic aquifer (Batu, 1998).

Assuming that the principal axes can be defined and the dispersion tensor can be transformed, only elements of the major diagonal remain, all others being zero. Under these conditions, the matrix representation of the tensor given by Eq. (2-32) becomes

$$[D_{ij}] = \begin{bmatrix} D_x & 0 & 0 \\ 0 & D_y & 0 \\ 0 & 0 & D_z \end{bmatrix} \tag{2-33}$$

As a result, the convective–dispersive flux components given by Eqs. (2-31) take, respectively, the following forms:

$$F_x = \varphi_e v_x C - \varphi_e D_x \frac{\partial C}{\partial x} \tag{2-34a}$$

$$F_y = \varphi_e v_y C - \varphi_e D_y \frac{\partial C}{\partial y} \tag{2-34b}$$

$$F_z = \varphi_e v_z C - \varphi_e D_z \frac{\partial C}{\partial z} \tag{2-34c}$$

where v_x, v_y, and v_z are the ground water velocity components in the x, y, and z directions, respectively, and are expressed in relation to the Darcy velocity components as

$$v_x = \frac{q_x}{\varphi_e}, \quad v_y = \frac{q_y}{\varphi_e}, \quad v_z = \frac{q_z}{\varphi_e} \tag{2-35}$$

and the Darcy velocity components are as follows:

$$q_x = K_x I_x, \quad q_y = K_y I_y, \quad q_z = K_z I_z \tag{2-36}$$

where I_x, I_y, and I_z are the hydraulic gradients in the x, y, and z directions, respectively. One should bear in mind that in Eqs. (2-34), D_x, D_y, and D_z are equivalent to D_{xx}, D_{yy}, and D_{zz}, respectively.

2.4.2.2 The Principal Axes of the Dispersion Tensor Can be Defined: the Porous Medium Is Homogeneous and Isotropic and Has a Unidirectional Flow

A homogeneous and isotropic aquifer is schematically defined in Figure 4-7. Consider two points in the aquifer, as shown in the figure, whose coordinates are (x_1, y_1, z_1) for point 1 and (x_2, y_2, z_2) for point 2. If, as shown in Figure 4-7, the hydraulic conductivities satisfy that $K_x(x_1, y_1, z_1) = K_y(x_1, y_1, z_1) = K_z(x_1, y_1, z_1) = K$ and $K_x(x_2, y_2, z_2) = K_y(x_2, y_2, z_2) = K_z(x_2, y_2, z_2) = K$ for points 1 and 2, respectively, the aquifer is said to be a homogeneous and isotropic aquifer.

For a unidirectional flow field, for example, in the x-coordinate direction,

$$I_x = I, \quad I_y = I_z = 0 \tag{2-37}$$

and

$$K_x = K_y = K_z = K \tag{2-38}$$

Eq. (2-36) takes the form

$$q_x = q = KI, \quad q_y = q_z = 0 \tag{2-39}$$

Then, Eq. (2-35) takes the form

$$v_x = U = \frac{q}{\varphi_e}, \quad v_y = v_z = 0 \tag{2-40}$$

Substitution of Eq. (2-40) into Eq. (2-34) gives

$$F_x = \varphi_e U C - \varphi_e D_x \frac{\partial C}{\partial x} \tag{2-41a}$$

$$F_y = -\varphi_e D_y \frac{\partial C}{\partial y} \tag{2-41b}$$

$$F_z = -\varphi_e D_z \frac{\partial C}{\partial z} \tag{2-41c}$$

2.5 DIFFERENTIAL EQUATIONS FOR SOLUTE TRANSPORT IN SATURATED POROUS MEDIA WITHOUT SORPTION AND DECAY

The differential equations for solute transport in Cartesian coordinates are presented in the following sections. Using the convective–dispersive flux components presented in the previous sections, the differential equations for solute transport are derived first for the general case. Equations for the other cases are presented as special cases.

2.5.1 GENERAL CASE: HOMOGENEOUS AND ANISOTROPIC DISPERSION

The differential element in Figure 2-6 is an infinitesimal parallelepiped and its dimensions are dx, dy, and dz. The concentration of a solute is defined as the mass of solute per unit volume of solution. The effective porosity of the medium is φ_e. The total mass of solute per unit time entering the element is

$$\text{entering mass per unit time} = F_x \, dy \, dz + F_y \, dx \, dz + F_z \, dx \, dy \tag{2-42}$$

The total mass of solute per unit time leaving the element is

$$\text{leaving mass per unit time} = \left(F_x + \frac{\partial F_x}{\partial x} \right) dy \, dz + \left(F_y + \frac{\partial F_y}{\partial y} \, dy \right) dx \, dz + \left(F_z + \frac{\partial F_z}{\partial z} \, dz \right) dx \, dy \tag{2-43}$$

where the partial terms indicate the spatial change of the solute mass in the specified direction. The difference of the mass rate entering and leaving the element is

$$\text{difference of mass per unit time} = \left(\frac{\partial F_x}{\partial x} + \frac{\partial F_y}{\partial y} + \frac{\partial F_z}{\partial z} \right) dx \, dy \, dz \tag{2-44}$$

Since the dissolved substance is assumed to be nonreactive, the difference between the flux into the element and the flux out of the element equals the amount of dissolved substance accumulated in the element. The rate of mass change in the element is

$$\text{mass change per unit time} = -\frac{\partial C}{\partial t} \, \varphi_e \, dx \, dy \, dz \tag{2-45}$$

Continuity or conservation of mass requires the per unit volume of the aquifer per unit time to be equal to the rate of solute concentration per unit volume of the aquifer. In other words, the difference of mass per unit time, as given by Eq. (2-44), must be equal to the mass change per unit time, as given by Eq. (2-45):

$$\frac{\partial F_x}{\partial x} + \frac{\partial F_y}{\partial y} + \frac{\partial F_z}{\partial z} = -\varphi_e \frac{\partial C}{\partial t} \tag{2-46}$$

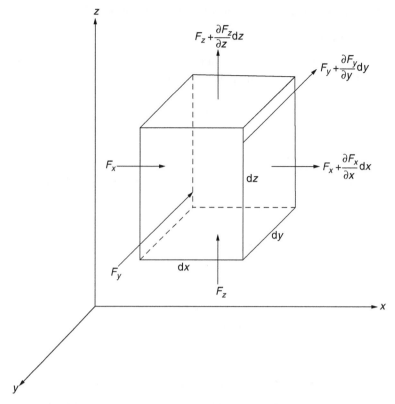

FIGURE 2-6 A differential parallelepiped element in a porous medium showing in and out solute fluxes.

The infinitesimal volume term, $dx\, dy\, dz$, exists in both Eqs. (2-44) and (2-45); therefore, these terms do not exist in Eq. (2-46), which means that Eq. (2-46) is valid for any aquifer volume. Substitution of the convective–dispersive flux components, as given by Eq. (2-31), into Eq. (2-46) gives

$$\varphi_e \frac{\partial C}{\partial t} = \frac{\partial}{\partial x}\left(\varphi_e D_{xx}\frac{\partial C}{\partial x} + \varphi_e D_{xy}\frac{\partial C}{\partial y} + \varphi_e D_{xz}\frac{\partial C}{\partial z}\right)$$

$$+ \frac{\partial}{\partial y}\left(\varphi_e D_{yx}\frac{\partial C}{\partial x} + \varphi_e D_{yy}\frac{\partial C}{\partial y} + \varphi_e D_{yz}\frac{\partial C}{\partial z}\right) \tag{2-47}$$

$$+ \frac{\partial}{\partial z}\left(\varphi_e D_{zx}\frac{\partial C}{\partial x} + \varphi_e D_{zy}\frac{\partial C}{\partial y} + \varphi_e D_{zz}\frac{\partial C}{\partial z}\right)$$

$$- \frac{\partial}{\partial z}\left(\varphi_e v_x C\right) - \frac{\partial}{\partial y}\left(\varphi_e v_y C\right) - \frac{\partial}{\partial z}\left(\varphi_e v_z C\right)$$

which is the most general case for solute transport in aquifers without sorption and decay.

2.5.2 Homogeneous and Isotropic Dispersion

For a homogeneous and isotropic medium, the convective–dispersive flux components are given by Eqs. (2-34). Substitution of these expressions into Eq. (2-46), assuming that effective porosity φ_e is not spatially variable, gives

$$\frac{\partial C}{\partial t} = D_x \frac{\partial^2 C}{\partial x^2} + D_y \frac{\partial^2 C}{\partial y^2} + D_z \frac{\partial^2 C}{\partial z^2} - v_x \frac{\partial C}{\partial x} - v_y \frac{\partial C}{\partial y} - v_z \frac{\partial C}{\partial z} \tag{2-48}$$

2.5.3 Homogeneous and Isotropic Medium with Uniform Flow

2.5.3.1 Three-Dimensional Case

In this case, the convective–dispersive flux components are given by Eq. (2-41). Substitution of these expressions into Eq. (2-46), assuming that porosity φ_e is not spatially variable, gives

$$\frac{\partial C}{\partial t} = D_x \frac{\partial^2 C}{\partial x^2} + D_y \frac{\partial^2 C}{\partial y^2} + D_z \frac{\partial^2 C}{\partial z^2} - U \frac{\partial C}{\partial x} \tag{2-49}$$

2.5.3.2 Two-Dimensional Case

In the case of two-dimensional solute transport, say in the x–y plane, Eq. (2-49) takes the form

$$\frac{\partial C}{\partial t} = D_x \frac{\partial^2 C}{\partial x^2} + D_y \frac{\partial^2 C}{\partial y^2} - U \frac{\partial C}{\partial x} \tag{2-50}$$

2.5.3.3 One-Dimensional Case

In the case of one-dimensional solute transport in the x-coordinate direction, which is also the unidirectional ground water flow direction, Eq. (2-50) takes the form

$$\frac{\partial C}{\partial t} = D_x \frac{\partial^2 C}{\partial x^2} - U \frac{\partial C}{\partial x} \tag{2-51}$$

2.6 DIFFERENTIAL EQUATIONS FOR SOLUTE TRANSPORT IN SATURATED POROUS MEDIA UNDER SORPTION AND DECAY CONDITIONS

The differential equations presented in Section 2.5 are based on the assumptions that (a) a mass transfer for a solute constituent does not exist between the liquid and solid portions of the groundwater system and (b) the constituent does not degrade during its transport. In the following sections, the differential equations given in Section 2.5 will be modified to include these effects.

2.6.1 Solute Transport Equations under Sorption Conditions

Sorption is the process of mass transfer between the liquid and solid portions of a ground water system. In order to account for sorption effects, modifications are necessary in the solute transport differential equations that are presented in Section 2.5. As will be shown below, under sorption conditions, only the temporal term $\partial C/\partial t$ in the above differential equations will be affected. In other words, its effects are the same in one-, two-, and three-dimensional solute transport cases. Therefore, inclusion of the sorption effects will be shown on the one-dimensional solute transport equation as given by Eq. (2-51).

2.6.1.1 One-Dimensional Solute Transport Equation with Sorption and Desorption

Definition of Sorption and Desorption and their Incorporation into the One-Dimensional Equation
The one-dimensional convective–dispersive solute transport equation for homogeneous media, as given by Eq. (2-51), modified to account for the effect of chemical reactions, can be written as (see, e.g., Gillham and Cherry, 1982b)

$$\frac{\partial C}{\partial t} = D_x \frac{\partial^2 C}{\partial x^2} - U \frac{\partial C}{\partial x} - G \tag{2-52}$$

where G ($M/L^3/T$) is a source–sink term that represents the rate at which the dissolved species is removed from solution. In other words, G represents the mass of solute per unit volume of solution per unit time.

When a porous medium is considered under sorption conditions, an additional term must be included in Eq. (2-51), as shown in Eq. (2-52), in order to account for the interaction between the solute constituent and the grains of the porous medium. This term is expressed as (see, e.g., van Genuchten, 1974; Gillham and Cherry, 1982b)

$$G = \frac{\rho_b}{\varphi} \frac{\partial S}{\partial t} \tag{2-53}$$

where S is the concentration of the solute in the solid phase (M/M), ρ_b (M/L^3) the dry bulk density of the porous medium, and φ the total porosity. Assuming that (1) the chemical processes occur rapidly relative to the flow rate and chemical equilibrium is achieved, and (2) under isothermal conditions the concentration of the solute in the solution is a function only of the concentration in the solid phase, it follows that

$$\frac{\partial S}{\partial t} = \frac{\partial S}{\partial C} \frac{\partial C}{\partial t} \tag{2-54}$$

Substitution of Eq. (2-54) into Eq. (2-53) yields

$$G = \frac{\rho_b}{\varphi} \frac{\partial S}{\partial C} \frac{\partial C}{\partial t} \tag{2-55}$$

and combining of Eqs. (2-52) and (2-55) gives

$$\left(1 + \frac{\rho_b}{\varphi} \frac{\partial S}{\partial C}\right) \frac{\partial C}{\partial t} = D_x \frac{\partial^2 C}{\partial x^2} - U \frac{\partial C}{\partial x} \tag{2-56}$$

For solute constituents that undergo transfer between the liquid and solid phases of a ground water system, the equilibrium partitioning of the species between the two phases is commonly measured by batch tests. In batch tests, a specified volume of solution is brought in contact with a known mass of soil for enough time to allow equilibrium partitioning of the solute species to occur between the liquid and solid phases. From these test results, graphs of the concentration in solution C against the concentration on the solids S can be obtained. These graphs are known as *isotherms*. In Eq. (2-56), $\partial S/\partial C$ is the slope of the isotherm. Various isotherm forms and their mathematical representations are described by Golubev and Garibyants (1971) and Smith (1970) (Gillham and Cherry, 1982b).

The reversal of the sorption process is called desorption. Solute molecules sorbed to a solid phase may be released back into solution as a result of changes in system conditions.

2.6.1.2 Freundlich Isotherm

There are a variety of isotherm equations, but most of the models that are used in practice are based on the Freundlich isotherm and are in the following form:

$$S = KC^N \tag{2-57}$$

where K and N are constants. Substitution of Eq. (2-57) into Eq. (2-56) allows the one-dimensional solute transport equation to be written in terms of one variable, C:

$$\frac{\rho_b}{\varphi} KN \frac{\partial C}{\partial t} C^{N-1} + \frac{\partial C}{\partial t} = D_x \frac{\partial^2 C}{\partial x^2} - U \frac{\partial C}{\partial x} \tag{2-58}$$

or

$$R_d \frac{\partial C}{\partial t} = D_x \frac{\partial^2 C}{\partial x^2} - U \frac{\partial C}{\partial x} \tag{2-59}$$

where

$$R_d = 1 + \frac{\rho_b}{\varphi} KNC^{N-1} \qquad (2\text{-}60)$$

2.6.1.3 Linear Sorption Isotherm ($N = 1$) and Distribution Coefficient: The K_d Approach

For most solute constituent cases, the isotherm, given by Eq. (2-57), is linear, which means that $N = 1$. For this special case, K_d is used instead of K and Eq. (2-57) takes the form

$$S = K_d C \qquad (2\text{-}61)$$

Therefore, Eq. (2-60) takes the form

$$R_d = 1 + \frac{\rho_b}{\varphi} K_d \qquad (2\text{-}62)$$

According to Eq. (2-61), the S vs. C data will be a straight line on an arithmetic plot (see Figure 2-9). From Eq. (2-61) we get

$$K_d = \frac{\Delta S}{\Delta C} \qquad (2\text{-}63)$$

which means that once the S vs. C data are plotted as shown in Figure 2-7, the value of K_d can be determined graphically for a given solute constituent.

The proportionality parameter K_d is known as the distribution coefficient and is widely used in ground water contamination studies. This parameter is a representation of the partitioning between the liquid and solid phases only if the reactions of the partitioning are fast and reversible and only the isotherm is linear. Fortunately, many contaminant constituents of interest in ground water studies satisfy these requirements. This approach will be called the K_d approach. Note that if $K_d = 0$ in Eq. (2-62), which means no sorption, R_d becomes equal to unity. Thus, Eq. (2-59) becomes the same as Eq. (2-51).

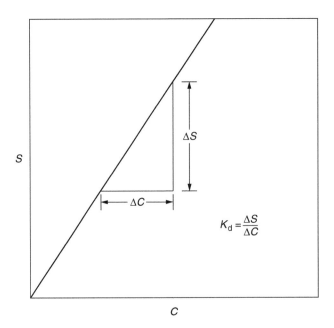

FIGURE 2-7 Linear sorption isotherm with C vs. S.

The distribution coefficient K_d can be expressed as

$$K_d = \frac{\text{mass of solute on the solid phase per unit mass of solid phase}}{\text{concentration of solute and solution}} \quad (2\text{-}64)$$

or in dimensional form,

$$[K_d] = \frac{\text{M/M}}{\text{M/L}^3} = \frac{\text{L}^3}{\text{M}} \quad (2\text{-}65)$$

In practice, the following units are used: the first one is

$$K_d = \frac{\mu g \text{ (sorbed solute mass)/g (mass of solid)}}{\mu g/\text{mL (aqueous concentration)}} = \text{mL/g} \quad (2\text{-}66)$$

and the second is

$$K_d = \frac{mg \text{ (sorbed solute mass)/kg (mass of solid)}}{mg/\text{L (aqueous concentration)}} = \text{L/kg} \quad (2\text{-}67)$$

The units for K_d, given by Eqs. (2-66) and (2-67), are widely used in practice. Apart from these, m^3/kg is also used as a unit for K_d.

2.6.1.4 Retardation Factor R_d

The term R_d, as given by Eq. (2-60) or (2-62), is known as the *retardation factor*. The term *retardation coefficient* is also used occasionally. The retardation equation in the form of Eq. (2-62) was introduced in the chemical literature by Mayer and Tompkins (1947) and was first incorporated into the convection–dispersion equation for use in chromatographic studies in chemical engineering by Vermeulen and Hiester (1952). Schematic illustration of the retardation concept is illustrated in Section 8.2.1. The K_d/K_{oc} approach to represent the transport of sorbing contaminants was first reported in the ground water literature by Higgins (1959), Nelson (1959), and Hashimoto (1964). For further details on estimation procedures of K_d, the reader should refer to Chapter 8.

2.6.1.5 Representative Parameter Values in the Retardation Factor (R_d) Equation

The dry bulk density ρ_b of the porous medium can also be expressed in terms of the density ρ_r of aquifer grains as

$$\rho_b = (1-\varphi)\rho_r \quad (2\text{-}68)$$

For unconsolidated granular deposits, the porosity φ, expressed as a fraction, generally ranges from 0.20 to 0.40. The average mass density ρ_r of minerals or the aquifer grains is approximately 2.65 g/cm^3 (Freeze and Cherry, 1979). Therefore, from Eq. (2-68), the range of bulk mass densities can be determined as

$$\rho_{b_{min}} = (1-\varphi_{max})\rho_r = (1-0.40)(2.65 \text{ g/cm}^3) = 1.59 \text{ g/cm}^3$$

$$\rho_{b_{max}} = (1-\varphi_{min})\rho_r = (1-0.20)(2.65 \text{ g/cm}^3) = 2.12 \text{ g/cm}^3$$

Using these values, from Eq. (2-62), the minimum and maximum retardation factor expression values can be written respectively as

$$R_{d_{min}} = 1 + \frac{\rho_{b_{min}}}{\varphi_{max}}K_d = 1 + \frac{1.59 \text{ g/cm}^3}{0.40}K_d = 1 + 3.98 \text{ (g/cm}^3)K_d \quad (2\text{-}69)$$

$$R_{d_{max}} = 1 + \frac{\rho_{b_{max}}}{\varphi_{min}}K_d = 1 + \frac{2.12 \text{ g/cm}^3}{0.20}K_d = 1 + 10.60 \text{ (g/cm}^3)K_d \quad (2\text{-}70)$$

EXAMPLE 2-2

Based on field tests, the K_d values of the three halogenated organic compounds in a sandy unconfined aquifer are given in Table 2-2 (Gillham et al., 1990). Based on these values, (a) determine the minimum and maximum retardation factor (R_d) values of each constituent and (b) discuss the results.

SOLUTION

(a) For tetrachloroethylene, substitution of its K_d value from Table 2-2 into Eq. (2-69) gives the minimum retardation factor value

$$R_{d_{min}} = 1 + (3.98 \text{ g/cm}^3) (0.20 \text{ cm}^3/\text{g}) = 1.80$$

Similarly, with the same K_d value, Eq. (2-70) gives

$$R_{d_{max}} = 1 + (10.60 \text{ g/cm}^3) (0.20 \text{ cm}^3/\text{g}) = 3.12$$

The values given above are listed in Table 2-3. The R_d values for 1,2-dichloroethylene and hexachloroethane can be calculated similarly and their values are listed in Table 2-3. (b) The R_d values listed in Table 2-3 should be viewed as values that are based on some extreme case scenarios. For example, as can be seen from Eq. (2-69), the minimum R_d value is based on the maximum porosity value and the maximum R_d value is based on the minimum porosity value (see Eq. [2-70]). The results in Table 2-3 show that the maximum R_d value of a constituent is almost twice the minimum R_d value of the same constituent.

2.6.1.6 Two- and Three-Dimensional Solute Transport Equations with Sorption and Desorption

As can be seen from Eq. (2-59), only its left term, $\partial C/\partial t$, changes to $R_d \, \partial C/\partial t$. Therefore, only this type of modification needs to be made in all two- and three-dimensional solute transport differential equations as given by Eqs. (2-47)–(2-50).

TABLE 2-2
Distribution Coefficient (K_d) Data for Example 2-2

Constituent	Distribution Coefficient K_d (cm^3/g)
Tetrachloroethylene	0.20
1,2-Dichlorobenzene	0.29
Hexachloroethane	0.31

TABLE 2-3
Minimum $R_{d_{min}}$ and Maximum $R_{d_{max}}$ Values for the Constituents in Example 2-2

Constituent	Distribution Coefficient K_d (cm^3/g)	$R_{d_{min}}$	$R_{d_{max}}$
Tetrachloroethylene	0.20	1.80	3.12
1,2-Dichlorobenzene	0.29	2.15	4.07
Hexachloroethane	0.31	2.23	4.29

2.6.2 Solute Transport Equations under Degradation Conditions

The term "degradation" is a general term and covers both radioactive decay and biodegradation. The general principles of radioactive decay and biodegradation as well as their incorporation into the solute transport differential equations are presented below.

2.6.2.1 Radioactive Decay

Radioactive Decay Equation
All pure radioactive substances decay in accordance with the same law. For a radioactive substance, experiments have shown that

$$\Delta C = -\nu C \Delta t \tag{2-71}$$

where C is the mass concentration (M/L^3) of material still undecayed, ΔC the mass concentration (M/L^3) that decays in a time interval Δt (T), and ν (T^{-1}) the decay constant. The decay constant is specific to each radioactive solute constituent. In the limiting case, i.e., $\Delta t \to 0$, Eq. (2-71) takes the form (Bear, 1972)

$$\frac{\partial C}{\partial t} = -\nu C \tag{2-72}$$

Decay Constant and Half-Life
The activity, which is the number of disintegrations per second of a radioactive substance, decreases with time. Each radioactive isotope has its own characteristic rate of decrease. If the logarithm of activity is plotted as a function of time, a straight line results, which shows that the activity is an exponential function of time. The rate of decomposition is proportional to the amount of radioactive substance present and is given by (Mercer et al., 1982)

$$N = N_0 e^{-\nu t} \tag{2-73}$$

where N is the number of radioactive atoms remaining after time t and N_0 is the original number at time $t = 0$. As defined above, the coefficient ν is the decay constant or disintegration constant and has units of T^{-1}.

Information for a radioactive constituent is usually given in terms of radioisotope half-lives, instead of ν. The relationship between half-life $t_{1/2}$ and decay constant ν is straightforward. The half-life $t_{1/2}$ is defined as the time at which the number of radioactive atoms remaining is just one half of the original number, i.e., $N = N_0/2$. Therefore, Eq. (2-73) can be evaluated as

$$\frac{N}{N_0} = \frac{1}{2} = e^{-\nu t_{1/2}} \tag{2-74}$$

or taking the natural logarithms (ln) of both sides gives

$$\nu = \frac{0.693}{t_{1/2}} \tag{2-75}$$

With the known value of half-life $t_{1/2}$ for a radioactive substance, the decay constant ν can be determined from Eq. (2-75).

Decay constants are generally reported in terms of half-lives, mostly in years. Table 2-4 gives half-lives of nuclides in Kocher (1981) as presented in Mercer et al. (1982).

EXAMPLE 2-3

The half-life of U-235 is given as $t_{1/2} = 7.0 \times 10^8$ years (see Table 2-4). Determine its decay constant v.

SOLUTION

From Eq. (2-75),

$$v = \frac{0.693}{7.0 \times 10^8 \text{ years}} = 9.9 \times 10^{-6} \text{ year}^{-1} = 2.71 \times 10^{-12} \text{ day}^{-1}$$

TABLE 2-4
Half-Lives and Radioactive Daughter Products (with Half-Lives Exceeding a Few Months) of Selected Radionuclides

Nuclide	Half-Life (years)	Daughter Radionuclide
C-14	5.7×10^3	—
Ni-59	7.5×10^4	—
Se-79	$\leq 6.5 \times 10^4$	—
Sr-90	29	—
Zr-93	1.5×10^{-6}	Nb-93m
Nb-93m	15	—
Tc-99	2.1×10^5	—
Pd-107	6.5×10^6	—
Sn-126	1.0×10^5	—
I-129	1.6×10^7	—
Cs-135	2.3×10^6	—
Cs-137	30	—
Sm-151	90	—
Eu-154	8.8	—
Pb-210	22	Po-210
Po-210	0.38	—
Ra-226	1.6×10^3	Pb-210
Ra-228	5.8	—
Ac-227	22	—
Th-229	7.3×10^3	—
Th-230	7.7×10^4	Ra-226
Th-232	1.4×10^{10}	Ra-228
Pa-231	3.3×10^4	Ac-227
U-233	1.6×10^5	Th-229
U-234	2.4×10^5	Th-230
U-235	7.0×10^8	Pa-231
U-238	4.5×10^9	U-234
Np-237	2.1×10^6	U-233
Pu-238	88	U-234
Pu-239	2.4×10^4	U-235
Pu-240	6.6×10^3	U-236
Am-241	4.3×10^2	Np-237
Am-243	7.4×10^3	Pu-239

Source: From Kocher, D.C., *Radioactive Decay Data Tables*, U.S. Department of Energy Technical Information Center, Oak Ridge, Tennessee, 1981. With permission; as presented in Mercer, J.W. et al., NUREG/CR-3066, Washington, DC, 1982.

2.6.2.2 Biodegradation

Biological transformation or biodegradation may influence the behavior of contaminants in ground water. A broad range of organic compounds such as pesticides, halogenated hydrocarbons, aromatic hydrocarbons, amines, alcohols, and others have shown biodegradation based on laboratory studies of soils, sediments, and waters (Cherry et al., 1984). The information available in the pertinent literature is generally limited to the biodegradation rates of these compounds with the exception of pesticides. Therefore, here only literature information on pesticides biodegradation will be presented.

Pesticide Biodegradation in Soils
Pesticides stand out as one of the major developments of the 20th century. There is no doubt that pesticides have improved longevity and quality of life, chiefly in the area of public health. Unfortunately, pesticides are poisons and can be particularly dangerous when misused. Most pesticides break down or degrade over time as a result of several chemical and microchemical reactions in soils. Sunlight breaks down some pesticides as well. Some pesticides produce intermediate substances, called "metabolites," as they degrade. The biological activity of these substances may also have environmental significance. Therefore, agricultural producers as well as environmental scientists are concerned with the fate of pesticides in agroecosystems (Rao et al., 1983). Some general principles of pesticide degradation are presented below based on the studies of Rao and Davidson (1980).

Three Important Phases of Pesticide Degradation
There are three important phases of pesticide degradation to be considered: (1) the causes of degradation; (2) the pathways of degradation; and (3) the rates of degradation. Photochemical, chemical, and microbiological transformations are the principal causes of pesticide degradation in soils. Among these causes, photodecomposition may have little practical significance for pesticides located below the ground surface. However, microbiological transformations are more important because they cause an extensive breakdown (to H_2O and CO_2) of the pesticide molecules (Rao and Davidson, 1980). Here, only the microbiological transformation mechanism is discussed.

Aerobic and Anaerobic Degradation
Degradation occurring in an oxygen-rich environment is called aerobic degradation or oxic degradation. If degradation occurs in an oxygen-depleted or -limited environment, it is known as anaerobic degradation or anoxic degradation. The most significant factors determining pesticide degradation rates are (1) soil type, (2) soil water content, (3) pH, (4) temperature, (5) clay, and (6) organic matter content. An increase in the soil pH will generally increase the rate of chemical degradation of many pesticides. The most effective and yet unpredictable influence on pesticide degradation is exerted by the microbiological populations of the soil and the environmental variables that control their activity. The effects of temperature and soil water content have been most intensely studied in this regard. Increased microbial activity and decreased sorption associated with higher temperatures generally enhance pesticide degradation. The half-lives of pesticides are known to increase with a decrease in the soil water content. At high soil water contents (approaching saturation), pesticide degradation rate is determined by the relative rates of decomposition under aerobic versus anaerobic conditions (Rao and Davidson, 1980).

Surface soil generally exists under aerobic conditions and in a highly oxidized state. However, as the soil water content approaches saturation or the soil surface becomes submerged under water, the availability of atmospheric oxygen becomes limited. Also, the oxygen trapped in the soil is quickly depleted by the metabolic activity of the aerobic soil microorganisms. As a result, within a short period of time, the soil environment becomes anaerobic. The rate of degradation and the

TABLE 2-5
Degradation Rate Coefficients and Half-Lives for Several Pesticides
Under Laboratory and Field Conditions

Pesticide	Test Condition	Rate Coefficient v (day $^{-1}$)		Half-Life $t_{1/2}$ (days)	
		Mean	% CV[a]	Mean	% CV[a]
(A) *Herbicides*					
2,4-D	Laboratory[b]	0.066	74.2	16	56.25
	Laboratory	0.051	23.5	15	33.3
	Field	3.6	83.3	5	100.0
2,4,5-T	Laboratory	0.029	51.7	33	66.7
	Laboratory[b]	0.035	82.9	16	68.8
Atrazine	Laboratory[b]	0.019	47.4	48	68.8
	Laboratory	0.0001	70.4	6900	71.5
	Field	0.042	33.3	20	50.0
Simazine	Laboratory[b]	0.014	71.4	75	73.3
	Field	0.022	95.5	64	93.8
Trifluralin	Laboratory[b]	0.008	65.5	132	82.6
	Laboratory[b] (Anaerobic)	0.025	—	28	—
	Laboratory (Chain)	0.0013	—	544	—
	Field	0.02	65.0	46	41.3
Bromacil	Laboratory[b]	0.0077	49.4	106	42.5
	Laboratory	0.0024	116.2	901	116.2
	Field	0.0038	100.0	349	76.8
Terbacil	Laboratory[b]	0.015	33.3	50	26.0
	Laboratory	0.0045	124.0	679	124.5
	Field	0.006	55.0	175	88.6
Linuron	Laboratory[b]	0.0096	19.8	75	18.7
	Field	0.0034	41.2	230	29.3
Diuron	Laboratory	—	—	—	—
	Field	0.0031	58.1	328	64.6
Dicamba	Laboratory[b]	0.022	80.2	14	85.7
	Laboratory (Ring)	0.0022	—	309	—
	Laboratory (Chain)	0.0044	—	147	—
	Field	0.093	16.1	8	12.5
Picloram	Laboratory[b]	0.0073	58.9	138	67.4
	Laboratory	0.0008	111.3	8600	184.2
	Field	0.033	51.5	31	77.4
Dalapon	Laboratory[b]	0.047	—	15	—
TCA	Laboratory[b]	0.059	103.4	46	119.6
	Field	0.073	—	22	—
Glyphosate	Laboratory[b]	0.1	121.0	38	139.5
	Laboratory	0.0086	93.0	903	191.8
Paraquat	Laboratory[b]	0.0016	—	487	—
	Field	0.00015	—	4747	—
(B) *Insecticides*					
Parathion	Laboratory[b]	0.029	48.3	35	82.3
	Field	0.057	101.8	18	44.4

(continued)

TABLE 2-5 (*Continued*)

Pesticide	Test Condition	Rate Coefficient v (day^{-1})		Half-Life $t_{1/2}$ (days)	
		Mean	% CV[a]	Mean	% CV[a]
Methyl	Laboratory[b]	0.16	—	4	—
Parathion	Field	0.046	—	15	—
Diazinon	Laboratory[b]	0.023	108.7	48	62.5
	Laboratory	0.022	—	32	—
Fonofos	Laboratory[b]	0.012	—	60	—
Malathion	Laboratory[b]	1.4	71.4	0.8	87.5
Phorate	Laboratory[b]	0.0084	—	82	—
	Field	0.01	30.0	7.5	24.0
Carbofuran	Laboratory[b]	0.047	87.2	37	94.6
	Laboratory	0.0013	—	535	—
	Laboratory[b] (Anaerobic)	0.026	50.0	44	95.4
	Field	0.016	87.5	68	61.8
Carbaryl	Laboratory[b]	0.037	56.8	22	40.9
	Laboratory (Chain)	0.0063	101.6	309	91.9
	Field	0.10	79.2	12	91.7
DDT	Laboratory[b]	0.00013	130.8	1657	98.3
	Laboratory[b] (Anaerobic)	0.0035	82.9	692	123.4
Aldrin and	Laboratory[b]	0.013	—	53	1
dieldrin	Field	0.0023	100.0	1237	198.4
Endrin	Laboratory[b] (anaerobic)	0.03	53.3	31	61.3
	Field (Aerobic)	0.0015	—	460	—
	Field (Anaerobic)	0.0053	—	130	—
Chlordane	Field	0.0024	104.2	1214	202.1
Heptachlor	Laboratory[b]	0.011	—	63	—
	Field	0.0046	119.6	426	82.6
Lindane	Laboratory[b]	0.0026	—	266	—
	Laboratory (Anaerobic)	0.0046	—	151	—
(C) *Fungicides*					
PCP	Laboratory[b]	0.02	60.0	48	60.4
	Laboratory (Anaerobic)	0.07	44.3	15	100.0
	Field	0.05	—	14	—
Captan	Field	0.231	—	3	—

[a] Coefficient of variation.

[b] These rates are based on the disappearance of solvent-extractable parent compound under aerobic incubation conditions, unless stated otherwise.

Source: From Ou, L.T. et al., *Retention and Transformation of Pesticides and Phosphorous in Soil–Water Systems: A Review of Data Base*, U.S. EPA, 1980. With permission; as presented in Rao, P.S.C. and Davidson, J.M., in *Environmental Impact of Nonpoint Source Pollution*, Overcash, M.R. and Davidson, J.M., Eds., Ann Arbor Science Publishers, Inc., Ann Arbor, MI, 1980, pp. 23–67.

degradation pathway of pesticides under anaerobic conditions may be different from that under aerobic conditions (Rao and Davidson, 1980).

Pesticide Degradation Equations
The two basic types of equations that were considered are as follows (Hamaker, 1972; Rao and Davidson, 1980):

$$\frac{\partial C}{\partial t} = -vC^n \tag{2-76}$$

$$\frac{\partial C}{\partial t} = -V_{max}\left(\frac{C}{\alpha + C}\right) \tag{2-77}$$

where C is the concentration (μg/mL), t the time (days), v the kinematic rate coefficient (day^{-1}), n the order of the reaction (dimensionless), V_{max} the maximum rate, and α a constant (μg/mL). It should be noted that when $n = 1$, Eq. (2-76) turns to be exactly the same as Eq. (2-72), which is a first-order kinetic equation.

Pesticide Degradation Rates Under Field Conditions
Ou et al. (1980) have conducted an extensive literature search to develop a database for first-order rate coefficients v and half-lives $t_{1/2}$ for pesticide degradation in soils. A summary of their data is listed in Table 2-5, which is also presented in Rao and Davidson (1980). The v and $t_{1/2}$ values computed using field data are based on the disappearance of the parent compound. Values of v and $t_{1/2}$ calculated on the basis of mineralization ($^{14}CO_2$ evolution) as well as parent compound disappearance under laboratory conditions are also included. Unless otherwise specified, the laboratory studies were performed under aerobic conditions and involved measurements of $^{14}CO_2$ evolution from ^{14}C-labeled compounds. In Table 2-5, the means and coefficient of variation values for v and $t_{1/2}$ are also shown. As can be seen from Table 2-5, in most cases, the half-lives of pesticides under field conditions are much smaller than those under laboratory conditions. This can be attributed to the fact that under field conditions, a number of factors and processes contribute to pesticide disappearance, while laboratory studies are generally designed to study only one of these processes. Table 2-5 shows that the coefficient of variation (%CV) values are less than 100 in a majority of the cases. As a result, the degradation rate can be estimated within a factor of 2 to 4 for most pesticides using the available data.

2.7 GENERALIZED SOLUTE TRANSPORT DIFFERENTIAL EQUATIONS FOR SATURATED POROUS MEDIA

In Section 2.5, multidimensional solute transport differential equations are presented without including sorption and decay effects. In this section, the generalized forms of the same equations are presented by including the sorption and decay effects.

2.7.1 GENERAL CASE: HOMOGENEOUS AND ANISOTROPIC DISPERSION

As mentioned in Section 2.6.1, under sorption conditions, only the term $\partial C/\partial t$ in Eq. (2-46) changes to $R_d \partial C/\partial t$ (see Eq. [2-59]). The effects of decay are represented by Eq. (2-72). Incorporating these into Eq. (2-46) gives

$$\frac{\partial F_x}{\partial x} + \frac{\partial F_y}{\partial y} + \frac{\partial F_z}{\partial z} = -\varphi_e R_d \frac{\partial C}{\partial t} - \varphi_e v R_d C \tag{2-78}$$

Substitution of the convective–dispersive flux components, as given by Eqs. (2-31a), (2-31b), and (2-31c) into Eq. (2-78) gives

$$\varphi_e R_d \frac{\partial C}{\partial t} = \frac{\partial}{\partial x}\left(\varphi_e D_{xx}\frac{\partial C}{\partial x} + \varphi_e D_{xy}\frac{\partial C}{\partial y} + \varphi_e D_{xz}\frac{\partial C}{\partial z}\right)$$

$$+ \frac{\partial}{\partial y}\left(\varphi_e D_{yx}\frac{\partial C}{\partial x} + \varphi_e D_{yy}\frac{\partial C}{\partial y} + \varphi_e D_{yz}\frac{\partial C}{\partial z}\right) \qquad (2\text{-}79)$$

$$+ \frac{\partial}{\partial z}\left(\varphi_e D_{zx}\frac{\partial C}{\partial x} + \varphi_e D_{zy}\frac{\partial C}{\partial y} + \varphi_e D_{zz}\frac{\partial C}{\partial z}\right)$$

$$- \frac{\partial}{\partial x}(\varphi_e v_x C) - \frac{\partial}{\partial y}(\varphi_e v_y C) - \frac{\partial}{\partial z}(\varphi_e v_z C) - \varphi_e v R_d C$$

which is the most general case for solute transport in aquifers.

2.7.2 HOMOGENEOUS AND ISOTROPIC DISPERSION

For a homogeneous and isotropic medium, the convective–dispersive flux components are given by Eqs. (2-34). Substitution of these expressions into Eq. (2-78), assuming that the effective porosity φ_e is not spatially variable, gives

$$R_d \frac{\partial C}{\partial t} = D_x \frac{\partial^2 C}{\partial x^2} + D_y \frac{\partial^2 C}{\partial y^2} + D_z \frac{\partial^2 C}{\partial z^2} - v_x \frac{\partial C}{\partial x} - v_y \frac{\partial C}{\partial y} - v_z \frac{\partial C}{\partial z} - v R_d C \qquad (2\text{-}80)$$

2.7.3 HOMOGENEOUS AND ISOTROPIC MEDIUM WITH UNIFORM FLOW FIELD

2.7.3.1 Three-Dimensional Case

In this case, the convective–dispersive flux components are given by Eq. (2-41). Substitution of these expressions into Eq. (2-78), assuming that porosity φ_e is not spatially variable, gives

$$R_d \frac{\partial C}{\partial t} = D_x \frac{\partial^2 C}{\partial x^2} + D_y \frac{\partial^2 C}{\partial y^2} + D_z \frac{\partial^2 C}{\partial z^2} - U \frac{\partial C}{\partial x} - v R_d C \qquad (2\text{-}81)$$

2.7.3.2 Two-Dimensional Case

In the case of two-dimensional solute transport, i.e., in the x–y plane, Eq. (2-81) takes the form

$$R_d \frac{\partial C}{\partial t} = D_x \frac{\partial^2 C}{\partial x^2} + D_y \frac{\partial^2 C}{\partial y^2} - U \frac{\partial C}{\partial x} - v R_d C \qquad (2\text{-}82)$$

2.7.3.3 One-Dimensional Case

In this case for one-dimensional solute transport in the x-coordinate direction, which is also the unidirectional ground water flow direction, Eq. (2-82) takes the form

$$R_d \frac{\partial C}{\partial t} = D_x \frac{\partial^2 C}{\partial x^2} - U \frac{\partial C}{\partial x} - v R_d C \qquad (2\text{-}83)$$

2.8 SOLUTE TRANSPORT DIFFERENTIAL EQUATIONS FOR UNSATURATED POROUS MEDIA UNDER UNIFORM FLOW CONDITIONS

Solute transport in unsaturated porous media under transient flow conditions is a relatively more complex phenomenon compared with that in saturated porous media, and solutions to this kind of problems can be found only by numerical techniques (see, e.g., Warrick et al., 1971; Bresler, 1973, 1975).

Here, the main focus is on the solute transport equations under uniform unsaturated flow conditions. It is shown that if the unsaturated flow field is assumed to be uniform, the same form of solute transport equations for saturated porous media can also be used for unsaturated porous media as well.

2.8.1 UNSATURATED POROUS MEDIA FLOW EQUATIONS UNDER UNIFORM FLOW CONDITIONS

For a three-dimensional unsaturated porous media flow, with the x-axis chosen positively downward, the Darcy velocity components in x-, y-, and z-coordinate directions can, respectively, be written as (e.g., Philip, 1969; Batu, 1982)

$$u = -K(h)\frac{\partial h}{\partial x} + K(h) \tag{2-84a}$$

$$v = -K(h)\frac{\partial h}{\partial y} \tag{2-84b}$$

$$w = -K(h)\frac{\partial h}{\partial z} \tag{2-84c}$$

where h is the soil water pressure head and $K(h)$ the unsaturated hydraulic conductivity. Let us assume that there is a uniform infiltration rate u_0 at the soil surface and

$$u_0 \leq K_s \tag{2-85}$$

where K_s is the saturated hydraulic conductivity of the unsaturated zone. Whenever the flow system reaches a steady state, the water content θ will be uniform throughout the unsaturated zone. Consequently, h will be constant and the same at every point in the unsaturated zone. This uniform soil water pressure head will be given by h_u. As a result, the partial derivatives of h with respect to x, y, and z will be zero. Therefore, from Eqs. (2-84a)–(2-84c) one can write

$$u = K(h_u), \quad v = 0, \quad w = 0 \tag{2-86}$$

Let θ_u be the water content corresponding to h_u. Therefore, from Eq. (2-86),

$$u_0 = K(\theta_u) \tag{2-87}$$

If u_0 (the infiltration rate at the soil surface) is given, θ_u can be determined from the $K = K(\theta)$ relationship. Then, the ground water velocity can be determined as

$$U = \frac{u_0}{\theta_u} = \frac{K(\theta_u)}{\theta_u} \tag{2-88}$$

2.8.2 SOLUTE TRANSPORT DIFFERENTIAL EQUATIONS FOR UNSATURATED POROUS MEDIA UNDER UNIFORM FLOW CONDITIONS

For the coordinate system and flow conditions described in Section 2.8.1, an equation of continuity for three-dimensional solute transport can be written as (e.g., Warrick et al., 1971; Bresler, 1975)

$$\frac{\partial(\theta C)}{\partial t} = \frac{\partial}{\partial x}\left(\theta D_x \frac{\partial C}{\partial x}\right) + \frac{\partial}{\partial y}\left(\theta D_y \frac{\partial C}{\partial y}\right) + \frac{\partial}{\partial z}\left(\theta D_z \frac{\partial C}{\partial z}\right) - \frac{\partial(uC)}{\partial x} \tag{2-89}$$

which is valid for one-dimensional flow in the x coordinate direction and three-dimensional solute transport. For $\theta = \theta_u$ and $u = u_0$, Eq. (2-89) takes the form

$$\frac{\partial C}{\partial t} = D_x \frac{\partial^2 C}{\partial x^2} + D_y \frac{\partial^2 C}{\partial y^2} + D_z \frac{\partial^2 C}{\partial z^2} - U \frac{\partial C}{\partial x} \tag{2-90}$$

which is exactly the same as Eq. (2-49) in the case of saturated solute transport. Equation (2-90) can similarly be extended to solute transport cases to include sorption and decay effects as in Eq. (2-81) under uniform unsaturated flow conditions. One must bear in mind that in the R_d equation (2-62), φ_e must be replaced by θ_u.

2.9 VOLUME- AND FLUX-AVERAGED CONCENTRATIONS UNDER UNIFORM FLOW CONDITIONS

2.9.1 RELATIONSHIP BETWEEN VOLUME- AND FLUX-AVERAGED CONCENTRATIONS UNDER UNIFORM FLOW CONDITIONS

Definitions of the volume-averaged (resident) concentration and flux-averaged (flowing) concentration are given in Section 2.2. Several researchers have discussed the importance of distinguishing between volume- (C_r) and flux-averaged (C_f) concentrations during one-dimensional (Brigham, 1974; Kreft and Zuber, 1978; Parker and van Genuchten, 1984) and two-dimensional (Batu and van Genuchten, 1990) solute transport. For a uniform ground water flow field, the convective–dispersive flux component in the x-coordinate direction is given by Eq. (2-41a) in terms of volume-average concentration C_r. Equation (2-41a) can also be written as

$$\frac{F_x}{\varphi_e U} = C_r - \frac{D_x}{U}\frac{\partial C_r}{\partial x} \tag{2-91}$$

and, using Eq. (2-17), it takes the form

$$C_f = C_r - \frac{D_x}{U}\frac{\partial C_r}{\partial x} \tag{2-92}$$

Equation (2-92) gives the flux-averaged concentration C_f in terms of the volume-averaged concentration C_r. This equation can be used immediately to derive the flux-averaged solution (first-type condition solution) from the volume-averaged solution (third-type condition solution) (refer Section 2.10 for the first- and third-type conditions). Alternatively, the third-type solution may be derived from the first-type solution by using the inverse of Eq. (2-92) (van Genuchten et al., 1984, Eqs. [74a] and [74b]):

$$C_r(x,y,z,t) = \frac{U}{D_x}\exp\left(\frac{Ux}{D_x}\right)\int_0^\infty \exp\left(-\frac{U\xi}{D_x}\right)C_f(\xi,y,z,t)\,\mathrm{d}\xi \tag{2-93}$$

It must be emphasized that Eqs. (2-92) and (2-93) are the relationships between the first-and third-type source conditions based on analytical solutions of Eq. (2-81). These relationships only hold for simplified solute transport problems, i.e., for unidirectional steady flow perpendicular to the input boundary at $x = 0$. In more general situations (including radial flow), additional terms will be generated when Eq. (2-92) is used to derive the flux-averaged concentration from the volume-averaged concentration (Sposito and Barry, 1987). As a result, first-type analytical solutions do not always represent flux-averaged concentrations in two- and three-dimensional solute transport domains.

2.9.2 DIFFERENTIAL EQUATION FOR FLUX-AVERAGED CONCENTRATIONS UNDER UNIFORM FLOW CONDITIONS

The transformation given by Eq. (2-92) is shown to leave the one-dimensional solute transport equation invariant, indicating that the one-dimensional solute transport equation can be formulated in terms of both C_r (volume-averaged concentration) and C_f (flux-averaged concentration) (Kreft and Zuber, 1979; Parker and van Genuchten, 1984). Equation (2-92) also applies to the two-dimensional case and this is shown by Batu and van Genuchten (1990). Equation (2-92) also applies to the three-dimensional case. This is shown below.

2.9.2.1 Invariance of C_r and C_f-Based Solute Transport Differential Equations

Equation (2-81), and its two- (Eq. [2-82]) and one-dimensional (Eq. [2-83]) forms, are all based on volume-averaged concentration, which is defined in Section 2.2.1. Equation (2-81) can also be expressed as

$$R_d \frac{\partial C_r}{\partial t} = D_x \frac{\partial^2 C_r}{\partial x^2} + D_y \frac{\partial^2 C_r}{\partial y^2} + D_z \frac{\partial^2 C_r}{\partial z^2} - U \frac{\partial C_r}{\partial x} - \nu R_d C_r \tag{2-94}$$

The invariance of Eq. (2-94) can be easily shown by rewriting the equation as

$$R_d \frac{\partial C_r}{\partial t} = D_x \frac{\partial^2 C_r}{\partial x^2} + D_y \frac{\partial^2 C_r}{\partial y^2} + D_z \frac{\partial^2 C_r}{\partial z^2} - U \frac{\partial C_r}{\partial x} - \nu R_d C_r$$

$$+ \frac{D_x}{U} \frac{\partial}{\partial x} \left(R_d \frac{\partial C_r}{\partial t} - D_x \frac{\partial^2 C_r}{\partial x^2} - D_y \frac{\partial^2 C_r}{\partial y^2} - D_z \frac{\partial^2 C_r}{\partial z^2} + U \frac{\partial C_r}{\partial x} + \nu R_d C_r \right) \tag{2-95}$$

Note that the last term (in brackets) is identical to zero as predicted by Eq. (2-94). Combining the terms in Eq. (2-95) gives

$$R_d \frac{\partial}{\partial t} \left(C_r - \frac{D_x}{U} \frac{\partial C_r}{\partial x} \right) = \left(D_x \frac{\partial^2 C_r}{\partial x^2} - \frac{D_x^2}{U} \frac{\partial^3 C_r}{\partial x^3} \right) + \left(D_y \frac{\partial^2 C_r}{\partial y^2} - \frac{D_x D_y}{U} \frac{\partial^3 C_r}{\partial x \partial y^2} \right)$$

$$+ \left(D_z \frac{\partial^2 C_r}{\partial z^2} - \frac{D_x D_z}{U} \frac{\partial^3 C_r}{\partial x \partial z^2} \right) - U \frac{\partial C_r}{\partial x} + D_x \frac{\partial^2 C_r}{\partial x^2} - \nu R_d C_r + \frac{\nu D_x R_d}{U} \frac{\partial C_r}{\partial x} \tag{2-96}$$

and with some rearranging,

$$R_d \frac{\partial}{\partial t} \left(C_r - \frac{D_x}{U} \frac{\partial C_r}{\partial x} \right) = D_x \frac{\partial^2}{\partial x^2} \left(C_r - \frac{D_x}{U} \frac{\partial C_r}{\partial x} \right)$$

$$+ D_y \frac{\partial^2}{\partial y^2} \left(C_r - \frac{D_x}{U} \frac{\partial C_r}{\partial x} \right) + D_z \frac{\partial^2}{\partial z^2} \left(C_r - \frac{D_x}{U} \frac{\partial C_r}{\partial x} \right) \tag{2-97}$$

$$- U \frac{\partial}{\partial x} \left(C_r - \frac{D_x}{U} \frac{\partial C_r}{\partial x} \right) - \nu R_d \left(C_r - \frac{D_x}{U} \frac{\partial C_r}{\partial x} \right)$$

and with Eq. (2-92),

$$R_d \frac{\partial C_f}{\partial t} = D_x \frac{\partial^2 C_f}{\partial x^2} + D_y \frac{\partial^2 C_f}{\partial y^2} + D_z \frac{\partial^2 C_f}{\partial z^2} - U \frac{\partial C_f}{\partial x} - \nu R_d C_f \tag{2-98}$$

takes the same form as Eq. (2-94).

Equations (2-94) and (2-98) describe the same physical processes, as D_x, D_y, D_z and U represent precisely the same physical quantities in both equations. This duality of interpretation of the physical meaning of the concentrations in the convection–dispersion equation imposes a need to stipulate carefully boundary conditions in keeping with the desired meaning. However, the identical mathematical form of Eqs. (2-94) and (2-98) must not be allowed to obscure the fundamental distinction between volume-averaged (resident) and flux-averaged (flowing) concentrations (Parker and van Genuchten, 1984).

2.10 INITIAL AND BOUNDARY CONDITIONS

2.10.1 INTRODUCTION

The type of differential equations presented by Eqs. (2-79)–(2-83) is commonly encountered in various branches of mathematical physics such as fluid dynamics, electricity, magnetism, heat flow, ground water flow, and others. Even though the aforementioned equations apply to solute transport in a flow field, they do not include any information regarding any specific case of flow or the shape and boundaries of the solute transport domain. Consequently, each of these equations has an infinite number of solutions and each corresponds to a particular solute transport case in a flow field.

The shape and boundaries as well as initial conditions for a particular solute transport problem correspond to one of the solutions of the corresponding differential equation for the aquifer. Therefore, the specification of appropriate initial and boundary conditions for a solute transport problem is a necessary step in determining the corresponding solution. Based on numerous initial and boundary conditions, a number of solutions were developed for the above-mentioned one-, two-, and three-dimensional solute transport differential equations during the past four decades. An understanding of the physical meaning of different initial and boundary conditions is important not only for the development of mathematical solutions, but also for the selection of a solution from a set of solutions to be used for a particular case faced in practice. The solution of a specific problem is one of the special solutions of solute transport differential equations. Since the specific solution is dependent on the boundary conditions, the problem is called the *boundary value problem*.

Based on the above-mentioned facts, in the following sections, the initial and boundary conditions typically encountered in solute transport problems will be presented with examples.

2.10.2 INITIAL CONDITIONS

Specification of initial conditions for a particular problem is only required for unsteady-state solute transport problems. For steady-state solute transport problems, initial conditions are not required. For unsteady-state solute transport problems, the distribution of concentration in the solute transport domain at a particular time is assumed to be known. Initial conditions for concentration are generally specified when time is zero ($t = 0$). The mathematical expression for this condition is

$$C(x,y,z,t = 0) = f(x,y,z) \qquad (2\text{-}99)$$

for all points inside the flow domain at $t = 0$. For analytical solute transport solutions, the initial concentration is generally taken as zero. Therefore, Eq. (2-99) takes the form

$$C(x,y,z,0) = 0 \qquad (2\text{-}100)$$

2.10.3 TYPES OF BOUNDARY CONDITIONS

After defining the solute transport problem for a particular case, the first step is to convert the physical problem into a mathematical problem. The second step is to establish the framework of the boundary value problem. In solute transport problems, the final solution is usually expressed in terms of concentration C, solute mass M, and solute flux F.

The selection of boundary conditions is a critically important step for conceptualizing and developing a model for a solute transport system. Even if one of the boundary conditions is inappropriate for a defined solute transport problem, it is most likely that the final expression will be erroneous or will produce ambiguous results.

2.10.3.1 Constant Concentration Boundary

The constant concentration boundary condition, is also known as the *Dirichlet boundary condition* or the *first-type boundary condition*. The general mathematical form of the steady constant concentration boundary condition is

$$C = f_1(x,y,z) \qquad (2\text{-}101)$$

If the boundary condition changes with time, the equation takes the form

$$C = f_2(x,y,z,t) \qquad (2\text{-}102)$$

Here, f_1 and f_2 are known functions. These boundary conditions may be valid for the whole flow domain or for a part of it. Examples for this type of boundary condition are presented in Chapters 3 to 5.

2.10.3.2 Flux Boundary

Continuous Flux Boundary Condition
The term *flux* means the mass of solute per unit time passing through a unit surface area. More specifically, values of the flux components, which are given by Eqs. (2-31a), to (2-31c), have specified values along the boundaries of the solute transport domain. Here, the main focus is the solute flux components, which are given by Eqs. (2-41a)–(2-41c), respectively, for the *x*-, *y*-, and *z*-coordinate directions in a unidirectional flow field. Based on these equations, the mathematical expression of the flux boundary is

$$F_n = f(x,y,z,t) \tag{2-103}$$

where F_n is the flux component normal to the boundary surface and $f(x, y, z, t)$ is a known function. Of these three flux components, the one that coincides with the direction of the uniform velocity U field in the *x*-coordinate direction, which is given by Eq. (2-41a), has a special importance. This is due to the fact that the analytical solutions, based on flux-type boundary conditions, specifically use Eq. (2-41a). For a flux boundary perpendicular to the *x*-coordinate direction (or with the direction of uniform velocity U), Eq. (2-41a) takes the following form:

$$\varphi_e UC - \varphi_e D_x \frac{\partial C}{\partial x} = F_M f(t), \quad x = 0 \tag{2-104}$$

which is also known the *third-type boundary condition* or *Cauchy boundary condition*. In Eq. (2-104), F_M is the constant solute flux (mass per unit area per unit time) and $f(t)$ is a function of time, which can take several distributions, such as a constant value in time (continuous feed solution) and an exponential increase or decrease in function with time. Examples for this type of boundary condition are presented in Chapters 3 to 5.

Instantaneous Flux Boundary Condition
Mathematical Expression of the Condition
If a known amount of mass M is instantaneously introduced in an area A, which is perpendicular to the *x*-coordinate direction, the corresponding mathematical expression of Eq. (2-104) for this type of boundary condition is

$$\frac{M}{A} \delta(t) = \varphi_e UC - \varphi_e D_x \frac{\partial C}{\partial x} \tag{2-105}$$

in which $\delta(t)$ is the Dirac delta function or impulse function (see, e.g., Spiegel, 1965).

Mathematical and Physical Bases of the Condition
The graphical representation of the Dirac delta function is shown in Figure 2-8. Consider the function $F_\varepsilon(t)$ such that

$$F_\varepsilon(t) = \frac{1}{\varepsilon}, \quad 0 \le t \le \varepsilon \tag{2-106a}$$

$$F_\varepsilon(t) = 0, \quad t > \varepsilon \tag{2-106b}$$

where $\varepsilon > 0$. It is geometrically evident that as $\varepsilon \to 0$, the height of the rectangle in Figure 2-8 increases indefinitely and the width decreases in such a way that the area is always equal to 1 and

$$\delta(t) = \lim_{\varepsilon \to 0} F_\varepsilon(t) \tag{2-107}$$

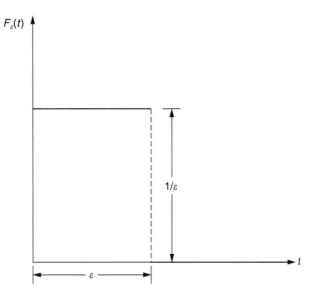

FIGURE 2-8   Graphical representation of the Dirac delta function.

This idea has led to a limiting function, denoted by $\delta(t)$, approached by $F_\varepsilon(t)$ as $\varepsilon \to 0$. This limiting function is called *unit impulse function* or *Dirac delta function*. Mathematically speaking, such a function does not exist. However, manipulations or operations with it can be rigorous. Equations (2-105) and (2-107) clearly show that the unit of $\delta(t)$ is T^{-1}. As a result, the dimensions on the left-hand side of Eq. (2-105), $M/A\delta(t)$, is $M/L^2/T$, which is the same as the dimension on its right-hand side.

2.10.3.3 No-Solute Flux Boundary

Here, as in Section 2.10.3.2, the main focus is the solute flux components, which are given by Eqs. (2-41a), (2-41b), and (2-41c), respectively, for the x-, y-, and z-coordinate directions in a unidirectional flow field. Based on these equations, the mathematical expression of the no-solute flux boundary is

$$F_n = 0 \qquad\qquad (2\text{-}108)$$

Under the framework of this expression, Eqs. (2-41a), (2-41b), and (2-41c), respectively, the following equations result:

$$UC - D_x \frac{\partial C}{\partial x} = 0 \qquad\qquad (2\text{-}109a)$$

$$\frac{\partial C}{\partial y} = 0 \qquad\qquad (2\text{-}109b)$$

and

$$\frac{\partial C}{\partial z} = 0 \qquad\qquad (2\text{-}109c)$$

Eq. (2-109a) is valid for boundaries perpendicular to the uniform flow velocity U and is a special form of Eqs. (2-104) or (2-105). Equations (2-109b) and (2-109c) represent impermeable boundaries that are parallel to the uniform flow velocity U field. Equations (2-109b) and (2-109c) are based on the fact that in Eqs. (2-41b) and (2-41c), the terms φD_y and φD_z cannot be zero. Therefore, in order to have $F_y = 0$ and $F_z = 0$, Eqs. (2-109b) and (2-109c), respectively, must be satisfied.

2.11 EQUATIONS FOR THE MECHANICAL DISPERSION COEFFICIENTS FOR SATURATED POROUS MEDIA

During the past four decades, research has been conducted to determine expressions for the components of the dispersion tensor, which is given by Eq. (2-32), and efforts toward this end are still in progress. In order to accomplish this, laboratory and field experiments, deterministic and stochastic approaches are being used. In this section, the state-of-the-art status for expressions of the dispersion coefficients will be presented.

2.11.1 LABORATORY EXPERIMENTS — LEAD EQUATIONS

Since the early 1960s, attempts have been made to correlate the dispersion coefficients (the components of the dispersion tensor given by Eq. [2-32]) with media characteristics such as grain-size distribution, permeability, and porosity of porous materials. Experimental investigations are based on mostly one-dimensional column experiments, although a limited number of two-dimensional as well as three-dimensional experiments are also available in the literature. Analytical solutions of Eqs. (2-49)–(2-51) are used in analyzing the experimental data.

2.11.1.1 Results from One-Dimensional Experiments

For a one-dimensional solute transport case, Eq. (2-27) can also be expressed as

$$D_x = D^* + D_{m_x} \qquad (2\text{-}110)$$

where D_x is the longitudinal dispersion coefficient, D^* the effective molecular diffusion coefficient, and D_{m_x} the mechanical dispersion coefficient. Rumer (1962) determined the values of D_x for one-dimensional solute transport in a uniform ground water flow field within a certain range of Reynolds numbers. The media used were quartz gravel and glass beads. The quartz gravel had a mean size of 1.65 mm; the glass beads had a mean size of 0.39 mm. The porosity of both materials were 0.39. The seepage velocity U for quartz gravel ranged from 0.04 to 0.4 cm/sec, whereas for glass beads it ranged from 0.01 to 0.2 cm/sec. In the experiments, NaCl was used as the tracer solution. As mentioned in Section 2.3.1, for relatively coarse materials, the effective molecular diffusion coefficient is negligible as compared with the mechanical dispersion coefficient. Therefore, the measured quantity is approximately equal to D_{m_x}. In other words, $D_{m_x} \cong D_x$. Rumer (1962) used Eq. (3-14) to analyze the results. The general equation that relates the mechanical dispersion coefficient to the seepage velocity is given by (Rumer, 1962)

$$D_{m_x} = \alpha_L U^n \qquad (2\text{-}111)$$

in which α_L denotes a constant for a given porous medium (expressed in cm when $n = 1$) and n is an exponent that is constant for a given porous medium in place. The terms α_L and n will vary for each medium and will also vary with porosity for a given medium. The correlation of n and α_L media characteristics is not completely known.

The values of α_L and n for quartz gravel and glass beads used by Rumer (1962) are given in Table 2-6. The factor α_L, which is measured in cm when $n = 1$, can be thought of as an intrinsic dispersivity. The *intrinsic dispersivity* is a parameter expressing the mixing ability or mass transport properties of the media. It is clear that α_L does not completely describe the dispersive properties ability of a medium when n is different from unity, but n has been found by other investigators to be always nearly equal to unity. Substitution of Eq. (2-111) into Eq. (2-110) gives

$$D_x = D^* + \alpha_L U^n \qquad (2\text{-}112)$$

TABLE 2-6
Parameters of the Mechanical Dispersion Coefficient Equation (2-111) Based on Rumer's Experiment

Material	α_L	n
Quartz gravel	0.2	1.083
Glass beads	0.027	1.105

Source: From Rumer, R.R., *J. Hydraul. Div.*, 88, 147–172, 1962. With permission.

2.11.1.2 Results from Two-Dimensional Experiments

Based on the previous investigations (Ebach and White, 1958; Raimondi et al., 1959; Rumer, 1962; Harleman et al., 1963), Harleman and Rumer (1963) generalized all related experiment-led longitudinal mechanical dispersion equations in the following form:

$$\frac{D_{mx}}{\nu} = \psi_1 R^{n_1} \tag{2-113}$$

where

$$R = \frac{Ud}{\nu} \tag{2-114}$$

in which ν is the kinematic viscosity, ψ_1 and n_1 are parameters characterizing the medium, R the Reynolds number based on the average grain size d as the characteristic length, and U the seepage velocity. Harleman and Rumer (1963) showed that ψ_1 is independent of particle size for nearly uniform media and is dependent on particle shape and possible size distribution.

Harleman and Rumer (1963) used a similar postulate; using Eq. (2-113) for the transverse dispersion coefficient as

$$\frac{D_{my}}{\nu} = \psi_2 R^{n_2} \tag{2-115}$$

the ratio of the longitudinal and transverse mechanical dispersion coefficients should be of the form

$$\frac{D_{mx}}{D_{my}} = \lambda R^n \tag{2-116}$$

where

$$\lambda = \frac{\psi_1}{\psi_2}, \; n = n_1 - n_2 \tag{2-117}$$

In accordance with the above discussion, λ and n should be functions only of the media characteristics.

Harleman and Rumer (1963) used a porous medium that was composed of Dow Chemical "Palespan 8" plastic spheres. The spheres had a mean size of 0.96 mm and their average hydraulic conductivity was 0.8 cm/sec. The parameters in Eqs. (2-113)–(2-117) are given in Table 2-7. Variation of D_{mx}/D_{my} with R based on Eq. (2-116) and values in Table 2-7 is shown in Table 2-8. Harleman and Rumer noted that the extension of Eq. (2-116) for R values greater than 10 is not justified. Substitution of Eq. (2-114) into Eq. (2-113) gives

$$D_{mx} = \lambda_1 U^{n_1} \tag{2-118}$$

where

$$\lambda_1 = \psi_1 \nu \left(\frac{d}{\nu}\right)^{n_1} \tag{2-119}$$

TABLE 2-7
Parameters for D_{m_x} and D_{m_y} for Dow Chemical "Palespan 8" Plastic Based on Harleman and Rumer's (1963) Experiments

Parameter	Value
ψ_1	0.66
n_1	1.2
ψ_2	0.036
n_2	0.7
$\lambda = \psi_1/\psi_2$	18.3
$n = n_1 - n_2$	0.5

TABLE 2-8
Variation of D_{m_x}/D_{m_y} with R for Dow Chemical "Palespan 8" Plastic Based on Harleman and Rumer's (1963) Experiments

R	D_{m_x}/D_{m_y}
0.05	4.09
0.10	5.79
0.20	8.18
0.30	10.02
0.40	11.57
0.50	12.94
0.60	14.18
0.70	15.31
0.80	16.37
0.90	17.36
1.00	18.30

Similarly, substitution of Eq. (2-114) into Eq. (2-115) gives

$$D_{m_y} = \lambda_2 U^{n_2} \tag{2-120}$$

where

$$\lambda_2 = \psi_2 \nu \left(\frac{d}{\nu}\right)^{n_2} \tag{2-121}$$

In Eqs. (2-118) and (2-120) λ_1, λ_2, n_1, and n_2 are experimentally determined constants for the medium. Since the 1960s, other researchers have conducted similar laboratory experiments on various porous materials and the equations they obtained have similar forms (see, e.g., Guvanasen and Volker, 1983). As can be seen from Tables 2-6 and 2-7, the exponents n_1 and n_2 are approximately equal to 1.0 and λ_1 and λ_2 depend on several factors and are denoted by α_L and α_T, respectively. Therefore, Eqs. (2-118) and (2-120), respectively, take the forms

$$D_{m_x} = D_{m_L} \cong \alpha_L U \tag{2-122}$$

$$D_{m_y} = D_{m_T} \cong \alpha_T U \tag{2-123}$$

where α_L and α_T are called the longitudinal and transverse dispersivities, respectively.

2.11.2 GENERALIZED EQUATIONS WITH THEORY AND POSTULATION

In Section 2.11.1, equations for mechanical dispersion coefficients based on laboratory experiments are presented. Equations 2-122 and 2-123 are the final, practical-oriented equations for mechanical dispersion equations. In the early 1960s, these equations were generalized based on theory and postulation (Bear, 1961; Scheidegger, 1961).

2.11.2.1 Fundamental Relationships for Dispersion Coefficients

The study of dispersion phenomena involves the application of probability theory for predicting the spatial distribution of tracer particles with time in a porous medium. De Josselin de Jong (1958) investigated tracer migration from a point source by using statistical methods, and showed that the plume of tracer particles will be normally distributed in three dimensions around a center traveling with the mean ground water velocity (Bear, 1972). The resulting surfaces of equal concentration form an ellipsoid of revolution about the center or an ellipse in a plane. Figure 2-9 schematically represents the distribution of tracer particles at successive times as they move within a porous medium as a result of a solute slug through a homogeneous porous medium. Initially (at $t = t_0$), the slug forms a sphere of revolution having its center at $x = x_0$. Gradually, the center of the slug moves further from its initial location within the porous medium by convection with the flow of the porous medium, and the plume expands due to the molecular and mechanical dispersion effects. As a result of these effects, the resulting surfaces of equal concentration form an ellipsoid of revolution about the center or ellipse in a plane. In Figure 2-9, the ellipses are the cross-section of the ellipsoids either in the horizontal or vertical plane. The spreading in the direction of flow is the longitudinal dispersion, while the spreading in the direction perpendicular to the flow is known as transverse dispersion. As a result, in three dimensions, there are three dispersion coefficients: the longitudinal dispersion coefficient D_x, the transverse horizontal dispersion coefficient D_y, and the transverse vertical dispersion coefficient D_z. The variances σ_x^2, σ_y^2, and σ_z^2 of tracer particles' distribution in the x, y, and z directions, respectively, have the following relations with the D_x, D_y, and D_z dispersion coefficients (Bear, 1972):

$$\sigma_x^2 = 2D_x t, \quad \sigma_y^2 = 2D_y t, \quad \sigma_z^2 = 2D_z t \tag{2-124}$$

In the case of one-dimensional molecular diffusion, the corresponding equation is derived in Eq. (6-20), Section 6.3.1.3.

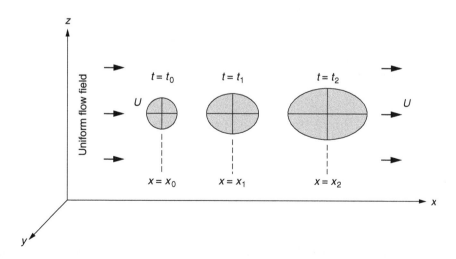

FIGURE 2-9 The areal variation in tracer concentration in a steady uniform flow field.

2.11.2.2 Mechanical Dispersion Coefficients

Bear (1961, 1972) analyzed the distribution of a solute plume generated from a point injection in a uniform flow field whose velocity components (v_x, v_y, v_z) are given by Eq. (2-35). Their corresponding displacements are L_x, L_y, and L_z, respectively. Bear observed that neither the velocity components nor their corresponding displacements play a role in the analysis of the plume. Bear reasoned that this is probably due to the nature of the porous medium with its interconnected channels distributed randomly in all directions, and that it is the true local velocities in the channels (neglecting molecular diffusion) that cause dispersion, and not the average velocity over the entire cross-section, whose components are v_x, v_y, and v_z. In an isotropic medium, the dispersion in the direction of the mean displacement $\mathbf{L}$ and perpendicular to it (see Figure 2-10) can be obtained from Eqs. (2-122) and (2-123). From Eqs. (2-122) and (2-123) one, respectively, can write

$$D_{m_L} = \alpha_T \frac{|\mathbf{L}|}{t} \tag{2-125}$$

$$D_{m_T} = \alpha_T \frac{|\mathbf{L}|}{t} \tag{2-126}$$

Here α_L and α_T are medium constants, but not the L_x and L_y components of $\mathbf{L}$ in the x–y coordinate system. Bear suggests resolving the components L_i further and taking their projections L_{ij} in the direction of the flow and perpendicular to it. This is shown in Figure 2-10. The components of the displacement vector $\mathbf{L}$ are $L_x = L \cos \beta$ and $L_y = L \sin \beta$ as shown in Figure 2-10a. In Figure 2-10b, the displacement on the flow direction is $AB = L \cos^2 \beta + L \sin^2 \beta = L (\cos^2 \beta + \sin^2 \beta) = L$, and the displacement perpendicular to the flow direction is $CD = L \cos \beta \sin \beta$. Thus, displacement is defined as a second-rank tensor in two dimensions and takes the form

$$[L_{ij}] = \begin{bmatrix} L\cos^2 \beta & L\sin \beta \cos \beta \\ L\sin \beta \cos \beta & L\sin^2 \beta \end{bmatrix} = L \begin{bmatrix} \cos^2 \beta & \sin \beta \cos \beta \\ \sin \beta \cos \beta & \sin^2 \beta \end{bmatrix} \tag{2-127}$$

In a three-dimensional space, $i, j = 1, 2, 3$. Bear (1972) generalized Eq. (2-124) as

$$(\sigma^2)_{kl} = 2(D_m)_{kl}\, t = 2\alpha_{ijkl}L_{ij} \tag{2-128}$$

where α_{ijkl}, which is a fourth-rank tensor, is the geometrical dispersivity of the porous medium, and for saturated porous media, flow is a property of the geometry of the solid matrix. This fourth-rank tensor has certain properties of symmetry (Bear, 1961; Scheidegger, 1961). In general, α_{ijkl} ($i, j, k, l = 1, 2, 3$) contains 81 components, of which all but those with four equal indices, or with two pairs of equal indices, are equal to zero (Bear, 1972).

For a uniform flow field with $L = vt$, Eq. (2-128) can be expressed as

$$[L_{ij}] = \begin{bmatrix} (v\cos \beta)^2 & v \sin \beta\, v \cos \beta \\ v \sin \beta . v \cos \beta & (v\sin \beta)^2 \end{bmatrix} = \begin{bmatrix} v_i v_j \\ v \end{bmatrix} \tag{2-129}$$

Hence, from Eqs. (2-128) and (2-129), one can write

$$(D_m)_{kl} = \alpha_{ijkl} \frac{v_i v_j}{v} t \tag{2-130}$$

where the summation convention is used. The general form of Eq. (2-130) has been obtained by Nikolaevskij (1959) from an analogy with the statistical theory of turbulence.

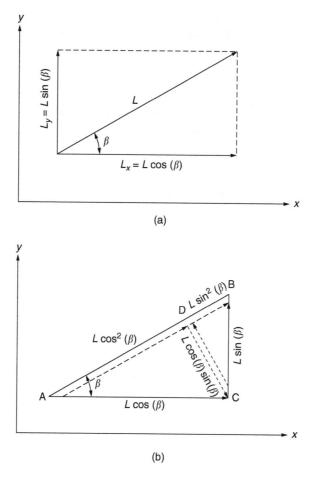

FIGURE 2-10   The resolution of the solute displacement L: (a) mean displacement and (b) components of the displacement tensor.

2.11.2.3   Dispersivities of the Medium

For an isotropic porous medium, Bear (1960, 1961, 1972) relates α_{ijkl} in Eq. (2-130) to α_L and α_T, which are, respectively, the longitudinal and the transverse dispersivity of the medium. In two dimensions,

$$\alpha_{1111} = \alpha_{2222} = \alpha_L, \quad \alpha_{1122} = \alpha_{2211} = \alpha_T$$

$$\alpha_{1112} = \alpha_{1121} = \alpha_{1211} = \alpha_{1222} = \alpha_{2111} = \alpha_{2122} = \alpha_{2212} = \alpha_{2221} = 0 \qquad (2\text{-}131)$$

$$\alpha_{1212} = \alpha_{1221} = \alpha_{2112} = \alpha_{2121} = \tfrac{1}{2}(\alpha_L - \alpha_T)$$

In three dimensions, similar terms must be added with subscript 3 replacing 2. Scheidegger (1961) expressed Eq. (2-131) in the general form as

$$\alpha_{ijkm} = \alpha_T \delta_{ij} \delta_{km} + \frac{(\alpha_L - \alpha_T)}{2} (\delta_{ik} \delta_{jm} + \delta_{im} \delta_{jk}) \qquad (2\text{-}132)$$

where δ_{ij} is the Kronecker delta.

Scheidegger (1961) analyzed the symmetrical properties of α_{ijkl}, which in a three-dimensional space includes 81 components, and extended the analysis to anisotropic media. Scheidegger, based on certain symmetry properties, showed that

$$\alpha_{ijkm} = \alpha_{ijmk}, \quad \alpha_{ijkm} = \alpha_{jikm}, \quad \alpha_{ijkm} = \alpha_{jimk}, \quad \alpha_{ijkm} = \alpha_{kmij} \tag{2-133}$$

Actually, Scheidegger (1961) leaves the last equality questionable, while Bear (1961) assumes that this symmetry exists in isotropic media. Even in an anisotropic medium, only 36 terms are nonzero. However, the number of constants in a specific type of anisotropy is not given. In an isotropic medium, 21 terms are nonzero (Bear, 1972).

2.11.2.4 Mechanical Dispersion Coefficients: General Case

Substitution of Eq. (2-132) into Eq. (2-130) gives (Bear, 1979)

$$D_{m_{ij}} = \alpha_T v \delta_{ij} + (\alpha_L - \alpha_T)\frac{v_i v_j}{v} \tag{2-134}$$

or, as in Eq. (2-32), the mechanical dispersion coefficient tensor takes the form

$$[D_{m_{ij}}] = \begin{bmatrix} D_{mxx} & D_{mxy} & D_{mxz} \\ D_{myx} & D_{myy} & D_{myz} \\ D_{mzx} & D_{mzy} & D_{mzz} \end{bmatrix} \tag{2-135}$$

and can be generalized from Eq. (2-27), using Eqs. (2-32) and (2-135), as

$$[D_{ij}] = [D_{m_{ij}}] + [D_{ij}^*] \tag{2-136}$$

Since the nine components of the D^*_{ij} tensor are equal, combining Eqs. (2-135) and (2-136) gives

$$[D_{ij}] = \begin{bmatrix} D_{mxx}+D^* & D_{mxy}+D^* & D_{mxz}+D^* \\ D_{myx}+D^* & D_{myy}+D^* & D_{myz}+D^* \\ D_{mzx}+D^* & D_{mzy}+D^* & D_{mzz}+D^* \end{bmatrix} \tag{2-137}$$

In the Cartesian coordinates, the velocity vector $\mathbf{v}$ (its components are v_x, v_y, and v_z) can be expressed as

$$|\mathbf{v}| = v = (v_x^2 + v_y^2 + v_z^2)^{1/2} \tag{2-138}$$

From Eqs. (2-134) and (2-138), the mechanical dispersion coefficient in the x-coordinate direction can be expressed as

$$D_{m_{xx}} = \frac{\alpha_T v^2 + (\alpha_L - \alpha_T)v_x^2}{v}$$

$$= \frac{\alpha_T v^2 + (\alpha_L - \alpha_T)v_x^2}{v} \tag{2-139}$$

$$= \frac{\alpha_T(v_x^2 + v_y^2 + v_z^2) + (\alpha_L - \alpha_T)v_x^2}{v}$$

$$= \alpha_L \frac{v_x^2}{v} + \alpha_T \frac{v_y^2}{v} + \alpha_T \frac{v_z^2}{v}$$

Similarly, the other mechanical dispersion coefficients tensor components in Eq. (2-137) can be written as

$$D_{m_{yy}} = \alpha_L \frac{v_y^2}{v} + \alpha_T \frac{v_x^2}{v} + \alpha_T \frac{v_z^2}{v} \tag{2-140}$$

$$D_{m_{zz}} = \alpha_L \frac{v_z^2}{v} + \alpha_T \frac{v_x^2}{v} + \alpha_T \frac{v_y^2}{v} \tag{2-141}$$

$$D_{m_{xy}} = D_{m_{yx}} = (\alpha_L - \alpha_T)\frac{v_x v_y}{v} \tag{2-142}$$

$$D_{m_{xz}} = D_{m_{zx}} = (\alpha_L - \alpha_T)\frac{v_x v_z}{v} \tag{2-143}$$

and

$$D_{m_{yz}} = D_{m_{zy}} = (\alpha_L - \alpha_T)\frac{v_y v_z}{v} \tag{2-144}$$

2.11.2.5 A Special Case for Mechanical Dispersion Coefficients: the Average Velocity (*v*) Coincides with the x-Coordinate Axis

Let us choose the x–y–z Cartesian coordinate system such that one of its axes, for example, x, coincides with the direction of the average velocity v. Therefore, $v_y = v_z = 0$. Then, Eq. (2-138) gives

$$|v| = v_x \equiv U \tag{2-145}$$

Under these conditions, using Eq. (2-145), Eq. (2-139) takes the form

$$D_{m_x} \equiv D_{m_{xx}} = \alpha_L \frac{v_x^2}{v_x} + 0 + 0 = \alpha_L U \tag{2-146}$$

Similarly, Eqs. (2-140) and (2-141), respectively, take the following forms:

$$D_{m_{yy}} = \alpha_T U \tag{2-147}$$

and

$$D_{m_{zz}} = \alpha_T U \tag{2-148}$$

and the remaining mechanical dispersion components, given by Eqs. (2-142), (2-143), and (2-144), respectively, take the following forms:

$$D_{m_{xy}} = D_{m_{yx}} = 0 \tag{2-149}$$

$$D_{m_{xz}} = D_{m_{zx}} = 0 \tag{2-150}$$

$$D_{m_{yz}} = D_{m_{zy}} = 0 \tag{2-151}$$

Therefore, for a uniform flow that is aligned with the x-axis, Eq. (2-135) takes the form

$$[D_{m_{ij}}] = \begin{bmatrix} \alpha_L U & 0 & 0 \\ 0 & \alpha_T U & 0 \\ 0 & 0 & \alpha_T U \end{bmatrix} \tag{2-152}$$

The axes of the coordinate system in which $D_{m_{ij}}$ is expressed by Eqs. (2-146)–(2-151), namely, in the direction of flow at a point and perpendicular to it, are called the principal axes of the dispersion. The coefficients, $D_{m_{xx}}$, $D_{m_{yy}}$, and $D_{m_{zz}}$ are called the principal values of the coefficient of mechanical dispersion. In this case, $D_{m_{xx}}$ is called the coefficient of longitudinal mechanical dispersion $D_{m_{yy}}$ and $D_{m_{zz}}$ and are called coefficients of transverse mechanical dispersion.

Generalization

The transverse dispersivities in the y- and z-coordinate directions may not be the same. Therefore, Eq. (2-152) can be generalized by distinguishing the transverse dispersivities in the y and z directions as

$$[D_{m_{ij}}] = \begin{bmatrix} \alpha_L U & 0 & 0 \\ 0 & \alpha_{TY} U & 0 \\ 0 & 0 & \alpha_{TZ} U \end{bmatrix} \tag{2-153}$$

where α_{TY} and α_{TZ} are the transverse dispersivities in the y and z directions, respectively. If the y direction corresponds to vertical direction and the z direction corresponds to horizontal direction, the dispersion coefficients in the x, y, and z directions can, respectively, be expressed as

$$D_x = D_{m_x} + D^* = \alpha_L U + D^* \tag{2-154}$$

$$D_y = D_{m_y} + D^* = \alpha_{TH} U + D^* \tag{2-155}$$

$$D_z = D_{m_z} + D^* = \alpha_{TV} + D^* \tag{2-156}$$

where α_{TH} and α_{TV} are the transverse horizontal dispersivity and transverse vertical dispersivity, respectively.

PROBLEMS

2.1   What is the main mechanism of quantifying solute transport in aquifers ?

2.2   In the case of groundwater, what is the solvent?

2.3   The concentration of benzene in groundwater is given as 15 mg/L. What are the equivalent units in ppm and g/m^3?

2.4   For a porous medium column shown in Figure 2-2, the given values are $F_x = 0.30$ mg/day/m², $\varphi_e = 0.25$, and $U = 0.15$ m/day. Determine the value of flux-averaged concentration in ppb.

2.5   The thickness of a liner is 1 m. The chloride concentration in the upper part is given as 45 mg/L. Under steady-state diffusive transport conditions, determine the maximum solute flux rate through the liner per square meter area (m²). Assume $\tau = 0.5$.

2.6   What is the base of the mechanical dispersive flux equation?

2.7   Write the complete form of the flux equation for one-dimensional case under hydrodynamic dispersion conditions. Explain the meaning of each term.

2.8   Hydrogeologic investigations for a sandy aquifer revealed that $K_x = 25$ m/day, $K_y = 15$ m/day, and $K_z = 2$ m/day. Based on these values, determine (a) the type of the aquifer and (b) the conditions that need to be satisfied in order for it to be homogeneous and isotropic.

2.9   Derive Eq. (2-49).

2.10 Evaluate the porosity terms in Eq. (2-59).

2.11 Show that the retardation factor R_d, as given by Eq. (2-62), is a dimensionless quantity.

3 Analytical Solute Transport Modeling in Aquifers

3.1 INTRODUCTION

The physical mechanisms of solute transport in porous media followed by the governing differential equations of solute transport as well as the associated potential solute source conditions, and initial and boundary conditions were presented in Chapter 2. These equations as well as the initial and boundary conditions are the foundations of solute transport modeling approaches. In this chapter, some analytical solute transport models, which have practical importance, are presented. Some important concepts regarding solute transport modeling are also included using some analytical solute transport models.

Determination of analytical solutions is a different task and in this chapter, with the exception of some simple cases, some derivational details of the solutions are included as example problems for interested readers and the focal points for the presentation of an analytical solute transport model is (1) initial and boundary conditions, (2) solution(s), (3) special cases, and (4) potential applications. As mentioned in Chapter 2, selection of a predictive solute transport model for a given real-world case requires a thorough knowledge about the physical mechanisms of solute transport in porous media, governing equations with the initial and boundary conditions, and the limitations of the candidate models for selection.

3.2 DETERMINISTIC VS. STOCHASTIC MODELING APPROACHES

In this chapter, some analytical solute transport models of practical importance in the deterministic sense is presented. Since the 1950s, a number of analytical and numerical solute transport models were developed by various investigators using the *deterministic approach*. All the analytical solute transport models presented in this chapter are based on the deterministic approach. The development and usage of this type of models have been intensified especially since the 1970s. Since the mid1970s, *stochastic approaches* started to appear in the literature mainly to evaluate and determine equations for dispersion coefficients by taking into account the spatial variability of flow and transport parameters. Chapter 6 is devoted to this subject.

In principle, stochastic approaches recognize that both flow and transport parameters of a porous medium are affected by uncertainty and treat them as random variables. This randomness leads to the definition of the flow and transport models in the stochastic sense, and predictions are made in terms of probabilities rather than in the traditional deterministic sense (Dagan, 1997). Although the uncertain nature of flow and transport parameters is widely recognized, especially since the mid-1970s, there is no general consensus on the best way to deal with it mathematically (Neuman, 1997). According to Freeze et al. (1990), there are three basic approaches used to propagate these uncertainties: (1) first-order analysis, (2) perturbation analysis, and (3) Monte-Carlo analysis. The *first-order analysis* (e.g., Dagan, 1989) is a simple and direct means of propagating uncertainty. This approach can be employed with either analytical or numerical solutions of the governing equations, but is limited to linear or nearly linear systems. Beyond this, this approach requires only the first two moments (the mean and the variance) of the input parameters, and it provides only the first two moments of the predicted output variables. In *pertubation analysis* (e.g., Gelhar et al., 1993), both the input parameters and the output variables are defined in terms of a mean and its perturbation about the mean. The relationship between input and output uncertainties can be developed using two general techniques. One of the techniques involves developing relationships in the spectral domain by using the theory of Fourier–Stieltjes integrals (see, e.g., Lumley and Panofsky, 1964) and an inverse Fourier transform. This approach is best suited to analytical solutions, an infinite flow domain, and a small coefficient of variation in input uncertainty. The foundations of the perturbation analysis, and based on this technique, analysis of dispersion coefficients for spatially variable media are covered in Chapter 6. The third basic approach, Monte-Carlo simulation, provides the most general approach to uncertainty propagation. With this approach, a large number of equally probable realizations of each parameter field are generated, and the hydrogeological simulation model is run for each realization.

Stochastic models as predictive tools for solute transport migration are on research level and extension of these efforts to real-world practice is fairly limited. Therefore, deterministic-based analytical and numerical modeling approaches still play an important role in the practice of solute transport analysis in aquifers.

3.3 SOME SELECTED ANALYTICAL SOLUTE TRANSPORT MODELS AND FUNDAMENTAL MODELING CONCEPTS

3.3.1 INTRODUCTION

Analytical solute transport models are generally limited to cases with steady and uniform flow fields and to problems that have relatively simple initial and boundary conditions. Despite these limitations, analytical solute transport models can be useful for determining the fate and transport of various contaminant constituents (e.g., dissolved chemicals and radionuclides) in soil and aquifer systems. Compared to numerical models, analytical models can provide more insight into the conceptual behavior of the system. Computer solutions of analytical models are usually less bulky, easier to operate, cheaper to run, and contain fewer parameters than numerical models.

3.3.1.1 Advantages of Analytical Solute Transport Models

The advantages of analytical models may be summarized as follows (van Genuchten, 1982):

1. They can form useful complements to numerical models, either by being part of numerical models or for their verification.
2. They often provide important initial estimates for a number of simplified cases. For example, the long-term vertical transport of chemicals during unsteady saturated–unsaturated flow can be described quite accurately with appropriate analytical models. The same is true for two- or three-dimensional transport from various subsurface point, line, and area sources.
3. They are important tools for estimating transport parameters from laboratory or field data.

Source conditions that are implemented in solute transport modeling applications are presented in Section 2.10.3. Based on these source conditions, in this section, some practically important one-, two-, and three-dimensional analytical solute transport models will be presented with their potential application areas. The following approach are applied followed in presenting these solutions: (1) problem statement and assumptions, (2) governing differential equation, (3) initial and boundary conditions, (4) solution for concentration distribution, (5) convective–dispersive flux components, (6) special solutions and (7) application areas. During this presentation process, details regarding the previous steps for the final solution will generally be skipped and the pertinent literature will be referred. However, for some cases, partial details regarding the final solution is presented with problem-type approach example.

3.3.2 FIRST-TYPE SOURCE ANALYTICAL MODELS

In this section, some practically important first-type source analytical models will be presented with their potential application areas. For this type of models the solute source itself is treated as the *first-type boundary condition*. The terms *Dirichlet boundary condition* and *constant concentration boundary condition* are also widely used for this type of boundary condition (see Section 2.10.3.1 for details).

3.3.2.1 One-Dimensional Analytical Solute Transport Models for the First-Type Sources

The one-dimensional analytical solutions for hydrodynamic dispersion in a semiinfinite porous medium having steady unidirectional flow field under the first-type source condition has a special importance in the literature as well as in practice. This is due to the fact that these models are the first ones in this field and are relatively simple such that they can be used as predictive tools with mostly by hand calculations using tables for *exponential function* and *error function*.

To the author's knowledge, one-dimensional analytical solutions for hydrodynamic dispersion for the first-type sources have been obtained by Glueckauf et al. (1949) and Gröbner and Hofreiter (1950); by Lapidus and Amundson (1952) and Ogata and Banks (1961) for a constant input concentration; and by Ebach and White (1958) for an input concentration that is a periodic function of time. Marino (1974, 1975) extended the one-dimensional solution for decaying solutes with exponentially varying time-dependent first-type source cases.

Problem Statement and Assumptions
The problem under consideration is unsteady-state one-dimensional solute transport in a unidirectional flow field as shown in Figure 3-1. The medium is considered of having a semiinfinite parallelepiped form. The assumptions are as follows:

1. The solute source is located at $x = 0$ and its concentration is given as an exponentially variable time-dependent quantity.

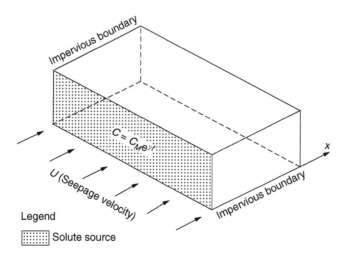

FIGURE 3-1 Schematic representation of one-dimensional solute transport with the first-type source condition in a unidirectional flow field.

2. The medium goes to infinity along the x-coordinate direction.
3. The flow is steady and has a uniform seepage (or ground water) velocity, U.
4. The source plane of the parallelepiped is perpendicular to the flow direction.

Governing Differential Equation
The governing differential equation for one-dimensional hydrodynamic dispersion equation is given by Eq. (2-83) as

$$R_d \frac{\partial C}{\partial t} = D_x \frac{\partial^2 C}{\partial x^2} - U \frac{\partial C}{\partial x} - v R_d C \tag{3-1}$$

where C is the concentration, x the coordinate in the flow direction, and t the time. The potential units of C that is widely used in practice are included in Section 2.2.1. In Eq. (3-1), R_d is the retardation factor (given by Eq. [2-62]) and v is the decay constant. (See Sections 2.6.2.1 and 2.6.2.2 for its meaning and range of values for various contaminant constituents.)

Initial and Boundary Conditions
The initial condition of $C(x, t)$ is

$$C(x, 0) = 0, \quad x \geq 0 \tag{3-2}$$

The boundary conditions for $C(x, t)$ are

$$C(0, t) = C_M e^{\gamma t}, \quad t \geq 0 \tag{3-3}$$

$$C(\infty, t) = 0, \quad t \geq 0 \tag{3-4}$$

$$\frac{\partial C(\infty, t)}{\partial x} = 0, \quad t \geq 0 \tag{3-5}$$

where $\gamma(T^{-1})$ is a parameter, C_M is the source concentration when $\gamma = 0$, and e = 2.7182818..., Equation (3-2) states that initially the concentration is zero everywhere in the transport domain. Equation (3-3) states that the source concentration is exponentially varying time-dependent quantity. Equations (3-4) and (3-5) state that both the concentration, C, and its gradient, $\partial C/\partial x$, with respect to x, approach zero when the x coordinate goes to infinity.

From Eq. (3-3), the function

$$\frac{C(0, t)}{C_M} = e^{\gamma t} \tag{3-6}$$

performs such that the source concentration can be represented as exponentially varying time-dependent quantity. If $\gamma < 0$, the source concentration decreases with time; if $\gamma = 0$, the source concentration is independent of time; and if $\gamma > 0$, the source concentration increases with time. Figure 3-2 shows the variation of $C(0, t)/C_M$ with time for different values of γ ranging from -0.0001 to -0.0100 day^{-1}. If observed data are available in accordance with Eq. (3-6) for time-dependent source concentration, the value of γ may be determined from the curve-fitting techniques.

Unsteady-State Solution for Concentration Distribution
General Case
The expression for the concentration distribution is determined as a special case from the two-dimensional solution of Batu (1989), which is presented in Eq. (3-77). The final solution is

$$C(x, t) = C_M P_1[\exp(-P_2) \, \text{erfc}(P_3 - P_4) + \exp(P_2) \, \text{erfc}(P_3 + P_4)] \tag{3-7}$$

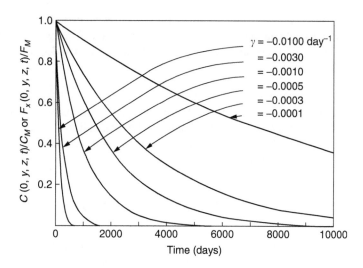

FIGURE 3-2 The variation of $C(0, t)/C_M$ with time for different values of γ. (Adapted from Batu, V., *Water Resour. Res.*, 29, 2881–2892, 1993.)

where

$$P_1 = \frac{1}{2}\exp\left(\gamma t + \frac{Ux}{2D_x}\right)$$ (3-8)

$$P_2 = \left(\frac{vR_d}{D_x} + \frac{U^2}{4D_x^2} + \frac{\gamma R_d}{D_x}\right)^{1/2} x$$ (3-9)

$$P_3 = \frac{R_d x}{2(D_x R_d t)^{1/2}}$$ (3-10)

$$P_4 = \left[\left(v + \frac{U^2}{4D_x R_d} + \gamma\right)t\right]^{1/2}$$ (3-11)

Equation (3-7) can also be expressed as

$$C(x, t) = \frac{1}{2}C_M \exp(\gamma t)\left\{\exp\left[\frac{x(U-P)}{2D_x}\right]\text{erfc}\left[\frac{R_d x - Pt}{2(D_x R_d t)^{1/2}}\right] + \exp\left[\frac{x(U+P)}{2D_x}\right]\text{erfc}\left[\frac{R_d x + Pt}{2(D_x R_d t)^{1/2}}\right]\right\}$$ (3-12)

where

$$P = [U^2 + 4D_x R_d(v + \gamma)]^{1/2}$$ (3-13)

Equation (3-12) is the same as the equation derived by Marino (1974, Eq. [35], p. 1016) for $R_d = 1$ with the exception of notation for different variables.

Special Case for $\gamma = 0$ and $v = 0$
If $\gamma = 0$ and $v = 0$, P, given by Eq. (3-13), becomes equal to U and Eq. (3-12) takes the form

$$C(x, t) = \frac{1}{2}C_M\left\{\text{erfc}\left[\frac{R_d x - Ut}{2(D_x R_d t)^{1/2}}\right] + \exp\left(\frac{Ux}{D_x}\right)\text{erfc}\left[\frac{R_d x + Ut}{2(D_x R_d t)^{1/2}}\right]\right\}$$ (3-14)

It appears that Eq. (3-14) was derived by several researchers independently. According to Marino (1974), Eq. (3-14) was first derived for $R_d = 1$ by Glueckauf et al. (1949). However, Bear (1972) states that Eq. (3-14) was first derived by Gröbner and Hofreiter (1949, 1950). On the other hand, according to van Genuchten and Parker (1984), Eq. (3-14) was first derived by Lapidus and Amundsen (1952). Finally, Ogata and Banks (1961) present another derivation of Eq. (3-14) for $R_d = 1$. With $R_d = 1$, Eq. (3-14) becomes equivalent to the equation included in Freeze and Cherry (1979, Eq. [9.5], p. 391).

Error and Complementary Error Functions
Equation (3-14) is widely used for both concentration prediction purposes as well as the determination of solute transport parameters under laboratory and field conditions. In Eq. (3-14), erfc(x) is the *complementary error function* and erf(x) is the *error function* and they are defined as (e.g., Carslaw and Jaeger, 1959; Spiegel, 1968)

$$\text{erfc}(x) = 1 - \text{erf}(x) = \frac{2}{\pi^{1/2}} \int_x^\infty e^{-u^2} du, \ \text{erf}(x) = \frac{2}{\pi^{1/2}} \int_0^x e^{-u^2} du \qquad (3\text{-}15)$$

The exponential function, exp(A), and the complementary error function, erfc(B), mostly appear as a product: exf(A,B) = exp(A) erfc(B) in most analytical solutions for solute transport. Two different approximations are used for exf(A,B) (Abramovitz and Stegun, 1970; van Genuchten and Alves, 1982). For $0 \leq B \leq 3$ (Abramowitz and Stegun, 1970, Eq. [7.1.26])

$$\text{exf}(A, B) = \exp(A) \, \text{erfc}(B) \cong \exp(A - B^2)(a_1\tau + a_2\tau^2 + a_3\tau^3 + a_4\tau^4 + a_5\tau^5) \qquad (3\text{-}16)$$

where

$$\tau = \frac{1}{1+0.3275911B} \qquad (3\text{-}17)$$

$a_1 = 0.2548296$, $a_2 = -0.2844967$, $a_3 = 1.421414$, $a_4 = -1.453152$, and $a_5 = 1.061405$, and for $B > 3$ (Abramowitz and Stegun, 1970, Eq. [7.1.14]):

$$\text{exf}(A, B) = \exp(A) \, \text{erfc}(B) \cong \frac{1}{\pi^{1/2}} \exp(A - B^2)/(B + 0.5)/(B + 1)/(B + 1.5)/(B + 2)/$$

$$(B + 2.5)/(B + 1)))))) \qquad (3\text{-}18)$$

For negative values of B, the following additional relation is used:

$$\text{exf}(A, B) = 2\exp(A) - \text{exf}(A, -B) \qquad (3\text{-}19)$$

The function exf(A, B) cannot be used for very small or very large values of its arguments A and B. The function returns zero for the following two conditions:

$$|A| > 170, \quad B \leq 0 \quad \text{or} \quad |A - B^2| > 170, \quad b > 0 \qquad (3\text{-}20)$$

Hand calculations can also be carried out by using tables for the error function and complementary error function. Table 3-1 presents values for these functions for a practical range of values. Table 3-1 was prepared using a FORTRAN program presented in van Genuchten and Alves (1982, p. 115, Table 4), which is based on Eqs. (3-15)–(3-20).

Graphical Form of Eq. (3-14) for $R_d = 1$
By defining the following variables

$$\xi = \frac{Ut}{x}, \quad \eta = \frac{D_x}{Ux} \qquad (3\text{-}21)$$

TABLE 3-1
Values of Error Function [erf(x)] and Complementary Error Function [erfc(x)] for Different x Values

x	erf(x)	erfc(x)	x	erf(x)	erfc(x)	x	erf(x)	erfc(x)
0.01	0.011284	0.988716	0.49	0.511668	0.488332	0.97	0.829870	0.170130
0.02	0.022565	0.977435	0.50	0.520500	0.479500	0.98	0.834231	0.165769
0.03	0.033841	0.966159	0.51	0.529244	0.470756	0.99	0.838508	0.161492
0.04	0.045111	0.954889	0.52	0.537899	0.462101	1.00	0.842701	0.157299
0.05	0.056372	0.943628	0.53	0.546464	0.453536	1.01	0.846810	0.153190
0.06	0.067622	0.932378	0.54	0.554939	0.445061	1.02	0.850838	0.149162
0.07	0.078858	0.921142	0.55	0.563323	0.436677	1.03	0.854784	0.145216
0.08	0.090078	0.909922	0.56	0.571616	0.428384	1.04	0.858650	0.141350
0.09	0.101281	0.898719	0.57	0.579816	0.420184	1.05	0.862436	0.137564
0.10	0.112463	0.887537	0.58	0.587923	0.412077	1.06	0.866144	0.133856
0.11	0.123623	0.876377	0.59	0.595937	0.404063	1.07	0.869773	0.130227
0.12	0.134758	0.865242	0.60	0.603856	0.396144	1.08	0.873326	0.126674
0.13	0.145867	0.854133	0.61	0.611681	0.388319	1.09	0.876803	0.123197
0.14	0.156947	0.843053	0.62	0.619412	0.380588	1.10	0.880205	0.119795
0.15	0.167996	0.832004	0.63	0.627046	0.372954	1.11	0.883533	0.116467
0.16	0.179012	0.820988	0.64	0.634586	0.365414	1.12	0.886788	0.113212
0.17	0.189992	0.810008	0.65	0.642029	0.357971	1.13	0.889971	0.110029
0.18	0.200936	0.799064	0.66	0.649377	0.350623	1.14	0.893082	0.106918
0.19	0.211840	0.788160	0.67	0.656628	0.343372	1.15	0.896124	0.103876
0.20	0.222702	0.777298	0.68	0.663782	0.336218	1.16	0.899096	0.100904
0.21	0.233522	0.766478	0.69	0.670840	0.329160	1.17	0.902000	0.098000
0.22	0.244296	0.755704	0.70	0.677801	0.322199	1.18	0.904837	0.095163
0.23	0.255022	0.744978	0.71	0.684666	0.315334	1.19	0.907608	0.092392
0.24	0.265700	0.734300	0.72	0.691433	0.308567	1.20	0.910314	0.089686
0.25	0.276326	0.723674	0.73	0.698104	0.301896	1.21	0.912956	0.087044
0.26	0.286900	0.713100	0.74	0.704678	0.295322	1.22	0.915534	0.084466
0.27	0.297418	0.702582	0.75	0.711156	0.288844	1.23	0.918050	0.081950
0.28	0.307880	0.692120	0.76	0.717537	0.282463	1.24	0.920505	0.079495
0.29	0.318283	0.681717	0.77	0.723822	0.276178	1.25	0.922900	0.077100
0.30	0.328627	0.671373	0.78	0.730010	0.269990	1.26	0.925236	0.074764
0.31	0.338908	0.661092	0.79	0.736103	0.263897	1.27	0.927514	0.072486
0.32	0.349126	0.650874	0.80	0.742101	0.257899	1.28	0.929734	0.070266
0.33	0.359279	0.640721	0.81	0.748003	0.251997	1.29	0.931899	0.068101
0.34	0.369365	0.630635	0.82	0.753811	0.246189	1.30	0.934008	0.065992
0.35	0.379382	0.620618	0.83	0.759524	0.240476	1.31	0.936063	0.063937
0.36	0.389330	0.610670	0.84	0.765143	0.234857	1.32	0.938065	0.061935
0.37	0.399206	0.600794	0.85	0.770668	0.229332	1.33	0.940015	0.059985
0.38	0.409009	0.590991	0.86	0.776100	0.223900	1.34	0.941914	0.058086
0.39	0.418739	0.581261	0.87	0.781440	0.218560	1.35	0.943762	0.056238
0.40	0.428392	0.571608	0.88	0.786687	0.213313	1.36	0.945562	0.054438
0.41	0.437969	0.562031	0.89	0.791843	0.208157	1.37	0.947313	0.052687
0.42	0.447468	0.552532	0.90	0.796908	0.203092	1.38	0.949016	0.050984
0.43	0.456887	0.543113	0.91	0.801883	0.198117	1.39	0.950673	0.049327
0.44	0.466225	0.533775	0.92	0.806768	0.193232	1.40	0.952285	0.047715
0.45	0.475482	0.524518	0.93	0.811563	0.188437	1.41	0.953853	0.046147
0.46	0.484655	0.515345	0.94	0.816271	0.183729	1.42	0.955376	0.044624
0.47	0.493745	0.506255	0.95	0.820891	0.179109	1.43	0.956857	0.043143
0.48	0.502750	0.497250	0.96	0.825424	0.174576	1.44	0.958297	0.041703

(Continued)

TABLE 3-1 (*Continued*)

x	erf(x)	erfc(x)	x	erf(x)	erfc(x)	x	erf(x)	erfc(x)
1.45	0.959695	0.040305	1.96	0.994426	0.005574	2.47	0.999522	0.000478
1.46	0.961054	0.038946	1.97	0.994664	0.005336	2.48	0.999547	0.000453
1.47	0.962373	0.037627	1.98	0.994892	0.005108	2.49	0.999571	0.000429
1.48	0.963654	0.036346	1.99	0.995111	0.004889	2.50	0.999593	0.000407
1.49	0.964898	0.035102	2.00	0.995322	0.004678	2.51	0.999614	0.000386
1.50	0.966105	0.033895	2.01	0.995525	0.004475	2.52	0.999634	0.000366
1.51	0.967277	0.032723	2.02	0.995719	0.004281	2.53	0.999654	0.000346
1.52	0.968414	0.031586	2.03	0.995906	0.004094	2.54	0.999672	0.000328
1.53	0.969516	0.030484	2.04	0.996086	0.003914	2.55	0.999689	0.000311
1.54	0.970586	0.029414	2.05	0.996258	0.003742	2.56	0.999706	0.000294
1.55	0.971623	0.028377	2.06	0.996423	0.003577	2.57	0.999721	0.000279
1.56	0.972628	0.027372	2.07	0.996582	0.003418	2.58	0.999736	0.000264
1.57	0.973603	0.026397	2.08	0.996734	0.003266	2.59	0.999750	0.000250
1.58	0.974547	0.025453	2.09	0.996880	0.003120	2.60	0.999764	0.000236
1.59	0.975462	0.024538	2.10	0.997020	0.002980	2.61	0.999777	0.000223
1.60	0.976348	0.023652	2.11	0.997155	0.002845	2.62	0.999789	0.000211
1.61	0.977207	0.022793	2.12	0.997283	0.002717	2.63	0.999800	0.000200
1.62	0.978038	0.021962	2.13	0.997407	0.002593	2.64	0.999811	0.000189
1.63	0.978843	0.021157	2.14	0.997525	0.002475	2.65	0.999821	0.000179
1.64	0.979622	0.020378	2.15	0.997638	0.002362	2.66	0.999831	0.000169
1.65	0.980376	0.019624	2.16	0.997747	0.002253	2.67	0.999841	0.000159
1.66	0.981105	0.018895	2.17	0.997851	0.002149	2.68	0.999849	0.000151
1.67	0.981810	0.018190	2.18	0.997951	0.002049	2.69	0.999858	0.000142
1.68	0.982493	0.017507	2.19	0.998046	0.001954	2.70	0.999866	0.000134
1.69	0.983153	0.016847	2.20	0.998137	0.001863	2.71	0.999873	0.000127
1.70	0.983790	0.016210	2.21	0.998224	0.001776	2.72	0.999880	0.000120
1.71	0.984407	0.015593	2.22	0.998308	0.001692	2.73	0.999887	0.000113
1.72	0.985003	0.014997	2.23	0.998388	0.001612	2.74	0.999893	0.000107
1.73	0.985578	0.014422	2.24	0.998464	0.001536	2.75	0.999899	0.000101
1.74	0.986135	0.013865	2.25	0.998537	0.001463	2.76	0.999905	0.000095
1.75	0.986672	0.013328	2.26	0.998607	0.001393	2.77	0.999910	0.000090
1.76	0.987190	0.012810	2.27	0.998674	0.001326	2.78	0.999916	0.000084
1.77	0.987691	0.012309	2.28	0.998738	0.001262	2.79	0.999920	0.000080
1.78	0.988174	0.011826	2.29	0.998798	0.001202	2.80	0.999925	0.000075
1.79	0.988641	0.011359	2.30	0.998857	0.001143	2.81	0.999929	0.000071
1.80	0.989090	0.010910	2.31	0.998912	0.001088	2.82	0.999933	0.000067
1.81	0.989524	0.010476	2.32	0.998965	0.001035	2.83	0.999937	0.000063
1.82	0.989943	0.010057	2.33	0.999016	0.000984	2.84	0.999941	0.000059
1.83	0.990347	0.009653	2.34	0.999064	0.000936	2.85	0.999944	0.000056
1.84	0.990736	0.009264	2.35	0.999111	0.000889	2.86	0.999948	0.000052
1.85	0.991111	0.008889	2.36	0.999155	0.000845	2.87	0.999951	0.000049
1.86	0.991472	0.008528	2.37	0.999197	0.000803	2.88	0.999954	0.000046
1.87	0.991821	0.008179	2.38	0.999237	0.000763	2.89	0.999956	0.000044
1.88	0.992156	0.007844	2.39	0.999275	0.000725	2.90	0.999959	0.000041
1.89	0.992479	0.007521	2.40	0.999311	0.000689	2.91	0.999961	0.000039
1.90	0.992790	0.007210	2.41	0.999346	0.000654	2.92	0.999964	0.000036
1.91	0.993090	0.006910	2.42	0.999379	0.000621	2.93	0.999966	0.000034
1.92	0.993378	0.006622	2.43	0.999411	0.000589	2.94	0.999968	0.000032
1.93	0.993656	0.006344	2.44	0.999441	0.000559	2.95	0.999970	0.000030
1.94	0.993922	0.006078	2.45	0.999469	0.000531	2.96	0.999972	0.000028
1.95	0.994179	0.005821	2.46	0.999497	0.000503	2.97	0.999973	0.000027

(Continued)

TABLE 3-1 (Continued)

x	erf(x)	erfc(x)	x	erf(x)	erfc(x)	x	erf(x)	erfc(x)
2.98	0.999975	0.000025	3.18	0.999993	0.000007	3.38	0.999998	0.000002
2.99	0.999976	0.000024	3.19	0.999994	0.000006	3.39	0.999998	0.000002
3.00	0.999978	0.000022	3.20	0.999994	0.000006	3.40	0.999998	0.000002
3.01	0.999979	0.000021	3.21	0.999994	0.000006	3.41	0.999999	0.000001
3.02	0.999981	0.000019	3.22	0.999995	0.000005	3.42	0.999999	0.000001
3.03	0.999982	0.000018	3.23	0.999995	0.000005	3.43	0.999999	0.000001
3.04	0.999983	0.000017	3.24	0.999995	0.000005	3.44	0.999999	0.000001
3.05	0.999984	0.000016	3.25	0.999996	0.000004	3.45	0.999999	0.000001
3.06	0.999985	0.000015	3.26	0.999996	0.000004	3.46	0.999999	0.000001
3.07	0.999986	0.000014	3.27	0.999996	0.000004	3.47	0.999999	0.000001
3.08	0.999987	0.000013	3.28	0.999996	0.000004	3.48	0.999999	0.000001
3.09	0.999988	0.000012	3.29	0.999997	0.000003	3.49	0.999999	0.000001
3.10	0.999988	0.000012	3.30	0.999997	0.000003	3.50	0.999999	0.000001
3.11	0.999989	0.000011	3.31	0.999997	0.000003	3.51	0.999999	0.000001
3.12	0.999990	0.000010	3.32	0.999997	0.000003	3.52	0.999999	0.000001
3.13	0.999990	0.000010	3.33	0.999998	0.000002	3.53	0.999999	0.000001
3.14	0.999991	0.000009	3.34	0.999998	0.000002	3.54	0.999999	0.000001
3.15	0.999992	0.000008	3.35	0.999998	0.000002	3.55	0.999999	0.000001
3.16	0.999992	0.000008	3.36	0.999998	0.000002	3.56	1.000000	0.000000
3.17	0.999993	0.000007	3.37	0.999998	0.000002			

Ogata and Banks (1961) expressed Eq. (3-14) for $R_d = 1$ as

$$\frac{C(\xi, \eta)}{C_M} = \frac{1}{2}\left\{\operatorname{erfc}\left[\frac{1-\xi}{2(\xi\eta)^{1/2}}\right] + \exp\left(\frac{1}{\eta}\right)\operatorname{erfc}\left[\frac{1+\xi}{2(\xi\eta)^{1/2}}\right]\right\} \qquad (3\text{-}22)$$

Curves based on Eq. (3-22) are shown in Figure 3-3 for various values of the dimensionless group $\eta = D_x/(Ux)$ for $C(\xi, \eta)/C_M$ vs. $\xi = Ut/x$. As can be seen from Figure 3-3, as η becomes small the concentration distribution becomes nearly symmetrical to the value of $\xi = 1$. However, for large values of η, asymmetrical concentration distributions become noticeable. This indicates that for large values of D_x or small values of x the contribution of the second term in Eq. (3-22) becomes significant as ξ approaches unity.

Approximated Form of Eq. (3-22)
According to Ogata and Banks (1961), the second term of Eq. (3-22) can be neglected with a maximum error of less than 3% for values of $\eta = D_x/(Ux) < 0.002$. Therefore, under these conditions, Eq. (3-22) becomes

$$\frac{C(\xi, \eta)}{C_M} \cong \frac{1}{2}\operatorname{erfc}\left[\frac{1-\xi}{2(\xi\eta)^{1/2}}\right] \qquad (3\text{-}23)$$

Unsteady-State Convective–Dispersive Flux Component
From Eq. (2-41a), the convective–dispersive flux in the x-coordinate direction is

$$F_x = \varphi_e UC - \varphi_e D_x \frac{\partial C}{\partial x} \qquad (3\text{-}24)$$

where F_x is the convective–dispersive flux component in the x direction and the other components, F_y and F_z, are all zero because the solute transport is one-dimensional. In Eq. (3-24), φ_e is the effective porosity, and all the other parameters are defined earlier. Since the dimensions of U and C are

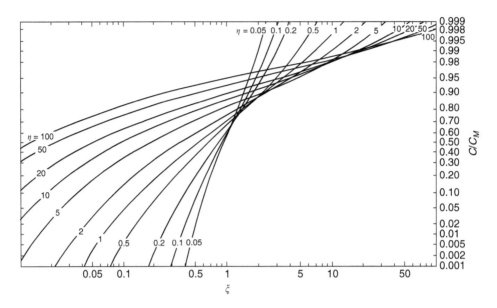

FIGURE 3-3   Graphical representation of Eq. (3-22). (Adapted from Ogata, A. and Banks, R.B., *Geological Survey Professional Paper 411-A*, United States Government Printing Office, Washington, DC, 1961, 7pp.)

[L/T] and [M/L^3], respectively, the dimensions of F_x is [M/L^2/T]. This physically means that F_x is the mass of solute entering per unit area per unit time.

From Eqs. (3-7) and (3-24), the expression for F_x can be determined by applying the common differentiation rules for exponential and error functions. For the error function given by Eq. (3-15), Leibnitz's rule for differentiating an integral (see, e.g., Spiegel, 1968, 1971) needs to be applied. The procedure outlined in Example 3-6 is used. The final expression for $F_x(x, t)$ is

$$F_x(x, t) = \varphi_e U C(x, t) - \frac{1}{2}\varphi_e C_M P_1 S_1 - \varphi_e C_M D_x P_1 (S_2 + S_3) \tag{3-25}$$

where $C(x, t)$ and P_1 are given by Eqs. (3-7) and (3-8), respectively, and

$$S_1 = \exp(-P_2)\,\mathrm{erfc}(P_3 - P_4) + \exp(P_2)\,\mathrm{erfc}(P_3 + P_4) \tag{3-26}$$

$$S_2 = -\,Q_1\exp(-P_2)\,\mathrm{erfc}(P_3 - P_4) - \exp(-P_2)P_{10}\exp[-(P_3 - P_4)^2] \tag{3-27}$$

$$S_3 = Q_1\exp(P_2)\,\mathrm{erfc}(P_3 + P_4) - \exp(P_2)P_{10}\exp[-(P_3 + P_4)^2] \tag{3-28}$$

$$Q_1 = \left(\frac{vR_d}{D_x} + \frac{U^2}{4D_x^2} + \frac{\gamma R_d}{D_x}\right)^{1/2} \tag{3-29}$$

and

$$P_{10} = \frac{R_d}{\pi(D_x R_d t)^{1/2}} \tag{3-30}$$

Steady-State Solution for Concentration Distribution
Under steady-state solute transport conditions, Eq. (3-1) takes the form

$$D_x\frac{\partial^2 C}{\partial x^2} - U\frac{\partial C}{\partial x} - R_d \nu C = 0 \tag{3-31}$$

and the initial condition given by Eq. (3-2) is not necessary. The solution for C will only be the function of x. Therefore, the boundary conditions given by Eqs. (3-3), (3-4), and (3-5) will have, respectively, the following forms for $C(x)$ with $\gamma = 0$:

$$C(0) = C_M \tag{3-32}$$

$$C(\infty) = 0 \tag{3-33}$$

$$\frac{\partial C(\infty)}{\partial x} = 0 \tag{3-34}$$

One must bear in mind that if $v = 0$ in Eq. (3-31), the solution of the boundary-value problem defined by Eqs. (3-31)–(3-34) is known, that is $C = C_M$. In other words, the concentration at every point of the x coordinate becomes equal to C_M. However, this is not the case if $v \geq 0$. The solution for this case be derived either by solving the above-defined boundary-value problem or using Eq. (3-7) with $\gamma = 0$ and making t to go infinity. Here, the second procedure is used with the following equations:

$$erfc(x) = 1 - erf(x) \tag{3-35}$$

$$erf(-x) = -erf(x) \tag{3-36}$$

$$erf(\infty) = 1 \tag{3-37}$$

Using these equations and, for $\gamma = 0$ and $t \to \infty$, Eq. (3-7) gives

$$C(x) = C_M \exp\left[\frac{Ux}{2D_x} - x\left(\frac{vR_d}{D_x} + \frac{U^2}{4D_x^2} \right)^{1/2} \right] \tag{3-38}$$

If $v = 0$ in Eq. (3-38), it can be easily seen that $C(x) = C_M$.

Steady-State Convective–Dispersive Flux Component
From Eqs. (3-24) and (3-38), the convective–dispersive flux component expression under steady-state solute transport conditions can be written as

$$F_x(x) = C(x)\left\{ \varphi_e U - \left[\frac{U}{2} - D_x\left(\frac{vR_d}{D_x} + \frac{U^2}{4D_x^2} \right)^{1/2} \right] \right\} \tag{3-39}$$

Applications of the One-Dimensional Analytical Models
One-dimensional analytical solute transport models potentially have a lot of applications for finding quick solutions to many problems faced in practice. These models can be applied for solute transport predictions in saturated zones and unsaturated zones as outlined in Sections 3.3.2.2 and 3.3.2.3. Here, some important concepts as well as examples regarding their application will be presented.

Theoretical Breakthrough Curves
From Eq. (3-14) with $R_d = 1$, the normalized concentration C/C_M vs. time t at a given distance from the source results in an S-shaped curve as shown in Figure 3-4. As can be seen from Figure 3-4, the concentration breakthrough curves arrive at points further from the source at later times and are more spread out. Figure 3-5 presents the normalized concentration vs. distance curves at various times. Figure 3-4 and Figure 3-5 show that U is the velocity of the $C/C_M = 0.5$ point on the concentration profile of the constituent. For a retarded constituent, U_c is the velocity of the $C/C_M = 0.5$

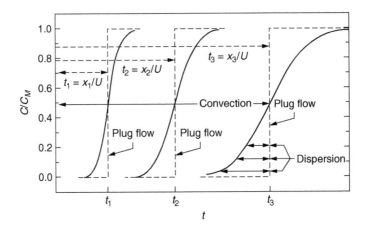

FIGURE 3-4 Normalized concentration, C/C_M, vs. time, t, curves derived from the one-dimensional convection-dispersion equation. (Adapted from Palmer, C.D. and Johnson, R.L., *Transport and Fate of contaminants in the subsurface* U.S. EPA/625/4-89/019, September 1989, pp. 5–22, Chap. 2.)

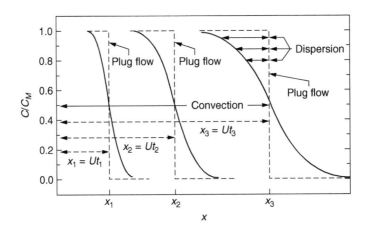

FIGURE 3-5 Normalized concentration, C/C_M, vs. distance, x, curves derived from the one-dimensional convection–dispersion equation. (Adapted from Palmer, C.D. and Johnson, R.L., U.S. EPA/625/4-89/019, September 1989, pp. 5–22, Chap. 2.)

point on the concentration profile of the retarded constituent and from Eq. (2-62) the expression for U_c is

$$U_c = \frac{U}{R_d} = \frac{U}{1 + (\rho_b/\varphi)K_d} \qquad (3\text{-}40)$$

As mentioned in Section 2.6.1, Eq. (2-62) or Eq. (3-40) is commonly known as the retardation factor equation.

Experimental Breakthrough Curves for Saturated Porous Media and Evaluation of the Model
Breakthrough curves reveal important features regarding the transport of solutes in saturated porous media. Some of these features are presented below (Gillham and Cherry, 1982b).

During the past four decades, a large number of laboratory experiments have been conducted to determine the validity of the convection-dispersion model. The majority of these studies have been one-dimensional miscible-displacement experiments in which the concentration history of the column effluent was compared with the concentrations predicted by the model. The one-dimensional convection–dispersion equation is given by Eq. (2-83) and Eq. (3-1). For an experiment in which a column of porous medium is initially saturated with solution having a tracer concentration of zero and the initial solution is displaced with solution of concentration C_0, the concentration history in the column effluent is generally represented by a breakthrough curve. This curve is a graph of dimensionless concentration, C/C_M vs. dimensionless time represented by the number of pore volumes, PV. Here C is the measured concentration in the effluent and PV is defined as

$$PV = \frac{\text{the volume of effluent}}{\text{total volume of voids in the porous medium sample}} \tag{3-41}$$

Equation (3-14), which is a special form of Eq. (3-7), is a well-known model in practice. Typical deviations from Eq. (3-14) with $R_d = 1$ are shown in Figure 3-6. When UL/D_x (L is the column length) is large, the breakthrough curve is symmetrical and passes through the point $C/C_M = 0.5$ and PV = 1.0.

For nonreactive tracers, experimental variability undoubtedly contributes to the observed differences. By considering a large number of experimental data, Scheidegger (1963) concluded that consistent discrepancies generally occur. These discrepancies have been attributed to the presence of dead-end pores (Deans, 1963; Coats and Smith, 1964; Fatt et al., 1966; Baker, 1977) and to solution–solid interface processes that are not accounted for in the spatial averaging process (Fried, 1975). Fried noted that discrepancies between measured and predicted results seldom exceed 2 to 3% and concluded that, for all practical purposes, the model is a good representation of the dispersion process.

Although the conclusion of Fried (1975) appears to be appropriate for coarse-grained geologic materials, there are a number of identified cases for which the convection–dispersion equation is not an adequate model. In fine-grained materials where an anion was used as the tracer, the early appearance of the breakthrough curve has been attributed to anion exclusion (Kemper and Rollins, 1966; Thomas and Swoboda, 1970; Appelt et al., 1975). In aggregated media, an extreme tailing of the breakthrough curve has been attributed to diffusion onto or out of the aggregates (Philip, 1968; Passioura, 1971; Passioura and Rose, 1971; Weeks et al., 1976), and similar effects observed for unsaturated media have been attributed to diffusion into relatively immobile zones of water (Orlob and Radhakrishna, 1958; Biggar and Nielsen, 1960). Typical deviations from the model are shown in Figure 3-6.

The Effect of Velocity on the Longitudinal Dispersion Coefficient
Laboratory experiments have provided considerable information on the dispersion process, such as the effect of velocity on the dispersion coefficient. Much of the available data was compiled by

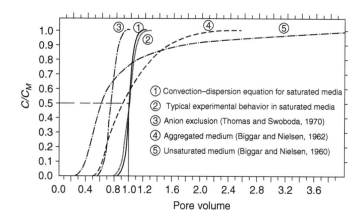

FIGURE 3-6 Typical deviations from the breakthrough curve in saturated porous media predicted by the one-dimensional first-type source model for $R_d = 1$ given by Eq. (3-14). (Adapted from Gillham, R.W., and Cherry, J.A., *Recent Trends in Hydrogeology*, The Geological Society of America, Special Paper 189, Boulder, CO, 1982b, pp. 31–62.)

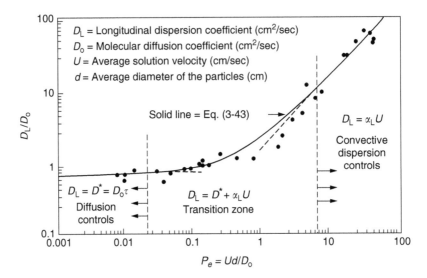

FIGURE 3-7 Dimensionless longitudinal dispersion coefficient, D_L/D_0, vs. Peclet number, $Pe = Ud/D_0$. (Adapted from Perkins, T.K. and Johnston, O.C., *Soc. Petrol. Eng. J.*, 3, 70–84, 1963.)

Perkins and Johnston (1963) and later by Rose (1973). Figure 3-7 presents a graph of dimensionless longitudinal dispersion coefficient D_L/D_0 (D_L is equivalent to D_x in Eq. [3-1]) vs. the *Peclet number* Pe, defined as

$$Pe = \frac{Ud}{D_0} \tag{3-42}$$

where D_0 is the molecular diffusion coefficient in pure water (see Eq. [2-22]) and d is a characteristic length, generally taken to be the mean grain diameter. As can be seen from Figure 3-7, the longitudinal dispersion coefficient is strongly dependent on the average solution velocity U as well as on the geometrical properties of the medium, represented by d. At low Peclet numbers, the dispersion coefficient has a constant value and is equal to the effective molecular diffusion coefficient of the tracer in porous medium. Figure 3-7 also shows that the longitudinal dispersion coefficient increases with increase in Peclet number and becomes a linear function of Pe at Peclet numbers in excess of approximately 1.0.A graph of the following empirical equation of Perkins and Johnston (1963) is included in Figure 3-7.

$$\frac{D_L}{D_0} = \frac{D^*}{D_0} + 1.75\frac{Ud}{D_0} \tag{3-43}$$

Figure 3-7 and Eq. (3-43) show that at low velocities (low *Pe*), the dispersion process is predominantly the result of molecular diffusion, whereas at high velocities the mechanical mixing process becomes dominant. At intermediate values of *Pe*, both diffusion and mechanical mixing play a significant role in the dispersion process.

Molecular Diffusion and Convection (Advection)-Dominated Solute Transport
As shown in Figure 3-7, mechanical mixing dominates the longitudinal dispersion coefficient only at high Peclet numbers. At low Peclet numbers, the longitudinal dispersion coefficient is approximately equal to the effective molecular diffusion coefficient. These features are shown in the following example.

EXAMPLE 3-1

This example is adapted from Gillham and Cherry (1982b). For a fine sand aquifer, the following values are given: $K_h = 1.0\times10^{-3}$ cm/sec (horizontal hydraulic conductivity), $d = 0.02$ cm (mean grain diameter), $\varphi_e = 0.35$ (effective porosity), and $I = 1.0 \times 10^{-2}$ cm/cm. In this aquifer, chloride migration is not known and $D_0 = 1.26 \times 10^{-9}$ m²/sec (molecular diffusion coefficient for chloride in pure water) (see Table 2-1). See the equations in Section 2.4.2.2 for definitions of these parameters. Using the values given above, determine (1) the value of Peclet number and (2) the status of the transport and the value of the longitudinal dispersion coefficient.

SOLUTION

1. From Eq. (2-39), the Darcy velocity is

$$q = K_h I = (1.0 \times 10^{-3} \text{cm/sec})(1.0 \times 10^{-2} \text{cm/cm}) = 1.0 \times 10^{-5} \text{cm/sec}$$

From the first expression of Eq. (2-40), we get

$$U = \frac{q}{\varphi_e} = \frac{1.0 \times 10^{-5} \text{cm/sec}}{0.35} = 2.86 \times 10^{-6} \text{cm/sec}$$

From Eq. (3-42), the Peclet number is

$$Pe = \frac{Ud}{D_0} = \frac{(2.86 \times 10^{-6} \text{ cm/sec})(0.02 \text{ cm})}{1.26 \times 10^{-5} \text{cm}^2/\text{sec}} = 4.54 \times 10^{-3}$$

2. From Figure 3-7 the corresponding value is $D_L/D_0 = 0.70$, i.e., the longitudinal dispersion coefficient $D_L = 0.7D_0 = D^*$. According to Eq. (2-22), the tortuosity factor τ is 0.7. Therefore,

$$D_L = D^* = 0.7D_0 = (0.7)(1.26 \times 10^{-5} \text{ cm}^2/\text{sec}) = 0.88 \times 10^{-5} \text{ cm}^2/\text{sec}$$

and thus the longitudinal dispersion coefficient is equal to the effective molecular diffusion coefficient. As a conclusion, it can be said that the transport is basically controlled by molecular diffusion. In other words, it is molecular diffusion-dominated solute transport.

A Computer Program (ST1A) for the One-Dimensional First-Type Source Model
A one-dimensional MS-DOS-based FORTRAN solute transport computer program, called solute transport one-dimensional first-type (ST1A), has been developed by the author. ST1A is a menu-driven program and calculates values of concentration and convective-dispersive flux components under unsteady state solute transport conditions [Eqs. (3-7) and (3-25)] and steady state solute transport conditions [Eqs. (3-38) and (3-39)]. This program can be run interactively in IBM compatible computers.

As mentioned above, the exponential term, $e^{\gamma t}$, increases the capability of the model such that exponentially varying solute source concentrations can be taken into account. With negative $\gamma (T^{-1})$ values, a decaying source concentration can be represented in the model. There are some requirements in the usage of this parameter. In Eqs. (3-9) and (3-11), γ is a square root term. Therefore, in order to take these square roots, some of the terms under each square root must be greater than zero. A relatively simple analysis can show that the necessary condition for this requirement is as follows:

$$v + \frac{U^2}{4D_x R_d} > |-\gamma| \tag{3-44}$$

For example, if $v = 0$, $U = 0.15$ m/day, $D_x = 3.195$ m²/day, and $R_d = 1$, the condition is 0.00176 day^{-1} > $|-\gamma|$. This means that for this transport field as long as the absolute values of $\gamma < 0.00176$ day^{-1}, decaying sources can be handled by the model.

Two examples based on the ST1A computer program are presented below.

EXAMPLE 3-2

For the ground water flow field of an aquifer, the following parameters are given: $C_M = 1$ mg/L = 1 g/m^3, $K_h = 10$ m/day, $I = 0.0025$ m/m, $\varphi_e = 0.25$, $D^* = 0$ m^2/day, $K_d = 0$ cm^3/g, $\rho_b = 1.8$ g/cm^3, $\gamma = 0$ day^{-1}, and $v = 0$ day^{-1}. Using the ST1A computer program, (1) determine the convective–dispersive flux variation at $x = 0$ for $\alpha_L = 0.1$, 1, and 10 m and (2) discuss the convective–dispersive flux variation at $x = 0$ with time for different α_L values.

SOLUTION

1. The convective–dispersive flux, $F_x(0, t)$, vs. time for $\alpha_L = 0.1$, 1, and 10 m is shown in Figure 3-8.
2. As can be seen from Figure 3-8, the flux is initially much higher; in fact, at zero time, its value is infinite at the source. From the same figure it can also be observed that the magnitude of the flux increases with the increase in longitudinal dispersivity α_L value. However, as time increases, the values of $F_x(0, t)$ approach to a constant value of 0.025 g/m^2/day. Figure 3-8 shows that as α_L becomes smaller, the approach to the constant flux value occurs earlier. As can be observed from Eq. (3-24), the longitudinal dispersion coefficient D_x is responsible for this situation. For $D_x = 0$ at the source, Eq. (3-24) gives $F_x = \varphi_e U C_M = K_h I C_M = (10$ m/day$)(0.0025$ m/m$)$ $(1$ g/m$^3) = 0.025$ g/m^2/day, which is exactly the same as the value in Figure 3-8 at greater t values. In conclusion, it can be said that for the first-type source cases, the convective–dispersive flux at the source is initially at a high value and with the increase in time it approaches a constant value. A comparative evaluation of this type of fluxes is presented in Section 3.3.3.1. A theoretical investigation of this type of fluxes is presented in Batu and van Genuchten (1990).

EXAMPLE 3-3

In the ground water flow field of an aquifer the following parameters are given: $C_M = 1$ mg/L = 1 g/m^3, $I = 0.0025$ m/m; $\varphi_e = 0.25$; $\alpha_L = 10$ m, $D^* = 0$ m^2/day, $K_d = 0$ cm^3/g, $\rho_b = 1.8$ g/cm^3, $\gamma = 0$ day^{-1}, and $v = 0$ day^{-1}. Using the ST1A computer program at $x = 10$ m from the source for $K_h = 1$, 10, and 100 m/day, (1) determine the concentration vs. time curves, (2) determine the convective–dispersive flux vs. time curves, and (3) discuss the results.

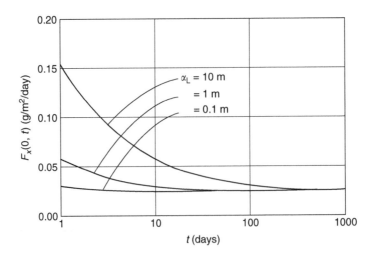

FIGURE 3-8   Convective–dispersive flux vs. time at $x = 0$ for $\alpha_L = 0.1$, 1, and 10 m in Example 3-2.

SOLUTION

1. For $K_h = 1$, 10, and 100 m/day with $\varphi_e = 0.25$, and $I = 0.0025$ m/m, $U = KI/\varphi_e$ (combination of Eqs. [2-39] and [2-40]), gives, respectively, $U = 0.01$, 0.1, and 1 m/day. The concentration vs. time curves at $x = 10$ m are shown in Figure 3-9.
2. The convective–dispersive fluxes at $x = 10$ m for $U = 0.01$, 0.1, and 1 m/day are shown in Figure 3-10.
3. Figure 3-9 shows that all the concentration vs. time curves (also called as breakthrough curves) for different U values all have similar shapes, but as the ground water velocity (U) decreases, the elapsed time increases for concentrations greater than zero. From $F_x = \varphi_e U C_M$ one gets $F_x = 0.0025$ g/m^2/day for $U = 0.01$ m/day, $F_x = 0.025$ g/m^2/day for $U = 0.1$ m/day, and $F_x = 0.25$ g/m^2/day for $U = 1$ m/day. Figure 3-10 shows that $F_x = 0.25$ g/m^2/day for $U = 1$ m/day is reached after approximately 100 days elapsed time. Figure 3-10 also shows that the corresponding fluxes for $U = 0.1$ and 0.01 m/day are reached in relatively longer time periods.

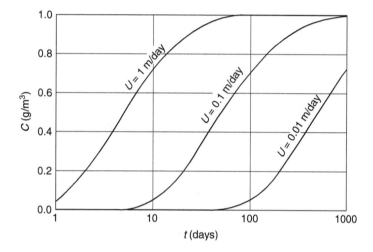

FIGURE 3-9 Concentration vs. time at $x = 10$ m for $U = 0.01$, 0.1, and 1 m/day in Example 3-3.

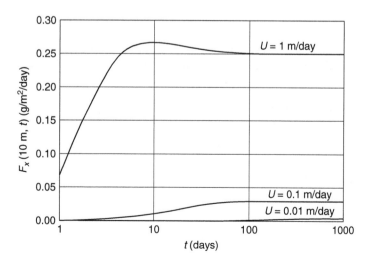

FIGURE 3-10 Convective–dispersive flux vs. time at $x = 10$ m for $U = 0.01$, 0.1, and 1 m/day in Example 3-3.

3.3.2.2 Two-Dimensional Analytical Solute Transport Models for the First-Type Strip Sources

There are a variety of analytical solute transport models in the literature based on the convective–dispersive solute transport partial differential equation published during the past half century. However, most of these analytical models started to appear in the literature for one-dimensional cases. The analytical solutions developed by Shen (1976), Cleary and Ungs (1978), and Batu (1983) take into account the dimension of the solute source that was assumed to be located in an infinite medium. Models for finite domains have more flexibility than the ones in an infinite medium in the sense that the effects of impervious horizontal and vertical planar boundaries can be taken into account. To the author's knowledge, the analytical model of Bruch and Street (1967) is the first two-dimensional analytical model in the literature for a first-type strip source located symmetrically between two impervious boundaries. Later, Batu (1989) developed a generalized two-dimensional first-type source model for bounded media and showed that the Bruch and Street model is one of its special forms. In this section, this model (Batu, 1989) as well as its applications will be presented.

Problem Statement and Assumptions
The problem under consideration is an unsteady-state two-dimensional solute transport from multiple strip first-type sources between two impervious planar boundaries in a unidirectional flow field as shown in Figure 3-11. The x- and z-axis are the planar coordinates either in a horizontal or vertical plane. The assumptions are as follows:

1. The solute sources are planar and are located irregularly along the z-axis in a unidirectional steady ground water velocity (U) field.
2. The solute sources are located at $x = 0$ and they are perpendicular to the velocity of the flow field.
3. The source concentrations are functions of the z coordinate and time (t) through an exponential function.
4. The medium goes to infinity in the x-coordinate direction and z varies between 0 and H_p.

Governing Differential Equation
The governing differential equations for solute transport in a unidirectional flow field are given in Section 2.7.3, and its two-dimensional form in the x and y coordinates is given by Eq. (2-82). Its form in the x and z coordinates is

$$\frac{\partial C}{\partial t} = \frac{D_x}{R_d}\frac{\partial^2 C}{\partial x^2} + \frac{D_z}{R_d}\frac{\partial^2 C}{\partial z^2} - \frac{U}{R_d}\frac{\partial C}{\partial x} - vC \qquad (3\text{-}45)$$

where C is the concentration, x and z are the planar coordinates, t the time, R_d the retardation factor (given by Eq. [2-62]) and v the decay constant. The potential units of C widely used in practice are included in Section 2.2.1.

Initial and Boundary Conditions
The initial condition for $C(x, z, t)$ is

$$C(x, z, 0) = 0 \qquad (3\text{-}46)$$

The boundary conditions for $C(x, z, t)$ at $x = 0$ are

$$C(0, z, t) = C_{M_1}(z)e^{\gamma t}, \quad 0 < z < H_1 \qquad (3\text{-}47)$$

$$C(0, z, t) = C_{M_{p-1}}(z)e^{\gamma t}, \quad H_{p-2} < z < H_{p-1} \qquad (3\text{-}48)$$

$$C(0, z, t) = C_{Mp}(z)e^{\gamma t}, \quad H_{p-1} < z < H_p \qquad (3\text{-}49)$$

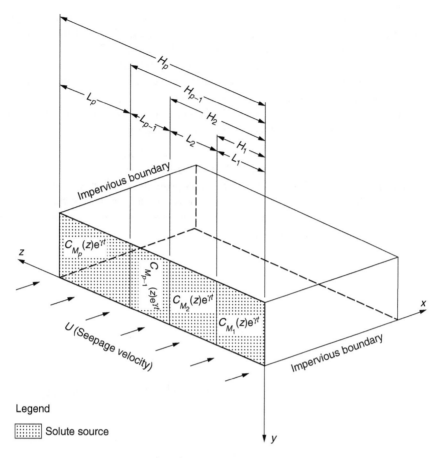

FIGURE 3-11 Schematic representation of two-dimensional solute transport from multiple strip sources with the first-type source condition in a unidirectional flow field. (Adapted from Batu, V., *Water Resour. Res.*, 25, 1125–1132, 1989.)

where

$$H_i = \sum_{j=1}^{i} L_j, \quad i = 1, 2, ..., p \tag{3-50}$$

The other boundary conditions at the impervious boundaries and infinity are as follows:

$$\frac{\partial C}{\partial z} = 0, \quad z = 0, \quad 0 < x < \infty \tag{3-51}$$

$$\frac{\partial C}{\partial z} = 0, \quad z = H_p, \quad 0 < x < \infty \tag{3-52}$$

$$C = 0, \quad x = \infty, \quad 0 < z < H_p \tag{3-53}$$

$$\frac{\partial C}{\partial x} = 0, \quad x = \infty, \quad 0 < x < H_p \tag{3-54}$$

Equation (3-46) states that initially the concentration is zero everywhere in the flow field. Equations (3-47)–(3-49) state that the concentration function $C(0, z, t)$ varies with the z coordinate and time t and is the first-type or Dirichlet boundary conditions. The function

$$\frac{C(0, z, t)}{C_{M_i}(z)} = e^{\gamma t} \tag{3-55}$$

performs such that the source concentration can be represented as exponentially varying time-dependent quantity. Equation (3-55) is a generalized version of Eq. (3-6) and the properties of the concentration function are described in Section 3.3.2.1. Figure 3-2 shows the variation of $C(0, z, t)/C_{M_i}$ with time for different values of γ ranging from -0.0001 to -0.0100 day^{-1}. If observed data are available for time-dependent source concentration, the value of γ may be determined from the curve-fitting technique. Equations (3-51) and (3-52) state that the planes at $z = 0$ and $z = H_p$ are impervious boundaries. Finally, Eqs. (3-53) and (3-54) state that both the concentration C and its gradient $(\partial C/\partial x)$ with respect to x, approach zero when the x distance goes to infinity.

Unsteady-State General Solution for Concentration Distribution
The boundary-value problem defined by Eqs. (3-45)–(3-54), was solved by the Laplace transform and Fourier analysis techniques (Batu, 1989). For this purpose, the solution was obtained first in the Laplace domain and then the concentration variation was obtained by taking the inverse Laplace transform of the Laplace domain solution. The final results are presented below and the intermediate steps are presented in the associated examples.

General Solution in the Laplace Domain
The general solution in the Laplace domain is (Batu, 1989)

$$u(x, z, s) = u_1(x, z, s) + u_2(x, z, s) \tag{3-56}$$

where

$$u_1(x, z, s) = \frac{1}{H_p(s-\gamma)}\left[\int_0^{H_p} C_M(z)dz\right]\exp\left(\frac{Ux}{2D_x} - p_0 x\right) \tag{3-57}$$

$$u_2(x, z, s) = \frac{2}{H_p(s-\gamma)}\sum_{n=1}^{\infty}\left[\int_0^{H_p} C_M(z)\cos(\lambda_n z)dz\right]\exp\left(\frac{Ux}{2D_x} - p_n x\right)\cos(\lambda_n z) \tag{3-58}$$

where

$$p_0 = \left[\frac{R_d}{D_x}(s+v) + \frac{U^2}{4D_x^2}\right]^{1/2} \tag{3-59}$$

$$p_n = \left[\frac{D_z}{D_x}\lambda_n^2 + \frac{R_d}{D_x}(s+v) + \frac{U^2}{4D_x^2}\right]^{1/2} \tag{3-60}$$

$$\lambda_n = \frac{n\pi}{H_p}, \quad n = 0, 1, 2, \ldots \tag{3-61}$$

and $u(x, z, s)$ is the Laplace transform of $C(x, z, t)$ and is defined as (e.g., Spiegel, 1965)

$$u(x, z, s) = f(s) = L[C(x, z, t)] = \int_0^{\infty} e^{-st}C(x, z, t)dt \tag{3-62}$$

Derivation of Eq. (3-56) is given in Example 3-4.

EXAMPLE 3-4

Solve the boundary-value problem defined by Eqs. (3-46)–(3-55) in the Laplace domain, i.e. derive Eq. (3-56).

SOLUTION

The Laplace transforms of each term in Eq. (3-45) with respect to time are as follows (e.g., Spiegel, 1965):

$$L\left[\frac{\partial C}{\partial t}\right] = su(x, z, s) - C(x, z, 0) \tag{E3-4-1}$$

$$L\left[\frac{\partial^2 C}{\partial x^2}\right] = \frac{\partial^2 u}{\partial x^2} \quad L\left[\frac{\partial^2 C}{\partial z^2}\right] = \frac{\partial^2 u}{\partial z^2} \quad L\left[\frac{\partial C}{\partial x}\right] = \frac{\partial u}{\partial x} \quad L\left[\frac{\partial C}{\partial z}\right] = \frac{\partial u}{\partial z} \qquad \text{(E3-4-2)}$$

where $u(x, z, s)$ is defined by Eq. (3-62). The substitution of these expressions into Eq. (3-45) gives

$$R_d(s + v)u = D_x\frac{\partial^2 u}{\partial x^2} + D_z\frac{\partial^2 u}{\partial z^2} - U\frac{\partial u}{\partial x} \qquad \text{(E3-4-3)}$$

which is the two-dimensional solute transport differential equation in the Laplace domain.

The *method of separation of variables* (e.g., Churchill, 1941; Spiegel, 1965; Batu, 1983) is applied to solve Eq. (E3-4-3). with

$$u(x, z, s) = X(x, s)Z(z, s) \qquad \text{(E3-4-4)}$$

where $X(x, s)$ and $Z(z, s)$ are some unknown functions. The substitution of Eq. (E3-4-4) into Eq. (E3-4-3) gives

$$\frac{D_x}{D_z}\frac{X''}{X'} - \frac{U}{D_z}\frac{X'}{X} - \frac{(s+v)R_d}{D_z} = -\frac{Z''}{Z} = \mu \qquad \text{(E3-4-5)}$$

where μ must be a constant. By examining the boundary conditions, it can be concluded that the constant value should be $\mu = \lambda^2 > 0$. Introducing this value into Eq. (E3-4-5), two ordinary differential equations can be obtained:

$$X'' - \frac{U}{D_x}X' - \frac{1}{D_x}[(s + v)R_d + \lambda^2 D_z]X = 0 \qquad \text{(E3-4-6)}$$

$$Z'' + \lambda^2 Z = 0 \qquad \text{(E3-4-7)}$$

Solution of Eq. (E3-4-6)

Equation (E3-4-6) is a linear, homogeneous second-order ordinary differential equation and can be written as (Spiegel, 1968, Eq. [18.7], p. 105)

$$\frac{d^2X}{dx^2} + a\frac{dX}{dx} + bX = 0 \qquad \text{(E3-4-8)}$$

where

$$a = -\frac{U}{D_x}, \quad b = -\frac{1}{D_x}[(s + v)R_d + \lambda^2 D_z] \qquad \text{(E3-4-9)}$$

The characteristic equation (E3-4-8) is

$$m^2 + am + b = 0 \qquad \text{(E3-4-10)}$$

with two roots m_1 and m_2. Three cases exists for these roots (Spiegel, 1968, p. 105), and from the inspection of the boundary conditions of the problem it can be observed that Case 1 satisfies the condition. For Case 1, m_1 and m_2 are real and distinct and the corresponding solution of Eq. (E3-4-8) is

$$X = A_1 \exp(m_1 x) + B_1 \exp(m_2 x) \qquad \text{(E3-4-11)}$$

where A_1 and B_1 are constants and

$$m_1 = -\frac{a}{2} - \left(\frac{a^2}{4} - b\right)^{1/2}, \quad m_2 = -\frac{a}{2} + \left(\frac{a^2}{4} - b\right)^{1/2} \qquad \text{(E3-4-12)}$$

By using Eq. (E3-4-9) they take the following forms:

$$m_1 = \frac{U}{2D_x} - p, \quad m_2 = \frac{U}{2D_x} + p \tag{E3-4-13}$$

where

$$p = \left[\frac{D_z}{D_x} \lambda^2 + \frac{R_d}{D_x}(s + v) + \frac{U^2}{4D_x^2} \right]^{1/2} \tag{E3-4-14}$$

Therefore, the substitution of Eq. (E3-4-13) into Eq. (E3-4-11) gives the solution of Eq. (E3-4-6) as

$$X(x, s) = A_1 \exp\left(\frac{Ux}{2D_x} - px \right) + B_1 \exp\left(\frac{Ux}{2D_x} + px \right) \tag{E3-4-15}$$

Solution of Eq. (E3-4-7)
Equation (E3-4-7) is a linear, homogeneous second-order ordinary differential equation and corresponds to Case 3 (Spiegel, 1968, Eq. [18.7], p. 105). By comparing Eq. (E3-4-7) with Eq. (E3-4-8), it can be observed that $a = 0$ and $b = \lambda^2$. Therefore, the characteristic equation given by Eq. (E3-4-10) becomes

$$m^2 + b = 0 \tag{E3-4-16}$$

and its roots are

$$m_1 = -\, i\lambda, \quad m_2 = +\, i\lambda \tag{E3-4-17}$$

and the final corresponding solution for Eq. (E3-4-16) is (Spiegel, 1968, Eq. [18-7], p. 105, Case 3)

$$Z(z, s) = A_2 \sin(\lambda z) + B_2 \cos(\lambda z) \tag{E3-4-18}$$

where A_2 and B_2 are constants.

Usage of Boundary Conditions at Infinity and Impermeable Boundaries
The substitution of Eqs. (E3-4-15) and (E3-4-18) into Eq. (E3-4-4) gives

$$u(x, z, s) = \left[A_1 \exp\left(\frac{Ux}{2D_x} - px \right) + B_1\left(\frac{Ux}{2D_x} + px \right) \right][A_2\sin(\lambda z) + B_2\cos(\lambda z)] \tag{E3-4-19}$$

and its derivative with respect to z is

$$\frac{\partial u}{\partial z} = \left[A_1 \exp\left(\frac{Ux}{2D_x} - px \right) + B_1\left(\frac{Ux}{2D_x} + px \right) \right][A_2\lambda \cos(\lambda z) - B_2\lambda \sin(\lambda z)] \tag{E3-4-20}$$

To satisfy the boundary conditions expressed by Eqs. (3-53) and (3-54), B_1 in Eq. (E3-4-19) must vanish. It can also be observed that A_2 in Eq. (E3-4-20) must vanish in order to satisfy the boundary condition given by Eq. (3-51). As a result, Eq. (E3-4-20) takes the form

$$\frac{\partial u}{\partial z} = -\, A_1 B_2 \exp\left(\frac{Ux}{2D_x} - px \right)\lambda \sin(\lambda z) \tag{E3-4-21}$$

By using the boundary condition given by Eq. (3-52) and the last expression of Eq. (E3-4-2), and from Eq. (E3-4-21), one can write

$$A_1 B_2 \exp\left(\frac{Ux}{2D_x} - px \right)\lambda \sin(\lambda H_p) = 0 \tag{E3-4-22}$$

In order to satisfy this condition, the following condition must be satisfied:

$$\sin(\lambda H_p) = 0 \tag{E3-4-23}$$

which results in Eq. (3-61). The substitution of $A_2 = 0$ and $B_1 = 0$, into Eq. (E3-4-19) and by applying the principle of superposition (since the partial differential equation and boundary conditions are linear), the following equation can be obtained:

$$u(x, z, s) = \sum_{n=0}^{\infty} A_n \exp\left(\frac{Ux}{2D_x} - p_n x\right) \cos(\lambda_n z) \qquad (E3\text{-}4\text{-}24)$$

where $A_n = (A_1 B_2)_n$, p_n is given by Eq. (3-60), and λ_n is given by Eq. (3-61).

Usage of Boundary Condition at $x = 0$ and Determination of A_n
The boundary conditions at $x = 0$, as given by Eqs. (3-47)–(3-49), need to be expressed in the Laplace domain. Their Laplace transform is

$$L[C(0, z, t)] = u(0, z, s) = L[C_M(z)e^{\gamma t}] = \frac{C_M(z)}{s-\gamma} \qquad (E3\text{-}4\text{-}25)$$

where

$$C_M(z) = C_{M_1}(z) + \cdots + C_{M_{p-1}}(z) + C_{M_p}(z) \qquad (E3\text{-}4\text{-}26)$$

For $x = 0$, by making use of Eq. (E3-4-26), Eq. (E3-4-24) becomes

$$\frac{C_M(z)}{s-\gamma} = \sum_{n=0}^{\infty} A_n \cos(\lambda_n z) \qquad (E3\text{-}4\text{-}27)$$

By considering the right-hand side of Eq. (E3-4-27) as a Fourier cosine series for the interval $z = 0$ to H_p yields (e.g., Churchill, 1941; Spiegel, 1971)

$$A_n = \frac{2}{H_p} \int_0^{H_p} \frac{C_M(z)}{s-\gamma} \cos(\lambda_n z)\,dz \qquad (E3\text{-}4\text{-}28)$$

The substitution of Eq. (E3-4-28) into Eq. (E3-4-24) gives

$$u(x, z, s) = \sum_{n=0}^{\infty}\left[\frac{2}{H_p}\int_0^{H_p}\frac{C_M(z)}{s-\gamma}\cos(\lambda_n z)dz\right]\exp\left(\frac{Ux}{2D_x}-p_n x\right)\cos(\lambda_n z) \qquad (E3\text{-}4\text{-}29)$$

or

$$u(x, z, s) = \left(\frac{1}{2}\right)\left[\frac{2}{H_p}\int_0^{H_p}\frac{C_M(z)}{s-\gamma}dz\right]\exp\left(\frac{Ux}{2D_x}-p_0 x\right) + \sum_{n=1}^{\infty}\left[\frac{2}{H_p}\int_0^{H_p}\frac{C_M(z)}{s-\gamma}\cos(\lambda_n z)dz\right]$$

$$\times \exp\left(\frac{Ux}{2D_x}-p_n x\right)\cos(\lambda_n z) \qquad (E3\text{-}4\text{-}30)$$

Note that the 1/2 in front of the first term on the right-hand side of Eq. (E3-4-30) is the coefficient of the first term of the Fourier series. It can be easily observed that Eq. (E3-4-30) is equivalent to Eq. (3-56).

Inverse Laplace Transform of Eq. (3-56)
Determination of the inverse Laplace transform is given in Example 3-5. After taking the inverse Laplace transform of Eq. (3-56), the general solution for concentration distribution can be expressed as (Batu, 1989)

$$C(x, z, t) = C_1(x, z, t) + C_2(x, z, t) \qquad (3\text{-}63)$$

The expression for $C_1(x, z, t)$ is

$$C_1(x, z, t) = \frac{I_0}{2H_p}\exp\left(\gamma t + \frac{Ux}{2D_x}\right)$$

$$\times \{\exp(-G_1^{1/2}x)\,\mathrm{erfc}[G_2 - (G_3 t)^{1/2}] + \exp(G_1^{1/2}x)\,\mathrm{erfc}[G_2 + (G_3 t)^{1/2}]\} \qquad (3\text{-}64)$$

where

$$I_0 = \int_0^{H_p} C_M(z)\, dz \tag{3-65}$$

$$G_1 = \frac{vR_d}{D_x} + \frac{U^2}{4D_x^2} + \frac{\gamma R_d}{D_x} \tag{3-66}$$

$$G_2 = \frac{R_d x}{2(D_x R_d t)^{1/2}} \tag{3-67}$$

and

$$G_3 = v + \frac{U^2}{4D_x R_d} + \gamma \tag{3-68}$$

The expression for $C_2(x, z, t)$ is

$$C_2(x, z, t) = \exp\left(\gamma t + \frac{Ux}{2D_x}\right) \sum_{n=1}^{\infty} I_n \cos(\lambda_n z) \tag{3-69}$$

$$\times \{\exp(-K_n^{1/2}x)\, \mathrm{erfc}[G_2 - (L_n t)^{1/2}] + \exp(K_n^{1/2}x)\, \mathrm{erfc}[G_2 + (L_n t)^{1/2}]\}$$

where

$$I_n = \frac{1}{H_p} \int_0^{H_p} C_M(z)\cos(\lambda_n z)dz \tag{3-70}$$

$$K_n = G_1 + \frac{D_z}{D_x} \lambda_n^2 \tag{3-71}$$

and

$$L_n = G_3 + \frac{D_z}{R_d} \lambda_n^2 \tag{3-72}$$

EXAMPLE 3-5

Determine the inverse Laplace transform of Eq. (3-56) to obtain the final solution for concentration distribution, $C(x, z, t)$, given by Eq. (3-63).

SOLUTION

The inverse Laplace transform of Eq. (3-56) is

$$L^{-1}[u(x, z, s)] = L^{-1}[u_1(x, z, s)] + L^{-1}[u_2(x, z, s)] \tag{E3-5-1}$$

where

$$L^{-1}[u(x, z, s)] = C(x, z, t), \quad L^{-1}[u_1(x, z, s)] = C_1(x, z, t), \quad L^{-1}[u_2(x, z, s)] = C_2(x, z, t) \tag{E3-5-2}$$

By combining Eqs. (E3-5-1) and (E3-5-2) one has Eq. (3-63). Therefore, the inverse Laplace transforms of $u_1(x, z, s)$ and $u_2(x, z, s)$ need to be taken, and they are given below.

Inverse Laplace Transform of $u_1(x, z, s)$
In order to determine the inverse Laplace transform of $u_1(x, z, s)$ as given by Eq. (3-57), according to the second expression of Eq. (E3-5-2), some manipulations need to be made. Equation (3-57) can also be expressed as

$$u_1(x, z, s - b) = \frac{1}{H_p(s+b-\gamma-b)}\left[\int_0^{H_p} C_M(z)dz\right]\exp\left(\frac{Ux}{2D_x}\right)\exp\left\{-\left[\frac{R_d}{D_x}(s - b)\right]^{1/2}x\right\} \tag{E3-5-3}$$

where

$$b = -v - \frac{U^2}{4D_x R_d} \qquad \text{(E3-5-4)}$$

By using the first translation or shifting property for the inverse Laplace transform (Spiegel, 1965, Theorem 2-3, p. 43), Eq. (E3-5-3) can be written as

$$u_1(x, z, s - b) = \frac{2}{H_p}\left[\int_0^{H_p} C_M(z)dz\right]\exp\left(\frac{Ux}{2D_x}\right)f_1(s - b) \qquad \text{(E3-5-5)}$$

where

$$f_1(s - b) = \frac{1}{(s+b-\gamma-b)}\exp\left\{-\left[\frac{R_d}{D_x}(s - b)\right]^{1/2}x\right\} \qquad \text{(E3-5-6)}$$

and

$$F_1(t) = L^{-1}[f_1(s - b)] = e^{bt}E_1(t) \qquad \text{(E3-5-7)}$$

$$L^{-1}[e_1(s)] = E_1(t), \quad e_1(s) = \frac{1}{s-(\gamma-b)}\exp\left[-\left(\frac{R_d}{D_x}s\right)^{1/2}x\right] \qquad \text{(E3-5-8)}$$

The inverse Laplace transform of $e_1(s)$ is (Carslaw and Jaeger, 1959, Eq. [19], p. 495)

$$E_1(t) = \frac{1}{2}\exp\left[\left(v + \frac{U^2}{4D_x R_d} + \gamma\right)t\right] \qquad \text{(E3-5-9)}$$

$$\times \{\exp(-G_1 x)\text{erfc}[G_2 - (G_3 t)^{1/2}] + \exp(G_1 x)\text{erfc}[G_2 + (G_3 t)^{1/2}]\}$$

where G_1, G_2, and G_3 are defined by Eqs. (3-66), (3-67), and (3-68), respectively. Therefore, the inverse Laplace transform of $u_1(x, z, s-b)$ in Eq. (E3-5-5) is

$$C_1(x, z, t) = L^{-1}[u_1(x, z, s - b)] = \frac{1}{H_p}\left[\int_0^{H_p} C_M(z)\,dz\right]\exp\left(\frac{Ux}{2D_x}\right)L^{-1}[f_1(s - b)] \qquad \text{(E3-5-10)}$$

or by using Eq. (E3-5-7),

$$C_1(x, z, t) = \frac{1}{H_p}\left[\int_0^{H_p} C_M(z)\,dz\right]\exp\left(\frac{Ux}{2D_x} - \frac{U^2 t}{4D_x R_d} - vt\right)E_1(t) \qquad \text{(E3-5-11)}$$

The substitution of Eqs. (E3-5-4) and (E3-5-9) into Eq. (E3-5-11) gives the final expression of $C_1(x, z, t)$ as given by Eq. (3-64).

Inverse Laplace transform of $u_2(x, z, s)$
Similarly, the inverse Laplace transform of Eq. (3-58) can be taken and the final result for $C_2(x, z, t)$ is given by Eq. (3-69).

Unsteady-State Special Solution for a Single-Strip Source
Concentration Distribution
The geometry for a single-strip source is shown in Figure 3-12, which is a special case of Figure 3-11. As can be seen from Figure 3-11, the source can be located at any place between the impervious boundaries. The source conditions expressed by Eqs. (3-47)–(3-49) take the forms

$$C(0, z, t) = \begin{cases} C_M e^{\gamma t} & D_1 < z < D_1 + 2B \qquad \text{(3-73)} \\ 0 & \text{otherwise} \qquad \text{(3-74)} \end{cases}$$

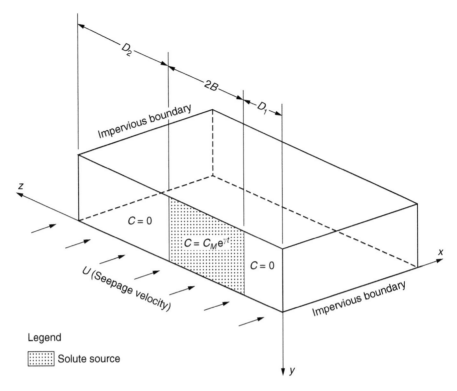

FIGURE 3-12   Schematic picture of a single-strip source in a bounded medium having unidirectional flow field. (Adapted from Batu, V., *Water Resour. Res.*, 25, 1125–1132, 1989.)

where C_M is the source concentration having a constant value. The substitution of $C_M(z) = C_M$ into Eq. (3-65) determines the expression for I_0 as

$$I_0 = 2BC_M \qquad (3\text{-}75)$$

Similarly, from Eq. (3-70) one can determine

$$I_n = \frac{C_M}{n\pi}\{\sin[\lambda_n(D_1 + 2B)] - \sin(\lambda_n D_1)\} \qquad (3\text{-}76)$$

By combination Eqs. (3-63), (3-64), (3-69), (3-75), and (3-76) one has the final expression for concentration distribution for a single-strip source:

$$C(x, z, t) = C_M P_1[\exp(-P_2)\,\text{erfc}(P_3 - P_4) + \exp(P_2)\,\text{erfc}(P_3 + P_4)]$$

$$+ \frac{C_M}{\pi}P_5 P_6 \sum_{n=1}^{\infty} P_{7_n}\,[\exp(-P_{8_n})\,\text{erfc}(P_3 - P_{9_n}) + \exp(P_{8_n})\,\text{erfc}(P_3 + P_{9_n})] \qquad (3\text{-}77)$$

where

$$P_1 = \frac{B}{D_1 + 2B + D_2}\exp\!\left(\gamma t + \frac{Ux}{2D_x}\right) \qquad (3\text{-}78)$$

$$P_2 = \left(\frac{vR_d}{D_x} + \frac{U^2}{4D_x^2} + \frac{\gamma R_d}{D_x}\right)^{1/2}x \qquad (3\text{-}79)$$

$$P_3 = \frac{R_d x}{2(D_x R_d t)^{1/2}} \tag{3-80}$$

$$P_4 = \left[\left(v + \frac{U^2}{4D_x R_d} + \gamma \right) t \right]^{1/2} \tag{3-81}$$

$$P_5 = \exp\left(\frac{Ux}{2D_x} \right), \quad P_6 = \exp(\gamma t) \tag{3-82}$$

$$P_{7_n} = \frac{1}{n} \{ \sin[\lambda_n(D_1 + 2B)] - \sin(\lambda_n D_1) \} \cos(\lambda_n z) \tag{3-83}$$

$$P_{8_n} = \left(\frac{vR_d}{D_x} + \frac{D_z}{D_x} \lambda_n^2 + \frac{U^2}{4D_x^2} + \frac{\gamma R_d}{D_x} \right)^{1/2} x \tag{3-84}$$

$$P_{9_n} = \left[\left(v + \frac{D_z}{R_d} \lambda_n^2 + \frac{U^2}{4D_x R_d} + \gamma \right) t \right]^{1/2} \tag{3-85}$$

and

$$\lambda_n = \frac{n\pi}{D_1 + 2B + D_2}, \quad n = 1, 2, \ldots \tag{3-86}$$

Special Solution for a Symmetrically Located Strip Source (Bruch and Street, 1967)
Bruch and Street (1967) developed an analytical solution for a symmetrically located strip source in Figure 3-12. Owing to symmetry, the authors developed a solution for one half of the transport domain (Bruch and Street, 1967, Figure 1b, p. 21). Therefore, in Figure 3-12, if $D_1 = 0$ and $D_2 = D$, and $2B$ is replaced by B, the geometry developed by Bruch and Street will be obtained. Moreover, the authors assumed that $R_d = 1$, $\gamma = 0$, and $v = 0$. By substituting these values into Eqs. (3-77)–(3-86), Eq. (3-77) takes the form

$$
\begin{aligned}
\frac{C(x, z, t)}{C_M} &= \frac{B}{2(B+D)} \operatorname{erfc}\left[\frac{x - Ut}{2(D_x t)^{1/2}} \right] + \frac{B}{2(B+D)} \exp\frac{Ux}{D_x} \operatorname{erfc}\left[\frac{x + Ut}{2(D_x t)^{1/2}} \right] \\
&+ \frac{1}{2} \sum_{n=1}^{\infty} F_n \cos\left(\frac{n\pi z}{B+D} \right) \exp\left[\frac{1}{2} \left(\frac{U}{D_x} - J_n \right) x \right] \operatorname{erfc}\left[\frac{x - J_n D_x t}{2(D_x t)^{1/2}} \right] \\
&+ \frac{1}{2} \sum_{n=1}^{\infty} F_n \cos\left(\frac{n\pi z}{B+D} \right) \exp\left[\frac{1}{2} \left(\frac{U}{D_x} + J_n \right) x \right] \operatorname{erfc}\left[\frac{x + J_n D_x t}{2(D_x t)^{1/2}} \right]
\end{aligned} \tag{3-87}
$$

where

$$F_n = \frac{2}{\pi} \frac{1}{n} \sin\left(\frac{n\pi B}{B+D} \right) \tag{3-88}$$

and

$$J_n = \left[\left(\frac{U}{D_x} \right)^2 + 4\lambda_n^2 \frac{D_z}{D_x} \right]^{1/2} \tag{3-89}$$

Equation (3-87) is exactly the same equation, with the exception of the notation, as obtained by Bruch and Street (1967, Eq. [44], p. 27).

Special Solution for the One-Dimensional Case
If $D_1 = 0$ and $D_2 = 0$ in Figure 3-12, the geometry turns out to be the case shown in Figure 3-1, which corresponds to the one-dimensional case and the corresponding solution is given in Section

3.3.2.1. By substituting these values into Eq. (3-86), one can write $\lambda_n = n\pi/(2B)$. Since $\sin(n\pi) = 0$ and $\sin(0) = 0$, substitution of $\lambda_n = n\pi/(2B)$ into Eq. (3-83) gives $P_{7_n} = 0$. Therefore, Eq. (3-77) turns out to be exactly the same as Eq. (3-7), which is the solution for concentration distribution for the one-dimensional case.

Convective–Dispersive Flux Components
From Eqs. (2-41a) and (2-41c), the convective–dispersive flux components for two-dimensional solute transport in x on z-coordinate directions, respectively, are

$$F_x = \varphi_e UC - \varphi_e D_x \frac{\partial C}{\partial x} \tag{3-90a}$$

$$F_z = - \varphi_e D_z \frac{\partial C}{\partial z} \tag{3-90b}$$

where F_x and F_z are the convective–dispersive flux components in the x direction and z direction, respectively, and the other flux component (F_y) is zero, because the solute transport is two-dimensional in the x–z plane. In Eq. (3-90), φ_e is the effective porosity and the other parameters are as defined previously. Since the dimensions of U and C are [L/T] and [M/L^3], respectively, the dimension of F_x and F_z is [M/L^2/T]. This physically means that both F_x and F_z are the masses of solute entering per unit area per unit time.
The expression for the convective–dispersive flux component in the x direction is

$$F_x(x, z, t) = \varphi_e UC(x, z, t) - \frac{1}{2} \varphi_e UC_M P_1 S_1 - \varphi_e C_M D_x P_1 (S_2 + S_3) \tag{3-91}$$

$$- \frac{1}{\pi} \varphi_e C_M D_x P_5 P_6 \sum_{n=1}^{\infty} P_{7_n} \left(\frac{U}{2D_x} S_{4_n} + S_{5_n} + S_{6_n} \right)$$

where S_1, S_2, and S_3 are defined by Eqs. (3-26), (3-27), and (3-28), respectively, and

$$S_{4_n} = \exp(-P_{8_n}) \operatorname{erfc}(P_3 - P_{9_n}) + \exp(P_{8_n}) \operatorname{erfc}(P_3 + P_{9_n}) \tag{3-92}$$

$$S_{5_n} = - P_3 \exp(-P_{8_n}) \operatorname{erfc}(P_3 - P_{9_n}) - \exp(-P_{8_n})P_{10} \exp[- (P_3 - P_{9_n})^2] \tag{3-93}$$

$$S_{6_n} = Q_{2_n} \exp(P_{8_n}) \operatorname{erfc}(P_3 + P_{9_n}) - P_{10} \exp[P_{8_n} - (P_3 + P_{9_n})^2] \tag{3-94}$$

$$Q_{2_n} = \left(\frac{vR_d}{D_x} + \frac{D_z}{D_x} \lambda_n^2 + \frac{U^2}{4D_x^2} + \frac{\gamma R_d}{D_x} \right)^{1/2} \tag{3-95}$$

P_{10} is defined by Eq. (3-30), and the rest are defined by Eqs. (3-78)–(3-86). The derivation of Eq. (3-91) is given in Example 3-6.
    The expression for the convective–dispersive flux component in the z direction can easily be determined from Eq. (3-90b) using Eq. (3-77). As can be seen from Eqs. (3-77) and (3-83), only P_{7_n} is the function of z. The result is

$$F_z(x, z, t) = \frac{1}{\pi} \varphi_e C_M D_z P_5 P_6 \sum_{n=1}^{\infty} Q_{3_n} [\exp(- P_{8_n})\operatorname{erfc}(P_3 - P_{9_n}) + \exp(P_{8_n})\operatorname{erfc}(P_3 + P_{9_n})] \tag{3-96}$$

where

$$Q_{3_n} = \frac{1}{n} \{\sin[\lambda_n(D_1 + 2B)] - \sin(\lambda_n D_1)\}\lambda_n \sin(\lambda_n Z) \tag{3-97}$$

and the rest are as defined previously.

EXAMPLE 3-6

Derive the convective–dispersive flux component, $F_x(x, z, t)$, in the x direction as given by Eq. (3-91).

SOLUTION

The convective–dispersive flux component in the x direction, $F_x(x, z, t)$, is derived by combining Eqs. (3-77) and (3-90a). As can be seen from Eq. (3-77), the expression of $C(x, z, t)$ includes complementary error functions [erfc( )] and their derivatives with respect to x that require some special attention. There are four erfc( ) terms in Eq. (3-77) and the first one is

$$F_1(x) = \text{erfc}(P_3 - P_4) = \text{erfc}(Ax - B) \tag{E3-6-1}$$

where

$$A = \frac{R_d}{2(D_x R_d t)^{1/2}} \quad B = \left[\left(v + \frac{U^2}{4D_x R_d} + \gamma \right) t \right]^{1/2} \tag{E3-6-2}$$

Equation (E3-6-1) can be expressed as (e.g., Spiegel, 1968, p. 183)

$$F_1(x) = 1 - \text{erf}(Ax - B) = 1 - \frac{2}{\pi^{1/2}} \int_0^{Ax-B} e^{-\xi^2} d\xi \tag{E3-6-3}$$

In order to determine the expression for $\partial F_1(x)/\partial x$, Leibnitz's rule for differentiation of integrals (e.g., Spiegel, 1968, Eq. [15.14], p. 95) as given by

$$\frac{\partial}{\partial x} \int_{u(x)}^{v(x)} f(\xi, x) \, d\xi = \int_{u(x)}^{v(x)} \frac{\partial}{\partial x} f(\xi, x) \, d\xi + f(v, x) \frac{\partial v}{\partial x} - f(u, x) \frac{\partial u}{\partial x} \tag{E3-6-4}$$

needs to be used. Comparing Eqs. (E3-6-3) and (E3-6-4) gives

$$u(x) = 0, \quad v(x) = Ax - B, \quad f(\xi, x) = \frac{2}{\pi^{1/2}} e^{-\xi^2} \tag{E3-6-5}$$

From these expressions one can write

$$\frac{\partial f(\xi, x)}{\partial x} = 0, \quad f(v, x) = \frac{2}{\pi^{1/2}} e^{-v^2} \tag{E3-6-6}$$

$$\frac{\partial v}{\partial x} = A, \quad \frac{\partial u}{\partial x} = 0 \tag{E3-6-7}$$

From Eq. (E3-6-3),

$$\frac{\partial F_1(x)}{\partial x} = \frac{\partial}{\partial x} \left[1 - \frac{2}{\pi^{1/2}} \int_0^{Ax-B} e^{-\xi^2} d\xi \right] \tag{E3-6-8}$$

By using Eq. (E3-6-4) with the expressions given by Eqs. (E3-6-6) and (E3-6-7), the following expression can be obtained:

$$\frac{\partial F_1(x)}{\partial x} = - \frac{2}{\pi^{1/2}} A \exp[-(Ax - B)^2] \tag{E3-6-9}$$

By applying the same procedure to the rest of the complementary error functions and taking the other derivatives with respect to x, and with some additional manipulations, Eq. (3-91) can be finally determined.

Unsteady-State Special Solutions for Other Sources

Equation (3-63) is the general solution for the resulting concentration distribution for any given $C_M(z)$ source concentration distribution. With the source concentration functions (Figure 3-11),

$C_M(z)$, the integrals I_0 and I_n given by Eqs. (3-65) and (3-70), respectively, need to be evaluated analytically or numerically depending on the type of the function for $C_M(z)$. Two examples are presented below.

Solution for Two-Strip Sources with Uniform Source Concentrations
The geometry of two-strip sources with uniform source concentrations is shown in Figure 3-13. The boundary conditions at $x = 0$ expressed by Eqs. (3-47)–(3-49) take the forms

$$C(0, z, t) = \begin{cases} C_{M_1}e^{\gamma t} & D_1 < z < D_1 + 2B_1 & \text{(3-98)} \\ C_{M_2}e^{\gamma t} & D_1 + 2B_1 + D_3 < z < D_1 + 2B_1 + D_3 + 2B_2 & \text{(3-99)} \\ 0 & \text{otherwise} & \text{(3-100)} \end{cases}$$

where C_{M_1} and C_{M_2} are the uniform source concentrations. Equation (3-63) is the general solution and for a special case, only the terms I_0 (Eq. [3-65]) and I_n (Eq. [3-70]) need to be evaluated using the source geometry. By using the above source conditions, from Eq. (3-65)

$$I_0 = \int_{D_1}^{D_1+2B_1} C_{M_1} dz + \int_{D_1+2B_1+D_3}^{D_1+2B_1+D_3+2B_2} C_{M_2} dz = 2B_1 C_{M_1} + 2B_2 C_{M_2} \quad \text{(3-101)}$$

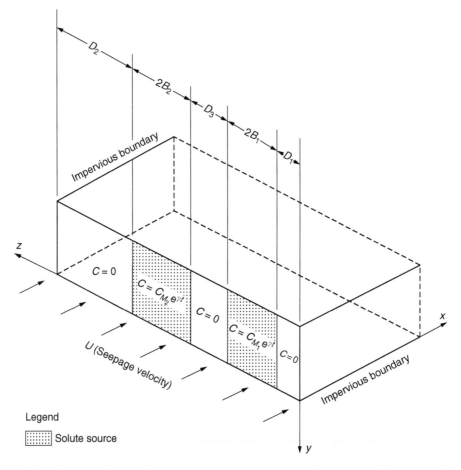

FIGURE 3-13 Schematic picture of two-strip sources in a bounded medium having unidirectional flow field.

Similarly, from Eq. (3-70), one can write

$$I_n = \frac{1}{H_p}\left[\int_{D_1}^{D_1+2B_1} C_{M_1}\cos(\lambda_n z)\,dz + \int_{D_1+2B_1+D_3}^{D_1+2B_1+D_3+2B_2} C_{M_2}\cos(\lambda_n z)\,dz\right] \quad (3\text{-}102)$$

or after some manipulation one gets

$$I_n = \frac{C_{M_1}}{H_p\lambda_n}\{\sin[\lambda_n(D_1+2B_1)] - \sin(\lambda_n D_1)\}$$

$$(3\text{-}103)$$

$$+ \frac{C_{M_2}}{H_p\lambda_n}\{\sin[\lambda_n(D_1+2B_1+D_3+2B_2)] - \sin[\lambda_n(D_1+2B_1+D_3)]\}$$

where $H_p = D_1 + 2B_1 + D_3 + 2B_2 + D_2$. Equations (3-61), (3-63), (3-64), (3-69), (3-101), and (3-103) form the concentration distribution solution for two-strip sources whose geometry is shown in Figure 3-13.

Solution for a Single-Strip Source with Linearly Varying Source Concentration
The geometry of a single-strip source with linearly varying source concentration is shown in Figure 3-14. The boundary conditions at $x = 0$ expressed by Eqs. (3-47)–(3-49) take the forms

$$C(0,z,t) = \begin{cases} C_M(z)e^{\gamma t} = \frac{1}{2}C_{M_{max}}\frac{z-D_1}{B}e^{\gamma t} & D_1 < z < D_1+2B \quad (3\text{-}104) \\ 0 & \text{otherwise} \quad (3\text{-}105) \end{cases}$$

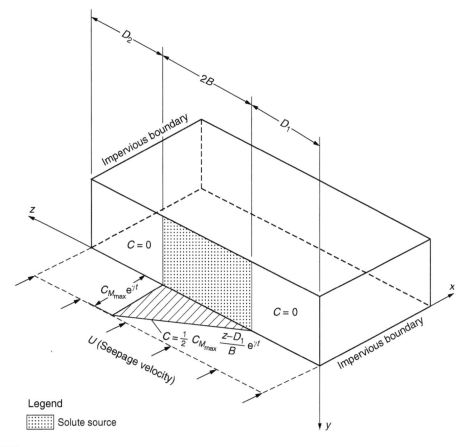

FIGURE 3-14 Schematic picture of a single-strip source having linearly varying source concentration.

Equation (3-104) implies that at $z = D_1$, $C(0, z=D_1, t) = 0$; and at $z = D_1 + 2B$, $C(0, z=D_1+2B, t) = e^n C_{M_{max}}$ where $C_{M_{max}}$ is the maximum concentration at the source. The variation of $C(0, z, t)$ is shown in Figure 3-14. Equation (3-63) is the general solution and for a special case, only the terms I_0 (Eq. [3-65]) and I_n (Eq. [3-70]) need to be evaluated using the source geometry. It can be observed from Eq. (3-104) and the source conditions given by Eqs. (3-47)–(3-49) that

$$C_M(z) = \frac{1}{2} C_{M_{max}} \frac{z-D_1}{B} \tag{3-106}$$

By substituting Eq. (3-106) into Eq. (3-65) and with some manipulations one can obtain

$$I_0 = \int_0^{H_p} C_M(z)\, dz = \frac{1}{2} C_{M_{max}} \int_{D_1}^{D_1+2B} \frac{z-D_1}{B}\, dz = \frac{1}{2} C_{M_{max}}(2B - D_1) \tag{3-107}$$

Similarly, by substituting Eq. (3-106) into Eq. (3-70) one can write

$$I_n = \frac{1}{H_p} \int_{D_1}^{D_1+2B} \frac{1}{2} C_{M_{max}} \frac{z-D_1}{B} \cos(\lambda_n z)\, dz \tag{3-108}$$

$$= \frac{1}{2} \frac{C_{M_{max}}}{H_p} \left[\int_{D_1}^{D_1+2B} \left(\frac{z}{B}\right) \cos(\lambda_n z)\, dz - \left(\frac{D_1}{B}\right) \int_{D_1}^{D_1+2B} \cos(\lambda_n z)\, dz \right]$$

By using the corresponding formula for the first integral in Eq. (3-108) from the integral tables (e.g., Spiegel, 1968, Eq. [14.370], p. 77) and after some manipulations, the following formula can be obtained:

$$I_n = \frac{1}{2} \frac{C_{M_{max}}}{(D_1+2B+D_2)B} \frac{1}{\lambda_n} \left\{ \frac{1}{\lambda_n} \cos[\lambda_n(D_1 + 2B)] + 2B \sin[\lambda_n(D_1 + 2B)] - \frac{1}{\lambda_n} \cos(\lambda_n D_1) \right\} \tag{3-109}$$

Equations (3-61), (3-63), (3-64), (3-69), (3-107), and (3-109) form the concentration distribution solution for a single-strip source with linearly variable source concentration and is shown in Figure 3-14.

Steady-State General Solution for Concentration Distribution
The solution for the steady- state solute transport case (see Figure 3-11 with $\gamma = 0$) can be obtained by using two different procedures. The first procedure is to use the corresponding differential equation under steady-state solute transport conditions with the associated boundary conditions. The second procedure to the solution developed for unsteady-state concentration distribution is obtained at the time t goes to infinity. Here, the first procedure is presented below and its details are presented in Example 3-7. The second procedure is presented in Example 3-8.

Under steady-state solute transport conditions, Eq. (3-45) takes the form

$$D_x \frac{\partial^2 C}{\partial x^2} + D_z \frac{\partial^2 C}{\partial z^2} - U \frac{\partial C}{\partial z} - R_d v C = 0 \tag{3-110}$$

For steady-state solute transport, the initial condition given by Eq. (3-46) is not necessary. The solution for the concentration distribution C will be only the function of x and z. Therefore, the boundary conditions given by Eqs. (3-47)–(3-49) will have the following forms:

$$C(0, z) = C_M(z), \quad 0 < z < H_1 \tag{3-111}$$

$$C(0, z) = C_{M_{p-1}}(z), \quad H_{p-2} < z < H_{p-1} \tag{3-112}$$

$$C(0, z) = C_{M_p}(z), \quad H_{p-1} < z < H_p \tag{3-113}$$

where H_i is given by Eq. (3-50). The other boundary conditions as given by Eqs. (3-51)–(3-54) remain the same.

The boundary-value problem defined by Eqs. (3-110)–(3-113) as well as Eqs. (3-51)–(3-54) was solved by the Fourier analysis technique. The final result for steady-state concentration distribution is given as (Batu, 1989)

$$C(x, z) = \frac{I_0}{H_p} \exp\left(\frac{Ux}{2D_x} - p_0 x\right) + 2\sum_{n=1}^{\infty} I_n \exp\left(\frac{Ux}{2D_x} - p_n x\right) \cos(\lambda_n z) \tag{3-114}$$

where

$$p_0 = \left(\frac{R_d}{D_x} v + \frac{U^2}{4D_x^2}\right)^{1/2} \tag{3-115}$$

$$p_n = \left(\frac{R_d}{D_x} v + \frac{D_z}{D_x} \lambda_n^2 + \frac{U^2}{4D_x^2}\right)^{1/2} \tag{3-116}$$

and I_0 and I_n are given by Eqs. (3-65) and (3-70), respectively. The derivation of Eq. (3-114) is given both in Examples 3-7 and 3-8. Note that the first term of the Fourier cosine series (for $n = 0$) in Eq. (E3-7-7) is divided by 2.

EXAMPLE 3-7

Derive the general steady-state concentration distribution expression as given by Eq. (3-114) by solving the corresponding boundary-value problem.

SOLUTION

The solution of this example is similar to the solution of Example 3-4 in many respects. The method of separation of variables (e.g., Churchill, 1941; Spiegel, 1965; Batu, 1983) is applied to solve the boundary-value problem defined by Eqs. (3-110)–(3-113). By following the same method in Example 3-4 with

$$C(x, z) = X(x)Z(z) \tag{E3-7-1}$$

Eq. (3-110) takes the form

$$\frac{D_x}{D_z} \frac{X''}{X'} - \frac{U}{D_z} \frac{X'}{X} - \frac{R_d}{D_z} v = -\frac{Z''}{Z} = \mu \tag{E3-7-2}$$

where μ must be a constant. Similarly, in Example 3-4, from Eq. (E3-7-2), two ordinary differential equations can be obtained. The solutions for $X(x)$ and $Z(z)$ are given by Eqs. (E3-4-15) and (E3-4-18), respectively. Therefore, $C(x, z)$ can be expressed as

$$C(x, z) = \left[A_1 \exp\left(\frac{Ux}{2D_x} - px\right) + B_1 \exp\left(\frac{Ux}{2D_x} + px\right)\right][A_2 \sin(\lambda z) + B_2 \cos(\lambda z)] \tag{E3-7-3}$$

where

$$p = \left(\frac{R_d}{D_x} v + \frac{D_z}{D_x} \lambda^2 + \frac{U^2}{4D_x^2}\right)^{1/2} \tag{E3-7-4}$$

Using the boundary conditions at the impervious boundaries and at infinity (Eqs. [3-51]–[3-54]), it can be shown that both A_2 and B_1 must be zero. Therefore, the solution can be expressed as (see Example 3-4)

$$C(x, z) = \sum_{n=0}^{\infty} A_n \exp\left(\frac{Ux}{2D_x} - p_n x\right) \cos(\lambda_n z) \tag{E3-7-5}$$

where p_n is given by Eq. (3-116). In order to determine the expression for A_n, the boundary conditions at $x = 0$ (Eqs. [3-111]–[3-113]) need to be used. For $x = 0$, Eq. (E3-7-5) gives

$$C_M(0, z) = \sum_{n=0}^{\infty} A_n \cos(\lambda_n z) \qquad \text{(E3-7-6)}$$

By considering the right-hand side of Eq. (E3-7-6) as a Fourier cosine series for the interval $z = 0$ to H_p, yields (e.g., Churchill, 1941; Spiegel, 1971)

$$A_n = \frac{2}{H_p} \int_0^{H_p} C_M(z) \cos(\lambda_n z) \, dz \qquad \text{(E3-7-7)}$$

By substituting Eq. (E3-7-7) into Eq. (E3-7-6) and after some manipulations (see Eqs. [E3-4-27]–[E3-4-30]), Eq. (3-114) can be finally obtained.

EXAMPLE 3-8

Derive the general steady-state concentration distribution expression, as given by Eq. (3-114), from the unsteady-state general solution given by Eq. (3-63).

SOLUTION

Equations (3-63), (3-64), and (3-69) form the general solution under unsteady-state solute transport conditions for the geometry shown in Figure 3-11. The corresponding steady-state solute transport solution will be obtained for $\gamma = 0$ and make the time t approach infinity. Substituting these values into Eq. (3-64) one can write

$$C_1(x, z) = \frac{I_0}{2H_p} \exp\left(\frac{Ux}{2D_x}\right) [\exp(-p_0 x)\, \mathrm{erfc}(-\infty) + \exp(p_0 x)\, \mathrm{erfc}(\infty)] \qquad \text{(E3-8-1)}$$

where p_0 is given Eq. (3-115). By using Eq. (3-35), Eq. (E3-8-1) takes the form

$$C_1(x, z) = \frac{I_0}{2H_p} \exp\left(\frac{Ux}{2D_x}\right) \{\exp(-p_0 x)[1 - \mathrm{erf}(-\infty)] + \exp(p_0 x)[1 - \mathrm{erf}(\infty)]\} \qquad \text{(E3-8-2)}$$

or by using Eqs. (3-36) and (3-37), it takes the following form:

$$C_1(x, z) = \frac{I_0}{H_p} \exp\left(\frac{Ux}{2D_x} - p_0 x\right) \qquad \text{(E3-8-3)}$$

Similarly, with Eq. (3-35), Eq. (3-69) takes the form

$$C_2(x, z) = \exp\left(\frac{Ux}{2D_x}\right) \sum_{n=1}^{\infty} I_n \cos(\lambda_n) \left\{ \exp\left[-\left(v\frac{R_d}{D_x} + \frac{D_z}{D_x}\lambda_n^2 + \frac{U^2}{4D_x^2}\right)^{1/2} x\right] \mathrm{erfc}(-\infty) \right.$$
$$\left. + \exp\left[\left(v\frac{R_d}{D_x} + \frac{D_z}{D_x}\lambda_n^2 + \frac{U^2}{4D_x^2}\right)^{1/2} x\right] \mathrm{erfc}(\infty) \right\} \qquad \text{(E3-8-4)}$$

Since $\mathrm{erfc}(-\infty) = 2$ and $\mathrm{erf}(\infty) = 0$, Eq. (E3-8-4) takes the form

$$C_2(x, z) = \exp\left(\frac{Ux}{2D_x}\right) \sum_{n=1}^{\infty} 2I_n \cos(\lambda_n z) \exp\left[-\left(v\frac{R_d}{D_x} + \frac{D_z}{D_x}\lambda_n^2 + \frac{U^2}{4D_x^2}\right)^{1/2} x\right] \qquad \text{(E3-8-5)}$$

Combination of Eqs. (3-63), (3-70), (3-116), (E3-8-3), and (E3-8-5) results in Eq. (3-114), which is the final equation for the steady-state concentration distribution.

Steady-State Special Solution for a Single-Strip Source
Concentration Distribution

The geometry of a single-strip source is shown in Figure 3-12 (with $\gamma = 0$), which is a special case of Figure 3-11 (with $\gamma = 0$). As can be seen from Figure 3-11, the source can be located at any place between the impervious boundaries. The source conditions expressed by Eqs. (3-111)–(3-113) take the forms

$$C(0, z) = \begin{cases} C_M & D_1 < z < D_1 + 2B \quad (3\text{-}117) \\ 0 & \text{otherwise} \quad (3\text{-}118) \end{cases}$$

where C_M is the source concentration having a constant value. Note that Eqs. (3-117) and (3-118) are, respectively, exactly the same as Eqs. (3-73) and (3-74) for $\gamma = 0$. By substituting Eqs. (3-117) and (3-118) into Eqs. (3-65) and (3-70), respectively, Eqs. (3-75) and (3-76) can be obtained. By introducing Eqs. (3-75) and (3-76) into Eq. (3-114), finally the following expression can be obtained for the concentration distribution under steady-state solute transport conditions:

$$C(x, z) = \frac{2BC_M}{D_1 + 2B + D_2} \exp\left[\left(\frac{U}{2D_x} - p_0\right)x\right]$$

$$+ \frac{2}{\pi} C_M \sum_{n=1}^{\infty} \frac{1}{n} \exp\left[\left(\frac{U}{2D_x} - p_n\right)x\right]\{\sin[\lambda_n(D_1 + 2B)] - \sin(\lambda_n D_1)]\} \cos(\lambda_n z) \quad (3\text{-}119)$$

where p_0 and p_n are defined by Eqs. (3-115) and (3-116), respectively and λ_n is given by Eq. (3-86).

Special Solution for a Symmetrically Located Strip Source (Bruch and Street, 1967)

Bruch and Street (1967) developed an analytical solution for a symmetrically located strip source (Figure 3-12). Due to the symmetry, the authors developed a solution for one half of the transport domain (Bruch and Street, 1967, p. 21, Figure 1b). Therefore, in Figure 3-12, if $D_1 = 0$, $D_2 = D$, and $2B$ is replaced by B, the geometry developed by Bruch and Street will be obtained. The corresponding solution for $R_d = 1$ and $v = 0$ in Eq. (3-119) is

$$\frac{C(x, z)}{C_M} = \frac{B}{B+D} + \frac{2}{\pi}\sum_{n=1}^{\infty}\frac{1}{n}\sin\left(\frac{n\pi B}{B+D}\right)\cos\left(\frac{n\pi z}{B+D}\right)\exp\left\{\frac{Ux}{2D_x} - \left[\frac{D_z}{D_x}\left(\frac{n\pi}{B+D}\right)^2 + \frac{U^2}{4D_x^2}\right]^{1/2}x\right\} \quad (3\text{-}120)$$

or

$$\frac{C(x, z)}{C_M} = \frac{B}{B+D} + \sum_{n=1}^{\infty} F_n \cos\left(\frac{n\pi z}{B+D}\right)\exp\left[\frac{1}{2}\left(\frac{U}{D_x} - J_n\right)x\right] \quad (3\text{-}121)$$

where

$$F_n = \frac{2}{\pi}\frac{1}{n}\sin\left(\frac{n\pi B}{B+D}\right), \quad n = 1, 2, \dots \quad (3\text{-}122)$$

$$J_n = \left[\left(\frac{U}{D_x}\right)^2 + \frac{4n^2\pi^2}{(B+D)^2}\frac{D_z}{D_x}\right]^{1/2} \quad (3\text{-}123)$$

Equation (3-121) is exactly the same equation, with the execption of the notation, as obtained by Bruch and Street (1967, Eq. [26], p. 24).

Special Solution for the One-Dimensional Case

If $D_1 = 0$ and $D_2 = 0$ in Figure 3-12, the geometry turns out to be the case shown in Figure 3-1, which corresponds to the one-dimensional case and the corresponding solution is given in Section 3.3.2.1. By substituting these values into Eq. (3-86), one can write $\lambda_n = n\pi/(2B)$. Since $\sin(n\pi) = 0$ and $\sin(0) = 0$, by substitution of $\lambda_n = n\pi/(2B)$ into Eq. (3-119), it can be seen that the second term on the right-hand side of Eq. (3-119) becomes zero. Therefore, Eq. (3-119) turns out to be exactly the same as Eq. (3-38), which is the solution for concentration distribution for the one-dimensional case under steady-state conditions.

Convective–Dispersive Flux Components
By using Eq. (3-119) in Eq. (3-90a), the expression for $F_x(x,z)$ can be determined as

$$F_x(x, z) = \varphi_e U C(x, z) - \frac{2BD_x\varphi_e C_M}{D_1+2B+D_2}\left(\frac{U}{2D_x} - p_0\right)\exp\left[\left(\frac{U}{2D_x} - p_0\right)x\right]$$

$$- \frac{2}{\pi}D_x\varphi_e C_M\sum_{n=1}^{\infty}R_{1_n}R_{2_n}\exp(R_{2_n}x)\cos(\lambda_n z) \tag{3-124}$$

where

$$R_{1_n} = \frac{1}{n}\{\sin[\lambda_n(D_1 + 2B)] - \sin(\lambda_n D_1)\} \tag{3-125}$$

and

$$R_{2_n} = \frac{U}{2D_x} - \left(\frac{R_d}{D_x}v + \frac{D_z}{D_x}\lambda_n^2 + \frac{U^2}{4D_x^2}\right)^{1/2} \tag{3-126}$$

and $C(x, z)$ is given by Eq. (3-119). Similarly, by using Eq. (3-119) in Eq. (3-90b), the expression for $F_z(x, z)$ can be determined as

$$F_z(x, z) = \frac{2}{\pi}\varphi_e D_z C_M\sum_{n=1}^{\infty}R_{1_n}\exp(R_{2_n}x)\lambda_n\sin(\lambda_n z) \tag{3-127}$$

Steady-State Special Solutions for Other Sources
Equation (3-114) is the general solution for the resulting concentration distribution for any given $C_M(z)$ source concentration distribution. With the source concentration function (Figure 3-11 with $\gamma = 0$), $C_M(z)$, the integrals I_0 and I_n given by Eqs. (3-64) and (3-70), respectively, need to be evaluated analytically or numerically depending on the type of the function $C_M(z)$. Two examples are presented below.

Solution for Two-Strip Sources with Uniform Source Concentrations
The geometry of two-strip sources with uniform source concentrations is shown in Figure 3-13 (with $\gamma = 0$). The boundary conditions at $x = 0$ are expressed by Eqs. (3-98)–(3-100) (with $\gamma = 0$). The expressions for I_0 and I_n are given by Eqs. (3-101) and (3-103), respectively. Therefore, Eqs. (3-61), (3-101), (3-103), and (3-114) form the solution for two-strip sources under steady-state solute transport conditions.

Solution for a Single-Strip Source with Linearly Varying Source Concentration
The geometry of a single-strip source with linearly varying source concentration is shown in Figure 3-14 (with $\gamma = 0$). The boundary conditions at $x = 0$ are expressed by Eqs. (3-104) and (3-105) (with $\gamma = 0$). The expressions for I_0 and I_n are given by Eqs. (3-107) and (3-109), respectively. Therefore, Eqs. (3-61), (3-107), (3-109), and (3-114) form the solution for a linearly varying source concentration under steady-state solute transport conditions.

Evaluation of the Models
The analytical models presented in the previous sections are evaluated from their verification as well as applicability limitation point of view and the results are presented below.

Comparisons with Numerical Models
The computer program ST2A, described in the following sections, was developed to predict spatially and temporally concentration and convective–dispersive flux components from an exponentially varying time-dependent solute strip source in a uniform flow field bounded by two impervious

boundaries (Figure 3-12). The normalized concentration results of ST2A are compared with two different finite element ground water flow and solute transport codes. The first is a finite element ground water flow and solute transport code, called GEOFLOW, developed by International Technology Corporation (IT, 1986). GEOFLOW is a quasi-three-dimensional code that can simulate the ground water flow and solute transport in a two-dimensional medium. The second code is PTC (Princeton Transport Code), which is a complete three-dimensional code, developed by the Princeton University (Babu et al., 1987). The modeling parameters used for the comparison of the two codes are shown in Figure 3-15 and Figure 3-16. The finite element grid covers an area of 75 m$\times$50 m. The geometrical values for the source area are $2B = 10$ m, $D_1 = 5$ m, and $D_2 = 35$ m. These values correspond to an asymmetrical source location with respect to the impervious boundaries. The uniform ground water velocity U is 0.1 m/day. The longitudinal and transverse dispersion coefficients are 1 m^2/day (D_x) and 0.1 m^2/day (D_z), respectively. A retardation factor (R_d) of 1 was selected for the comparisons. Both γ and v were assigned values of zero. The time step (Δt) for the numerical codes and total simulation time for the solute transport example were selected to be 1 and 100 days, respectively. The Courant number C controls oscillations of the numerical solution arising from the temporal discretization. The resulting criterion is $C = U\Delta t/\Delta x < 1$ (Daus and Frind, 1985), where Δx is the node spacing in the x direction. The values of U, Δt, and Δx satisfy this condition well.

Figure 3-15 presents the longitudinal normalized concentration distribution as a function of the x coordinate for $z = 10$ and 16.25 m. Figure 3-16 presents the normalized concentration distributions as a function of the z coordinate for $x = 5$ and 20 m. As can be seen from Figure 3-15 and Figure 3-16, ST2A and the two different finite element codes show a good agreement in their results.

Representation of Exponentially Variable Time-Dependent Source Concentrations
As mentioned in Section 3.3.2.1, the exponential term, $e^{\gamma t}$, increases the capability of the model in a way that exponentially varying solute source concentrations can be taken into account. With

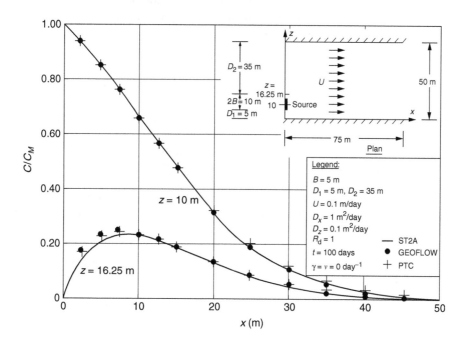

FIGURE 3-15 Comparison of normalized concentrations of ST2A, GEOFLOW, and PTC along the longitudinal direction for $z = 10$ and 16.25 m. (Adapted from Batu, V., *Water Resour. Res.*, 25, 1125–1132, 1989.)

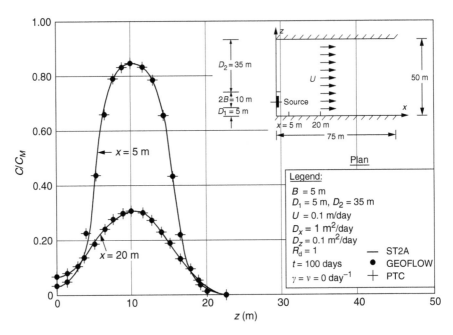

FIGURE 3-16 Comparison of normalized concentrations of ST2A, GEOFLOW, and PTC along the trans-verse direction for x = 5 and 20 m. (Adapted from Batu, V., *Water Resour. Res.*, 25, 1125–1132, 1989.)

negative $\gamma(T^{-1})$ values a decaying source concentration can be represented in the model. There are some requirements in the usage of this parameter. In Eqs. (3-79), (3-81), (3-84), and (3-85) γ is under the square root. Therefore, in order to be able to take these square roots, some of the terms under each square root must be greater than zero. A relatively simple analysis can show that the nec-essary condition for this requirement is the expression given by Eq. (3-44). An example is presented in the paragraph following Eq. (3-44).

Number of Terms Selection for the Infinite Series in the Solutions
The number of terms N in the infinite series of the unsteady-state solution (Eq. [3-77]) and the steady-state solution (Eq. [3-119]) play a key role in determining accurate results. Numerical experiments with the ST2A program showed that the area close to the source requires more terms than the other areas in order to determine accurate results. The line at which the source is located (x = 0) is the most sensitive location. Based on the boundary condition expressed by Eqs. (3-73) and (3-74) or Eqs. (3-117) and (3-118), the normalized concentration variation at x = 0 is known. Here, the results of a series of computer runs with different N values are pre-sented for the normalized concentration (C/C_M) variation at x = 0. The following values are used for these examples: $2B$ = 100 m, D_1 = 1,000 m, D_2 = 1,000 m, φ_e = 0.20, K = 1 m/day, and I = 0.02 m/m (U = 0.1 m/day). The normalized concentration variation for N = 200, 400, 1,000, 3,000, 5,000, and 10,000 are shown in Figure 3-17–Figure 3-22, respectively. These figures clearly show that oscillations decrease with increase in N values. Generally, N = 3,000 produces reasonably accurate results in the solute transport domain with the exception of the domain close to the source.

Application of the Two-Dimensional Analytical Solute Transport Models to Ground Water Flow Systems
The general model and its derivative models under unsteady- and steady-state solute transport con-ditions presented above can be applied to solute transport predictions in aquifers if the ground water velocity can be assumed to be uniform. The models can also be applied to unsaturated zones with some additional assumptions. These are all discussed below.

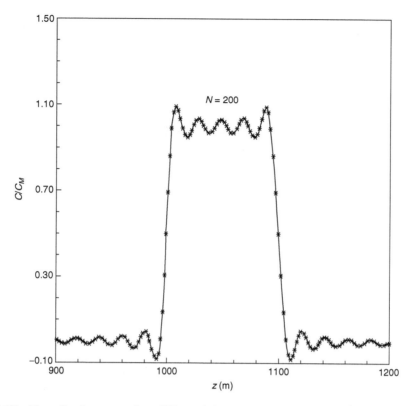

FIGURE 3-17   Normalized concentration, C/C_M, variation around the source $x = 0$ for $N = 200$.

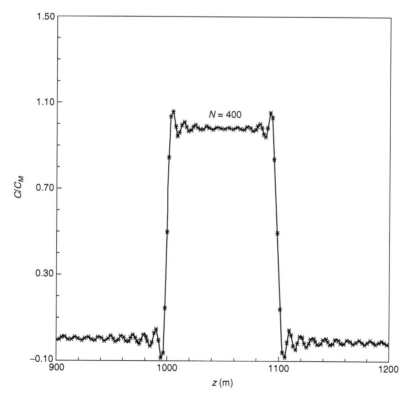

FIGURE 3-18   Normalized concentration, C/C_M, variation around the source $x = 0$ for $N = 400$.

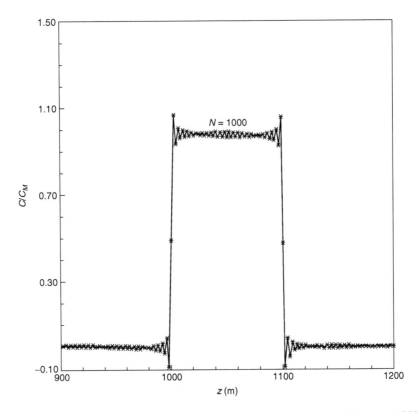

FIGURE 3-19   Normalized concentration, C/C_M, variation around the source $x = 0$ for $N = 1000$.

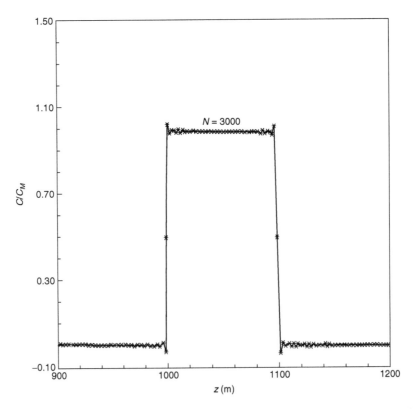

FIGURE 3-20   Normalized concentration, C/C_M, variation around the source $x = 0$ for $N = 3000$.

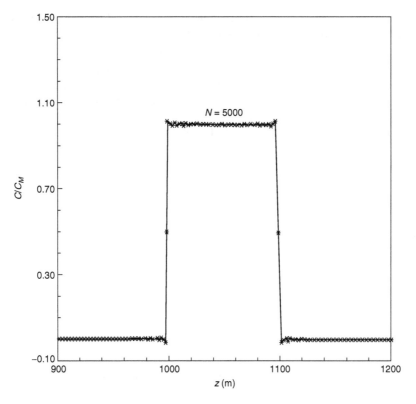

FIGURE 3-21   Normalized concentration, C/C_M, variation around the source $x = 0$ for $N = 5000$.

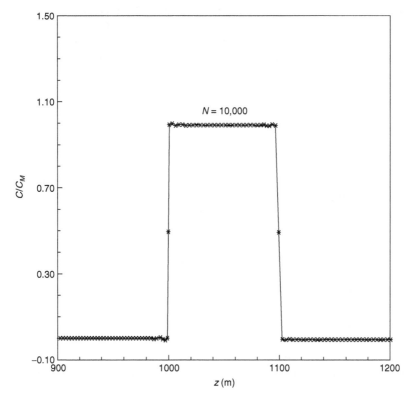

FIGURE 3-22   Normalized concentration, C/C_M, variation around the source $x = 0$ for $N = 10,000$.

In this section, a computer program for a single-strip source, called ST2A, is described. Then, the application of this program is presented with examples.

A Computer Program (ST2A) for the Two-Dimensional First-Type Strip Source Model

A two-dimensional MS-DOS-based FORTRAN solute transport computer program, called solute transport two-dimensional first-type (ST2A), has been developed by the author. ST2A is a menu-driven program and calculates values of concentration and convective–dispersive flux components under unsteady-state solute transport conditions (Eqs. [3-77], [3-91], and [3-96]) and steady state solute transport conditions (Eqs. [3-119], [3-124], and [3-127]). This program can be run interactively in IBM compatible computers.

Application to Saturated Zones

The two-dimensional analytical solute transport model whose geometry is shown in Figure 3-12, can be used for different purposes. The impervious boundaries around the solute transport domain increase the potential usage areas of the model. For ground water flow cases, the effects of impervious boundaries such as bedrock outcrops and clay zones surrounding the main ground water flow zone can be taken into account. If the aquifer goes to infinity from both sides of the source, its effect can be taken into account by assigning large values to D_1 and D_2 as compared with the source width $2B$. The model can also be used along a cross-section of a confined or unconfined aquifer such that the $z = 0$ plane is the bottom of the aquifer and $z = D_1 + 2B + D_2$ plane is the upper boundary of the aquifer. For a confined aquifer, $z = D_1 + 2B + D_2$ plane is the bottom of the confining layer and for an unconfined aquifer is the water table of the unconfined aquifer. For these cases, the source will be infinitely long in the direction of the y-coordinate axis. For two-dimensional areal model case in the x–z plane, as shown in Figure 3-12, it is assumed that the source extends from the bottom of a confined or unconfined aquifer to the top of the confined aquifer or to the water table of the unconfined aquifer.

EXAMPLE 3-9

For the source and aquifer geometry shown in Figure 3-12 the following parameters are given: $K_h = 13.875$ m/day, $I = 0.0027027$ m/m, $\varphi_e = 0.25$ ($U = 0.15$ m/day), $K_d = 0$ cm^3/g, $\rho_b = 1.78$ cm^3/g ($R_d = 1$), $D^* = 0$ m^2/day, $C_M = 1$ unit, $2B = 10$ m, $D_1 = 1,000$ m, $D_2 = 1,000$ m, $\alpha_L = 21.3$ m, and $\alpha_T = 4.3$ m. With these parameters, using the ST2A program, (1) determine the normalized isoconcentration curves after 10 and 20 years elapsed times for constant source concentration case, and the contours for the convective–dispersive flux components in the x-coordinate direction after 10 years and (2) determine the normalized isoconcentration curves after 10 and 20 years elapsed times for a decaying source concentration case using $\gamma = 0.001$ day^{-1}.

SOLUTION

By using the given data, the ST2A program was run to generate the coordinates, normalized concentration, and convective–dispersive flux components data at a given time. The output of ST2A was used to generate contours using the SURFER^{TM} contouring program (Golden Software, Inc., 1988).

1. In Figure 3-23 and Figure 3-24 the normalized concentration contours after 10 and 20 years, respectively, are presented. Figure 3-25 presents the contours for the convective–dispersive flux components after 10 years in the x-coordinate direction.
2. Using the given data with $\gamma = 0.001$ day^{-1}, the ST2A program was run for the decaying source concentration case. Figure 3-26 presents the temporal variation of the decaying source function for the given value of γ. Figure 3-27 and Figure 3-28 present the normalized concentration contours for 10 and 20 years, respectively. Notice from Figure 3-26 that the normalized solute source concentration becomes approximately zero

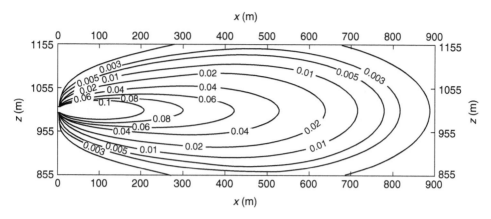

FIGURE 3-23 Normalized isoconcentration curves for a constant source concentration after 10 years for Example 3-9, produced from the ST2A program.

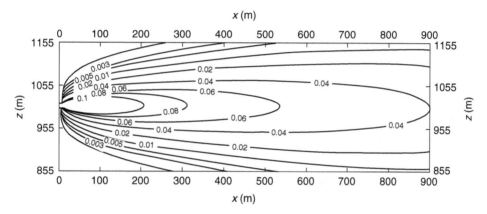

FIGURE 3-24 Normalized isoconcentration curves for a constant source concentration after 20 years for Example 3-9 produced from the ST2A program.

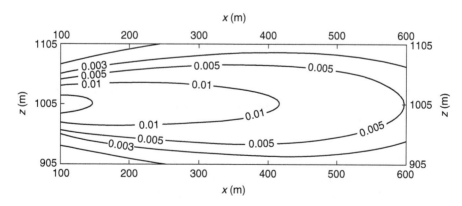

FIGURE 3-25 Contours for the convective–dispersive flux component in the x-coordinate direction $[F_x/(\varphi_e C_M)]$ for a constant source concentration after 10 years for Example 3-9, produced from the ST2A program.

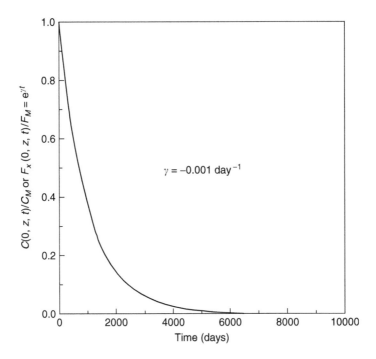

FIGURE 3-26 $C(0, z, t)/C_M$ vs. time for $\gamma = -0.001$ day^{-1} for Example 3-9.

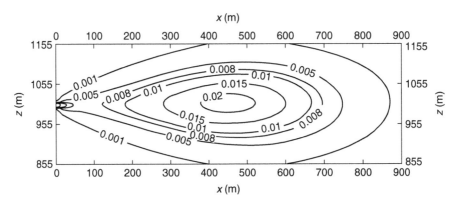

FIGURE 3-27   Normalized isoconcentration curves for a decaying source concentration after 10 years for Example 3-9, produced from the ST2A program.

after approximately 6000 days (16.4 years). As can be seen from these figures, the center of the solute plume is shifted approximately 450 m from the source after 10 years (Figure 3-27) and 1000 m after 20 years (Figure 3-28).

Application to Unsaturated Zones
The schematic diagram in Figure 3-29 shows the applicability of the two-dimensional ST2A model for a single constant concentration source. The associated governing equations for flow and solute transport in unsaturated porous media under uniform infiltration rate conditions are given in Section 2.8.1. Some important points for the application and assumptions are given below:

1.   It is assumed that there is a uniform infiltration rate (v_0) at the soil surface. As shown in Section 2.8.1, this infiltration rate creates a constant ground water velocity in the

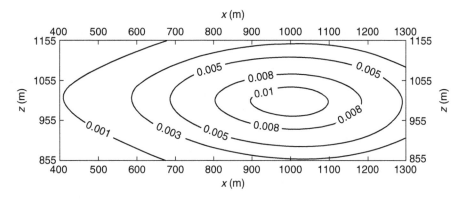

FIGURE 3-28   Normalized isoconcentration curves for a decaying source concentration after 20 years for Example 3-9, produced from the ST2A program.

unsaturated zone. Also in the same section it is shown that the ground water velocity U can be determined using the unsaturated hydraulic conductivity vs. water content relationship for a given soil, i.e., $K = K(\theta)$.

2. Liquid waste from a surface contaminant source is being discharged to the unsaturated zone. Landfills, some units of factories, or other industrial complexes may be considered as surface contaminant sources.

3. The length L of the source in practice is finite, but in the model it is infinite (in the y-coordinate axis).

4. It is assumed that the waste liquid seeping from the bottom of the source ditch reaches to the top of the unsaturated zone and creates a constant concentration C_M of a certain solute species in the area beneath the ditch (the dotted zones in Figure 3-29).

5. The source concentration C_M can also be taken into account as an exponentially varying time-dependent quantity according to Eq. (3-73).

3.3.2.3   Three-Dimensional Analytical Solute Transport Models for the First-Type Rectangular Sources

As mentioned in Section 3.3.2.2, during the past two decades, a number of two-dimensional analytical solute transport models of the transport and fate of contaminants in subsurface flow systems have been developed. However, the number of three-dimensional analytical solute transport models is fairly limited (Sagar, 1982; Yeh, 1981; Domenico and Robbins, 1985; Huyakorn et al., 1987; Leij et al., 1991; Batu, 1996). These analytical models are based on infinite or semiinfinite medium assumption with the exception of Yeh (1981), Huyakorn et al. (1987), and Batu (1996). Yeh (1981) and Huyakorn et al. (1987) presented three-dimensional analytical solute transport models from partially penetrating finite sources by taking into account the thickness of aquifer. The final solutions of the aforementioned models are in integral form and they require numerical integration. As mentioned by Batu (1996), the following are the three limitations of these models: (1) the finite source is adjacent to the top boundary of aquifer, (2) the source location along the aquifer thickness cannot be changed, and (3) the source concentration is assumed to be constant spatially and temporally. Also these models assume that the aquifer boundaries approach infinity horizontally. The model developed by Batu (1996) is a fairly general three-dimensional analytical solute transport model from multiple time- and space-dependent rectangular solute sources in a unidirectional bounded parallelepiped-type flow field with the first-type source condition. Since this model is more general than the other

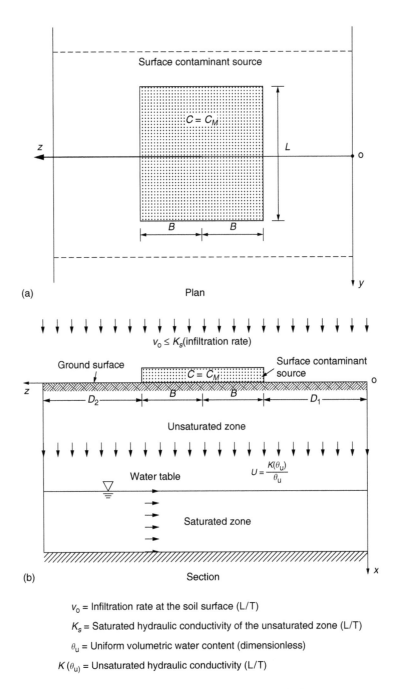

FIGURE 3-29 Schematic representation for the applicability of the ST2A program to unsaturated flow zones. (Adapted from Batu, V., *Ground Water*, 26, 71–77, 1988.)

ones, in this section, only the model developed by Batu (1996) will be presented with its potential application areas.

Problem Statement and Assumptions
The geometry of the generalized three-dimensional analytical solute transport model is illustrated in Figure 3-30. All solute sources are located on the *yOz* plane and the four other planes

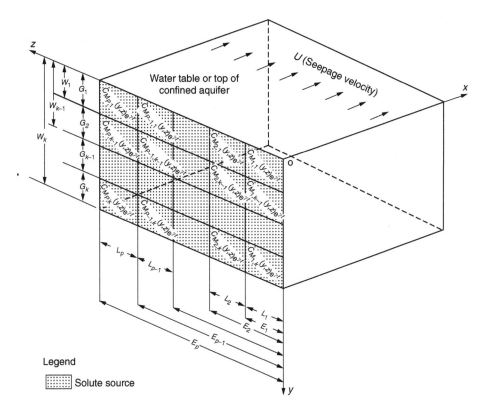

FIGURE 3-30   Schematic representation of the three-dimensional solute transport model with the first-type source inputs in a bounded medium having unidirectional flow field. (Adapted from Batu, V., *J. Hydrol.*, 174, 57–82, 1996.)

that are perpendicular to the yOz plane are assumed to be impervious. All five planes defined above form a parallelepiped, which goes to infinity in the x-coordinate direction. The multiple rectangular sources are located irregularly along the yOz plane as shown in Figure 3-30 and they have space- and time-dependent source concentrations. The uniform flow field is assumed to exist along the x-coordinate direction and the system approaches infinity along the same coordinate direction. The upper boundary of the model (xOz plane) from a practical standpoint corresponds to the water table of an unconfined aquifer or the upper boundary of a confined aquifer. The lower boundary, which is parallel to the xOz plane at $y = W_k$ distance (see Figure 3-30), may represent a bedrock or an impervious boundary or the lower boundary of an unconfined or confined aquifer.

Under the framework of the above description, the problem is an unsteady-state three-dimensional solute transport from multiple rectangular sources between four impervious planar boundaries in a unidirectional flow field as shown in Figure 3-30. The x- and z-axis are the planar coordinates and the y axis is the vertical coordinate. The assumptions are as follows:

1. The solute sources are planar and located irregularly on the y–z plane in a unidirectional steady ground water velocity U field.
2. The solute sources are located at $x = 0$ plane and they are perpendicular to the velocity of the flow field.
3. The source concentrations are functions of the y and z coordinates and time t through an exponential function.
4. The medium goes to infinity in the x-coordinate direction and y varies between 0 and W_k and z varies between 0 and E_p.

Governing Differential Equation

The governing differential equations for solute transport in unidirectional flow field are given in Section 2.7.3 and its three-dimensional form in the x, y, and z coordinates is given by Eq. (2-81), which can also be written as

$$\frac{\partial C}{\partial t} = \frac{D_x}{R_d}\frac{\partial^2 C}{\partial x^2} + \frac{D_y}{R_d}\frac{\partial^2 C}{\partial y^2} + \frac{D_z}{R_d}\frac{\partial^2 C}{\partial z^2} - \frac{U}{R_d}\frac{\partial C}{\partial x} - vC \tag{3-128}$$

where C is the concentration, x, y, and z are the coordinates, and t is the time. The potential units of C widely used in practice are included in Section 2.2.1.2. In Eq. (3-128), R_d is the retardation factor (given by Eq. [2-62]) and v the decay constant.

Initial and Boundary Conditions

The initial condition for $C(x, z, t)$ is

$$C(x, y, z, 0) = 0 \tag{3-129}$$

The values of concentration at the sources are specified. Therefore, these are the first-type or Dirichlet boundary conditions. The concentrations in the first row of rectangles on the yOz plane are given as follows:

$$C(0, y, z, t) = C_{M_{1,1}}(y, z)e^{\gamma t}, \quad 0 < y < W_1 \quad 0 < z < E_1 \tag{3-130}$$

$$C(0, y, z, t) = C_{M_{p-1,1}}(y, z)e^{\gamma t}, \quad 0 < y < W_1 \quad E_{p-2} < z < E_{p-1} \tag{3-131}$$

$$C(0, y, z, t) = C_{M_{p,1}}(y, z)e^{\gamma t}, \quad 0 < y < W_1 \quad E_{p-1} < z < E_p \tag{3-132}$$

Similarly, expressions for 2nd, ..., (k–1), and kth row can be written. The expressions for the kth row are as follows:

$$C(0, y, z, t) = C_{M_{1,k}}(y, z)e^{\gamma t}, \quad W_{k-1} < y < W_k \quad 0 < z < E_1 \tag{3-133}$$

$$C(0, y, z, t) = C_{M_{p-1,k}}(y, z)e^{\gamma t}, \quad W_{k-1} < y < W_k \quad E_{p-2} < z < E_{p-1} \tag{3-134}$$

$$C(0, y, z, t) = C_{M_{p,k}}(y, z)e^{\gamma t}, \quad W_{k-1} < y < W_k \quad E_{p-1} < z < E_p \tag{3-135}$$

where $\gamma\,(\mathrm{T}^{-1})$ is a constant, and

$$E_i = \sum_{j=1}^{i} L_j \quad i = 1, 2, ..., p \tag{3-136}$$

$$W_i = \sum_{j=1}^{i} G_j \quad i = 1, 2, ..., k \tag{3-137}$$

The boundary conditions at the impervious planes and infinite x distance are

$$\frac{\partial C}{\partial y} = 0, \quad y = 0, \quad 0 < x < \infty, \quad 0 < z < E_p \tag{3-138}$$

$$\frac{\partial C}{\partial y} = 0, \quad y = F_k, \quad 0 < x < \infty, \quad 0 < z < E_p \tag{3-139}$$

$$\frac{\partial C}{\partial z} = 0, \quad z = 0, \quad 0 < x < \infty, \quad 0 < y < W_k \tag{3-140}$$

$$\frac{\partial C}{\partial z} = 0, \quad z = E_p, \quad 0 < x < \infty, \quad 0 < y < W_k \tag{3-141}$$

$$C = 0, \quad x = \infty, \quad 0 < y < W_k, \quad 0 < z < E_p \tag{3-142}$$

$$\frac{\partial C}{\partial x} = 0, \quad x = \infty, \quad 0 < y < W_k, \quad 0 < z < E_p \tag{3-143}$$

Equation (3-129) states that initially the concentration is zero everywhere in the flow field. Equations (3-130)–(3-135) state that the concentration $C(0, y, z, t)$ varies with the y and z coordinates and time t and are the first-type or Dirichlet boundary conditions. The function

$$\frac{C(0, y, z, t)}{C_{M_i}(y, z)} = e^{\gamma t} \tag{3-144}$$

performs such that the source concentration can be represented as exponentially varying time-dependent quantity. Equation (3-144) is a generalized version of Eqs. (3-6) and (3-55) and the properties of this function are described in Sections 3.3.2.1 and 3.3.2.2. Figure 3-2 shows the variation of $C(0, y, z, t)/C_M$ with time for different values of γ ranging from -0.0001 to -0.0100 day^{-1}. If observed data are available for time-dependent source concentration, the value of γ may be determined from the curve-fitting techniques. Equations (3-138) and (3-139) state that the solute fluxes perpendicular to the xOz plane ($y = 0$) and the plane parallel to the xOz plane at a distance $y = F_k$ are zero. Therefore, they are impervious boundaries. Equations (3-140) and (3-141) state that the solute fluxes perpendicular to the xOy ($z = 0$) plane and to the plane parallel to the xOy plane at a distance $z = E_p$ are zero. Therefore, they are impervious boundaries. Finally, Eqs. (3-142) and (3-143) state that both the concentration C and its gradient with respect to $x(\partial C/\partial x)$ approach zero when the x goes to infinity.

Unsteady-State General Solution for Concentration Distribution
The boundary-value problem defined by Eqs. (3-128)–(3-144) is solved by the Laplace transform and Fourier analysis techniques. For this purpose, the solution is obtained first in the Laplace domain and then the concentration variation is determined by taking the inverse Laplace transform of the Laplace domain solution. In the following sections, the intermediate steps and final solutions are presented based on Batu (1996).

General Solution in the Laplace Domain
The general solution in the Laplace domain is (Batu, 1996)

$$u(x, y, z, s) = \sum_{m=0}^{\infty} \sum_{n=0}^{\infty} B_{mn} \cos(\eta_m y) \cos(\lambda_n z) \exp\left[\left(\frac{U}{2D_x} - K_{mn}\right)x\right] \tag{3-145}$$

where B_{mn} is a constant,

$$\lambda_n = \frac{n\pi}{E_p}, \quad n = 0, 1, 2, \ldots \tag{3-146}$$

$$\eta_m = \frac{m\pi}{W_k}, \quad m = 0, 1, 2, \ldots \tag{3-147}$$

and

$$K_{mn} = \left[\frac{D_z}{D_x}\lambda_n^2 + \frac{D_y}{D_x}\eta_m^2 + \frac{R_d}{D_x}(s + v) + \frac{U^2}{4D_x^2}\right]^{1/2} \tag{3-148}$$

and $u(x, y, z, s)$ is the Laplace transform of $C(x, y, z, t)$ and is defined as (e.g., Spiegel, 1965)

$$u(x, y, z, s) = f(s) = L\{C(x, y, z, t)\} = \int_0^\infty e^{-st} C(x, y, z, t)\, dt \tag{3-149}$$

Derivation of Eq. (3-145) is given in Example 3-10.

The general solution in the Laplace domain given by Eq. (3-145) is expressed as (Batu, 1996)

$$u(x, y, z, s) = u_1(x, y, z, s) + u_2(x, y, z, s) + u_3(x, y, z, s) + u_4(x, y, z, s) \tag{3-150}$$

where

$$u_1(x, y, z, s) = \left(\frac{1}{2}\right)\left(\frac{1}{2}\right) B_{00} \exp\left[\left(\frac{U}{2D_x} - K_{00}\right)x\right] \tag{3-151}$$

$$u_2(x, y, z, s) = \frac{1}{2}\sum_{n=1}^\infty B_{0n}\cos(\lambda_n z)\exp\left[\left(\frac{U}{2D_x} - K_{0n}\right)x\right] \tag{3-152}$$

$$u_3(x, y, z, s) = \frac{1}{2}\sum_{m=1}^\infty B_{m0}\cos(\eta_m y)\exp\left[\left(\frac{U}{2D_x} - K_{m0}\right)x\right] \tag{3-153}$$

and

$$u_4(x, y, z, s) = \sum_{m=1}^\infty\sum_{n=1}^\infty B_{mn}\cos(\eta_m y)\cos(\lambda_n z)\exp\left[\left(\frac{U}{2D_x} - K_{mn}\right)x\right] \tag{3-154}$$

where

$$K_{00} = \left[\frac{R_d}{D_x}(s+v) + \frac{U^2}{4D_x^2}\right]^{1/2} \tag{3-155}$$

$$K_{0n} = \left[\frac{D_z}{D_x}\lambda_n^2 + \frac{R_d}{D_x}(s+v) + \frac{U^2}{4D_x^2}\right]^{1/2} \tag{3-156}$$

and

$$K_{m0} = \left[\frac{D_y}{D_x}\eta_m^2 + \frac{R_d}{D_x}(s+v) + \frac{U^2}{4D_x^2}\right]^{1/2} \tag{3-157}$$

with K_{mn} given by Eq. (3-148). Equations (3-150)–(3-157) are based on the fact that B_{0n} and B_{m0} are one-half and B_{00} is one-quarter (Carslaw and Jaeger, 1959, p. 182).

The expression for B_{mn} in Eq. (3-145), which is derived in Example 3-10, is as follows:

$$B_{mn} = \frac{4}{E_p W_k}\int_0^{W_k}\int_0^{E_p}\frac{C_M(y, z)}{s-\gamma}\cos(\eta_m y)\cos(\lambda_n z)\, dy\, dz \tag{3-158}$$

From Eq. (3-158), B_{00}, B_{0n}, and B_{m0} can, respectively, be expressed as

$$B_{00} = \frac{4}{E_p W_k}\int_0^{W_k}\int_0^{E_p}\frac{C_M(y, z)}{s-\gamma}\, dy\, dz \tag{3-159}$$

$$B_{0n} = \frac{4}{E_p W_k}\int_0^{W_k}\int_0^{E_p}\frac{C_M(y, z)}{s-\gamma}\cos(\lambda_n z)\, dy\, dz \tag{3-160}$$

and

$$B_{m0} = \frac{4}{E_p W_k}\int_0^{W_k}\int_0^{E_p}\frac{C_M(y, z)}{s-\gamma}\cos(\eta_m y)\, dy\, dz \tag{3-161}$$

Now, the general solution in the Laplace domain is complete.

EXAMPLE 3-10

Solve the boundary-value problem defined by Eqs. (3-128)–(3-144) in the Laplace domain, i.e., derive Eq. (3-150).

SOLUTION

The Laplace transform of each term in Eq. (3-128) with respect to time are as follows (see, e.g., Spiegel, 1965):

$$L\left[\frac{\partial C}{\partial t}\right] = su(x, y, z, s) - C(x, y, z, t) \tag{E3-10-1}$$

$$L\left[\frac{\partial^2 C}{\partial x^2}\right] = \frac{\partial^2 u}{\partial x^2} \quad L\left[\frac{\partial^2 C}{\partial y^2}\right] = \frac{\partial^2 u}{\partial y^2} \quad L\left[\frac{\partial^2 C}{\partial z^2}\right] = \frac{\partial^2 u}{\partial z^2}$$

$$L\left[\frac{\partial C}{\partial x}\right] = \frac{\partial u}{\partial x} \quad L\left[\frac{\partial C}{\partial y}\right] = \frac{\partial u}{\partial y} \quad L\left[\frac{\partial C}{\partial z}\right] = \frac{\partial u}{\partial z} \tag{E3-10-2}$$

where $u(x, y, z, s)$ is defined as (e.g., Spiegel, 1965)

$$u(x, y, z, s) = f(s) = L[C(x, y, z, t)] = \int_0^\infty e^{-st} C(x, y, z, t)\, dt \tag{E3-10-3}$$

The substitution of these expressions into Eq. (3-128) gives

$$R_d(s + v)u = D_x \frac{\partial^2 u}{\partial x^2} + D_y \frac{\partial^2 u}{\partial y^2} + D_z \frac{\partial^2 u}{\partial y^2} - U \frac{\partial u}{\partial x} \tag{E3-10-4}$$

which is the three-dimensional solute transport differential equation in the Laplace domain.

The method of separation of variables (e.g., Churchill, 1941; Spiegel, 1965; Batu, 1983) is applied to solve Eq. (E3-10-4) with

$$u(x, y, z, s) = X(x, s)Y(y, s)Z(z, s) \tag{E3-10-5}$$

where $X(x, s)$, $Y(y, s)$, and $Z(z, s)$ are some unknown functions. The substitution of Eq. (E3-10-5) into Eq. (E3-10-4) gives

$$-\frac{Z''}{Z} = \frac{D_x}{D_z}\frac{X''}{X} + \frac{D_y}{D_z}\frac{Y''}{Y} - \frac{U}{D_z}\frac{X'}{X} - \frac{R_d}{D_z}(s + v) = \mu \tag{E3-10-6}$$

where μ must be a constant (Churchill, 1941). By examining the boundary conditions, it can be concluded that the constant value should be as $\mu = \lambda^2 > 0$. By introducing this value into Eq. (E3-10-6), the following ordinary and partial differential equations, respectively, can be obtained:

$$Z'' + \lambda^2 Z = 0 \tag{E3-10-7}$$

$$\frac{D_x}{D_z}\frac{X''}{X} + \frac{D_y}{D_z}\frac{Y''}{Y} - \frac{U}{D_z}\frac{X'}{X} - \frac{R_d}{D_z}(s + v) = \lambda^2 \tag{E3-10-8}$$

Equation (E3-10-8) can be rearranged as

$$-\frac{Y''}{Y} = \frac{D_x}{D_y}\frac{X''}{X} - \frac{U}{D_y}\frac{X'}{X} - \frac{R_d}{D_y}(s + v) - \lambda^2 \frac{D_z}{D_y} = \eta^2 \tag{E3-10-9}$$

Again, from Eq. (E3-10-9), the following two ordinary differential equations can be obtained:

$$Y'' + \eta^2 Y = 0 \qquad\qquad\qquad\text{(E3-10-10)}$$

$$X'' - \frac{U}{D_x} X' - \frac{1}{D_x}[(s + v)R_d + \lambda^2 D_z + \eta^2 D_y]X = 0 \qquad\qquad\text{(E3-10-11)}$$

Solutions of Eqs. (E3-10-7), (E3-10-10), and (E3-10-11)
Equations (E3-10-7) and (E3-10-10) have the same forms as Eq. (E3-4-7). Therefore, by using the same procedure, the following solutions for $Z(z)$ and $Y(y)$, respectively, can be obtained:

$$Z(z) = A_1 \sin(\lambda z) + B_1 \cos(\lambda z) \qquad\qquad\qquad\text{(E3-10-12)}$$

$$Y(y) = A_2 \sin(\eta y) + B_2 \cos(\eta y) \qquad\qquad\qquad\text{(E3-10-13)}$$

where A_1, B_1, A_2, and B_2 are constants. Equation (E3-10-11) has the similar form of Eq. (E3-4-6). Therefore, by using the same procedure, its solution can be written as

$$X(x) = A_3 \exp\left[\left(\frac{U}{2D_x} + K\right)x\right] + B_3 \exp\left[\left(\frac{U}{2D_x} - K\right)x\right] \qquad\text{(E3-10-14)}$$

where A_3 and B_3 are constants and

$$K = \left[\frac{D_z}{D_x}\lambda^2 + \frac{D_y}{D_x}\eta^2 + \frac{R_d}{D_x}(s + v) + \frac{U^2}{4D_x^2}\right]^{1/2} \qquad\text{(E3-10-15)}$$

Usage of the Boundary Conditions at Infinity and at the Impermeable Boundaries
The substitution of Eqs. (E3-10-12)–(E3-10-14) into Eq. (E3-10-5) gives

$$u(x, y, z, s) = [A_1 \sin(\lambda z) + B_1 \cos(\lambda z)][A_2 \sin(\eta y) + B_2 \cos(\eta y)]$$

$$\times \left\{A_3 \exp\left[\left(\frac{U}{2D_x} + K\right)x\right] + B_3 \exp\left[\left(\frac{U}{2D_x} - K\right)x\right]\right\} \qquad\text{(E3-10-16)}$$

To satisfy the boundary conditions expressed by Eqs. (3-142) and (3-143), A_3 in Eq. (E3-10-16) must vanish i.e., $A_3 = 0$. From Eq. (E3-10-16) the derivative of u with respect to y is

$$\frac{\partial u}{\partial y} = B_3[A_2\eta \cos(\eta y) - B_2\eta \sin(\eta y)][A_1 \sin(\lambda z) + B_1 \cos(\lambda z)] \exp\left[\left(\frac{U}{2D_x} - K\right)x\right] \quad\text{(E3-10-17)}$$

The derivative of u with respect to z is

$$\frac{\partial u}{\partial z} = B_3[A_2 \sin(\eta y) + B_2 \cos(\eta y)][A_1\lambda \cos(\lambda z) - B_1 \lambda \sin(\lambda z)] \exp\left[\left(\frac{U}{2D_x} - K\right)x\right]\text{(E3-10-18)}$$

It can be observed that from Eqs. (3-140) and (E3-10-18), A_1 must vanish in Eq. (E3-10-16); similarly, from Eqs. (3-138) and (E3-10-17), A_2 must vanish in Eq. (E3-10-16), that is $A_1 = 0$ and $A_2 = 0$. As a result, Eqs. (E3-10-17) and (E3-10-18), respectively, take the forms

$$\frac{\partial u}{\partial y} = - B_1 B_2 B_3\eta \sin(\eta y)\cos(\lambda z) \exp\left[\left(\frac{U}{2D_x} - K\right)x\right] \qquad\text{(E3-10-19)}$$

$$\frac{\partial u}{\partial z} = - B_1 B_2 B_3 \lambda \cos(\eta y)\sin(\lambda z) \exp\left[\left(\frac{U}{2D_x} - K\right)x\right] \qquad\text{(E3-10-20)}$$

Using the boundary condition given by Eq. (3-141) and the last expression of Eq. (E3-10-2) one can write

$$B_1 B_2 B_3 \; \lambda \; \cos(\eta y) \sin(\lambda E_p) \exp\left[\left(\frac{U}{2D_x} - K\right)x\right] = 0 \tag{E3-10-21}$$

In order to satisfy the above-mentioned condition, the following condition must be satisfied:

$$\sin(\lambda E_p) = 0 \tag{E3-10-22}$$

which results in Eq. (3-146). Similarly, using the boundary condition given by Eq. (3-139), and the fifth expression of Eq. (E3-10-2), one can write

$$B_1 B_2 B_3 \eta \sin(\eta W_k) \cos(\lambda z) \exp\left[\left(\frac{U}{2D_x} - K\right)x\right] = 0 \tag{E3-10-23}$$

In order to satisfy the above-mentioned condition, the following condition must be satisfied:

$$\sin(\eta W_k) = 0 \tag{E3-10-24}$$

which results in Eq. (3-147). By substituting Eqs. (3-146) and (3-147) and $A_1 =$ and $A_2 = 0$ into Eq. (E3-10-5) and applying the principle of superposition (since the partial differential equation and boundary conditions are linear), Eq. (3-145) can be obtained.

Usage of Boundary Conditions at x = 0 and Determination of B_{mn}
The boundary conditions at $x = 0$, as given by Eqs. (3-130)–(3-135) need to be expressed in the Laplace domain. Their Laplace transform is

$$L[C(0, y, z, t)] = u(0, y, z, s) = L[C_M(0, y, z)e^{\gamma t}] = \frac{C_M(0, y, z)}{s - \gamma} \tag{E3-10-25}$$

where

$$C_M(0, y, z) = C_{M_1}(0, y, z) + \cdots + C_{M_{p-1}}(0, y, z) + C_{M_p}(0, y, z) \tag{E3-10-26}$$

For $x = 0$, by using Eq. (E3-10-25), Eq. (3-145) becomes

$$\frac{C_M(0, y, z)}{s - \gamma} = \sum_{m=0}^{\infty} \sum_{n=0}^{\infty} B_{mn} \cos(\eta_m y) \cos(\lambda_n z) \tag{E3-10-27}$$

The coefficients of B_{mn} are the coefficients of the double Fourier cosine series expansion of $C_M(0, y, z)/(s-\gamma)$. The substitution of Eqs. (3-146) and (3-147) into Eq. (E3-10-27) gives

$$\frac{C_M(0, y, z)}{s - \gamma} = \sum_{m=0}^{\infty} \sum_{n=0}^{\infty} B_{mn} \cos\left(\frac{m\pi y}{W_k}\right) \cos\left(\frac{n\pi z}{E_p}\right) \tag{E3-10-28}$$

B_{mn} can be obtained by the method given below (Hildebrand, 1976, p. 455). If both sides of Eq. (E3-10-28) are multiplied by $\cos(p\pi y/W_k) \cos(q\pi z/E_p)$ where p and q are arbitrary positive integers, and if the results are integrated over the whole area at $x = 0$,

$$\int_0^{W_k} \int_0^{E_p} \frac{C_M(0, y, z)}{s - \gamma} \; \cos\left(\frac{p\pi y}{W_k}\right) \cos\left(\frac{p\pi z}{E_p}\right) dy \; dz$$

$$= \sum_{m=0}^{\infty} \sum_{n=0}^{\infty} B_{mn} \int_0^{W_k} \int_0^{E_p} \cos\left(\frac{p\pi y}{W_k}\right) \cos\left(\frac{q\pi z}{E_p}\right) \cos\left(\frac{m\pi y}{W_k}\right) \cos\left(\frac{n\pi z}{E_p}\right) dy \; dz \tag{E3-10-29}$$

Equation (E3-10-29) can also be written as

$$\int_0^{W_k} \int_0^{E_p} \frac{C_M(0, y, z)}{s-\gamma} \cos\left(\frac{p\pi y}{W_k}\right) \cos\left(\frac{p\pi z}{E_p}\right) dy\, dz$$

$$= \sum_{m=0}^{\infty} \sum_{n=0}^{\infty} B_{mn} \left[\int_0^{W_k} \cos\left(\frac{p\pi y}{W_k}\right) \cos\left(\frac{m\pi y}{W_k}\right) dy \right] \left[\int_0^{E_p} \cos\left(\frac{q\pi z}{E_p}\right) \cos\left(\frac{n\pi z}{E_p}\right) dz \right] \qquad \text{(E3-10-30)}$$

and hence, based on the following relation (Hildebrand, 1976, Eq. [146], p. 217)

$$\int_0^l \cos\left(\frac{m\pi x}{l}\right) \cos\left(\frac{n\pi x}{l}\right) dx = 0, \quad m \neq n \qquad \text{(E3-10-31)}$$

the product vanishes unless $p = m$ and $q = n$, in which case it has the value

$$\left[\int_0^{W_k} \cos^2\left(\frac{m\pi y}{W_k}\right) dy \right] \left[\int_0^{E_p} \cos^2\left(\frac{n\pi z}{E_p}\right) dz \right]$$

$$= \left(\frac{W_k}{2}\right)\left(\frac{E_p}{2}\right) \quad \text{for } n = 1, 2, ..., \quad m = 1, 2, ... \qquad \text{(E3-10-32)}$$

$$= (W_k)(E_p) \quad \text{for } n = 0, \quad m = 0$$

Thus, the double series on the right-hand side of Eq. (E3-10-29) reduces to a single term, for which $m = p$ and $n = q$, and there follows the expression for B_{mn} given by Eq. (3-158). Therefore, Eqs. (3-145) and (3-158) form the solution of the boundary-value problem in the Laplace domain.

Inverse Laplace Transform of Eq. (3-150)
After taking the inverse Laplace transform of Eq. (3-150), the general solution for concentration distribution can be expressed as (Batu, 1996)

$$C(x, y, z, t) = C_1(x, y, z, t) + C_2(x, y, z, t) + C_3(x, y, z, t) + C_4(x, y, z, t) \qquad \text{(3-162)}$$

This solution is derived in Example 3-11. The expression for $C_1(x, y, z, t)$ is

$$C_1(x, y, z, t) = \tfrac{1}{2} I_0 F_1 F_2 [\exp(-F_3)\, \text{erfc}(F_4 - F_5) + \exp(F_3)\, \text{erfc}(F_4 + F_5)] \qquad \text{(3-163)}$$

where

$$I_0 = \frac{1}{E_p W_k} \int_0^{W_k} \int_0^{E_p} C_M(0, y, z)\, dy\, dz \qquad \text{(3-164)}$$

$$F_1 = \exp\left(\frac{U}{2D_x}\right), \quad F_2 = \exp(\gamma t) \qquad \text{(3-165)}$$

$$F_3 = \left(\frac{vR_d}{D_x} + \frac{U^2}{4D_x^2} + \frac{\gamma R_d}{D_x}\right)^{1/2} x \qquad \text{(3-166)}$$

$$F_4 = \frac{R_d x}{2(D_x R_d t)^{1/2}} \qquad \text{(3-167)}$$

and

$$F_5 = \left[\left(\nu + \frac{U^2}{4D_x R_d} + \gamma \right) t \right]^{1/2} \tag{3-168}$$

The expression for $C_2(x, y, z, t)$ is

$$C_2(x, y, z, t) = 2F_1 F_2 \sum_{m=1}^{\infty} \sum_{n=1}^{\infty} I_{mn} \cos(\eta_m y) \cos(\lambda_n z)$$

$$\times [\exp(-F_{6_{mn}}) \operatorname{erfc}(F_4 - F_{7_{mn}}) + \exp(F_{6_{mn}}) \operatorname{erfc}(F_4 + F_{7_{mn}})] \tag{3-169}$$

where

$$I_{mn} = \frac{1}{E_p W_k} \int_0^{W_k} \int_0^{E_p} C_M(0, y, z) \cos(\eta_m y) \cos(\lambda_n z) \, dy \, dz \tag{3-170}$$

$$F_{6_{mn}} = x \left(\frac{\nu R_d}{D_x} + \frac{D_z}{D_x} \lambda_n^2 + \frac{D_y}{D_x} \eta_m^2 + \frac{U^2}{4D_x^2} + \frac{\gamma R_d}{D_x} \right)^{1/2} \tag{3-171}$$

$$F_{7_{mn}} = \left[\left(\nu + \frac{D_z}{R_d} \lambda_n^2 + \frac{D_y}{R_d} \eta_m^2 + \frac{U^2}{4D_x R_d} + \gamma \right) t \right]^{1/2} \tag{3-172}$$

and F_1 and F_2 are defined by Eq. (3-165), and F_4 is defined by Eq. (3-167).

The expression for $C_3(x, y, z, t)$ is

$$C_3(x, y, z, t) = F_1 F_2 \sum_{n=1}^{\infty} I_n \cos(\lambda_n z)[\exp(-F_{8_n}) \operatorname{erfc}(F_4 - F_{9_n}) + \exp(F_{8_n}) \operatorname{erfc}(F_4 + F_{9_n})] \tag{3-173}$$

where

$$I_n = \frac{1}{E_p W_k} \int_0^{W_k} \int_0^{E_p} C_M(0, y, z) \cos(\lambda_n z) \, dy \, dz \tag{3-174}$$

$$F_{8_n} = x \left(\frac{\nu R_d}{D_x} + \frac{D_z}{D_x} \lambda_n^2 + \frac{U^2}{4D_x^2} + \frac{\gamma R_d}{D_x} \right)^{1/2} \tag{3-175}$$

$$F_{9_n} = \left[\left(\nu + \frac{D_z}{R_d} \lambda_n^2 + \frac{U^2}{4D_x^2} + \gamma \right) t \right]^{1/2} \tag{3-176}$$

and F_1 and F_2 are defined by Eq. (3-165), and F_4 is defined by Eq. (3-167).

The expression for $C_4(x, y, z, t)$ is

$$C_4(x, y, z, t) = F_1 F_2 \sum_{m=1}^{\infty} I_m \cos(\eta_m y)[\exp(-F_{10_m}) \operatorname{erfc}(F_4 - F_{11_m}) + \exp(F_{10_m}) \operatorname{erfc}(F_4 + F_{11_m})] \tag{3-177}$$

where

$$I_m = \frac{1}{E_p W_k} \int_0^{W_k} \int_0^{E_p} C_M(0, y, z) \cos(\eta_m y) \, dy \, dz \tag{3-178}$$

$$F_{10_m} = x \left(\frac{\nu R_d}{D_x} + \frac{D_z}{D_x} \eta_m^2 + \frac{U^2}{4D_x^2} + \frac{\gamma R_d}{D_x} \right)^{1/2} \tag{3-179}$$

$$F_{11_m} = \left[\left(\nu + \frac{D_y}{R_d} \eta_m^2 + \frac{U^2}{4D_x R_d} + \gamma \right) t \right]^{1/2} \tag{3-180}$$

and F_1 and F_2 are defined by Eq. (3-165) and F_4 is defined by Eq. (3-167).

EXAMPLE 3-11

Determine the inverse Laplace transform of Eq. (3-150) to obtain the final solution for concentration distribution, $C(x, y, z, t)$, given by Eq. (3-162).

SOLUTION

The inverse Laplace transform of Eq. (3-150) is

$$L^{-1}[u(x, y, z, s)] = L^{-1}[u_1(x, y, z, s)] + L^{-1}[u_2(x, y, z, s)]$$

$$+ L^{-1}[u_3(x, y, z, s)] + L^{-1}[u_4(x, y, z, s)] \qquad \text{(E3-11-1)}$$

where

$$L^{-1}[u(x, y, z, s)] = C(x, y, z, t)$$

$$L^{-1}[u_1(x, y, z, s)] = C_1(x, y, z, t), \quad L^{-1}[u_2(x, y, z, s)] = C_2(x, y, z, t) \qquad \text{(E3-11-2)}$$

$$L^{-1}[u_3(x, y, z, s)] = C_3(x, y, z, t), \quad L^{-1}[u_4(x, y, z, s)] = C_4(x, y, z, t)$$

Combination of Eqs. (E3-11-1) and (E3-11-2) gives Eq. (3-162). Therefore, the inverse Laplace transform of, $u_1(x, y, z, s)$, $u_2(x, y, z, s)$, $u_3(x, y, z, s)$, and $u_4(x, y, z, s)$, need to be taken and they are given below.

Inverse Laplace Transform of $u_1(x, y, z, s)$
In order to determine the inverse Laplace transform of $u_1(x,y,z,s)$ as given by Eq. (3-151), some manipulations need to be made first. From Eqs. (3-151) and (3-159), the following expression can be written:

$$u_1(x, y, z, s - b) = \frac{F_1}{E_p W_k(s-\gamma)} \exp\left\{-\left[\frac{R_d}{D_x}(s-b)\right]^{1/2} x\right\}\left[\int_0^{W_k}\int_0^{E_p} C_M(0, y, z)\, dy\, dz\right] \qquad \text{(E3-11-3)}$$

where

$$b = -v - \frac{U^2}{4D_x R_d} \qquad \text{(E3-11-4)}$$

From the first translation or shifting property for the inverse Laplace transform (e.g., Spiegel, 1965, Theorem 2-3, p. 43), Eq. (E3-11-3) can be written as

$$u_1(x, y, z, s - b) = \frac{1}{E_p W_k} F_1 f_1(s - b)\left[\int_0^{W_k}\int_0^{E_p} C_M(0, y, z)\, dy\, dz\right] \qquad \text{(E3-11-5)}$$

where

$$f_1(s - b) = \frac{1}{(s+b-\gamma-b)} \exp\left\{-\left[\frac{R_d}{D_x}(s-b)\right]^{1/2}\right\} \qquad \text{(E3-11-6)}$$

and

$$F_1(t) = L^{-1}[f_1(s - b)] = e^{bt} E_1(t) \qquad \text{(E3-11-7)}$$

$$L^{-1}[e_1(s)] = E_1(t), \quad e_1(s) = \frac{1}{s-(\gamma-b)} \exp\left[-\left(\frac{R_d}{D_x}s\right)^{1/2} x\right] \qquad \text{(E3-11-8)}$$

The inverse Laplace transform of $e_1(s)$ is (Carslaw and Jaeger, 1959, Eq. [19], p. 495)

$$E_1(t) = \tfrac{1}{2}\exp(F_5)[\exp(-F_3)\,\text{erfc}(F_4-F_5) + \exp(F_3)\,\text{erfc}(F_4+F_5)] \qquad \text{(E3-11-9)}$$

where F_3, F_4, and F_5 are given by Eqs. (3-166), (3-167), and (3-168), respectively. Therefore, the inverse Laplace transform of Eq. (E3-11-5) is

$$C_1(x, y, z, t) = L^{-1}[u_1(x, y, z, s - b)] = \frac{1}{E_p W_k} F_1 \exp(bt) E_1(t) \left[\int_0^{W_k} \int_0^{E_p} C_M(0, y, z) \, dy \, dz \right] \text{(E3-11-10)}$$

The substitution of Eq. (E3-11-9) into Eq. (E3-11-10) and after simplification, Eq. (3-163), which is the expression for $C_1(x,y,z,t)$, can be obtained.

Inverse Laplace Transform of $u_2(x, y, z, s)$
Similarly the inverse Laplace transform of Eq. (3-152) can be taken and the final result for $C_2(x, y, z, t)$ is given by Eq. (3-169).

Inverse Laplace Transform of $u_3(x, y, z, s)$
Similarly the inverse Laplace transform of Eq. (3-153) can be taken and the final result for $C_3(x, y, z, t)$ is given by Eq. (3-173).

Inverse Laplace Transform of $u_4(x, y, z, s)$
Similarly the inverse Laplace transform of Eq. (3-154) can be taken and the final result for $C_4(x, y, z, t)$ is given by Eq. (3-177).

Unsteady-State Special Solution for a Single Rectangular Source
Concentration Distribution
The geometry of a single rectangular source is shown in Figure 3-31, which is a special case of Figure 3-30. The rectangular source is located at any place along the yOz plane and its width and height are $2B$ and $2H$, respectively. The distances from the right and left edges of the source to the vertical impervious boundaries are D_1 and D_2, respectively. The source conditions expressed by Eqs. (3-130)–(3-135) take the forms

$$C(0, y, z, t) = \begin{cases} C_M e^{\gamma t} & D_1 < z < D_1 + 2B, \quad H_1 < y < H_1 + 2H \qquad \text{(3-181)} \\ \\ 0 & \text{otherwise} \qquad\qquad\qquad\qquad\qquad\qquad\quad \text{(3-182)} \end{cases}$$

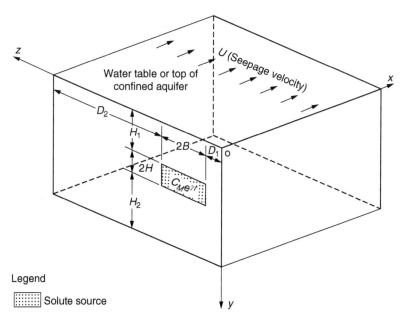

FIGURE 3-31 The geometry of a single rectangular source with uniform solute source concentration. (Adapted from Batu, V., *J. Hydrol.*, 174, 57–82, 1996.)

where C_M is the source concentration having a constant value. The substitution of Eqs. (3-181) and (3-182) into Eq. (3-164) with $C_M(0, y, z) = C_M$ gives

$$I_0 = \frac{4BHC_M}{(H_1+2H+H_2)(D_1+2B+D_2)}$$

(3-183)

Similarly, the expression for I_{mn} can be obtained. The substitution of Eqs. (3-181) and (3-182) into Eq. (3-170) gives

$$I_{mn} = \frac{C_M}{\pi^2} \frac{S_n}{n} \frac{S_m}{m}$$

(3-184)

where

$$S_n = \sin[\lambda_n(D_1 + 2B)] - \sin(\lambda_n D_1)$$

(3-185)

$$S_m = \sin[\eta_m(H_1 + 2H)] - \sin(\eta_m H_1)$$

(3-186)

The substitution of Eqs. (3-181) and (3-182) into Eq. (3-174) gives the expression for I_n as;

$$I_n = \frac{C_M}{\pi} \frac{2H}{H_1+2H+H_2} \frac{S_n}{n}$$

(3-187)

Finally, the expression for I_m can be similarly obtained by substitution of Eqs. (3-181) and (3-182) into Eq. (3-178) as

$$I_m = \frac{C_M}{\pi} \frac{2B}{D_1+2B+D_2} \frac{S_m}{m}$$

(3-188)

Therefore, with Eqs. (3-183), (3-184), (3-187), and (3-188), Eqs. (3-163), (3-169), (3-173), and (3-177) form the solution for concentration distribution for a single rectangular source as shown in Figure 3-31. In the above-mentioned expressions, λ_n and η_m are given by Eqs. (3-146) and (3-147), respectively, with $E_p = D_1 + 2B + D_2$ and $W_k = H_1 + 2H + H_2$.

Special Solution for Two-Dimensional Cases

Two special solutions corresponding to the two-dimensional special cases can be obtained from the three-dimensional solution for a rectangular source whose geometry is shown in Figure 3-31. The first two-dimensional special case is shown in Figure 3-32, which shows that the source fully penetrates the entire thickness of the aquifer in the y-coordinate direction. In other words, the source becomes a vertical strip. This corresponds to Figure 3-12, which is the geometry of the ST2A model. The second two-dimensional special case is shown in Figure 3-33, which shows that the source fully penetrates the entire length of the aquifer in the z-coordinate direction. The special solutions corresponding to these cases are derived in Example 3-12.

EXAMPLE 3-12

Derive the two-dimensional unsteady-state special solutions from the three-dimensional solution for a rectangular source as given by Eqs. (3-163), (3-169), (3-173), (3-177), (3-183), (3-184), (3-187), and (3-188).

SOLUTION

First Special Case: The geometry of the first special case is shown in Figure 3-32 where $H_1 = 0$ and $H_2 = 0$. The substitution of these values into Eqs. (3-163) and (3-183) gives

$$C_1(x, z, t) = \frac{BC_M F_1 F_2}{D_1+2B+D_2} [\exp(-F_3) \text{ erfc}(F_4-F_5) + \exp(F_3) \text{ erfc}(F_4+F_5)] \quad \text{(E3-12-1)}$$

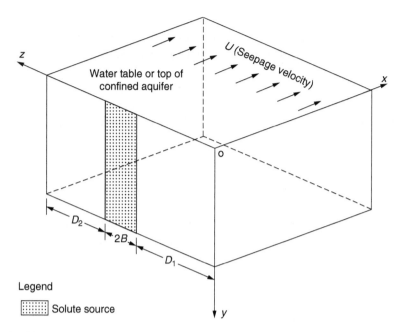

FIGURE 3-32   First two-dimensional special solution of the three-dimensional solution for a rectangular source: vertical strip source.

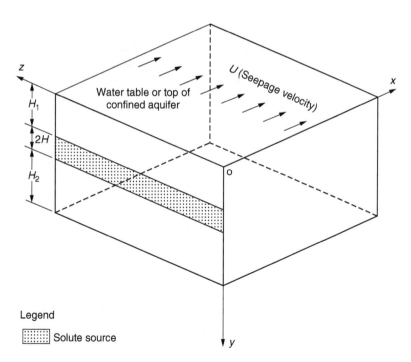

FIGURE 3-33   Second two-dimensional special solution of the three-dimensional solution for a rectangular source: horizontal strip source.

Since $H_1 = 0$, Eq. (3-147) gives $\eta_m = m\pi/(2H)$ and the sine terms in Eq. (3-186) become zero, resulting in $S_m = 0$. As a result, $I_{mn} = 0$ and thereby $C_2(x, y, z, t) = 0$ according to Eq. (3-169). Further substitution of $H_1 = 0$ and $H_2 = 0$ into Eq. (3-187) gives

$$I_n = \frac{C_M}{\pi} \frac{S_n}{n} \qquad\qquad (E3\text{-}12\text{-}2)$$

The substitution of Eq. (E3-12-2) into Eq. (3-173) gives the following equation:

$$C_3(x, z, t) = \frac{C_M}{\pi} F_1 F_2 \sum_{n=1}^{\infty} \frac{S_n}{n} \cos(\lambda_n z)$$

$$\times [\exp(-F_{8_n}) \operatorname{erfc}(F_4 - F_{9_n}) + \exp(F_{8_n}) \operatorname{erfc}(F_4 + F_{9_n})] \qquad (E3\text{-}12\text{-}3)$$

Since $S_m = 0$, $I_m = 0$ according to Eq. (3-188). Therefore, Eq. (3-177) results in $C_4(x,y,z,t) = 0$. Combining Eqs. (E3-12-1) and (E3-12-3) and substituting it into Eq. (3-162) with $C_2(x,y,z,t) = 0$ and $C_4(x, y, z, t) = 0$ gives

$$C(x, z, t) = \frac{BC_M}{D_1 + 2B + D_2} F_1 F_2 [\exp(-F_3) \operatorname{erfc}(F_4 - F_5) + \exp(F_3) \operatorname{erfc}(F_4 + F_5)] \qquad (E3\text{-}12\text{-}4)$$

$$+ \frac{C_M}{\pi} F_1 F_2 \sum_{n=1}^{\infty} \frac{S_n}{n} \cos(\lambda_n z)[\exp(- F_{8_n}) \operatorname{erfc}(F_4 - F_{9_n}) + \exp(F_{8_n}) \operatorname{erfc}(F_4 + F_{9_n})]$$

The following observations can be made by comparing Eq. (E3-12-4) with Eq. (3-77): The product $BC_M/(D_1 + 2B + D_2)F_1 F_2$ in Eq. (E3-12-4) is exactly the same as that of the P_1 as given by Eq. (3-78). The expression for F_3 (given by Eq. [3-166]) in Eq. (E3-12-4) is exactly the same as the expression for P_2 (given by Eq. [3-79]) in Eq. (3-77); the expression for F_4 (given by Eq. [3-167]) in Eq. (E3-12-4) is exactly the same as the expression for P_3 (given by Eq. [3-80]) in Eq. (3-77); the expression for F_5 (given by Eq. [3-168]) in Eq. (E3-12-4) is exactly the same as the expression for P_4 (given by Eq. [3-81]) in Eq. (3-77); the product of $C_M F_1 F_2/\pi$ in Eq. (E3-12-4) is exactly the same as $C_M P_5 P_6/\pi$ in Eq. (3-77), where P_5 and P_6 are given by the expressions in Eq. (3-82); in Eq. (E3-12-4) S_n/n (S_n is given by Eq. [3-185]) is exactly the same as P_{7n} (given by Eq. [3-83]) in Eq. (3-77); the expression for F_{8n} (given by Eq. [3-175]) in Eq. (E3-12-4) is exactly the same as the expression for P_{8n} (given by Eq. [3-84]) in Eq. (3-77); the expression for F_{9n} (given by Eq. [3-176]) in Eq. (E3-12-4) is exactly the same as the expression for P_{9n} (given by Eq. [3-85]) in Eq. (3-77). Thus, Eq. (E3-12-4) is exactly the same as Eq. (3-77).

Second Special Case: The geometry of the second special case corresponds to $D_1 = 0$ and $D_2 = 0$ (see Figure 3-33). This case is equivalent to the first special case with the exception that the strip source is parallel to the z-coordinate axis. Introducing these expressions into Eqs. (3-163), (3-169), (3-173), (3-177), (3-183), (3-184), (3-187), and (3-188), it can be shown that $C_2(x, y, z, t)$ and $C_3(x, y, z, t)$ both become zero and the sum of $C_1(x, y, z, t)$ and $C_4(x, y, z, t)$ is exactly the same as the solution for a single-strip source as given by Eq. (3-77), with the exception that H, H_1, and H_2 define the source geometry instead of B, D_1, and D_2.

Special Solution for the One-Dimensional Case
In Section 3.3.2.2, the one-dimensional solution is obtained as a special solution from the two-dimensional solution as given by Eq. (3-77). As shown above, Eq. (3-77) is a special case of the three-dimensional solution. Therefore, it can be concluded that the same one-dimensional solution is a special case of the three-dimensional solution as well.

Convective–Dispersive Flux Components
From Eqs. (2-41a)–(2-41c), the convective–dispersive flux components in the x-, y-, and z-coordinate directions, respectively, can be easily determined using the same procedure for the

convective–dispersive flux components for the two-dimensional case in Section 3.3.2.2. Since the mathematical expressions are too lengthy they are not presented in this chapter.

Unsteady-State Special Solutions for Other Sources
Equation (3-162) is the general solution for the resulting concentration distribution for any given $C_M(0, y, z)$ source concentration distribution. With the source concentration functions (see Figure 3-30), $C_M(0, y, z)$, the integrals I_0, I_{mn}, I_n, and I_m given by Eqs. (3-164), (3-170), (3-174), and (3-178), respectively, need to be evaluated analytically or numerically depending on the type of the function $C_M(0, y, z)$. One example is presented below.

Solution for Two Rectangular Sources with Uniform Source Concentrations
The geometry of two rectangular sources with uniform source concentrations is shown in Figure 3-34. The boundary conditions at $x = 0$, expressed by Eqs. (3-130)–(3-135), take the following forms:

$$C(0, y, z, t) = C_{M_1} e^{\gamma t},$$

$$D_1 < z < D_1 + 2B_1, \quad A_1 + 2H_1 + A_3 < y < A_1 + 2H_1 + A_3 + 2H_2 \qquad (3\text{-}189)$$

$$C(0, y, z, t) = C_{M_2} e^{\gamma t},$$

$$D_1 + 2B_1 + D_3 < z < D_1 + 2B_1 + D_3 + 2B_2, \quad A_1 < y < A_1 + 2H_1 \qquad (3\text{-}190)$$

where C_{M_1} and C_{M_2} are the source concentrations of the rectangular sources. As mentioned earlier, in order to determine the $C(x, y, z, t)$ concentration distribution, only I_0, I_{mn}, I_n, and I_m given by

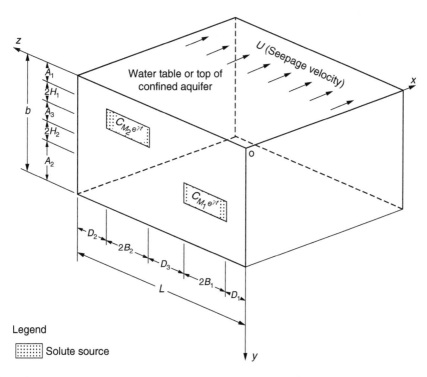

FIGURE 3-34 The geometry of two rectangular sources with uniform solute sources concentrations. (Adapted from Batu, V., *J. Hydrol.*, 174, 57–82, 1996.)

Eqs. (3-164), (3-170), (3-174), and (3-178), respectively, need to be evaluated. With the source conditions given by Eqs. (3-189) and (3-190), Eq. (3-164) gives

$$
I_0 = \frac{1}{Lb} \int_{y=A_1+2H_1+A_3}^{A_1+2H_1+A_3+2H_2} C_{M_1} dy \int_{z=D_1}^{D_1+2B_1} dz
$$

$$
+ \frac{1}{Lb} \int_{y=A_1}^{A_1+2H_1} C_{M_2} dy \int_{z=D_1+2B_1+D_3}^{D_1+2B_1+D_3+2B_2} dz
\tag{3-191}
$$

where

$$
L = D_1 + 2B_1 + D_3 + 2B_2 + D_2, \qquad b = A_1 + 2H_1 + A_3 + 2H_2 + A_2
\tag{3-192}
$$

From Eq. (3-191) one gets

$$
I_0 = \frac{4}{Lb} (B_1 H_2 C_{M_1} + B_2 H_1 C_{M_2})
\tag{3-193}
$$

From Eq. (3-170)

$$
I_{mn} = \frac{1}{Lb} \int_{y=A_1+2H_1+A_3}^{A_1+2H_1+A_3+2H_2} C_{M_1} \cos(\eta_m y) dy \int_{z=D_1}^{D_1+2B_1} \cos(\lambda_n z) \, dz
$$

$$
+ \frac{1}{Lb} \int_{y=A_1}^{A_1+2H_1} C_{M_2} \cos(\eta_m y) \, dy \int_{z=D_1+2B_1+D_3}^{D_1+2B_1+D_3+2B_2} \cos(\lambda_n z) \, dz
\tag{3-194}
$$

where

$$
\lambda_n = \frac{n\pi}{L}, \qquad \eta_m = \frac{m\pi}{b}
\tag{3-195}
$$

From Eqs. (3-146) and (3-147), after evaluation, Eq. (3-194) gives

$$
I_{mn} = \frac{C_{M_1}}{\pi^2} \frac{1}{m} \frac{1}{n} \left\{ \sin\left[\frac{m\pi}{b}(b - A_2) \right] - \sin\left[\frac{m\pi}{b}(A_1 + 2H_1 + A_3) \right] \right\}
$$

$$
\times \left\{ \sin\left[\frac{n\pi}{L}(D_1 + 2B) \right] - \sin\left(\frac{n\pi D_1}{L} \right) \right\}
\tag{3-196}
$$

$$
+ \frac{C_{M_2}}{\pi^2} \frac{1}{m} \frac{1}{n} \left\{ \sin\left[\frac{m\pi}{b}(A_1 + 2H_1) \right] - \sin\left(\frac{m\pi A_1}{b} \right) \right\}
$$

$$
\times \left\{ \sin\left[\frac{n\pi}{L}(L - D_2) \right] - \sin\left[\frac{n\pi}{L}(D_1 + 2B_1 + D_3) \right] \right\}
$$

From Eq. (3-174),

$$
I_n = \frac{1}{Lb} \int_{y=A_1+2H_1+A_3}^{A_1+2H_1+A_3+2H_2} C_{M_1} dy \int_{z=D_1}^{D_1+2B_1} \cos(\lambda_n z) \, dz
$$

$$
+ \frac{1}{Lb} \int_{y=A_1}^{A_1+2H_1} C_{M_2} dy \int_{z=D_1+2B_1+D_3}^{D_1+2B_1+D_3+2B_2} \cos(\lambda_n z) \, dz
\tag{3-197}
$$

and on integration

$$
I_n = C_{M_1} \frac{2H_2}{Lb} \frac{1}{\lambda_n} \left\{ \sin[\lambda_n(D_1 + 2B_1)] - \sin(\lambda_n D_1) \right\}
$$

$$
+ C_{M_2} \frac{2H_1}{Lb} \frac{1}{\lambda_n} \left\{ \sin[\lambda_n(L - D_2)] - \sin(\lambda_n(D_1 + 2B_1 + D_3)) \right\}
\tag{3-198}
$$

Finally, from Eq. (3-178),

$$I_m = \frac{1}{Lb} \int_{y=A_1+2H_1+A_3}^{A_1+2H_1+A_3+2H_2} C_{M_1} \cos(\eta_m y) \, dy \int_{z=D_1}^{D_1+2B_1} dz$$

$$+ \frac{1}{Lb} \int_{y=A_1}^{A_1+2H_1} C_{M_2} \cos(\eta_m y) \, dy \int_{z=D_1+2B_1+D_3}^{D_1+2B_1+D_3+2B_2} dz \qquad (3\text{-}199)$$

and on integration

$$I_m = C_{M_1} \frac{2B_1}{Lb} \frac{1}{\eta_m} \left\{ \sin[\eta_m(b-A_2)] - \sin(\eta_m(A_1+2H_1+A_3)] \right\}$$

$$+ C_{M_2} \frac{2B_2}{Lb} \frac{1}{\eta_m} \left\{ \sin[\eta_m(A_1+2H_1)] - \sin(\eta_m A_1) \right\} \qquad (3\text{-}200)$$

Therefore, with Eqs. (3-193), (3-196), (3-198), and (3-200), Eqs. (3-162), (3-163), (3-169), (3-173), and (3-177) form the solution for two rectangular sources.

Steady-State General Solution for Concentration Distribution
The solution for the steady-state solute transport case (see Figure 3-30 with $\gamma = 0$) can be obtained by using two different procedures as is used for the two-dimensional solution in Section 3.3.2.2. The first procedure is to use the corresponding differential equation under steady-state solute transport conditions with the associated boundary conditions. The second procedure is to use the solution developed for unsteady-state concentration distribution as the time t approaches infinity. The first procedure can be used as in Example 3-7.

From Eq. (3-128), the differential equation under steady-state solute transport can be written as

$$D_x \frac{\partial^2 C}{\partial x^2} + D_y \frac{\partial^2 C}{\partial y^2} + D_z \frac{\partial^2 C}{\partial z^2} - U \frac{\partial C}{\partial x} - R_d v C = 0 \qquad (3\text{-}201)$$

For steady-state solute transport, the initial condition given by Eq. (3-129) is not necessary. The source conditions under steady-state solute transport conditions can be written from the corresponding unsteady-state solute transport source conditions, which are given by Eqs. (3-130)–(3-135) with $\gamma = 0$. Therefore, the solution for the concentration distribution C will only be the function of x, y, and z. The values of concentration at the sources are specified. Likewise, these are the first-type or Dirichlet boundary conditions. The concentrations in the first row of rectangles on the yOz plane are given as follows:

$$C(0, y, z) = C_{M_{1,1}}(0, y, z), \quad 0 < y < W_1, \quad 0 < z < E_1 \qquad (3\text{-}202)$$

$$C(0, y, z) = C_{M_{p-1,1}}(0, y, z), \quad 0 < y < W_1, \quad E_{p-2} < z < E_{p-1} \qquad (3\text{-}203)$$

$$C(0, y, z) = C_{M_{p,1}}(0, y, z), \quad 0 < y < W_1, \quad E_{p-1} < z < E_p \qquad (3\text{-}204)$$

Similarly, expressions for 2nd, ..., $(k-1)$th, kth row can be written. The expressions for the kth row are as follows:

$$C(0, y, z) = C_{M_{1,k}}(0, y, z), \quad W_{k-1} < y < W_k, \quad 0 < z < E_1 \qquad (3\text{-}205)$$

$$C(0, y, z) = C_{M_{p-1,k}}(0, y, z), \quad W_{k-1} < y < W_k, \quad E_{p-2} < z < E_{p-1} \qquad (3\text{-}206)$$

$$C(0, y, z) = C_{M_{p,k}}(0, y, z), \quad W_{k-1} < y < W_k, \quad E_{p-1} < z < E_p \qquad (3\text{-}207)$$

where E_i and W_i are defined by Eqs. (3-136) and (3-137), respectively. The other boundary conditions as given by Eqs. (3-138)–(3-143) remain the same.

Determination of the General Steady-State Solution from the General Unsteady-State Solution
As mentioned above, the solution developed for unsteady-state concentration distribution will be obtained when the time t approaches infinity. Equations (3-162), (3-163), (3-169), (3-173), and (3-177) form the general solution under unsteady-state solute transport conditions for the geometry shown in Figure (3-30). The corresponding steady-state solute transport solution will be obtained for $\gamma = 0$ when the time approaches infinity.

From Eqs. (3-163) and (3-165), the expression for $C_1(x, y, z)$ takes the form

$$C_1(x, y, z) = \frac{1}{2} I_0 \exp\left(\frac{Ux}{2D_x}\right) \exp(0) \tag{3-208}$$

$$\times \left\{ \exp\left[-\left(\frac{\nu R_d}{D_x} + \frac{U^2}{4D_x^2} \right)^{1/2} x \right] \text{erfc}(0 - \infty) + \exp\left[\left(\frac{\nu R_d}{D_x} + \frac{U^2}{4D_x^2} \right)^{1/2} x \right] \text{erfc}(0 + \infty) \right\}$$

or by using Eqs. (3-35),

$$C_1(x, y, z) = \frac{1}{2} I_0 \exp\left(\frac{Ux}{2D_x}\right) \tag{3-209}$$

$$\times \left\{ \exp\left[-\left(\frac{\nu R_d}{D_x} + \frac{U^2}{4D_x^2} \right)^{1/2} x \right] [1 - \text{erf}(-\infty)] + \exp\left[\left(\frac{\nu R_d}{D_x} + \frac{U^2}{4D_x^2} \right)^{1/2} x \right] [1 - \text{erf}(\infty)] \right\}$$

and by using Eqs. (3-36) and (3-37), finally it takes the form

$$C_1(x, y, z) = I_0 \exp\left[\frac{Ux}{2D_x} - \left(\frac{\nu R_d}{D_x} + \frac{U^2}{4D_x^2} \right)^{1/2} x \right] \tag{3-210}$$

Similarly, from Eqs. (3-169), (3-173), and (3-177) one can write, for $C_2(x, y, z)$, $C_3(x, y, z)$, and $C_4(x, y, z)$, respectively, as

$$C_2(x, y, z) = 4 \exp\left(\frac{Ux}{2D_x}\right) \sum_{m=1}^{\infty} \sum_{n=1}^{\infty} I_{mn} P_{mn} \cos(\eta_m y) \cos(\lambda_n z) \tag{3-211}$$

$$C_3(x, y, z) = 2\exp\left(\frac{Ux}{2D_x}\right) \sum_{n=1}^{\infty} I_n P_n \cos(\lambda_n z) \tag{3-212}$$

$$C_4(x, y, z) = 2\exp\left(\frac{Ux}{2D_x}\right) \sum_{m=1}^{\infty} I_m P_m \cos(\eta_m y) \tag{3-213}$$

where

$$P_{mn} = \exp\left[-\left(\frac{\nu R_d}{D_x} + \frac{D_z}{D_x}\lambda_n^2 + \frac{D_y}{D_x}\eta_m^2 + \frac{U^2}{4D_x^2} \right)^{1/2} x \right] \tag{3-214}$$

$$P_n = \exp\left[-\left(\frac{\nu R_d}{D_x} + \frac{D_z}{D_x}\lambda_n^2 + \frac{U^2}{4D_x^2} \right)^{1/2} x \right] \tag{3-215}$$

and

$$P_m = \exp\left[-\left(\frac{\nu R_d}{D_x} + \frac{D_y}{D_x}\eta_m^2 + \frac{U^2}{4D_x^2} \right)^{1/2} x \right] \tag{3-216}$$

In the above-mentioned expressions, λ_n and η_m are given by Eqs. (3-146) and (3-147), respectively, and I_{mn}, I_n, and I_m are given by Eqs. (3-170), (3-174), and (3-178), respectively. The sum of the four components given above forms the general solution for steady-state solute transport conditions:

$$C(x, y, z) = C_1(x, y, z) + C_2(x, y, z) + C_3(x, y, z) + C_4(x, y, z) \tag{3-217}$$

Steady-State Special Solution for a Single Rectangular Source
Concentration Distribution
The geometry of a single rectangular source is shown in Figure 3-31 with $\gamma = 0$, which is a special case of Figure 3-30. The rectangular source is located at any place along the yOz plane and its width and height are $2B$ and $2H$, respectively. The source conditions expressed by Eqs. (3-202)–(3-207) take the forms

$$C(0, y, z) = \begin{cases} C_M & D_1 < z < D_1 + 2B, \quad H_1 < y < H_1 + 2H \quad \text{(3-218)} \\ 0 & \text{otherwise} \quad\quad\quad\quad\quad\quad\quad\quad\quad\quad\quad\quad\quad\quad \text{(3-219)} \end{cases}$$

where C_M is the source concentration having a constant value. The expressions for I_0, I_{mn}, I_n, and I_m for a rectangular source are given by Eqs. (3-183), (3-184), (3-187), and (3-188), respectively. By substitution of these expressions into Eqs. (3-210), (3-211), (3-212), and (3-213), and using Eq. (3-217), gives

$$C(x, y, z) = \frac{4BHC_M}{(D_1+2B+D_2)(H_1+2H+H_2)} \exp\left[\frac{Ux}{2D_x} - \left(\frac{vR_d}{D_x} + \frac{U^2}{4D_x^2} \right)^{1/2} x \right]$$

$$+ \frac{4C_M}{2D_x} \exp\left(\frac{Ux}{2D_x} \right) \sum_{m=1}^{\infty} \sum_{n=1}^{\infty} \frac{S_n}{n} \frac{S_m}{m} P_{mn} \cos(\eta_m y) \cos(\lambda_n z) \quad \text{(3-220)}$$

$$+ \frac{4HC_M}{\pi(H_1+2H+H_2)} \exp\left(\frac{Ux}{2D_x} \right) \sum_{n=1}^{\infty} \frac{S_n}{n} P_n \cos(\lambda_n z)$$

$$+ \frac{4BC_M}{\pi(D_1+2B+D_2)} \exp\left(\frac{Ux}{2D_x} \right) \sum_{m=1}^{\infty} \frac{S_m}{m} P_m \cos(\eta_m y)$$

where S_n and S_m are given by Eqs. (3-185) and (3-186), respectively. The expressions for P_{mn}, P_n, and P_m are given by Eqs. (3-214), (3-215), and (3-216), respectively, and the expressions for λ_n and η_m are given by Eqs. (3-146) and (3-147), respectively, with $E_p = D_1+2B+D_2$ and $W_k = H_1+2H+H_2$.

Special Solution for the Two-Dimensional Cases
Two special solutions corresponding to the two-dimensional special cases can be obtained by using the three-dimensional solution for a rectangular source, which is given by Eq. (3-220).

First Special Case: The geometry of the first special case, which corresponds to $H_1 = 0$ and $H_2 = 0$, is shown in Figure 3-32. By substituting these values into Eq. (3-220) one gets

$$C(x, z) = \frac{2BC_M}{D_1+2B+D_2} \exp\left[\frac{Ux}{2D_x} - \left(\frac{vR_d}{D_x} + \frac{U^2}{4D_x^2} \right)^{1/2} x \right] + \frac{2}{\pi} \exp\left(\frac{Ux}{2D_x} \right) \sum_{n=1}^{\infty} \frac{S_n}{n} \cos(\lambda_n z) P_n \quad \text{(3-221)}$$

where S_n and P_n are given by Eqs. (3-185) and (3-215), respectively. Note that the second and fourth terms on the right-hand side of Eq. (3-220) become zero due to the same reasons as in deriving Eq. (E3-12-4). The corresponding equation to Eq. (3-221) is Eq. (3-114). By substituting Eqs. (3-115) and (3-116) into Eq. (3-119) it can be observed that the resulting expression becomes exactly the same as Eq. (3-221).

Second Special Case: The geometry of the first special case, which corresponds to $D_1 = 0$ and $D_2 = 0$, is shown in Figure 3-33. This is equivalent to the first special case with the exception that the source is parallel to the z-coordinate axis. Introducing these values into Eq. (3-220), it can be shown that the second and third terms on the right-hand side of Eq. (3-220) become zero and the sum of the rest of two terms become exactly the same as the solution for a single-strip source as given by Eq. (3-119), with the exception that H, H_1, and H_2 define the source geometry instead of D, D_1, and D_2.

Special Solution for the One-Dimensional Case
In Section 3.3.2.2, the one-dimensional solution is obtained as a special solution from the two-dimensional solution as given by Eq. (3-119). As shown above, Eq. (3-119) is a special case of the three-dimensional solution. Therefore, it can be concluded that the same one-dimensional solution is a special case of the three-dimensional solution as well.

Convective–Dispersive Flux Components
From Eqs. (2-41a)–(2-41c), the convective–dispersive flux components in the x-, y-, and z-coordinate directions, respectively, can be easily determined using the same procedure for the convective–dispersive flux components for the two-dimensional case in Section 3.3.2.2. Since the mathematical expressions are too lengthy they are not presented in this chapter.

Steady-State Special Solutions for Other Sources
Equation (3-217) is the general solution for the resulting concentration distribution for any given $C_M(0, y, z)$. With the source concentration functions (Figure 3-30), $C_M(0, y, z)$, the integrals I_0, I_{mn}, I_n, and I_m given by Eqs. (3-164), (3-170), (3-174), and (3-178), respectively, need to be evaluated analytically or numerically depending on the type of the function $C_M(0, y, z)$. An example is presented below.

Solution for Two Rectangular Sources with Uniform Source Concentrations
The geometry of two rectangular sources with uniform source concentrations is shown in Figure 3-34. The boundary conditions at $x = 0$ expressed by Eqs. (3-202)–(3-207) take the following forms:

$$C_M(0, y, z) = C_{M_1},$$

$$D_1 < z < D_1 + 2B_1, \quad A_1 + 2H_1 + A_3 < y < A_1 + 2H_1 + A_3 + 2H_3 \quad (3\text{-}222)$$

$$C_M(0, y, z) = C_{M_2},$$

$$D_1 + 2B_1 + D_3 < z < D_1 + 2B_1 + D_3 + 2B_2, \quad A_1 < y < A_1 + A_1 < 2H_1 \quad (3\text{-}223)$$

where C_{M_1} and C_{M_2} are the source concentrations of the rectangular sources. As mentioned earlier, in order to determine the $C(x,y,z)$ concentration distribution, only I_0, I_{mn}, I_n, and I_m given by Eqs. (3-164), (3-170), (3-174), and (3-178), respectively, need to be evaluated. Their evaluated forms for the conditions given by Eqs. (3-222) and (3-223) are determined previously and given by Eqs. (3-193), (3-196), (3-198), and (3-200), respectively. Therefore, with Eqs. (3-193), (3-196), (3-198), and (3-200), Eqs. (3-210)–(3-213), and (3-217) form the solution for two rectangular sources.

Evaluation of the Models
The analytical models presented in the previous sections are evaluated from their verification as well as applicability limitation point of view and the results are presented below (Batu, 1996).

Special Features of the Three-Dimensional Analytical Models
The solutions for unsteady- and steady-state cases are fairly general such that solutions for any combination of rectangular sources located along a vertical plane (yOz plane in Figure 3-30) perpendicular to the flow direction can be generated by using source geometry. The source concentration on every rectangle can be space-dependent, i.e., the function of y and z coordinates. The exponential term, $e^{\gamma t}$, increases the potential usage of the general model such that the exponentially varying time-dependent source concentrations can be simulated. Variation of γ with the source strength (normalized concentration or normalized flux) and with time for decreasing sources of various γ values is illustrated in Figure 3-2. If historical data are available for time-dependent source concentrations, the value of γ can be determined by using curve-fitting techniques.

One of the main advantages of the models is that the source locations can be changed along the yOz plane. Specifically, this situation can be observed in Figure 3-31 for a single-source case and Figure 3-34 for a two-source case. Beyond this flexibility, another important advantage of these models over the models of Yeh (1981) and Huyakorn et al. (1987) is that they do not require numerical integration. In the models of Yeh (1981) and Huyakorn et al. (1987), the source is adjacent to the water table of unconfined aquifer (or the top boundary of confined aquifer), and therefore, its location cannot be changed. Moreover, all these models require numerical integration.

The single rectangular source has a special practical importance; therefore, this solution is given in detail. The impervious boundaries around the solute source increase the potential usage of the model. If the the aquifer approaches infinity from both sides of the source, its effect can be taken into account by assigning large values to D_1 and D_2 as compared with the source width $2B$. One of the important features of this solution is that the source can be located at any place in the yOz plane that is perpendicular to the flow direction. Assigning $H_1 = 0$ corresponds to the case that was developed by Huyakorn et al. (1987), in which the source is adjacent to the water table or to the top boundary of a confined aquifer.

Special Solutions
Some verifications have already been done for the three-dimensional analytical models. As mentioned above in detail, it is shown that the one-dimensional analytical models under unsteady- and steady-state solute transport conditions (Section 3.3.2.1) and two-dimensional analytical models under unsteady- and steady-state solute transport conditions (Section 3.3.2.2) are exact special cases of the corresponding three-dimensional unsteady- and steady-state solute transport models.

Comparisons with the Analytical Model of Huyakorn et al. (1987)
A computer program ST3A, described in the following sections, was developed to predict spatial and temporal concentration and convective–dispersive flux components from an exponentially varying time-dependent rectangular solute source in a uniform flow field bounded by four impermeable planes (Figure 3-31). The results of this model are compared with the results of a three-dimensional analytical solute transport model developed by Huyakorn et al. (1987). Huyakorn and others presented comparisons between their model and other two models developed by Yeh (1981) and Domenico and Robbins (1985) for transient and steady-state conditions. Only steady-state solute transport comparisons based on Figure 7 and Figure 9 of Huyakorn et al. (1987) are presented. The reasons of not presenting unsteady-state comparisons are presented in Batu (1996). The results of the ST3A model are compared with Cases 1 and 2 of Huyakorn et al. (1987). Case 1 corresponds to a partially penetrating source and the corresponding input data are listed in Table 3-2. Case 2 corresponds to a fully penetrating source and the corresponding input data are listed in Table 3-2. The comparisons between the present model and that of Huyakorn et al. (1982) for Cases 1 and 2 are shown in Figure 3-35 and Figure 3-36, respectively. As seen from these figures, a good correlation exists between the results; the minor differences may be attributed to the algorithms used in the numerical integration and perhaps time discretization of the Huyakorn et al. (1987) model. As mentioned above, the present model (ST3A) does not require numerical integration and all the solutions are in the form of infinite single and double series.

Representation of Exponentially Variable Time-Dependent Source Concentrations
As mentioned in Section 3.3.2.1, the exponential term, $e^{\gamma t}$, increases the capability of the model such that exponentially varying solute source concentrations can be taken into account. With negative $\gamma(T^{-1})$ values, a decaying source concentration can be represented in the model. There are some requirements for the usage of this parameter. In Eqs. (3-166), (3-168), (3-171), (3-172), (3-175), (3-176), (3-179), and (3-180), γ is under the square root. Therefore, in order to take these square roots, some of the terms under each square root must be greater than zero. A relatively simple analysis can show that the necessary condition for this requirement is the expression given by Eq. (3-44). An example is presented in the paragraph following Eq. (3-44).

Number of Terms Selection for the Infinite Series in the Solutions
As can be seen from the final expressions, the mathematics of the three-dimensional concentration distribution is basically composed of single and double infinite series. Therefore, the number of terms, N, in the single infinite series plays a key role in the accuracy of results. The higher the values of N, the

TABLE 3-2

Input Data for the Single Rectangular Source Model (ST3A) for Figure 3-35

Parameter	Value
Source concentration, C_M	1.0 unit
Half-width of the rectangular source, B	121.90 m
First boundary distance, D_1	10000.0 m
Second boundary distance, D_2	10000.0 m
Half height of the rectangular source, H	5.0 m
Distance to the upper boundary, H_1	0.0 m
Distance to the bottom, H_2	30.0 m
Initial value of the x coordinate	0.0 m
Maximum value of the x coordinate	800.0 m
Increment of the x coordinate	50.0 m
Initial value of the z coordinate	10121.9 m
Maximum value of the z coordinate	10121.9 m
Increment of the z coordinate	10.0 m
Initial value of the y coordinate	0.0 m
Maximum value of the y coordinate	0.0 m
Increment of the y coordinate	10.0 m
Effective porosity, φ_e	0.20
Total porosity, φ	0.40
Hydraulic conductivity, K	0.20 m/day
Hydraulic gradient, I	0.082 m/m
Longitudinal dispersivity, α_L	15.40 m
Transverse horizontal dispersivity, α_{TH}	1.54 m
Transverse vertical dispersivity, α_{TV}	1.54 m
Effective molecular diffusion coefficient, D^*	0.0 m²/day
Number of terms parameter in the series, N	500

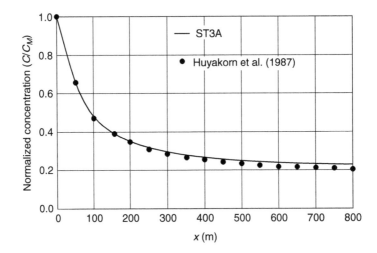

FIGURE 3-35 Comparison of the single rectangular source results with those of Huyakorn et al. (1987) for partially penetrated source case. (Adapted from Batu, V., *J. Hydrol.*, 174, 57–82, 1996.)

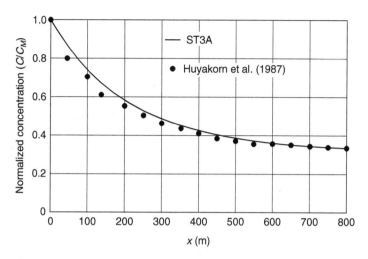

FIGURE 3-36 Comparison of the single rectangular source results with those of Huyakorn et al. (1987) for fully penetrated source case. (Adapted from Batu, V., *J. Hydrol.*, 174, 57–82, 1996.)

more accurate the results. The execution time of the double series is significantly longer than that for the single series. For example, if $N = 500$, 500 terms are required for one of the single series; whereas for one of the double series $N^2 = 250,000$ terms are required. It is clear that the higher values of N will increase the accuracy of results, but will adversely affect the execution time. Owing to this, an analysis was carried out by Batu (1996) to establish some guidelines for the selection of N values to produce accurate results and reduce the execution time. The results of this analysis are presented below.

The analysis was carried out on a numerical example whose geometry is shown in Figure 3-37. The major input data are listed in Table 3-3. Computer runs with the ST3A program, which is described below, were carried out with the data in Table 3-3 for different time periods and z values, and the results for a time period of 100 days are presented in Figure 3-38, which shows that the normalized concentration C/C_M vs. distance x for various N values after 100 days for $y = 0$ m and $z = 505$ m, which correspond to the symmetry plane of the source at the xOz plane (see Figure 3-37). As can be seen from Figure 3-38, the curves tend to approach each other with increase in N values. Owing to the source conditions described by Eqs. (3-181) and (3-182), at $x = 0$. the value of C/C_M must be equal to 1 and $N = 500$ establishes the closest value among others to this normalized concentration value. The results in Figure 3-38 show that higher N values are required to improve the accuracy of results around the source area and $N = 50$ and 100 produce results that are significantly lower than the expected ones. Figure 3-38 also shows that $N = 200$, 400, and 500 produce approximately the same results for x values greater than 10 m. The following conclusions can be drawn from the analysis presented above: (1) convergence increases with the increase in values of N; and (2) higher N values are required to improve the accuracy around the source zone. Based on the numerical results, this zone may approximately be specified as $x < 2B$ and $N = 400$ to 500. For other zones, i.e., when $x > 2B$, smaller N values, approximately 200, can produce reasonably accurate results.

Application of the Three-Dimensional Analytical Solute Transport Models to Ground Water Flow Systems

The general model and its derivative models under unsteady- and steady-state solute transport conditions can be applied to solute transport predictions in aquifers if the ground water velocity can be assumed to be uniform. The models can also be applied to unsaturated zones with some additional assumptions. All these are discussed in the following paragraphs.

In this section, a computer program ST3A, for a single rectangular source is described. Then, application of this program is presented with examples.

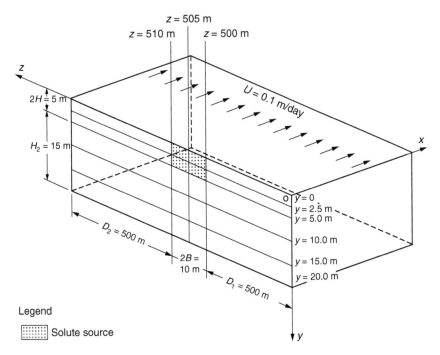

FIGURE 3-37   The geometry of the single rectangular source for Example 3-13.

TABLE 3-3
Input Data for the Single Rectangular Source Model (ST3A) for Figure 3-36

Parameter	Value
Source concentration, C_M	1.0 unit
Half width of the rectangular source, B	19.29 m
First boundary distance, D_1	10000.0 m
Second boundary distance, D_2	10000.0 m
Hafh height of the rectangular source, H	5.0 m
Distance to the upper boundary, H_1	0.0 m
Distance to the bottom, H_2	0.0 m
Initial value of the x coordinate	0.0 m
Maximum value of the x coordinate	800.0 m
Increment of the x coordinate	50.0 m
Initial value of the z coordinate	10019.29 m
Maximum value of the z coordinate	10019.29 m
Increment of the z coordinate	10.0 m
Initial value of the y coordinate	0.0 m
Maximum value of the y coordinate	0.0 m
Increment of the y coordinate	1.0 m
Effective porosity, φ_e	0.20
Total porosity, φ	0.40
Hydraulic conductivity, K	0.20 m/day
Hydraulic gradient, I	0.082 m/m
Longitudinal dispersivity, α_L	15.40 m
Transverse horizontal dispersivity, α_{TH}	1.54 m
Transverse vertical dispersivity, α_{TV}	1.54 m
Effective molecular diffusion coefficient, D^*	0.0 m²/day
Number of terms parameter in the series, N	500

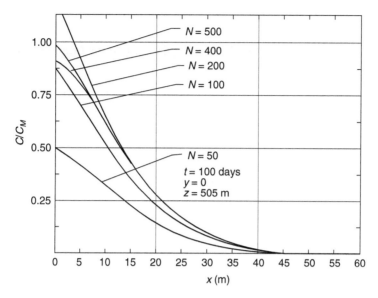

FIGURE 3-38   Normalized concentration, C/C_M, vs. x coordinate for various N values and $y = 0$ and $z = 505$ m for 100 days for the geometry in Figure 3-37. (Adapted from Batu, V., *J. Hydrol.*, 174, 57–82, 1996.)

A Computer Program (ST3A) for the Three-Dimensional First-Type Rectangular Source Model
A three-dimensional MS-DOS-based FORTRAN solute transport program, called solute transport three-dimensional first-type (ST3A), has been developed by the author. ST3A is a menu-driven program and calculates values of concentration and convective–dispersive flux components under unsteady-state solute transport conditions (Eq. [3-162] for concentration distribution) and steady state solute transport conditions (Eqs. [3-220] for concentration distribution).This program can be run interactively in IBM compatible computers.

Application to Saturated Zones
The three-dimensional analytical solute transport model for a rectangular source (see Figure 3-31) can be used for different purposes. The impervious boundaries around the solute transport domain increase the potential usage areas of the model. For ground water flow cases, the effects of impervious boundaries, such as bedrock outcrops and clay zones surrounding the main ground water flow zone, can be taken into account. If the aquifer boundaries approaches infinity horizontally, its effect can be taken into account by assigning large values to D_1 and D_2 as compared with the source width $2B$. The rectangular source can be at any location in the cross-section of an aquifer. Assigning $H_1 = 0$ is the case that the rectangular source starts at the water table of an unconfined aquifer or at the top of a confined aquifer, which corresponds to the Huyakorn et al. (1987) model.

EXAMPLE 3-13

For the input data given in Table 3-4 (see also Figure 3-37) and using the ST3A program, after 1 year, determine the normalized concentration contours (a) at $y = 0$ horizontal plane; (b) at $y = 20$ m horizontal plane; (c) at $z = 495$ m vertical plane; and (d) at $x = 10$ m vertical plane.

SOLUTION

Using the given data, the ST3A program was run to generate the coordinates and normalized concentrations at the given time. The output of ST3A was used to generate contours using the

TABLE 3-4
Input Data for a Single Rectangular Source for Example 3-13

Parameter	Value
Darcy velocity, q	0.025 m/day
Effective porosity, φ_e	0.25
Longitudinal dispersivity, α_L	10 m
Transverse horizontal dispersivity, α_{TH}	1 m
Transverse vertical dispersivity, α_{TV}	1 m
Distribution coefficient, K_d	0
Effective molecular diffusion coefficient, D^*	0
Source concentration parameter, γ	0
Decay constant, v	0
Source width, $2B$	10 m
Distance to the right boundary, D_1	500 m
Distance to the right boundary, D_2	500 m
Source height, $2H$	5 m
Distance to the upper boundary, H_1	0
Distance to the lower boundary, H_2	15 m
Source concentration, CM	1 unit

Adapted from Batu, V., *J. Hydrol.*, 174, 57–82, 1996.

SURFER™ contouring package (Golden Software, Inc., 1988):

1. The normalized concentration contours at $y = 0$ horizontal plane after 1 year are presented in Figure 3-39.
2. The normalized concentration contours at $y = 20$ m horizontal plane after 1 year are presented in Figure 3-40.
3. The normalized concentration contours at $z = 495$ m vertical plane after 1 year are presented in Figure 3-41.
4. The normalized concentration contours at $x = 10$ m vertical plane after 1 year are presented in Figure 3-42.

Application to Unsaturated Zones
Application of the ST3A model to unsaturated zones can similarly be made as in the case for ST2A, which is presented in Section 3.2.2.2. The conceptual model for the ST3A model is the same as in Figure 3-29 with the exception that $L = 2H$. The associated governing equations for flow and solute transport in unsaturated porous media under uniform infiltration rate conditions are given in Section 2.8. The five important points mentioned for the ST2A in Section 3.3.2.2 are also valid for the three-dimensional case.

3.3.3 Third-Type Source Analytical Solute Transport Models

In this section, some well-known third-type source analytical solute transport models are presented with their potential application areas. For this type of models, the solute source itself is treated as the *third-type boundary condition*. The other well-known terms for this type of boundary condition are *flux boundary condition* or *Cauchy boundary condition* (for more details see Section 2.10.3.2).

3.3.3.1 One-Dimensional Analytical Solute Transport Models for the Third-Type Sources

Similarly, the first-type source one-dimensional analytical solute transport model (Section 3.3.2.1), the one-dimensional analytical model under the third-type (or flux-type) source condition has a

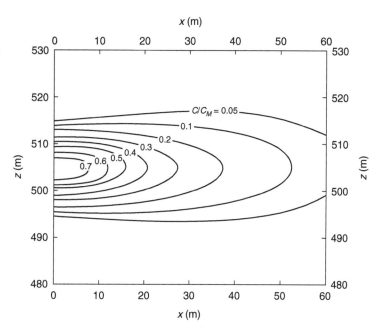

FIGURE 3-39 Normalized concentration , C/C_M, contours at $y = 0$ horizontal plane after 1 year for Example 3-13. (Adapted from Batu, V., *J. Hydrol.*, 174, 57–82, 1996.)

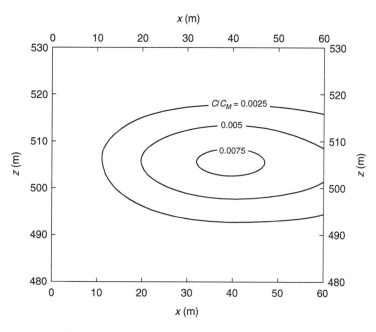

FIGURE 3-40 Normalized concentration , C/C_M, contours at $y = 20$ m horizontal plane after 1 year for Example 3-13. (Adapted from Batu, V., *J. Hydrol.*, 174, 57–82, 1996.)

special importance in the literature. The solution was first obtained by Lindstrom et al. (1967) in similar form of the first-type source condition model in Section 3.3.2.1. This solution is also presented as a special case of Batu's two-dimensional general third-type source model (Batu, 1993),

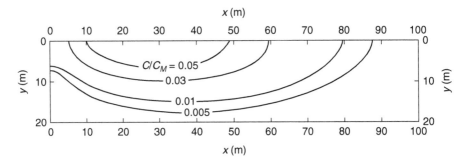

FIGURE 3-41 Normalized concentration, C/C_M, contours at $z = 495$ m vertical plane after 1 year for Example 3-13. (Adapted from Batu, V., *J. Hydrol.*, 174, 57–82, 1996.)

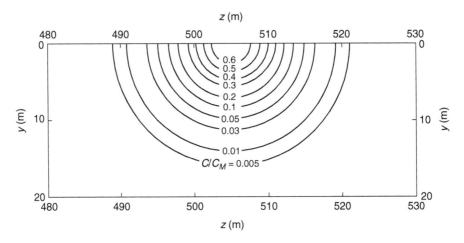

FIGURE 3-42 Normalized concentration, C/C_M, contours at $x = 10$ m vertical plane after 1 year for Example 3-13. (Adapted from Batu, V., *J. Hydrol.*, 174, 57–82, 1996.)

which is presented in detail in Section 3.3.3.2. In this section, one-dimensional analytical model followed by the solution of Lindstrom et al. (1967) is presented.

Problem Statement and Assumptions
The problem under consideration is unsteady-state one-dimensional solute transport in a unidirectional flow field as shown in Figure 3-43. The medium is considered of having a semiinfinite parallelepiped form. The solute source is located at $x = 0$ and its flux is exponentially variable time-dependent quantity. The assumptions listed in (2)–(4) in Section 3.3.2.1 for the first-type source case are also valid for the third-type source case.

Governing Differential Equation
The governing differential equation is given by Eq. (3-1).

Initial and Boundary Conditions
The initial condition for $C(x, t)$ is

$$C(x, 0) = 0, \quad x \geq 0 \tag{3-224}$$

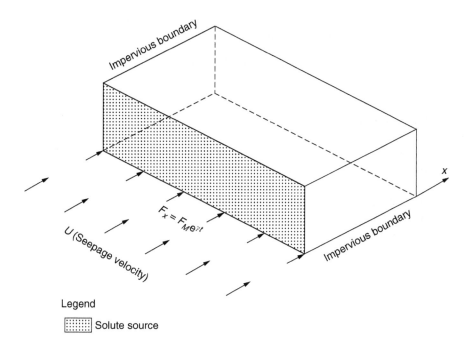

Legend

:::::: Solute source

FIGURE 3-43   Schematic representation of one-dimensional solute transport with the third-type source condition in a unidirectional flow field.

The boundary conditions for $C(x, t)$ are

$$F_x(0,t) = F_M e^{\gamma t}, \quad t \geq 0 \tag{3-225}$$

$$C(\infty, t) = 0, \quad t \geq 0 \tag{3-226}$$

$$\frac{\partial C(\infty, t)}{\partial x} = 0, \quad t \geq 0 \tag{3-227}$$

where $F_x(0, t)$ (M/L²/T) is the solute input flux at $x = 0$ (mass per unit area per unit time), γ (T⁻¹) is a parameter, F_M (M/L²/T) is the solute flux when $\gamma = 0$ at $x = 0$, and e = 2.7182818... .

Equation (3-224) states that initially the concentration is zero everywhere in the transport domain. Equation (3-225) states that the solute flux, which is time-dependent, has a specified value at $x = 0$. Therefore, it is the *third-type* or *flux-type source condition*. This type of condition is also called *Cauchy condition* (see Section 2.10.3.2). Equations (3-226) and (3-227) state that both the concentration C and its gradient with respect to x ($\partial C/\partial x$) approach to zero when the x distance tends to infinity.

From Eq. (3-225), the function

$$\frac{F_x(0, t)}{F_M} = e^{\gamma t} \tag{3-228}$$

performs such that the source flux can be represented as exponentially varying time-dependent quantity. If $\gamma < 0$, the source flux decreases with time; if $\gamma = 0$, the source flux is independent of time; and if $\gamma > 0$, the source concentration increases with time. Figure 3-2 shows the variation of $F_x(0, t)/F_M$ with time for different values of γ ranging from −0.0001 to −0.0100 day⁻¹. If observed data are available for time-dependent source flux, the value of γ may be determined from the curve-fitting techniques.

The general form of the flux-type source condition is given by Eq. (2-104) and for the exponentially varying source flux case it takes the form

$$\varphi_e UC - \varphi_e D_x \frac{\partial C}{\partial x} = F_M e^{\gamma t} \tag{3-229}$$

where $f(t) = e^{\gamma t}$.

Unsteady-State Solution for Concentration Distribution
General Case
The solution can be obtained by solving the above-mentioned boundary-value problem defined by Eqs. (3-1) and (3-224)–(3-229). The solution can also be obtained as a special case from the single-strip source solution (see Eq. [3-271]). Because it is simpler, the second method will be used. The expression for the concentration distribution is determined as a special case from the two-dimensional model of Batu (1993), which is presented in Eq. (3-271). From this solution, the solution for concentration distribution can be written as

$$C(x, t) = \frac{F_M}{\varphi_e} \frac{1}{D_x} \int_0^t F_1(\xi)[F_2(\xi) - F_3(\xi)F_4(\xi)] \quad \exp[\gamma(t - \xi)]d\xi \tag{3-230}$$

where

$$F_1(\xi) = \exp\left(\frac{Ux}{2D_x} - v\xi - \frac{U^2\xi}{4D_x R_d}\right) \tag{3-231}$$

$$F_2(\xi) = \left(\frac{D_x}{\pi R_d \xi}\right)^{1/2} \exp\left(-\frac{R_d x^2}{4D_x \xi}\right) \tag{3-232}$$

$$F_3(\xi) = \frac{U}{2R_d} \exp\left(\frac{Ux}{2D_x} + \frac{U^2\xi}{4R_d D_x}\right) \tag{3-233}$$

and

$$F_4(\xi) = \text{erfc}\left[\frac{R_d x}{2(D_x R_d \xi)^{1/2}} + \frac{U}{2}\left(\frac{\xi}{R_d D_x}\right)^{1/2}\right] \tag{3-334}$$

Special Case for $\gamma = 0$ and $v = 0$
If $\gamma = 0$ and $v = 0$, Eq. (3-230) takes the form

$$C(x, t) = \frac{F_M}{\varphi_e} \frac{1}{D_x} \int_0^t \exp\left(\frac{Ux}{2D_x} - \frac{U^2\xi}{4D_x R_d}\right)\left\{\left(\frac{D_x}{\pi R_d \xi}\right)^{1/2} \exp\left(-\frac{R_d x^2}{4D_x \xi}\right)\right.$$
$$\left. - \frac{U}{2R_d} \exp\left(\frac{Ux}{2D_x} + \frac{U^2\xi}{4R_d D_x}\right)\text{erfc}\left[\frac{R_d x}{2(D_x R_d \xi)^{1/2}} + \frac{U}{2}\left(\frac{\xi}{R_d D_x}\right)^{1/2}\right]\right\}d\xi \tag{3-235}$$

If in Eqs. (3-230) and (3-235)

$$F_M = \varphi_e U C_M \tag{3-236}$$

the concentration $C(x, t)$ becomes *volume-averaged (resident) concentration* and is denoted by $C_r(x, t)$ (see Sections 2.2.2 and 2.9.1 for definitions). With Eq. (3-236), Eq. (3-235) takes the form

$$\frac{C_r(x, t)}{C_M} = \frac{U}{D_x} \int_0^t \exp\left(\frac{Ux}{2D_x} - \frac{U^2\xi}{4D_x R_d}\right)\left\{\left(\frac{D_x}{\pi R_d \xi}\right)^{1/2} \exp\left(-\frac{R_d x^2}{4D_x \xi}\right)\right. \tag{3-237}$$
$$\left. - \frac{U}{2R_d} \exp\left(\frac{Ux}{2D_x} + \frac{U^2\xi}{4R_d D_x}\right)\text{erfc}\left[\frac{R_d x}{2(D_x R_d \xi)^{1/2}} + \frac{U}{2}\left(\frac{\xi}{R_d D_x}\right)^{1/2}\right]\right\}d\xi$$

Solution of Lindstrom et al. (1967) for $R_d = 1$, $\varphi = 0$, and $v = 0$
Lidstrom et al. (1967) solved Eq. (3-1) for $R_d = 1$ and $v = 0$ with the initial condition given by Eq. (3-224), and boundary conditions given by Eqs. (3-226) and (3-227). They used the following source condition instead of Eq. (3-229) for $\varphi = 0$:

$$UC - D_x \frac{\partial C}{\partial x} = \frac{F_M}{\varphi_e} = UC_M \qquad (3\text{-}238)$$

where $F_M = \varphi_e UC_M$. The boundary-value problem defined above was solved by Lidstrom et al. (1967) and the following solution was obtained (see, e.g., van Genuchten and Parker [1984], Eq. 5, p.704):

$$\frac{C_r(x, t)}{C_M} = \frac{1}{2}\text{erfc}\left[\frac{R_d x - Ut}{2(D_x R_d t)^{1/2}}\right] + \left(\frac{U^2 t}{\pi D_x R_d}\right)^{1/2}\exp\left[-\frac{(R_d x - Ut)^2}{4 D_x R_d t}\right] \qquad (3\text{-}239)$$

$$- \frac{1}{2}\left(1 + \frac{Ux}{D_x} + \frac{U^2 t}{D_x R_d}\right)\exp\left(\frac{Ux}{D_x}\right)\text{erfc}\left[\frac{R_d x + Ut}{2(D_x R_d t)^{1/2}}\right]$$

Although Eqs. (3-237) and (3-239) appear to be different, they produce the same numerical results as listed in Table 3-5.

Unsteady-State Convective–Dispersive Flux Component
The convective–dispersive flux component in the x-coordinate direction (F_x) can be determined by substituting Eq. (3-230) into Eq. (3-24). However, the expression for F_x can be determined as a special case from the two-dimensional F_x expression of Batu (1993), which is presented in Eq. (3-272). From this expression, F_x can be written as

$$F_x = \varphi_e UC - \varphi_e D_x V \qquad (3\text{-}240)$$

where C is given by Eq. (3-230) and

$$V = \frac{F_M}{\varphi_e}\frac{1}{D_x}\int_0^t [F_{1x}(F_2 - F_3 F_4) + F_1(F_{2x} - F_3 F_{4x} - F_4 F_{3x})]\exp[\gamma(t - \xi)]\,d\xi \qquad (3\text{-}241)$$

in which F_1, F_2, F_3, and F_4 are given by Eqs. (3-231)–(3-234), respectively. The other expressions are as follows:

$$F_{1x} = \frac{\partial F_1}{\partial x} = \frac{U}{2D_x}\exp\left(\frac{Ux}{2D_x} - v\xi - \frac{U^2 \xi}{4 D_x R_d}\right) \qquad (3\text{-}242)$$

$$F_{2x} = \frac{\partial F_2}{\partial x} = -\frac{R_d x}{2 D_x \xi}\left(\frac{D_x}{\pi R_d \xi}\right)^{1/2}\exp\left(-\frac{R_d x^2}{4 D_x \xi}\right) \qquad (3\text{-}243)$$

$$F_{3x} = \frac{\partial F_3}{\partial x} = \frac{U^2}{4 R_d D_x}\exp\left(\frac{Ux}{2D_x} + \frac{U^2 \xi}{4 R_d D_x}\right) \qquad (3\text{-}244)$$

and

$$F_{4x} = \frac{\partial F_4}{\partial x} = -\frac{R_d}{(\pi D_x R_d \xi)^{1/2}}\exp\left\{-\left[\frac{R_d x}{2(D_x R_d \xi)^{1/2}} + \frac{U}{2}\left(\frac{\xi}{R_d D_x}\right)^{1/2}\right]^2\right\} \qquad (3\text{-}245)$$

Steady-State Solution for Concentration Distribution
Under steady-state solute transport conditions, Eq. (3-31) is the governing differential equation. For steady-state solute transport, the initial condition given by Eq. (3-224) is not necessary. The

solution for C will be only the function of x. Therefore, the boundary condition at the source for $F_x(x)$ given by Eqs. (3-225) and (3-229), with $\gamma = 0$, will have the following form:

$$F_x(0) = F_M = \varphi_e UC - \varphi_e D_x \frac{\partial C}{\partial x} \tag{3-246}$$

The other boundary conditions are given by Eqs. (3-33) and (3-34). The expression for the concentration distribution is determined as a special case from the two-dimensional model of Batu (1993), which is presented in Eq. (3-294). From this solution, the solution for concentration distribution can be written as

$$C(x) = \frac{F_M}{\varphi_e(U/2 + D_x p_0)} \exp\left(\frac{Ux}{2D_x} - p_0 x \right) \tag{3-247}$$

where

$$p_0 = \left(\frac{R_d}{D_x} v + \frac{U^2}{4D_x^2} \right)^{1/2} \tag{3-248}$$

If $v = 0$ in Eq. (3-247), it can easily be seen that $C(x) = F_M/(\varphi_e U)$. If $F_M = \varphi_e UC_M$, then $C(x) = C_M$. In other words, the concentration at every point of the x coordinate becomes equal to C_M.

Steady-State Convective–Dispersive Flux
From Eqs. (3-24) and (3-247), the convective–dispersive flux expression under steady-state solute transport conditions can be written as

$$F_x(x) = \varphi_e UC(x) - \frac{D_x F_M}{(U/2 + D_x p_0)} \left(\frac{U}{2D_x} - p_0 \right) \exp\left(\frac{Ux}{2D_x} - p_0 x \right) \tag{3-249}$$

where $C(x)$ is given by Eq. (3-247).

Evaluation of the Models
Equation (3-230) is obtained as a special solution from a more general two-dimensional solute transport model as given by Eq. (3-271). The two-dimensional model is verified well against a finite-difference solute transport code with excellent comparison (see Section 3.3.3.2). As mentioned above, the Lindstrom et al. (1967) model corresponds to $\gamma = 0$, $v = 0$, and $F_M = \varphi_e UC_M$ in Eq. (3-230) and the end expression is given by Eq. (3-239). Both models are based on the volume-averaged concentration C_r. Equation (3-237) corresponds to Eq. (3-239), which is the model of Lindstrom et al. (1967). As can be seen from Eqs. (3-237) and (3-239), their mathematical expressions are different, but as listed in Table 3-5, their results are exactly the same. Table 3-5 presents comparison of the results of these two models for $U = 1$ m/day, $C_M = 1$ unit, $D_x = 4$ m²/day, and $t = 5$ days (Numerical results for the Lindstrom et al. model are taken from van Genuchten and Alves [1984, Table 6, p. 118]). The numerical results for Eq. (3-237) are determined by running the computer program, ST1B, which is written for Eq. (3-230). This program is described in the following sections. For further evaluation regarding the numerical integration and Gauss points of Eq. (3-230), see Sections 3.3.3.2 and 3.3.3.3.

Applications of the One-Dimensional Analytical Solute Transport Models
As mentioned in Section 3.3.2.1, one-dimensional analytical solute transport models potentially have a lot of applications for finding quick solutions for many problems faced in practice. These models can be applied to solute transport predictions in saturated and unsaturated zones as outlined in Sections 3.3.2.2 and 3.3.2.3. In this chapter, some important concepts as well as examples regarding their application are presented.

A Computer Program for the One-Dimensional Third-Type Source Model (ST1B)
A one-dimensional MS-DOS-based FORTRAN solute transport computer program, solute transport one-dimensional third-type (ST1B), has been developed by the author. ST1B is a menu-driven program

TABLE 3-5
Comparison of the Results of the One-Dimensional Model with Lindstrom et al. (1967) Model for $C_M = 1$ unit, $U = 1$ m/day, $D_x = 4$ m²/day, and $t = 5$ days. (Numerical results for the Lindstrom and other models are taken from van Genuchten and Alves [1984, Table 6, p. 118].)

Distance x(m)	Eq. (3-237)[a]	Eq. (3-239)[b]
	C/C_M	
0.0	0.7619	0.7640
2.0	0.6376	0.6376
4.0	0.5023	0.5023
6.0	0.3712	0.3712
8.0	0.2559	0.2559
10.0	0.1638	0.1638
12.0	0.0970	0.0970
14.0	0.0530	0.0530
16.0	0.0266	0.0266
18.0	0.0123	0.0123
20.0	0.0052	0.0052

[a] Based on Batu (1993).
[b] Based on Lindstrom et al. (1967).

and calculates values of concentration and convective–dispersive flux components under unsteady-state solute transport conditions (Eqs. [3-230] and [3-239]) and steady state solute transport conditions (Eqs. [3-247] and [3-249]). This program can be run interactively in IBM compatible computers.

In the ST1B program, the time t must be greater than zero. A zero value gives runtime error. This is due to the fact that the time t is in the denominator in some expressions of Eqs. (3-230) and (3-240). This is the case for almost all analytical solute transport models. The model of Lindstrom et al. (1967), as given by Eq. (3-239), has the same situation.

As mentioned above, the exponential term, $e^{\gamma t}$, increases the capability of the model such that exponentially varying solute source fluxes can be taken into account. With negative $\gamma(T^{-1})$ values a decaying source flux can be represented in the model. Unlike the first-type source model in Section 3.3.2, there are no restrictive requirements for γ for the third-type source model as Eq. (3-44) for the first-type source model.

Two examples based on the ST1B computer program are presented below.

EXAMPLE 3-14

For the aquifer whose flow and transport parameters are given in Example 3-2, (1) determine the solute input flux at $x = 0$ and discuss its status and (2) using the ST1B computer program, determine the concentration vs. time at the source ($x = 0$) for $\alpha_L = 0.1$, $\alpha_L = 1$, and $\alpha_L = 10$ m and discuss the results.

SOLUTION

(1) From Eqs. (2-39) and (2-40), the ground-water velocity can be found as

$$U = \frac{1}{\varphi_e} KI = \frac{1}{0.25}(10 \text{ m/day})(0.0025 \text{ m/m}) = 0.1 \text{ m/day}$$

and from Eq. (3-236) the solute input flux is

$$F_M = \varphi_e U C_M = (0.25)(0.1 \text{ m/day})(1 \text{ g/m}^3 = 0.025 \text{ g/day/m}^2$$

As mentioned above, if the solute source flux has the form of Eq. (3-236), $C(x, t)$ becomes volume-averaged (resident) concentration and is denoted by $C_r(x, t)$ (see Sections 2.2 and 2.9 for definitions). Unlike the first-type source case, the solute source flux is a given quantity; whereas; for the first-type source case, the source concentration, C_M, is the given quantity and the solute source flux at the source varies with time (see Example 3-2).

(2) The volume-averaged concentration, C_r, vs. time variation at the source is shown in Figure 3-44. Figure 3-44 shows that C_r increases with time and after some elapsed time its value reaches to the concentration in the solute flux ($C_M = 1 \text{ g/m}^3$). Figure 3-44 clearly shows that the elapsed time period depends on the value of α_L. As can be seen from Figure 3-44, for $\alpha_L = 0.1$ m, it takes approximately 15 days, for $\alpha_L = 1$ m it takes approximately 100 days, and for $\alpha_L = 10$ m it takes longer time period. This is not unexpected due to the fact that higher α_L values cause more dispersion, which delays the buildup period of solute at the source.

EXAMPLE 3-15

Using the ST1B computer program for the aquifer whose flow and transport parameters are given in Example 3-3 (1) determine the volume-averaged concentration C_r vs. time curves at $x = 10$ m and compare them with the ones for the first-type source in Example 3-3; (2) determine the convective–dispersive flux vs. time curves; and (3) discuss the results.

SOLUTION

(1) For $K_h = 1, 10, 100$ m/day with $\varphi_e = 0.25$, and $I = 0.0025$ m/m, $U = KI/\varphi_e$ (a combination of Eqs. [2-39] and [2-40]), gives respectively, $U = 0.01, 0.1$, and 1 m/day. The substitution of these values into Eq. (3-236) with $C_M = 1 \text{ g/m}^3$ and $U = 0.01, 0.1$, and 1 m/day, respectively, gives $F_M = 0.0025, 0.025, 0.25 \text{ g/day/m}^2$. The concentration vs. time curves for three different U values are given in Figure 3-45. For comparison, the first-type (or flux-averaged) concentrations for the same parameters given in Figure 3-9 are also shown.

(2) The convective–dispersive fluxes at $x = 10$ m for $U = 0.01, 0.1$, and 1 m/day are given in Figure 3-46.

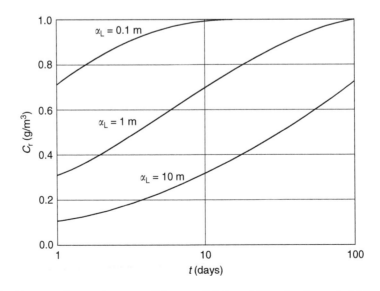

FIGURE 3-44   Concentration vs. time at $x = 0$ for $\alpha_L = 0.1, 1$, and 10 m for Example 3-14.

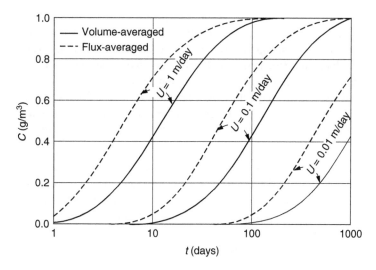

FIGURE 3-45   Concentration vs. time at $x = 10$ m for $U = 0.01, 0.1,$ and 1 m/day for Example 3-15.

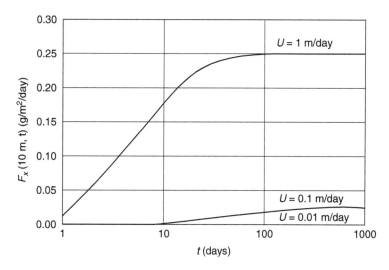

FIGURE 3-46   Convective–dispersive flux at $x = 10$ m for $U = 0.01, 0.1,$ and 1 m/day for Example 3-15.

(3) Figure 3-45 shows that the flux-averaged (first-type source) concentrations are higher than the volume-averaged (third-type) concentrations and with increase in time t the discrepancies become smaller, and for large time they become approximately the same. For example, for $U = 0.1$ m/day after 100 days, the first-type source condition predicts aproximately 1.7 times higher concentration than the third-type source condition does. The principal reason for this situation is that the first-type source condition corresponds to a situation that the solute flux at the source decreases with time (see Figure 3-8), whereas the solute flux for the third-type source condition is constant. Figure 3-46 shows that the solute fluxes at $x = 10$ m increases with time and approach constant values at large times as the case in Figure 3-10 for the first-type source case.

Comparative Evaluation of the First- and Third-Type Source Conditions
Proper formulation of solute transport source conditions is critically important for the interpretation of observed data from column displacement experiments as well as subsequent extrapolation of the experimental results to solute transport problems in the field. Unfortunately, the most appropriate source conditions are not always intuitively apparent. As a result, in many instances, vastly different solutions have

been applied to physically identical problems. In order to provide answers to these potential problems, van Genuchten and Parker (1984) presented a detailed analysis of several analytical solute transport models that have previously been applied to column tracer studies. Using basic principles of mass conservation, van Genuchten and Parker showed that the first-type source condition at the inlet boundary incorrectly predicts the volume-averaged concentration in semi-infinite systems. The authors also showed that the third-type source condition preserve mass in semi-infinite systems. The authors also presented evaluations for the first- and third-type source conditions for finite systems. In this section the results of van Genuchten and Parker (1984) are presented for semi-infinite systems.

Mass Balance Constraints
The first- and third-type source solutions (Eqs. [3-14] and [3-239]) were evaluated by van Genuchten and Parker (1984) and their evaluation is presented in this section. Assuming no decay ($v = 0$) and time-independent source concentration ($\gamma = 0$), mass balance considerations lead to the following equality for the solute transport situation depicted in Figure 3-43:

$$\varphi_e U C_M t = \varphi_e R_d \int_0^\infty C_r(x, t)\, dt \tag{3-250}$$

van Genuchten and Parker (1984) showed that by substitution of Eq. (3-239), the volume-averaged concentration model, on the right-hand side of Eq. (3-250) and performing the integration results exactly in the same expression on the left-hand side of Eq. (3-250). Physically, this means that whatever is added at $x = 0$ (term on the left-hand side in Eq. [3-250]) must be found inside the domain (term on the right-hand side in Eq. [3-250]).

Mass Balance Error for the First-Type Source Condition Model
The first-type source condition given by Eq. (3-3) (with $\gamma = 0$) assumes that the concentration itself can be specified at $x = 0$, a situation that usually is not possible in practice. As a result, van Genuchten and Parker (1984) stated that Eq. (3-3) is incorrect for displacement experiments where tracer solution is added at a specified rate. By substituting Eq. (3-14) into Eq. (3-250), van Genuchten and Parker showed that the mass balance as given by Eq. (3-250) is not satisfied. The authors defined the relative error E as

$$E = \frac{1}{U C_M t}\left[R_d \int_0^\infty C_r(x, t)\, dx - U C_M t \right] \tag{3-251}$$

and after integration they obtained the following expression:

$$E(\zeta) = \frac{1}{(\pi\zeta)^{1/2}}\exp\left(-\frac{\zeta}{4}\right) + \frac{1}{\zeta} - \left(\frac{1}{2} + \frac{1}{\zeta}\right)\mathrm{erfc}\left(\frac{1}{2}\zeta^{1/2}\right) \tag{3-252}$$

where

$$\zeta = \frac{U^2 t}{D_x R_d} \tag{3-253}$$

Equation (3-252) can also be derived by considering the actual solute flux F_x at $x = 0$ as given by Eq. (3-24). Therefore, its special form at $x = 0$ becomes

$$F_x(0, t) = \varphi_e U C_M - \varphi_e D_x \frac{\partial C_r(0, t)}{\partial x} \tag{3-254}$$

Equation (3-24) indicates that in addition to the convective solute mass flux $\varphi_e U C_M$, a dispersive solute flux component is operative at $x = 0$. This dispersive flux is responsible for the observed mass balance deviations, as can be shown by defining the relative error E in terms of F_x:

$$E = \frac{1}{\varphi_e U t}\left[\int_0^t F_x(0, \tau)\, d\tau - \varphi_e U C_M t \right] = -\frac{D_x}{U C_M t}\int_0^t \frac{\partial C_r(0, t)}{\partial x}\, dt \tag{3-255}$$

van Genuchten and Parker (1984) showed that by substituting Eq. (3-14) into Eq. (3-255) also leads to Eq. (3-252). Figure 3-47 shows that the relative error in the mass balance can be extremely large for small values of the dimensionless parameter ζ, i.e., at early times, for solute transport domains where dispersive transport is large as compared with convective transport, and for domains where the solute is strongly adsorbed by the solid phase (large R_d). Based on the aforementioned analysis, van Genuchten and Parker (1984) concluded that the first-type source condition given by Eq. (3-3) (with $\gamma = 0$) fails to satisfy the mass balance constraints and the discrepancy is not always negligible.

Relationship between the First- and Third-Type Source Models
See Section 3.3.3.2.

3.3.3.2   Two-Dimensional Analytical Solute Transport Models for the Third-Type Strip Sources

As mentioned in Section 3.3.2.2, there are a variety of two-dimensional analytical solute transport models in the literature published during the past half century for the first-type sources. However, the number of two-dimensional analytical solute transport models for the third-type sources are fairly limited and they are based on point source approach in infinite media (Hunt, 1978; Wilson and Miller, 1978) and they do not take into account the size of the source. The two-dimensional analytical model developed by Batu (1993), is based on finite strip sources in a bounded medium. As far as the author is aware, this is the only two-dimensional analytical model dealing with third-type strip sources. In this section, this model as well as its applications are presented.

Problem Statement and Assumptions
The problem under consideration is unsteady-state two-dimensional solute transport from multiple third-type (or flux-type) strip sources between two impervious planar boundaries in a unidirectional flow field as shown in Figure 3-48. The x- and z-coordinate axis are the planar coordinates either in a horizontal or vertical plane. The solute fluxes are functions of the z coordinate and time t through an exponential function. The assumptions listed in (1), (2), and (4) in Section 3.3.2.2 for the first-type source case are also valid for the third-type source case.

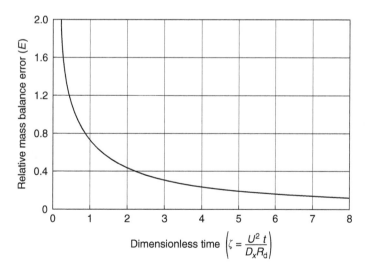

FIGURE 3-47   Relative mass balance error, E, vs. dimensionless time, ζ, for the first-type analytical solution given by Eq. (3-14). (Adapted from van Genuchten, M.T. and Parker, J.C., *Soil Sci. Soc. Am. J.*, 48, 703–708, 1984.)

Governing Differential Equation
The governing differential equation is given by Eq. (3-45).

Initial and Boundary Conditions
The initial condition for $C(x, z, t)$ is given by Eq. (3-46). The boundary conditions for $C(x, z, t)$ at $x = 0$ are

$$F_x(0, z, t) = F_{M_1}(z)e^{\gamma t}, \quad 0 < z < H_1 \tag{3-256}$$

$$F_x(0, z, t) = F_{M_{p-1}}(z)e^{\gamma t}, \quad H_{p-2} < z < H_{p-1} \tag{3-257}$$

$$F_x(0, z, t) = F_{M_p}(z)e^{\gamma t}, \quad H_{p-1} < z < H_p \tag{3-258}$$

where, $F_x(0, z, t)$ (M/L^2/T) is the solute input flux at $x = 0$ (mass per unit area per unit time), γ(T^{-1}) is a parameter, F_{M_i} (M/L^2/T) is the solute input flux when $\gamma = 0$ at $x = 0$, and e = 2.7182818..., and H_i is given by Eq. (3-50). The other boundary conditions at the impervious boundaries and infinity are given by Eqs. (3-51)–(3-54). Equations (3-256)–(3-258) state that the flux $F_x(0, z, t)$, which is time-dependent, has specified values at $x = 0$ and these values vary with the z coordinate. Therefore, they are the third-type or flux-type source conditions. These type of conditions are also called Cauchy condition (see Section 2.10.3.2). The physical meaning of the rest of conditions are given in Section 3.3.2.2.
   From Eqs. (3-256)–(3-258), the function

$$\frac{F_x(0, z, t)}{F_{M_i}(z)} = e^{\gamma t} \tag{3-259}$$

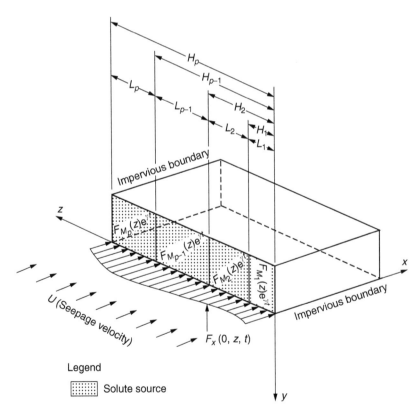

FIGURE 3-48 Schematic representation of two-dimensional solute transport from multiple strip sources with the third-type source condition in a unidirectional flow field. (Adapted from Batu, V., *Water Resour. Res.*, 29, 2881–2892, 1993.)

performs such that the source flux can be represented as exponentially varying time-dependent quantity. If $\gamma < 0$, the source flux decreases with time; if $\gamma = 0$, the source flux is independent of time; and if $\gamma > 0$, the source concentration increases with time. Figure 3-2 shows the variation of the function given by Eq. (3-259) for different values of γ ranging from -0.0001 to -0.0100 day^{-1}. If observed data are available for time-dependent flux sources, the value of γ may be determined from the curve-fitting techniques.

Unsteady-State General Solution for Concentration Distribution
The boundary-value problem defined by Eqs. (3-45), (3-46), (3-51)–(3-54), and (3-256)–(3-258) was solved by the Laplace transform and Fourier analysis techniques (Batu, 1993). For this purpose, the solution was obtained first in the Laplace domain and then the concentration variation was obtained by taking the inverse Laplace transform of the Laplace domain solution. The final results are presented in the following sections and the intermediate steps are presented in the associated examples.

General Solution in the Laplace Domain
As indicated above, the governing differential equation, and the initial and boundary conditions given by Eq. (3-45), (3-46) and (3-51)–(3-54) for the two-dimensional first-type solution case (Section 3.3.2.2) are exactly the same as the third-type source solution case. Using these equations, in Example 3-4, Eq. (E3-4-24) is obtained as the solution with A_n as unknown and this solution is also valid for the third-type source case as well. Using the source conditions at $x = 0$, given by Eqs. (3-256)–(3-258) and (E3-4-24), the general solution in the Laplace domain is expressed in the form of Eq. (3-56), and $u_1(x, z, s)$ and $u_2(x, z, s)$ are as follows (Batu, 1993):

$$u_1(x, z, s) = \frac{1}{H_p(s - \gamma)} \frac{\exp\left(\dfrac{Ux}{2D_x} - p_0 x\right)}{\left(\dfrac{U}{2} + D_x p_0\right)} I_0(z) \tag{3-260}$$

$$u_2(x, z, s) = \frac{2}{H_p(s - \gamma)} \sum_{n=1}^{\infty} \frac{\exp\left(\dfrac{Ux}{2D_x} - p_n x\right)}{\left(\dfrac{U}{2} + D_x p_n\right)} \cos(\lambda_n z) I_n(z) \tag{3-261}$$

where p_0 and p_n are given by Eqs. (3-59) and (3-60), respectively, λ_n is given by Eq. (3-61), and

$$I_0 = \frac{1}{\varphi_e} \int_0^{H_p} F_M(z)\, dz \tag{3-262}$$

$$I_n(z) = \frac{1}{\varphi_e} \int_0^{H_p} F_M(z)\cos(\lambda_n z)\, dz \tag{3-263}$$

Determination of A_n in Eq. (E3-4-24) from the usage of Eqs. (3-255)–(3-257) is shown in Example 3-16.

EXAMPLE 3-16

Using the boundary conditions at $x = 0$ by Eqs. (3-256)–(3-258), and (E3-4-24), determine $u_1(x, y, s)$ and $u_2(x, y, s)$ in Eq. (3-56) for the third-type source.

SOLUTION

The boundary conditions at the source, given by Eqs. (3-256)–(3-258) need to be expressed in the Laplace domain. The convective–dispersive flux in the x direction (F_x) is given by Eq. (3-90a).

Therefore, taking the Laplace transform of the combined form of Eqs. (3-90a), (3-256)–(3-258) gives

$$\frac{F_M(z)}{\varphi_e(s - \gamma)} = Uu - D_x \frac{\partial u}{\partial x}$$

(E3-16-1)

From Eq. (E3-4-24)

$$\frac{\partial u}{\partial x} = \sum_{n=0}^{\infty} A_n \left(\frac{U}{2D_x} - P_n \right) \exp \left(\frac{Ux}{2D_x} - P_n x \right) \cos(\lambda_n z)$$

(E3-16-2)

and at $x = 0$ it gives

$$\frac{\partial u(0, z, s)}{\partial x} = \sum_{n=0}^{\infty} A_n \left(\frac{U}{2D_x} - P_n \right) \cos(\lambda_n z)$$

(E3-16-3)

For $x = 0$, Eq. (E3-4-24) gives

$$u(0, z, s) = \sum_{n=0}^{\infty} A_n \cos(\lambda_n z)$$

(E3-16-4)

Since Eq. (E3-16-1) is only valid at $x = 0$, substitution of Eqs. (E3-16-3) and (E3-16-4) into Eq. (E3-16-1) gives

$$\frac{F_M(z)}{\varphi_e(s - \gamma)} = \sum_{n=0}^{\infty} B_n \cos(\lambda_n z)$$

(E3-16-5)

where

$$B_n = A_n \left(\frac{U}{2} + D_x P_n \right)$$

(E3-16-6)

In Eq. (E3-16-6), B_n can be determined as done for Eq. (E3-4-27) in Example 3-4. By considering the right-hand side of Eq. (E3-16-5) as a Fourier cosine series for the interval $z = 0$ to H_m, yields (e.g., Churchill, 1941, Spiegel, 1971)

$$B_n = \frac{2}{H_p} \int_0^{H_p} \frac{F_M(z)}{\varphi_e(s - \gamma)} \cos(\lambda_n z) \, dz$$

(E3-16-7)

The substitution of Eq. (E3-16-7) into Eq. (E3-16-6) and solving for A_n gives

$$A_n = \frac{1}{\left(\dfrac{U}{2} + D_x P_n \right)} \frac{2}{H_p} \int_0^{H_p} \frac{F_M(z)}{\varphi_e(s - \gamma)} \cos(\lambda_n z) \, dz$$

(E3-16-8)

and substitution of Eq. (E3-16-8) into Eq. (E3-4-24) results in the following equation:

$$u(x, z, s) = \sum_{n=0}^{\infty} \left[\frac{1}{\left(\dfrac{U}{2} + D_x P_n \right)} \frac{2}{H_p} \int_0^{H_p} \frac{F_M(z)}{\varphi_e(s - \gamma)} \cos(\lambda_n z) \, dz \right]$$

(E3-16-9)

$$\times \exp \left(\frac{Ux}{2D_x} - P_n x \right) \cos(\lambda_n z)$$

or after some manipulations, Eq. (E3-16-9) results in the form of Eq. (3-56) where $u_1(x, z, s)$ and $u_2(x, z, s)$ are given by Eqs. (3-260) and (3-261), respectively. It must be remembered that in writing Eq. (3-260) there is a $(1/2)$ coefficient for the first term of the Fourier cosine series.

Inverse Laplace Transforms of Eqs. (3-260) and (3-261)
Determination of the inverse Laplace transforms of Eqs. (3-260) and (3-261) are given in Example 3-17. The sum of Eqs. (3-260) and (3-261) forms the Laplace domain solution according to

Eq. (3-56) and its inverse Laplace transform is the general solution (Batu, 1993):

$$C(x, z, t) = C_1(x, z, t) + C_2(x, z, t) \qquad (3\text{-}264)$$

where

$$C_1(x, z, t) = \frac{1}{H_p} \frac{I_0}{D_x} \int_0^t F_1(\xi)[F_2(\xi) - F_3(\xi)F_4(\xi)]\exp[\gamma(t - \xi)] \, d\xi \qquad (3\text{-}265)$$

and

$$C_2(x, z, t) = \frac{2}{H_p} \frac{1}{D_x} \sum_{n=1}^{\infty} I_n(z)\cos(\lambda_n z) \int_0^t F_1(\xi) \qquad (3\text{-}266)$$

$$\times \exp\left(-\frac{D_z}{R_d} \lambda_n^2 \xi \right)[F_2(\xi) - F_3(\xi)F_4(\xi)]\exp[\gamma(t - \xi)] \, d\xi$$

where $F_1(\xi)$, $F_2(\xi)$, $F_3(\xi)$, and $F_4(\xi)$ are given by Eqs. (3-231)–(3-234), respectively.

EXAMPLE 3-17

Derive Eq. (3-264) by determining the inverse Laplace transforms of Eqs. (3-260) and (3-261).

SOLUTION

The Laplace domain solution is given by Eq. (3-56) and its inverse Laplace transform is given by Eqs. (E3-5-1) and (E3-5-2) and combining them gives Eq. (3-264). Therefore, the inverse Laplace transforms of $u_1(x, y, s)$ and $u_2(x, y, z)$ need to be taken, and they are given below.

Inverse Laplace Transform of $u_1(x, z, s)$
In order to determine the inverse Laplace transform of $u_1(x, z, s)$ as given by Eq. (3-260), according to the second expression of Eq. (E3-5-2), some manipulations need to be made. Equation (3-260) can also be expressed as

$$u_1(x, z, s - b) = \frac{1}{H_p} I_0(z)e_1(s)e_2(s - b)\exp\left(\frac{Ux}{2D_x} \right) \qquad (E3\text{-}17\text{-}1)$$

where

$$e_1(s) = \frac{1}{s - \gamma} \qquad (E3\text{-}17\text{-}2)$$

$$e_2(s - b) = \frac{\exp\left\{ -\left[\frac{R_d}{D_x}(s - b) \right]^{1/2} x \right\}}{\frac{U}{2} + D_x\left[\frac{R_d}{D_x}(s - b) \right]^{1/2}} \qquad (E3\text{-}17\text{-}3)$$

and b is given by Eq. (E3-5-4). The inverse Laplace transform of Eq. (E3-17-2) is

$$L^{-1}[e_1(s)] = E_1(t) = \exp(\gamma t) \qquad (E3\text{-}17\text{-}4)$$

According to the first translation or shifting property for the inverse Laplace transform, Eq. (E3-17-3) can be expressed as (Spiegel, 1965, Theorem 2-3, p. 43)

$$L^{-1}[e_2(s - b)] = E_2(t) = \exp(bt)G_1(t) \qquad (E3\text{-}17\text{-}5)$$

where

$$L^{-1}[g_1(s)] = G_1(t) \qquad (E3\text{-}17\text{-}6)$$

and

$$g_1(s) = \frac{\exp[-(R_d/D_x s)^{1/2}x]}{D_x[U/(2D_x) + (R_d/D_x s)^{1/2}]} \tag{E3-17-7}$$

The inverse Laplace transform of Eq. (E3-17-7) is (Carslaw and Jaeger, 1959, Eq. [12], p. 494)

$$G_1(t) = \frac{1}{D_x}\left(\frac{D_x}{\pi R_d t}\right)^{1/2}\exp\left(-\frac{R_d x^2}{4D_x t}\right) - \frac{U}{2D_x R_d} \tag{E3-17-8}$$

$$\times \exp\left[\frac{Ux}{2D_x} + \frac{D_x t}{R_d}\left(\frac{U}{2D_x}\right)^2\right]\mathrm{erfc}\left[\frac{R_d x}{2(D_x R_d t)^{1/2}} + \frac{U}{2}\left(\frac{t}{R_d D_x}\right)^{1/2}\right]$$

According to the *convolution theorem* or *Faltung's theorem* (e.g., Spiegel, 1965, Theorem 2-10, p. 45), from the above-mentioned expressions, the inverse Laplace transform of Eq. (E3-17-1) can be expressed as

$$C_1(x, z, t) = \frac{1}{H_p}I_0(z)\exp\left(\frac{Ux}{2D_x}\right)\int_0^t E_1(t-\xi)E_2(\xi)\,d\xi \tag{E3-17-9}$$

where ξ is a dummy variable.

Letting $t - \xi = \eta$ or $\xi = t - \eta$, Eq. (E3-17-9) takes the form

$$C_1(x, z, t) = \frac{I_0(z)}{H_p}\int_0^t E_1(\eta)E_2(t-\eta)\,d\eta \tag{E3-17-10}$$

This shows that the convolution of F and G obeys the *cumulative law* of algebra (e.g., Spiegel, 1965, Problem 21, p. 45; 56).

The substitution of Eqs. (E3-17-4) and (E3-17-5) along with Eq. (E3-17-8) into Eq. (E3-17-9) results in Eq. (3-265).

Inverse Laplace Transform of $u_2(x, z, s)$
Similarly, the inverse Laplace transform of Eq. (3-261) can be taken and the final result for $C_2(x,z,t)$ is given by Eq. (3-266).

Unsteady-State Special Solution for a Single-Strip Source
Concentration Distribution
The geometry of a single-strip source is shown in Figure 3-49, which is a special case of Figure 3-48. As can be seen from Figure 3-49, the source can be located at any place between the impervious boundaries. The source conditions expressed by Eqs. (3-256)–(3-258) take the forms

$$F_x(0, z, t)=\begin{cases} F_M e^{\gamma t} & D_1 < z < D_1 + 2B \tag{3-267} \\ \\ 0 & \text{otherwise} \tag{3-268} \end{cases}$$

where F_M(M/L^2/T) is the solute source flux when time is zero. The substitution of Eqs. (3-267) and (3-268) into Eq. (3-262), i.e., $F_M(z) = F_M$, gives

$$I_0 = 2B\frac{F_M}{\varphi_e} \tag{3-269}$$

Similarly, from Eq. (3-263), one can determine

$$I_n = \frac{F_M}{\varphi_e}\frac{1}{\lambda_n}\{\sin[\lambda_n(D_1 + 2B)] - \sin(\lambda_n D_1)\} \tag{3-270}$$

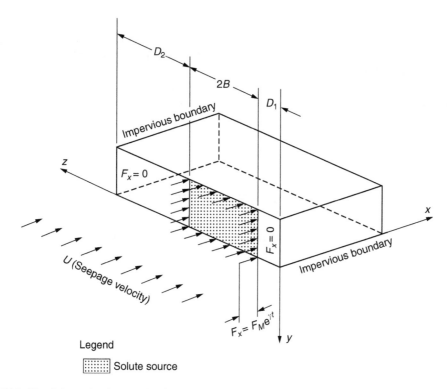

FIGURE 3-49 Schematic picture of a single-strip source in a bounded medium having unidirectional flow field. (Adapted from Batu, V., *Water Resour. Res.*, 29, 2881–2892, 1993.)

where λ_n is given by Eq. (3-86). Equations (3-86), (3-264)–(3-266), (3-269), and (3-270) form the solution for single-strip source having uniform solute source flux. Therefore, from the aforementioned equations, the final solution takes the form (Batu, 1993)

$$C(x, z, t) = \frac{F_M}{\varphi_e} \frac{2B}{(D_1 + 2B + D_2)} \frac{1}{D_x} \int_0^t F_1(\xi)[F_2(\xi) - F_3(\xi)F_4(\xi)]\exp[\gamma(t - \xi)] \, d\xi$$

$$+ \frac{2}{\pi} \frac{F_M}{\varphi_e} \frac{1}{D_x} \sum_{n=1}^{\infty} \frac{1}{n} \left\{\sin[\lambda_n(D_1 + 2B)] - \sin(\lambda_n D_1)\right\} \cos(\lambda_n z) \qquad (3\text{-}271)$$

$$\times \int_0^t F_1(\xi)\exp\left(-\frac{D_z}{R_d} \lambda_n^2 \xi\right)[F_2(\xi) - F_3(\xi)F_4(\xi)] \exp[\gamma(t - \xi)] \, d\xi$$

where $F_1(\xi)$, $F_2(\xi)$, $F_3(\xi)$, and $F_4(\xi)$ are given by Eqs. (3-231)–(3-234),- respectively.

Special Solution for the One-Dimensional Case
If $D_1 = 0$ and $D_2 = 0$ in Figure 3-49, the geometry turns out to be the case shown in Figure 3-43 in Section 3.3.3.1, which corresponds to the one-dimensional case and the corresponding solution is given in Section 3.3.3.1. By substituting these values into Eq. (3-86) one can write $\lambda_n = n\pi/(2B)$. Since $\sin(n\pi) = 0$ and $\sin(0) = 0$, by substituting of $\lambda_n = n\pi/(2B)$ into Eq. (3-271), it can be observed that the second part of the right-hand side of Eq. (3-271) becomes zero, and the rest of its parts turn out to be exactly the same as Eq. (3-230), which is the one-dimensional solution.

Convective–Dispersive Flux Components
The convective–dispersive flux components for two-dimensional solute transport in the x- and z-coordinate direction are given by Eqs. (3-90a) and (3-90b), respectively. The physical meaning and dimensions of these fluxes are presented in Section 3.3.2.2.

Following the same procedure as outlined in Example 3-6, from Eqs. (3-90a) and (3-271), the flux component in the x-coordinate direction can be expressed as

$$F_x = \varphi_e UC - \varphi_e D_x(V_1 + V_2) \tag{3-272}$$

where

$$V_1 = \frac{F_M}{\varphi_e} \frac{2B}{(D_1 + 2B + D_2)} \frac{1}{D_x} \int_0^t [F_{1x}(F_2 - F_3 F_4) \tag{3-273}$$

$$+ F_1(F_{2x} - F_3 F_{4x} - F_4 F_{3x})]\exp[\gamma(t - \xi)]\,d\xi$$

and

$$V_2 = \frac{2}{\pi} \frac{F_M}{\varphi_e} \frac{1}{D_x} \sum_{n=1}^{\infty} \frac{1}{n} \left\{\sin[\lambda_n(D_1 + 2B)] - \sin(\lambda_n D_1)\right\} \cos(\lambda_n z) \tag{3-274}$$

$$\times \int_0^t \exp\left(-\frac{D_z}{R_d}\lambda_n^2 \xi\right)[F_{1x}(F_2 - F_3 F_4) + F_1(F_{2x} - F_3 F_{4x} - F_4 F_{3x})]\exp[\gamma(t - \xi)]\,d\xi$$

in which $F_1(\xi)$, $F_2(\xi)$, $F_3(\xi)$, and $F_4(\xi)$ are given by Eqs. (3-231)–(3-234), respectively, and F_{1x}, F_{2x}, F_{3x}, and F_{4x} are given by Eqs. (3-242)–(3-245), respectively. The expression for the convective–dispersive flux component in the direction can easily be determined from Eq. (3-90b) using Eq. (3-271) as

$$F_z = -\varphi_e D_z V_3 \tag{3-275}$$

where

$$V_3 = -\frac{2}{\pi} \frac{F_M}{\varphi_e} \frac{1}{D_x} \sum_{n=1}^{\infty} \frac{1}{n} \left\{\sin[\lambda_n(D_1 + 2B)] - \sin(\lambda_n D_1)\right\} \lambda_n \sin(\lambda_n z) \tag{3-276}$$

$$\times \int_0^t F_1(\xi)\exp\left(-\frac{D_z}{R_d}\lambda_n^2 \xi\right)(F_2 - F_3 F_4)\exp[\gamma(t - \xi)]\,d\xi$$

Unsteady-State Special Solutions for Other Sources
Equation (3-264) along with Eqs. (3-265) and (3-266) is the general solution for the resulting concentration distribution for any given $F_M(z)$ source flux distribution. With the source flux functions (Figure 3-48), $F_M(z)$, the integrals I_0 and I_n given by Eqs. (3-262) and (3-263), respectively, need to be evaluated analytically or numerically depending on the type of the function for $F_M(z)$.

Solution for Two Strip Sources with Uniform Solute Source Fluxes
The geometry of two strip sources with uniform source concentrations is shown in Figure 3-50. The boundary conditions at $x = 0$ expressed by Eqs. (3-256)–(3-258) take the forms

$$F_{x_1}(0, z, t) = F_{M_1} e^{\gamma t}, \quad D_1 < z < D_1 + 2B \tag{3-277}$$

$$F_{x_2}(0, z, t) = F_{M_2} e^{\gamma t}, \quad D_1 + 2B_1 + D_3 < z < D_1 + 2B_1 + D_3 + 2B_2 \tag{3-278}$$

where F_{M_1} and F_{M_2} (units are $M/L^2/T$) are the uniform solute source fluxes. Equation (3-264) is the general solution and for a special case, the integrals I_0 and I_n given by Eqs. (3-262) and (3-263),

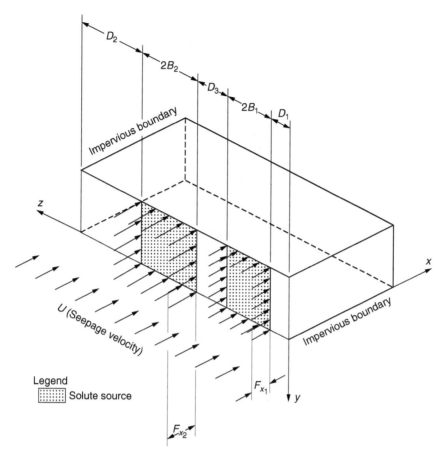

FIGURE 3-50 Schematic representation of two strip sources with uniform solute source fluxes in a bounded medium having unidirectional flow field.

respectively, need to be evaluated using the source geometry. Using the above-mentioned source conditions, from Eq. (3-262),

$$I_0 = \int_{D_1}^{D_1+2B_1} \frac{F_{M_1}}{\varphi_e}\, dz + \int_{D_1+2B_1+D_3}^{D_1+2B_1+D_3+2B_2} \frac{F_{M_2}}{\varphi_e}\, dz = 2B_1 \frac{F_{M_1}}{\varphi_e} + 2B_2 \frac{F_{M_2}}{\varphi_e} \qquad (3\text{-}279)$$

Similarly, from Eq. (3-263), one can write

$$I_n = \int_{D_1}^{D_1+2B_1} \frac{F_{M_1}}{\varphi_e} \cos(\lambda_n z)\, dz + \int_{D_1+2B_1+D_3}^{D_1+2B_1+D_3+2B_2} \frac{F_{M_2}}{\varphi_e} \cos(\lambda_n z)\, dz \qquad (3\text{-}280)$$

or after some manipulations one gets

$$I_n = \frac{F_{M_1}}{\varphi_e \lambda_n} \{\sin[\lambda_n(D_1 + 2B_1)] - \sin(\lambda_n D_1)\}$$

$$+ \frac{F_{M_2}}{\varphi_e \lambda_n} \{\sin[\lambda_n(D_1 + 2B_1 + D_3 + 2B_2)] - \sin[\lambda_n(D_1 + 2B_1 + D_3)]\} \qquad (3\text{-}281)$$

Equations (3-264)–(3-266), (3-279), and (3-281) form the solution for two strip sources with uniform solute source fluxes.

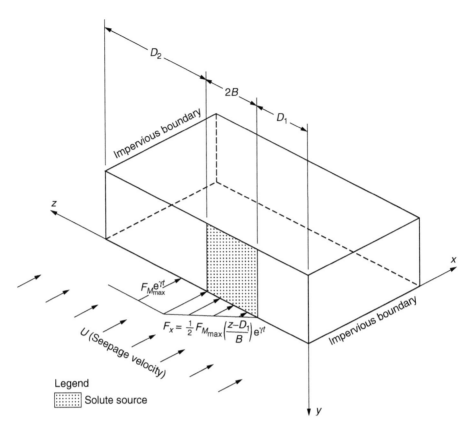

FIGURE 3-51 Schematic representation of a single-strip source having linearly varying solute source flux.

Solution for a Single-Strip Source with Linearly Varying Solute Source Flux
The geometry of a single-strip source with linearly varying solute source flux is shown in Figure
3-51. The boundary conditions at $x = 0$ expressed by Eqs. (3-256)–(3-258) take the forms

$$F_x(0, z, t) = \begin{cases} \dfrac{1}{2} F_{M_{max}} \dfrac{z - D_1}{B} e^{\gamma t} & D_1 < z < D_1 + 2B \quad (3\text{-}282) \\ 0 & \text{otherwise} \quad (3\text{-}283) \end{cases}$$

Equation (3-282) means that at $z = D_1$, $F_x(0, z = D_1, t) = 0$; and at $z = D_1 + 2B$, $F_x(0, z = D_1+2B,$
$t) = F_{M_{max}}$. The variation of $F_x(0, z, t)$ is shown in Figure 3-51. Equation (3-264) is the general solution and for a special case, the integrals I_0 and I_n given by Eqs. (3-262) and (3-263), respectively,
need to be evaluated by using the source geometry. It can be observed from Eq. (3-282) and the
source conditions given by Eqs. (3-256)–(3-258) that

$$F_M(z) = \frac{1}{2} F_{M_{max}} \left(\frac{z - D_1}{B} \right) \quad (3\text{-}284)$$

By substituting Eq. (3-284) into Eq. (3-262) and with some manipulations one can obtain

$$I_0 = \int_0^{H_p} \frac{F_M(z)}{\varphi_e} \, dz = \frac{1}{\varphi_e} \int_{D_1}^{D_1+2B} \frac{1}{2} F_{M_{max}} \frac{z - D_1}{B} \, dz = \frac{1}{2} \frac{F_{M_{max}}}{\varphi_e} (2B - D_1) \quad (3\text{-}285)$$

Similarly, by substituting Eq. (3-284) into Eq. (3-263) one can write

$$I_n = \frac{1}{\varphi_e} \frac{1}{2} F_{M_{max}} \int_{D_1}^{D_1+2B} \frac{z - D_1}{B} \cos(\lambda_n z) \, dz \quad (3\text{-}286)$$

which is similar to the integral in Eq. (3-108). Therefore, the final result can be written as

$$I_n = \frac{1}{2}\frac{F_{M_{max}}}{B}\frac{1}{\lambda_n}\left\{\frac{1}{\lambda_n}\cos[\lambda_n(D_1+2B)] + 2B\sin[\lambda_n(D_1+2B)] - \frac{1}{\lambda_n}\cos(\lambda_n D_1)\right\} \quad (3\text{-}287)$$

Equations (3-61), (3-265), (3-266), (3-285), and (3-287) form the concentration distribution solution for singlestrip source with linearly variable source flux shown in Figure 3-51.

Steady-State General Solution for Concentration Distribution
The solution for steady-state solute transport case (Figure 3-48 with $\gamma = 0$) can be obtained by using two different procedures. The first procedure is to use the corresponding differential equation under steady-state solute transport conditions with the associated boundary conditions. The second procedure is to use the solution developed for unsteady-state concentration distribution as the time t approaches infinity. Here, the first procedure is presented below and its details are presented in Example 3-18 (Batu, 1993).

Under steady-state solute transport conditions the governing equation is given by Eq. (3-110). The solution for the concentration distribution, C, will be only the function of x and z and the initial condition given by Eq. (3-46) is not necessary. The boundary conditions given by Eqs. (3-256)–(3-258) will have the following forms:

$$F_x(0, z, t) = F_{M_1}(z), \quad 0 < z < H_1 \quad (3\text{-}288)$$

$$F_x(0, z, t) = F_{M_{p-1}}(z), \quad H_{p-2} < z < H_{p-1} \quad (3\text{-}289)$$

$$F_x(0, z, t) = F_{M_p}(z), \quad H_{p-1} < z < H_p \quad (3\text{-}290)$$

where H_i is given by Eq. (3-50). The other boundary conditions, given by Eqs. (3-51)–(3-54), remain the same.

The boundary-value problem for steady-state two-dimensional solute transport conditions defined above is exactly the same as the corresponding one in Section 3.3.2.2 with the exception of the boundary conditions at the sources. The solution of the same problem is given in Example 3-7 and Eq. (E3-7-5) is the solution and p_n is given by Eq. (3-116). In order to determine the expression for A_n in Eq. (E3-7-5), the boundary conditions at $x = 0$ (Eqs. [3-288]–[3-290]) need to be used. The final result for steady-state concentration distribution is given as

$$C(x, z) = \frac{I_0(z)}{H_p\left(\frac{U}{2}+D_x p_0\right)}\exp\left(\frac{Ux}{2D_x}-p_0 x\right) + \frac{2}{H_p}\sum_{n=1}^{\infty}\frac{I_n(z)}{\left(\frac{U}{2}+D_x p_n\right)}\exp\left(\frac{Ux}{2D_x}-p_n x\right)\cos(\lambda_n z) \quad (3\text{-}291)$$

where λ_n, p_0, p_n, $I_0(z)$, and $I_n(z)$ are given by Eqs. (3-61), (3-115), (3-116), (3-262), and (3-263), respectively. The derivation of Eq. (3-291) is given in Example 3-18.

EXAMPLE 3-18

Derive Eq. (3-291) using the general solution given by Eq. (E3-7-5) with the source boundary conditions given by Eqs. (3-288)–(3-290).

SOLUTION

At the sources ($x = 0$), from Eq. (3-90a), one can write

$$F_x(0, z) = \varphi_e U C(0, z) - \varphi_e D_x \frac{\partial C(0, z)}{\partial x} \quad (E3\text{-}18\text{-}1)$$

From Eq. (E3-7-5),

$$\frac{\partial C}{\partial x} = \sum_{n=0}^{\infty} A_n \left(\frac{U}{2D_x} - p_n \right) \exp\left(\frac{Ux}{2D_x} - p_n x \right) \cos(\lambda_n z) \qquad \text{(E3-18-2)}$$

For $x = 0$, Eqs. (E3-7-5) and (E3-18-2) give

$$C(0, z) = \sum_{n=0}^{\infty} A_n \cos(\lambda_n z) \qquad \text{(E3-18-3)}$$

$$\frac{\partial C(0, z)}{\partial x} = \sum_{n=0}^{\infty} A_n \left(\frac{U}{2D_x} - p_n \right) \cos(\lambda_n z) \qquad \text{(E3-18-4)}$$

The substitution of Eqs. (E3-18-3) and (E3-18-4) into Eq. (E3-18-1) gives

$$\frac{F_x(0, z)}{\varphi_e} = \frac{F_M(z)}{\varphi_e} = \sum_{n=0}^{\infty} B_n \cos(\lambda_n z) \qquad \text{(E3-18-5)}$$

where

$$B_n = A_n \left(\frac{U}{2} + D_x p_n \right) \qquad \text{(E3-18-6)}$$

Considering the right-hand side of Eq. (E3-18-5) as a Fourier cosine series for the interval $z = 0$ to H_p yields (e.g., Churchill, 1941; Spiegel, 1971)

$$B_n = \frac{2}{H_p} \int_0^{H_p} \frac{F_M(z)}{\varphi_e} \cos(\lambda_n z)\, dz \qquad \text{(E3-18-7)}$$

The substitution of Eq. (E3-18-7) into Eq. (E3-18-6) and solving it for A_n yields

$$A_n = \frac{1}{\left(\dfrac{U}{2} + D_x p_n \right)} \frac{2}{H_p} \int_0^{H_p} \frac{F_M(z)}{\varphi_e} \cos(\lambda_n z)\, dz \qquad \text{(E3-18-8)}$$

Finally, introducing Eq. (E3-18-8) into Eq. (E3-7-5) results in

$$C(x, z) = \frac{2}{H_p} \sum_{n=0}^{\infty} \frac{1}{\left(\dfrac{U}{2} + D_x p_n \right)} \left[\int_0^{H_p} \frac{F_M(z)}{\varphi_e} \cos(\lambda_n z)\, dz \right] \exp\left(\frac{Ux}{2D_x} - p_n x \right) \cos(\lambda_n z) \qquad \text{(E3-18-9)}$$

Equation (13-18-9) is the general solution and is valid for any solute flux distribution at $x = 0$. By evaluating the first term of the series for $n = 0$, Eq. (E3-18-9) takes the form of Eq. (3-291) as given above. It must be remembered that the first term in the Fourier cosine series (for $n = 0$) is divided by 2 as a requirement for the Fourier cosine series.

Steady-State Special Solution for a Single-Strip Source

Concentration Distribution
The geometry of a single-strip source is shown in Figure 3-49 (with $\gamma = 0$), which is a special case of Figure 3-48 (with $\gamma = 0$). As can be seen from Figure 3-49, the source can be located at any place between the impervious boundaries. The source conditions given by Eqs. (3-256)–(3-258) take the following forms:

$$F_x(0, z) = \begin{cases} F_M & D_1 < z < D_1 + 2B & \text{(3-292)} \\[2mm] 0 & \text{otherwise} & \text{(3-293)} \end{cases}$$

where $F_M (M/L^2/T)$ is the solute source flux. The general solution is given by Eq. (3-291) and for a single-strip source case, $I_0(z)$ and $I_n(z)$ are given by Eqs. (3-269) and (3-270), respectively. The substitution of these expressions into Eq. (3-291) gives the following expression for concentration distribution:

$$C(x, z) = \frac{2BF_M}{\varphi_e\left(\dfrac{U}{2} + D_x p_0\right)(D_1 + 2B + D_2)} \exp\left(\frac{Ux}{2D_x} - p_0 x\right) \tag{3-294}$$

$$+ \frac{2}{\pi} \frac{F_M}{\varphi_e} \sum_{n=1}^{\infty} R_{1_n} \cos(\lambda_n z) \frac{\exp(R_{2_n} x)}{\left[\dfrac{U}{2} + (R_d D_x v + D_x D_z \lambda_n^2 + U^2/4)^{1/2}\right]}$$

where λ_n, p_0, p_n, R_{1_n}, and R_{2_n}, are given by Eqs. (3-61), (3-115), (3-116), (3-125), and (3-126), respectively.

Special Solution for the One-Dimensional Case
If $D_1 = 0$ and $D_2 = 0$ in Figure 3-49, the geometry turns out to be the case in Figure 3-43, which corresponds to the one-dimensional case and the corresponding solution is given in Section 3.3.3.1. By substituting these values in Section 3.3.3.2, one can write $\lambda_n = n\pi/2B$. Since $\sin(n\pi) = 0$ and $\sin(0) = 0$, substitution of $\lambda_n = n\pi/2B$ into Eq. (3-294), it can be observed that the second part of the right-hand side of Eq. (3-294) becomes zero, and the rest of parts turn out to be exactly the same as Eq. (3-247), which is the one-dimensional solution.

Convective–Dispersive Flux Components
Using Eq. (3-294) in Eq. (3-90a), the expression for $F_x(x,z)$ can be determined as

$$F_x = \varphi_e U C(x, z) - \frac{2BD_x F_M}{(U/2 + D_x p_0)(D_1 + 2B + D_2)} \exp\left(\frac{U}{2D_x} - p_0\right) \exp\left(\frac{Ux}{2D_x} - p_0 x\right) \tag{3-295}$$

$$- \frac{2}{\pi} D_x F_M \sum_{n=1}^{\infty} R_{1_n} \cos(\lambda_n z) R_{2_n} \frac{\exp(R_{2_n} x)}{[U/2 + (R_d D_x v + D_x D_z \lambda_n^2 + U^2/4)^{1/2}]}$$

where λ_n, p_0, R_{1_n}, and R_{2_n} are given by Eqs. (3-61), (3-115), (3-125), and (3-126), respectively.
Similarly, using Eq. (3-294) in Eq. (3-90b) the expression for $F_z(x, z)$ can be determined as

$$F_z = D_z F_M \sum_{n=1}^{\infty} R_{1_n} \lambda_n \sin(\lambda_n z) \frac{\exp(R_{2_n} x)}{[U/2 + (R_d D_x v + D_x D_z \lambda_n^2 + U^2/4)^{1/2}]} \tag{3-296}$$

Steady-State Solutions for Other Sources
Equation (3-291) is the general solution for the resulting concentration distribution for any given $F_M(z)$ solute source flux distribution. With the solute source flux functions (Figure 3-48 with $\gamma = 0$), $F_M(z)$, the integrals I_0 and I_n given by Eqs. (3-262) and (3-263), respectively, need to be evaluated analytically or numerically depending on the function $F_M(z)$. Two examples are presented below.

Solution for Two-Strip Sources with Uniform Source Concentrations
The geometry of two-strip sources with uniform solute source fluxes is shown in Figure 3-50 (with $\gamma = 0$). The boundary conditions at $x = 0$ are expressed by Eqs. (3-288)–(3-290) (with $\gamma = 0$). The expressions for I_0 and I_n are given by Eqs. (3-279) and (3-281), respectively. Therefore, Eqs. (3-61), (3-279), (3-281), and (3-291) form the solution for two-strip sources having constant source fluxes under steady-state solute transport conditions.

Solution for a Single-Strip Source with Linearly Varying Source Concentration
The geometry of a single-strip source with linearly varying solute source fluxes is shown in Figure 3-51 (with $\gamma = 0$). The boundary conditions at $x = 0$ are expressed by Eqs. (3-288)–(3-290) (with $\gamma = 0$). The expressions for I_0 and I_n are given by Eqs. (3-285) and (3-287), respectively. Therefore, Eqs. (3-61), (3-285), (3-287), and (3-291) form the solution for a single-strip source having linearly varying source flux under steady-state solute transport conditions.

Evaluation of the Models

The general unsteady-state solution for concentration distribution given by Eq. (3-264) was evaluated for a single-strip source case. Therefore, the unsteady-state solution for a single-strip source, which is given by Eq. (3-271), was used for model evaluation. The results of this evaluation, based on Batu (1993) are presented below.

As mentioned in Section 3.3.3.1, if in Eq. (3-271), F_M is given by Eq. (3-236), the concentration $C(x, z, t)$ becomes volume-averaged (resident) concentration and is denoted by $C_r(x, z, t)$ (see Sections 2.2.2 and 2.9.1 for definitions). With $F_M = \varphi_e U C_M$, the results of Eq. (3-271) were compared with two different numerical codes. As can be seen from Eq. (3-271), the concentration solution includes integrals as well as infinite series. Notice that for the second part of Eq. (3-271) integration must be performed for all n values of the infinite series. Since currently no analytical method exists to determine these integrals, numerical integration procedure was used for their evaluation. For these integrals, it was shown that the Gauss method is powerful and provides accurate results (Batu, 1993). Since the integrals have limits 0 and t, whereas the Gauss method requires limits of -1 and $+1$, the following change of variable is performed:

$$\xi = \frac{t}{2}(\xi' + 1) \qquad (3\text{-}297)$$

Thus, when ξ' is -1, ξ is 0; when $\xi' = 1$, $\xi = t$; and ξ' becomes the variable in the Gauss integration. Moreover

$$d\xi = \frac{t}{2}d\xi' \qquad (3\text{-}298)$$

ξ and $d\xi$ are substituted into Eq. (3-271) and based on the modified form of Eq. (3-271) a computer program, ST2B, has been developed by the author. This program is described below.

The results of ST2B model are compared with other available analytical and numerical models. The results are presented below.

Comparison with the One-Dimensional Analytical Model

As mentioned in Section 3.3.3.1, the one-dimensional analytical model for the third-type solute source, as given by Eq. (3-230), is a special form of the more general two-dimensional model given by Eq. (3-271). Therefore, the verification study conducted for the one-dimensional model case is valid also for the two-dimensional model case.

Comparisons with Numerical Models

The numerical model codes that were selected are called FTWORK (Faust et al., 1990) and SWIFT III (Reeves et al., 1986; Ward, 1988) and the comparison details are included in Batu (1993). Both FTWORK and SWIFT III are complete block-centered, three-dimensional finite-difference ground water flow and solute transport codes. Here, some details regarding the comparison with FTWORK are presented. For the numerical model, a finite modeling domain had to be chosen rather than a domain that extends to infinity in the x direction of the analytical model. The finite-difference grid that was selected for comparison is a modified form of the grid in the FTWORK user manual (Faust et al., 1990, Figure 4.20). Ground water flow and solute transport parameters were selected to be the same. The modeling area was chosen to be 185×53 m, which contained 19 columns and 39 rows (Figure 3-52). As can be seen from Figure 3-52, the grid dimensions range from 0.25 to 10 m with finer grid dimensions around the source. The solute source is located at the upgradient boundary asymmetrically, and its geometrical values are $2B = 5$ m, $D_1 = 45$ m, and $D_2 = 3$ m. Spatial increments of the finite-difference grid satisfied the appropriate criterion, i.e., the Peclet number should be lesser than 2 to eliminate numerical dispersion (Huyakorn and Pinder, 1983). The length of the modeling domain was selected such that the solute front with the minimum retardation factor and assigned hydrogeological parameters would not reach the boundary at $x = 185$ m at the end of 180 days, which is the total simulation time. Finer grid dimensions were used around the source area in

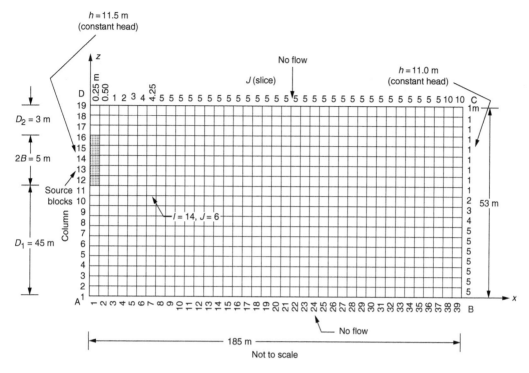

FIGURE 3-52 The finite-difference grid used for verification of the two-dimensional third-type solute source model. (Adapted from Batu, V., *Water Resour. Res.*, 29, 2881–2892, 1993.)

order to increase accuracy for comparison. The boundaries at $x = 0$ and $x = 185$ m (AD and BC in Figure 3-52) were simulated as constant head for the finite-difference model (FTWORK). The other two boundaries which are parallel to the main ground water flow direction (AB and CD) were simulated as no-flow boundaries. The constant-head boundary values are 11.5 and 11.0 m for the upgradient and downgradient boundaries (AD and BC), respectively. These boundary conditions define a one-dimensional ground water flow field. The hydraulic conductivity, $K = K_x = K_y = K_z$, and effective porosity, φ_e, were selected to be 13.875 m/day and 0.25, respectively. With the hydraulic gradient $I = (0.5\ \text{m})/(185\ \text{m}) = 0.0027027$ m/m, the corresponding ground water velocity is $U = 0.15$ m/day. The longitudinal dispersivity, α_L and transverse dispersivity, α_T, and retardation factor, R_d values are 21.3 m, 4.3 m, and 1, respectively. The effective molecular diffusion coefficient, D^*, is zero and the solute source concentration is $C_M = 1$ g/m^3. The solute source flux was selected according to Eq. (3-236) and with the above given values $F_M = (0.25)(0.15\ \text{m/day})(1\ \text{g/m}^3) = 0.0375$ g/day/m^2. Therefore, the predicted concentrations are volume-averaged (resident) concentrations denoted by C_r (see Sections 2.2.2 and 2.9.1 for definitions).

The simulated volume-averaged concentrations as a function of distance and time are compared for the analytical model of a single-strip source, called ST2B, and finite-difference numerical model (FTWORK). Longitudinal and transverse normalized concentration vs. distance comparisons at $x = 22.5$ m and $z = 51.5$ m are shown in Figure 3-53 and Figure 3-54, respectively, for a time period of 180-days. Figure 3-55 presents the volume-averaged concentration vs. time at a point (for the analytical model) and at the cell center (for the numerical model) shown in Figure 3-52. Review of Figures 3-53–3-55 indicates that good correlation exits among the results of the analytical model (ST2B) and numerical model (FTWORK).

Numerical experiments were carried out in the FTWORK runs to determine the maximum time step producing numerically stable results. The results showed that $\Delta t = 1$ day was appropriate and this value was used for all comparisons.

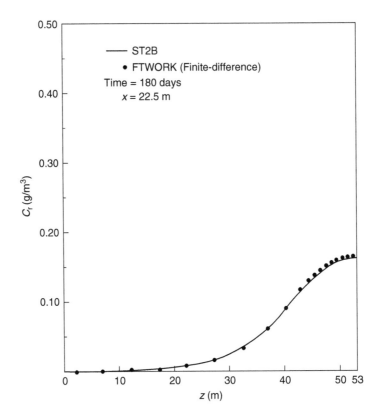

FIGURE 3-53   Concentration vs. transverse distance comparison at $x = 22.5$ m between the third-type source two-dimensional analytical model and a finite-difference numerical model after 180 days. (Adapted from Batu, V., *Water Resour. Res.*, 29, 2881–2892, 1993.)

As in the previous comparison, the results between the analytical model (ST2B) and the other numerical model, SWIFT III (Reeves et al., 1986; Ward, 1988), were very close and are not presented here.

Representation of Exponentially Variable Time-Dependent Source Concentrations
As mentioned previously, the solute source flux can be represented as exponentially varying time-dependent quantity. If $\gamma < 0$, the source flux decreases with time; if $\gamma = 0$, the source flux is independent of time; and if $\gamma > 0$, the source concentration increases with time. Figure 3-2 shows the variation of the function given by Eq. (3-259) for different values of γ ranging from -0.0001 to -0.0100 day^{-1}. If observed data are available for time-dependent flux sources, the value of γ may be determined from the curve-fitting techniques.

Selection of Number of Quadrature Points in the Gaussian Integration
As mentioned previously, the integrals in Eq. (3-271) were evaluated numerically based on the transformation expressions given by Eqs. (3-297) and (3-298) by means of Gaussian integration procedure with 4, 5, 10, 20, 60, 104, or 256 quadrature points. The same method was applied for the integrals in the convective–dispersive flux components as given by Eqs. (3-272) and (3-275). In the ST2B computer program, which is presented later, the user has options to use the above-mentioned quadrature points. Quadrature points higher than 20 produce accurate results.

Selection of Number of Terms for the Infinite Series in the Solutions
Apart from the number of Gauss points, the number of terms in the infinite series plays a key role in determining accurate results. Numerical experiments showed that at the source, the number of

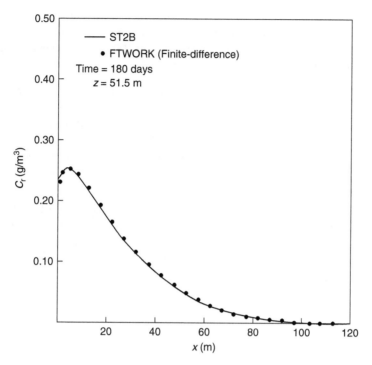

FIGURE 3-54 Concentration vs. longitudinal distance comparison at $z = 51.5$ m between the third-type source two-dimensional analytical model and a finite-difference numerical model after 180 days. (Adapted from Batu, V., *Water Resour. Res.*, 29, 2881–2892, 1993.)

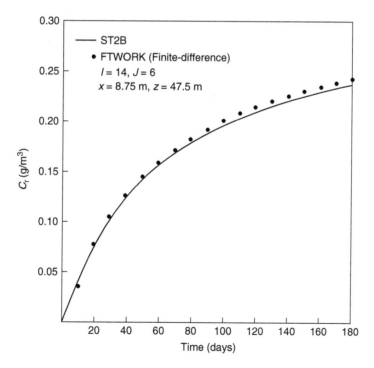

FIGURE 3-55 Concentration vs. time comparison at the point $(x = 8.75$ m, $z = 47.5$ m) between the third-type source two-dimensional analytical model and a finite-difference numerical model after 180 days. (Adapted from Batu, V., *Water Resour. Res.*, 29, 2881–2892, 1993.)

terms are the highest for converged results. The required number of terms decreases with the increasing distance from the source. Generally, terms around 400 produce reasonably accurate results. In most cases, the number of terms between 50 and 100 produce acceptable results with the exception of the source location and its close proximity.

Application of the Two-Dimensional Analytical Models to Ground Water Flow Systems

The general model and its derivative models under unsteady- and steady-state solute transport conditions presented above can be applied to solute transport predictions in aquifers if the ground water velocity can be assumed to be uniform. The models can also be applied to unsaturated zones with some additional assumptions. All these are all discussed in the following sections.

In this section, first, a computer program ST2B for a single-strip source, is described. Then, application of this program is presented with examples.

A Computer Program (ST2B) for the Two-Dimensional Third-Type Strip Source Model

A two-dimensional MS-DOS-based FORTRAN solute transport computer program, solute transport two-dimensional third-type (ST2B), has been developed by the author. ST2B is a menu-driven program and calculates values of concentration and convective–dispersive flux components under unsteady-state solute transport conditions (Eqs. [3-271], [3-272], and [3-275]) and steady state solute transport conditions (Eqs. [3-294], [3-295], and [3-296]). This program can be run interactively in IBM compatible computers.

Application to Saturated Zones

The two-dimensional analytical solute transport model whose geometry is shown in Figure 3-49, can be used for different purposes. The impervious boundaries around the solute transport domain increase the potential usage areas of the model. For ground water flow cases, the effects of impervious boundaries, such as bedrock outcrops and clay zones surrounding the main ground water flow zone, can be taken into account. If the aquifer approaches infinity from both sides of the source, its effect can be taken into account by assigning large values to D_1 and D_2 as compared with the source width $2B$. The model can also be used along a cross-section of a confined or unconfined aquifer such that the $z = 0$ plane is the bottom of the aquifer and $z = D_1 + 2B + D_2$ plane the upper boundary of the aquifer. For a confined aquifer, the $z = D_1 + 2B + D_2$ plane is the bottom of the confining layer and for an unconfined aquifer is the water table of the unconfined aquifer. For these cases, the source will be infinitely long in the direction of the y-coordinate axis. For two-dimensional area model case in the x–z plane, as shown in Figure 3-49, it is assumed that the source extends from the bottom of a confined or unconfined aquifer to the top of the confined aquifer or to the water table of the unconfined aquifer.

EXAMPLE 3-19

For a decaying source and aquifer geometry shown in Figure 3-49, the following parameters are given: $K_h = 13.875$ m/day; $I = (0.5 \text{ m})/(185 \text{ m}) = 0.0027027$ m/m; $\varphi_e = 0.25$ ($U = 0.15$ m/day); $K_d = 0$ cm^3/g; $\rho_b = 1.78$ g/cm^3 ($R_d = 1$); $D^* = 0$ m^2/day; $C_M = 1$ g/m^3; $F_M = 0.0375$ g/day/m^2; $2B = 10$ m; $D_1 = 1000$ m; $D_2 = 1000$ m; $v = 0$ day^{-1}; $\gamma = 0.001$ day^{-1}; $\alpha_L = 21.3$ m; and $\alpha_T = 4.3$ m. With these parameters, using the ST2B program, determine the isoconcentration curves after 5, 10, and 20 years elapsed times.

SOLUTION

By using the given data, the ST2B program was run to generate the coordinates and concentration data at a given time. The output of ST2B was used to generate contours using the SURFER^{TM} contouring program (Golden Software, Inc., 1988).

Using the given data with $\gamma = 0.001$ day^{-1}, the ST2B program was run for the decaying source flux case. Figure 3-26 presents the temporal variation of the decaying source function for the given value of γ. Figures 3-56, 3-57, and 3-58 present the concentration contours after 5, 10, and 20 years,

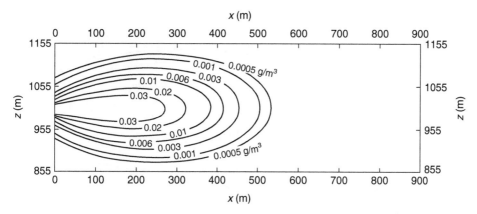

FIGURE 3-56 Concentration contours for a decaying source after 5 years for Example 3-19. (Adapted from Batu, V., *Water Resour. Res.*, 29, 2881–2892, 1993.)

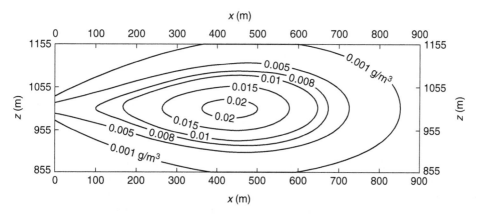

FIGURE 3-57 Concentration contours for a decaying source after 10 years for Example 3-1.9 (Adapted from Batu, V., *Water Resour. Res.*, 29, 2881–2892, 1993.)

respectively. Notice from Figure 3-26 that the normalized solute source flux becomes approximately zero after approximately 6000 days (16.4 years). It can be seen from these figures that the center of the solute plume is shifted approximately 180 m from the source after 5 years (Figure 3-56), 440 m after 10 years (Figure 3-57), and 980 m after 20 years (Figure 3-58).

Application to Unsaturated Zones

The schematic diagram in Figure 3-59 shows the applicability of the two-dimensional ST2B model for a single constant source flux. The associated governing equations for flow and solute transport in unsaturated porous media under uniform infiltration rate conditions are given in Section 2.8. Some important points for the application and assumptions are given below:

1. It is assumed that there is a uniform infiltration rate, v_0, at the soil surface. As shown in Section 2.8.1, this infiltration rate creates a constant ground water velocity in the unsaturated zone. Also in the same section it is shown that the ground water velocity, U, can be determined using the unsaturated hydraulic conductivity vs. water content relationship for a given soil, i.e., $K = K(\theta)$.

2. Liquid waste from a surface contaminant source is being discharged to the unsaturated zone. Landfills, some units of factories, or other industrial complexes may be considered as surface contaminant sources.

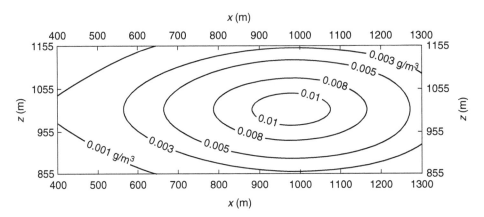

FIGURE 3-58 Concentration contours for a decaying source after 20 years for Example 3-19. (Adapted from Batu, V., *Water Resour. Res.*, 29, 2881–2892, 1993.)

3. The length L of the source in practice is finite, but in the model it is infinite (in the y-coordinate axis).
4. It is assumed that the waste liquid seeping from the bottom of the source ditch reaches to the top of the unsaturated zone and creates a constant source flux, F_M of a certain solute species in the area beneath the ditch.
5. The source flux, F_M can also be taken into account as a exponentially varying time-dependent quantity according to Eq. (3-267).

Comparative Evaluation of the First- and Third-Type Source Models
As mentioned in Section 3.3.3.1, proper formulation of solute transport conditions is critically important for the interpretation of observed data from column displacement experiments. As a result, in Section 3.3.3.1, the analysis of van Genuchten and Parker (1984) regarding the evaluation of the first- and third-type source conditions for one-dimensional analytical models are presented. Later, Batu and van Genuchten (1990), and Batu (1993) extended this evaluation process to two-dimensional analytical models. In this section, the evaluation process is presented for two-dimensional analytical models.

Mass Balance Constraints
The first- and third-type source conditions were evaluated for two-dimensional models in van Genuchten and Batu (1990) in semiinfinite medium based on the analytical models of Batu (1983), and Batu and van Genuchten (1990). Here, the two-dimensional evaluation process will be presented based on the above-mentioned analytical model presented in the earlier sections based on Batu (1993).

   Mass balance evaluation will be done for one of the special cases. A single-strip source case with constant solute source flux given by Eq. (3-236) will be used (Figure 3-49). As mentioned in Section 3.3.3.1, if in Eq. (3-271) F_M is given by Eq. (3-236), the concentration $C(x, z, t)$ becomes volume-averaged (resident) concentration and is denoted by $C_r(x, z, t)$ (see Sections 2.2.2 and 2.9.1 for definitions). With $F_M = \varphi_e U C_M$ and assuming no decay ($v = 0$) and time-independent solute flux ($\gamma = 0$), mass balance requirement for solute transport is

$$2BF_M t = 2B\varphi_e U C_M t = \varphi_e R_d \int_{z=0}^{H_p} \int_{x=0}^{\infty} C_r(x, z, t)\, \mathrm{d}x\, \mathrm{d}z \qquad (3\text{-}299)$$

The left-hand side of Eq. (3-299) gives the mass of solute entering into the medium through the source between 0 and t. The thickness of the source perpendicular to the x–z plane is considered to

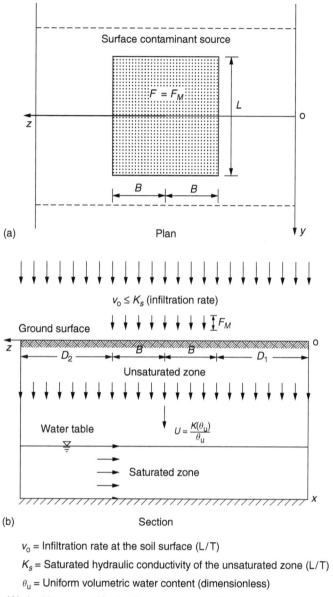

FIGURE 3-59 Schematic representation for the applicability of the ST2B program to unsaturated flow zones. (Adapted from Batu, V., *Ground Water*, 26, 71–77, 1988.)

be of unit length. The right-hand side of Eq. (3-299) gives the total mass of the solute recovered in the system during the same time period. In other words, in order to satisfy the mass balance constraint, whatever is added to the source (term on the left) must be found inside the solute transport domain (term on the right). Taking the Laplace transform of Eq. (3-299) gives

$$2B\varphi_e U C_M \frac{1}{s^2} = \varphi_e R_d \int_{z=0}^{H_p} \int_{x=0}^{\infty} u_r(x, z, s) \, dx \, dz \qquad (3\text{-}300)$$

It is much easier to work with Eq. (3-300) than Eq. (3-299). By substituting Eq. (3-56) with Eqs. (3-260) and (3-261) into Eq. (3-300) by using Eqs. (3-262) and (3-263) with Eq. (3-236), it can be shown that the right-hand side of Eq. (3-300) results in exactly the same form as its left-hand side. This means that the mass balance requirement is satisfied. This is shown in Example 3-20.

EXAMPLE 3-20

Show that substitution of $u_r(x, z, s)$, which is the sum of Eqs. (3-260) and (3-261), into Eq. (3-300) by using Eqs. (3-262) and (3-263) with Eq. (3-236), results in exactly the same form as the left-hand side of Eq. (3-300).

SOLUTION

The substitution of Eq. (3-236) into Eqs. (3-262) and (3-263), respectively, gives

$$I_0 = 2BUC_M \tag{E3-20-1}$$

and

$$I_n = UC_M \frac{1}{\lambda_n} \left\{ \sin[\lambda_n(D_1 + 2B)] - \sin(\lambda_n D_1) \right\} \tag{E3-20-2}$$

where λ_n is given by Eq. (3-61). With $\gamma = 0$, $v = 0$, and $H_p = D_1 + 2B + D_2$, and using Eqs. (E3-20-1) and (E3-20-2), Eqs. (3-260) and (3-261), respectively, take the following forms:

$$u_{r_1} = \frac{2BUC_M}{(D_1 + 2B + D_2)} \frac{\exp\left(\dfrac{Ux}{2D_x} - p_0 x\right)}{\left(\dfrac{U}{2} + D_x p_0\right)} \frac{1}{s} \tag{E3-20-3}$$

and

$$u_{r_2}(x, z, s) = \frac{2}{\pi} UC_M \sum_{n=1}^{\infty} \frac{1}{n} \left\{ \sin[\lambda_n(D_1 + 2B)] - \sin(\lambda_n D_1) \right\}$$

$$\times \cos(\lambda_n z) \frac{\exp\left(\dfrac{Ux}{2D_x} - p_n x\right)}{\left(\dfrac{U}{2} + D_x p_n\right)} \frac{1}{s} \tag{E3-20-4}$$

From Eqs. (3-59) and (3-60) with $v = 0$, p_0 and p_n, respectively, take the following forms

$$p_0 = \left(\frac{R_d}{D_x} s + \frac{U^2}{4D_x^2} \right)^{1/2} \tag{E3-20-5}$$

and

$$p_n = \left(\frac{D_z}{D_x} \lambda_n^2 + \frac{R_d}{D_x} s + \frac{U^2}{4D_x^2} \right)^{1/2} \tag{E3-20-6}$$

Also, from Eq. (3-56),

$$u_r(x, z, s) = u_{r_1}(x, z, s) + u_{r_2}(x, z, s) \tag{E3-20-7}$$

The substitution of Eq. (E3-20-7) into Eq. (3-300) gives

$$2B\varphi_e UC_M \frac{1}{s^2} = I_1 + I_2 \tag{E3-20-8}$$

where

$$I_1 = \varphi_e R_d \frac{2BUC_M}{(D_1 + 2B + D_2)} \frac{1}{\left(\dfrac{U}{2} + D_x p_0\right)} \frac{1}{s} \int_{z=0}^{D_1+2B+D_2} \int_{x=0}^{\infty} \exp\left(\frac{Ux}{2D_x} - p_0 x \right) dx \, dz \tag{E3-20-9}$$

and

$$I_2 = \frac{2}{\pi} U C_M \varphi_e R_d \frac{1}{s} \int_{z=0}^{D_1+2B+D_2} \int_{x=0}^{\infty} \sum_{n=1}^{\infty} \frac{1}{n} \left\{ \sin[\lambda_n(D_1 + 2B)] - \sin(\lambda_n D_1) \right\}$$

$$\times \cos(\lambda_n z) \frac{\exp\left(\dfrac{Ux}{2D_x} - p_n x \right)}{\left(\dfrac{U}{2} + D_x p_n \right)} \, dx \, dz \qquad (E3\text{-}20\text{-}10)$$

By performing the double integration of Eqs. (E3-20-9) and (E3-20-10) one can obtain

$$I_1 = 2BU\varphi_e C_M \frac{1}{s^2} \qquad (E3\text{-}20\text{-}11)$$

and

$$I_2 = 0 \qquad (E3\text{-}20\text{-}12)$$

The substitution of Eqs. (E3-20-1) and (E3-20-2) into Eq. (E3-20-8) results in exactly the same form of the left-hand side of Eq. (E3-20-8). This means that the mass balance constraint as expressed by Eq. (3-299) is satisfied.

Relationship between the First- and Third-Type Source Models
Eq (2-92) gives the flux-averaged concentration C_f in terms of volume-averaged concentration C_r. The equation can be used immediately to derive the first-type solution from the third-type. As mentioned in Section 2.9.1, alternatively, the third-type solution may be derived from the first-type by using the inverse of Eq. (2-92), which is given as Eq. (2-93) and its two-dimensional form is

$$C_r(x, z, t) = \frac{U}{D_x} \exp\left(\frac{Ux}{D_x} \right) \int_x^{\infty} \exp\left(-\frac{U\xi}{D_x} \right) C_f(\xi, z, t) \, d\xi \qquad (3\text{-}301)$$

The first-type solution for $C_f(x, z, t)$ for a strip source is given by Eq. (3-77) and substituting it into Eq. (3-301) gives the solution for $C_r(x, z, t)$ for the same solute transport problem. However, this procedure may not be efficient due to the fact that complicated integrals may potentially be found in the $C_r(x, z, t)$ expression. Instead, it is relatively easier to work in the Laplace domain form of Eq. (3-301). For example, Batu and van Genuchten (1990) determined their Laplace domain third-type source solution (Batu and van Genuchten, 1990, Eq. [16]) for a strip source in a semiinfinite medium, with its exact form, from the corresponding Laplace domain first-type source solution of Batu (1983, Eq. [33]). In this section, the same procedure will be used to show that $C_r(x, z, t)$ for a strip source in a finite medium can be determined from the corresponding $C_f(x, z, t)$ solution. This is shown in Example 3-21.

EXAMPLE 3-21

Determine the third-type source solution for volume-averaged concentration distribution (Eq. [3-271]) from the first-type source solution for flux-averaged concentration distribution (Eq. [3-77]).

SOLUTION

Since it is relatively easier, the third-type solution will be obtained in the Laplace domain. Taking the Laplace transform of Eq. (3-301) gives

$$u_r(x, z, s) = \frac{U}{D_x} \exp\left(\frac{Ux}{D_x} \right) \int_x^{\infty} \exp\left(-\frac{U\xi}{D_x} \right) u_f(\xi, z, s) \, d\xi \qquad (E3\text{-}21\text{-}1)$$

With $x = \xi$, from Eq. (E3-4-30), $u_f(x, z, s)$ takes the form

$$u_f(\xi, z, s) = \frac{I_0}{H_p(s-\gamma)} \exp\left(\frac{U\xi}{2D_x} - p_0\xi \right) + \frac{2}{(s-\gamma)} \sum_{n=1}^{\infty} \exp\left(\frac{U\xi}{2D_x} - p_n\xi \right) \cos(\lambda_n z) \qquad (E3\text{-}21\text{-}2)$$

The substitution of Eq. (E3-21-2) into Eq. (E3-21-1) gives

$$u_r(x, z, s) = \frac{U}{D_x}\exp\left(\frac{Ux}{D_x}\right)\frac{I_0}{H_p(s - \gamma)}I_1 + 2\frac{U}{D_x}\exp\left(\frac{Ux}{D_x}\right)\frac{1}{(s - \gamma)}\sum_{n=1}^{\infty}I_n\cos(\lambda_n z)I_{2n} \qquad \text{(E3-21-3)}$$

where the expressions for the integrals I_1 and I_{2n} and their evaluated forms, respectively, are

$$I_1 = \int_x^{\infty}\exp\left(-\frac{U\xi}{2D_x} - p_0\xi\right)d\xi = \frac{\exp\left(-\dfrac{Ux}{2D_x} - p_0x\right)}{\left(\dfrac{U}{2D_x} + p_0\right)} \qquad \text{(E3-21-4)}$$

and

$$I_{2n} = \int_x^{\infty}\exp\left(-\frac{U\xi}{2D_x} - p_n\xi\right)d\xi = \frac{\exp\left(-\dfrac{Ux}{2D_x} - p_nx\right)}{\left(\dfrac{U}{2D_x} + p_n\right)} \qquad \text{(E3-21-5)}$$

The substitution of Eqs. (E3-21-4) and (E3-21-5) into Eq. (E3-21-3) gives

$$u_r(x, z, s) = \frac{U}{D_x}\exp\left(\frac{Ux}{D_x}\right)\frac{I_0}{H_p(s - \gamma)}\frac{\exp\left(-\dfrac{Ux}{2D_x} - p_0x\right)}{\left(\dfrac{U}{2D_x} + p_0\right)}$$

$$+ 2\frac{U}{D_x}\exp\left(\frac{Ux}{D_x}\right)\frac{1}{(s - \gamma)}\sum_{n=1}^{\infty}I_n\cos(\lambda_n z)\frac{\exp\left(-\dfrac{Ux}{2D_x} - p_nx\right)}{\left(\dfrac{U}{2D_x} + p_n\right)} \qquad \text{(E3-21-6)}$$

For the first-type source case I_0 and I_n are given by Eqs. (3-65) and (3-70), respectively, and for a strip source case the corresponding expressions for I_0 and I_n are given by Eqs. (3-75) and (3-76), respectively. The substitution of Eqs. (3-75) and (3-76) into Eq. (E3-21-6) finally gives the Laplace domain solution for the third-type strip source:

$$u_r(x, z, s) = \frac{2BUC_M}{H_p(s - \gamma)}\frac{\exp\left(\dfrac{Ux}{2D_x} - p_0x\right)}{\left(\dfrac{U}{2} + D_xp_0\right)} + \frac{2UC_M}{\pi}\frac{1}{(s - \gamma)}\sum_{n=1}^{\infty}\frac{1}{n}$$

$$\times\{\sin[\lambda_n(D_1 + 2B)] - \sin(\lambda_nD_1)\}\cos(\lambda_nz)\frac{\exp\left(\dfrac{Ux}{2D_x} - p_nx\right)}{\left(\dfrac{U}{2} + D_xp_n\right)} \qquad \text{(E3-21-7)}$$

It can be easily verified that combination of Eqs. (3-56), (3-260), (3-261), (3-269), and (3-270) gives exactly the same form of Eq. (E3-21-7) with $F_M = \varphi_e UC_M$ as given by Eq. (3-236). Taking the inverse Laplace transform of Eq. (E3-21-7) results in exactly the same form of Eq. (3-271) with $F_M = \varphi_e UC_M$.

Convective–Dispersive Flux Components Comparison at the Source for the First- and Third-Type Sources
As shown in Example 3-2, the solute flux at the source ($x = 0$) varies with time depending on the value of longitudinal dispersivity, α_L (see Figure 3-8). Batu and van Genuchten (1990) present results for convective–dispersive flux component in the x-coordinate direction (F_x) at the source ($x = 0$) using two-dimensional volume-averaged analytical solute transport model for a strip source in a semiinfinite medium (Batu and van Genuchten, 1990, Eq. [30], p. 342) with an example. The only difference between this model and its corresponding form, as given above by Eq. (3-271), is that Eq. (3-271) is valid in a finite domain (see Figure 3-49). By assigning large values to D_1 and D_2 the results of these two models virtually become the same.

In Figure 3-60 the convective–dispersive flux component for the first-type source condition along the z-axis, $F_x/(\varphi_e C_M)$, is presented at $x = 0$ based on the model of Batu and van Genuchten

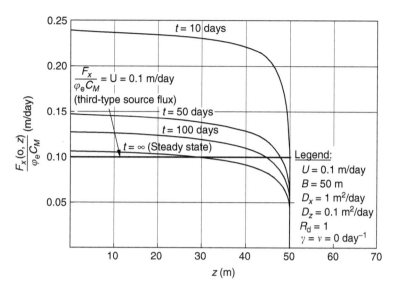

FIGURE 3-60 Temporal and spatial variations of $F_x(0,z)/(\varphi_e C_M)$ at $x = 0$ for the first-type source condition. (Adapted from Batu, V. and van Genuchten, M.T., *Water Resour. Res.*, 26, 339–350, 1990.)

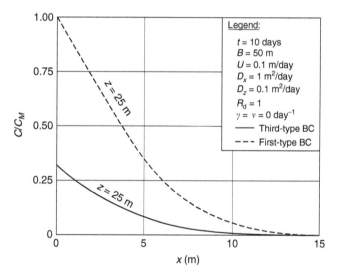

FIGURE 3-61 Comparisons of the normalized concentrations for the first- and third-type conditions for $t = 10$ days. (Adapted from Batu, V. and van Genuchten, M.T., *Water Resour. Res.*, 26, 339–350, 1990.)

(1990). The parameters used are $B = 50$ m, $U = 0.1$ m/day, $D_x = 1$ m^2/day, $D_z = 0.1$ m^2/day, $R_d = 1$, $\gamma = 0$, and $v = 0$, and are shown in Figure 3-60. Figure 3-60 clearly shows that the flux decreases with the increase in time and becomes constant when the system reaches steady-state. This is in contrast to similar results for the third-type source condition whose flux component at $x = 0$ remains constant at $F_x/(\varphi_e C_M) = U$ (= 0.1 m/day) according to Eq. (3-236).

Concentration Comparisons for the First- and Third-Type Sources
The concentration comparisons for the first- and third-type sources are presented for the numerical example described above. In Figure 3-61–3-63, the concentration distributions for the first- and third-type source conditions for $t = 10$, 100, and ∞ (steady-state) days are compared. The concentration

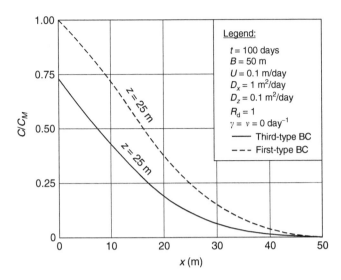

FIGURE 3-62 Comparisons of the normalized concentrations for the first- and third-type conditions for $t = 100$ days. (Adapted from Batu, V. and van Genuchten, M.T., *Water Resour. Res.*, 26, 339–350, 1990.)

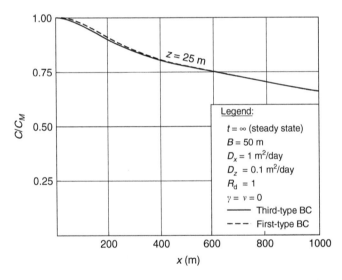

FIGURE 3-63 Comparisons of the normalized concentrations for the first- and third-type conditions for $t = \infty$ (steady-state). (Adapted from Batu, V. and van Genuchten, M.T., *Water Resour. Res.*, 26, 339–350, 1990.)

profiles correspond to $z = 25$ m and all other parameters are included above. The first-type source results were generated using the analytical two-dimensional model of Batu (1983) and the third-type source results were generated using the two-dimensional analytical model of Batu and van Genuchten (1990), which is described above. Figure 3-61–3-63 clearly show that the first-type source condition predicts normalized concentrations with significant discrepancies with the third-type source condition during unsteady-state solute transport conditions, especially near the input boundary. The principal reason for this situation is that the first-type source condition corresponds to the situation that the solute flux at the source decreases with time (see Figure 3-60), whereas the solute flux for the third-type source condition case is constant. For large times and steady-state transport conditions the results become approximately the same. For example, the first-type source condition in this example predicts more than three times higher concentrations at $t = 10$ days than does the third-type source condition.

3.3.3.3 Three-Dimensional Analytical Solute Transport Models for the Third-Type Rectangular Sources

As mentioned in Section 3.3.2.3, during the past two decades a number of three-dimensional analytical solute transport models of the transport and fate of contaminants in subsurface flow systems have been developed. However, the number of three-dimensional analytical solute transport models is fairly limited (Hunt, 1978; Yeh, 1981; Leij et al., 1991). These analytical models are based on infinite or semiinfinite medium assumption with the exception of Yeh (1981), which is based on partially penetrating finite sources by taking into account the thickness of aquifer. In the Yeh (1981) model, the finite source is adjacent to the top boundary of aquifer and its location along the aquifer thickness cannot be changed. The model presented in Batu (1997a) is fairly general three-dimensional analytical solute transport model from multiple time- and space-dependent rectangular solute sources in a unidirectional bounded parallelepiped-type flow field with the third-type source condition. Since this model is more general than the other ones mentioned above, in this section, only Batu (1997a) model will be presented with its potential application areas.

Problem Statement and Assumptions

The geometry of the generalized three-dimensional third-type sources analytical solute transport model is illustrated in Figure 3-64. All solute sources are located on the yOz plane and the four other planes that are perpendicular to the yOz plane are assumed to be impervious. All five planes defined above form a parallelepiped that goes to infinity in the x-coordinate direction. The multiple rectangular sources are located irregularly along the yOz plane as shown in Figure 3-64 and have space- and time-dependent source fluxes. The uniform flow field is assumed to exist along the x-coordinate direction and the system goes to infinity along the same coordinate direction. The upper boundary of the model (xOz plane) from a practical standpoint corresponds to the water table

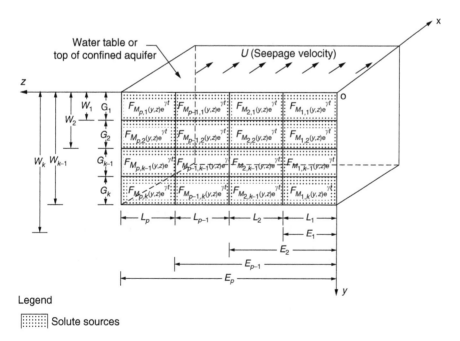

FIGURE 3-64 Schematic representation of the three-dimensional solute transport model with the third-type source inputs in a bounded medium having unidirectional flow field. (Adapted from Batu, V., unpublished paper, 1997a.)

of an unconfined aquifer or the upper boundary of a confined aquifer. The lower boundary, which is parallel to the xOz plane at $y = W_k$ distance (see Figure 3-64), may represent a bedrock or an impervious boundary, or the lower boundary of an unconfined or confined aquifer.

Under the framework of the above-mentioned description, the problem is unsteady-state three-dimensional solute transport from multiple rectangular sources between four impervious planar boundaries in a unidirectional flow field as shown in Figure 3-64. The x- and z-axis are the planar coordinates and the y-axis is the vertical coordinate. The assumptions are as follows:

1. The solute sources are planar rectangles and located irregularly on the y–z plane in a uni-directional steady ground water velocity (U) field.
2. The solute sources are located at $x = 0$ plane and they are perpendicular to the velocity of the flow field.
3. The source fluxes are functions of the y and z coordinates and time (t) through an expo-nential function and
4. The medium goes to infinity in the x-coordinate direction and y varies between 0 and W_k and z varies between 0 and E_p.

Governing Differential Equation
The governing differential equation is given in Eq. (3-128).

Initial and Boundary Conditions
The initial condition for $C(x, z, t)$ is given by Eq. (3-129). The values of fluxes at the sources are specified. Therefore, these are the third- or flux-type boundary conditions. The fluxes in the first row of rectangles on the yOz plane are given as follows:

$$F(0, y, z, t) = F_{M_{1,1}}(y, z)e^{\gamma t}, \quad 0 < y < W_1, \quad 0 < z < E_1 \tag{3-302}$$

$$F(0, y, z, t) = F_{M_{p-1,1}}(y, z)e^{\gamma t}, \quad 0 < y < W_1, \quad E_{p-2} < z < E_{p-1} \tag{3-303}$$

$$F(0, y, z, t) = F_{M_{p,1}}(y, z)e^{\gamma t}, \quad 0 < y < W_1, \quad E_{p-1} < z < E_p \tag{3-304}$$

Similarly, expressions for 2nd,...,$(k-1)$, kth row can be written. The expressions for the kth row are as follows:

$$F(0, y, z, t) = F_{M_{1,k}}(y, z)e^{\gamma t}, \quad W_{k-1} < y < W_k \quad 0 < z < E_1 \tag{3-305}$$

$$F(0, y, z, t) = F_{M_{p-1,k}}(y, z)e^{\gamma t}, \quad W_{k-1} < y < W_k \quad E_{p-2} < z < E_{p-1} \tag{3-306}$$

$$F(0, y, z, t) = F_{M_{p,k}}(y, z)e^{\gamma t}, \quad W_{k-1} < y < W_k \quad E_{p-1} < z < E_p \tag{3-307}$$

where $F_x(0, y, z, t)$ (M/L²/T) is the solute input flux at $x = 0$ (mass per unit area per unit time), γ(T⁻¹) is a parameter, $F_{M_{i,j}}$ (M/L²/T) is the solute input flux when $\gamma = 0$ at $x = 0$, and e = 2.7182818 ..., and E_i and W_i are given by Eqs. (3-136) and (3-137), respectively. The other boundary conditions at the impervious boundaries and infinity are given by Eqs. (3-138)–(3-143). Equations (3-302)–(3-307) state that the flux $F_x(0, y, z, t)$, which is time-dependent, has specified values at $x = 0$ and these values vary with the y and z coordinates. Therefore, they are the third-type or flux-type source conditions. These type of conditions are also called Cauchy condition (see Section 2.10.3.2). The physical meaning of the rest of boundary conditions is given in Section 3.3.2.3.

The function

$$\frac{F_x(0, y, z, t)}{F_M(y, z)} = e^{\gamma t} \tag{3-308}$$

performs such that the source flux can be represented as exponentially varying time-dependent quantity. Equation (3-308) is a generalized version of Eqs. (3-228) and (3-259) and the properties of this function are described in Section 3.3.3.1. Figure 3-2 shows the variation of $F(0, y, z, t)/F_M$ with time for different values of γ ranging from -0.0001 to -0.0100 day^{-1}. If observed data are available for time-dependent source concentration, the value of γ may be determined from the curve-fitting techniques. The physical meaning of the rest of boundary conditions is given in Section 3.3.2.3.

Unsteady-State Solution for Concentration Distribution
The boundary-value problem, defined by Eqs. (3-128) and (3-129), Equations (3-302)–(3-307) and Eqs. (3-138)–(3-143), is solved by the Laplace transform and Fourier analysis techniques. For this purpose, the solution is first obtained in the Laplace domain and then the concentration variation is determined by taking the inverse Laplace transform of the Laplace domain solution. It can be observed that the aforementioned equations are the same as the first-type three-dimensional analytical model, which is presented in Section 3.3.2.3, with the exception of the conditions at $x = 0$. Therefore, there are common equations in the solution process of the three-dimensional analytical solutions. As a result, the solution of the third-type three-dimensional analytical model will be presented by using the common equations as appropriate.

General Solution in the Laplace Domain
As can be seen from Section 3.3.2.3, the Laplace domain solution for the first-type source case is derived by using the initial condition and all boundary conditions with the exception of the ones at $x = 0$. Therefore, Eq. (3-145) is valid for the third-type source conditions as well, and derivation of Eq. (3-145) is presented in Example 3-10. The final form of the general solution in the Laplace domain is in the form of Eq. (3-150), using the boundary conditions at $x = 0$, and the expressions for $u_1(x, y, z, s)$, $u_2(x, y, z, s)$, $u_3(x, y, z, s)$, and $u_4(x, y, z, s)$ are as follows:

$$u_1(x, y, z, s) = \left(\frac{1}{2}\right)\left(\frac{1}{2}\right)B_{00}\exp\left[\left(\frac{U}{2D_x} - K_{00}\right)x\right] \tag{3-309}$$

$$u_2(x, y, z, s) = \frac{1}{2}\sum_{n=1}^{\infty} B_{0n}\cos(\lambda_n z)\exp\left[\left(\frac{U}{2D_x} - K_{0n}\right)x\right] \tag{3-310}$$

$$u_3(x, y, z, s) = \frac{1}{2}\sum_{m=1}^{\infty} B_{m0}\cos(\eta_m y)\exp\left[\left(\frac{U}{2D_x} - K_{m0}\right)x\right] \tag{3-311}$$

and

$$u_4(x, y, z, s) = \sum_{m=1}^{\infty}\sum_{n=1}^{\infty} B_{mn}\cos(\eta_m y)\cos(\lambda_n z)\exp\left[\left(\frac{U}{2D_x} - K_{mn}\right)x\right] \tag{3-312}$$

where K_{mn}, K_{00}, K_{0n}, and K_{m0} are given by Eqs. (3-148), (3-155), (3-156), and (3-157), respectively. Equation (3-150) and Eqs. (3-309)–(3-312) are based on the fact that in Eq. (3-145), B_{0n} and B_{m0} are one-half and B_{00} is one-quarter (Carslaw and Jaeger, 1959, p. 182). The expression for B_{mn}, which is derived in Example 3-22, is as follows:

$$B_{mn} = \frac{1}{\left(\dfrac{U}{2} + D_x K_{mn}\right)}\frac{4}{E_p W_k}\int_0^{W_k}\int_0^{E_p}\frac{F_M(y, z)}{\varphi_e(s-\gamma)}\cos(\eta_m y)\cos(\lambda_n z)\, dy\, dz \tag{3-313}$$

From Eq. (3-313), B_{00}, B_{0n}, and B_{m0}, can, respectively, be expressed as

$$B_{00} = \frac{1}{\left(\dfrac{U}{2} + D_x K_{00}\right)}\frac{4}{E_p W_k}\int_0^{W_k}\int_0^{E_p}\frac{F_M(y, z)}{\varphi_e(s-\gamma)}\, dy\, dz \tag{3-314}$$

$$B_{0n} = \frac{1}{\left(\dfrac{U}{2} + D_x K_{0n}\right)}\frac{4}{E_p W_k}\int_0^{W_k}\int_0^{E_p}\frac{F_M(y, z)}{\varphi_e(s-\gamma)}\cos(\lambda_n z)\, dy\, dz \tag{3-315}$$

and

$$B_{m0} = \frac{1}{\left(\dfrac{U}{2} + D_x K_{m0}\right)} \frac{4}{E_p W_k} \int_0^{W_k} \int_0^{E_p} \frac{F_M(y, z)}{\varphi_e(s-\gamma)} \cos(\eta_m y)\, dy\, dz \qquad (3\text{-}316)$$

Now, the general solution in the Laplace domain is complete. Determination procedure of B_{mn} is given in Example 3-22.

EXAMPLE 3-22

Determine the expression for B_{mn} in Eq. (3-145) by using the boundary conditions at $x = 0$ given by Eqs. (3-302)–(3-307).

SOLUTION

In order to determine the expression for B_{mn} the boundary conditions at $x = 0$ given by Eqs. (3-302)–(3-307) need to be used. The procedure is similar to that used in Example 3-16. The boundary conditions at $x = 0$ need to be expressed in the Laplace domain. The convective–dispersive flux in the x direction (F_x) is given by Eq. (3-90a). Therefore, taking the Laplace transform of the combined form of Eqs. (3-90a), (3-302)–(3-307) gives

$$\frac{F_M(y, z)}{\varphi_e(s-\gamma)} = Uu - D_x \frac{\partial u}{\partial x} \qquad (E3\text{-}22\text{-}1)$$

From Eq. (3-145) one can write

$$\frac{\partial u}{\partial x} = \sum_{m=0}^{\infty}\sum_{m=0}^{\infty} B_{mn}\left(\frac{U}{2D_x} - K_{mn}\right)\cos(\eta_m y)\cos(\lambda_n z)\exp\left[\left(\frac{U}{2D_x} - K_{mn}\right)x\right] \qquad (E3\text{-}22\text{-}2)$$

For $x = 0$, Eqs. (3-145) gives

$$u(0, y, z, s) = \sum_{m=0}^{\infty}\sum_{n=0}^{\infty} B_{mn}\cos(\eta_m y)\cos(\lambda_n z) \qquad (E3\text{-}22\text{-}3)$$

For $x = 0$, Eq. (E3-22-2) gives

$$\frac{\partial u(0, y, z, s)}{\partial x} = \sum_{m=0}^{\infty}\sum_{n=0}^{\infty} B_{mn}\left(\frac{U}{2D_x} - K_{mn}\right)\cos(\eta_m y)\cos(\lambda_n z) \qquad (E3\text{-}22\text{-}4)$$

By substituting Eqs. (E3-22-3) and (E3-22-4) into Eq. (E3-22-1) and after some manipulation, one can write

$$\frac{F_M(y, z)}{\varphi_e(s-\gamma)} = \sum_{m=0}^{\infty}\sum_{n=0}^{\infty} E_{mn}\cos(\eta_m y)\cos(\lambda_n z) \qquad (E3\text{-}22\text{-}5)$$

where

$$E_{mn} = B_{mn}\left(\frac{U}{2} + D_x K_{mn}\right) \qquad (E3\text{-}22\text{-}6)$$

Equation (E3-22-5) is in the form of Eq. (E3-10-27). Therefore, the determination procedure for E_{mn} presented through Eqs. (E3-10-28) through (E3-10-32) can be used. As a result, the expression for E_{mn} can be written as

$$E_{mn} = \frac{4}{W_k E_p} \int_0^{W_k} \int_0^{E_p} \frac{F_M(y, z)}{\varphi_e(s-\gamma)} \cos(\eta_m y)\cos(\lambda_n z)\, dy\, dz \qquad (E3\text{-}22\text{-}7)$$

Combining Eqs. (E3-22-6) and (E3-22-7) gives Eq. (3-313).

Inverse Laplace Transforms of Eqs. (3-309)–(3-312)
The sum of the inverse Laplace transforms of Eqs. (3-309)–(3-312) according to Eq. (3-150) is

$$C(x, y, z, t) = C_1(x, y, z, t) + C_2(x, y, z, t) + C_3(x, y, z, t) + C_4(x, y, z, t) \qquad (3\text{-}317)$$

Upon inspection of Eqs. (3-309)–(3-312) it can be concluded that their inverse Laplace transforms are similar to that of Eq. (3-260), and details of its inverse Laplace transform are given in Example 3-17. By following the same procedure, the inverse Laplace transform of $u_1(x, y, z, s)$, as given by Eq. (3-309), can be obtained as

$$C_1(x, y, z, t) = \frac{I_0}{D_x} \int_0^t F_1(\xi)[F_2(\xi) - F_3(\xi)F_4(\xi)]\exp[\gamma(t - \xi)]\, d\xi \qquad (3\text{-}318)$$

where $F_1(\xi)$, $F_2(\xi)$, $F_3(\xi)$, and $F_4(\xi)$ are given by Eqs. (3-231)–(3-234), respectively, and

$$I_0 = \frac{1}{W_k E_p} \frac{1}{\varphi_e} \int_0^{W_k} \int_0^{E_p} F_M(y, z)\, dy\, dz \qquad (3\text{-}319)$$

Similarly, the inverse Laplace transform of $u_2(x, y, z, s)$, as given by Eq. (3-310), can be written as

$$C_2(x, y, z, t) = \frac{2}{D_x} \sum_{n=1}^{\infty} I_n \cos(\lambda_n z) \qquad (3\text{-}320)$$

$$\times \int_0^t F_1(\xi)\exp\left(-\frac{D_z}{R_d}\lambda_n^2\xi\right)[F_2(\xi) - F_3(\xi)F_4(\xi)]\exp[\gamma(t - \xi)]\, d\xi$$

where $F_1(\xi)$, $F_2(\xi)$, $F_3(\xi)$, and $F_4(\xi)$ are given by Eqs. (3-231)–(3-234), respectively, and

$$I_n = \frac{1}{W_k E_p} \frac{1}{\varphi_e} \int_0^{W_k} \int_0^{E_p} F_M(y, z)\cos(\lambda_n z)\, dy\, dz \qquad (3\text{-}321)$$

Similarly, the inverse Laplace transform of $u_3(x, y, z, s)$, as given by Eq. (3-311), can be written as

$$C_3(x, y, z, t) = \frac{2}{D_x} \sum_{m=1}^{\infty} I_m \cos(\eta_m y)$$

$$\times \int_0^t F_1(\xi)\exp\left(-\frac{D_y}{R_d}\eta_m^2\xi\right)[F_2(\xi) - F_3(\xi)F_4(\xi)]\exp[\gamma(t - \xi)]\, d\xi \qquad (3\text{-}322)$$

where $F_1(\xi)$, $F_2(\xi)$, $F_3(\xi)$, and $F_4(\xi)$ are given by Eqs. (3-231)–(3-234), respectively, and

$$I_m = \frac{1}{W_k E_p} \frac{1}{\varphi_e} \int_0^{W_k} \int_0^{E_p} F_M(y, z)\cos(\eta_m z)\, dy\, dz \qquad (3\text{-}323)$$

Similarly, the inverse Laplace transform of $u_4(x, y, z, s)$, as given by Eq. (3-312), can be written as

$$C_4(x, y, z, t) = \frac{4}{D_x} \sum_{m=1}^{\infty} \sum_{n=1}^{\infty} I_{mn} \cos(\eta_m y)\cos(\lambda_n z)$$

$$\times \int_0^t F_1(\xi)\exp\left(-\frac{D_y}{R_d}\eta_m^2\xi - \frac{D_z}{R_d}\lambda_n^2\xi\right)[F_2(\xi) - F_3(\xi)F_4(\xi)]\exp[\gamma(t - \xi)]\, d\xi \qquad (3\text{-}324)$$

where $F_1(\xi)$, $F_2(\xi)$, $F_3(\xi)$, and $F_4(\xi)$ are given by Eqs. (3-231) (3-234), respectively, and

$$I_{mn} = \frac{1}{W_k E_p} \frac{1}{\varphi_e} \int_0^{W_k} \int_0^{E_p} F_M(y, z)\cos(\eta_m z)\cos(\lambda_n z)\, dy\, dz \qquad (3\text{-}325)$$

With Eqs. (3-317)–(3-325), the general solution for $C(x, y, z, t)$ is now complete.

Unsteady-State-Special Solution for a Single Rectangular Source
Concentration Distribution
The geometry of a single rectangular source is shown in Figure 3-65, which is a special form of Figure 3-64 with $\gamma = 0$. The rectangular source is located at any place along the yOz plane and its width and height are $2B$ and $2H$, respectively. The distances from the right and left edges of the source to the vertical impervious boundaries are D_1 and D_2, respectively. The source conditions expressed by Eqs. (3-302)–(3-307) take the forms

$$F_x(0, y, z, t) = \begin{cases} F_M e^{\gamma t} & D_1 < z < D_1 + 2B, \quad H_1 < y < H_1 + 2H \quad \text{(3-326)} \\ 0 & \text{otherwise} \quad\quad\quad\quad\quad\quad\quad\quad\quad\quad\text{(3-327)} \end{cases}$$

where F_M ($\text{M L}^{-2}\text{T}^{-1}$) is solute source flux when $t = 0$. The substitution of Eqs. (3-326) and (3-327) into Eq. (3-319) gives

$$I_0 = \frac{4BH}{(H_1 + 2H + H_2)(D_1 + 2B + D_2)} \frac{F_M}{\varphi_e} \quad\quad (3\text{-}328)$$

Similarly, the expression for I_{mn} can be obtained. The substitution of Eqs. (3-326) and (3-327) into Eq. (3-325) gives

$$I_{mn} = \frac{F_M}{\varphi_e} \frac{1}{\pi^2} \frac{S_n}{n} \frac{S_m}{m} \qu\quad\quad (3\text{-}329)$$

where S_n and S_m are given by Eqs. (3-185) and (3-186), respectively.
    The substitution of Eqs. (3-326) and (3-327) into Eq. (3-321) gives the expression for I_n as

$$I_n = \frac{F_M}{\varphi_e} \frac{1}{\pi} \frac{2H}{(H_1 + 2H + H_2)} \frac{S_n}{n} \qu\quad\quad (3\text{-}330)$$

where S_n is given by Eq. (3-185).

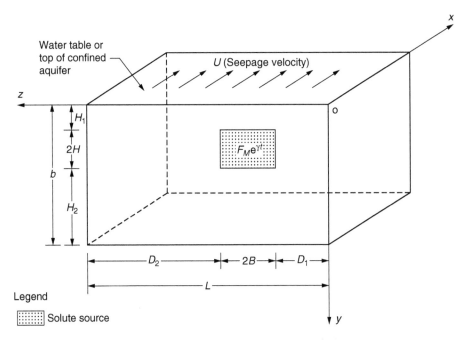

FIGURE 3-65 The geometry of a single rectangular source with uniform solute source flux. (Adapted from Batu, V., unpublished paper, 1997a.)

Finally, the expression for I_m can be similarly obtained by substituting of Eqs. (3-326) and (3-327) into Eq. (3-323) as

$$I_m = \frac{F_M}{\varphi_e} \frac{2B}{(D_1 + 2B + D_2)} \frac{S_m}{m} \tag{3-331}$$

where S_m is given by Eq. (3-186).

Therefore, with Eqs. (3-328)–(3-331), Eqs. (3-317), (3-318), (3-320), (3-322), and (3-324) form the solution for concentration distribution for a single rectangular source shown in Figure 3-65. In the above-mentioned expressions, λ_n and η_m are given by Eqs. (3-146) and (3-147), respectively, with $E_p = D_1 + 2B + D_2$ and $W_k = H_1 + 2H + H_2$.

Special Solution for Two-Dimensional Cases

Two special solutions corresponding to the two-dimensional special cases can be obtained from the three-dimensional third-type source solution for a rectangular source whose geometry is shown in Figure 3-65. The procedure is similar to the two-dimensional special solutions for the three-dimensional first-type source solution presented in Section 3.3.2.3. The first two-dimensional special case is shown in Figure 3-32, which shows that the source fully penetrates the entire thickness of the aquifer in the y-coordinate direction. In other words, the source becomes a vertical strip. This corresponds to Figure 3-49, which is the geometry of the ST2B model. The second two-dimensional special case is shown in Figure 3-33, which shows that the source fully penetrates the entire length of the aquifer in the z-coordinate direction. The special solutions corresponding to these cases are derived in Example 3-23, which is presented below.

EXAMPLE 3-23

Derive the two-dimensional unsteady-state special solutions from the three-dimensional solution for a rectangular source, which is the combination of Eqs. (3-317), (3-318), (3-320), (3-322), (3-324), and (3-328)–(3-331).

SOLUTION

First Special Case: The geometry of the first special case is shown in Figure 3-32 where $H_1 = 0$ and $H_2 = 0$. The substitution of these values into Eq. (3-328) gives

$$I_0 = \frac{F_M}{\varphi_e} \frac{2B}{(D_1 + 2B + D_2)} \tag{E3-23-1}$$

The substitution of Eq. (E3-23-1) into Eq. (3-318) results in exactly the same form as of the first part of the right-hand side of Eq. (3-271), which is the two-dimensional third-type strip source model. Since $H_1 = 0$, Eq. (3-147) gives $\eta_m = m\pi/2H$ and the sine terms in Eq. (3-186) become zero, resulting in $S_m = 0$. As a result, $I_{mn} = 0$ and thus $C_4(x, y, z, t) = 0$ according to Eq. (3-324). Further substitution of $H_1 = 0$ and $H_2 = 0$ into Eq. (3-330) gives

$$I_n = \frac{F_M}{\varphi_e} \frac{1}{\pi} \frac{S_n}{n} \tag{E3-23-2}$$

The substitution of Eq. (E3-23-2) along with Eq. (3-185) into Eq. (3-320) results in exactly the same form as the second part of the right-hand side of Eq. (3-271), which is the two-dimensional third-type strip source model. Since $S_m = 0$, $I_m = 0$ according to Eq. (3-188). Therefore, Eq. (3-322) results $C_3(x, y, z, t) = 0$. Now the proof is complete, i.e., Eq. (3-271) is a special solution of the three-dimensional third-type rectangular source model.

Second Special Case: The geometry of the second special case corresponds to $D_1 = 0$ and $D_2 = 0$ (see Figure 3-33). This case is equivalent to the first special case with the exception that the strip source is parallel to the z-coordinate axis. By introducing these values into Eqs. (3-317),

(3-318), (3-320), (3-322), (3-324), (3-328), (3-329), (3-330), and (3-331), it can be shown that $C_2(x, y, z, t)$ and $C_4(x, y, z, t)$ become zero and the sum of $C_1(x, y, z, t)$ and $C_3(x, y, z, t)$ is exactly the same as the solution for a single-strip source as given by Eq. (3-271), with the exception that H, H_1, and H_2 define the source geometry instead of B, D_1, and D_2.

Special Solution for the One-Dimensional Case
In Section 3.3.3.2, the one-dimensional solution, as given by Eq. (3-230), is obtained as a special solution from the two-dimensional solution. As shown above, Eq. (3-271) is a special case of the three-dimensional solution. Therefore, it can be concluded that the same one-dimensional solution is a special case of the three-dimensional solution as well.

Convective–Dispersive Flux Components
From Eqs. (2-41a)–(2-41c), the convective–dispersive flux components in the x-, y-, and z-coordinate directions, respectively, can be easily determined using the same procedure as in the convective–dispersive flux components for the two-dimensional case in Section 3.3.3.2. Since the mathematical expressions are too lengthy they are not presented in this chapter.

Unsteady-State Special Solutions for Other Sources
Equation (3-317) is the general solution for the resulting concentration distribution for any given $F_M(y, z)$ source concentration distribution. With the source concentration functions (Figure 3-64), $F_M(y, z)$, the integrals I_0, I_n, I_m, and I_{mn} given by Eqs. (3-319), (3-321), (3-323), and (3-325), respectively, need to be evaluated analytically or numerically depending on the type of the function $F_M(y, z)$. One example is presented below.

Solution for Two Rectangular Sources with Uniform Solute Source Fluxes
The geometry of two rectangular sources with uniform source fluxes is shown in Figure 3-66. The boundary conditions at $x = 0$ expressed by Eqs. (3-302)–(3-307) take the

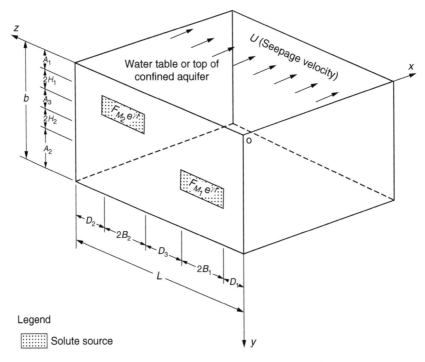

FIGURE 3-66 The geometry of two rectangular sources with uniform solute source fluxes. (Adapted from Batu, V., unpublished paper, 1997a.)

following forms:

$$F_x(0, y, z, t) = F_{M_1}e^{\gamma t},$$
$$D_1 < z < D_1 + 2B_1, \quad A_1 + 2H_1 + A_3 < y < A_1 + 2H_1 + A_3 + 2H_2 \qquad (3\text{-}332)$$

$$F_x(0, y, z, t) = F_{M_2}e^{\gamma t}$$
$$D_1 + 2B_1 + D_3 < z < D_1 + 2B_1 + D_3 + 2B_2, \quad A_1 < y < A_1 + 2H_1 \qquad (3\text{-}333)$$

where F_{M_1} and F_{M_2} are the source fluxes of the rectangular sources. As mentioned earlier, in order to determine the $C(x, y, z, t)$ concentration distribution, only I_0, I_n, I_m, and I_{mn}, given by Eqs. (3-319), (3-321), (3-323), and (3-325), respectively, need to be evaluated. With the source conditions given by Eqs. (3-332) and (3-333), Eq. (3-319) gives

$$I_0 = \frac{1}{Lb}\int_{y=A_1+2H_1+A_3}^{A_1+2H_1+A_3+2H_2} F_{M_1}\,dy\int_{z=D_1}^{D_1+2B_1}dz$$
$$+ \frac{1}{Lb}\int_{y=A_1}^{A_1+2H_1} F_{M_2}\,dy\int_{z=D_1+2B_1+D_3}^{D_1+2B_1+D_3+2B_2}dz \qquad (3\text{-}334)$$

where

$$L = D_1 + 2B_1 + D_3 + 2B_2 + D_2, \quad b = A_1 + 2H_1 + A_3 + 2H_2 + A_2 \qquad (3\text{-}335)$$

From Eq. (3-334) one gets

$$I_0 = \frac{4}{Lb}(B_1H_2F_{M_1} + B_2H_1F_{M_2}) \qquad (3\text{-}336)$$

From Eq. (3-321)

$$I_n = \frac{1}{Lb}\int_{y=A_1+2H_1+A_3}^{A_1+2H_1+A_3+2H_2} F_{M_1}\,dy\int_{z=D_1}^{D_1+2B_1}\cos(\lambda_n z)\,dz \qquad (3\text{-}337)$$
$$+ \frac{1}{Lb}\int_{y=A_1}^{A_1+2H_1} F_{M_2}\,dy\int_{z=D_1+2B_1+D_3}^{D_1+2B_1+D_3+2B_2}\cos(\lambda_n z)\,dz$$

where, from Eq. (3-146) and the first expression of Eq. (3-335),

$$\lambda_n = \frac{n\pi}{L} \qquad (3\text{-}338)$$

and on integration one gets

$$I_n = \frac{2H_2}{Lb}F_{M_1}\frac{1}{\lambda_n}\{\sin[\lambda_n(D_1 + 2B_1)] - \sin(\lambda_n D_1)\}$$
$$+ \frac{2H_1}{Lb}F_{M_2}\frac{1}{\lambda_n}\{\sin[\lambda_n(L - D_2)] - \sin[\lambda_n(D_1 + 2B_1 + D_3)]\} \qquad (3\text{-}339)$$

From Eq. (3-323)

$$I_m = \frac{1}{Lb}\int_{y=A_1+2H_1+A_3}^{A_1+2H_1+A_3+2H_2} F_{M_1}\cos(\eta_m y)\,dy\int_{z=D_1}^{D_1+2B_1}dz$$
$$+ \frac{1}{Lb}\int_{y=A_1}^{A_1+2H_1} F_{M_2}\cos(\eta_m y)\,dy\int_{z=D_1+2B_1+D_3}^{D_1+2B_1+D_3+2B_2}dz \qquad (3\text{-}340)$$

where, from Eq. (3-147), and the second expression of Eq. (3-335),

$$\eta_m = \frac{m\pi}{b} \tag{3-341}$$

and on integration

$$I_m = \frac{2B_1}{Lb} F_{M_1} \frac{1}{\eta_m} \{\sin[\eta_m(b - A_2)] - \sin[\eta_m(A_1 + 2H_1 + A_3)]\} \tag{3-342}$$

$$+ \frac{2B_2}{Lb} F_{M_2} \frac{1}{\eta_m} \{\sin[\eta_m(A_1 + 2H_1)] - \sin(\eta_m A_1)\}$$

From Eq. (3-325),

$$I_{mn} = \frac{1}{Lb} \int_{y=A_1+2H_1+A_3}^{A_1+2H_1+A_3+2H_2} F_{M_1} \cos(\eta_m y)\, dy \int_{z=D_1}^{D_1+2B_1} \cos(\lambda_n z)\, dz \tag{3-343}$$

$$+ \frac{1}{Lb} \int_{y=A_1}^{A_1+2H_1} F_{M_2} \cos(\eta_m y)\, dy \int_{z=D_1+2B_1+D_3}^{D_1+2B_1+D_3+2B_2} \cos(\lambda_n z)\, dz$$

After evaluation, Eq. (3-343) gives

$$I_{mn} = \frac{F_{M_1}}{\pi^2} \frac{1}{m} \frac{1}{n} \left\{ \sin\left[\frac{m\pi}{b}(b - A_2)\right] - \sin\left[\frac{m\pi}{b}(A_1 + 2H_1 + A_3)\right] \right\}$$

$$\times \left\{ \sin\left[\frac{n\pi}{L}(D_1 + 2B)\right] - \sin\left(\frac{n\pi D_1}{L}\right) \right\}$$

$$+ \frac{F_{M_2}}{\pi^2} \frac{1}{m} \frac{1}{n} \left\{ \sin\left[\frac{m\pi}{b}(A_1 + 2H_1)\right] - \sin\left(\frac{m\pi A_1}{b}\right) \right\} \tag{3-344}$$

$$\times \left\{ \sin\left[\frac{n\pi}{L}(L - D_2)\right] - \sin\left[\frac{n\pi}{L}(D_1 + 2B_1 + D_3)\right] \right\}$$

Therefore, with Eqs. (3-336), (3-339), (3-342), and (3-344), Eqs. (3-317), (3-318), (3-320), (3-322), and (3-324) form the solution for two rectangular sources having constant solute source fluxes.

Steady-State General Solution for Concentration Distribution
The solution for steady-state solute transport case (Figure 3-64 with $\gamma = 0$) can be obtained by using two different procedures. The first procedure is to use the corresponding differential equation under steady-state solute transport conditions with the associated boundary conditions. The second procedure is to use the solution developed for unsteady-state concentration distribution as the time (t) approaches infinity. Here, the first procedure is presented below and its details are presented in Example 3-24 (Batu, 1997a).

Under steady-state solute transport conditions the governing equation is given by Eq. (3-201). The solution for the concentration distribution C will be the only function of x, y, and z coordinates, and the initial condition given by Eq. (3-129) is not necessary. The fluxes in the first row of rectangles on the yOz plane are given as follows:

$$F(0, y, z) = F_{M_{1,1}}(y, z), \quad 0 < y < W_1, \quad 0 < z < E_1 \tag{3-345}$$

$$F(0, y, z) = F_{M_{p-1,1}}(y, z), \quad 0 < y < W_1, \quad E_{p-2} < z < E_{p-1} \tag{3-346}$$

$$F(0, y, z) = F_{M_{p,1}}(y, z), \quad 0 < y < W_1, \quad E_{p-1} < z < E_p \tag{3-347}$$

Similarly, expressions for 2nd, ..., $(k-1)$, kth row can be written. The expressions for the kth row are as follows:

$$F(0, y, z) = F_{M_{1,k}}(y, z), \quad W_{k-1} < y < W_k, \quad 0 < z < E_1 \tag{3-348}$$

$$F(0, y, z) = F_{M_{p-1,k}}(y, z), \quad W_{k-1} < y < W_k, \quad E_{p-2} < z < E_{p-1} \tag{3-349}$$

$$F(0, y, z) = F_{M_{p,k}}(y, z), \quad W_{k-1} < y < W_k, \quad E_{p-1} < z < E_p \tag{3-350}$$

where E_i and W_i are given by Eqs. (3-136) and (3-137), respectively. The other boundary conditions at the impervious boundaries and infinity as given by Eqs. (3-138)–(3-143) remain the same.

The solution of Eq. (3-201) with the boundary conditions given Eqs. (3-138)–(3-143) can easily be obtained using the solution in the Laplace domain. Equation (3-201) becomes equivalent to Eq. (E3-10-4) for $s = 0$ and replacing u by C. Therefore, the general solution is as follows:

$$C(x, y, z) = \sum_{m=0}^{\infty} \sum_{n=0}^{\infty} B_{mn} \cos(\eta_m y) \cos(\lambda_n z) \exp\left[\left(\frac{U}{2D_x} - K_{mn}\right)x\right] \tag{3-351}$$

where K_{mn}, which corresponds to $s = 0$ in Eq. (3-148), is

$$K_{mn} = \left(\frac{R_d}{D_x}v + \frac{D_z}{D_x}\lambda_n^2 + \frac{D_y}{D_x}\eta_m^2 + \frac{U^2}{4D_x^2}\right)^{1/2} \tag{3-352}$$

where λ_n and η_m are given by Eqs. (3-146) and (3-147), respectively.

By using the boundary conditions at the source, as given by Eqs. (3-345)–(3-350), the final form of the general solution is (details of the derivational process are given below in Example 3-24)

$$C(x, y, z) = C_1(x, y, z) + C_2(x, y, z) + C_3(x, y, z) + C_4(x, y, z) \tag{3-353}$$

The expression for $C_1(x, y, z)$ is

$$C_1(x, y, z) = I_0 \frac{\exp[(U/(2D_x) - K_{00})x]}{(U/2 + D_x K_{00})} \tag{3-354}$$

where I_0 is given by Eq. (3-319) and

$$K_{00} = \left(\frac{R_d}{D_x}v + \frac{U^2}{4D_x^2}\right)^{1/2} \tag{3-355}$$

The expression for $C_2(x, y, z)$ is

$$C_2(x, y, z) = 2\sum_{n=1}^{\infty} I_n \cos(\lambda_n z) \frac{\exp[(U/(2D_x) - K_{0n})x]}{(U/2 + D_x K_{0n})} \tag{3-356}$$

where

$$K_{0n} = \left(\frac{R_d}{D_x}v + \frac{D_z}{D_x}\lambda_n^2 + \frac{U^2}{4D_x^2}\right)^{1/2} \tag{3-357}$$

and I_n is given by Eq. (3-321).

The expression for $C_3(x, y, z)$ is

$$C_3(x, y, z) = 2\sum_{m=1}^{\infty} I_m \cos(\eta_m y) \frac{\exp[(U/(2D_x) - K_{m0})x]}{(U/2 + D_x K_{m0})} \tag{3-358}$$

where

$$K_{m0} = \left(\frac{R_d}{D_x}v + \frac{D_y}{D_x}\eta_m^2 + \frac{U^2}{4D_x^2}\right)^{1/2} \tag{3-359}$$

and I_m is given by Eq. (3-323).

The expression for $C_4(x, y, z)$ is

$$C_4(x, y, z) = 4\sum_{m=1}^{\infty}\sum_{n=1}^{\infty} I_{mn} \cos(\eta_m y) \cos(\lambda_n z) \frac{\exp[(U/(2D_x)) - K_{mn})x]}{(U/2 + D_x K_{mn})} \tag{3-360}$$

where I_{mn} and K_{mn} are given by Eqs.(3-325) and (3-352), respectively.

EXAMPLE 3-24

Derive the steady-state general solution as given by Eq. (3-353) from Eq. (3-351) using the source conditions given by Eqs. (3-345)–(3-350).

SOLUTION

At the sources ($x = 0$), from Eq. (3-90a) one can write

$$F_x(0, y, z) = \varphi_e U C(0, y, z) - \varphi_e D_x \frac{\partial C(0, y, z)}{\partial x} \tag{E3-24-1}$$

From Eq. (3-351)

$$\frac{\partial C}{\partial x} = \sum_{m=0}^{\infty}\sum_{n=0}^{\infty} B_{mn}\left(\frac{U}{2D_x} - K_{mn}\right) \exp[((U/(2D_x)) - K_{mn})x] \cos(\eta_m y) \cos(\lambda_n z) \tag{E3-24-2}$$

For $x=0$, Eq. (3-351) gives

$$C(0, y, z) = \sum_{m=0}^{\infty}\sum_{n=0}^{\infty} B_{mn} \cos(\eta_m y) \cos(\lambda_n z) \tag{E3-24-3}$$

For $x=0$, Eq. (E3-24-2) gives

$$\frac{\partial C(0, y, z)}{\partial x} = \sum_{m=0}^{\infty}\sum_{n=0}^{\infty} B_{mn}\left(\frac{U}{2D_x} - K_{mn}\right) \cos(\eta_m y) \cos(\lambda_n z) \tag{E3-24-4}$$

The substitution of Eqs. (E3-24-3) and (E3-24-4) into Eq. (E3-24-1) gives

$$\frac{F_x(0, y, z)}{\varphi_e} = \sum_{m=0}^{\infty}\sum_{n=0}^{\infty} E_{mn} \cos(\eta_m y) \cos(\lambda_n z) \tag{E3-24-5}$$

where

$$E_{mn} = B_{mn}(U/2 + D_x K_{mn}) \tag{E3-24-6}$$

Equation (E3-24-5) is in the form of Eq. (E3-10-27). Therefore, the determination procedure for E_{mn} presented by Eqs. (E3-10-28)–(E3-10-32) can be used. As a result, the expression for E_{mn} can be written as

$$E_{mn} = \frac{4}{W_k E_p} \int_0^{W_k}\int_0^{E_p} \frac{F_M(y, z)}{\varphi_e} \cos(\eta_m y) \cos(\lambda_n z) \, dy \, dz \tag{E3-24-7}$$

The combination of Eqs. (E3-24-6) and (E3-24-7) gives

$$B_{mn} = \frac{1}{\left(\dfrac{U}{2} + D_x K_{mn}\right)} \frac{4}{E_p W_k} \int_0^{W_k}\int_0^{E_p} \frac{F_M(y, z)}{\varphi_e} \cos(\eta_m y) \cos(\lambda_n z) \, dy \, dz \tag{E3-24-8}$$

and substitution of Eq. (E3-24-8) into Eq. (3-351) gives the following expression:

$$C(x, y, z) = 4\sum_{m=0}^{\infty}\sum_{n=0}^{\infty} I_{mn} \cos(\eta_m y) \cos(\lambda_n z) \frac{\exp\left[\left(\dfrac{U}{2D_x} - K_{mn}\right)x\right]}{\left(\dfrac{U}{2} + D_x K_{mn}\right)} \tag{E3-24-9}$$

where I_{mn} and K_{mn} are given by Eqs.(3-325) and (3-352), respectively. The components of $C(x, y, z)$ are given by Eqs. (3-354), (3-356), (3-358), and (3-360) for $C_1(x, y, z)$, $C_2(x, y, z)$, $C_3(x, y, z)$, $C_4(x, y, z)$, respectively. These expressions are based on the fact that B_{0n} and B_{m0} are one-half and B_{00} is one-quarter in Eq. (3-351) (Carslaw and Jaeger, 1959, p. 182).

Steady-State Special Solution for a Single Rectangular Source
Concentration Distribution
The geometry of a single rectangular source is shown in Figure 3-65, which is a special form of Figure 3-64 with $\gamma = 0$. The rectangular source is located at any place along the yOz plane and its width and height are $2B$ and $2H$, respectively. The distances from the right and left edges of the source to the vertical impervious boundaries are D_1 and D_2, respectively. The source conditions, expressed by Eqs. (3-345)–(3-350), take the forms

$$F_x(0, y, z) = \begin{cases} F_M & D_1 < z < D_1 + 2B, \quad H_1 < y < H_1 + 2H & (3\text{-}361) \\ 0 & \text{otherwise} & (3\text{-}362) \end{cases}$$

where F_M (M/L²/T) is solute source flux. The expressions for I_0, I_{mn}, I_n, and I_m for a rectangular source are given by Eqs. (3-328), (3-329), (3-330), and (3-331), respectively. The substitution of Eq. (3-328) into Eq. (3-354) gives

$$C_1(x, y, z) = \frac{F_M}{\varphi_e} \frac{4BH}{(H_1 + 2H + H_2)(D_1 + 2B + D_2)} \frac{\exp\left[\left(\dfrac{U}{2D_x} - K_{00}\right)x\right]}{\left(\dfrac{U}{2} + D_x K_{00}\right)} \tag{3-363}$$

and substitution of Eq. (3-330) into Eq. (3-356) gives

$$C_2(x, y, z) = \frac{F_M}{\varphi_e} \frac{4H}{(H_1 + 2H + H_2)} \frac{1}{\pi} \sum_{n=1}^{\infty} \frac{S_n}{n} \frac{\exp\left[\left(\dfrac{U}{2D_x} - K_{0n}\right)x\right]}{\left(\dfrac{U}{2} + D_x K_{0n}\right)} \cos(\lambda_n z) \tag{3-364}$$

where S_n is given by Eq. (3-185). The substitution of Eq. (3-331) into Eq. (3-358) gives

$$C_3(x, y, z) = \frac{F_M}{\varphi_e} \frac{4B}{(D_1 + 2B + D_2)} \frac{1}{\pi} \sum_{m=1}^{\infty} \frac{S_m}{m} \frac{\exp\left[\left(\dfrac{U}{2D_x} - K_{m0}\right)x\right]}{\left(\dfrac{U}{2} + D_x K_{m0}\right)} \cos(\eta_m y) \tag{3-365}$$

where S_m is given by Eq. (3-186). The substitution of Eq. (3-329) in Eq. (3-360) gives

$$C_4(x, y, z) = \frac{F_M}{\varphi_e} \frac{4}{\pi^2} \sum_{m=1}^{\infty} \sum_{n=1}^{\infty} \frac{S_n}{n} \frac{S_m}{m} \frac{\exp\left[\left(\dfrac{U}{2D_x} - K_{mn}\right)x\right]}{\left(\dfrac{U}{2} + D_x K_{mn}\right)} \cos(\eta_m y) \cos(\lambda_n z) \tag{3-366}$$

Equations (3-353), (3-363), (3-364), (3-365), and (3-366) form the solution for a single rectangular source under steady-state solute transport conditions.

Special Solutions for the Two-Dimensional Cases
Two special solutions corresponding to the two-dimensional special cases can be obtained from the three-dimensional third-type source solution for a rectangular source whose geometry is shown in Figure 3-65. The procedure is similar to the two-dimensional special solutions for the three-dimensional first-type source solution presented in Section 3.3.2.3. The first two-dimensional special case is shown in Figure 3-32, which shows that the source fully penetrates the entire thickness of the aquifer in the y-coordinate direction. In other words, the source becomes a vertical strip. This corresponds to Figure 3-49, which is the geometry of the ST2B model. The second two-dimensional special case is shown in Figure 3-33, which shows that the source fully penetrates the entire length

of the aquifer in the z-coordinate direction. The special cases corresponding to these special cases are derived in Example 3-25, which is presented below.

EXAMPLE 3-25

Derive the two-dimensional steady-state special solutions from the three-dimensional solution for a rectangular source which is the combination of Eqs. (3-328)–(3-331), (3-353), (3-363)–(3-366).

SOLUTION

First Special Case: The geometry of the first special case is shown in Figure 3-32 for which $H_1 = 0$ and $H_2 = 0$. The substitution of these values into Eq. (3-328) gives Eq. (E3-23-1). The substitution of Eq. (E3-23-1) into Eq. (3-354) results in exactly the same form as the first part of the right-hand side of Eq. (3-294), which is the two-dimensional third-type strip source model. Since $H_1 = 0$, Eq. (3-147) gives $\eta_m = m\pi/2H$ and the sine terms in Eq. (3-186) become zero, resulting in $S_m = 0$. As a result, $C_4(x, y, z) = 0$ according to Eq. (3-366). Further substitution of $H_1 = 0$ and $H_2 = 0$ into Eq. (3-330) gives Eq. (E3-23-2). The substitution of Eq. (E3-23-2) along with Eq. (3-185) into Eq. (3-364) results in exactly the same form of the second part of the right-hand side of Eq. (3-294) in, which is the two-dimensional third-type strip source model. Since $S_m = 0$, $I_m = 0$ according to Eq. (3-188). Therefore, Eq. (3-365) results $C_4(x, y, z) = 0$. Now the proof is complete, i.e., Eq. (3-294) is a special solution of the three-dimensional third-type rectangular source model.

    Second Special Case: The geometry of the second special case corresponds to $D_1 = 0$ and $D_2 = 0$ (see Figure 3-33). This case is equivalent to the first special case with the exception that the strip source is parallel to the z-coordinate axis. Introducing these values into Eqs. (3-328)–(3-331), (3-353), (3-363)–(3-366) it can be shown that $C_2(x, y, z)$ and $C_4(x, y, z)$ both become zero and the sum of $C_1(x, y, z)$ and $C_3(x, y, z)$ turn out to be exactly the same as the solution for a single-strip source as given by Eq. (3-294), with the exception that H, H_1, and H_2 define the source geometry instead of B, D_1, and D_2.

Special Solution for the One-Dimensional Case
In Section 3.3.3.2, the one-dimensional solution, as given by Eq. (3-247), is obtained as a special solution from the two-dimensional solution. As shown above, Eq. (3-294) is a special case of the three-dimensional solution. Therefore, it can be concluded that the same one-dimensional solution is a special case of the three-dimensional solution as well.

Convective–Dispersive Flux Components
From Eqs. (2-41a)–(2-41c), the convective–dispersive flux components in the x-, y-, and z-coordinate directions, respectively, can easily be determined using the same procedure for the convective–dispersive flux components for the two-dimensional case in Section 3.3.3.2. Since the mathematical expressions are too lengthy they are not presented in this chapter.

Steady-State Special Solutions for Other Sources
Eq. (3-353) is the general solution for the resulting concentration distribution for any given $F_M(y, z)$ source concentration distribution. With the source concentration functions (Figure 3-64 with $\gamma = 0$), $F_M(y, z)$, the integrals I_0, I_n, I_m, and I_{mn} given by Eqs. (3-319), (3-321), (3-323), and (3-325), respectively, need to be evaluated analytically or numerically depending on the type of the function $F_M(y, z)$. An example is presented below.

Solution for Two Rectangular Sources with Uniform Solute Source Fluxes
The geometry of two rectangular sources with uniform source concentrations is shown in Figure 3-66 with $\gamma = 0$. The boundary conditions at $x = 0$, expressed by Eqs. (3-345)–(3-350), take the following forms:

$$F_x(0, y, z) = F_{M_1}, \quad D_1 < z < D_1 + 2B_1, \quad A_1 + 2H_1 + A_3 < y < A_1 + 2H_1 + A_3 + 2H_2 \quad (3\text{-}367)$$

$$F_x(0, y, z) = F_{M_2}, \quad D_1 + 2B_1 + D_3 < z < D_1 + 2B_1 + D_3 + 2B_2, \quad A_1 < y < A_1 + 2H_1 \quad (3\text{-}368)$$

where F_{M_1} and F_{M_2} are the source fluxes of the rectangular sources. As mentioned earlier, in order to determine the $C(x,y,z)$ concentration distribution, only I_0, I_n, I_m, and I_{mn} which are given by Eqs. (3-319), (3-321), (3-323), and (3-325), respectively, need to be evaluated. For the source conditions given by Eqs. (3-367) and (3-368), the expressions for I_0, I_n, I_m, and I_{mn} are given by Eqs. (3-336), (3-339), (3-342), and (3-344), respectively. Therefore, with Eqs. (3-336), (3-339), (3-342), and (3-344), Eqs. (3-353), (3-354), (3-356), (3-358), and (3-360) form the solution for two rectangular sources having constant solute source fluxes.

Evaluation of the Models
The three-dimensional third-type source analytical solute transport models presented in the previous sections are evaluated from their verification as well as applicability limitations point of view and the results, based on Batu (1997a), are presented below.

As mentioned in Section 3.3.3.1, if in Eq. (3-317), F_M is given by Eq. (3-236), the concentration $C(x, y, z, t)$ becomes volume-averaged (resident) concentration and is denoted by $C_r(x, y, z, t)$ (see Sections 2.2.2 and 2.9.1 for definitions). As can be seen from the components of the solution, the concentration solution includes integrals as well as infinite series. Note that for $C_2(x, y, z, t)$, $C_3(x, y, z, t)$, and $C_4(x, y, z, t)$ expressions integration must be performed for every n and m values in the infinite series. These integrals are similar to those expressions for the two-dimensional third-type solution, as given by Eq. (3-271) and the Gauss numerical integration method is appropriate for their evaluation (Batu, 1997a). The Gauss method was applied according to Eqs. (3-297) and (3-298). Equations (3-297) and (3-298) are substituted in the single rectangular source solution, which is the combination of Eqs. (3-317), (3-318), (3-320), (3-322), (3-324), and (3-328)–(3-331). Based on the aforementioned expressions, a computer program, ST3B, has been developed by the author. This program is described in the following sections.

Special Features of the Three-Dimensional Analytical Solute Transport Models
The solutions for unsteady- and steady-state solute transport cases are fairly general such that solutions for any combinations of rectangular sources located along a vertical plane (yOz plane in Figure 3-64) perpendicular to the flow direction can be generated using the source geometry. The source flux on every rectangle can be space-dependent, i.e., the function of y and z coordinates. The exponential term, $e^{\gamma t}$, increases the potential usage of the general model such that exponentially varying time-dependent source concentrations can be simulated. Variation of γ with the source strength (normalized concentration or normalized flux) with time is illustrated in Figure 3-2 for various decreasing γ value sources. If historical data are available for time-dependent source fluxes, the variation of γ can be determined by using curve-fitting techniques.

The single rectangular source has a special importance. Therefore, this solution will be elaborated in detail. The impervious boundaries around the solute source increase the potential usage of the model. If the aquifer approaches infinity from both sides of the source, its effect can be taken into account by assigning large values to D_1 and D_2 as compared with the source width $2B$. One of the important features of this solution is that the source can be located at any place in the yOz plane that is perpendicular to the flow direction. Assigning $H_1 = 0$ corresponds to the case that the source is adjacent to the water table or to the top boundary of a confined aquifer.

Special Solutions
Some verifications have already been done for the three-dimensional analytical models. As mentioned above in detail, it is shown that the one-dimensional analytical models under unsteady- and steady-state solute transport conditions (Section 3.3.3.1) and two-dimensional analytical models under unsteady- and steady-state solute transport conditions (Section 3.3.3.2) are exact special cases of the corresponding three-dimensional unsteady- and steady-state solute transport models.

Comparisons with Numerical Models
The numerical model codes that were selected are called FTWORK (Faust et al., 1990) and MODFLOWT (Duffield, 1996). MODFLOWT is a MODFLOW (McDonald and Harbaugh, 1984)

family code. Both FTWORK and MODFLOWT are complete block-centered, three-dimensional finite-difference ground water flow and solute transport codes. The three-dimensional finite-difference grid used for comparison for FTWORK is shown in Figure 3-67. The same grid is also valid for MODFLOWT with the exception that the columns are numbered from left to right, i.e., opposite to FTWORK column numbers. As can be seen from Figure 3-67, the thickness b of the aquifer is 5 m and a total of 10 modeling layers with each 0.5 m thick were considered for the corresponding numerical models. The horizontal grid of Figure 3-67 is shown in Figure 3-52. The boundary conditions described in Section 3.3.3.2 are the same for all ten layers in Figure 3-67. As in Figure 3-52, constant head boundary values (11.5 m at the yOz lane and 11.0 m at the downgradient plane of the parallelepiped) were assigned in Figure 3-67. These boundary conditions define a one-dimensional ground water flow field in the x-coordinate direction. Similar to the two-dimensional comparison in Section 3.3.3.2, the hydraulic conductivity, $K = K_x = K_y = K_z$, and effective porosity, φ_e, were selected to be 13.875 m/day and 0.25, respectively. With the hydraulic gradient $I = 0.5$ m/185 m $= 0.0027027$ m/m, the corresponding ground water velocity, U is 0.15 m/day. The longitudinal dispersivity, α_L, transverse horizontal dispersivity, α_{TH}, transverse vertical dispersivity, α_{TV}, and retardation factor, R_d, values are 21.3 m, 4.3 m, 4.3 m and 1, respectively. The effective molecular diffusion coefficient, D^*, is zero and the solute source concentration is $C_M = 1$ g/m^3.

The solute source flux was selected according to Eq. (3-236) and with the above given values $F_M = (0.25)(0.15$ m/day$)(1$ g/m$^3) = 0.0375$ g/day/m^2. Therefore, the predicted concentrations are volume-averaged (resident) concentrations denoted by C_r (see Sections 2.2.2 and 2.9.1 for definitions). In ST3B code, only F_M and not C_M is the input. Therefore, $F_M = 0.0375$ g/day/m^2 as used for the present rectangular source model. This value was assigned on the $2B \times 2H$ source area in Figure 3-67. As can be seen from Figure 3-67, $2B = 5$ m and $2H = 0.5$ m.

The numerical codes, FTWORK and MODFLOWT, require different forms of data than each other and the analytical model. Since the FTWORK code is designed to handle solute sources with only the third-type source condition, assigning $C_M = 1$ g/m^3 to each shaded block in Figure 3-67 is sufficient. In MODFLOWT, the appropriate way is to use injection wells according to $Q = KIA$

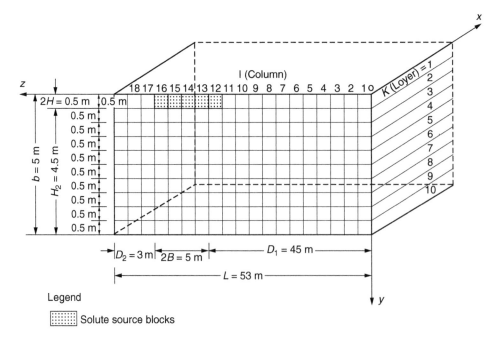

FIGURE 3-67 The finite-difference grid used for verification of the three-dimensional third-type solute source model. (Adapted from Batu, V., unpublished paper, 1997a.)

where Q is the injection rate to each source block, I is the hydraulic gradient, and A is the shaded area of each block face in Figure 3-67. With these values, Q = (13.875 m/day)(0.0027027 m/m)[(1.0 m)(0.5 m)] = 0.01875 m³/day. This value was assigned to each source block in Figure 3-67 for the MODFLOWT model.

Numerical experiments were carried out in the FTWORK and MODFLOWT runs to determine the maximum time step producing numerically stable results. The results showed that Δt = 1 day was appropriate for both models and this value was used in all comparisons.

Comparison between the results of the analytical model for a single rectangular source, called ST3B, and the two three-dimensional numerical models, FTWORK and MODFLOWT, are presented in Figure 3-68–3-71. Figure 3-68 presents C_r vs. the z coordinate for x = 22.5 m and y = 1.25 m for the analytical model (the corresponding grid coordinates for FTWORK are I = 1–19, J = 9, and K = 3; and for MODFLOWT I = 1–19, J = 9, and K = 3) after 180 days elapsed time. Figure 3-69 presents C_r vs. the x coordinate for y = 1.25 m and z = 51.5 m for the analytical model (the corresponding grid coordinates for FTWORK are I = 18, J = 1–39, and K = 3; and for MODFLOWT I = 2, J = 1–39, and K = 3) after 180 days elapsed time. Figure 3-70 presents C_r vs. the y coordinate for x = 8.75 m and z = 47.5 m for the analytical model (the corresponding grid coordinates for FTWORK are I = 14, J = 6, and K = 1–10; and for MODFLOWT I = 6, J = 6, and K = 1–10) after 180 days elapsed time. Figure 3-71 presents C_r vs. time at a point having the coordinates x = 8.75 m, y = 0.25 m, and z = 47.5 m for the analytical model (the corresponding grid coordinates for FTWORK are I = 14, J = 6, and K = 1; and for MODFLOWT I = 6, J = 6, and K = 1). Review of Figure 3-68–3-71 indicates that good agreement exists among the results of the analytical and the two numerical models. Minor discrepancies can be attributed to the roundoff errors, horizontal and vertical grid spacings, and algorithms used both in the analytical and numerical models.

Representation of Exponentially Variable Time-Dependent Source Concentrations
As mentioned previously, the solute source flux can be represented as exponentially varying time-dependent quantity. If $\gamma < 0$, the source flux decreases with time; if $\gamma = 0$, the source flux is independent of time; and if $\gamma > 0$, the source concentration increases with time. Figure 3-2 shows the

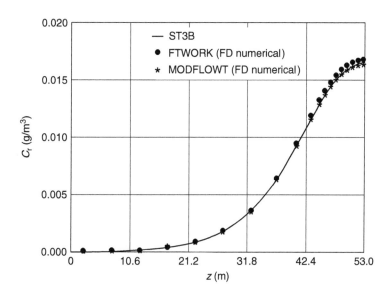

FIGURE 3-68   Concentration vs. transverse distance (z) comparison at x = 22.5 m and y = 1.25 m between the third-type source three-dimensional analytical model and two different finite-difference numerical models after 180 days. (Adapted from Batu, V., unpublished paper, 1997a.)

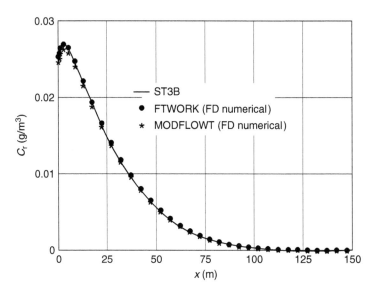

FIGURE 3-69 Concentration vs. longitudinal distance (x) comparison at $y = 1.25$ m and $z = 51.5$ m between the third-type source three-dimensional analytical model and two different finite-difference numerical models after 180 days. (Adapted from Batu, V., unpublished paper, 1997a.)

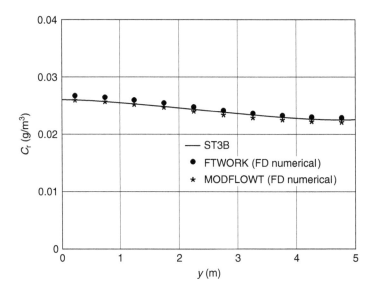

FIGURE 3-70 Concentration vs. vertical distance, y, comparison at $x = 8.75$ m and $z = 47.5$ m between the third-type source three-dimensional analytical model and two different finite-difference numerical models after 180 days (Adapted from Batu, V., unpublished paper, 1997a.)

variation of the function given by Eq. (3-308) for different values of γ ranging from -0.0001 to -0.0100 day^{-1}. If observed data are available for time-dependent flux sources, the value of γ may be determined from the curve-fitting techniques.

Number of Quadrature Points Selection in the Gaussian Integration
As mentioned previously, the integrals in Eq. (3-317) were evaluated numerically based on the transformation expressions given by Eqs. (3-297) and (3-298) by means of Gaussian integration procedure with 4, 5, 10, 20, 60, 104, or 256 quadrature points. The same method was applied for the

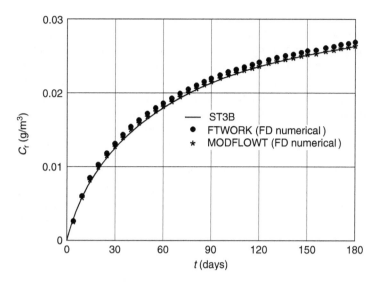

FIGURE 3-71 Concentration vs. time (t) comparison at point ($x = 22.5$ m, $y = 0.25$ m, $z = 47.5$ m) 1.25 m between the third-type source three-dimensional analytical model and two different finite-difference numerical models after 180 days. (Adapted from Batu, V., unpublished paper, 1997a.)

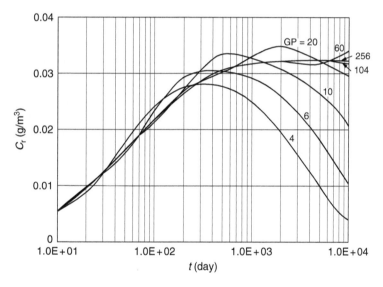

FIGURE 3-72 Concentration vs. time curves generated from the three-dimensional analytical model for different number of Gauss points (GP) for $N = 50$. (Adapted from Batu, V., unpublished paper, 1997a.)

integrals in the convective–dispersive flux components in the ST3B program. For longer time periods, more Gauss points (GP) are required in order to have accurate results. Figure 3-72 presents concentration vs. time curves for different number of GP for a point having the coordinates $x = 8.75$ m ($J = 4$), and $y = 0.25$ m ($K = 1$) (refer Figure 3-52 for the coordinates). As can be seen from Figure 3-72, the accuracy increases as the number of GP increases. Obviously, for the total time period of 10,000 days (27.40 years) the results corresponding to the lower GP values (GP = 4, 6, 10, and 20) are not acceptable. However, for small time periods (up to 100 days) their results may be treated as approximate values. In the ST3B program, which is presented later, the user has options to use the above-mentioned quadrature points. Quadrature points higher than 20 produce accurate results.

Number of Terms Selection for the Infinite Series in the Solutions
Apart from the number of Gauss Points (GP), the number of terms in the infinite series (N) plays a key role in determining accurate results. Numerical experiments showed that at the source, the number of terms are the highest for converged results. The required number of terms decreases with the increasing distance from the source. Generally, the values of N between 50 and 100 around the source area produce reasonably accurate results. As the x coordinate increases, the value of N can be as low as 10 and yet the results have acceptable accuracy. This situation can be observed in Figure 3-73 which presents C_r vs. time t at a point having the coordinates $x = 8.75$ m ($J = 4$), $z = 47.5$ m ($I = 14$), and $y = 0.25$ m ($K = 1$) for $N = 10$ and 50 (see Figure 3-52 for the coordinates).

Application of the Three-Dimensional Analytical Models to Ground Water Flow Systems
The general model and its derivative models under unsteady- and steady-state solute transport conditions can be applied to solute transport predictions in aquifers if the ground water velocity can be assumed to be uniform. The models can also be applied to unsaturated zones with some additional assumptions. These are all discussed in the following paragraphs.
    In this section first, a computer program ST3B, for a single rectangular source, is described. Then, application of this program is presented with examples.

A Computer Program (ST3B) for the Three-Dimensional Third-Type Rectangular Source Model
A three-dimensional MS-DOS-based FORTRAN solute transport program, called solute transport three-dimensional third-type (ST3B), has been developed by the author. ST3B is a menu-driven program and calculates values of concentration distribution and convective–dispersive flux components under unsteady-state solute transport conditions (Eqs. [3-317], [3-318], [3-320], [3-322], [3-324], [3-328]–[3-331] for concentration distribution) and steady state solute transport conditions (Eqs. [3-28]–[3-29], [3-30], [3-31], [3-53], [3-54], [3-56], [3-58], and [3-60] for concentration distribution). This program can be run interactively in IBM compatible computers.

Application to Saturated Zones
The three-dimensional analytical solute transport model for a rectangular source (see Figure 3-65) can be used for different purposes. The impervious boundaries around the solute transport domain increase the potential usage areas of the model. For ground water flow cases, the effects of impervious boundaries, such as bedrock outcrops and clay zones surrounding the main ground water flow

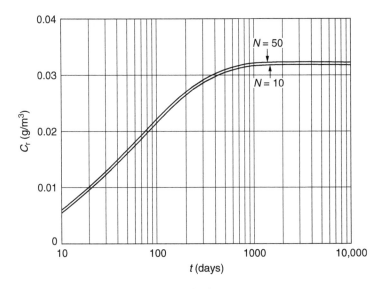

FIGURE 3-73   Concentration vs. time curves generated from the three-dimensional analytical model for $N = 10$ and 50 using GP (Gauss points) = 256. (Adapted from Batu, V., unpublished paper, 1997a.)

zone, can be taken into account. If the aquifer boundaries approach infinity horizontally, its effect can be taken into account by assigning large values to D_1 and D_2 as compared with the source width ($2B$). The rectangular source can be at any location in the cross-section of an aquifer. Assigning $H_1 = 0$ is the case that the rectangular source starts at the water table of an unconfined aquifer or at the top of a confined aquifer.

Application to Unsaturated Zones
Application of the ST3B model to unsaturated zones can similarly be made as in the case for ST2A, which is presented in Section 3.2.2.2. The conceptual model for the ST3B model is the same as in Figure 3-59 with the exception that $L = 2H$. The associated governing equations for flow and solute transport in unsaturated porous media under uniform infiltration rate conditions are given in Section 2.8. The five important points mentioned for the ST2B in Section 3.3.3.2 are also valid for the third-type three-dimensional source case.

Comparative Evaluation of the First- and Third-Type Source Models
This subject is extensively elaborated in Section 3.3.3.2 for two-dimensional third-type source case model. The mass balance constraints given by Eqs. (3-299) and (3-300) can easily be extended to the three-dimensional cases by adding the y coordinate to the expressions. Example 3-20 can easily be extended to the three-dimensional case. Similarly, Eq. (2-93) can be used to obtain the third-type three-dimensional solution from the first-type three-dimensional solution given in Section 3.3.2.3.

3.3.4 Instantaneous Source Analytical Solute Transport Models

There are some limited number of solutions generally adapted from the heat conduction literature for instantaneous sources based on the assumption that the boundaries of the solute transport domain approaches infinity in all directions for point sources (Hunt, 1978; Wilson and Miller, 1978) and for parallelepiped sources (Domenico and Robbins, 1985). As a result, these solutions do not take into account the aquifer thickness and, consequently, the relative position of the source along the aquifer thickness. In this section, in accordance with the geometry of the models presented in Sections 3.3.2 and 3.3.3, one-, two-, and three-dimensional analytical solutions will be presented for areal instantaneous sources based on Batu (1997b, 1997c). These solutions are based on the *instantaneous flux boundary condition*. The mathematical as well as physical bases of this condition are included in Section 2.10.3.2.

3.3.4.1 One-Dimensional Analytical Solute Transport Models for Instantaneous Sources

This model was obtained as a special solution from the general three-dimensional solution developed by Batu (1997b) and, to the author's knowledge , this model has not been published in the literature yet. Similar to the first-type source one-dimensional analytical model (Section 3.3.2.1) and third-type source one-dimensional analytical model (Section 3.3.3.1), this model will probably have a special importance in the literature. In the following sections, the one-dimensional instantaneous source model will be presented as a special case from the more general two- and three-dimensional instantaneous source analytical models.

Problem Statement and Assumptions
The problem under consideration is one-dimensional solute transport from an instantaneous planar source in a unidirectional flow field as shown in Figure 3-74. The medium is considered of having a semiinfinite parallelepiped form. The solute mass M is assumed to be entered in the column at $x = 0$ and the medium approaches infinity along the x-coordinate direction. The assumptions listed in (2)–(4) in Section 3.3.2.1 for the first-type source case are also valid for the instantaneous source case as well.

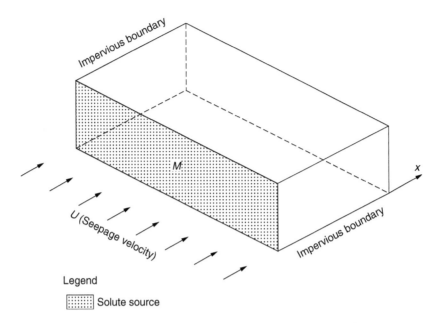

FIGURE 3-74   Schematic representation of the one-dimensional solute transport for an instantaneous planar source in a unidirectional flow field. (Adapted from Batu, V., unpublished paper, 1997b.)

Governing Differential Equation
The governing differential equation is given by Eq. (3-1).

Initial and Boundary Conditions
The initial condition for $C(x, t)$ is

$$C(x, t) = 0, \quad x \geq 0 \tag{3-369}$$

The boundary conditions for $C(x, t)$ are

$$\frac{M}{V\varphi_e}\delta(t) = \frac{M}{(1\,\text{unit})(1\,\text{unit})(1\,\text{unit})\varphi_e}\delta(t) = C_M\delta(t) = \frac{F_x(0)}{\varphi_e} \quad t \geq 0 \tag{3-370}$$

$$C(\infty, t) = 0, \quad t \geq 0 \tag{3-371}$$

$$\frac{\partial C(\infty, t)}{\partial x} = 0, \quad t \geq 0 \tag{3-372}$$

where M (its unit is M) is the mass of the solute introduced instantaneously at $x = 0$ (see Figure 3-74) at an unit volume V [(1 unit)(1 unit)(1 unit)], $\delta(t)$ is *Dirac delta function* or *impulse function*, $C_M(\text{M/L}^3)$ is the instantaneous source concentration, and $F_x(0)$ is the instantaneous solute source flux at $x = 0$.
Equation (3-370) states that a mass M of solute is introduced instantaneously at $x = 0$ shown in Figure 3-74. The mathematical and physical bases of Eq. (3-370) are given in more detail in Section 2.10.3.2. Since the unit of $\delta(t)$ is T^{-1}, the dimensions of each side of Eq. (3-370) are $\text{M/L}^2/\text{T}$. Equations (3-371) and (3-372) state that both the concentration, C, and its gradient $(\partial C/\partial x)$ approach zero when the x distance approaches infinity.

Solution for Concentration Distribution

The solution for concentration distribution can be obtained by solving the above-mentioned boundary-value problem defined by Eqs. (3-1) and (3-369)–(3-372). The solution can also be obtained as a special case from the single-strip source solution (Eq. [3-404]). Because it is simpler, here, the second method for the determination of the solution will be presented.

The substitution of $D_1 = D_2 = 0$ in Eqs. (3-401) and (3-403), respectively, gives

$$C_1(x, t) = \frac{M}{2B\varphi_e} \frac{1}{D_x} F_1(t)[F_2(t) - F_3(t)F_4(t)] \tag{3-373}$$

and

$$C_2(x, t) = 0 \tag{3-374}$$

Equation (3-374) is due to the fact that $\sin[\lambda_n(D_1+2B)] = \sin[n\pi/(D_1+2B+D_2)] = \sin(n\pi) = 0$ and $\sin(\lambda_n D_1) = 0$. The substitution of Eqs. (3-373) and (3-374) into Eq. (3-395) and keeping in mind that $2B$ is unit, the solution takes the form

$$C(x, t) = \frac{M}{\varphi_e} \frac{1}{D_x} F_1(t)[F_2(t) - F_3(t)F_4(t)] \tag{3-375}$$

where

$$F_1(t) = \exp\left(\frac{Ux}{2D_x} - vt - \frac{U^2t}{4D_x R_d} \right) \tag{3-376}$$

$$F_2(t) = \left(\frac{D_x}{\pi R_d t} \right)^{1/2} \exp\left(-\frac{R_d x^2}{4D_x t} \right) \tag{3-377}$$

$$F_3(t) = \frac{U}{2R_d} \exp\left(\frac{Ux}{2D_x} + \frac{U^2t}{4R_d D_x} \right) \tag{3-378}$$

$$F_4(t) = \operatorname{erfc}\left[\frac{R_d x}{2(D_x R_d t)^{1/2}} + \frac{U}{2}\left(\frac{t}{R_d D_x} \right)^{1/2} \right] \tag{3-379}$$

Equation (3-375) can also be written as

$$C(x, t) = C_M \frac{1}{D_x} F_1(t)[F_2(t) - F_3(t)F_4(t)] \tag{3-380}$$

where C_M is the instantaneous solute source concentration and is defined as

$$C_M = \frac{M}{\varphi_e} \tag{3-381}$$

Equation (3-380) is a special form of Eq. (3-404) for $2B = 1$, $D_1 = 0$, and $D_2 = 0$.

Convective–Dispersive Flux Component

The convective–dispersive flux component, F_x, in the x-coordinate direction can be determined from Eq. (3-405). Since $C_2 = 0$, Eq. (3-405) gives

$$F_x = \varphi_e UC - \varphi_e D_x V \tag{3-382}$$

where C is given by Eq. (3-375) and

$$V = \frac{M}{\varphi_e} \frac{1}{D_x}\left[\frac{\partial F_1}{\partial x}(F_2 - F_3 F_4) + F_1\left(\frac{\partial F_2}{\partial x} - F_3 \frac{\partial F_4}{\partial x} - F_4 \frac{\partial F_3}{\partial x} \right) \right] \tag{3-383}$$

in which

$$\frac{\partial F_1}{\partial x} = \frac{U}{2D_x} \exp\left(\frac{Ux}{2D_x} - vt - \frac{U^2 t}{4D_x R_d} \right) \tag{3-384}$$

$$\frac{\partial F_2}{\partial x} = -\frac{R_d x}{2D_x t} \left(\frac{D_x}{\pi R_d t} \right)^{1/2} \exp\left(-\frac{R_d x^2}{4D_x t} \right) \tag{3-385}$$

$$\frac{\partial F_3}{\partial x} = \frac{U^2}{4R_d D_x} \exp\left(\frac{Ux}{2D_x} + \frac{U^2 t}{4R_d D_x} \right) \tag{3-386}$$

$$\frac{\partial F_4}{\partial x} = -\frac{R_d}{(\pi D_x R_d t)^{1/2}} \exp\left\{ -\left[\frac{R_d x}{2(D_x R_d t)^{1/2}} + \frac{U}{2}\left(\frac{t}{R_d D_x} \right)^{1/2} \right]^2 \right\} \tag{3-387}$$

and F_1, F_2, F_3, and F_4 are given by Eqs. (3-376), (3-377), (3-378), and (3-379), respectively.

Evaluation of the Model
Equation (3-375) is obtained as a special solution from a more general two-dimensional solute transport model as given by Eq. (3-404). The two-dimensional solute transport model is obtained from a more general three-dimensional solute transport model as given by Eq. (3-429). As can be seen from Section 3.3.4.3, there is good agreement between one of the special forms for a rectangular source of the three-dimensional analytical model and a three-dimensional finite-difference numerical model. Since the one-dimensional analytical model is a special form of the three-dimensional analytical model, the verification is also valid for the one-dimensional analytical model as well.

Applications of the One-Dimensional Analytical Solute Transport Model
As mentioned in Sections 3.3.2.1 and 3.3.3.1, one-dimensional analytical solute transport models potentially have a lot of applications for finding quick solutions for many problems faced in practice. These models can be applied to solute transport predictions in saturated and unsaturated zones as outlined in Sections 3.3.2.2 and 3.3.2.3. In this chapter, an example is presented for the one-dimensional instantaneous solute transport model.

A Computer Program for the One-Dimensional Instantaneous Source Model (ST1C)
A one-dimensional MS-DOS-based FORTRAN solute transport computer program, called solute transport one-dimensional instantaneous type (ST1C), has been developed by the author. ST1C is a menu-driven program and calculates values of concentration and convective–dispersive flux component (Eqs. [3-375] and [3-382]). This program can be run interactively in IBP compatible computers.

In the ST1C program the time t must be greater than zero. A zero value gives runtime error. This is due to the fact that the time t is in the denominator in some expressions of Eqs. (3-375) and (3-382). This is the case for almost all analytical solute transport models.

One example is presented below based on the ST1C computer program.

EXAMPLE 3-26

Using the ST1C program, for the aquifer whose flow and transport parameters are given in Example 3-2 for $M = 1000$ g instantaneous solute mass, (1) determine the concentration vs. distance curves after 100 days for $\alpha_L = 0.1$, 1, and 10 m; (2) determine the concentration vs. time curve at $x = 0$ for $\alpha_L = 1$ m; and (3) determine the convective–dispersive flux vs. distance curve after 100 days for $\alpha_L = 1$ m.

SOLUTION

(1) The concentration vs. distance curves after 100 days for $\alpha_L = 0.1$, 1, and 10 m are given in Figure 3-75.

(2) The concentration vs. time curve at $x = 0$ for $\alpha_L = 1$ m is given in Figure 3-76.

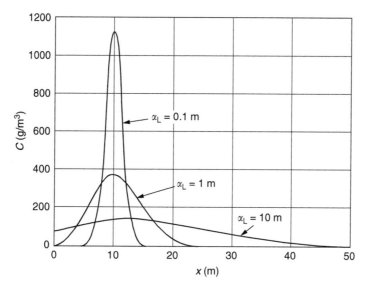

FIGURE 3-75   Concentration vs. distance after 100 days for α_L = 0.1, 1, and 10 m for Example 3-25.

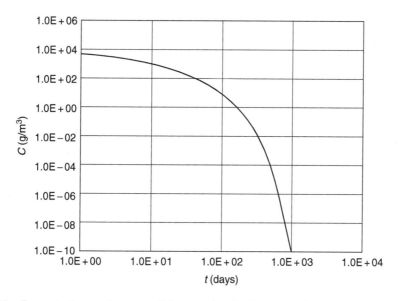

FIGURE 3-76   Concentration vs. time at $x = 0$ for α_L = 1 m for Example 3-25.

(3) The convective–dispersive flux vs. distance curve after 100 days for $\alpha_L = 1$ m is given in Figure 3-77.

3.3.4.2   Two-Dimensional Analytical Solute Transport Models for Instantaneous Strip Sources

This model was obtained as a special case from the general three-dimensional solution developed by Batu (1997b), and to the author's knowledge, this model has not been published in the literature yet. Similar to the first-type source two-dimensional analytical model (Section 3.3.2.2) and third-type source two-dimensional analytical model (Section 3.3.3.2), this model will probably have a special importance in the literature. In the following sections, the two-dimensional instantaneous

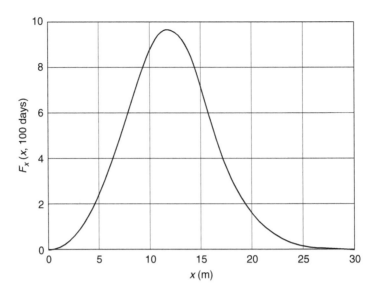

FIGURE 3-77   Convective–dispersive flux after 100 days for $\alpha_L = 1$ m for Example 3-25.

source model will be presented as a special case from the more general three-dimensional instantaneous source analytical model (Batu, 1997c).

Problem Statement and Assumptions
The problem under consideration is two-dimensional solute transport from multiple instantaneous strip sources between two impervious planar boundaries in a unidirectional flow field as shown in Figure 3-78. The x- and z-coordinate axis are the planar coordinates either in a horizontal or vertical plane. The solute masses are functions of the z coordinate. The assumptions listed in (1), (2), and (4) in Section 3.3.2.2 for the first-type source case are also valid for the instantaneous source case as well. It is also assumed that the corresponding mass to each strip source is uniformly mixed in the source.

Governing Differential Equation
The governing differential equation is given by Eq. (3-45).

Initial and Boundary Conditions
The initial condition for $C(x, z, t)$ is given by Eq. (3-46). The boundary conditions for $C(x, z, t)$ at $x = 0$ are

$$\frac{M_1}{V_1 \varphi_e} \delta(t) = \frac{M_1}{(L_1)(1\text{unit})(1\text{unit})\varphi_e} \delta(t) = C_{M_1} \delta(t) = \frac{F_x(0, z)}{\varphi_e}, \quad 0 < z < H_1 \quad \text{(3-388)}$$

$$\frac{M_{p-1}}{V_{p-1} \varphi_e} \delta(t) = \frac{M_{p-1}}{(L_{p-1})(1\text{unit})(1\text{unit})\varphi_e} \delta(t) = C_{M_{p-1}} \delta(t) = \frac{F_x(0, z)}{\varphi_e}, \quad H_{p-2} < z < H_{p-1} \quad \text{(3-389)}$$

$$\frac{M_p}{V_p \varphi_e} \delta(t) = \frac{M_p}{(L_p)(1\text{unit})(1\text{unit})\varphi_e} \delta(t) = C_{M_p} \delta(t) = \frac{F_x(0, z)}{\varphi_e}, \quad H_{p-1} < z < H_p \quad \text{(3-390)}$$

where M_i (its unit is M) is the mass of solute introduced instantaneously at the L_i distance interval at $x = 0$ (see Figure 3-78) at a volume V_i [$(L_i)(1$ unit$)(1$ unit$)$], $\delta(t)$ is Dirac delta function or impulse

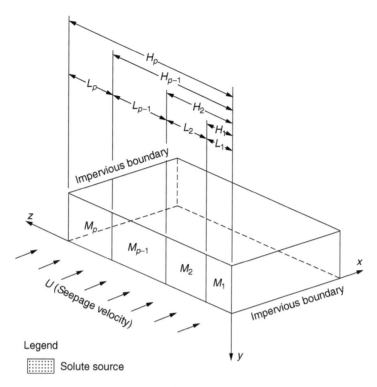

FIGURE 3-78 Schematic representation of the two-dimensional solute transport from multiple instantaneous strip sources in a bounded unidirectional flow field. (Adapted from Batu, V., unpublished paper, 1997b.)

function, C_{M_i} (M/L^3) is the instantaneous source concentration, and $F_x(0, z)$ (M/L^2/T) is the instantaneous solute source flux at $x = 0$; H_i is given by Eq. (3-50). The other boundary conditions at the impervious boundaries and infinity are given by Eqs. (3-51)–(3-54). Equations (3-388)–(3-390) state that a mass of solute M_i is introduced instantaneously at the V_i volume whose dimensions are L_i, 1 unit, and 1 unit. The mathematical and physical bases of Eqs. (3-388)–(3-390) are given in more detail in Section 2.10.3.2. Since the unit of $\delta(t)$ is T^{-1}, the dimensions of each side of Eqs. (3-388)–(3-390) are M/L^2/T. The physical meaning of the rest of boundary conditions is given in Section 3.3.2.2.

Solution for Concentration Distribution
The boundary-value problem defined by Eqs. (3-45), (3-46), (3-51)–(3-54), and (3-388) through (3-390) was solved by the Laplace transform and Fourier analysis techniques (Batu, 1997b). For this purpose, the solution was obtained first in the Laplace domain and then the concentration variation was obtained by taking the inverse Laplace transform of the Laplace domain solution. The final results are presented in the following paragraphs and the intermediate steps are presented in the associated examples.

General Solution in the Laplace Domain
As indicated above, the governing differential equation, and the initial and boundary conditions given by Eqs. (3-45), (3-46), (3-51)–(3-54) for the two-dimensional first-type solution case (Section 3.3.2.2) are exactly the same as the instantaneous source solution case. Using these equations, in Example 3-4, Eq. (E3-4-24) is obtained as the solution with A_n as unknown and this solution is also valid for the instantaneous source case as well. Using the source conditions at $x = 0$, given by

Eqs. (3-388)–(3-390) and (E3-4-24), the general solution in the Laplace domain is expressed in the form of Eq. (3-56), and $u_1(x, z, s)$ and $u_2(x, z, s)$ are as follows (Batu, 1997b):

$$u_1(x, z, s) = \frac{1}{H_p} \frac{\exp(Ux/(2D_x) - p_0 x)}{(U/2 + D_x p_0)} I_0(z) \tag{3-391}$$

$$u_2(x, z, s) = \frac{2}{H_p} \sum_{n=1}^{\infty} \frac{\exp(Ux/(2D_x) - p_n x)}{(U/2 + D_x p_n)} \cos(\lambda_n z) I_n(z) \tag{3-392}$$

where p_0, p_n, and λ_n are given by Eqs. (3-59), (3-60), and (3-61), respectively, and

$$I_0(z) = \int_0^{H_p} C_M \, dz \tag{3-393}$$

$$I_n(z) = \int_0^{H_p} C_M \cos(\lambda_n z) \, dz \tag{3-394}$$

Determination of A_n in Eq. (E3-4-24) from the usage of Eqs. (3-388)–(3-390) is shown in Example 3-26.

EXAMPLE 3-26

Using the boundary conditions at $x = 0$ by Eqs. (3-388)–(3-390) and (E3-4-24), determine $u_1(x, z, s)$ and $u_2(x, z, s)$ in Eq. (3-56) for the instantaneous-type source.

SOLUTION

The boundary conditions at the sources, given by Eqs. (3-388)–(3-390) need to be expressed in the Laplace domain. The convective–dispersive flux, F_x, in the x direction is given by Eq. (3-90a). Therefore, taking the Laplace transform of the combined form of Eqs. (3-90a) and (3-388)–(3-390) gives

$$C_M(z) = Uu - D_x \frac{\partial u}{\partial x} \tag{E3-26-1}$$

From Eq. (E3-4-24), the expressions for $\partial u(0,y,s)/\partial x$ and $u(0,y,s)$, which are the expressions at $x = 0$, are given by Eqs.(E3-16-3) and (E3-16-4), respectively. Since Eq. (E3-26-1) is only valid at $x = 0$, the substitution of Eqs. (E3-16-3) and (E3-16-4) into Eq. (E3-26-1) gives

$$C_M(z) = \sum_{n=0}^{\infty} B_n \cos(\lambda_n z) \tag{E3-26-2}$$

where B_n is given by Eq. (E3-16-6). In Eq. (E3-26-2), B_n can be determined as in Eq. (E3-4-27). Considering the right-hand side of Eq. (E3-26-2) as a Fourier cosine series for the interval $z = 0$ to H_m yields (e.g., Churchill, 1941; Spiegel, 1971)

$$B_n = \frac{2}{H_p} \int_0^{H_p} C_M \cos(\lambda_n z) \, dz \tag{E3-26-3}$$

The substitution of Eq. (E3-26-3) into Eq. (E3-16-6) and solving for A_n gives

$$A_n \frac{1}{\left(\dfrac{U}{2} + D_x p_n\right)} \frac{2}{H_p} \int_0^{H_p} C_M \cos(\lambda_n z) \, dz \tag{E3-26-4}$$

and the substitution of Eq. (E3-26-4) into Eq. (E3-4-24) results in the following expression:

$$u(x, z, s) = \sum_{n=0}^{\infty} \left[\frac{1}{\left(\dfrac{U}{2} + D_x p_n\right)} \frac{2}{H_p} \int_0^{H_p} C_M \cos(\lambda_n z) \, dz \right] \tag{E3-26-5}$$

$$\times \exp(Ux/(2D_x) - p_n x) \cos(\lambda_n z)$$

which is the general solution in the Laplace domain and is valid for any solute instantaneous mass distribution at $x = 0$. After some manipulations, Eq. (E3-26-5) results in the form of Eq. (3-56), in which $u_1(x, z, s)$ and $u_2(x, z, s)$ are given by Eqs. (3-391) and (3-392), respectively. It must be remembered that in writing Eq. (3-391) there is a 1/2 coefficient for the first term of the Fourier series.

Inverse Laplace Transforms of Eqs. (3-391) and (3-392)
Determination of the inverse Laplace transforms of Eqs. (3-391) and (3-392) is given in Example 3-27. The sum of Eqs. (3-391) and (3-391) forms the Laplace domain solution according to Eq. (3-56), and its inverse Laplace transform is the general solution (Batu, 1997b)

$$C(x, z, t) = C_1(x, z, t) + C_2(x, z, t) \tag{3-395}$$

where

$$C_1(x, z, t) = \frac{1}{H_p} \frac{I_0(z)}{D_x} F_1(t)[F_2(t) - F_3(t)F_4(t)] \tag{3-396}$$

and

$$C_2(x, z, t) = \frac{2}{H_p} \frac{1}{D_x} F_1(t)[F_2(t) - F_3(t)F_4(t)] \tag{3-397}$$

$$\times \sum_{n=1}^{\infty} I_n(z) \cos(\lambda_n z) \exp\left(-\frac{D_z}{R_d} \lambda_n^2 t\right)$$

in which $F_1(t)$, $F_2(t)$, $F_3(t)$, and $F_4(t)$ are given by Eqs. (3-376), (3-377), (3-378), and (3-379), respectively.

EXAMPLE 3-27

Derive Eq. (3-395) by determining the inverse Laplace transforms of Eqs. (3-391) and (3-392).

SOLUTION

It is known that the inverse Laplace transforms of $u_1(x, z, s)$ and $u_2(x, z, s)$ are $C_1(x, z, t)$ and $C_2(x, z, t)$, respectively, and are presented below.

Inverse Laplace Transform of $u_1(x, z, s)$
In order to determine the inverse Laplace transform of $u_1(x, z, s)$, as given by Eq. (3-391), according to the second expression of Eq. (E3-5-2), some manipulations need to be made. Equation (3-391) can also be expressed as

$$u_1(x, z, s - b) = \frac{1}{H_p} I_0(z) \exp\left(\frac{Ux}{2D_x}\right) e_2(s-b) \tag{E3-27-1}$$

where $I_0(z)$ is given by Eq. (3-393), $e_2(s-b)$ is given by Eq. (E3-17-3), and the expression for b is given by Eq. (E3-5-4). As a result, the inverse Laplace transform of $u_1(x, z, s - b)$ is the same as the inverse Laplace transform of Eq. (E3-17-3). Therefore, the inverse Laplace transform of $u_1(x, z, s)$ can be expressed by Eq. (3-396) as given earlier.

Inverse Laplace Transform of $u_2(x, z, s)$
The inverse Laplace transform procedure of $u_2(x, z, s)$ is similar to that for $u_1(x, z, s)$ as presented earlier. Therefore, the inverse Laplace transform for $u_2(x, z, s)$ is given by Eq. (3-397).

Special Solution for a Single Instantaneous Strip Source
Concentration Distribution
The geometry of a single-strip source is shown in Figure 3-79, which is a special case of Figure 3-78. As can be seen from Figure 3-79, the source can be located at any place between the impervious

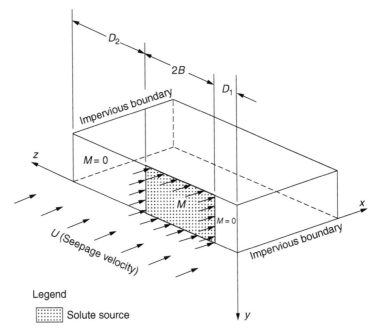

FIGURE 3-79 Schematic picture of a single instantaneous strip source in a bounded unidirectional flow field. (Adapted from Batu, V., unpublished paper, 1997b.)

boundaries. The source conditions expressed by Eqs. (3-388)–(3-390) take the forms

$$\frac{M}{V\varphi_e}\delta(t) = \begin{cases} \dfrac{M}{(2B)(1\,\text{unit})(1\,\text{unit})\varphi_e}\delta(t) = C_M\delta(t) = \dfrac{F_x(0, z)}{\varphi_e} & D_1 < z < D_1 + 2B \quad (3\text{-}398) \\[2mm] 0 & \text{otherwise} \quad (3\text{-}399) \end{cases}$$

where M (its unit is M) is the mass of the solute introduced instantaneously at $x = 0$ (see Figure 3-79) at a volume V [$(2B)(1\,\text{unit})(1\,\text{unit})$], $\delta(t)$ is Dirac delta function or impulse function, C_M (M/L^3) is the instantaneous source concentration, and $F_x(0, z)$ is the instantaneous solute source flux at $x = 0$. The mathematical and physical bases of Eqs. (3-398) and (3-399) are given in more detail in Section 2.10.3.2. Since the unit of $\delta(t)$ is T^{-1}, the dimensions of each side of Eqs. (3-398) and (3-399) are M/L^2/T.

Introducing $C_M = M/(2B\varphi_e)$ into Eq. (3-393),

$$I_0 = \frac{M}{\varphi_e} \tag{3-400}$$

The substitution of Eq. (3-400) into Eq. (3-396) with $H_p = D_1 + 2B + D_2$ gives

$$C_1(x, z, t) = \frac{M}{2B\varphi_e} \frac{2B}{(D_1 + 2B + D_2)} \frac{1}{D_x} F_1(t)[F_2(t) - F_3(t)F_4(t)] \tag{3-401}$$

Similarly, the substitution of Eqs. (3-398) and (3-399) into Eq. (3-394) gives

$$I_n = \frac{M}{2B\varphi_e} \frac{1}{\lambda_n} \{\sin[\lambda_n(D_1 + 2B)] - \sin(\lambda_n D_1)\} \tag{3-402}$$

where λ_n is given by Eq. (3-61). The substitution of Eq. (3-402) into Eq. (3-397) gives

$$C_2(x, z, t) = \frac{2}{\pi} \frac{M}{2B\varphi_e} \frac{1}{D_x} F_1(t)[F_2(t) - F_3(t)F_4(t)]$$

$$(3\text{-}403)$$

$$\times \sum_{n=1}^{\infty} \frac{1}{n} \{\sin[\lambda_n(D_1 + 2B)] - \sin(\lambda_n D_1)\} \cos(\lambda_n z) \exp\left(-\frac{D_z}{R_d} \lambda_n^2 t\right)$$

According to Eq. (3-395), the sum of $C_1(x, z, t)$ and $C_2(x, z, t)$, which are given by Eqs. (3-401) and (3-403), respectively, is the solution for $C(x, z, t)$:

$$C(x, z, t) = \frac{M}{2B\varphi_e} \frac{2B}{(D_1 + 2B + D_2)} \frac{1}{D_x} F_1(t)[F_2(t) - F_3(t)F_4(t)]$$

$$+ \frac{2}{\pi} \frac{M}{2B\varphi_e} \frac{1}{D_x} F_1(t)[F_2(t) - F_3(t)F_4(t)] \qquad (3\text{-}404)$$

$$\times \sum_{n=1}^{\infty} \frac{1}{n} \{\sin[\lambda_n(D_1 + 2B)] - \sin(\lambda_n D_1)\} \cos(\lambda_n z) \exp\left(-\frac{D_z}{R_d} \lambda_n^2 t\right)$$

where $F_1(t)$, $F_2(t)$, $F_3(t)$, and $F_4(t)$ are given by Eqs. (3-376), (3-377), (3-378), and (3-379), respectively.

Special Solution for the One-Dimensional Case
If $D_1 = 0$ and $D_2 = 0$ in Figure 3-79, the geometry turns out to be the case in Figure 3-74, which corresponds to the one-dimensional case and the corresponding solution is given in Section 3.3.4.1. By substituting these values into Eq. (3-86) one can write $\lambda_n = n\pi/(2B)$. Since $\sin(n\pi) = 0$, substitution of $\lambda_n = n\pi/(2B)$ into Eq. (3-404) it can be seen that the second term on the right-hand side of Eq. (3-404) becomes zero. Therefore, Eq. (3-404) turns out to be exactly the same as Eq. (3-373), which is the solution for concentration distribution for the one-dimensional case.

Convective–Dispersive Flux Components
The convective–dispersive flux components for two-dimensional solute transport in the x- and z-coordinate directions are given by Eqs. (3-90a) and (3-90b), respectively. The physical meaning and dimensions of these fluxes are presented in Section 3.3.2.2.

Following the same procedure as outlined in Example 3-6, from Eqs. (3-90a), (3-401), and (3-403), the flux component in the x-coordinate direction can be expressed as

$$F_x = \varphi_e U(C_1 + C_2) - \varphi_e D_x\left(\frac{\partial C_1}{\partial x} + \frac{\partial C_2}{\partial x}\right) \qquad (3\text{-}405)$$

where C_1 and C_2 are given Eqs. (3-401) and (3-403), respectively. Using the same procedure in Section 3.3.2.2, the following expression for $\partial C_1/\partial x$ can be written:

$$\frac{\partial C_1}{\partial x} = \frac{M}{\varphi_e} \frac{1}{(D_1 + 2B + D_2)} \frac{1}{D_x}\left[\frac{\partial F_1}{\partial x}(F_2 - F_3 F_4) + F_1\left(\frac{\partial F_2}{\partial x} - F_3\frac{\partial F_4}{\partial x} - F_4\frac{\partial F_3}{\partial x}\right)\right] \quad (3\text{-}406)$$

where F_1, F_2, F_3, and F_4 are given by Eqs. (3-376), (3-377), (3-378), and (3-379), respectively, and from these equations, the expressions for $\partial F_1/\partial x$, $\partial F_2/\partial x$, $\partial F_3/\partial x$, and $\partial F_4/\partial x$ are given by Eqs. (3-384), (3-385), (3-386), and (3-387), respectively. The procedure used in Example 3-6 is used in determining Eq. (3-406). Similarly, from Eq. (3-403), one can write

$$\frac{\partial C_2}{\partial x} = \frac{M}{2B\varphi_e} \frac{2}{\pi} \frac{1}{D_x}\left[\frac{\partial F_1}{\partial x}(F_2 - F_3 F_4) + F_1\left(\frac{\partial F_2}{\partial x} - F_3\frac{\partial F_4}{\partial x} - F_4\frac{\partial F_3}{\partial x}\right)\right]$$

$$\times \sum_{n=1}^{\infty} \frac{1}{n} \{\sin[\lambda_n(D_1 + 2B)] - \sin(\lambda_n D_1)\} \cos(\lambda_n z) \exp\left(-\frac{D_z}{R_d} \lambda_n^2 t\right) \quad (3\text{-}407)$$

The expression for the convective–dispersive flux component in the z direction can easily be determined from Equations (3-90b) and (3-403) as

$$F_z = -\varphi_e D_z \frac{\partial C_2}{\partial z} \tag{3-408}$$

since C_1, as given by Eq. (3-401), is not a function of z, $\partial C_1/\partial x = 0$, and from Eq. (3-403) one can write

$$\frac{\partial C_2}{\partial z} = -\frac{2}{\pi} \frac{M}{2B\varphi_e} \frac{1}{D_x} F_1(F_2 - F_3 F_4)$$

$$\times \sum_{n=1}^{\infty} \frac{1}{n} \{\sin[\lambda_n(D_1 + 2B)] - \sin(\lambda_n D_1)\} \lambda_n \sin(\lambda_n z) \tag{3-409}$$

Special Solutions for Other Sources

Equations (3-395), along with Eqs. (3-396) and (3-397), is the general solution for the resulting concentration distribution for any given $C_M(z)$ solute mass distribution which appears only in Eqs. (3-393) and (3-394). With the solute mass distribution functions (Figure 3-78), the integrals $I_0(z)$ and $I_n(z)$ given by Eqs. (3-393) and (3-394), respectively, need to be evaluated analytically or numerically depending on the type of the solute mass distribution functions. One example is given below.

Solution for Two Instantaneous Strip Sources

Two geometry of two-strip sources is shown in Figure 3-80. The boundary conditions at $x = 0$, expressed by Eqs. (3-388)–(3-390), take the forms

$$\frac{M_1}{V_1 \varphi_e} = \frac{M_1}{(2B)(1\text{unit})(1\text{unit})\varphi_e} \delta(t) = C_{M_1} \delta(t) = \frac{F_x(0, z)}{\varphi_e}, \tag{3-410}$$

$$D_1 < z < D_1 + 2B$$

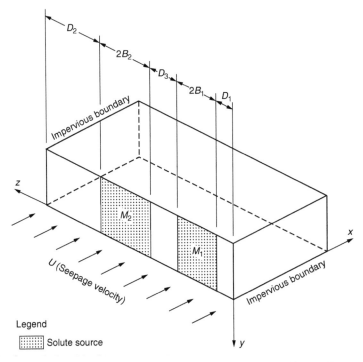

FIGURE 3-80 Schematic picture of two instantaneous strip sources with uniform solute source masses in a bounded medium having unidirectional flow field. (Adapted from Batu, V., unpublished paper, 1997b.)

$$\frac{M_2}{V_2\varphi_e} = \frac{M_2}{(2B_2)(1\text{unit})(1\text{unit})\varphi_e}\delta(t) = C_{M_2}\delta(t) = \frac{F_x(0, z)}{\varphi_e},$$

(3-411)

$$D_1 + 2B_1 + D_3 < z < D_1 + 2B_1 + D_3 + 2B_2$$

where M_1 and M_2 (units are M) are the solute source masses. Equation (3-395) is the general solution, and for a special case the integrals I_0 and I_n, given by Eqs. (3-393) and (3-394), respectively, need to be evaluated using the source geometry. Using the above-mentioned source conditions, from Eq. (3-393),

$$I_0 = \int_{D_1}^{D_1+2B_1}\frac{M_1}{2B_1\varphi_e}\,dz + \int_{D_1+2B_1+D_3}^{D_1+2B_1+D_3+2B_2}\frac{M_2}{2B_2\varphi_e}\,dz = \frac{M_1+M_2}{\varphi_e}$$

(3-412)

Similarly, from Eq. (3-394), one can write

$$I_n = \int_{D_1}^{D_1+2B_1}\frac{M_1}{2B_1\varphi_e}\cos(\lambda_n z)\,dz + \int_{D_1+2B_1+D_3}^{D_1+2B_1+D_3+2B_2}\frac{M_2}{2B_2\varphi_e}\cos(\lambda_n z)\,dz$$

(3-413)

or after some manipulations one gets

$$I_n = \frac{M_1}{2B_1\varphi_e}\frac{1}{\lambda_n}\{\sin[\lambda_n(D_1 + 2B_1)] - \sin(\lambda_n D_1)\}$$

(3-414)

$$+ \frac{M_2}{2B_2\varphi_e}\frac{1}{\lambda_n}\{\sin[\lambda_n(D_1 + 2B_1 + D_3 + 2B_2)] - \sin[\lambda_n(D_1 + 2B_1 + D_3)]\}$$

Equations (3-395)–(3-397), (3-412), and (3-414) form the solution for two–strip sources with instantaneous uniform solute source masses.

Evaluation of the Models
Comparison with Numerical Models
Equation (3-395) is obtained as a special solution from a more general three-dimensional solute transport model as given by Eq. (3-429). As can be seen from Section 3.3.4.3, there is good agreement between one of the special forms for a rectangular source of the three-dimensional analytical model and a three-dimensional finite-difference numerical model. Since the two-dimensional analytical model is a special form of the three-dimensional analytical model, the verification is also valid for the two-dimensional analytical model as well.

Number of Terms Selection for the Infinite Series in the Solutions
The number of terms, N, in the infinite series of the solution plays a key role in determining accurate results. Infinite series only exist in $C_2(x, z, t)$ component of the solution (Eq. [3-397] for the general solution and Eq. [3-403] for the strip source solution). The exponential term in the aforementioned equations acts such that the converged results can be obtained with a relatively less number of terms (N). Numerical experiments showed that N having values around 250 produces fairly accurate results.

Application of the Two-Dimensional Analytical Models to Ground Water Flow Systems
The general model and its derivative models presented above can be applied to solute transport predictions in aquifers if the ground water velocity can be assumed to be uniform. The models can also be applied to unsaturated zones with some additional assumptions. These are all discussed in the following sections.

In this section, first, a computer program, called ST2C, for a single-strip source is described. Then, application of this program is presented with an example.

A Computer Program (ST2C) for the Two-Dimensional Instantaneous Strip Source Model
A two-dimensional MS-DOS-based FORTRAN solute transport computer program, called solute transport two-dimensional instantaneous-type (ST2C), has been developed by the author. ST2C is

a menu-driven program and calculates values of concentration and convective–dispersive flux components (Eqs. [3-395], [3-401], [3-403], [3-405], and [3-408]). This program can be run interactively in IBM compatible computers.

Application to Saturated Zones

The two-dimensional analytical solute transport model, whose geometry is shown in Figure 3-79, can be used for different purposes. The impervious boundaries around the solute transport domain increase the potential usage areas of the model. For ground water flow cases, the effects of impervious boundaries such as bedrock outcrops and clay zones surrounding the main ground water flow zone can be taken into account. If the aquifer approaches infinity from both sides of the source, its effect can be taken into account by assigning large values to D_1 and D_2 as compared with the source width $2B$. The model can also be used along a cross-section of a confined or unconfined aquifer such that the $z = 0$ plane is the bottom of the aquifer and $z = D_1 + 2B + D_2$ plane is the upper boundary of the aquifer. For a confined aquifer, $z = D_1 + 2B + D_2$ plane is the bottom of the confining layer and for an unconfined aquifer is the water table of the unconfined aquifer. For these cases the source will be infinitely long in the direction of the y-coordinate axis. For two-dimensional areal model case in the x–z plane, as shown in Figure 3-79, it is assumed that the source extends from the bottom of a confined or unconfined aquifer to the top of the confined aquifer or to the water table of the unconfined aquifer.

EXAMPLE 3-28

For a instantaneous strip source and aquifer geometry shown in Figure 3-79, the following parameters are given: $K_h = 13.875$ m/day; $I = (0.5 \text{ m}/(185 \text{ m}) = 0.0027027$ m/m; $\varphi_e = 0.25$ ($U = 0.15$ m/day); $K_d = 0$ cm^3/g; $\rho_b = 1.78$ cm^3/g ($R_d = 1$); $D^* = 0$ m^2/day; $M = 1,000$ g; $2B = 10$ m; $D_1 = 1,000$ m; $D_2 = 1,000$ m; v $= 0$ day^{-1}; $\alpha_L = 21.3$ m; and $\alpha_T = 4.3$ m. With these parameters, using the ST2C program for 5, 10, and 20 years elapsed times, (1) determine the contours for $C(x, z, t)$ and discuss the results; (2) determine the contours for $F_x(x, z, t)$ and discuss the results; and (3) determine the contours for $F_z(x, z, t)$ and discuss the results.

SOLUTION

Using the given data, the ST2C program was run to generate the coordinates and concentration data at a given time. The output of ST2C was used to generate contours using the SURFER™ contouring program (Golden Software, Inc., 1988).

(1) The concentration, $C(x, z, t)$ (g/m^3) contours after 5, 10, and 20 years are given in Figure 3-81, 3-82, and 3-83, respectively. Figure 3-81 shows that after 5 years elapsed time the center of the plume is approximately 300 m away from the original source location in the x-coordinate direction.

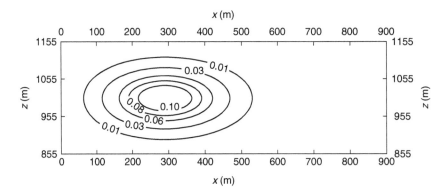

FIGURE 3-81 Concentration (g/m^3) contours after 5 years elapsed time for an instantaneous strip source for Example 3-28. (Adapted from Batu, V., unpublished paper, 1997b.)

After 10 years elapsed time, the center of the plume is approximately 560 m away from its original location in the x-coordinate direction (Figure 3-82). After 20 years elapsed time, the center of the plume is approximately 1110 m away from its original location in the x-coordinate direction (Figure 3-83).

(2) The contours for $F_x(x, z, t)$ (g/day/m²) after 5, 10, and 20 years elapsed times are given in Figure 3-84, 3-85, and 3-86, respectively.

(3) The contours for $F_z(x, z, t)$ (g/day/m²) after 5, 10, and 20 years elapsed times are given in Figure 3-87, 3-88, and 3-89, respectively. Note that the contours at both sides of the $z = 1000$ m line geometrically are symmetrical, but the corresponding contour values have the opposite signs. This is not unexpected due to the fact that the z-coordinate axis has positive sign in the upward direction and the symmetry line is located at $z = 1000$ m. Notice also that along this symmetry line $F_z = 0$, as expected.

Application to Unsaturated Zones

The schematic diagram in Figure 3-90 shows the applicability of the two-dimensional ST2C model for a single instantaneous source. The associated governing equations for flow and solute transport in unsaturated porous media under uniform infiltration rate conditions are given in Section 2.8. Some important points for the application and assumptions are given below:

1. It is assumed that there is a uniform infiltration rate, v_0, at the soil surface. As shown in Section 2.8.1, this infiltration rate creates a constant ground water velocity in the unsaturated zone. Also in the same section it is shown that the ground water velocity, U, can

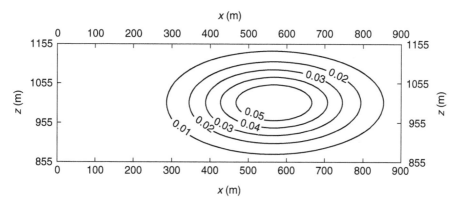

FIGURE 3-82　Concentration(g/m³) contours after 10 years elapsed time for an instantaneous strip source for Example 3-28. (Adapted from Batu, V., unpublished paper, 1997b.)

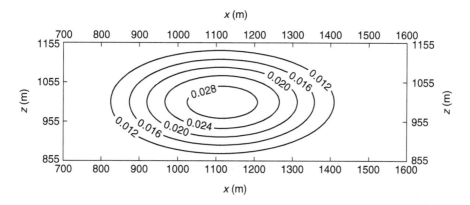

FIGURE 3-83　Concentration(g/m³) contours after 20 years elapsed time for an instantaneous strip source for Example 3-28. (Adapted from Batu, V., unpublished paper, 1997b.)

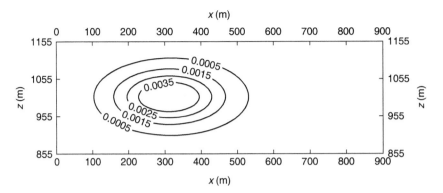

FIGURE 3-84 Contours for $F_x(x, z, t)$ (g/day/m²) after 5 years elapsed time for an instantaneous strip source for Example 3-28. (Adapted from Batu, V., unpublished paper, 1997b.)

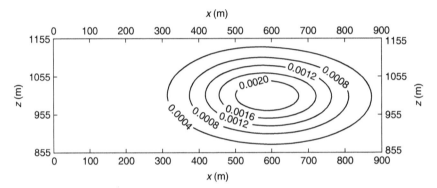

FIGURE 3-85 Contours for $F_x(x, z, t)$ (g/day/m²) after 10 years elapsed time for an instantaneous strip source for Example 3-28. (Adapted from Batu, V., unpublished paper, 1997b.)

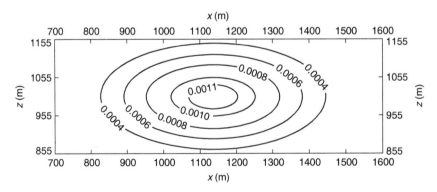

FIGURE 3-86 Contours for $F_x(x, z, t)$ (g/day/m²) after 20 years elapsed time for an instantaneous strip source for Example 3-28. (Adapted from Batu, V., unpublished paper, 1997b.)

be determined using the unsaturated hydraulic conductivity vs. water content relationship for a given soil, i.e., $K = K(\theta)$.

2. Liquid waste from a surface contaminant source is being discharged to the unsaturated zone. Landfills, some units of factories, or other industrial complexes may be considered as surface contaminant sources.

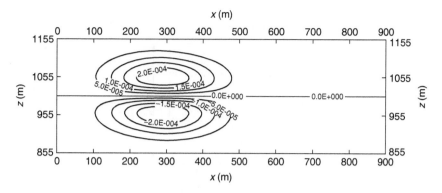

FIGURE 3-87　Contours for $F_z(x, z, t)$ (g/day/m²) after 5 years elapsed time for an instantaneous strip source for Example 3-28. (Adapted from Batu, V., unpublished paper, 1997b.)

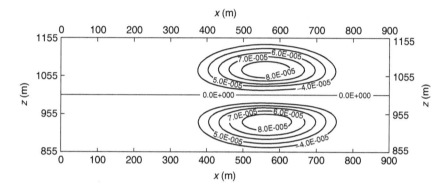

FIGURE 3-88　Contours for $F_z(x, z, t)$ (g/day/m²) after 10 years elapsed time for an instantaneous strip source for Example 3-28. (Adapted from Batu, V., unpublished paper, 1997b.)

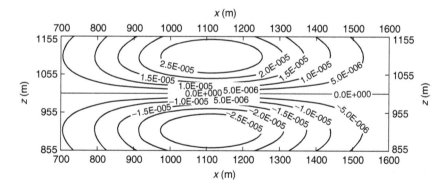

FIGURE 3-89　Contours for $F_z(x, z, t)$ (units g/day/m²) after 20 years elapsed time for an instantaneous strip source for Example 3-28. (Adapted from Batu, V., unpublished paper, 1997b.)

3.　The length of the source, L, in practice is finite, but in the model it is infinite (in the y-coordinate axis).
4.　It is assumed that the waste liquid seeping from the bottom of the source ditch reaches to the top of the unsaturated zone and creates an instantaneous source (M) of a certain solute species in the area beneath the ditch.

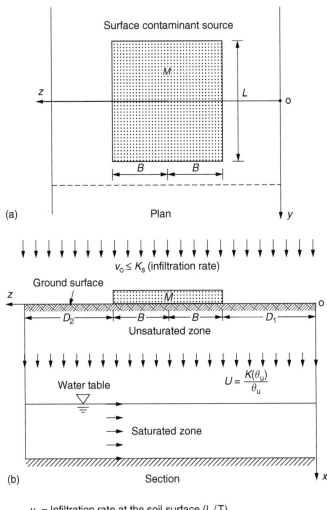

v_0 = Infiltration rate at the soil surface (L/T)

K_s = Saturated hydraulic conductivity of the unsaturated zone (L/T)

θ_u = Uniform volumetric water content (dimensionless)

$K(\theta_u)$ = Unsaturated hydraulic conductivity (L/T)

FIGURE 3-90 Schematic representation for the applicability of the ST2C program to unsaturated

3.3.4.3 Three-Dimensional Analytical Solute Transport Models for Instantaneous Rectangular Sources

This model was developed as a general three-dimensional analytical solution by Batu (1997c), and to the author's knowledge, this model has not been published in the literature yet. Similar to the first-type source three-dimensional analytical model (Section 3.3.2.3) and third-type source three-dimensional analytical model (Section 3.3.3.3), this model will probably have a special importance in the literature. In the following sections, the three-dimensional instantaneous source model will be presented with its potential application areas.

Problem Statement and Assumptions
The geometry of the generalized three-dimensional analytical solute transport model for instantaneous rectangular sources is illustrated in Figure 3-91. All solute sources are located on the yOz

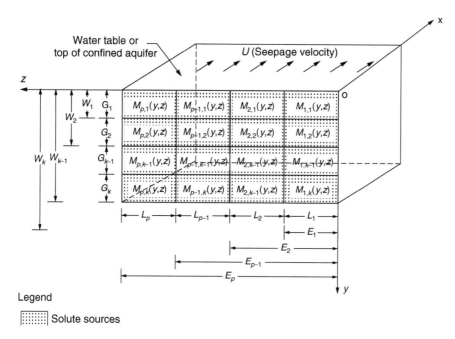

FIGURE 3-91 Schematic representation of the three-dimensional solute transport model with instantaneous rectangular sources in a bounded unidirectional flow field. (Adapted from Batu, V., unpublished paper, 1997c.)

plane and the four other planes that are perpendicular to the yOz plane are assumed to be impervious. All five planes defined above form a parallelepiped which goes to infinity in the x-coordinate direction. The multiple instantaneous rectangular sources are located irregularly along the yOz plane as shown in Figure 3-91. The uniform flow field is assumed to exist along the x-coordinate direction and the system goes to infinity along the same coordinate direction. The upper boundary of the model (xOz plane) from a practical standpoint corresponds to the water table of an unconfined aquifer or the upper boundary of a confined aquifer. The lower boundary, which is parallel to the xOz plane at $y = W_k$ distance (see Figure 3-91), may represent a bedrock or an impervious boundary or the lower boundary of an unconfined or confined aquifer.

Under the framework of the above-mentioned description, the problem is a three-dimensional solute transport from instantaneous multiple rectangular sources between four impervious planar boundaries in a unidirectional flow field as shown in Figure 3-91. The x- and z-axis are the planar coordinates and the y axis is the vertical coordinate. The assumptions are as follows:

1. The solute sources are planar and located irregularly on the y–z plane in a unidirectional steady ground water velocity, U, field.
2. The solute sources are located at $x = 0$ plane and they are perpendicular to the velocity of the flow field.
3. The corresponding solute mass to each rectangle is uniformly mixed.
4. The medium goes to infinity in the x-coordinate direction and y varies between 0 and W_k and z varies between 0 and E_p.

Governing Differential Equation
The governing differential equation is given by Eq. (3-128).

Initial and Boundary Conditions

The initial condition for $C(x, z, t)$ is given by Eq. (3-129). The boundary conditions in the first row of rectangles on the yOz at $x = 0$ are

$$\frac{M_{1,1}}{V_{1,1}\varphi_e}\delta(t) = \frac{M_{1,1}}{(L_1)(G_1)(1\text{unit})\varphi_e}\delta(t) = C_{M1,1}\delta(t) = \frac{F_x(0, y, z)}{\varphi_e} \quad 0<y<W_1 \quad 0<z<E_1 \quad (3\text{-}415)$$

$$\frac{M_{p-1,1}}{V_{p-1,1}\varphi_e}\delta(t) = \frac{M_{p-1,1}}{(L_{p-1})(G_1)(1\text{unit})\varphi_e}\delta(t) = C_{p-1,1}\delta(t) = \frac{F_x(0, y, z)}{\varphi_e} \quad 0<y<W_1 \quad E_{p-2}<z<E_{p-1} \quad (3\text{-}416)$$

$$\frac{M_{p,1}}{V_{p,1}\varphi_e}\delta(t) = \frac{M_{p,1}}{(L_p)(G_1)(1\text{unit})\varphi_e}\delta(t) = C_{p,1}\delta(t) = \frac{F_x(0, y, z)}{\varphi_e} \quad 0<y<W_1 \quad E_{p-1}<z<E_p \quad (3\text{-}417)$$

Similarly, expressions for 2nd, ..., $(k-1)$, kth row can be written. The expressions for the kth row are as follows:

$$\frac{M_{1,k}}{V_{1,k}\varphi_e}\delta(t) = \frac{M_{1,k}}{(L_1)(G_k)(1\text{unit})\varphi_e}\delta(t) = C_{M1,k}\delta(t) = \frac{F_x(0, y, z)}{\varphi_e} \quad W_{k-1}<y<W_k \quad 0<z<E_1 \quad (3\text{-}418)$$

$$\frac{M_{p-1,k}}{V_{p-1,k}\varphi_e}\delta(t) = \frac{M_{p-1,k}}{(L_{p-1})(G_k)(1\text{unit})\varphi_e}\delta(t) = C_{p-1,k}\delta(t) = \frac{F_x(0, y, z)}{\varphi_e} \quad W_{k-1}<y<W_k \quad E_{p-2}<z<E_{p-1} \quad (3\text{-}419)$$

$$\frac{M_{p,k}}{V_{p,k}\varphi_e}\delta(t) = \frac{M_{p,k}}{(L_p)(G_k)(1\text{unit})\varphi_e}\delta(t) = C_{p,k}\delta(t) = \frac{F_x(0, y, z)}{\varphi_e} \quad W_{k-1}<y<W_k \quad E_{p-1}<z<E_p \quad (3\text{-}420)$$

where $M_{i,j}$ (its unit is M) is the mass of solute introduced instantaneously at the $V_{i,j}$ area, whose dimensions are L_i, G_j, and 1 unit; $\delta(t)$ is Dirac delta function or impulse function, $C_{M_{i,j}}$ (M/L^3) the instantaneous source concentration in the $V_{i,j}$ volume, and $F_x(0, y, z)$ (M/L^2/T) the instantaneous solute source flux at $x = 0$; E_i and W_i are given by Eqs. (3-136) and (3-137), respectively. The other boundary conditions at the impervious boundaries and infinity are given by Eqs. (3-138)–(3-143). Equations (3-415)–(3-420) state that a mass of solute $M_{i,j}$ is introduced instantaneously at the area $A_{i,j}$. The mathematical and physical bases of Eqs. (3-415)–(3-420) are given in more detail in Section 2.10.3.2. Since the unit of $\delta(t)$ is T^{-1}, the dimensions of each side of Eqs. (3-415)–(3-420) are M/L^2/T. The physical meaning of the rest of boundary conditions is given in Section 3.3.2.3.

Solution for Concentration Distribution

The boundary-value problem defined by Eqs. (3-128), (3-129), (3-415)–(3-420), (3-138)–(3-143) is solved by the Laplace transform and Fourier analysis techniques. For this purpose, the solution is first obtained in the Laplace domain and then the concentration variation is determined by taking the inverse Laplace transform of the Laplace domain solution. It can be observed that the afore-mentioned equations are the same as the first-type three-dimensional analytical model (Section 3.3.2.3) as well as the third-type three-dimensional analytical model (Section 3.3.3.3), with the exception of the conditions at $x = 0$. Therefore, there are common equations in the solution process of the three-dimensional analytical solutions. As a result, the solution of the instantaneous three-dimensional analytical model will be presented by using the common equations as appropriate.

General Solution in the Laplace Domain

As can be seen from Section 3.3.2.3, the Laplace domain solution for the first-type source case is derived by using the initial condition and all boundary conditions with the exception of the ones at $x = 0$. Therefore, Eq. (3-145) is valid for the third-type source conditions as well, and derivation of Eq. (3-145) is presented in Example 3-10. The final form of the general solution in the Laplace domain is in the form of Eq. (3-150); and using the boundary conditions at $x = 0$, the expressions for $u_1(x, y, z, s)$, $u_2(x, y, z, s)$, $u_3(x, y, z, s)$, and $u_4(x, y, z, s)$ are as follows:

$$u_1(x, y, z, s) = \left(\frac{1}{2}\right)\left(\frac{1}{2}\right)B_{00} \exp\left[\left(\frac{U}{2D_x} - K_{00}\right)x\right] \tag{3-421}$$

$$u_2(x, y, z, s) = \frac{1}{2}\sum_{n=1}^{\infty} B_{0n} \exp\left[\left(\frac{U}{2D_x} - K_{0n}\right)x\right] \tag{3-422}$$

$$u_3(x, y, z, s) = \frac{1}{2}\sum_{m=1}^{\infty} B_{m0} \cos(\eta_m y) \exp\left[\left(\frac{U}{2D_x} - K_{m0}\right)x\right] \tag{3-423}$$

and

$$u_4(x, y, z, s) = \sum_{m=1}^{\infty}\sum_{n=1}^{\infty} B_{mn}\cos(\eta_m y)\cos(\lambda_n z)\exp\left[\left(\frac{U}{2D_x} - K_{mn}\right)x\right] \tag{3-424}$$

where K_{mn}, K_{00}, K_{0n}, and K_{m0} are given by Eqs. (3-148), (3-155), (3-156), and (3-157), respectively. In the above-mentioned equations, λ_n and η_m are given by Eqs. (3-146) and (3-147), respectively. Equation (3-150) and (3-309)–(3-312) are based on the fact that in Eq. (3-145), B_{0n} and B_{m0} are one-half and B_{00} is one-quarter (Carslaw and Jaeger, 1959, p. 182). The expression for B_{mn}, which is derived in Example 3-29, is as follows:

$$B_{mn} = \frac{1}{\left(\dfrac{U}{2} + D_x K_{mn}\right)} \frac{4}{W_k E_p}\int_0^{W_k}\int_0^{E_p} C_M\cos(\eta_m y)\cos(\lambda_n z)\ dy\ dz \tag{3-425}$$

From Eq. (3-425), B_{00}, B_{0n}, and B_{m0}, can, respectively, be expressed as

$$B_{00} = \frac{1}{\left(\dfrac{U}{2} + D_x K_{00}\right)} \frac{4}{W_k E_p}\int_0^{W_k}\int_0^{E_p} C_M dy\ dz \tag{3-426}$$

$$B_{0n} = \frac{1}{\left(\dfrac{U}{2} + D_x K_{0n}\right)} \frac{4}{W_k E_p}\int_0^{W_k}\int_0^{E_p} C_M\cos(\lambda_n z)dy\ dz \tag{3-427}$$

and

$$B_{m0} = \frac{1}{\left(\dfrac{U}{2} + D_x K_{m0}\right)} \frac{4}{W_k E_p}\int_0^{W_k}\int_0^{E_p} C_M\cos(\eta_m y)dy\ dz \tag{3-428}$$

Now, the general solution in the Laplace domain is complete. The procedure for determining B_{mn} is given in Example 3-29 below.

EXAMPLE 3-29

Determine the expression for B_{mn} in Eq. (3-145) by using the boundary conditions at $x = 0$ given by Eqs. (3-415)–(3-420).

SOLUTION

In order to determine the expression for B_{mn} the boundary conditions at $x = 0$ given by Eqs. (3-415)–(3-420) need to be used. The procedure is similar to that used in Example 3-22. The boundary conditions at $x = 0$ need to be expressed in the Laplace domain. The convective–dispersive flux, F_x, in the x direction is given by Eq. (3-90a). Therefore, taking the Laplace transform of the combined form of Eqs. (3-90a), and (3-315)–(3-320) gives

$$C_M = Uu - D_x\frac{\partial u}{\partial x} \tag{E3-29-1}$$

From Eq. (3-145), one can write

$$\frac{\partial u}{\partial x} = \sum_{m=0}^{\infty} \sum_{m=0}^{\infty} B_{mn}\left(\frac{U}{2D_x} - K_{mn}\right)\cos(\eta_m y)\cos(\lambda_n z)\exp\left[\left(\frac{U}{2D_x} - K_{mn}\right)x\right] \qquad \text{(E3-29-2)}$$

For $x = 0$, Eqs. (3-145) gives

$$u(0, y, z, s) = \sum_{m=0}^{\infty} \sum_{n=0}^{\infty} B_{mn}\cos(\eta_m y)\cos(\lambda_n z) \qquad \text{(E3-29-3)}$$

For $x = 0$, Eq. (E3-29-2) gives

$$\frac{\partial u(0, y, z, s)}{\partial x} = \sum_{m=0}^{\infty} \sum_{n=0}^{\infty} B_{mn}\left(\frac{U}{2D_x} - K_{mn}\right)\cos(\eta_m y)\cos(\lambda_n z) \qquad \text{(E3-29-4)}$$

The substitution of Eqs. (E3-29-3) and (E3-29-4) into Eq. (E3-29-1) and after some manipulation gives

$$C_M = \sum_{m=0}^{\infty} \sum_{n=0}^{\infty} E_{mn}\cos(\eta_m y)\cos(\lambda_n z) \qquad \text{(E3-29-5)}$$

where

$$E_{mn} = B_{mn}\left(\frac{U}{2} + D_x K_{mn}\right) \qquad \text{(E3-29-6)}$$

Equation (E3-29-5) is in the form of Eq. (E3-10-27). Therefore, the determination procedure for E_{mn} presented through Eqs. (E3-10-28)–(E3-10-32) can be used. As a result, the expression for E_{mn} can be written as

$$E_{mn} = \frac{4}{W_k E_p} \int_0^{W_k} \int_0^{E_p} C_M \cos(\eta_m y)\cos(\lambda_n z)dydz \qquad \text{(E3-29-7)}$$

Combination of Eqs. (E3-29-6) and (E3-29-7) gives Eq. (3-425).

Inverse Laplace Transforms of Eqs. (3-421), (3-422), (3-423), and (3-424)
Determination of the inverse Laplace transforms of Eqs. (3-421)–(3-424) are given in Example 3-30. The sum of the inverse Laplace transforms of Eqs. (3-421)–(3-424) according to Eq. (3-150) is

$$C(x, y, z, t) = C_1(x, y, z, t) + C_2(x, y, z, t) + C_3(x, y, z, t) + C_4(x, y, z, t) \qquad \text{(3-429)}$$

The expression for $C_1(x, y, z, t)$ is

$$C_1(x, y, z, t) = \frac{I_0(y, z)}{D_x}F_1(t)[F_2(t) - F_3(t)F_4(t)] \qquad \text{(3-430)}$$

where

$$I_0(y, z) = \frac{1}{W_k E_p} \int_0^{W_k} \int_0^{E_p} C_M dydz \qquad \text{(3-431)}$$

The expression for $C_2(x, y, z, t)$ is

$$C_2(x, y, z, t) = \frac{2}{D_x}F_1(t)[F_2(t) - F_3(t)F_4(t)] \qquad \text{(3-432)}$$

$$\times \sum_{n=1}^{\infty} I_n(y, z)\cos(\lambda_n z)\exp\left(-\frac{D_z}{R_d}\lambda_n^2 t\right)$$

where

$$I_n(y, z) = \frac{1}{W_k E_p} \int_0^{W_k} \int_0^{E_p} C_M \cos(\lambda_n z) \, dy \, dz \qquad (3\text{-}433)$$

The expression for $C_3(x, y, z, t)$ is

$$C_3(x, y, z, t) = \frac{2}{D_x} F_1(t)[F_2(t) - F_3(t)F_4(t)]$$

$$\times \sum_{m=1}^{\infty} I_m(y, z) \cos(\eta_m y) \exp\left(-\frac{D_y}{R_d} \eta_m^2 t\right) \qquad (3\text{-}434)$$

where

$$I_m(y, z) = \frac{1}{W_k E_p} \int_0^{W_k} \int_0^{E_p} C_M \cos(\eta_m y) \, dy \, dz \qquad (3\text{-}435)$$

The expression for $C_4(x, y, z, t)$ is

$$C_4(x, y, z, t) = \frac{4}{D_x} F_1(t)[F_2(t) - F_3(t)F_4(t)]$$

$$\times \sum_{m=1}^{\infty} \sum_{n=1}^{\infty} I_{mn}(y, z) \cos(\eta_m y)\cos(\lambda_n z) \exp\left(-\frac{D_y}{R_d} \eta_m^2 t - \frac{D_z}{R_d} \lambda_n^2 t\right) \qquad (3\text{-}436)$$

where

$$I_{mn}(y, z) = \frac{1}{W_k E_p} \int_0^{W_k} \int_0^{E_p} C_M \cos(\eta_m y) \cos(\lambda_n z) \, dy \, dz \qquad (3\text{-}437)$$

In Eqs. (3-430), (3-432), (3-434), and (3-436), the expressions for $F_1(t)$, $F_2(t)$, $F_3(t)$, and $F_4(t)$ are given by Eqs. (3-376), (3-377), (3-378), and (3-379), respectively.

EXAMPLE 3-30

Derive Eq. (3-429) by determining the inverse Laplace transform of Eqs. (3-421)–(3-424).

SOLUTION

It is known that the inverse Laplace transforms of $u_1(x, z, s)$, $u_2(x, z, s)$, $u_3(x, z, s)$, and $u_4(x, z, s)$ are $C_1(x, y, z, t)$, $C_2(x, y, z, t)$, $C_3(x, y, z, t)$, and $C_4(x, y, z, t)$, respectively, and are presented below.

Inverse Laplace Transform of $u_1(x, y, z, s)$
In order to determine the inverse Laplace transform of $u_1(x, y, z, s)$, as given by Eq. (3-421), according to the second expression of Eq. (E3-5-2), some manipulations need to be made. Equation (3-421) can also be expressed as

$$u_1(x, y, z, s - b) = \frac{1}{H_p} I_0(y, z) \exp\left(\frac{Ux}{2D_x}\right) e_2(s - b) \qquad (E3\text{-}30\text{-}1)$$

where $I_0(y, z)$ is given by Eq. (3-431) and $e_2(s-b)$ is given by Eq. (E3-17-3), and the expression for b is given by Eq. (E3-5-4). As a result, the procedure for taking the inverse Laplace transform of $u_1(x, y, z, s-b)$ is the same as the inverse Laplace transform of Eq. (E3-17-3). Therefore, the inverse Laplace transform of $u_1(x, y, z, s)$ can be expressed by Eq. (3-430) as given earlier.

Inverse Laplace Transform of $u_2(x, y, z, s)$
The inverse Laplace transform procedure of $u_2(x, y, z, s)$ is similar to that for $u_1(x, y, z, s)$ as presented earlier. Therefore, the inverse Laplace transform for $u_2(x, y, z, s)$ is given by Eq. (3-432).

Inverse Laplace Transform of $u_3(x, y, z, s)$
The inverse Laplace transform procedure of $u_3(x, y, z, s)$ is similar to that for $u_1(x, y, z, s)$ as presented earlier. Therefore, the inverse Laplace transform for $u_3(x, y, z, s)$ is given by Eq. (3-434).

Inverse Laplace Transform of $u_4(x, y, z, s)$
The inverse Laplace transform procedure of $u_4(x, y, z, s)$ is similar to that for $u_1(x, y, z, s)$ as presented earlier. Therefore, the inverse Laplace transform for $u_4(x, y, z, s)$ is given by Eq. (3-436).

Special Solution for a Single Rectangular Source
Concentration Distribution
The geometry of a single rectangular source is shown in Figure 3-92, which is a special form of Figure 3-91. The rectangular source is located at any place along the yOz plane and its width and height are $2B$ and $2H$, respectively. The distances from the right and left edges of the source to the vertical impervious boundaries are D_1 and D_2, respectively. The source conditions expressed by Eqs. (3-415)– (3-420) are in the forms

$$\frac{M}{V\varphi_e}\delta(t) = \begin{cases} \dfrac{M}{(2B)(2H)(1\ \text{unit})\varphi_e}\delta(t)=C_M\delta(t)=\dfrac{F_x(0, y, z)}{\varphi_e} & D_1<z<D_1+2B,\ \ H_1<y<H_1+2H \quad (3\text{-}438) \\[2ex] 0 & \text{otherwise} \hspace{4.5em} (3\text{-}439) \end{cases}$$

where M (its unit is M) is the mass of solute introduced instantaneously at $x = 0$ (see Figure 3-92) in a volume V [$(2B)(2H)(1\ \text{unit})$], $\delta(t)$ is Dirac delta function or impulse function, C_M (M/L^3) is the instantaneous source concentration, and $F_x(0,y,z)$ is the instantaneous solute flux at $x = 0$. The mathematical and physical bases of Eqs. (3-438) and (3-439) are given in more detail in Section 2.10.3.2. Since the unit of $\delta(t)$ is T^{-1}, the dimensions of each side of Eqs. (3-438) and (3-439) are M/L^2/T.

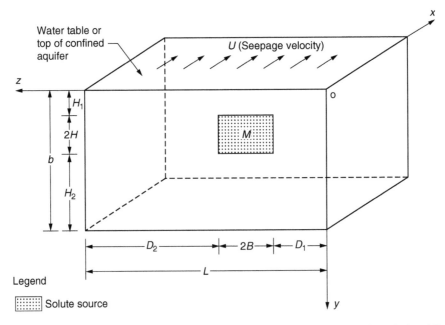

FIGURE 3-92 The geometry of a single instantaneous rectangular source within a bounded unidirectional flow field (Adapted from Batu, V., unpublished paper, 1997c.)

The substitution of Eqs. (3-438) and (3-439) into Eq. (3-431) gives

$$I_0 = \frac{M}{\varphi_e} \frac{1}{(H_1 + 2H + H_2)(D_1 + 2B + D_2)} \tag{3-440}$$

Similarly, substitution of Eqs. (3-438) and (3-439) into Eq. (3-437) gives

$$I_{mn} = \frac{M}{(2B)(2H)\varphi_e} \frac{1}{\pi^2} \frac{S_n}{n} \frac{S_m}{m} \tag{3-441}$$

where S_n and S_m are given by Eqs. (3-185) and (3-186), respectively.

Similarly, substitution Eqs. (3-438) and (3-439) into Eq. (3-433) gives

$$I_n = \frac{M}{(2B)(2H)\varphi_e} \frac{2H}{(H_1 + 2H + H_2)} \frac{1}{\pi} \frac{S_n}{n} \tag{3-442}$$

where S_n is given by Eq. (3-185).

Finally, substitution of Eqs. (3-438) and (3-439) into Eq. (3-435) one gets

$$I_m = \frac{M}{(2B)(2H)\varphi_e} \frac{2B}{(D_1 + 2B + D_2)} \frac{1}{\pi} \frac{S_m}{m} \tag{3-443}$$

where S_m is given by Eq. (3-186).

Therefore, with Eqs. (3-440)–(3-443), Eqs. (3-429), (3-430), (3-432), (3-434), and (3-436) form the solution for concentration distribution for a single rectangular source shown in Figure 3-92. In the above-mentioned expressions, λ_n and η_m are given by Eqs. (3-146) and (3-147), respectively, with $E_p = D_1 + 2B + D_2$ and $W_k = H_1 + 2H + H_2$.

Special Solution for the Two-Dimensional Cases

Two special solutions corresponding to the two-dimensional special cases can be obtained from the three-dimensional instantaneous source solution for a rectangular source whose geometry is shown in Figure 3-92. The procedure is similar to those two-dimensional special solutions obtained from the three-dimensional first-type source solution (Section 3.3.2.3) and three-dimensional third-type source solution (Section 3.3.3.3). The first two-dimensional special case is shown in Figure 3-32, which shows that the source fully penetrates the entire thickness of the aquifer in the y-coordinate direction. In other words, the source becomes a vertical strip. This corresponds to Figure 3-92, which is the geometry of the ST2C model. The second two-dimensional special case is shown in Figure 3-33, which shows that the source fully penetrates the entire length of the aquifer in the z-coordinate direction. The special cases corresponding to these special cases are derived in Example 3-31, which is presented below.

EXAMPLE 3-31

Derive the two-dimensional instantaneous strip source solution from the three-dimensional instantaneous rectangular source solution which is the combination of Eqs. (3-429), (3-430), (3-432), (3-434), (3-436), and (3-440)–(3-443).

SOLUTION

First Special Case: The geometry of the first-special case is shown in Figure 3-32 for which $H_1 = 0$ and $H_2 = 0$. For the two- and three-dimensional cases the instantaneous source concentrations are defined as

$$C_{M_{2D}} = \frac{M}{(2B)(1\text{unit})(1\text{unit})\varphi_e}, \quad C_{M_{3D}} = \frac{M}{(2B)(2H)(1\text{unit})\varphi_e} \tag{E3-31-1}$$

where the first expression is due to Eq. (3-398) and the second expression is due to Eq. (3-438). If $H_1 = 0$ and $H_2 = 0$ and using the second expression in Eq. (E3-31-1), the expression for $C_1(x, y, z, t)$, which is the combination of Eqs. (3-430) and (3-440), takes the form

$$C_1(x, y, z, t) = \frac{C_{M_{3D}}(2B)(2H)\varphi_e}{\varphi_e} \frac{1}{(2H)(D_1 + 2B + D_2)} \frac{1}{D_x} F_1(t)[F_2(t) - F_3(t)F_4(t)] \quad \text{(E3-31-2)}$$

After cancellation of the two $2H$ terms in the numerator and denominator in Eq. (E3-31-1) and using the first expression in Eq. (E3-31-1), it can be observed that Eq. (E3-31-2) reduces to the exact form of Eq. (3-401). Since $H_1 = 0$, Eq. (3-147) gives $\eta_m = m\pi/(2H)$ and the sine terms in Eq. (3-186) become zero, resulting in $S_m = 0$. As a result, $I_{mn} = 0$ and thereby $C_4(x, y, z, t) = 0$ according to the combination of Eqs. (3-436) and (3-441). Again substitution of $H_1 = 0$ and $H_2 = 0$ into the combination of Eqs. (3-432) and (3-442) gives exactly the same form of Eq. (3-403). Since $S_m = 0$, $I_m = 0$ according to Eq. (3-188). Therefore, $C_3(x, y, z, t) = 0$, which is the combination of Eqs. (3-434) and (3-443). Now the proof is complete, i.e., Eq. (3-404) is a special solution of the three-dimensional instantaneous rectangular source model.

Second Special Case: The geometry of the second special case corresponds to $D_1 = 0$ and $D_2 = 0$ (see Figure 3-33). This is equivalent to the first special case with the exception that the strip source is parallel to the z-coordinate axis. Introducing these values into Eqs. (3-429), (3-430), (3-432), (3-434), (3-436), and (3-440)–(3-443), it can be shown that $C_2(x, y, z, t)$ and $C_4(x, y, z, t)$ both become zero and the sum of $C_1(x, y, z, t)$ and $C_3(x, y, z, t)$ turns out to be exactly the same as the solution for a single-strip source as given by Eq. (3-404), with the exception that H, H_1, and H_2 define the source geometry instead of B, D_1, and D_2.

Special solution for the One-Dimensional Case
In Section 3.3.4.2, the one-dimensional solution as given by Eq. (3-375) is obtained as a special solution from the two-dimensional solution. As shown above, Eq. (3-404) is a special case of the three-dimensional solution. Therefore, it can be concluded that the same one-dimensional solution is a special case of the three-dimensional solution as well.

Convective–Dispersive Flux Components
From Eqs. (2-41a)–(2-41c), the convective–dispersive flux components in the x-, y-, and z-coordinate directions, respectively, can easily be determined using the same procedure for the convective–dispersive flux components for the two-dimensional case in Section 3.3.4.2. Since the mathematical expressions are too lengthy they are not presented in this chapter.

Special Solutions for Other Sources
Equation (3-429) is the general solution for the resulting concentration distribution for any given number of instantaneous rectangular sources. With the source masses (Figure 3-91), the integrals I_0, I_n, I_m, and I_{mn} given by Eqs. (3-431), (3-433), (3-435), and (3-437), respectively, need to be evaluated. An example is presented below.

Solution for Two Instantaneous Rectangular Sources
The geometry of two rectangular sources with uniform source concentrations is shown in Figure 3-93. The boundary conditions at $x = 0$ expressed by Eqs. (3-415)–(3-420) take the following forms:

$$\frac{M_1}{V_1\varphi_e}\delta(t) = \frac{M_1}{(2B_1)(2H_1)(1\text{unit})\varphi_e}\delta(t) = C_{M_1}\delta(t) = \frac{F_x(0, y, z)}{\varphi_e} \quad \text{(3-444)}$$

$$D_1 < z < D_1 + 2B_1, \quad A_1 + 2H_1 + A_3 < y < A_1 + 2H_1 + A_3 + 2H_2$$

$$\frac{M_2}{V_2\varphi_e}\delta(t) = \frac{M_2}{(2B_2)(2H_2)(1\text{unit})\varphi_e}\delta(t) = C_{M_2}\delta(t) = \frac{F_x(0, y, z)}{\varphi_e} \quad \text{(3-445)}$$

$$D_1 + 2B_1 + D_3 < z < D_1 + 2B_1 + D_3 + 2B_2, \quad A_1 < y < A_1 + 2H_1$$

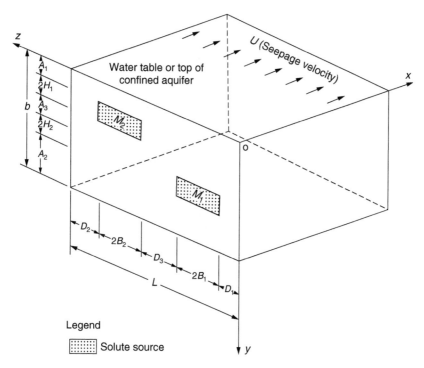

FIGURE 3-93 The geometry of two rectangular sources with instantaneous source masses (Adapted from Batu, V., unpublished paper, 1997b.)

where M_1 and M_2 are the instantaneous masses at the rectangular sources. As mentioned previously, in order to determine the $C(x, y, z, t)$ concentration distribution, only I_0, I_n, I_m, and I_{mn}, which are given by Eqs. (3-431), (3-433), (3-435), and (3-437), respectively, need to be evaluated. With the source conditions given by Eqs. (3-444) and (3-445), Eq. (3-431) gives

$$I_0 = \frac{1}{Lb} \int_{y=A_1+2H_1+A_3}^{A_1+2H_1+A_3+2H_2} M_1 dy \int_{z=D_1}^{D_1+2B_1} dz \qquad (3-446)$$

$$+ \frac{1}{Lb} \int_{y=A_1}^{A_1+2H_1} M_2 dy \int_{z=D_1+2B_1+D_3}^{D_1+2B_1+D_3+2B_2} dz$$

where

$$L = D_1 + 2B_3 + D_3 + 2B_2 + D_2, \quad b = A_1 + 2H_1 + A_3 + 2H_2 + A_2 \qquad (3-447)$$

From Eq. (3-446)

$$I_0 = \frac{4}{Lb}(B_1 H_2 M_1 + B_2 H_1 M_2) \qquad (3-448)$$

From Eq. (3-433)

$$I_n = \frac{1}{Lb} \int_{y=A_1+2H_1+A_3}^{A_1+2H_1+A_3+2H_2} M_1 dy \int_{z=D_1}^{D_1+2B_1} \cos(\lambda_n z)\, dz \qquad (3-449)$$

$$+ \frac{1}{Lb} \int_{y=A_1}^{A_1+2H_1} M_2 dy \int_{z=D_1+2B_1+D_3}^{D_1+2B_1+D_3+2B_2} \cos(\lambda_n z)\, dz$$

where from Eq. (3-146) and the first expression of Eq. (3-447),

$$\lambda_n = \frac{n\pi}{L} \tag{3-450}$$

and on integration,

$$
\begin{aligned}
I_n &= \frac{2H_2}{Lb} M_1 \frac{1}{\lambda_n} \{\sin[\lambda_n(D_1 + 2B_1)] - \sin(\lambda_n D_1)\} \\[2mm]
&\quad + \frac{2H_1}{Lb} M_2 \frac{1}{\lambda_n} \{\sin[\lambda_n(L - D_2)] - \sin[\lambda_n(D_1 + 2B_1 + D_3)]\}
\end{aligned}
\tag{3-451}
$$

From Eq. (3-435)

$$
\begin{aligned}
I_m &= \frac{1}{Lb} \int_{y=A_1+2H_1+A_3}^{A_1+2H_1+A_3+2H_2} M_1\cos(\eta_m y)dy \int_{z=D_1}^{D_1+2B_1} dz \\[2mm]
&\quad + \frac{1}{Lb} \int_{y=A_1}^{A_1+2H_1} M_2\cos(\eta_m y)dy \int_{z=D_1+2B_1+D_3}^{D_1+2B_1+D_3+2B_2} dz
\end{aligned}
\tag{3-452}
$$

where from Eq. (3-147) and the second expression of Eq. (3-447),

$$\eta_m = \frac{m\pi}{b} \tag{3-453}$$

and on integration

$$
\begin{aligned}
I_m &= \frac{2B_1}{Lb} M_1 \frac{1}{\eta_m} \{\sin[\eta_m(b - A_2)] - \sin[\eta_m(A_1 + 2H_1 + A_3)]\} \\[2mm]
&\quad + \frac{2B_2}{Lb} M_2 \frac{1}{\eta_m} \{\sin[\eta_m(A_1 + 2H_1)] - \sin(\eta_m A_1)\}
\end{aligned}
\tag{3-454}
$$

From Eq. (3-437)

$$
\begin{aligned}
I_{mn} &= \frac{1}{Lb} \int_{y=A_1+2H_1+A_3}^{A_1+2H_1+A_3+2H_2} M_1\cos(\eta_m y)dy \int_{z=D_1}^{D_1+2B_1} \cos(\lambda_n z)dz \\[2mm]
&\quad + \frac{1}{Lb} \int_{y=A_1}^{A_1+2H_1} M_2\cos(\eta_m y)dy \int_{z=D_1+2B_1+D_3}^{D_1+2B_1+D_3+2B_2} \cos(\lambda_n z)dz
\end{aligned}
\tag{3-455}
$$

After evaluation, Eq. (3-455) gives

$$
\begin{aligned}
I_{mn} &= \frac{M_1}{\pi^2} \frac{1}{m} \frac{1}{n} \left\{ \sin\left[\frac{m\pi}{b}(b - A_2) \right] - \sin\left[\frac{m\pi}{b}(A_1 + 2H_1 + A_3) \right] \right\} \\[2mm]
&\quad \times \left\{ \sin\left[\frac{n\pi}{L}(D_1 + 2B) \right] - \sin\left(\frac{n\pi D_1}{L} \right) \right\} \\[2mm]
&\quad + \frac{M_2}{\pi^2} \frac{1}{m} \frac{1}{n} \left\{ \sin\left[\frac{m\pi}{b}(A_1 + 2H_1) \right] - \sin\left(\frac{m\pi A_1}{b} \right) \right\} \\[2mm]
&\quad \times \left\{ \sin\left[\frac{n\pi}{L}(L - D_2) \right] - \sin\left[\frac{n\pi}{L}(D_1 + 2B_1 + D_3) \right] \right\}
\end{aligned}
\tag{3-456}
$$

Therefore, with Eqs. (3-448), (3-451), (3-454), and (3-456), Eqs. (3-429), (3-432), (3-434), and (3-436) form the solution for two instantaneous rectangular sources.

Evaluation of the Models
The three-dimensional instantaneous source analytical solute transport models presented in the previous sections are evaluated from their verification as well as applicability limitations point of view and the results, based on Batu (1997c), are presented below.

As can be seen from the components of the solution, the concentration solution includes only infinite series. Equations (3-429), (3-430), (3-432), (3-434), (3-436), and (3-440)–(3-443) form the solution for concentration distribution for a single rectangular source shown in Figure 3-92. Based on the aforementioned expressions, a computer program ST3C has been developed by the author. This program is described in the following sections.

Special Features of the Three-Dimensional Analytical Models
The solutions are fairly general such that solutions for any combinations of rectangular sources located along a vertical plane (yOz plane in Figure 3-91) perpendicular to the flow direction can be generated using the source geometry. The source flux on every rectangle can be space-dependent, i.e., the function of y and z coordinates.

The single rectangular source has a special importance. Therefore, this solution will be elaborated in detail. The impervious boundaries around the solute source increase the potential usage of the model. If the aquifer approaches infinity from both sides of the source, its effect can be taken into account by assigning large values to D_1 and D_2 as compared with the source width $2B$. One of the important features of this solution is that the source can be located at any place in the yOz plane that is perpendicular to the flow direction. Assigning $H_1 = 0$ corresponds to the case that the source is adjacent to the water table or to the top boundary of a confined aquifer.

Special Solutions
Some verifications have already been done for the three-dimensional analytical models. As mentioned above in detail, it is shown that the one-dimensional instantaneous source analytical model (Section 3.3.4.1) and two-dimensional instantaneous source analytical model (Section 3.3.4.2) are exact special cases of the corresponding three-dimensional instantaneous source solute transport model.

Comparisons with a Numerical Model
The numerical model code that was selected was called FTWORK (Faust et al., 1990), which is complete block-centered, three-dimensional finite-difference ground water flow and solute transport code. The finite-difference grid, shown in Figure 3-67 for comparison of results between ST3B and FTWORK, was used for comparison of results between ST3C and FTWORK as well. As can be seen from Figure 3-67, the thickness of the aquifer, b, is 5 m and a total of 10 modeling layers with each 0.5 m thick were considered for the FTWORK model. The horizontal grid of Figure 3-67 is shown in Figure 3-52. The boundary conditions described in Section 3.3.3.2 are the same for all ten layers in Figure 3-67. As in Figure 3-52, constant head boundary values (11.5 m at the yOz lane and 11.0 m at the down gradientplane of the parallelepiped) were assigned in Figure 3-67. These boundary conditions define a one-dimensional ground water flow field in the x-coordinate direction. Similar to the two-dimensional comparison in Section 3.3.3.2, the hydraulic conductivity, $K = K_x = K_y = K_z$, and effective porosity, φ_e, were selected to be 13.875 m/day and 0.25, respectively. With the hydraulic gradient $I = 0.5$ m/185 m $= 0.0027027$ m/m, the corresponding ground water velocity, U, is 0.15 m/day. The longitudinal dispersivity, α_L, transverse horizontal dispersivity, α_{TH}, transverse verticsl dispersivity, α_{TV}, and retardation factor, R_d values are 21.3 m, 4.3 m, 4.3 m, and 1, respectively. The effective molecular diffusion coefficient, D^*, was selected to be zero.

As mentioned previously, in the analytical model (ST3C) the mass M is assumed to be uniformly distributed over $2B \times 2H$ area whose thickness is considered to be unity (see Figure 3-92). This is due to the fact that the analytical model represents the source as a rectangle only. In the comparisons, $M = 0.625$ g of mass is considered. As can be seen from Figure 3-67, $2B = 5$ m and $2H$

= 0.5 m. Therefore, from Eq. (3-438), the instantaneous source concentration over the parallelepiped having the volume $V = (2B)(2H)(1 \text{ unit})$ with unit height is

$$C_M = \frac{M}{V\varphi_e} = \frac{M}{(2B)(2H)(1\text{unit})\varphi_e} = \frac{0.625\text{g}}{(5\text{m})(0.5\text{m})(1\text{m})(0.25)} = 1.0 \text{ g/m}^3$$

For ST3C, the instantaneous solute mass M is the input, whereas for FTWORK the corresponding instantaneous solute concentration is the input. As shown in Figure 3-52, the height of the source block is $\Delta x_1 = 0.25$ m. Therefore, as in the ST3C case, for FTWORK the instantaneous source concentration over the parallelepiped having the volume $V = (2B)(2H)(\Delta x_1)$ with Δx_1 height is

$$C_M = \frac{M}{(2B)(2H)(\Delta x_1)\varphi_e} = \frac{0.625\text{g}}{(5\text{m})(0.5\text{m})(0.25\text{m})0.25} = 4.0 \text{ g/m}^3$$

Notice that the value of C_M is dependent on the Δx_1 dimension of the source block. For FTWORK this value is assigned to the shaded source blocks in Figure 3-52, and to all remaining blocks initial concentration values were assigned to be zero.

Numerical experiments were carried out in the FTWORK runs to determine the maximum time step providing numerically converged results. The results showed that $\Delta t = 0.25$ days was appropriate in producing converged results and for all comparisons this value was used. For example, the value ($\Delta t = 1$ day) used for ST3B comparison in Section 3.3.3.3 was not appropriate in the sense that the numerical results were not converged. In other words, there were systematic discrepancies between the results of ST3C and FTWORK.

Comparison between the results of the analytical model for a single rectangular source, ST3C, and the three-dimensional numerical models, FTWORK, are presented in Figures 3-94 through 3-97. Figure 3-94 presents C vs. the z coordinate for $x = 22.5$ m and $y = 1.25$ m for the analytical model (the corresponding grid coordinates for FTWORK are $I = 1$ through 19, $J = 9$, and $K = 3$) after 180 days elapsed time. Figure 3-95 presents C vs. the x coordinate for $y = 1.25$ m and $z = 51.5$ m for the analytical model (the corresponding grid coordinates for FTWORK are $I = 18$, $J = 1$ through 39, and $K = 3$) after 180 days elapsed time. Figure 3-96 presents C vs. the y coordinate for $x = 8.75$ m and $z = 47.5$ m for the analytical model (the corresponding grid coordinates for FTWORK are $I = 14$, $J = 6$, and $K = 1$ through 10) after 180 days elapsed time. Figure 3-97 presents C vs. time at a point having the coordinates $x = 8.75$ m, $y = 0.25$ m, and $z = 47.5$ m for the

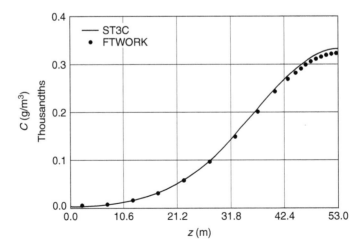

FIGURE 3-94 Concentration vs. transverse distance (z) comparison at $x = 22.5$ m and $y = 1.25$ m between the instantaneous source three-dimensional analytical model and the finite-difference numerical model after 180 days (Adapted from Batu, V., unpublished paper, 1997c.)

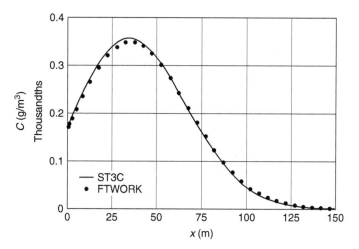

FIGURE 3-95 Concentration vs. longitudinal distance (x) comparison at $y = 1.25$ m and $z = 51.5$ m between the instantaneous source three-dimensional analytical model the finite-difference numerical model after 180 days. (Adapted from Batu, V., unpublished paper, 1997c.)

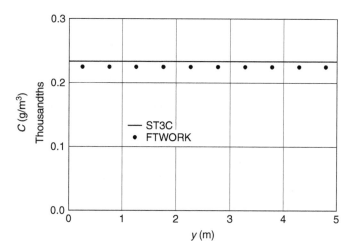

FIGURE 3-96 Concentration vs. vertical distance (y) comparison at $x = 8.75$ m and $z = 47.5$ m between the instantaneous source three-dimensional analytical model and the finite-difference numerical model after 180 days. (Adapted from Batu, V., unpublished paper, 1997c.)

analytical model (the corresponding grid coordinates for FTWORK are $I = 14$, $J = 6$, and $K = 1$). Review of Figures 3-94–3-97 indicates that good agreement exists among the results of the analytical and the numerical models. Minor discrepancies can be attributed to the roundoff errors, horizontal and vertical grid spacings, and algorithms used both in the analytical and numerical models.

Number of Terms Selection for the Infinite Series in the Solutions
The number of terms in the infinite series, N, plays a key role in determining accurate results. Numerical experiments showed that at the source the number of terms are the highest for converged results. The required number of terms decreases with the increasing distance from the source. Generally, values of N between 250 and 400 produce reasonably accurate results in all regions.

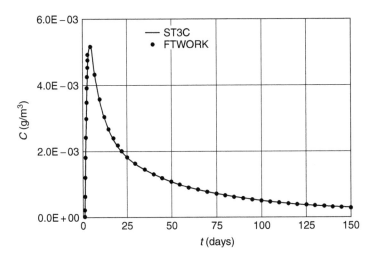

FIGURE 3-97 Concentration vs. time (t) comparison at point ($x = 8.75$ m, y $= 0.25$ m, $z = 47.5$ m) between the instantaneous source three-dimensional analytical model and finite-difference numerical model after 180 days. (Adapted from Batu, V., unpublished paper, 1997c.)

Application of the Three-Dimensional Analytical Models to Ground Water Flow Systems

The general model and its derivative models can be applied to solute transport predictions in aquifers if the ground water velocity can be assumed to be uniform. The models can also be applied to unsaturated zones with some additional assumptions. These are all discussed in the following sections.

In this section first, a computer program for a single rectangular source, ST3C, is described. Then, application of this program is presented with examples.

A Computer Program (ST3C) for the Three-Dimensional Third-Type Rectangular Source Model

A three-dimensional MS-DOS-based FORTRAN solute transport program, called solute transport three-dimensional instantaneous type (ST3C), has been developed by the author. ST3C is a menu-driven program and calculates values of concentration and convective–dispersive flux components (Eqs. [3-429], [3-430], [3-432], [3-434], [3-436] and [3-440]–[3-443]). This program can be run interactively in IBM compatible computers.

Application to Saturated Zones

The three-dimensional analytical solute transport model for a rectangular source (see Figure 3-92) can be used for different purposes. The impervious boundaries around the solute transport domain increase the potential usage areas of the model. For ground water flow cases, the effects of impervious boundaries such as bedrock outcrops and clay zones surrounding the main ground water flow zone can be taken into account. If the aquifer boundaries approach infinity horizontally, its effect can be taken into account by assigning large values to D_1 and D_2 as compared with the source width $2H$. The rectangular source can be at any location in the cross-section of an aquifer. Assigning $H_1 = 0$ is the case that the rectangular source starts at the water table of an unconfined aquifer or at the top of a confined aquifer.

Application to Unsaturated Zones

Application of the ST3C model to unsaturated zones can similarly be made as in the case for ST2A, which is presented in Section 3.2.2.2. The conceptual model for the ST3C model is the same as in Figure 3-59 with the exception that $L = 2H$. The associated governing equations for flow and solute transport in unsaturated porous media under uniform infiltration rate conditions are given in Section 2.8. The four important points mentioned for ST2C in Section 3.3.3.2 are also valid for the third-type source three-dimensional case.

3.3.4.4   Three-Dimensional Analytical Model For Instantaneous Parallelepiped Sources

Problem Statement and Assumptions
The problem under consideration is three-dimensional instantaneous parallelepiped source shown in Figure 3-98. The initial dimensions of the source in the x, y, and z directions are X, Y, and Z, respectively. The concentration within the parallelepiped at the beginning of migration ($t = 0$) is C_0. As shown in Figure 3-98, the origin of the Cartesian coordinate system is taken at the center of mass of the parallelepiped. The aquifer is of infinite extent in any of the directions of the coordinate system. Initially, the concentration outside of the parallelepiped is assumed to be zero.

Under the framework of the above-mentioned description, the problem is three-dimensional solute transport from instantaneous parallelepiped source shown in Figure 3-98. The x- and y-axis are the planar coordinates and the z-axis is the vertical coordinate. The assumptions are as follows:

1. The solute source is a parallelepiped source in a unidirectional steady ground water velocity (U) field.
2. The origin of the coordinates is at the center of the parallelepiped and the y-coordinate axis is perpendicular to the velocity of the flow field.
3. The solute mass in the parallelepiped is uniformly mixed.
4. The medium approaches infinity in all directions.

Governing Differential Equations
The governing differential equation with no decay ($v = 0$) is given by Eq. (3-128):

$$\frac{\partial C}{\partial t} = \frac{D_x}{R_d}\frac{\partial^2 C}{\partial x^2} + \frac{D_y}{R_d}\frac{\partial^2 C}{\partial y^2} + \frac{D_z}{R_d}\frac{\partial^2 C}{\partial z^2} - \frac{U}{R_d}\frac{\partial C}{\partial x} \tag{3-457}$$

A transformation with

$$\xi = x - \frac{U}{R_d}t, \quad \tau = t \tag{3-458}$$

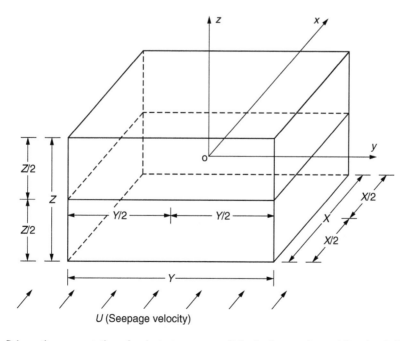

FIGURE 3-98   Schematic representation of an instantaneous parallelepiped source in a unidirectional flow field.

Eq. (3-457) takes the form

$$\frac{\partial C}{\partial \tau} = \frac{D_x}{R_d} \frac{\partial^2 C}{\partial \xi^2} + \frac{D_y}{R_d} \frac{\partial^2 C}{\partial y^2} + \frac{D_z}{R_d} \frac{\partial^2 C}{\partial z^2} \tag{3-459}$$

Derivation of Eq. (3-459) is shown below in Example 3-32. With the transformations

$$\bar{\xi} = \frac{\xi}{\left(\dfrac{D_x}{R_d}\right)^{1/2}}, \quad \bar{y} = \frac{y}{\left(\dfrac{D_y}{R_d}\right)^{1/2}}, \quad \bar{z} = \frac{z}{\left(\dfrac{D_z}{R_d}\right)^{1/2}} \tag{3-460}$$

Eq. (3-459) takes the form

$$\frac{\partial C}{\partial \tau} = \frac{\partial^2 C}{\partial \bar{\xi}^2} + \frac{\partial^2 C}{\partial \bar{y}^2} + \frac{\partial^2 C}{\partial \bar{z}^2} \tag{3-461}$$

where

$$\bar{\xi} = \left(\frac{R_d}{D_x}\right)^{1/2}\left(x - \frac{U}{R_d}t\right), \quad \bar{y} = \left(\frac{R_d}{D_y}\right)^{1/2} y, \quad \bar{z} = \left(\frac{R_d}{D_z}\right)^{1/2} z \tag{3-462}$$

EXAMPLE 3-32

Derive Eq. (3-459) from Eq. (3-457) using the transformation given by Eq. (3-458).

SOLUTION

Using the chain rule of partial differentiation (Ogata, 1970, p. 17), the expressions in Eq. (3-458) are substituted in Eq. (3-457). For example, the transformation of $\partial C/\partial x$ is given by

$$\frac{\partial C}{\partial x} = \frac{\partial C}{\partial \xi}\frac{\partial \xi}{\partial x} + \frac{\partial C}{\partial \tau}\frac{\partial \tau}{\partial x} = \frac{\partial C}{\partial \xi}(1) + \frac{\partial C}{\partial \tau}(0) = \frac{\partial C}{\partial \xi} \tag{E3-32-1}$$

From this, one can write

$$\frac{\partial^2 C}{\partial x^2} = \frac{\partial^2 C}{\partial \xi^2} \tag{E3-32-2}$$

The time differentiation follows the same rule and is given by

$$\frac{\partial C}{\partial t} = \frac{\partial C}{\partial \xi}\frac{\partial \xi}{\partial t} + \frac{\partial C}{\partial \tau}\frac{\partial \tau}{\partial t} = \frac{\partial C}{\partial \xi}\left(-\frac{U}{R_d}\right) + \frac{\partial C}{\partial \tau}(1) = \frac{\partial C}{\partial \tau} - \frac{U}{R_d}\frac{\partial C}{\partial \xi} \tag{E3-32-3}$$

The substitution of Eqs. (E3-32-1) and (E3-32-3) into Eq. (3-457) gives Eq. (3-459).

Initial and Boundary Conditions
The concentration $C(x, y, z, t)$ within the parallelepiped at the beginning ($t = 0$) is C_0. Therefore, the initial condition at the source is

$$C(x, y, z, 0) = C_0 \quad \text{for} \quad -\frac{X}{2} \leq x \leq \frac{X}{2}, \quad -\frac{Y}{2} \leq x \leq \frac{Y}{2}, \quad -\frac{Z}{2} \leq x \leq \frac{Z}{2} \tag{3-463}$$

Initially, the concentration outside of the parallelepiped is zero:

$$C(x, y, z, 0) = 0 \quad \text{for} \quad |x| > \frac{X}{2}, \quad |y| > \frac{Y}{2}, \quad |z| > \frac{Z}{2} \tag{3-464}$$

The boundary condition is that the concentration $C(x, y, z, t)$ approaches zero as the coordinates x, y, and z approach infinity:

$$C(x, y, z, t) = 0, \quad (x^2 + y^2 + z^2)^{1/2} \rightarrow \infty \tag{3-465}$$

Solution
With the transform variables given by Eq. (3-462), Eqs. (3-463) and (3-464), respectively, take the following forms:

$$C(\bar{\xi}, \bar{y}, \bar{z}, \tau = 0) = C_0 \text{ for } -\frac{\bar{\xi}_0}{2} \leq \bar{\xi} \leq \frac{\bar{\xi}_0}{2}, \quad -\frac{\bar{y}_0}{2} \leq \bar{y} \leq \frac{\bar{y}_0}{2}, \quad -\frac{\bar{z}_0}{2} \leq \bar{z} \leq \frac{\bar{z}_0}{2} \tag{3-466}$$

$$C(\bar{x}, \bar{y}, \bar{z}, \tau = 0) = 0 \quad \text{for} \quad |\bar{\xi}| > \frac{\bar{\xi}_0}{2}, \quad |\bar{y}| > \frac{\bar{y}_0}{2}, \quad |\bar{z}| > \frac{\bar{z}}{2} \tag{3-467}$$

where, from Eq. (3-462), for $t = 0$

$$\frac{\bar{\xi}_0}{2} = \left(\frac{R_d}{D_x}\right)^{1/2}\frac{X}{2}, \quad \frac{\bar{y}_0}{2} = \left(\frac{R_d}{D_y}\right)^{1/2}\frac{Y}{2}, \quad \frac{\bar{z}_0}{2} = \left(\frac{R_d}{D_z}\right)^{1/2}\frac{Z}{2} \tag{3-468}$$

And the boundary condition given by Eq. (3-465) reduces to

$$C(\bar{\xi}, \bar{y}, \bar{z}, \tau) = 0, \quad (\bar{\xi}^2 + \bar{y}^2 + \bar{z}^2)^{1/2} \rightarrow \infty \tag{3-469}$$

Now the solution of the problem is reduced to the solution of Eqs. (3-461), (3-466), (3-467), and (3-469) as (Carslaw and Jaeger, 1959;Eq. [4], p. 9 and; Eq. [10], p. 56)

$$C(x, y, z, t) = \frac{C_0}{8}\left[\text{erf}\left(\frac{\frac{\bar{\xi}_0}{2}-\bar{\xi}}{2t^{1/2}}\right) + \text{erf}\left(\frac{\frac{\bar{\xi}_0}{2}+\bar{\xi}}{2t^{1/2}}\right)\right]\left[\text{erf}\left(\frac{\frac{\bar{y}_0}{2}-\bar{y}}{2t^{1/2}}\right) + \text{erf}\left(\frac{\frac{\bar{y}_0}{2}+\bar{y}}{2t^{1/2}}\right)\right]$$

$$\times \left[\text{erf}\left(\frac{\frac{\bar{z}_0}{2}-\bar{z}}{2t^{1/2}}\right) + \text{erf}\left(\frac{\frac{\bar{z}_0}{2}+\bar{z}}{2t^{1/2}}\right)\right] \tag{3-470}$$

where erf(x) is the error function defined by Eq. (3-15) and is tabulated in Table 3-1. Using the appropriate expressions in Eqs. (3-462) and (3-468), the first term in the first bracket of the right-hand side of Eq. (3-470) can be expressed as

$$\text{erf}\left(\frac{\frac{\bar{\xi}_0}{2}-\bar{\xi}}{2t^{1/2}}\right) = \text{erf}\left[\frac{\frac{X}{2}-(x-\frac{U}{R_d}t)}{2\left(\frac{D_x t}{R_d}\right)^{1/2}}\right] \tag{3-471}$$

Since

$$\text{erf}(0) = 0, \quad \text{erf}(\infty) = 1, \quad \text{erf}(-x) = -\text{erf}(x) \tag{3-472}$$

Eq. (3-471) takes the form

$$\text{erf}\left(\frac{\frac{\bar{\xi}_0}{2}-\bar{\xi}}{2t^{1/2}}\right) = -\text{erf}\left[\frac{x-\frac{U}{R_d}t-\frac{X}{2}}{2\left(\frac{D_x t}{R_d}\right)^{1/2}}\right] \tag{3-473}$$

By writing similar expressions for the rest of error functions in Eq. (3-470), finally the following solution can be obtained:

$$C(x, y, z, t) = \frac{C_0}{8} \left\{ \text{erf}\left[\frac{x - \frac{U}{R_d}t + \frac{X}{2}}{2\left(\frac{D_x t}{R_d}\right)^{1/2}} \right] - \text{erf}\left[\frac{x - \frac{U}{R_d}t - \frac{X}{2}}{2\left(\frac{D_x t}{R_d}\right)^{1/2}} \right] \right\}$$

$$\left\{ \text{erf}\left[\frac{y + \frac{Y}{2}}{2\left(\frac{D_y t}{R_d}\right)^{1/2}} \right] - \text{erf}\left[\frac{y - \frac{Y}{2}}{2\left(\frac{D_y t}{R_d}\right)^{1/2}} \right] \right\} \left\{ \text{erf}\left[\frac{z + \frac{Z}{2}}{2\left(\frac{D_z t}{R_d}\right)^{1/2}} \right] - \text{erf}\left[\frac{z - \frac{Z}{2}}{2\left(\frac{D_z t}{R_d}\right)^{1/2}} \right] \right\} \quad (3\text{-}474)$$

With $R_d = 1$, Eq. (3-474) is exactly the same as, with exception of the defined variables, the equation derived by Hunt (1978, Eq. [33], pp. 82–83). The same solution for $R_d = 1$ is also included in Domenico and Robbins (1985, Eq. [11], p. 478) by referring Hunt's solution. Also, Moltyaner and Killey (1988, Eq. [5], p. 1616) give the same equation for $R_d = 1$ by referring the aforementioned solution in Carslaw and Jaeger (1959).

Special Solutions
Two-Dimensional Solutions (Instantaneous Rectangular Source Cases)
If $X = 0$ in Figure 3-98, the instantaneous parallelepiped source becomes a rectangle in the y–z plane where Y and Z are the dimensions of the rectangle. If $X = 0$ is introduced in Eq. (3-474), the first two bracketed error functions cancel each other and $C(x, y, z, t)$ becomes zero. This means that the solution is not valid for $X = 0$. In other words, X must be greater than zero, namely $|X| > 0$; two-dimensional solutions can be obtained from Eq. (3-474). The first one corresponds to the case where Z approaches infinity. This defines a parallelepiped whose boundaries along the z-coordinate direction approaches infinity. For this case, the last two error functions in Eq. (3-474), by using the equalities in Eq. (3-472), become

$$\text{erf}\left[\frac{z + \infty}{2\left(\frac{D_z t}{R_d}\right)^{1/2}} \right] - \text{erf}\left[\frac{z - \infty}{2\left(\frac{D_z t}{R_d}\right)^{1/2}} \right] = \text{erf}(\infty) - \text{erf}(-\infty) = 1 + 1 = 2 \quad (3\text{-}475)$$

and with this, Eq. (3-474) takes the form

$$C(x, y, t) = \frac{C_0}{4} \left\{ \text{erf}\left[\frac{x - \frac{U}{R_d}t + \frac{X}{2}}{2\left(\frac{D_x t}{R_d}\right)^{1/2}} \right] - \text{erf}\left[\frac{x - \frac{U}{R_d}t - \frac{X}{2}}{2\left(\frac{D_x t}{R_d}\right)^{1/2}} \right] \right\} \left\{ \text{erf}\left[\frac{y + \frac{Y}{2}}{2\left(\frac{D_y t}{R_d}\right)^{1/2}} \right] - \text{erf}\left[\frac{y - \frac{Y}{2}}{2\left(\frac{D_y t}{R_d}\right)^{1/2}} \right] \right\} \quad (3\text{-}476)$$

The second solution corresponds to the case where Y goes to infinity. For this case, the boundaries of the parallelepiped along the y-coordinate direction goes to infinity. Similar to the previous case, Eq. (3-474) takes the form

$$C(x, z, t) = \frac{C_0}{4} \left\{ \text{erf}\left[\frac{x - \frac{U}{R_d}t + \frac{X}{2}}{2\left(\frac{D_x t}{R_d}\right)^{1/2}} \right] - \text{erf}\left[\frac{x - \frac{U}{R_d}t - \frac{X}{2}}{2\left(\frac{D_x t}{R_d}\right)^{1/2}} \right] \right\} \left\{ \text{erf}\left[\frac{z + \frac{Z}{2}}{2\left(\frac{D_z t}{R_d}\right)^{1/2}} \right] - \text{erf}\left[\frac{z - \frac{Z}{2}}{2\left(\frac{D_z t}{R_d}\right)^{1/2}} \right] \right\} \quad (3\text{-}477)$$

One-Dimensional Solution

If in Eq. (3-476), Y approaches infinity, by treating the two error functions in the second bracket as in the two-dimensional cases, Eq. (3-476) becomes

$$C(x, t) = \frac{C_0}{2} \left\{ \text{erf}\left[\frac{x - \frac{U}{R_d}t + \frac{X}{2}}{2\left(\frac{D_x t}{R_d}\right)^{1/2}} \right] - \text{erf}\left[\frac{x - \frac{U}{R_d}t - \frac{X}{2}}{2\left(\frac{D_x t}{R_d}\right)^{1/2}} \right] \right\} \tag{3-478}$$

which is also reported in Moltyaner and Killey (1988, Eq. [6], p. 1616). Equation (3-478) can also be determined from Eq. (3-477) in the same manner. For this case, the parallelepiped source becomes infinite in the y- and z-coordinate directions.

PROBLEMS

3.1 Describe the mathematical foundations of the one-dimensional analytical solute transport model expressed by Eq. (3-14).

3.2 Under steady-state solute transport conditions for $v = 0$, show that the one-dimensional solution of Eq. (3-31) with the associated boundary conditions is $C(x) = C_M$.

3.3 For a contaminant source case shown in Figure 3-1, the concentration from a constant TCE (trichloroethylene) source at 20 m distance was measured as 25 μg/L after 180 days. The medium is a sandy aquifer and its hydrogeologic parameters are given as follows: $K = 5$ m/day, $I = 0.005$ m/m, and $\varphi_e = 0.25$. Solute transport parameters are given as follows: $\alpha_L = 2$ m, $D^* = 2.0 \times 10^{-9}$ m^2/sec, and $R_d = 2.5$. Determine the source concentration C_M).

3.4 The general solution for concentration distribution for the sources shown in Figure 3-11 is given by Eq. (3-63). Determine a special solution from this solution for a given source configuration?

3.5 The boundary conditions at $z = 0$ and $z = H_p$ (see Figure 3-11) are given by Eqs. (3-51) and (3-52). How the boundary conditions can be derived?

3.6 (1) determine the concentration distribution for a single-strip source (Eq. [3-77]) from the general solution (Eq. [3-63]). Also (2) derive the expressions for I_0 and I_n.

3.7 The two-dimensional solution for a strip source (Figure 3-12) is given by Eq. (3-77). Using this solution, derive the one-dimensional solute transport solution as given by Eq. (3-14).

3.8 Which equations represent the unsteady-state solution for multiple first-type rectangular source solute transport? What are the key characteristics of the solution?

3.9 Determine the concentration distribution solution for a single first-type rectangular source from the general solution?

3.10 For a third- or flux-type source case, the source condition is given by Eq. (3-225) and the source geometry for the one-dimensional solute transport case is given by Figure 3-43. With these, (1) explain the physical meaning of the third-type source condition and (2) compare it with the first-type source condition.

3.11 Evaluate the first- and third-type source conditions from the mass balance constraints points of view. Assume no decay and time-independent source concentration.

3.12 The boundary conditions at the impervious boundaries for the three-dimensional first-type rectangular sources are expressed by Eqs. (3-138)–(3-141). Explain why $\partial C/\partial y = 0$ and $\partial C/\partial z = 0$ on these boundaries?

3.13 Consider that in Figure 3-31, the areal extent of the aquifer is relatively large as compared with the width, $2B$, of the rectangular source. How can this situation be simulated?

4 Numerical Flow and Solute Transport Modeling in Aquifers

4.1 INTRODUCTION

Although analytical solute transport models are important for solute transport evaluation in ground water, they are not capable of handling all kinds of solute transport problems encountered in practice. This is mainly due to the fact that almost all analytical solute transport models in the literature are based on unidirectional flow fields as well as other restrictive assumptions. Note that analytical and numerical flow and solute transport modeling approaches should not be considered as two different ways of analysis of ground water problems, but analytical models should be considered as supplementary parts of numerical models. The other important point is that modeling applications for both flow and solute transport problems have started with relatively simple and mostly one-dimensional models. Also, many concepts were established through their applications in a variety of practical problems. For example, the most simple one-dimensional analytical solute transport model (Chapter 3, Section 3.3.2.1, Eq. [3.14]) is an important tool not only from its application point of view, but also because it is an evaluation tool for numerical solute transport models. As presented in detail in Chapter 3, most important concepts for the mechanism of solute transport and solute transport modeling can be realized much more effectively by studying both the foundations as well as applications of analytical solute transport models; this is true for analytical ground water flow models too.

The Darcy equation is the most simple ground water flow model and can be generated by simple hand calculations. For example, the Darcy velocity q (Chapter 2, Section 2.4.2.2, Eq. [2-39]) can be determined using the values of hydraulic gradient I and hydraulic conductivity K. Therefore, a sophisticated method is not needed. This model is the simple flow model of the analytical solute transport models presented in Chapter 3. In the case of numerical solute transport modeling, the flow field may not be uniform, and especially under various stresses applied to an aquifer (such as extraction wells, injection wells, drains, etc.), the flow field may potentially become significantly nonuniform. In order to conduct solute transport modeling in this type of flow field, the flow field needs to be determined first. This field can only be determined through a numerical ground water flow model. Therefore, before conducting a solute transport modeling study on a given subsurface flow system, first a ground water flow modeling study needs to be carried out. As a result, in this chapter, the foundations of both ground water and solute transport modeling will be presented.

The purpose of this chapter is to present the foundational features of numerical models based on the finite-difference method by emphasizing the application of flow and solute transport models to real-world projects, and not a detailed presentation of mathematical developments and computer programing aspects for numerical flow and transport models, which can be found in other well-known books (e.g., Bear, 1972, 1979; Wang and Anderson, 1982; Huyakorn and Pinder, 1983; Zheng and Bennett, 1995; and others) and some publicly available user's manuals for many computer codes (e.g., Prickett and Lonnquist, 1971; Prickett et al., 1981; McDonald and Harbaugh, 1984; Zheng, 1990; Faust et al., 1990; Duffield, 1996; and others).

4.2 NUMERICAL FLOW AND SOLUTE TRANSPORT MODELING APPROACHES

Numerical modeling techniques generally are based on (a) finite-difference methods; (b) finite-element methods; (c) collocation methods; (d) method of characteristics; and (e) boundary element methods. It is important to note that these methods are closely related. The finite-difference, finite element, and collocation methods essentially yield the same approximation. The method of characteristics is a variant of the finite-difference method and is particularly suitable for solving hyperbolic differential equations. Finally, the boundary element method, a variant of the conventional finite element method, is especially useful for the solution of elliptic differential equations (Huyakorn and Pinder, 1983).

Of the five methods mentioned above, the finite-difference method is the one that is mostly used for numerical ground water modeling applications. The majority of publicly available model codes for ground water flow and transport modeling applications in the North America and some other countries are based on the finite-difference method. Therefore, in this book, the main focus will be on the finite-difference-based modeling applications. Note that the general principles of flow and transport modeling applications are also applicable to computer codes based on the other four potential methods mentioned above. As a result, one should note that the modeling principles presented in the subsequent sections have generic characteristics, i.e., they can be applied to those computer codes based on methods other than the finite-difference method.

4.2.1 REASONS FOR THE POPULARITY OF THE FINITE-DIFFERENCE METHOD

For ground water flow and solute transport problems, the most popular approach in solving differential equations numerically is the finite-difference method; there are several reasons regarding this popularity. The fundamental principles of the finite-difference method are readily understood and do not require advanced training in applied mathematics. The finite-difference methods have firm theoretical foundations. Also, owing to the form and algebraic simplicity of the equations resulting from the finite-difference approximations, development of solution algorithms are relatively easier (Huyakorn and Pinder, 1983).

4.2.2 HISTORICAL BACKGROUND OF THE FINITE-DIFFERENCE METHOD

The usage of finite differences in subsurface flow modeling began with Peaceman and Rachford (1955). It appears that the first application of the finite difference theory to ground water flow was by Remson et al. (1965). In the following decade, additional publications started to appear in the ground water literature (Freeze and Witherspoon, 1966; Bittinger et al., 1967; Pinder and Bredehoeft, 1968; Prickett and Lonnquist, 1971). In the mid-1970s, two publicly available finite-difference computer codes (Trescott, 1975; Trescott et al., 1976) for ground water flow modeling were released by the United States Geological Survey. In the last quarter of the 20th century, the number of publications and computer codes for ground water flow and ground water solute transport modeling increased significantly. However, the application of finite-difference method to solute transport in ground water was relatively new; the first publication appears to have been that of Tanji et al. (1967).

4.3 FINITE-DIFFERENCE MODELING APPROACHES FOR GROUND WATER FLOW

4.3.1 FUNDAMENTALS OF FLOW IN AQUIFERS

Hydraulic head; drawdown; porosity; discharge velocity and ground water velocity; hydraulic conductivity, permeability, and transmissivity; and capillarity and capillary fringe are the key hydraulic and hydrogeological quantities to characterize aquifers. These quantities are used for almost every aquifer problem. In the following sections, definitions as well as some key aspects of these quantities are presented.

4.3.1.1 Hydraulic Head and Drawdown

Hydraulic Head

Hydraulic head, which is also called as head or piezometric head, is defined as the fluid pressure of water formation produced by the height above a given point. Geometrically, the hydraulic head at a point of an unconfined aquifer and a confined aquifer is shown in Figure 4-1 and Figure 4-2, respectively. The water level in a piezometer, which is a small tube that is open only at its bottom, in an unconfined aquifer is the same as the watertable elevation (see Figure 4-1). However, in the case of a confined aquifer, the water level in a piezometer is the same as the piezometric surface elevation of the aquifer at the same point (see Figure 4-2). As shown in Figure 4-1 and Figure 4-2, the hydraulic head $h(x, y, z, t)$ at point $A(x, y, z)$ consists of the pressure head $p(x, y, z, t)/\gamma$ and the elevation of the bottom of the pizometer z above an arbitrarily chosen datum elevation. The pressure head $p(x, y, z, t)/\gamma$ is the height of the column of water in the piezometer above its bottom or above the point $A(x, y, z)$. As a result, the hydraulic head is defined as (Hantush, 1964)

$$h(x, y, z, t) = \frac{p(x, y, z, t)}{\gamma} + z + f \tag{4-1}$$

where x, y, and z are the Cartesian coordinates, γ is the unit weight of water, f the elevation of the x–y plane above an arbitrarily chosen datum elevation, and t the time observation from a reference time.

The dimensions of hydraulic head is length (L) and is generally expressed as "meters of water" or "feet of water."

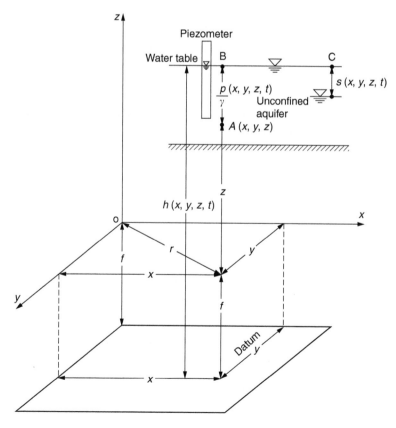

FIGURE 4-1 Hydraulic head in an unconfined aquifer. (Adapted from Batu, V., *Aquifer Hydraulics: A Comprehensive Guide to Hydrogeologic Data Analysis*, John Wiley & Sons, Inc., 1998, p. 26, Figure 2-2.)

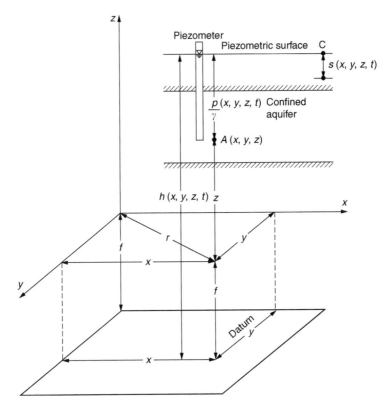

FIGURE 4-2 Hydraulic head in a confined aquifer. (Adapted from Batu, V., *Aquifer Hydraulics: A Comprehensive Guide to Hydrogeologic Data Analysis*, John Wiley & Sons, Inc., 1998, p. 26, Figure 2-3.)

Drawdown

Under extraction or injection conditions from a well, the position of the initial hydraulic head at a point in an aquifer changes, and the vertical distance between the initial hydraulic head at a point in the aquifer and the lowered position of the hydraulic head at the same point is called drawdown. For an unconfined aquifer, the water level at point C on the water table drops by a vertical distance $s(x, y, z, t)$, which is the drawdown at the same point (see Figure 4-1). For a confined aquifer case, the drawdown is defined with respect to the piezometric surface (see Figure 4-2).

The dimensions of drawdown is length (L), and is generally expressed as "meters of water" or "feet of water."

4.3.1.2 Porosity and Effective Porosity

Porosity of an aquifer, or porous medium φ, is defined as the percentage of void spaces:

$$\varphi = \frac{\text{volume of voids of a sample}}{\text{total volume of the sample}} \tag{4-2}$$

The *effective porosity* is the portion of pore space in a saturated porous material in which water flow occurs. In other words, it is the volume of the interconnected voids of a sample. This definition is based on the fact that all the pore space of a porous medium filled with water is not open for water flow. Hence, effective porosity φ_e is defined as

$$\varphi_e = \frac{\text{volume of the interconnected voids of a sample}}{\text{total volume of the sample}} \tag{4-3}$$

Obviously, the effective porosity of a porous medium is smaller than its porosity.

Table 4-1, based on the original data of Morris and Johnson (1967), presents representative porosity (φ_e) ranges as well as arithmetic mean (φ_{av}) for various aquifer materials. In general, clays have smaller pores than sandy materials but may have much greater porosity values.

4.3.1.3 Discharge Velocity and Ground Water Velocity

The *discharge velocity* is defined as the quantity of fluid that passes through a unit of total area of the porous medium in a unit time. In order to define these velocities better, let us consider a volume element in a porous medium as shown in Figure 4-3. The dimensions of the parallelepiped volume element are Δx, Δy, and Δz in the x, y, and z directions, respectively. Now, let us consider a horizontal section at z elevation from the x–y plane. The total area and the area of the pores of the horizontal section are A and A_p, respectively. Define m as

$$m = \frac{\text{the area of pores}}{\text{the total area}} = \frac{A_P}{A} \tag{4-4}$$

and let Q_z be the vertical component of Q discharge passing the area A_p. Therefore,

$$Q_z = mAv \tag{4-5}$$

is the vertical component of the Q discharge vector. Here, mv and v are the discharge velocity and ground water velocity (seepage velocity), respectively.

In general, the area of pores changes with the z coordinate. Therefore, Eq. (4-4) takes the form

$$m_z = \frac{A_p(z)}{A} \tag{4-6}$$

The average value of m over the parallelepiped of height Δz gives

$$m = \frac{1}{\Delta z} \int_0^{\Delta z} m(z)\, dz = \frac{1}{(\Delta x)(\Delta y)(\Delta z)} \int_0^{\Delta z} A_p(z)\, dz \tag{4-7}$$

TABLE 4-1
Porosity of Various Porous Materials

Group	Porous Material	No. of Analyses	Range of Porosity φ	Arithmetic Mean Average Porosity φ_{av}
Igneous Rocks	Weathered granite	8	0.34–0.57	0.45
	Weathered gabbro	4	0.42–0.45	0.43
	Basalt	94	0.03–0.35	0.17
Sedimentary Materials	Sandstone	65	0.14–0.49	0.34
	Siltstone	7	0.21–0.41	0.35
	Sand (fine)	243	0.26–0.53	0.43
	Sand (coarse)	26	0.31–0.46	0.39
	Gravel (fine)	38	0.25–0.38	0.34
	Gravel (coarse)	15	0.24–0.36	0.28
	Silt	281	0.34–0.61	0.46
	Clay	74	0.35–0.57	0.42
	Limestone	74	0.07–0.56	0.30
Metamorphic Rocks	Schist	18	0.04–0.49	0.38

Adapted from Morris, D.A. and Johnson, A.I., *U.S. Geological Survey Water Supply Paper 1839-D*, 1967; and Mercer, J.W. et al., U.S. Nuclear Regulatory Commission, NUREG/CR-3066, Washington, DC, 1982.

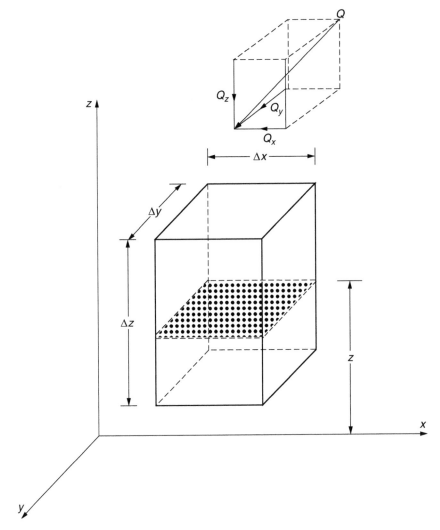

FIGURE 4-3   Schematic illustration for the derivation of the relationship between the discharge and ground water velocity. (Adapted from Batu, V., *Aquifer Hydraulics: A Comprehensive Guide to Hydrogeologic Data Analysis*, John Wiley & Sons, Inc., 1998, p. 29, Figure 2-4.)

or

$$m = \frac{1}{V} \int_0^{\Delta z} A_p(z)\, dz \qquad (4\text{-}8)$$

where V is the total volume and the integral is the volume of voids. As a result, the average value of m is the volume-based effective porosity φ_e. Therefore, Eq. (4-5) takes the form

$$Q_z = \varphi_e A v = qA \qquad (4\text{-}9)$$

and

$$q = \varphi_e v \qquad (4\text{-}10)$$

is the relationship between the discharge velocity and ground water velocity. The term seepage velocity is also used for ground water velocity.

4.3.1.4 Darcy's Law

Darcy's law was established by Henry Darcy in 1856. Based on the experimental study for flow of water through sands (Darcy, 1856). Here, the one-dimensional form of Darcy's law will be presented in the light of the fundamental principles of hydromechanics. The concepts regarding the components of hydraulic head is presented prior to Darcy's equation.

The schematic representation of the experimental apparatus shown in Figure 4-4 is equivalent to that used by Darcy in his original experiments with the exception that Darcy's apparatus was vertical. The pipe shown in Figure 4-4 is filled with a porous material and water flow occurs in the direction indicated by the arrow. Two piezometers are located at points 1 and 2 in the pipe. The geometric elevations of these two points are z_1 and z_2 with respect to a datum. During water flow, head losses occur between points 1 and 2. As a result, the hydraulic head at point 2, h_2, is less than the hydraulic head at point 1, h_1, by Δh. The pressure heads at points 1 and 2 are p_1/γ and p_2/γ, respectively.

For an incompressible inviscid (nonviscous) fluid flowing in a pipe, horizontal, vertical, or inclined, with smooth walls, the steady-state Bernoulli equation is

$$\frac{p}{\rho g} + z + \frac{v^2}{2g} = \text{constant} = h \tag{4-11}$$

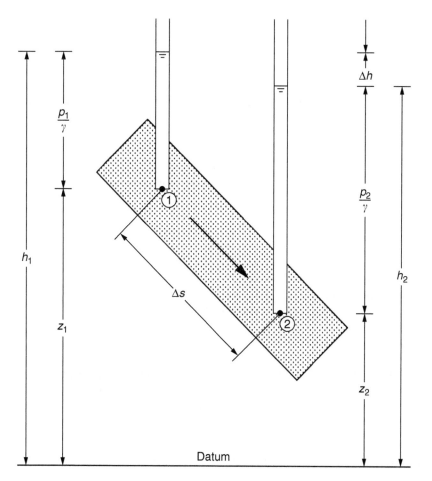

FIGURE 4-4 Schematic illustration of Darcy's experiment. (Adapted from Batu, V., *Aquifer Hydraulics: A Comprehensive Guide to Hydrogeologic Data Analysis*, John Wiley & Sons, Inc., 1998, p. 30, Figure 2-5.)

where ρ is the density of the fluid, g the acceleration due to gravity, p the pressure, z the geometric elevation from the datum, v the ground water (or seepage) velocity, and h the total head. Also

$$\frac{p}{\rho g} = \frac{p}{\gamma} = \text{piezometric head}$$

and

$$\frac{v^2}{2g} = \text{velocity head}$$

where γ is the unit weight of the fluid (water). The Bernoulli equation, Eq. (4-11), states that for all points of the pipe, the sum of the piezometric head p/γ, geometric elevation z, and velocity head $v^2/2g$ remains a constant value.

Now, let us examine the development of the flow in a porous medium. As shown in Figure 4-4, consider two points, 1 and 2, along the axis of the pipe, that are separated by a distance Δs from each other. Piezometers are inserted at these points and the water levels in the piezometers are at h_1 and h_2 elevations from the datum. In the ground water flow, due to resistance against the flow within the pores, the Bernoulli equation takes the following form for points 1 and 2:

$$\frac{p_1}{\gamma} + z_1 + \frac{v_1^2}{2g} = \frac{p_2}{\gamma} + z_2 + \frac{v_2^2}{2g} + \Delta h \tag{4-12}$$

where Δh is the head loss over the distance Δs.

The hydraulic gradient I or the gradient of piezometric head is defined as

$$I = -\lim_{\Delta s \to 0} \frac{\Delta h}{\Delta s} = -\frac{dh}{ds} \tag{4-13}$$

In porous media flow, velocities are relatively low. As a result, the velocity heads in Eq. (4-12) may be neglected without appreciable error. Therefore, from Eq. (4-12), one can write

$$\Delta h = \left(\frac{p_1}{\gamma} + z_1 \right) - \left(\frac{p_2}{\gamma} + z_2 \right) = h_1 - h_2 \tag{4-14}$$

which means that the expression for the hydraulic head at any point in the flow domain is

$$h = \frac{P}{\gamma} + z \tag{4-15}$$

Equation (4-15) is one of the key expressions in ground water hydraulics, whereas Eq. (4-11) is the hydraulic head expression for surface-water flows such as rivers, streams, canals, etc. In a river or stream, the average velocity is around 1 m/sec. However, in most aquifers, even 1 m/day is a very high velocity. In some tight formations, the velocity of water is mostly less than 1 m/year. Therefore, the velocity heads in porous media flow are extremely low and this is shown in the following example.

EXAMPLE 4-1

The average velocity in a river is around 1.5 m/sec. Based on the hydrogeologic investigations, the average ground water velocity in a nearby aquifer is around 1 m/day. Determine the velocity heads for each flow case and make a comparative evaluation.

SOLUTION

In the case of a river flow, the velocity head in Eq. (4-11) has the value

$$h_{swv} = \frac{v^2}{2g} = \frac{(1.5 \text{ m})^2}{2(9.81 \text{ m/sec}^2)} = 0.11 \text{ m}$$

The ground water velocity in the aquifer is

$$v = 1 \text{ m/day} = 1.16\times10^{-5} \text{ m/sec}$$

Therefore, the velocity head for the ground water flow is

$$h_{gwv} = \frac{v^2}{2g} = \frac{(1.16\times10^{-5} \text{ m/sec})^2}{2(9.81 \text{ m/sec}^2)} = 6.83\times10^{-12} \text{ m}$$

and the ratio between these velocity heads is

$$\frac{h_{gwv}}{h_{swv}} = \frac{6.83\times10^{-12} \text{ m}}{0.11 \text{ m}} = 6.21\times10^{-11}$$

The above values clearly show that the ground water velocity head is extremely small as compared with the other hydraulic head components. Therefore, the velocity head term in Eq. (4-11) can safely be neglected for flow in porous media.

Darcy Experiment

In order to generate the flow equation in porous media, a relationship is needed between the ground water velocity and hydraulic gradient. In 1856, Henry Darcy experimentally found that the discharge Q is proportional to the difference of water levels Δh, and inversely proportional to Δs (refer Figure 4-4):

$$Q \propto \Delta h, \qquad Q \propto \frac{1}{\Delta s} \tag{4-16}$$

By introducing a proportionality constant K, we get to the equation

$$Q = -KA\frac{h_2-h_1}{\Delta h} = -KA\frac{\Delta h}{\Delta s} \tag{4-17}$$

or

$$Q = -KA\frac{dh}{ds} \tag{4-18}$$

where A is the cross-sectional area of the column. From Eq. (4-18), the discharge velocity can be expressed as

$$q = \frac{Q}{A} = -K\frac{dh}{ds} = -KI \tag{4-19}$$

where q is the discharge velocity or Darcy velocity (also known as Darcy flux and specific discharge), K the hydraulic conductivity, and $I(=dh/ds)$ the hydraulic gradient. Equation (4-19) is commonly known as Darcy's law, there is a linear relationship between the hydraulic gradient I and the discharge velocity q. The negative sign in Eq. (4-19) indicates that the flow of water is in the direction of decreasing head. It must be emphasized that Darcy's law does not describe the state within individual pores. It represents the statistical macroscopic equivalent of the Navier–Stokes equations of motion for viscous flow of water in porous media. Equation (4-19) is the simple form of Darcy's law and its two- and three-dimensional forms are presented in Section 4.3.2.

The head loss Δh is independent of the inclination of the column. Figure 4-5 presents two schemes of experiments to check Darcy's law. In Figure 4-5a, the column is horizontal and head difference Δh occurs over the horizontal distance Δs in the porous medium. In Figure 4-5b, the column is vertical and Δh occurs over Δs in the porous medium. Different forms of Darcy's law as given by Eqs. (4-17)–(4-19), are also valid for these experimental cases.

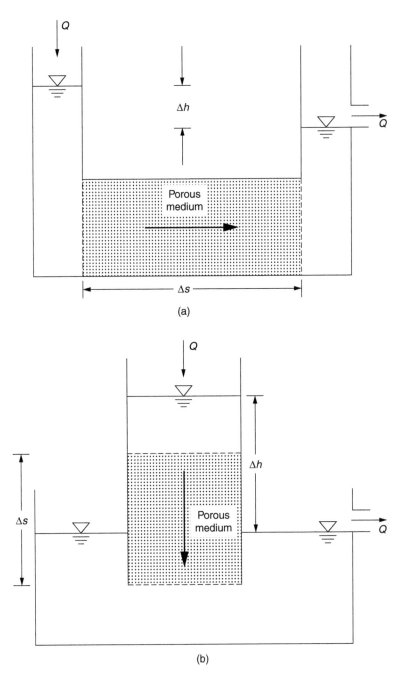

FIGURE 4-5   (a) Darcy experiment on a horizontal column and (b) Darcy experiment on a vertical column. (Adapted from Batu, V., *Aquifer Hydraulics: A Comprehensive Guide to Hydrogeologic Data Analysis*, John Wiley & Sons, Inc., 1998, p. 33, Figure 2-6.)

Validity of Darcy's Law
According to Eq. (4-19), Darcy's law states that the discharge velocity is proportional to the first power of the hydraulic gradient. Since the velocity in laminar flow is also proportional to the first power of the hydraulic gradient, it can be inferred that the flow in a porous medium must be laminar for the Darcy's law to be valid. In hydrodynamics, the usual criterion to determine whether the flow is laminar or turbulent is the Reynolds number. Experiments have shown that Darcy's

equation has definite limitations. For homogeneous aquifer materials, Darcy's law is valid only under the condition of the inequality for the Reynolds number:

$$N_R = \frac{\rho q d}{\mu} = \frac{q d}{v} \leq N_L \tag{4-20}$$

where ρ is the density of water, μ the dynamic viscosity, q the discharge velocity, d the average diameter of soil particles, v the kinematic viscosity of water, and N_L a number varying between 3 and 10 (Polubarinova-Kochina, 1962; Bear, 1979). With $v = 0.018$ cm^2/sec, under average temperature conditions, Eq. (4-20) takes the form

$$q d \leq 0.054 - 0.180 \tag{4-21}$$

where q is expressed in cm/sec and d in cm. For large grained materials or relatively fast water flow in aquifers, other relationships between q and I have been established based on experimental results. For most turbulent flows in porous media, a polynomial of the second degree has been considered (Jaeger, 1956):

$$I = A q + B q^2 \tag{4-22}$$

If turbulence is fully developed, the term Aq in Eq. (4-22) is negligible, and the constants A and B are determined experimentally.

4.3.1.5 Hydraulic Conductivity, Transmissivity, and Permeability

These quantities are widely used in characterizing aquifers. Their definitions with the units used in practice are presented below.

Hydraulic Conductivity
The definition of hydraulic conductivity is derived from Darcy's law. By solving Eq. (4-19) for K, one obtains

$$K = \frac{q}{I} = - \frac{Q}{A(dh/ds)} \tag{4-23}$$

It is apparent from Eq. (4-23) that K has the dimensions of "length/time" (L/T) or velocity. This coefficient is called *hydraulic conductivity* or *saturated hydraulic conductivity* and is a function of properties of both the porous medium and the fluid flowing through it. The saturated hydraulic conductivity of a given aquifer material can be determined in the laboratory by being one of the schematical devices shown in Figure 4-5. Tables and nomographs regarding saturated hydraulic conductivity values for different formations are presented in a number of publications, especially during the past century. It is important to note that these tables mostly use the general term "hydraulic conductivity" rather than making it specific for horizontal and vertical hydraulic conductivity values. By default, these values should be viewed as tables for "horizontal hydraulic conductivity," unless stated otherwise. It is a well-known fact that vertical hydraulic conductivity may be 1 to 3 orders of magnitude less than horizontal hydraulic conductivity, depending on the type of formation. For practical usage, Table 4-2 and 4-3 are of two different forms that are commonly used in some well-known literatures. Table 4-2 presents average ranges of the saturated hydraulic conductivity for a few encountered soils. From another perspective, Table 4-3 presents the range of values of hydraulic conductivity and permeability in a nomographic format. Different units are used in practice for hydraulic conductivity and their conversion factors are included in Table 4-4.

In the geotechnical engineering literature, the term coefficient of permeability is mostly used instead of hydraulic conductivity. Based on the aforementioned expressions, *hydraulic conductivity* can be defined as the rate of discharge of water under laminar flow conditions through a unit

TABLE 4-2
Hydraulic Conductivity of Porous Materials

Group	Porous Material	No. of Analysis	Range of Hydraulic Conductivity K (cm/sec)	Arithmetic Mean Average Hydraulic Conductivity K_{av} (cm/sec)
Igneous Rocks	Weathered granite	7	$(3.3–52)\times10^{-4}$	1.65×10^{-3}
	Weathered gabbro	4	$(0.5–3.8)\times10^{-4}$	1.98×10^{-4}
	Basalt	93	$(0.2–4250)\times10^{-6}$	9.45×10^{-6}
Sedimentary	Sandstone (fine)	20	$(0.5–2270)\times10^{-6}$	3.31×10^{-4}
Materials	Siltstone	8	$(0.1–142)\times10^{-8}$	1.9×10^{-7}
	Sand (fine)	159	$(0.2–189)\times10^{-4}$	2.28×10^{-3}
	Sand (medium)	255	$(0.9–567)\times10^{-4}$	1.65×10^{-2}
	Sand (coarse)	158	$(0.9–6610)\times10^{-4}$	5.20×10^{-2}
	Gravel	40	$(0.3–31.2)\times10^{-1}$	4.03×10^{-1}
	Silt	39	$(0.09–7090)\times10^{-7}$	2.83×10^{-5}
	Clay	19	$(0.1–47)\times10^{-8}$	9.00×10^{-8}
Metamorphic Rocks	Schist	17	$(0.002–1130)\times10^{-6}$	1.90×10^{-4}

Adapted from Morris, D.A. and Johnson, A.I., *U.S. Geological Survey Water Supply Paper 1839-D*, 1967; and Mercer, J.W. et al., U.S. Nuclear Regulatory Commission, NUREG/CR-3066, Washington, DC, 1982.

cross-sectional area of a porous medium under a unit hydraulic gradient and standard temperature conditions (usually 20°C).

Transmissivity
For a confined aquifer having thickness b, the *transmissivity* (or *transmissibility*) T is defined as

$$T = Kb \qquad (4\text{-}24)$$

This concept can be used for unconfined aquifers as well. In Eq. (4-24), b is the saturated thickness of the aquifer or the height of the watertable above the underlying aquitard or impermeable boundary. Different units are used in practice for transmissivity and its conversion factors are included in Table 4-5.

Permeability
As mentioned above, the term coefficient of permeability is used instead of hydraulic conductivity, especially in the geotechnical engineering literature. This is inappropriate because K not only depends on the permeability of the porous medium but also on the properties of the fluid flowing through the porous medium. The fluid properties that affect the flow are dynamic viscosity μ and γ. The dynamic viscosity μ of the fluid is a measure of the resistive force within the pores of the porous medium, and the specific weight γ may be considered as the driving force exerted by the gravity. On the other hand, the size of a typical pore is proportional to the mean grain diameter d. By using the method of dimensional analysis, the relationship between K, μ, γ, and d is determined as (Jacob, 1950)

$$K = \frac{Cd^2\gamma}{\mu} = k\frac{\gamma}{\mu} = k\frac{\rho g}{\mu} = k\frac{g}{v} \qquad (4\text{-}25)$$

where

$$K = Cd^2 \qquad (4\text{-}26)$$

TABLE 4-3

Hydraulic Conductivity and Permeability Range of Values in Different Units

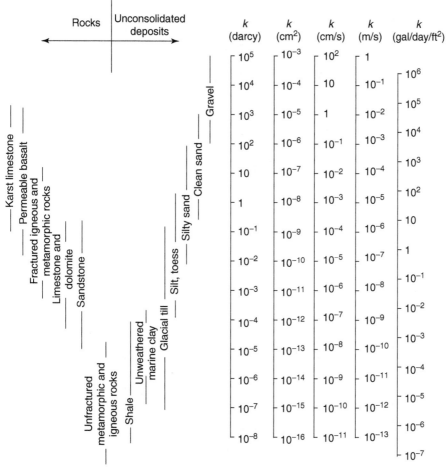

Adapted from Freeze, R.A. and Cherry, J.A., *Ground water*, Prentice Hall, Englewood Cliffs, NJ, 1979, 604 pp.

and

$$v = \frac{\mu}{\rho}$$
(4-27)

is the kinematic viscosity (L^2/T) and k depends on the character of only the porous medium. The shape factor C depends on the porosity of the medium, the grain sizes distribution, the shape of the grains, and their orientation and arrangement. The dimension of k is (L^2). Since k depends on only the character of the porous medium, it is called as the coefficient of permeability or simply permeability. The term intrinsic permeability is also used by some authors.

In Eq. (4-25), both γ and μ vary with temperature; therefore, the ratio γ/μ depends on the temperature as well. However, in many cases of aquifer problems, the temperature does not vary significantly and the hydraulic conductivity K may be then considered as constant in the associated equations.

TABLE 4-4
Conversion Factors for Hydraulic Conductivity (K)

Unit	Unit						
	cm/sec	m/sec	m/day	m/year	ft/sec	ft/day	ft/year	gpd/ft²
cm/sec	1	1.000E−2	8.640E+02	3.154E+02	3.281E−02	2.835E+03	1.035E+06	2.120E+04
m/sec	1.000E+02	1	8.640E+04	3.154E+07	3.281	2.835E+05	1.035E+08	2.120E+06
m/day	1.157E−03	1.157E−05	1	365	3.797E−05	3.281	1.198E+03	2.454E+01
m/year	3.171E−06	3.171E−08	2.740E−03	1	1.040E−07	8.989E−03	3.281	6.273E−02
ft/sec	3.048E+01	0.3048	2.633E+04	9.612E+06	1	8.640E+04	3.154E+07	6.463E+05
ft/day	3.528E−04	3.528E−06	0.3048	1.113E+02	1.157E−05	1	3.650E+02	7.48
ft/year	9.665E−07	9.665E−09	8.351E−04	9.612E+06	3.171E−06	2.740E−03	1	2.049E−02
gpd/ft²	4.717E−05	4.717E−07	4.075E−02	1.487E−01	1.547E−06	1.337E−01	4.880E+01	1

Notes: (1) Enter table at the first column with the given unit. Move right to the right column of the unit to be derived. Read the conversion factor as a multiplier

(2) Conversion factors are given in scientific format. For example, 1.157E−05 = 1.157×10^{-5}.

(3) Example: To convert 16.7 ft/day in to cm/sec: $16.7 \times 3.528E−04$ cm/sec = 5.891E−03 cm/sec.

TABLE 4-5
Conversion Factors for Transmissivity (T)

Unit	Unit						
	m²/sec	m²/day	m²/year	ft²/sec	ft²/day	ft²/year	gpd/ft
m²/sec	1	8.640E+04	3.154E+07	1.076E+01	9.300E+05	3.395E+08	6.956E+06
m²/day	1.157E−05	1	365	1.246E−04	1.076E+01	3.929E+03	8.051E+01
m²/year	3.171E−08	2.238E−03	1	3.413E−07	2.949E−02	1.076E+01	2.206E−01
ft²/sec	9.290E−02	8.027E+03	2.930E+06	1	8.640E+04	3.154E+07	6.463E+05
ft²/day	1.075E−06	9.290E−02	3.391E+01	1.157E−05	1	3.65	7.48
ft²/year	2.946E−09	2.545E−04	9.290E−02	3.171E−08	2.740E−03	1	2.049E−02
gpd/ft	1.438E−07	1.242E−02	4.533	1.547E−06	1.337E−01	4.880E+01	1

Notes: (1) Enter table at the first column with the given unit. Move right to the right column of the unit to be derived. Read the conversion factor as a multiplier.
(2) Conversion factors are given in scientific format. For example, 1.075E−06 = 1.075×10^{-6}.
(3) Example: To convert 5.4 m²/sec into gpd/ft: $5.4 \times 3.528E{-}04$ m²/sec = 30.753E+0.6 gpd/ft.

The units of permeability k can be cm² or ft². In the petroleum industry, the unit of permeability is darcy, according to the relationship

$$k = \frac{\mu Q/A}{\Delta p/\Delta s} \tag{4-28}$$

which is obtained from Eqs. (4-17) and (4-25) by replacing $h = p/\gamma$ where p is the pressure. Based on Eq. (4-28), 1 darcy is defined by using $\mu = 1$ centipoise = 0.01 dyn sec/cm², $Q = 1$ cm³/sec, $A = 1$ cm², $\Delta p = 1$ atmosphere = 1.0132×10^6 dyn/cm², and $\Delta s = 1$ cm. Therefore, substitution of the above values into Eq. (4-28) gives the definition of darcy:

$$1 \text{ darcy} = \frac{(1 \text{ centipoise})(1 \text{ cm}^3/\text{sec})/(1 \text{ cm}^2)}{(1 \text{ atmosphere})/(1 \text{ cm})} = 9.87 \times 10^{-9} \text{ cm}^2$$

EXAMPLE 4-2

The kinematic viscosity and permeability of an aquifer are given as $v = 1.12$ cm²/sec and $k = 7.5$ Darcy. Determine the value of the hydraulic conductivity of the aquifer.

SOLUTION

From Eq. (4-25),

$$K = k\frac{g}{v} = (7.5 \text{ darcy} \times 9.87 \times 10^{-9} \text{ cm}^2/\text{darcy})\frac{(981 \text{ cm/sec}^2)}{(1.14 \text{ cm}^2/\text{sec})}$$

$$K = 6.48 \times 10^{-4} \text{ cm/sec}$$

4.3.1.6 Classification of Aquifers with Respect to Hydraulic Conductivity

As seen in Section 4.3.1.4, hydraulic conductivity of a porous medium is a measure of its fluid-transmitting capability. In aquifers, hydraulic conductivity values usually vary from location to location; in other words, it is a spatially variable quantity. Also, hydraulic conductivity values show variations from direction to direction at a point or location. As a result, the flow rate and discharge velocity will vary from point to point as well as from direction to direction at a point. Aquifer problems associated with these situations, obviously cannot be handled with the simplistic form of the hydraulic conductivity as defined by the original form of Darcy's law as given by Eq. (4-19). For this reason, aquifers are categorized with respect to the variation of their hydraulic conductivity; the definitions of some cases are presented in the following sections.

Basic Definitions
Isotropic and Anisotropic Aquifers
If the hydraulic conductivity K at a point in an aquifer is the same in every direction, the aquifer is called an *isotropic aquifer*. On the contrary, if K varies with the direction, the aquifer is called an *anisotropic aquifer*.

Principal Directions of Anisotropy
For simplicity, the principal directions of anisotropy is explained in a plane perpendicular to the z-axis, i.e., the x–y plane (see Figure 4-6). Let θ be the angle between the x-axis and the axis at which its value is determined, i.e., the ξ-axis. Since K changes with θ on the x–y plane, the value of K varies between its minimum and maximum values from $\theta = 0$ to 2π rad. The axes on which the minimum and maximum values are perpendicular to each other are called the *principal directions of anisotropy*. For example, if these K values are located on the ξ- and η-axis in Figure 4-6, these axes are the principal directions of anisotropy. For the special case in Figure 4-6, the third axis, the ζ-axis, coincides with the z-axis. If one considers a plane perpendicular either to the ξ- or η-axis, the principal directions of anisotropy in that plane can be defined similarly.

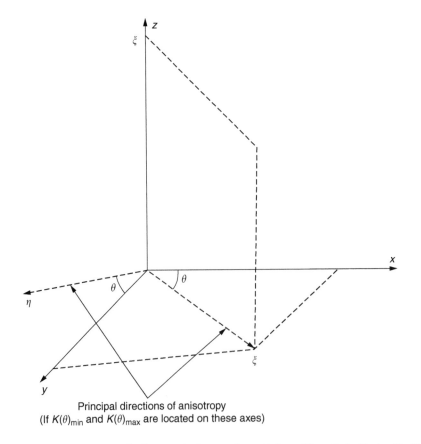

Principal directions of anisotropy
(If $K(\theta)_{min}$ and $K(\theta)_{max}$ are located on these axes)

FIGURE 4-6 Principal directions of anisotropy in the x–y plane. (Adapted from Batu, V., *Aquifer Hydraulics: A Comprehensive Guide to Hydrogeologic Data Analysis*, John Wiley & Sons, Inc., 1998, p. 47, Figure 2-9).

Classifications of Aquifers

Consider that the principal directions of anisotropy of an aquifer are in the directions of the x, y, and z coordinates in a Cartesian coordinate system. Four cases can be defined for heterogeneity and anisotropy (Freeze and Cherry, 1979). These cases are defined below.

Homogeneous and Isotropic Aquifer

This is the most simple case. A homogeneous and isotropic aquifer is schematically defined in Figure 4-7. As shown in Figure 4-7, consider two points in the aquifer whose coordinates are (x_1, y_1, z_1) for point 1 and (x_2, y_2, z_2) for point 2. If, as shown in Figure 4-7, the hydraulic conductivities satisfy the condition

$$K_x(x_1, y_1, z_1) = K_y(x_1, y_1, z_1) = K_z(x_1, y_1, z_1)$$
$$= K_x(x_2, y_z, z_2) = K_y(x_2, y_2, z_2) = K_z(x_2, y_2, z_2) = K \qquad (4\text{-}29)$$

for points 1 and 2 the aquifer is said to be homogeneous and isotropic aquifer. The expressions in Eq. (4-29) mean that the hydraulic conductivity does not vary from point to point as well as with direction.

Homogeneous and Anisotropic Aquifer

A homogeneous and anisotropic aquifer is schematically defined in Figure 4-8. Consider points 1 and 2 in the aquifer whose coordinates are (x_1, y_1, z_1) and (x_2, y_2, z_2), respectively. If the hydraulic conductivities satisfy the conditions

$$K_x(x_1, y_1, z_1) = K_x(x_2, y_2, z_2) = K_x$$
$$K_y(x_1, y_1, z_1) = K_y(x_2, y_2, z_2) = K_y \qquad (4\text{-}30)$$
$$K_z(x_1, y_1, z_1) = K_z(x_z, y_2, z_2) = K_z$$

for points 1 and 2, the aquifer is said to be homogeneous and anisotropic aquifer.

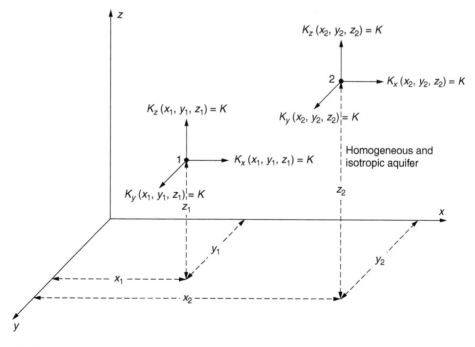

FIGURE 4-7 Homogeneous and isotropic aquifer. (Adapted from Batu, V., *Aquifer Hydraulics: A Comprehensive Guide to Hydrogeologic Data Analysis*, John Wiley & Sons, Inc., 1998, p. 48, Figure 2-10).

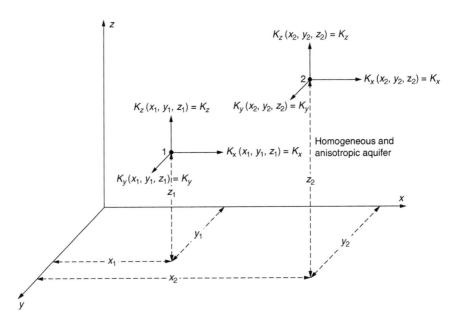

FIGURE 4-8 Homogeneous and anisotropic aquifer. (Adapted from Batu, V., *Aquifer Hydraulics: A Comprehensive Guide to Hydrogeologic Data Analysis*, John Wiley & Sons, Inc., 1998, p. 49, Figure 2-11).

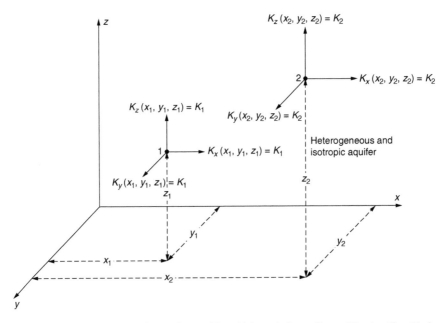

FIGURE 4-9 Heteregeneous and isotropic aquifer. (Adapted from Batu, V., *Aquifer Hydraulics: A Comprehensive Guide to Hydrogeologic Data Analysis*, John Wiley & Sons, Inc., 1998, p. 49, Figure 2-12).

Heterogeneous and Isotropic Aquifer

A heterogeneous and isotropic aquifer is schematically defined in Figure 4-9. Let the coordinates of points 1 and 2 be (x_1, y_1, z_1) and (x_2, y_2, z_2), respectively. If the hydraulic conductivities satisfy the conditions

$$K_x(x_1, y_1, z_1) = K_y(x_1, y_1, z_1) = K_z(x_1, y_1, z_1) = K_1$$

$$K_x(x_2, y_2, z_2) = K_y(x_2, y_2, z_2) = K_z(x_2, y_2, z_2) = K_2 \qquad (4\text{-}31)$$

for points 1 and 2, the aquifer is said to be heterogeneous and isotropic aquifer. If, as a special case of Eq. (4-31),

$$K_x(x_1, y_1, z_1) = K_y(x_1, y_1, z_1) = K_x(x_2, y_2, z_2) = K_y(x_2, y_2, z_2) = K_h$$
$$K_z(x_1, y_1, z_1) = K_z(x_2, y_2, z_2) = K_z = K_v \quad (4\text{-}32)$$

the aquifer is said to be transversely isotropic aquifer. Although the case defined by the expressions in Eq. (4-32) is relatively in a simplified form, it is widely used in deriving analytical solutions for well hydraulics under fully and partially penetrating well conditions (see e.g., Batu, 1998).

Heterogeneous and Anisotropic Aquifer
This is the most general case as shown in Figure 4-10. Consider points 1 and 2 whose coordinates are (x_1, y_1, z_1) and (x_2, y_2, z_2), respectively. If the hydraulic conductivities satisfy the conditions,

$$K_x(x_1, y_1, z_1) \neq K_y(x_1, y_1, z_1) \neq K_z(x_1, y_1, z_1)$$
$$K_x(x_2, y_2, z_2) \neq K_y(x_2, y_2, z_2) \neq K_z(x_2, y_2, z_2) \quad (4\text{-}33)$$

for points 1 and 2, the aquifer is said to be heterogeneous and anisotropic aquifer.

4.3.1.7 Storage Coefficient, Specific Yield, and Specific Retention

Storage Coefficient
The *storage coefficient* (or *storativity*) S of a saturated confined aquifer or confined layer of thickness b is defined as

$$S = S_s b \quad (4\text{-}34)$$

where S_s is given by (see, e.g., Batu, 1998, Eq. [2-63] p. 58)

$$S_s = \frac{\gamma}{E_s} + \frac{\varphi\gamma}{E_w} \quad (4\text{-}35)$$

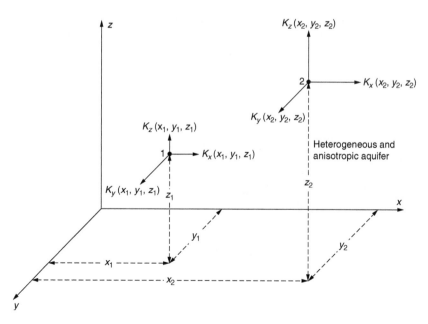

FIGURE 4-10 Heterogeneous and anisotropic aquifer. (Adapted from Batu, V., *Aquifer Hydraulics: A Comprehensive Guide to Hydrogeologic Data Analysis*, John Wiley & Sons, Inc., 1998, p. 50, Figure 2-13).

where S_s is the specific storage, dimension of L^{-1}, γ specific weight of water, E_s the modulus of elasticity of soil skeleton, E_w the modulus of elasticity of water, and φ the porosity of soil. The specific storage can be defined as the volume of water that is released from or taken into storage per unit volume of a confined aquifer or a confined aquifer layer per unit change in hydraulic head. Substitution of Eq. (4-35) into Eq. (4-34) gives

$$S = \frac{\gamma b}{E_s} + \frac{n\gamma b}{E_w} = S_k + S_w \tag{4-36}$$

S_k is the storage coefficient of soil and is defined as

$$S_k = \frac{\gamma b}{E_s} \tag{4-37}$$

and S_w is the storage coefficient of water and is defined as

$$S_w = \frac{\varphi\gamma b}{E_w} \tag{4-38}$$

Unlike S_s, the storage coefficient is dimensionless.

The definition of storage coefficient is further clarified on a confined aquifer (Figure 4-11), which has uniform thickness and its upper and lower boundaries are horizontal for convenience (Ferris et al., 1962). If the hydraulic head in the confined aquifer is decreased, some volume of water will be released proportionally to the change in hydraulic head. Let us consider a representative parallelepiped extending from the base of the aquifer to its piezometric surface as shown in Figure 4-11. Storage coefficient is the ratio between the volume of water released from storage in the parallelepiped, ΔV and $(\Delta A)(\Delta h)$:

$$S = \frac{\Delta V}{(\Delta A)(\Delta h)} \tag{4-39}$$

where ΔA and Δh are the horizontal cross-sectional area of the parallelepiped and the change in hydraulic head, respectively. Although the storage coefficient is defined for a confined aquifer, the concept can be applied to unconfined aquifers as well.

It is clear from Eq. (4-36) that, besides the compressibilities of soil and water, the storage coefficient S is a function of formation thickness b, which is a site-specific quantity. For most aquifers, confined or unconfined, the storage coefficient values are generally fall in the range of 0.00005 to 0.005 (Ferris et al., 1962; Freeze and Cherry, 1979). The storage coefficient of an aquifer can be measured by pumping and slug tests.

Specific Yield

The *specific yield* S_y is the measure of water yielded by gravity drainage when the watertable of an unconfined aquifer declines, in general terms, it is defined as the volume of water released from or taken into a unit area of an unconfined aquifer when the watertable moves by one unit of height (Meinzer, 1923; Ferris et al., 1962).

The definition of specific yield is further illuminated on an unconfined aquifer (Figure 4-12), which has uniform thickness and its lower boundary is horizontal for convenience (Ferris et al., 1962). When the hydraulic head in the unconfined aquifer is decreased, some volume of water will be released by gravity drainage. Let us consider a representative parallelepiped extending from the base of the aquifer to its watertable as shown in Figure 4-12. The specific yield is the ratio between the volume of water involved in the gravity drainage or refilling in the parallelepiped, ΔV_g and $(\Delta A)(\Delta h)$:

$$S_y = \frac{\Delta V_g}{(\Delta A)(\Delta h)} \tag{4-40}$$

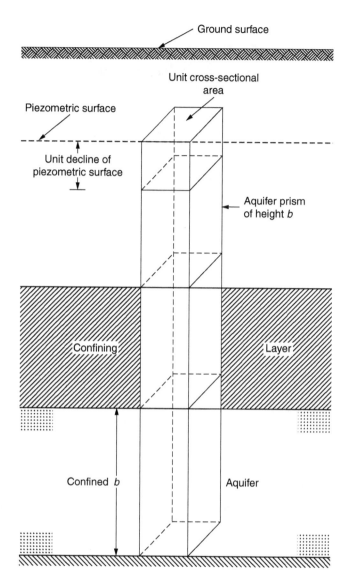

FIGURE 4-11 Diagram for storage coefficient for a confined aquifer. (Adapted from Ferris, J.G., et al., *U.S. Geological Survey Water Supply Paper 1536-E*, 1962, pp. 69–174.)

where ΔA and Δh are the horizontal cross-sectional area of the parallelepiped and the change in watertable elevation, respectively. Similar to the storage coefficient, the specific yield is a dimensionless quantity as well.

The specific yields of unconfined aquifers S_y are much higher than the storage coefficients S of confined aquifers. The specific yields of unconfined aquifers are generally in the range of 0.01 to 0.30, depending on the type of porous materials. For fine-grained porous materials, the specific yield values are significantly different from their porosity values and for coarse-grained materials the specific yield values are very close to their porosity values. Typical values for various porous materials are given in Table 4-6. Methods for the determination of specific yield include the analysis of time versus drawdown data from pumping tests in unconfined aquifers (see, e.g., Batu, 1998).

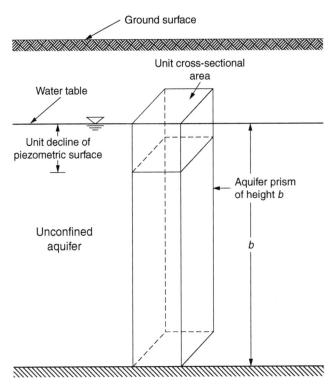

FIGURE 4-12 Diagram for specific yield for an unconfined aquifer. (Adapted from Ferris, J.G., et al., *U.S. Geological Survey Water Supply Paper 1536-E*, 1962, pp. 69–174.)

TABLE 4-6
Specific Yield of Various Porous Materials

Group	Porous Material	No. of Analysis	Range of Specific Yield S_y	Arithmetic Mean Average Specific Yield S_{yav}
Sedimentary Materials	Sandstone (fine)	47	0.02–0.40	0.21
	Sandstone (medium)	10	0.12–0.41	0.27
	Siltstone	13	0.01–0.33	0.12
	Sand (fine)	287	0.01–0.46	0.33
	Sand (medium)	297	0.16–0.46	0.32
	Sand (coarse)	143	0.18–0.43	0.30
	Gravel (fine)	33	0.13–0.40	0.28
	Gravel (medium)	13	0.17–0.44	0.24
	Gravel (coarse)	9	0.13–0.25	0.21
	Silt	299	0.01–0.39	0.20
	Clay	27	0.01–0.18	0.06
	Limestone	32	0–0.36	0.14
Wind-Laid Materials	Loess	5	0.14–0.22	0.18
	Eolian sand	14	0.32–0.47	0.38
	Tuff	90	0.02–0.47	0.21
Metamorphic Rocks	Schist	11	0.22–0.33	0.26

Adapted from Morris, D.A. and Johnson, A.I., *U.S. Geological Survey Water Supply Paper 1839-D*, 1967; and Mercer, J.W., et al., U.S. Nuclear Regulatory Commission, NUREG/CR-3066, Washington, DC, 1982.

Specific Retention

The *specific retention*, S_r, is the measure of water retained in the soil against gravity by capillarity and hygroscopic forces when the watertable of an unconfined aquifer drops. It is defined as the volume of water retained against gravity in a unit area of an unconfined aquifer when the watertable drops by one unit of height (Meinzer, 1923; Lohman, 1972).

The definition of specific retention can further be illuminated on an unconfined aquifer (Figure 4-12), which has uniform thickness and its lower boundary is horizontal for convenience. As mentioned above, when the hydraulic head in the unconfined aquifer is decreased, some volume of water will be retained by capillarity and hygroscopic forces against gravity. Let us consider a representative parallelepiped extending from the base of the aquifer to its water table as shown in Figure 4-12. The specific retention is the ratio between the volume of water involved in the retention in the parallelepiped, ΔV_r and $(\Delta A)(\Delta h)$:

$$S_r = \frac{\Delta V_r}{(\Delta A)(\Delta h)} = \varphi - S_y \qquad (4\text{-}41)$$

in which ΔA and Δh are the horizontal cross-sectional area of the parallelepiped and the change in watertable elevation, respectively.

Similar to storage coefficient S and specific yield S_y, the specific retention is a dimensionless quantity. In order to determine the values of specific retention S_r, the values given in Table 4-6 can be subtracted from the porosity values given in Table 4-1 according to Eq. (4-41).

4.3.2 EQUATIONS OF MOTION IN AQUIFERS: GENERALIZATION OF DARCY'S LAW

4.3.2.1 Equations of Motion: Principal Directions of Anisotropy Coincide with the Directions of Coordinate Axes

Both Cartesian and cylindrical coordinates have wide applications in solving aquifer hydraulics problems. In order to derive the governing partial differential equations of flow in aquifers as well as to solve these equations, expressions of Darcy velocities in these coordinates system are required. These equations are presented in the following sections.

Darcy Velocities in Two- and Three-Dimensional Cartesian Coordinates

Equations of motion is given in the Cartesian coordinates for the special case where the principal directions of anisotropy coincide with x-, y-, and z-coordinate axes. In other words, the Darcy velocity vector has three components that are aligned with the x-, y-, and z-coordinate axis. These components are shown on an infinitesimal parallelepiped volume that represents a point in the flow field (Figure 4-13).

Equation (4-19) is the one-dimensional differential form of Darcy's law and it states that the flow rate through porous materials in any direction is proportional to the negative rate of change in the hydraulic head in that direction. The negative sign in the equation implies that the fluid move in the direction of decreasing hydraulic head.

Darcy's law was extended to three-dimensional cases (e.g., Polubarinova-Kochina, 1962; Hantush, 1964; Bear, 1972, 1979; Freeze and Cherry, 1979). The general form of Darcy's law in a nonhomogeneous and anisotropic porous medium, with the principal axes of the hydraulic conductivity tensor parallel to the Cartesian coordinate axes in vector form is

$$\mathbf{q} = q_x \mathbf{i} + q_y \mathbf{j} + q_z \mathbf{k} \qquad (4\text{-}42)$$

where $\mathbf{q}$ is the Darcy velocity vector, and q_x, q_y, and q_z and $\mathbf{i}$, $\mathbf{j}$, and $\mathbf{k}$ are the Darcy velocity components and the unit vectors in the x, y, and z directions, respectively. The general, spatially

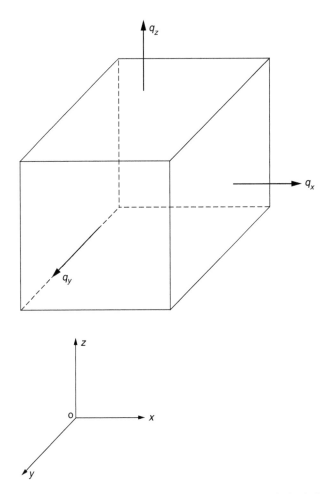

FIGURE 4-13   Darcy velocity components under the condition that the principal directions of anisotropy concide with x, y, and z directions of the coordinate axes.

and temporally variable forms of Darcy velocity components in the x, y, and z, directions, respectively, are

$$q_x = q_x(x, y, z, t) = - K_x(x, y, z)\frac{\partial h(x, y, z, t)}{\partial x} \tag{4-43a}$$

$$q_y = q_y(x, y, z, t) = - K_y(x, y, z)\frac{\partial h(x, y, z, t)}{\partial y} \tag{4-43b}$$

$$q_z = q_z(x, y, z, t) = - K_z(x, y, z)\frac{\partial h(x, y, z, t)}{\partial z} \tag{4-43c}$$

In the two-dimensional case, Eqs. (4-43a), (4-43b), and (4-43c), respectively, take the forms

$$q_x = q_x(x, y, t) = - K_x(x, y)\frac{\partial h(x, y, t)}{\partial x} \tag{4-44a}$$

$$q_y = q_y(x, y, t) = - K_x(x, y)\frac{\partial h(x, y, t)}{\partial x} \tag{4-44b}$$

$$q_z = 0 \tag{4-44c}$$

In reality, the hydraulic conductivity components of an aquifer are spatially variable quantities as indicated in the above equations. As a result, the Darcy velocity components will be spatially variable quantities as well. Aquifer test analysis methods are based on analytical and semianalytical solutions, which assume that the principal hydraulic conductivities do not vary with the coordinates. This assumption generally produces practically reasonable results for most cases encountered in various aquifers. Recent studies based on finite-difference and finite element numerical methods of solving ground water problems showed that spatial variability of hydraulic conductivity exhibits less level of spatially variable Darcy velocity components. In other words, the variability of Darcy velocity components are significantly less than the variability of hydraulic conductivity. For this purpose, Batu (1984) developed a finite element method to determine Darcy velocities in any kind of nonhomogeneous and anisotropic aquifers and the results on some hypothetical space-dependent hydraulic conductivity functions confirmed the above statements (see Figure 4 of Batu [1984]).

With the assumption of constant principal hydraulic conductivities, in aquifer hydraulics, it is further assumed that the principal hydraulic conductivities in the horizontal plane are the same, i.e., $K_x(x, y, z) = K_y(x, y, z) = K_h$ and $K_z(x, y, z) = K_z$, and Eqs. (4-43a), (4-43b), and (4-43c), respectively, take the following forms:

$$q_x = q_x(x, y, z, t) = - K_h \frac{\partial h(x, y, z, t)}{\partial x} \tag{4-45a}$$

$$q_y = q_y(x, y, z, t) = - K_h \frac{\partial h(x, y, z, t)}{\partial y} \tag{4-45b}$$

$$q_z = q_z(x, y, z, t) = - K_z \frac{\partial h(x, y, z, t)}{\partial z} \tag{4-45c}$$

In Eq. (4-45c), K_v is often used instead of K_z.

Darcy Velocities in Two- and Three-Dimensional Cylindrical Coordinates
Cylindrical coordinate system has wide applications in deriving analytical solutions for circular, vertical well hydraulics problems owing to the geometric shape of wells. A circular well generally has a finite radius and it is assumed that the central axis of the well coincides with the z-coordinate axis. The radial distance in a horizontal plane is measured from the vertical axis. Both the Cartesian and cylindrical coordinates are shown in Figure 4-14. Under extraction or injection condition from or to a circular well, a radial flow to or from the well occurs. Under this condition, the hydraulic conductivity K_r is the only horizontal hydraulic conductivity instead of K_x and K_y in the x and y directions, respectively. The vertical hydraulic conductivity K_z remains the same. Therefore, the equivalent forms of Eqs. (4-45) take the following forms:

$$q_r = q_r(r, z, t) = - K_r \frac{\partial h(r, z, t)}{\partial r} \tag{4-46a}$$

$$q_z = q_z(r, z, t) = - K_z \frac{\partial h(r, z, t)}{\partial z} \tag{4-46b}$$

It is apparent that the equivalent forms of Eq. (4-46a) in the Cartesian coordinates are Eqs. (4-45a) and (4-45b). Equation (4-46a) is based on the assumption that the principal direction of the horizontal hydraulic conductivity is in the radial direction K_r and its value is the same in all directions around the well. The direction of the vertical hydraulic conductivity is in the direction of the z coordinate.

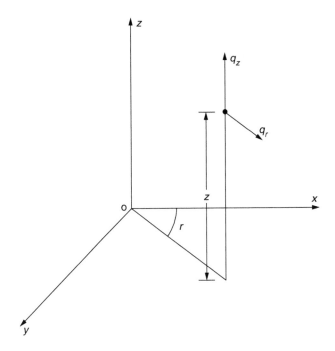

FIGURE 4-14 Darcy velocity components for the cylindrical coordinates system.

4.3.2.2 Equations of Motion: Principal Directions of Anisotropy Do Not Coincide with the Directions of Coordinate Axes

Equations of motion will be written for the general case where the principal directions of anisotropy do not coincide with x-, y-, and z-coordinate axes. These components are shown on an infinitesimal parallelepiped volume, which represents a point in the flow field shown in Figure 4-15. The Darcy velocity components of the vector $\mathbf{q}$ given by Eq. (4-42) are (see, e.g., Bear, 1972; Domenico and Schwartz, 1990)

$$q_{xx} = -K_{xx}\frac{\partial h}{\partial x} - K_{xy}\frac{\partial h}{\partial y} - K_{xz}\frac{\partial h}{\partial z} \tag{4-47a}$$

$$q_{yy} = -K_{yx}\frac{\partial h}{\partial x} - K_{yy}\frac{\partial h}{\partial y} - K_{yz}\frac{\partial h}{\partial z} \tag{4-47b}$$

$$q_{zz} = -K_{zx}\frac{\partial h}{\partial x} - K_{zy}\frac{\partial h}{\partial y} - K_{zz}\frac{\partial h}{\partial z} \tag{4-47c}$$

where x, y, and z are the Cartesian coordinates and K_{xx}, K_{xy}, ..., K_{zz} are the nine constant components of the hydraulic conductivity tensor in the most general case. The nine components in the matrix form display a second-rank symmetric tensor known as the hydraulic conductivity tensor (Bear, 1972). In Eqs. (4-47), the first subscript indicates the direction perpendicular to the plane upon which the Darcy velocity vector acts and the second subscript indicates the direction of the Darcy velocity vector in that plane. If the principal directions of anisotropy coincide with x, y, and z directions of the coordinate axes, the six components K_{xy}, K_{xz}, K_{yx}, K_{yz}, K_{zx}, and K_{zy} become equal to zero, and Eqs. (4-47) reduces to Eqs. (4-43) for $K_x(x, y, z) = K_x$, $K_y (x, y, z) = K_y$, and $K_z (x, y, z) = K_z$.

4.3.3 Principal Hydraulic Conductivities

As mentioned in Section 4.3.2.1, if the principal directions of anisotropy coincide with x, y, and z directions of the coordinate axes, the Darcy velocity vector has three components expressed by Eqs. (4-45),

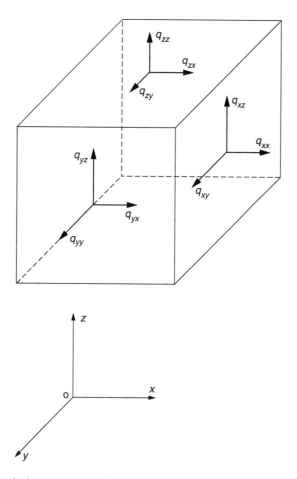

FIGURE 4-15 Darcy velocity components for the general case.

which are aligned with the x-, y-, and z-coordinate axis. In other words, the six components K_{xy}, K_{xz}, K_{yx}, K_{yz}, K_{zx}, and K_{zy} of the hydraulic conductivity tensor all become equal to zero. Under the conditions, K_{xx}, K_{yy}, and K_{zz} are called *principal hydraulic conductivities*, and the x-, y-, and z-coordinate directions are called *principal directions*. Now, let us consider the situation with Eqs. (4-47) and Figure 4-15, for which the aforementioned six components are not equal to zero. The question is: what is the orientation of the principal directions of anisotropy axes corresponding to Eqs. (4-47) The answer is that there are three principal directions of anisotropy (shown by ξ, η, and ζ) on which the aforementioned six hydraulic conductivity components are zero. The ξ, η, and ζ coordinates have the same origin with the x, y, and z coordinates, but their orientation as well as the corresponding hydraulic conductivities on these axes are not known. The solution of the two-dimensional version of this problem has significant practical importance and its details are presented below.

4.3.3.1 Principal Hydraulic Conductivities and Transmissivities for Two-Dimensional Case

Consider the two-dimensional form of Eqs. (4-47). For two-dimensional flow, Eqs. (4-43) takes the form

$$q_x = - K_{xx}\frac{\partial h}{\partial x} - K_{xy}\frac{\partial h}{\partial y} \tag{4-48a}$$

$$q_y = - K_{yx}\frac{\partial h}{\partial x} - K_{yy}\frac{\partial h}{\partial y} \tag{4-48b}$$

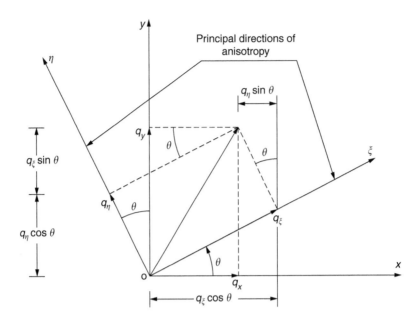

FIGURE 4-16 Rotation of two-dimensional coordinate axes. (Adapted from Batu, V., *Aquifer Hydraulics: A Comprehensive Guide to Hydrogeologic Data Analysis*, John Wiley & Sons, Inc., 1998, p. 85, Figure 2-28.)

The x–y and ξ–η coordinates are shown in Figure 4-16. The orientation angle θ and hydraulic conductivities or transmissivities on the ξ and η coordinates are not known. Now, we will derive the corresponding expressions in terms of K_{xx}, K_{yy}, K_{xy}, and K_{yx} regarding these quantities.

As can be seen from Figure 4-16, the Darcy velocity vector is the same in x–y and ξ–η coordinates systems. Therefore, one can write

$$q_x = q_\xi \cos\theta - q_\eta \sin\theta \tag{4-49a}$$

$$q_y = q_\eta \cos\theta + q_\eta \sin\theta \tag{4-49b}$$

From the geometry in Figure 4-16, the coordinates ξ–η are related to x–y coordinates by

$$x = \xi\cos\theta - \eta\sin\theta \tag{4-50a}$$

$$y = \eta\cos\theta + \xi\sin\theta \tag{4-50b}$$

On the other hand, from the rules of derivatives and Eqs. (4-50), the following equations can be written:

$$\frac{\partial h}{\partial \xi} = \frac{\partial h}{\partial x}\frac{\partial x}{\partial \xi} + \frac{\partial h}{\partial y}\frac{\partial y}{\partial \xi} = \frac{\partial h}{\partial x}\cos\theta + \frac{\partial h}{\partial y}\sin\theta \tag{4-51a}$$

$$\frac{\partial h}{\partial \eta} = \frac{\partial h}{\partial x}\frac{\partial x}{\partial \eta} + \frac{\partial h}{\partial y}\frac{\partial y}{\partial \eta} = -\frac{\partial h}{\partial x}\sin\theta + \frac{\partial h}{\partial y}\cos\theta \tag{4-51b}$$

From the definition of principal directions of anisotropy, the Darcy velocity components on the ξ and η coordinates in Figure 4-16, respectively are

$$q_\xi = -K_{\xi\xi}\frac{\partial y}{\partial \xi} \tag{4-52a}$$

$$q_\eta = -K_{\eta\eta}\frac{\partial h}{\partial \eta} \tag{4-52b}$$

From Eqs. (4-49), (4-51), and (4-52), the Darcy velocity components on the x and y coordinate directions can be written, respectively, as

$$q_x = -\,(K_{\xi\xi}\cos^2\theta + K_{\eta\eta}\sin^2\theta)\frac{\partial h}{\partial x} + (K_{\eta\eta} - K_{\xi\xi})\sin\theta\cos\theta\,\frac{\partial h}{\partial y} \qquad (4\text{-}53a)$$

$$q_y = -\,(K_{\xi\xi}\sin^2\theta + K_{\eta\eta}\cos^2\theta)\frac{\partial h}{\partial y} + (K_{\eta\eta} - K_{\xi\xi})\sin\theta\cos\theta\,\frac{\partial h}{\partial x} \qquad (4\text{-}53b)$$

From Eqs. (4-48a) and (4-53a) one can write

$$K_{xx} = K_{\xi\xi}\cos^2\theta + K_{\eta\eta}\sin^2\theta \qquad (4\text{-}54a)$$

$$K_{xy} = (K_{\xi\xi} - K_{\eta\eta})\sin\theta\cos\theta \qquad (4\text{-}54b)$$

Similarly, from Eqs. (4-48b) and (4-53b) one has

$$K_{yy} = K_{\xi\xi}\sin^2\theta + K_{\eta\eta}\cos^2\theta \qquad (4\text{-}55a)$$

$$K_{yx} = (K_{\xi\xi} - K_{\eta\eta})\sin\theta\cos\theta \qquad (4\text{-}55b)$$

It is interesting to see that $K_{xy} = K_{yx}$ from Eqs. (4-54b) and (4-55b). Finally, substitution of the equivalents of $\sin^2\theta$, $\cos^2\theta$, and $\sin\theta\cos\theta$ into Eqs. (4-54) and (4-55) gives

$$K_{xx} = \tfrac{1}{2}(K_{\xi\xi} + K_{\eta\eta}) + \tfrac{1}{2}(K_{\xi\xi} - K_{\eta\eta})\cos(2\theta) \qquad (4\text{-}56)$$

$$K_{yy} = \tfrac{1}{2}(K_{\xi\xi} + K_{\eta\eta}) - \tfrac{1}{2}(K_{\xi\xi} - K_{\eta\eta})\cos(2\theta) \qquad (4\text{-}57)$$

and

$$K_{xy} = K_{yx} = \tfrac{1}{2}(K_{\xi\xi} - K_{\eta\eta})\sin(2\theta) \qquad (4\text{-}58)$$

Since, according to Eq. (4-24), the transmissivity is a product of the hydraulic conductivity and aquifer thickness, by multiplying both sides of Eqs. (4-56)–(4-58) by the aquifer thickness b, the corresponding equations for transmissivity can be obtained as

$$T_{xx} = \tfrac{1}{2}(T_{\xi\xi} + T_{\eta\eta}) + \tfrac{1}{2}(T_{\xi\xi} - T_{\eta\eta})\cos(2\theta) \qquad (4\text{-}59)$$

$$T_{yy} = \tfrac{1}{2}(T_{\xi\xi} + T_{\eta\eta}) - \tfrac{1}{2}(T_{\xi\xi} - T_{\eta\eta})\cos(2\theta) \qquad (4\text{-}60)$$

$$T_{xy} = T_{yx} = \tfrac{1}{2}(T_{\xi\xi} - T_{\eta\eta})\sin(2\theta) \qquad (4\text{-}61)$$

4.3.3.2 Mohr's Circle for Principal Hydraulic Conductivities and Transmissivities for Two-Dimensional Case

In the mechanics of materials, Mohr's circle is a graphical interpretation of the formulas used to determine the principal stresses and the maximum shearing stresses at a point in a stressed rigid body. A German engineer, Otto Mohr (1835 to 1918), developed a useful graphical interpretation of the stress equations. The particular case for plane stress was first presented by another German engineer, K. Culmann (1829 to 1881) about 16 years before Mohr's paper. This method, commonly known as Mohr's circle in solid mechanics, involves the construction of a circle such that the coordinates of each point on the circle represent the normal and shearing stresses on one plane through the stressed point, and the angular position of the radius to the point gives the orientation of the plane. Mohr's circle is also used as a graphical interpretation of the formulas used to determine the maximum and minimum second moments of areas and maximum products of inertia (see, e.g., Terzaghi, 1943; Higdon et al., 1976). Also, since the 1980s, the Mohr's circle method is being used to determine the principal macrodispersivities for solute transport in aquifers (see Section 6.4.6.8).

The Mohr's circle approach is also a convenient way to interpret Eqs. (4-56), (4-57), and (4-58) to determine the principal hydraulic conductivities and their orientation angle (see, e.g., Verruijt, 1970; Bear, 1972). The Mohr's circle corresponding to Eqs. (4-56)–(4-58) is shown in Figure 4-17 with K_{xx} and K_{yy} plotted as horizontal coordinates and K_{xy} and K_{yx} as vertical coordinates. The circle is centered on the K_{xx}- and K_{yy}- axis at a distance $(K_{xx} + K_{yy})/2$ from K_{yx}- and $-K_{xy}$-axis and the radius of the circle is given by

$$r = \left[\left(\frac{K_{xx}-K_{yy}}{2}\right)^2 + K_{xy}^2\right]^{1/2} \qquad (4\text{-}62)$$

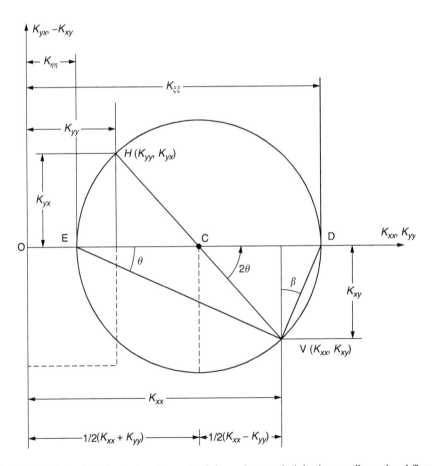

FIGURE 4-17 Mohr's circle for hydraulic conductivity and transmissivity in two-dimensional flow cases in aquifers. (Adapted from Batu, V., *Aquifer Hydraulics: A Comprehensive Guide to Hydrogeologic Data Analysis*, John Wiley & Sons, Inc., 1998, p. 88, Figure 2-29.)

From the horizontal coordinates of points E and D in Figure 4-17, the expressions for $K_{\xi\xi}$ and $K_{\eta\eta}$ can be determined as

$$K_{\xi\xi} = \tfrac{1}{2}(K_{xx} + K_{yy}) + \left[\tfrac{1}{4}(K_{xx} - K_{yy})^2 + K_{xy}^2 \right]^{1/2} \tag{4-63}$$

$$K_{\eta\eta} = \tfrac{1}{2}(K_{xx} + K_{yy}) - \left[\tfrac{1}{4}(K_{xx} - K_{yy})^2 + K_{xy}^2 \right]^{1/2} \tag{4-64}$$

The angle 2θ from CV to CD is counterclockwise or positive, and from Figure 4-17,

$$\tan(2\theta) = \frac{2K_{xy}}{K_{xx} - K_{yy}} \tag{4-65}$$

Since $\beta = \theta$ in Figure 4-17, the following equation can also be written:

$$\tan\theta = \tan\beta = \frac{K_{\xi\xi} - K_{xx}}{K_{xy}} \tag{4-66}$$

According to Eq. (4-24), the transmissivity is a product of the hydraulic conductivity and aquifer thickness. Therefore, by multiplying both sides of Eqs. (4-63) and (4-64) by the aquifer thickness b, the expressions for $T_{\xi\xi}$ and $T_{\eta\eta}$ can be obtained as

$$T_{\xi\xi} = \tfrac{1}{2}(T_{xx} + T_{yy}) + \left[\tfrac{1}{4}(T_{xx} - T_{yy})^2 + T_{xy}^2 \right]^{1/2} \tag{4-67}$$

$$T_{\eta\eta} = \tfrac{1}{2}(T_{xx} + T_{yy}) - \left[\tfrac{1}{2}(T_{xx} - T_{yy})^2 + T_{xy}^2 \right]^{1/2} \tag{4-68}$$

Similarly, Eqs. (4-65) and (4-66) can, respectively, be written as

$$\tan(2\theta) = \frac{2T_{xy}}{T_{xx} - T_{yy}} \tag{4-69}$$

$$\tan\theta = \tan\beta = \frac{T_{\xi\xi} - T_{xx}}{T_{xy}} \tag{4-70}$$

The above equations are widely used to determine the aquifer parameters under fully penetrating pumping wells in homogeneous and anisotropic confined nonleaky aquifers (see, e.g., Batu, 1998).

4.3.4 Directional Hydraulic Conductivities

In anisotropic aquifers, hydraulic conductivity varies from direction to direction. In Section 4.3.3, for a two-dimensional anisotropic aquifer case, the orientation of the principal directions as well as the expressions of hydraulic conductivities ($K_{\xi\xi}$ and $K_{\eta\eta}$) on these axes in terms of the hydraulic conductivity components (K_{xx}, K_{yy}, K_{xy}, and K_{yx}) are given. As shown in Figure 4-18, consider a two-dimensional flow system in which the (ξ, η) coordinate axes are the principal axes on which the principal hydraulic conductivities ($K_{\xi\xi}$ and $K_{\eta\eta}$) are known. The problem is to find the hydraulic conductivity expression in the r direction, which makes an angle θ with the ξ-coordinate axis. The corresponding expressions for three-dimensional flow systems are also presented.

4.3.4.1 Directional Hydraulic Conductivities and Transmissivities for Two-Dimensional Cases

Consider a two-dimensional homogeneous and anisotropic flow system in which the principal hydraulic conductivities ($K_{\xi\xi}$ and $K_{\eta\eta}$) or transmissivities ($T_{\xi\xi}$ and $T_{\eta\eta}$) are aligned with the

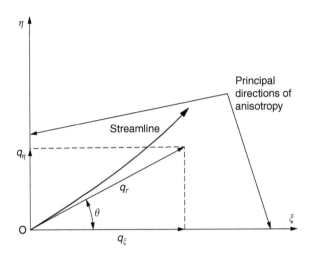

FIGURE 4-18   Flow line in a two-dimensional anisotropic aquifer.

ξ- and η-axis. The tangent to the flow line passing through the origin makes an angle θ with the ξ-axis. Therefore, the Darcy velocity along the streamline is

$$q_r = - K_r \frac{\partial h}{\partial r} \tag{4-71}$$

and its components in the ξ and η directions, respectively, are

$$q_\xi = - K_{\xi\xi} \frac{\partial h}{\partial \xi} = q_r \cos\theta \tag{4-72}$$

$$q_\eta = - K_{\eta\eta} \frac{\partial h}{\partial \eta} = q_r \sin\theta \tag{4-73}$$

Since $h = h\,(\xi, \eta)$ and $\xi = \xi(r)$, $\eta = \eta(r)$

$$\frac{\partial h}{\partial r} = \frac{\partial h}{\partial \xi} \frac{\partial \xi}{\partial r} + \frac{\partial h}{\partial \eta} \frac{\partial \eta}{\partial r} \tag{4-74}$$

and from Figure 4-18, one can write

$$\frac{\partial \xi}{\partial r} = \cos\theta, \qquad \frac{\partial \eta}{\partial r} = \sin\theta \tag{4-75}$$

Substitution of Eqs. (4-71)–(4-73), and (4-75) into Eq. (4-74) and after simplification, the following equation can be obtained:

$$\frac{1}{K_r} = \frac{\cos^2\theta}{K_{\xi\xi}} + \frac{\sin^2\theta}{K_{\eta\eta}} \tag{4-76}$$

By solving Eq. (4-76) for K_r, one gets

$$K_r = \frac{K_{\xi\xi}}{\cos^2\theta + (K_{\xi\xi}/K_{\eta\eta})\sin^2\theta} \tag{4-77}$$

According to Eq. (4-24), the transmissivity is a product of the hydraulic conductivity and aquifer thickness. Therefore, the principal transmissivities become

$$T_{\xi\xi} = K_{\xi\xi}\,b, \qquad T_{\eta\eta} = K_{\eta\eta}\,b \tag{4-78}$$

and by substituting these into Eq. (4-77), the transmissivity ($T_r = K_r b$) in the direction of s is

$$T_r = \frac{T_{\xi\xi}}{\cos^2\theta + (T_{\xi\xi}/T_{\eta\eta})\sin^2\theta}$$ (4-79)

Equations (4-77) and (4-79) give the directional hydraulic conductivity K_r and directional transmissivity T_r, respectively, in any angular direction θ.

Hydraulic Conductivity and Transmissivity Ellipses
Substitution of $\xi = r\cos\theta$ and $\eta = r\sin\theta$ into Eq. (4-76), one has

$$\frac{r^2}{K_r} = \frac{\xi^2}{K_{\xi\xi}} + \frac{\eta^2}{K_{\eta\eta}}$$ (4-80)

By drawing a segment of length $r = (K_r)^{1/2}$ in the direction of q_r, Eq. (4-80) takes the form (Maasland, 1957; Bear, 1972)

$$\frac{\xi^2}{K_{\xi\xi}} + \frac{\eta^2}{K_{\eta\eta}} = 1$$ (4-81)

Equation (4-81) is the hydraulic conductivity ellipse in the ξ- and η-coordinate system. The semi-axes of the ellipse in the ξ and η directions are $(K_{\xi\xi})^{1/2}$ and $(K_{\eta\eta})^{1/2}$, respectively (see Figure 4-19). The ellipse gives the directional hydraulic conductivity in the direction of the flow. The directional hydraulic conductivity K_r at angular direction θ can also be determined from Eq. (4-76) by using the known values of $K_{\xi\xi}$ and $K_{\eta\eta}$ on the principal axes.

The equations presented above can also be applied to directional transmissivities and transmissivity ellipse using Eqs. (4-78) with $T_r = K_r b$.

4.3.4.2 Directional Hydraulic Conductivities for Three-Dimensional Cases

For three-dimensional cases, equations similar to those in Section 4.3.4.1 can be developed (Massland, 1957; Bear, 1972). The corresponding equation to Eq. (4-76) is

$$\frac{1}{K_r} = \frac{\cos^2\alpha_x}{K_{\xi\xi}} + \frac{\cos^2\alpha_y}{K_{\eta\eta}} + \frac{\cos^2\alpha_z}{K_{\zeta\zeta}}$$ (4-82)

where α_x, α_y, and α_z are the angles between the direction of q_r and the three principal axes ξ, η, and ζ.

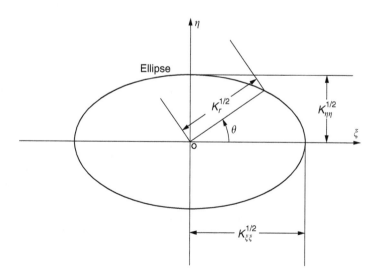

FIGURE 4-19 Hydraulic conductivity ellipse.

Hydraulic Conductivity Ellipsoid
With $\xi = r \cos \alpha_x$, $\eta = r \cos \alpha_y$, and $\zeta = r \cos \alpha_z$, Eq. (4-82) takes the form

$$\frac{r^2}{K_r} = \frac{\xi^2}{K_{\xi\xi}} + \frac{\eta^2}{K_{\eta\eta}} \frac{\zeta^2}{K_{\zeta\zeta}} \qquad (4\text{-}83)$$

By drawing a segment of length $r = (K_r)^{1/2}$ in the direction of q_r, Eq. (4-83) takes the form (Maasland, 1957; Bear, 1972)

$$\frac{\xi^2}{K_{\xi\xi}} + \frac{\eta^2}{K_{\eta\eta}} + \frac{\zeta^2}{K_{\zeta\zeta}} = 1 \qquad (4\text{-}84)$$

Equation (4-84) is the hydraulic conductivity ellipsoid in the ξ-, η-, and ζ-coordinate system. The semiaxes of the ellipse in the ξ, η, and ζ directions are $(K_{\xi\xi})^{1/2}$, $(K_{\eta\eta})^{1/2}$, and $(K_{\zeta\zeta})^{1/2}$, respectively. Figure 4-19 can be viewed as the vertical cross-section of the hydraulic conductivity ellipsoid. The ellipse gives the directional hydraulic conductivity in the direction of the flow. The directional hydraulic conductivity K_r at a direction defined by α_x, α_y, and α_z angles can also be determined from Eq. (4-82) using the known values of $K_{\xi\xi}$, $K_{\eta\eta}$, and $K_{\zeta\zeta}$ on the principal axes.

4.3.5 DERIVATION OF THE GOVERNING DIFFERENTIAL EQUATION IN RECTANGULAR COORDINATES

Analytical as well as numerical solution of an aquifer problem is based on its governing differential equation and the corresponding initial and boundary conditions. Initial and boundary conditions typically encountered in most aquifer problems are included in Section 4.3.9. In this section, the governing equations of flow in aquifers are presented both in rectangular (Cartesian) and cylindrical coordinates (see, e.g., Hantush, 1964; Bear, 1972, 1979).

4.3.5.1 General Case: The Principal Directions of Anisotropy Are Not the Coordinate Axes

Three-Dimensional Flow Equation
Consider a volume element ΔV in an aquifer whose mass is ΔM and W (unit T^{-1}) is the volumetric flux per unit volume and represents sources or sinks of water in this volume. According to the conservation of mass principle, the sum of the net inward flux through ΔV and the amount of mass generated within the element per unit time must be equal to the rate of mass accumulating within ΔV. Then

$$\text{rate of change of } \Delta M \text{ per unit time per unit volume} = \frac{1}{\Delta V} \frac{\partial(\Delta M)}{\Delta V} \qquad (4\text{-}85)$$

If $\mathbf{q}$ is the bulk velocity vector and ρ is the density of the fluid, the mass flow per unit area will be $\rho\mathbf{q}$. The mathematical expression of the aforementioned conservation of mass principle is (see, e.g., Hantush, 1964)

$$-\iint(\rho\mathbf{q}).d\mathbf{S} - W = \iiint\left[\frac{1}{\Delta V}\frac{\partial(\Delta M)}{\partial t}\right]dV \qquad (4\text{-}86)$$

where the left-hand side of the equation is the surface integral taken over the closed surface of the volume of ΔV and the right-hand side is the volume integral over the volume element. The dot on the left-hand side means that the scalar product or dot product of the two vector quantities $\rho\mathbf{q}$ and $d\mathbf{S}$ is taken (see, e.g., Spiegel, 1971). In order to proceed further, a relationship is needed between ΔM and ΔV. This relationship is derived in the ground water literature of the specific storage (S_s) equation and its expression is (see, e.g., Hantush, 1964, p. 295; Batu, 1998, p. 58)

$$\frac{\partial(\Delta M)}{\Delta V} = \rho S_s \partial h \qquad (4\text{-}87)$$

The surface integral on the left-hand side of Eq. (4-86) can be converted into a volume integral by applying the divergence theorem (see, e.g., Spiegel, 1971). With this theorem and by substituting Eq. (4-87) into Eq. (4-86) we get

$$- \iiint \left[\nabla \cdot (\rho \mathbf{q}) + W + \rho S_s \frac{\partial h}{\partial t} \right] dV = 0 \tag{4-88}$$

where ∇ is the gradient operator and defined as (e.g., Spiegel, 1971)

$$\nabla = \frac{\partial}{\partial x} \mathbf{i} + \frac{\partial}{\partial y} \mathbf{j} + \frac{\partial}{\partial z} \mathbf{k} \tag{4-89}$$

in which $\mathbf{i}$, $\mathbf{j}$, and $\mathbf{k}$ are the unit vectors in the x, y, and z directions, respectively. Since the volume element is arbitrary, the integral in Eq. (4-88) is valid for any volume element. As a result, with the assumption that the density of water ρ is constant, the following equation can be written:

$$- \nabla \cdot \mathbf{q} - W = S_s \frac{\partial h}{\partial t} \tag{4-90}$$

where $\mathbf{q}$ is the Darcy velocity in the general form and its components are given by Eqs. (4-47). By using tensor notation, Eqs. (4-47) can be expressed in a concised form as

$$\mathbf{q} = - K \nabla h \tag{4-91}$$

where K is the *hydraulic conductivity tensor* and is defined as

$$[K] = \begin{bmatrix} K_{xx} & K_{xy} & K_{xz} \\ K_{yx} & K_{yy} & K_{yz} \\ K_{zx} & K_{zy} & K_{zz} \end{bmatrix} \tag{4-92}$$

Substitution of Eq. (4-91) into Eq. (4-90) gives

$$\nabla \cdot (K \nabla h) - W = S_s \frac{\partial h}{\partial t} \tag{4-93}$$

Equation (4-93) can also be expressed in a scalar form by using Eqs. (4-89) and (4-92). Since its scalar form is too lengthy, only its two-dimensional form under some special conditions is presented below.

Two-Dimensional Flow Equation under Special Conditions
The two-dimensional form of Eq. (4-92) for $K_{xy} = K_{yx}$ is

$$[K] = \begin{vmatrix} K_{xx} & K_{xy} \\ K_{xy} & K_{yy} \end{vmatrix} \tag{4-94}$$

which represents a symmetric hydraulic conductivity tensor. By combining Eqs. (4-48), (4-90), and (4-94), one has

$$K_{xx} \frac{\partial^2 h}{\partial x^2} + 2K_{xy} \frac{\partial^2 h}{\partial x \partial y} + K_{yy} \frac{\partial^2 h}{\partial y^2} - W = S_s \frac{\partial h}{\partial t} \tag{4-95}$$

4.3.5.2   Special Cases: The Principal Directions of Anisotropy Are the Coordinate Axes

As mentioned in Section 4.3.2.1, if the principal directions of anisotropy coincide with x, y, and z directions of the coordinate axes, the Darcy velocity vector has three components expressed by Eqs. (4-43), which are aligned with the x-, y-, and z-coordinate axes. In other words, the six components K_{xy}, K_{xz}, K_{yx}, K_{yz}, K_{zx}, and K_{zy} of the hydraulic conductivity tensor all become equal to zero. Under this condition, Eq. (4-92) takes the form

$$[K] = \begin{bmatrix} K_{xx} & 0 & 0 \\ 0 & K_{yy} & 0 \\ 0 & 0 & K_{zz} \end{bmatrix} \tag{4-96}$$

and the corresponding forms of Darcy velocity components are given by Eqs. (4-43). The differential equations of several cases in accordance to Section 4.3.1.6 is presented below.

Homogeneous and Isotropic Aquifer
This is the most simple case and its definition is given in Figure 4-7. By substituting Eqs. (4-29) into Eqs. (4-43), Eq. (4-93) takes the form

$$\frac{\partial^2 h}{\partial x^2} + \frac{\partial^2 h}{\partial y^2} + \frac{\partial^2 h}{\partial z^2} - W = \frac{S_s}{K}\frac{\partial h}{\partial t} \tag{4-97}$$

which is the differential equation of flow in a homogeneous and isotropic aquifer.

Homogeneous and Anisotropic Aquifer
A homogeneous and anisotropic aquifer is schematically represented in Figure 4-8. Substitution of Eqs. (4-30) into Eqs. (4-43) and then into Eq. (4-93) gives

$$K_x\frac{\partial^2 h}{\partial x^2} + K_y\frac{\partial^2 h}{\partial y^2} + K_z\frac{\partial^2 h}{\partial z^2} - W = S_s\frac{\partial h}{\partial t} \tag{4-98}$$

which is the differential equation of flow in a homogeneous and anisotropic aquifer.

Heterogeneous and Isotropic Aquifer
A heterogeneous and isotropic aquifer is schematically represented in Figure 4-9. According to Eqs. (4-31), the three principal hydraulic conductivities at a point in the aquifer are equal, but their values vary from point to point. Therefore, with Eqs. (4-43), Eq. (4-93) takes the form

$$\frac{\partial}{\partial x}\left[K(x, y, z)\frac{\partial h}{\partial x}\right] + \frac{\partial}{\partial y}\left[K(x, y, z)\frac{\partial h}{\partial y}\right] + \frac{\partial}{\partial z}\left[K(x, y, z)\frac{\partial h}{\partial z}\right] - W = S_s\frac{\partial h}{\partial t} \tag{4-99}$$

which is the differential equation of flow in a heterogeneous and isotropic aquifer.

Transversely Isotropic Aquifer
From Eqs. (4-32), (4-43), and (4-93) one gets

$$K_h\left(\frac{\partial^2 h}{\partial x^2} + \frac{\partial^2 h}{\partial y^2}\right) + K_z\frac{\partial^2 h}{\partial z^2} - W = S_s\frac{\partial h}{\partial t} \tag{4-100}$$

which is the differential equation of flow in a transversely isotropic aquifer.

Heterogeneous and Anisotropic Aquifer
A heterogeneous and anisotropic aquifer is schematically represented in Figure 4-10. From the conditions expressed by Eqs. (4-33) and combining Eqs. (4-43) and (4-93) gives the corresponding differential equation of flow as

$$\frac{\partial}{\partial x}\left[K_x(x, y, z)\frac{\partial h}{\partial x}\right] + \frac{\partial}{\partial y}\left[K_y(x, y, z)\frac{\partial h}{\partial y}\right] + \frac{\partial}{\partial z}\left[K_z(x, y, z)\frac{\partial h}{\partial z}\right] - W = S_s\frac{\partial h}{\partial t} \tag{4-101}$$

The conditions for Eqs. (4-99) and (4-101) are that the hydraulic conductivity functions must be continuous and have continuous first derivative everywhere in the flow domain. The one-, two-, and three-dimensional forms of Eqs. (4-95), (4-97), (4-98), and (4-100) are widely used in developing analytical solutions for aquifer problems. Equations (4-99) and (4-101) can only be solved numerically with the exception of some special hydraulic conductivity distribution functions.

4.3.6 DERIVATION OF THE GOVERNING DIFFERENTIAL EQUATION IN CYLINDRICAL COORDINATES

A vertical extraction or injection well causes radial flow in the aquifer toward the well or from the well. It is relatively easier to work with partial differential equations of flow in aquifers, written in polar coordinates. Both rectangular and polar coordinates are shown in Figure 4-20, in which the well is located at the origin of the coordinate system and the axis of the well is perpendicular to the *x*–*y* plane. This implicitly means that only the *x* and *y* rectangular coordinates must be converted into polar coordinates, since the vertical coordinate *z* remains the same. Since the flow is radial, hydraulic conductivity in the radial direction or radial hydraulic conductivity K_r is used instead of K_x and K_y. This concept is based on the assumption that the hydraulic conductivity as well as hydraulic head do not change with the angular direction θ. Hydraulic head changes only with the radial distance *r* from the well. In the following sections, the differential equations of ground water flow in three- and two-dimensional cases will be presented.

4.3.6.1 Differential Equation in Three Dimensions for a Transversely Isotropic Aquifer Case

Definition of a transversely isotropic aquifer is given in Section 4.3.1.6 and the corresponding partial differential equation in rectangular coordinates is given by Eq. (4-100). Owing to the reasons mentioned above, only the derivatives of *h* with respect to *x* and *y* must be converted into polar coordinates. From the geometry of Figure 4-20 one can write

$$r = r\cos\theta \qquad y = r\sin\theta$$
$$r = (x^2 + y^2)^{1/2} \qquad \theta = \text{arc}\tan(y/x) \tag{4-102}$$

By applying the chain rule of differentiation to the derivatives of $h = h(x, y, t)$, the differential equation in cylindrical coordinates can be obtained as (the derivational details are given in Example 4-3)

$$K_r \frac{\partial^2 h}{\partial r^2} + \frac{K_r}{r}\frac{\partial h}{\partial r} + K_z \frac{\partial^2 h}{\partial z^2} - W = S_s \frac{\partial h}{\partial t} \tag{4-103}$$

Equation (4-103) has wide application in a circular fully and partially penetrating wells in aquifers.

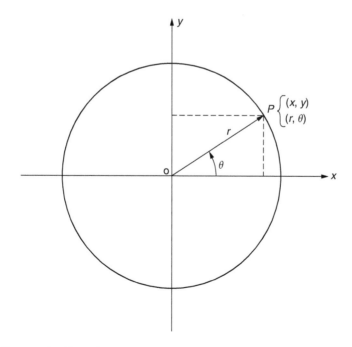

FIGURE 4-20   Rectangular (Cartesian) and polar coordinates.

EXAMPLE 4-3

Derive Eq. (4-103).

SOLUTION

By applying the chain rule of differentiation to the derivatives of $h = h(x, y, t)$, one can write

$$\frac{\partial h}{\partial x} = \frac{\partial h}{\partial r}\frac{\partial r}{\partial x} + \frac{\partial h}{\partial \theta}\frac{\partial \theta}{\partial x} \tag{E4-3-1}$$

By differentiating Eq. (E4-3-1) with respect to x one gets

$$\frac{\partial^2 h}{\partial x^2} = \frac{\partial}{\partial x}\left(\frac{\partial h}{\partial r}\frac{\partial r}{\partial x}\right) + \frac{\partial}{\partial x}\left(\frac{\partial h}{\partial \theta}\frac{\partial \theta}{\partial x}\right) = \frac{\partial}{\partial x}\left(\frac{\partial h}{\partial r}\right)\frac{\partial r}{\partial x} + \frac{\partial h}{\partial r}\frac{\partial^2 r}{\partial x^2}$$

$$+ \frac{\partial}{\partial x}\left(\frac{\partial h}{\partial \theta}\right)\frac{\partial \theta}{\partial x} + \frac{\partial h}{\partial \theta}\frac{\partial^2 \theta}{\partial x^2} \tag{E4-3-2}$$

Now, by applying the chain rule again, one finds

$$\frac{\partial}{\partial x}\left(\frac{\partial h}{\partial r}\right) = \frac{\partial}{\partial r}\left(\frac{\partial h}{\partial r}\right)\frac{\partial r}{\partial x} + \frac{\partial}{\partial \theta}\left(\frac{\partial h}{\partial r}\right)\frac{\partial \theta}{\partial x} = \frac{\partial^2 h}{\partial r^2}\frac{\partial r}{\partial x} + \frac{\partial}{\partial \theta}\left(\frac{\partial h}{\partial r}\right)\frac{\partial \theta}{\partial x} \tag{E4-3-3}$$

and

$$\frac{\partial}{\partial x}\left(\frac{\partial h}{\partial \theta}\right) = \frac{\partial}{\partial r}\left(\frac{\partial h}{\partial \theta}\right)\frac{\partial r}{\partial x} + \frac{\partial}{\partial \theta}\left(\frac{\partial h}{\partial r}\right)\frac{\partial \theta}{\partial x} = \frac{\partial}{\partial r}\left(\frac{\partial h}{\partial \theta}\right)\frac{\partial r}{\partial x} + \frac{\partial^2 h}{\partial \theta^2}\frac{\partial \theta}{\partial x} \tag{E4-3-4}$$

To determine the partial derivatives of $\partial r/\partial x$ and $\partial \theta/\partial x$, r and θ in Eqs. (4-102) have to be determined as

$$\frac{\partial r}{\partial x} = \tfrac{1}{2}(2x)(x^2 + y^2)^{1/2} = \frac{x}{(x^2+y^2)^{1/2}} = \frac{x}{r} \tag{E4-3-5}$$

and

$$\frac{\partial \theta}{\partial x} = \frac{y}{1+(y/x)^2}\left(-\frac{y}{x^2}\right) = -\frac{y}{x^2+y^2} = -\frac{y}{r^2} \tag{E4-3-6}$$

By differentiating Eqs. (E4-3-5) and (E4-3-6), the following expressions can be obtained:

$$\frac{\partial^2 r}{\partial x^2} = \frac{r - x(\partial r/\partial x)}{r^2} = \frac{1}{r} - \frac{x^2}{r^3} = \frac{r^2-x^2}{r^3} = \frac{y^2}{r^3} \tag{E4-3-7}$$

and

$$\frac{\partial^2 \theta}{\partial x^2} = -y\left(-\frac{2}{r^3}\right)\frac{\partial r}{\partial x} = \frac{2y}{r^3}\frac{x}{r} = \frac{2xy}{r^4} \tag{E4-3-8}$$

By substituting Eqs. (E4-3-3)–(E4-3-8) into Eq. (E4-3-2) gives

$$\frac{\partial^2 h}{\partial x^2} = \left[\frac{\partial^2 h}{\partial x^2}\frac{\partial r}{\partial x} + \frac{\partial}{\partial \theta}\left(\frac{\partial h}{\partial r}\right)\frac{\partial \theta}{\partial x}\right]\frac{\partial r}{\partial x} + \frac{\partial h}{\partial r}\frac{\partial^2 r}{\partial x^2} + \left[\frac{\partial}{\partial r}\left(\frac{\partial h}{\partial \theta}\right)\frac{\partial r}{\partial x}\right. \tag{E4-3-9}$$

$$\left. + \frac{\partial^2 h}{\partial x^2}\frac{\partial \theta}{\partial x}\right]\frac{\partial \theta}{\partial x} + \frac{\partial h}{\partial \theta}\frac{\partial^2 \theta}{\partial x^2}$$

or

$$\frac{\partial^2 h}{\partial x^2} = \frac{x^2}{r^2}\frac{\partial^2 h}{\partial r^2} - \frac{xy}{r^2}\frac{\partial}{\partial \theta}\left(\frac{\partial h}{\partial r}\right) + \frac{y^2}{r^3}\frac{\partial h}{\partial r} - \frac{xy}{r^2}\frac{\partial}{\partial r}\left(\frac{\partial h}{\partial \theta}\right)$$

$$+ \frac{y^2}{r^4}\frac{\partial^2 h}{\partial \theta^2} + 2\frac{xy}{r^4}\frac{\partial h}{\partial \theta} \tag{E4-3-10}$$

By using

$$\frac{\partial}{\partial \theta}\left(\frac{\partial h}{\partial r}\right) = \frac{\partial}{\partial r}\left(\frac{\partial h}{\partial \theta}\right) \tag{E4-3-11}$$

Eq. (E4-3-10) takes the form

$$\frac{\partial^2 h}{\partial x^2} = \frac{x^2}{r^2}\frac{\partial^2 h}{\partial r^2} - 2\frac{xy}{r^2}\frac{\partial}{\partial r}\left(\frac{\partial h}{\partial \theta}\right) + \frac{y^2}{r^3}\frac{\partial h}{\partial r} + \frac{y^2}{r^4}\frac{\partial^2 h}{\partial \theta^2} + 2\frac{xy}{r^4}\frac{\partial h}{\partial \theta} \tag{E4-3-12}$$

Similarly, the following equation can also be obtained:

$$\frac{\partial^2 h}{\partial y^2} = \frac{y^2}{r^2}\frac{\partial^2 h}{\partial r^2} + 2\frac{xy}{r^2}\frac{\partial}{\partial r}\left(\frac{\partial h}{\partial \theta}\right) + \frac{x^2}{r^3}\frac{\partial h}{\partial r} + \frac{x^2}{r^4}\frac{\partial^2 h}{\partial \theta^2} - 2\frac{xy}{r^4}\frac{\partial h}{\partial \theta} \tag{E4-3-13}$$

Substitution of Eqs. (E4-3-12) and (E4-3-13) into Eq. (4-100), replacing K_h by K_r and remembering that h does not change with θ, after simplification, one determines Eq. (4-103).

4.3.6.2 Differential Equation in Two Dimensions

In the case of two-dimensional radial flow, the third term on the left-hand side becomes zero and it is appropriate to use transmissivity instead of hydraulic conductivity by using the aquifer thickness b. Therefore, using Eqs. (4-24) and (4-29), when $W = 0$, the differential equation in two dimensions from Eq. (4-103) takes the form

$$\frac{\partial^2 h}{\partial r^2} + \frac{1}{r}\frac{\partial h}{\partial r} = \frac{S}{T}\frac{\partial h}{\partial t} \tag{4-104}$$

where $T = K_r b$.

4.3.7 Differential Equations under Steady-State Conditions

The term $S_s \partial h/(t$ or $(S/T)\partial h/(t$ in Eqs. (4-90)–(4-104) is the one that makes the cases transient. Therefore, if this term is zero, all of those equations correspond to steady-state case. For example, under this condition, Eq. (4-98), which corresponds to a homogeneous and anisotropic aquifer, takes the form

$$K_x\frac{\partial^2 h}{\partial x^2} + K_y\frac{\partial^2 h}{\partial y^2} + K_z\frac{\partial^2 h}{\partial z^2} - W = 0 \tag{4-105}$$

For a homogeneous and isotropic aquifer, when $W = 0$ Eq. (4-97) reduces to the well-known Laplace equation

$$\nabla^2 h = \frac{\partial^2 h}{\partial x^2} + \frac{\partial^2 h}{\partial y^2} + \frac{\partial^2 h}{\partial z^2} = 0 \tag{4-106}$$

where ∇^2 is the Laplacian operator.

4.3.8 GOVERNING DIFFERENTIAL EQUATION FOR THE FINITE-DIFFERENCE MODELING APPROACH

The finite-difference mathematical approach relates specifically to the widely used ground water flow modeling code MODFLOW (McDonald and Harbaugh, 1984) and one of its sister codes MODFLOWT (Duffield, 1996). As a result, the mathematical background presented here and in the subsequent sections are compatible with the mathematical backgrounds of MODFLOW and MOD-FLOWT.

The finite-difference approach is applied to ground water flow in heterogeneous and an isotropic aquifers and the governing differential equation is derived in Section 4.3.5.2 (Eq. [4-101]). Therefore, the governing differential equation for ground water flow in heterogeneous and anisotropic aquifers in the MODFLOW family codes is given by the following equation:

$$\frac{\partial}{\partial x}\left(K_{xx}\frac{\partial h}{\partial x}\right)+\frac{\partial}{\partial y}\left(K_{yy}\frac{\partial h}{\partial y}\right)+\frac{\partial}{\partial z}\left(K_{zz}\frac{\partial h}{\partial z}\right)-W=S_s\frac{\partial h}{\partial t} \qquad (4\text{-}107)$$

where x, y, and z (units are L) are the principal coordinate axes of the system. In other words, they are assumed to be aligned along the principal hydraulic conductivities K_{xx}, K_{yy}, and K_{zz} (units L/T), respectively (See Section 4.3.3 for principal hydraulic conductivities). The rest of six components (K_{xy}, K_{xz}, K_{yx}, K_{yz}, K_{zx}, and K_{zy}) in the hydraulic conductivity tensor (Eq. [4-92]) all become equal to zero. Therefore, the corresponding hydraulic conductivity tensor for Eq. (4-107) is the tensor given by Eq. (4-96). In Eq. (4-107), h is the hydraulic head (unit L) and is defined by Eq. (4-1), W (unit 1/T) the volumetric flux per unit volume and represents sources or sinks of water, S_s (unit 1/L) the specific storage and is defined by Eq. (4-35), and t (T) the time. In general, S_s, K_{xx}, K_{yy}, and K_{zz} may be functions of space and written as

$$S_s \equiv S_s(x, y, z), \quad K_{xx} \equiv K_x(x, y, z), K_{yy} \equiv K_{yy}(x, y, z), K_{zz} \equiv K_{zz}(x, y, z) \qquad (4\text{-}108)$$

and h and W may be functions of space and time and written as

$$h \equiv h(x, y, z, t), \qquad W \equiv W(x, y, z, t) \qquad (4\text{-}109)$$

Equation (4-107) describes ground water flow under nonequilibrium (unsteady) conditions in a heterogeneous and anisotropic porous medium. The derivation of Eq. (4-107) is given in Section 4.3.5 as Eq. (4-101), and in the light of Eq. (4-108), Eq. (4-101) is exactly the same as Eq. (4-107).

Equation (4-107), together with the specified flow and head conditions at the boundaries of an aquifer system and specified initial head conditions, forms a mathematical model of ground water flow. A solution of Eq. (4-107) for a ground water flow problem with its corresponding initial and boundary conditions is an algebraic expression for h (x, y, z, t) in analytical sense, i.e., when the derivatives of h (x, y, z, t) with respect to space (x, y, z) and time (t) are substituted into Eq. (4-107), the differential equation and its initial and boundary conditions are satisfied. A space- and time-varying head distribution of this nature characterizes the ground water flow system in which it measures both the energy of flow and the volume of water in storage and can be used to determine directions and rates of movement.

Analytical solutions of Eq. (4-107) can be obtained under some special conditions such as hydraulic conductivity variation and initial and boundary conditions (see, e.g., Hantush, 1964; Bear, 1979; Batu, 1998). For spatially variable hydraulic conductivity and complex initial and boundary condition cases, numerical methods must be used to obtain approximate solutions. As mentioned in Section 4.2, the finite-difference method is one of the potential numerical methods use to solve Eq. (4-107) and is popular for ground water modeling applications. In the finite-difference approach, Eq. (4-107) is replaced by a finite set of discrete points in space and time, and the partial derivatives are replaced by differences between functional values at these points. This process leads to systems of simultaneous linear algebraic difference equations and their solution yields values of h (x, y, z, t). The values of h (x, y, z, t) form an approximation to the space- and time-varying head distribution that is given by an analytical solution of the same partial differential equation, i.e., Eq. (4-107).

4.3.9 FINITE-DIFFERENCE GRID AND DISCRETIZATION CONVENTION

4.3.9.1 Grid Coordinates and Cartesian Coordinates

For MODFLOW, a three-dimensional finite-difference grid system is formed by parallelepipeds or rectangular blocks (or simply blocks). Figure 4-21 shows a spatial discretization of an aquifer system that is formed by blocks. The grid coordinates of a block are shown by i (row number), j (column number), and k (layer number). Therefore, an (i, j, k) coordinate system, which is a compatible computer array conventions, is used. For an aquifer system consisting of "$nrow$" (number of rows), "$ncol$" (number of columns), and "$nlay$" (number of layers), i is the row index, $i = 1, 2, ..., nrow$; j is the column index, $j = 1, 2, ..., ncol$; and k is the layer index, $k = 1, 2, ..., nlay$. For example, Figure 4-21 shows an aquifer system with $nrow = 5$, $ncol = 9$, and $nlay = 5$. The origin of the aquifer system ($i = 1, j = 1, k = 1$) is the upper-left corner of the topmost layer. This convention is adopted for the numerical models described in this book. Unless otherwise stated, the origin of the

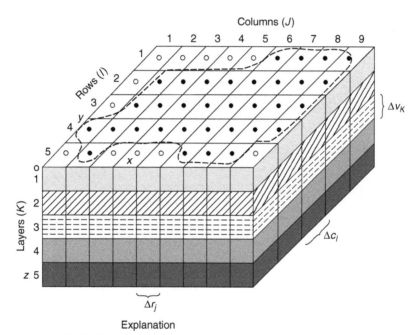

Explanation

- - - - Aquifer boundary

•   Active cell

o   Inactive cell

Δr_J Dimension of cell along the row direction, subscript (J) indicates the number of the column

Δc_I Dimension of cell along the column direction, subscript (I) indicates the number of the row

Δv_K Dimension of cell along the vertical direction, subscript (K) indicates the number of the layer

FIGURE 4-21 A discretized aquifer system. (Adapted from McDonald, M.G. and Harbaugh, A.W., *Open File Report 83–875*, U.S. Geological Survey, National Center, Reston, Virginia, 1984.)

rectangular (Cartesian) coordinates is the lower-left corner of the topmost layer. Therefore, points along a row form a line parallel to the x-axis, points along a column form a line parallel to the y-axis, and points along the vertical form a line parallel to the z-axis.

In Figure 4-21, the width of cells along the rows are designated as Δr_j for the jth column, the width of cells along the columns as Δc_i for the ith column; and the thickness of layers in the vertical direction as Δv_k for the kth layer. For example, the block with coordinates of $i = 5, j = 2$, and $k = 4$ has a volume of $(\Delta r_2)(\Delta c_5)(\Delta v_4)$.

4.3.9.2 Block- and Point-Centered Grid Discretizations

In the finite-difference method there are basically two conventions regarding the location of nodes. They are block-centered formulation and point-centered formulation, which are shown in Figure 4-22. As shown in Figure 4-22, in these two conventions, the aquifer is divided with two sets of parallel lines

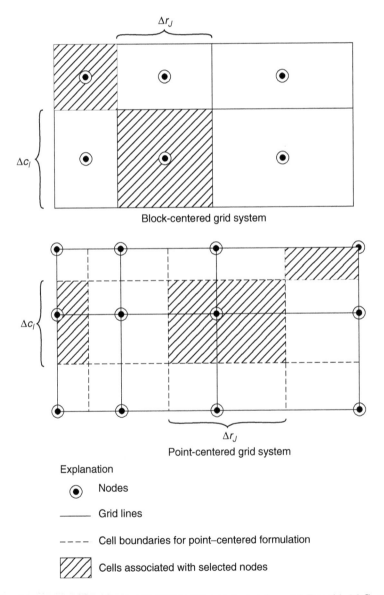

FIGURE 4-22   Block-centered and point-centered grids. (Adapted from McDonald, M.G. and Harbaugh, A.W., *Open File Report 83–875*, U.S. Geological Survey, National Center, Reston, Virginia, 1984.)

that are perpendicular to each other. In the three-dimensional block-centered formulation, the blocks are formed by the sets of parallel planes (parallel lines in two-dimensional cases) and the nodes are the center of the blocks. In the three-dimensional point-centered formulation, the nodes are at the intersection points of the sets of parallel planes and blocks are formed around the nodes with faces halfway between nodes. In either case, spacing of nodes should be such that the hydraulic properties are uniform in the whole volume of a block. The finite-difference formulation presented in the following sections is valid for the two formulations. Although the finite-difference model can accept the two formulations, only the block-centered formulation is included in the MODFLOW family codes.

4.3.9.3   Discretization with Time

As can be seen from Eq. (4-107), the head $h = h(x, y, z, t)$ is a function of space (x, y, z) and time t. Therefore, apart from the space discretization, in the finite-difference formulation, discretization of the continuous time domain is also required.

4.3.10   DERIVATION OF THE FINITE-DIFFERENCE EQUATION

Equation (4-107) is the governing equation of ground water flow in heterogeneous and anisotropic aquifers and is derived in Section 4.3.5 by applying the conservation of mass principle. Although Eq. (4-107) can be used in developing the finite-difference equation, in this section, the approach presented in the MODFLOW manual (McDonald and Harbaugh, 1984) will be followed. Similarly, the finite-difference equation for ground water flow can also be derived by the application of the conservation of mass principle. This principle is applied to a finite-difference cell shown in Figure 4-23. According to this principle, the sum of all flows into and out of the cell must be equal to the rate of

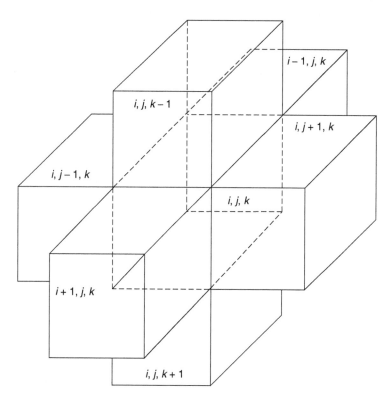

FIGURE 4-23   Cell (i, j, k) and indices for the six adjacent cells. (Adapted from McDonald, M.G. and Harbaugh, A.W., *Open File Report 83–875*, U.S. Geological Survey, National Center, Reston, Virginia, 1984.)

change in storage within the cell. By assuming that the density of ground water is constant, application of the conservation of mass principle for time period Δt for a cell gives

$$(\textstyle\sum Q_i)\Delta t = S_s \Delta h \Delta V = S \Delta V \tag{4-110}$$

where Q_i (L^3/T)is the ith flow rate into or out of the cell, S_s (L^{-1}) the specific storage (see Section 4.3.1.7), ΔV (L^3) the volume of the cell, and Δh (L) the change in head over a time interval of length Δt (T). According to Eq. (4-34), the product $(S_s \Delta h)$ is the storage coefficient S of the cell. Therefore, the right-hand sides of Eq. (4-110) represent the volume of water in the cell. As a result, Eq. (4-110) states that the net volume of water entering into the cell during time period Δt is equal to the storage of water in the cell. Equation (4-110) can also be expressed as

$$\textstyle\sum Q_i = S_s \frac{\Delta h}{\Delta t} \Delta V \tag{4-111}$$

4.3.10.1   Flow Rates into Cell (i, j, k) From the Six Adjacent Cells

Figure 4-23 shows an aquifer cell (i, j, k) and six adjacent cells $(i - 1, j, k)$, $(i + 1, j, k)$, $(i, j - 1, k)$, $(i, j + 1, k)$, $(i, j, k - 1)$, and $(i, j, k + 1)$. The flow rate into cell (i, j, k) in the row direction from cell $(i, j - 1, k)$ according to Darcy's law, can be expressed as

$$q_{i,j-1/2,k} = KR_{i,j-1/2,k} \Delta c_i \Delta v_k \frac{(h_{i,j-1,k} - h_{i,j,k})}{\Delta r_{j-1/2}} \tag{4-112}$$

where $q_{i,j-1/2,k}$ (L^3/T) is the flow rate through the face between cells (i, j, k) and $(i, j - 1, k)$, $KR_{i,j-1/2,k}$ (L/T) is the hydraulic conductivity along the row between nodes (i, j, k) and $(i, j - 1, k)$, and $\Delta r_{j-1/2}$ (L) is the distance between nodes (i, j, k) and $(i, j - 1, k)$. The index $(j - 1/2)$ (L) is used to indicate the space between nodes. It does not indicate a point exactly halfway between nodes (Figure 4-24). For example, $KR_{i,j-1/2,k}$ represents hydraulic conductivity in the entire region between nodes (i, j, k) and $(i, j - 1, k)$.

Similar expressions can be written for the flow rates into or out of the cell through the remaining five faces. The flow rate into the row direction through the face between cells (i, j, k) and $(i, j + 1, k)$ is

$$q_{i,j+1/2,k} = KR_{i,j+1/2,k} \Delta c_i \Delta v_k \frac{(h_{i,j+1,k} - h_{i,j,k})}{\Delta r_{j+1/2}} \tag{4-113}$$

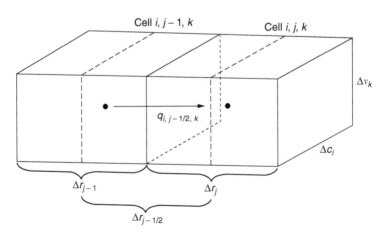

FIGURE 4-24   Flow into cell (i, j, k) from cell $(i, j - 1, k)$. (Adapted from McDonald, M.G. and Harbaugh, A.W., *Open File Report 83–875*, U.S. Geological Survey, National Center, Reston, Virginia, 1984.)

For the column direction, the flow rate through the forward face of the block is given by

$$q_{i+1/2,j,k} = KR_{i+1/2,j,k}\Delta r_j \Delta v_k \frac{(h_{i+1,j,k}-h_{i,j,k})}{\Delta c_{j+1/2}} \tag{4-114}$$

For the column direction, the flow rate through the rear face of the block is given by

$$q_{i-1/2,j,k} = KR_{i-1/2,j,k}\Delta r_j \Delta v_k \frac{(h_{i+1,j,k}-h_{i,j,k})}{\Delta c_{i-1/2}} \tag{4-115}$$

For the vertical direction, flow through the bottom face is given by

$$q_{i,j,k+1/2} = KV_{i,j,k+1/2}\Delta r_j \Delta c_i \frac{(h_{i,j,k+1}-h_{i,j,k})}{\Delta v_{k+1/2}} \tag{4-116}$$

For the vertical direction, flow through the upper face is given by

$$q_{i,j,k-1/2} = KV_{i,j,k-1/2}\Delta r_j \Delta c_i \frac{(h_{i,j,k-1}-h_{i,j,k})}{\Delta v_{k-1/2}} \tag{4-117}$$

Equations (4-112)–(4-117) express flow rates through the faces of cell (i, j, k) in terms of heads, grid dimensions, and hydraulic conductivity. Since grid dimensions and hydraulic conductivity remain the same throughout the solution process, the notation can be simplified by combining the unchanged quantities into a single constant, which multiplies the head, called hydraulic conductance or simply conductance. For example, in Eq. (4-112),

$$CR_{i,j-1/2,k} = KR_{i,j-1/2,k}\frac{\Delta c_i \, \Delta v_k}{\Delta r_{j-1/2}} \tag{4-118}$$

where $CR_{i, j-1/2, k}$ (L^2/T) is the conductance in row i and layer k between nodes $(i, j-1/2, k)$ and (i, j, k). According to Eq. (4-118), conductance is the product of hydraulic conductivity and cross-sectional area of flow divided by the length of the flow path, which is the distance between the nodes. Substitution of Eq. (4-118) into Eq. (4-112) gives

$$q_{i,j-1/2,k} = CR_{i,j-1/2,k}(h_{i,j-1,k} - h_{i,j,k}) \tag{4-119}$$

Similarly, Eqs. (4-113)–(4-117) can be written in the following forms:

$$q_{i,j+1/2,k} = CR_{i,j+1/2,k}(h_{i,j+1,k} - h_{i,j,k}) \tag{4-120}$$

$$q_{i-1/2,j,k} = CC_{i-1/2,j,k}(h_{i-1,j,k} - h_{i,j,k}) \tag{4-121}$$

$$q_{i+1/2,j,k} = CC_{i+1/2,j,k}(h_{i+1,j,k} - h_{i,j,k}) \tag{4-122}$$

$$q_{i,j,k-1/2} = CV_{i,j,k-1/2}(h_{i,j,k-1} - h_{i,j,k}) \tag{4-123}$$

$$q_{i,j,k+1/2} = CV_{i,j,k+1/2}(h_{i,j,k+1} - h_{i,j,k}) \tag{4-124}$$

where conductances are defined analogous to $CR_{i, j-1/2, k}$ in Eq. (4-118).

4.3.10.2 Flow Rates into the Cell From Outside the Aquifer

Equations (4-119)–(4-124) account for the flow into cell (i, j, k) from the six adjacent cells. Additional terms are required for accounting flows into the cell from outside the aquifer (seepage through stream beds, drains, areal recharge, evapotranspiration, and wells). These flows may be dependent on the head in the receiving cell, but independent of all other heads in the aquifer, or they may be entirely independent of head in the receiving cell. Flow from outside the aquifer may be represented by the expression

$$a_{i,j,k,n} = p_{i,j,k,n} h_{i,j,k} + q_{i,j,k,n} \tag{4-125}$$

where $a_{i,j,k,n}$ represents flow rate from the nth external source into cell (i, j, k) (L³/T), and $p_{i,j,k,n}$ (L³/T) and $q_{i,j,k,n}$ (L³/T) are constants. For example, consider that a cell is receiving flow from two sources, i.e., recharge from a well and seepage through a riverbed. For the first source ($n = 1$), since the flow from the well is assumed to be independent of head, $p_{i,j,k,1}$ is zero and $q_{i,j,k,1}$ is the recharge rate for the well. For this case

$$a_{i,j,k,1} = q_{i,j,k,1} \tag{4-126}$$

For the second source ($n = 2$), the seepage is proportional to the head difference between river stage and head in the cell (i, j, k) (Figure 4-25). Thus

$$a_{i,j,k,2} = CRIV_{i,j,k,2}(R_{i,j,k} - h_{i,j,k}) \tag{4-127}$$

where $CRIV_{i,j,k,2}$ (L²/T) is the conductance of the riverbed (Figure 4-25) in the cell (i, j, k) and $R_{i,j,k}$ (L) is the head in the river. Equation (4-127) can be rewritten as

$$a_{i,j,k,2} = - CRIV_{i,j,k,2} h_{i,j,k} + CRIV_{i,j,k,2} R_{i,j,k} \tag{4-128}$$

The conductance $CR_{i,j,k,2}$ corresponds to $p_{i,j,k,2}$ and the term $CRIV_{i,j,k,2} R_{i,j,k}$ corresponds to $q_{i,j,k,2}$. Similarly, all other external sources or stresses can be represented by an expression of the form of Eq. (4-128). In general, if there are N external sources or stresses affecting a single cell, the combined flow rate expression is

$$QS_{i,j,k} = \sum_{n=1}^{N} a_{i,j,k,n} = \sum_{n=1}^{N} p_{i,j,k,n} h_{i,j,k} + \sum_{n=1}^{N} q_{i,j,k,n} \sum_{n=1}^{N} \tag{4-129}$$

With $P_{i,j,k}$ and $Q_{i,j,k}$ defined as

$$P_{i,j,k} = \sum_{n=1}^{N} p_{i,j,k,n} \tag{4-130}$$

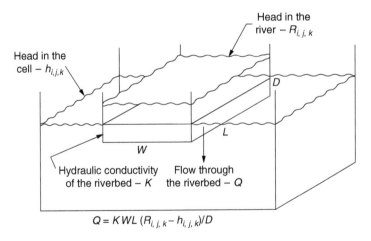

FIGURE 4-25   Leakage through a riverbed into a cell. (Adapted from McDonald, M.G. and Harbaugh, A.W., *Open File Report 83–875*, U.S. Geological Survey, National Center, Reston, Virginia, 1984.)

and

$$Q_{i,j,k} = \sum_{n=1}^{N} q_{i,j,k,n} \tag{4-131}$$

from Eq. (4-129), the general external flow term for cell (i, j, k) is

$$QS_{i,j,k} = P_{i,j,k}h_{i,j,k} + Q_{i,j,k} \tag{4-132}$$

4.3.10.3 Whole Continuity Equation

In the continuity equation, (4-111), ΣQ_i represents the flow rates between node (i, j, k) and the six adjacent nodes (Eqs. [4-112]–[4-117]), and the external flow rate QS (Eq. [4-132]).

The continuity equation (4-111) including the flow rates between node (i, j, k), the six adjacent nodes, and the external flow rate QS yields

$$q_{i,j-1/2,k} + q_{i,j+1/2,k} + q_{i-1/2,j,k} + q_{j+1/2,j,k} + q_{i,j,k-1/2}$$

$$+ q_{i,j,k+1/2} + QS_{i,j,k} = SS_{i,j,k}\frac{\Delta h_{i,j,k}}{\Delta t}\Delta r_j\Delta c_i\Delta v_k \tag{4-133}$$

where $\Delta h_{i,j,k}/\Delta t$ (L/T) is a finite-difference approximation for head change with respect to time, $SS_{i,j,k}$ (L^{-1}) is the specific storage of cell (i, j, k), and $\Delta r_j \Delta c_i \Delta v_k$ (L^3) is the volume of cell (i, j, k). Substitution of Eqs. (4-119)–(4-124) and Eq. (4-132) into Eq. (4-133) gives the finite-difference approximation for cell (i, j, k) as

$$CR_{i,j-1/2,k}(h_{i,j-1,k} - h_{i,j,k}) + CR_{i,j+1/2,k}(h_{i,j+1,k} - h_{i,j,k})$$

$$+ CC_{i-1/2,j,k}(h_{i-1,j,k} - h_{i,j,k}) + CC_{i+1/2,j,k}(h_{i+1,j,k} - h_{i,j,k}) \tag{4-134}$$

$$+ CV_{i,j,k-1/2}(h_{i,j,k-1} - h_{i,j,k}) + CV_{i,j,k+1/2}(h_{i,j,k+1} - h_{i,j,k})$$

$$+ P_{i,j,k}h_{i,j,k} + Q_{i,j,k} = SS_{i,j,k}(\Delta r_j\Delta c_i\Delta v_k)\frac{\Delta h_{i,j,k}}{\Delta t}$$

4.3.10.4 Approximation for Head Change with Respect to Time

The head difference $\Delta h_{i,j,k}$ in Eq. (4-134) must be expressed in terms of specific head values, which are related to the head values for determining flows into and out of the cell. In Figure 4-26, in the head versus time graph, for cell (i, j, k), two time values on the horizontal axis (t_m and t_{m-1}) and their corresponding head values on the vertical axis ($h_{i,j,k}^m$ and $h_{i,j,k}^{m-1}$) are shown. The geometric slope of the dotted line is $\Delta h_{i,j,k}^m / \Delta t_m$. The method utilized by MODFLOW is that the flow terms of Eq. (4-134) are evaluated at the more advanced time t_m, and the geometric slope $(\Delta h/\Delta t)$ of the head versus time curve (Figure 4-26) is evaluated as

$$\frac{\Delta^m h_{i,j,k}}{\Delta t_m} = \frac{h_{i,j,k}^m - h_{i,j,k}^{m-1}}{t_m - t_{m-1}} \tag{4-135}$$

Equation (4-135) means that the slope of head versus time, or time derivative, is approximated with the head change at the node over a time interval which precedes, and ends with, the time at which is evaluated. This approach is called a backward finite-difference approach, which means that $\Delta h/\Delta t$ is determined over a time interval extending backward in time from t_m, the time at which the flow terms are evaluated. There are other ways of approximating $\Delta h/\Delta t$, such as approximation over a time interval that begins at the time of flow evaluation and extends to some later time or is centered at the time of flow evaluation, both forward and backward from it. However, these alternatives cause numerical instability problems, which means if heads are calculated at successive times, and if for

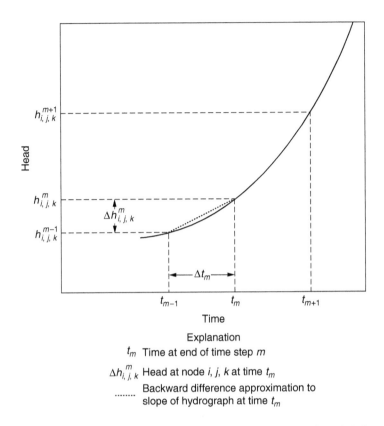

FIGURE 4-26   Head versus time for cell (i, j, k). (Adapted from McDonald, M.G. and Harbaugh, A.W., *Open File Report 83–875*, U.S. Geological Survey, National Center, Reston, Virginia, 1984.)

any reason errors enter the calculation at a particular time, these errors will increase at each succeeding time as the calculation progresses until finally they completely dominate the result. It is important to note that, by contrast, the backward finite-difference approach is always numerically stable. This means that errors introduced at any time diminish progressively at succeeding times. Therefore, the backward finite-difference approach is preferred over the other alternatives, despite the fact that it leads to large systems of equations, that must be solved simultaneously for each time at which heads are to be determined.

Equation (4-134) can be rewritten in backward finite-difference form by specifying the flow terms at t_m, the end of the time interval, and approximating the time derivative of head over the interval t_{m-1} to t_m gives

$$
\begin{aligned}
CR_{i,j-1/2,k}(h_{i,j-k}^m - h_{i,j,k}^m) &+ CR_{i,j+1/2,k}(h_{i,j+1,k}^m - h_{i,j,k}^m) \\
+ CC_{i-1/2,j,k}(h_{i-1,j,k}^m - h_{i,j,k}^m) &+ CC_{i+1/2,j,k}(h_{i+1,j,k}^m - h_{i,j,k}^m) \\
+ CV_{i,j,k-1/2}(h_{i,j,k-1}^m - h_{i,j,k}^m) &+ CV_{i,j,k+1/2}(h_{i,j,k+1}^m - h_{i,j,k}^m) \\
+ P_{i,j,k}h_{i,j,k}^m + Q_{i,j,k} &= SS_{i,j,k}(\Delta r_j \Delta c_i \Delta v_k)\frac{h_{i,j,k}^{m-1} - h_{i,j,k}^m}{t_m - t_{m-1}}
\end{aligned}
\tag{4-136}
$$

In Eq. (4-136), head, at the beginning of the time step, $h_{i,j,k}^m$, and all conductances and coefficients related to the node (i, j, k) are known. The seven heads at time t_m, the end of the time step, are unknown and these are part of the head distribution that are to be predicted. Equation (4-136) cannot be solved independently as it represents a single equation in seven unknowns. However, this type of equation can be written for each of the "n" number of cells in the system. Since there is only one unknown head for each cell, the system has "n" equations in "n" unknowns. This kind of system of equations can be solved simultaneously.

In most models, heads at some cells are known during the model construction process. These cells are called constant head cells and are further discussed in Sections 4.3.11.1 and 4.3.15.1. Since heads are constant with time at constant head cells, they do not need equations. At some cells there is no flow and they are called no-flow cells (see Section 4.3.15.1). No-flow cells are those to which there is no flow from adjacent cells. The rest of the cells are called variable head cells at which heads may vary with time. Therefore, the number of equations for a ground water system is equal to the number of variable head cells and an equation similar to Eq. (4-136) is required for each of them.

4.3.10.5 Transient Simulation

The objective of transient simulation is to determine head versus time variation at all variable head cells under the condition that the initial head distribution and boundary conditions are given. The total time period is divided discrete time steps starting from time t_1 at which a value of $h^1_{i,j,k}$ at each node is given. The first step in the solution process is to determine heads $h^2_{i,j,k}$ after the first time step (Δt_1) at $t_2 = t_1 + \Delta t_1$, which is the end of the first time step. Therefore, in Eq. (4-136), the subscript m is taken as 2 and the subscript $m - 1$, which appears in only one head term, is taken as 1. Therefore, Eq. (4-136) becomes

$$
\begin{aligned}
& CR_{i,j-1/2,k}(h^2_{i,j-1,k} - h^2_{i,j,k}) + CR_{i,j+1/2,k}(h^2_{i,j+1,k} - h^2_{i,j,k}) \\
& + CC_{i-1/2,j,k}(h^2_{i-1,j,k} - h^2_{i,j,k}) + CC_{i+1/2,j,k}(h^2_{i+1,j,k} - h^2_{i,j,k}) \qquad (4\text{-}137) \\
& + CV_{i,j,k-1/2}(h^2_{i,j,k-1} - h^2_{i,j,k}) + CV_{i,j,k+1/2}(h^2_{i,j,k+1} - h^2_{i,j,k}) \\
& + P_{i,j,k}h^2_{i,j,k} + Q_{i,j,k} = SS_{i,j,k}(\Delta r_j \Delta c_i \Delta v_k)\frac{h^2_{i,j,k} - h^1_{i,j,k}}{t_2 - t_1}
\end{aligned}
$$

With this process, a system of n equations in n unknowns is formulated where the unknowns are the heads at t_2. The heads for time t_2 can be determined by solving n number of equations simultaneously. Then, the process is repeated to determine heads at time $t_3 = t_2 + \Delta t_2$, the end of the second time step. To do this, again Eq. (4-136) is applied using $m - 1 = 2$ and $m = 3$. Similarly, a system of n equations in n unknowns is formulated where the unknowns are the heads at t_3. The heads for time t_3 can be determined by solving n number of equations simultaneously. This solution process is continued for as many time steps as necessary to cover the targeted entire time period.

4.3.10.6 Determination of Head Distribution

Different techniques are available to solve a set of finite-difference equations for flow to determine the head distribution at each time and some of the techniques used in ground water applications are described in Section 4.3.16. In this section, some details regarding these techniques especially the ones based on iteration will be presented.

As mentioned in Section 4.3.16, the solution of a set of finite-difference ground water flow equations is generally based on direct methods and iterative methods. The direct methods are also called direct algebraic methods and these methods are generally less desirable than iterative methods in ground water applications due to the existence of numerical problems.

An iterative method starts with a set of assigned head values to each cell and these heads are called initial trial solution. The assigned values are the best guess and a uniform value to a modeling layer or even to the whole modeling domain can be assigned. One should bear in mind that closer the values to the final solution, lesser the time taken for execution to derive the final results. The next step is to initiate the calculation process by using the initial trial solution to generate an interim solution, which more nearly satisfies the system of equations. The interim solution then becomes the new trial solution and the same process is repeated. Each repetition is called an itera-tion. The process is repeated until it "closes;" which means that the trial solution and the interim solution are "nearly" equal. The trial solution and interim solution are called to be nearly equal if, for each node, the difference between the trial head value and the interim head value is less than some arbitrarily established value. This value is usually called closure criterion. Once the closure criterion is satisfied, the interim solution is considered as a good approximation to the solution of the system of equations. During a time step, arrays of interim head values are generated in succes-sion with each array containing one interim head value for each node. In Figure 4-27, these arrays are represented by three-dimensional lattice symbols with a superscript used to indicate the level of iteration. Thus $h_{i,j,k}^{m,0}$ represents the initial trial value chosen for head at node (i, j, k), and $h_{i,j,k}^{m,1}$ is the interim head calculated during iteration 1 and the trial value used for iteration 2. Similarly, $h_{i,j,k}^{m,2}$ is the interim solution from iteration 2 and the trial value for iteration 3.

4.3.10.7   Final Form of the Finite-Difference Equation

As mentioned above, ground water flow can be simulated by writing the continuity equation for each cell [Eq. (4-136)], and solving the resulting systems of algebraic equations for head at each node. It is convenient to rearrange Eq. (4-136) such that all terms including heads at the end of the current time step are grouped on the left-hand side of the equation and all terms that are independent of head at the end of the current time step are on the right-hand side of the equation. Therefore, Eq. (4-136) gives

$$CV_{i,j,k-1/2}h^m_{i,j,k-1} + CC_{i-1/2,j,k}h^m_{i-1/2,j,k} + CR_{i,j-1/2,k}h^m_{i,j-1,k} \qquad (4\text{-}138)$$
$$+ (-CV_{i,j,k-1/2} - CC_{i-1/2,j,k} - CR_{i,j-1/2,k} - CR_{i,j+1/2,k}$$
$$- CC_{i+1/2,j,k} - CV_{i,j,k+1/2} + \text{HCOF}_{i,j,k})h^m_{i,j,k} + CR_{i,j+1/2,k}h^m_{i,j+1,k}$$
$$+ CC_{i+1/2,j,k}h^m_{i+1,j,k} + CV_{i,j,k+1/2}h^m_{i,j,k+1} = \text{RHS}_{i,j,k}$$

where

$$\text{HCOF}_{i,j,k} = P_{i,j,k} - \frac{SC1_{i,j,k}}{t_m - t_{m-1}} \qquad (4\text{-}139)$$

$$\text{RHS}_{i,j,k} = -Q_{i,j,k} - \frac{SC1_{i,j,k}h_{i,j,k}^{m-1}}{t_m - t_{m-1}} \qquad (4\text{-}140)$$

and

$$SC1_{i,j,k} = SS_{i,j,k}\,\Delta r_j \Delta c_i \Delta v_k \qquad (4\text{-}141)$$

The units of HCOF$_{i,j,k}$, RHS$_{i,j,k}$, and SC1$_{i,j,k}$ are (L²/T), (L³/T), and (L²), respectively.

Equation (4-138) is the finite-difference equation for the development of the system of linear equations from which spatial and temporal head distribution is determined and is the basis of the ground water flow model used in MODFLOW (McDonald and Harbaugh, 1984).

4.3.10.8   Approximate Solution and Truncation Error

Based on the foregoing analysis, it can be said that the iterative solution procedure yields only an approximation to the solution of the system of finite-difference equations for each time step. The accuracy of this approximation depends upon the closure criterion that is assigned for the numerical process. However, it is important to point out that even if exact solutions to the set of finite-differ-ence equations were obtained at each time step, these exact solutions would themselves be only an approximation to the solution of the differential equation of flow as given by Eq. (4-107). The dis-crepancy between the head ($h_{i,j,k}^m$) given by the solution to the system of finite-difference equations

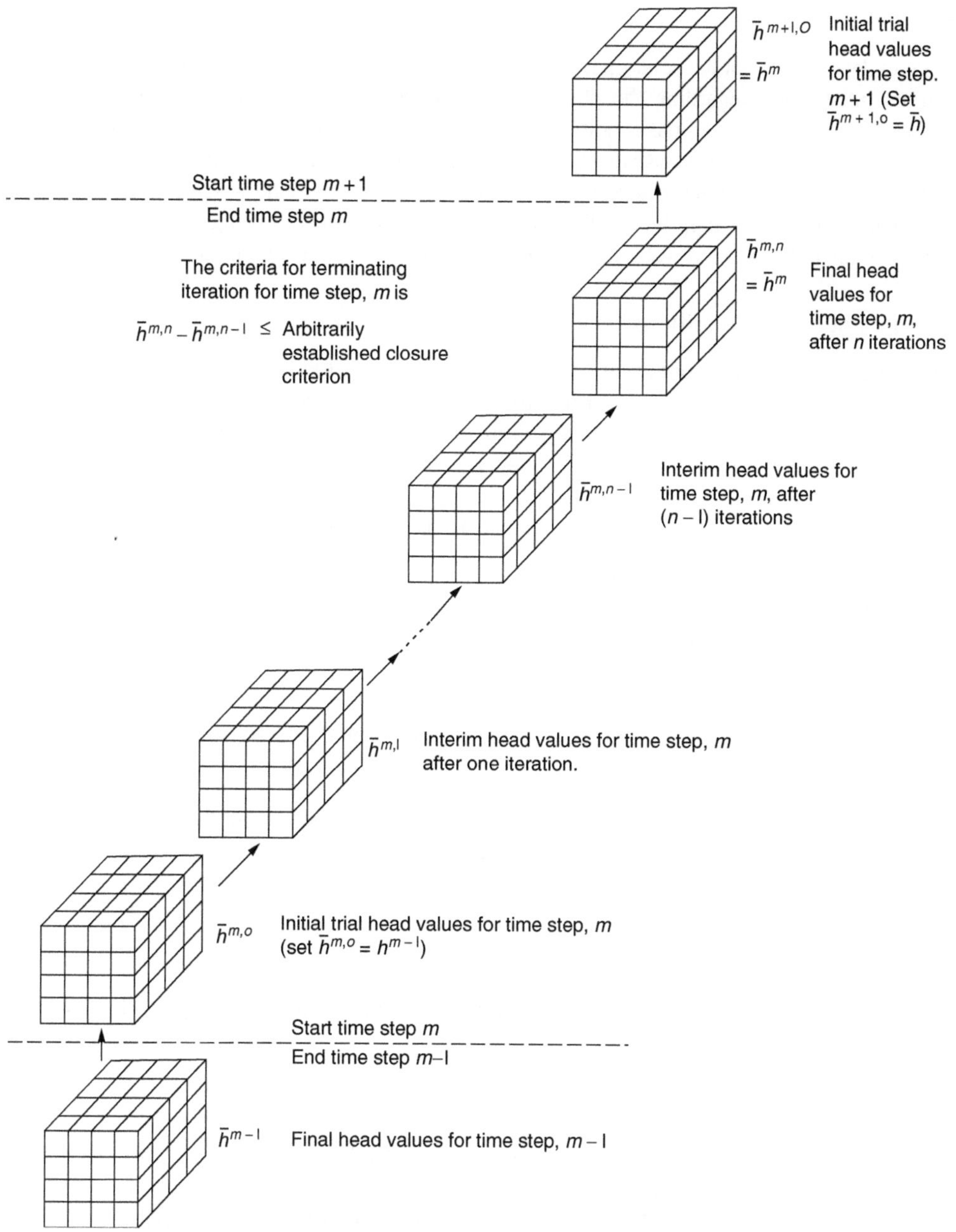
Initial trial head values for time step. m + 1 (Set h̄m+1,o = h̄)
h̄m+1,o = h̄m
Start time step m + 1
End time step m
The criteria for terminating iteration for time step, m is
h̄m,n − h̄m,n−1 ≤ Arbitrarily established closure criterion
h̄m,n = h̄m
Final head values for time step, m, after n iterations
h̄m,n−1
Interim head values for time step, m, after (n − 1) iterations
h̄m,1
Interim head values for time step, m after one iteration.
h̄m,o
Initial trial head values for time step, m (set h̄m,o = hm−1)
Start time step m
End time step m−1
h̄m−1
Final head values for time step, m − 1

FIGURE 4-27 Determination of head distribution iteratively. (Adapted from McDonald, M.G. and Harbaugh, A.W., *Open File Report 83–875*, U.S. Geological Survey, National Center, Reston, Virginia, 1984.)

for a given node and time and the head $[h\ (x_i, y_j, z_k, t_m)]$, which would be given by the formal solution of the differential equation for the corresponding point and time, is termed as truncation error. In general, the truncation error magnifies with the increase in grid spacing and time step length. Also, it must be noted that even if a formal solution of the differential equation can be determined, it would normally be only an approximation to conditions in the field, i.e., hydraulic conductivity and specific storage variation are seldom known with accuracy. Beyond this, uncertainties with regard to hydrologic boundaries, formation thicknesses are generally present.

4.3.11 BOUNDARY AND INITIAL CONDITIONS

In order to determine a unique solution from the finite-difference equations presented in Section 4.3.10, additional information regarding the physical state of the process is required. This information is described by boundary and initial conditions. For steady-state problems only boundary conditions are required, whereas for unsteady-state problems both boundary and initial conditions are required. Detailed information about the boundary conditions encountered in ground water modeling applications are given in several books (e.g., Bear, 1972, 1979; Batu, 1998). In the following sections, boundary conditions encountered in ground water modeling applications, as well as initial conditions are described.

4.3.11.1 Boundary Conditions

Mathematically, boundary conditions include the geometry of the boundary and the values of the dependent variable or its derivative normal to the boundary. The selection of boundary conditions is a critically important step for conceptualizating and developing a model for ground water system. Even if one of the boundary conditions is inappropriate for a defined aquifer problem, it is most likely that the final model will be an error or in an ambiguous form. The boundary conditions that are often used in solving ground water problems are described below.

Constant Head Boundary
This boundary condition is also known as the Dirichlet boundary condition or the first-type boundary condition. The frequently used term in the ground water literature is constant head boundary condition. According to this boundary condition, values of head are specified along the boundary. For example, if in Figure 4-21, the first row in the first layer ($I = 1$, $J = 1$–9, $K = 1$) is part of a surface water body (river, lake, pond, etc.), it can be represented as a constant head boundary.

Specified Head Boundary
The specified head boundary condition is used especially in numerical models. It is also known as prescribed head boundary condition. The general mathematical form of the steady-state constant head boundary condition at a cell is

$$h = h_C \tag{4-142}$$

where h_C(L) is the head value at that cell and constant all time. If the boundary head value changes with time, Eq. (4-142) takes the form

$$h = h_C(t) \tag{4-143}$$

where $h_C(t)$ (L) is the head versus time function at that cell. The common practice is to assign the known boundary head values, which are determined by field observations to the corresponding boundary segments of the model. In reality, the constant head boundary segments act as portions of the adjacent surface water bodies. Therefore, special care is required when using this type of boundary condition. The main criterion in using this type of boundary condition is to check the head values near the boundaries and make sure that they are not affected under different stress conditions (extraction, injection, etc.) in the ground water system. For example, as long as the cone of depression of an extraction or injection well does not reach the specified head boundary segment, they can be used safely. If the cone of depression of an extraction well reaches the specified head boundary segments, they act as if some real surface water bodies exists there. As a result, the solution will be an error.

Specified Flux Boundary
The term flux means the volume of fluid per unit time passing through a unit surface area. If the flux normal on a given part of the boundary surface can be specified as a function of space and time, that boundary is called a specified flux boundary. This type of boundary condition is also known as

Neuman boundary condition or second-type boundary condition. The simplest form of the specified flux boundary is the constant or spatially variable net precipitation recharge. The flux can also be specified using Darcy's equation (see Sections 4.3.1.4 and 4.3.2.1) as

$$q_n = -K_n \frac{\partial h}{\partial n} \tag{4-144}$$

where $q_n(L^2/T)$ is the flux, n represents the direction perpendicular to boundary, $K_n(L/T)$ the hydraulic conductivity in the direction perpendicular to the boundary, and $h(L)$ the hydraulic head.

No-Flow Boundary
There are several forms of the no-flow boundary condition. These are described below.

Impermable Boundary Layer
Relatively less pervious layers or formations act as impermeable boundary layers. For example, the top of a bedrock on which an aquifer rests acts as an impermeable layer. The Darcy velocity (see Sections 4.3.1.4 and 4.3.2.1) normal to this kind of boundaries is equal to zero. Therefore, from Eq. (4-144), one can write

$$q_n = -K_n \frac{\partial h}{\partial n} = 0 \tag{4-145}$$

Since K_n cannot be zero, the no-flow boundary condition takes the form

$$\frac{\partial h}{\partial n} = 0 \tag{4-146}$$

which is a special form of the Neuman boundary condition.

Water Divide
The water divide boundary is a special case of the no-flow boundary and generally occurs around a line formed by ground water moundings or river beds at which a line can be defined such that the ground water velocities perpendicular to that line are zero. This type of boundaries may change with seasons and care must be taken in defining them.

Seepage Face Boundary
An example of this type of boundary is the interface between an unconfined aquifer and an adjacent water body such that the watertable ends at a point that is above the free surface of the water body. The segment of the boundary above the water surface and below the watertable of the unconfined aquifer is called *seepage face*. Another example is a tunnel in an aquifer under saturated conditions through which there is water flow. Since water flows into the tunnel under atmospheric conditions, the internal surface of the tunnel acts as a seepage face as well.

The mathematical expression of a seepage face boundary can easily be written from Eq. (4-1) by taking $f = 0$ as

$$h = z \tag{4-147}$$

which is the boundary condition at the seepage face.

Head-Dependent Flux Boundary
If the flow rate is related to both the normal derivative and the head value, the boundary is called head-dependent flux boundary. For example, the volumetric flow rate per unit area of water is related to the normal derivative of head and head itself by

$$-K_n \frac{\partial h}{\partial n} = q_n(h_n) \tag{4-148}$$

where q_n is some function that describes the boundary flow rate given the head at the boundary h_b.

4.3.11.2 Initial Conditions

Specification of initial conditions for a particular problem is required only for unsteady-state ground water problems. For steady-state problems, initial conditions are not required. For unsteady-state problems, the distribution of hydraulic head [$h(x, y, z, t)$] for the flow system at a particular time is assumed to be known. Initial conditions are generally specified when time is zero ($t = 0$). The mathematical expression for this condition is

$$h = h(x, y, z, 0) \tag{4-149}$$

where $h(x, y, z, 0)$ is the initial head distribution function. The common practice for a finite-difference model is that an initial head value is assigned to each cell.

4.3.12 FINITE-DIFFERENCE GRID DISCRETIZATION OF SPACE

In ground water modeling applications, finite-difference grid discretization is one of the most important steps. The other step, which is equally important, is the orientation of the finite-difference grid system in an aquifer that exists in nature. Grid discretization is carried out both vertically (vertical grid discretization) and horizontally (horizontal grid discretization). For a two-dimensional model only horizontal grid discretization is required, whereas for a three-dimensional model both horizontal and vertical discretization are required. In the following sections, the principles of finite-difference grid system orientation and the principles of finite-difference grid discretization are described (McDonald and Harbaugh, 1984).

4.3.12.1 Finite-Difference Grid System Orientation

In practice, a ground water modeling study is generally conducted for a relatively small area of an existing aquifer or system of aquifers in nature. For a proper orientation, the hydrogeologic model of the natural system needs to be available. One must bear in mind that this kind of model is based on field data and assumptions, and is seldom complete. For the sake of discussion, it is assumed that the aforementioned information is available to the modeler(s). Under this condition, the main task is to separate the model region from the rest of the aquifer or system of aquifers. The boundary conditions account for the influence of flow conditions outside the selected domain. As seen in Section 4.3.11, later in Section 4.3.15, boundary conditions are often difficult to define along the edges of the modeling domain because the heads or inflow and outflow rates are poorly defined. One way of to overcoming this problem is to locate model boundaries along the natural hydrogeologic boundaries or parallel to the main flow direction. For example, if in Figure 4-21, the main ground water flow direction is approximately parallel to the direction of rows (the x-coordinate direction), the orientation of the grid system is said to be appropriate. This means that the principal directions of anisotropy are the coordinate axes (see Section 4.3.5.2), and the hydraulic conductivity tensor is given by Eq. (4-96). Therefore, the main partial differential equation of ground water flow is Eq. (4-101) or Eq. (4-107), on which the MODFLOW and its family codes are based. One must keep in mind that a main flow direction may not be defined properly in every modeling study. Therefore, the rule of thumb is to make one of the horizontal coordinate axes (x or y) parallel to the defined main ground water flow direction.

4.3.12.2 Vertical Grid Discretization

Vertical Discretization of Permeable Materials
In the finite-difference method, a rectilinear grid is used to divide the modeling region to be studied into rows, columns, and layers, which form parallelepiped cells with rectangular faces. The properties of the cells, which are assumed to be homogeneous, are used to formulate the coefficients

of the finite-difference equations given in Section 4.3.10. Generally, the grid is superimposed on a ground water system consisting of a sequence of stratigraphic units that are not quite horizontal (Figure 4-28). Thus, some cells may represent two very different material types, making the specification of physical properties difficult. Therefore, it is convenient to deform the grid such that grid layers follow the contours of the stratigraphic units. The most common methodology is to make the model layers conform with the identifiable geologic units. Consider the simple example shown in Figure 4-28a with a coarse sand, silt, and sand and gravel layer. For these three different formations, vertical grid discretization is accomplished using three model layers that coincide with these three formations as shown in Figure 4-28b. In the finite-difference method employed in the MODFLOW family codes, all cells are limited to being parallelepipeds with six rectangular faces. Therefore, in Figure 4-28b, each formational cell should be approximated by the nearest parallelepiped cell. In order to achieve this, only the upper and lower surfaces of a stratigraphic cell need to be adjusted to the nearest upper and lower planes of a model cell. With this approach, the thickness of a modeling layer will change from cell to cell.

Vertical Discretization of Less Permeable Materials (Aquitards)
Simulation Procedure of Systems Having Aquifers and Aquitards
There are some unique features of flow through low permeable formations (aquitards) separating aquifers that have considerably higher hydraulic conductivities. Studies conducted by some investigators during the second half of the 20th century have made significant contributions toward understanding these unique features. These are discussed below.

In the process of developing analytical ground water flow models for problems involving well hydraulics in a system of aquifers separated by confining layers, an important assumption is made regarding the direction of flow in the confining layers. More specifically, the analytical solutions developed for two highly permeable aquifers separated by a less permeable confining layer (or aquitard) have generated important insights for simplifying the modeling process. The assumption made is that *the flow is vertical through the confining layers and horizontal in the aquifers*. The

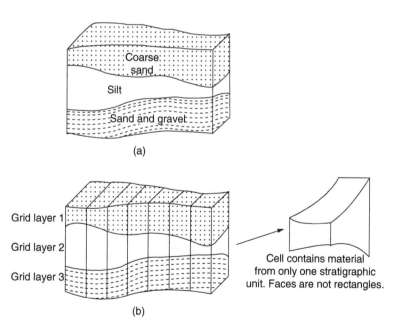

FIGURE 4-28 Modeling representation of a natural system of aquifers: (a) Aquifer cross-section; and (b) Aquifer cross-section with deformed grid superimposed. (Adapted from McDonald, M.G. and Harbaugh, A.W., *Open File Report 83–875*, U.S. Geological Survey, National Center, Reston, Virginia, 1984.)

validity of this assumption has been investigated by Neuman and Witherspoon (1969a, 1969b, 1969c), based on the finite element numerical modeling method. These authors found that when the hydraulic conductivities of the aquifers are two orders of magnitude greater than that of the aquitard, the errors introduced by this assumption are usually less than 5%. According to Neuman and Witherspoon (1969b), these errors increase with time, decrease with radial distance from the pumping well, and are smallest in the pumped aquifer and greatest in the aquitard. Since the hydraulic conductivity contrast between aquifer and aquitard is often greater than three orders of magnitude, it would appear that these assumptions can be used safely.

Since the flow is almost vertical in a low permeable formation, such as clay separated by two aquifers, the flow lines in the low permeable formation tend to become almost vertical, as shown in Figure 4-29. Since the flow in the aquifers is nearly horizontal, each of these units can be approximated accurately with just one or two layers. In the low permeable unit, the equipotential lines are nearly horizontal. Therefore, many layers are required to represent the change in head across the unit. Figure 4-30 shows a grid that represents head variation in the clay. In this example, the clay unit is represented by six grid layers. Therefore, the flow system shown in Figure 4-30 is modeled with eight layers, one for each sand unit and six for the clay unit.

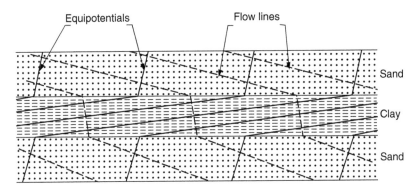

FIGURE 4-29   Flow net in a system of aquifers cross section having two high permeable units separated by a low permeable unit. (Adapted from McDonald, M.G. and Harbaugh, A.W., *Open File Report 83–875*, U.S. Geological Survey, National Center, Reston, Virginia, 1984.)

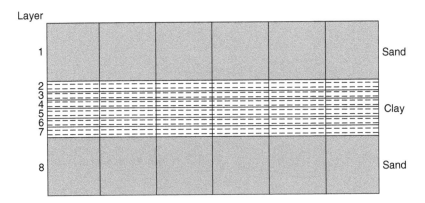

FIGURE 4-30   A system of aquifers cross-section having a low permeable unit represented by six modeling layers. (Adapted from McDonald, M.G. and Harbaugh, A.W., *Open File Report 83–875*, U.S. Geological Survey, National Center, Reston, Virginia, 1984.)

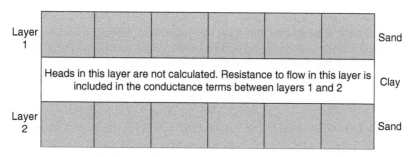

FIGURE 4-31   A system of aquifers cross-section in which a low permeable unit is not represented by a modeling layer. (Adapted from McDonald, M.G. and Harbaugh, A.W., *Open File Report 83–875*, U.S. Geological Survey, National Center, Reston, Virginia, 1984.)

An Alternative Simulation Procedure for the Effects of an Aquitard
The system of aquifers shown in Figure 4-30 is modeled with eight layers, one for each high permeable unit and six for the low permeable unit. If the heads in the high permeable units with the effects of the aquitard on the flow are needed, there is an alternative modeling procedure in MODFLOW. As shown in Figure 4-31, the upper sand unit would be Layer 1 and the lower sand unit would be Layer 2 in the model. The clay would not be modeled, namely, heads in the clay layer would not be calculated. However, the properties of the clay would be used to calculate conductance between the sand layers. The MODFLOW code has been specifically designed to handle this situation.

4.3.12.3   Horizontal Grid Discretization

The finite-difference equations, which correspond to the differential equation of ground water flow given by Eq. (4-107), presented in Section 4.3.10, include the grid dimensions Δr, Δc, and Δv, and time step Δt as parameters. The solution of the finite-difference equations must converge to Eq. (4-107) when the finite increments Δr, Δc, Δv, and Δt decrease in size. There are several guidelines regarding the process of selecting these quantities. In this section, the guidelines regarding the grid dimensions (Δr, Δc, and Δv) will be presented. The guidelines for time step Δt will be presented in Section 4.3.13.

The Rule of Thumb for Adjacent Grid Sizing
As mentioned previously, the finite-difference method requires that the selected modeling domain be discretized horizontally and vertically. The grid system is referenced in terms of a row, column, and layer-numbering scheme with block-centered nodes, as shown in Figure 4-21. The dimensions of each cell can be varied. Thus, relatively dense nodes can be selected in areas of interest and coarse nodes in areas of less importance. Overall, the variable grid sizing reduces the number of nodes in a model grid system. However, one must be careful in changing cell sizes gradually. The *rule of thumb is that the dimensions of adjacent cells in a given direction should not differ by more than a factor of 1.5.* More specifically, for example, if the grid size of a cell in a given direction is 50 m, the size of the adjacent grid in the same direction should not be more than 75 m. This approach is, however, on the conservative side. Experience has shown that in most cases, a factor of up to 2.0 can still generate numerically stable results.

Peclet Number Requirement for Grid Sizing
Requirements for solute transport modeling grids are more restrictive than for flow modeling grids. If a model grid is likely to be used for solute transport simulations, certain requirements must be met in order to generate stable results for solute transport. This is due to the fact that the convective–dispersive solute transport differential equation (Eq. [4-196]) is more difficult to solve

numerically than the flow equation. Theoretical investigation (Price et al., 1966) and experience (Huyakorn and Pinder, 1983) indicate that if the hydrodynamic dispersion coefficient is greater than zero, numerical oscillations may occur depending on the grid dimensions. The local Peclet number Pe (unitless), which is defined as

$$Pe = \frac{U \Delta l}{D} \qquad (4\text{-}150)$$

is a measure of the level of oscillations. In Eq. (4-150), U(L/T) is the ground water velocity, Δl (L) is the characteristic length in the ground water flow direction, and D (L²/T) is the hydrodynamic dispersion coefficient. Equation (4-150) is defined for one-dimensional solute transport cases based on uniform ground water flow U, which is given by Eq. (2-40). From Eq. (2-154), the expression for D can be written as

$$D = D_m + D^* = \alpha_L U + D^* \qquad (4\text{-}151)$$

where α_L(L) is the longitudinal dispersion coefficient, D_m(L²/T) is the mechanical dispersion coefficient, and D^* (L²/T) is the effective molecular diffusion coefficient. Since D^* is a negligibly small fraction of D_m, D^* can be neglected and Eq. (4-150) takes the form

$$Pe \cong \frac{\Delta l}{\alpha_L} \qquad (4\text{-}152)$$

It has been suggested by different investigators (e.g., Huyakorn and Pinder, 1983) that *grid spacing should be selected such that the local Peclet number does not exceed 2*. In most cases involving nonuniform flow, acceptable numerical solutions with very mild oscillations are achieved even when the local Pellet number is as high as 10 (Huyakorn and Pinder, 1983).

For one-dimensional cases, Δl corresponds to Δx. For two-dimensional cases, Δl can be selected as the maximum value among the values of Δx (Δr) and (Δc). Equation (4-152) shows that the potential ranges of longitudinal dispersivity (α_L) values in a modeling study including planned sensitivity analysis values must be established before constructing the finite-difference grid. The reader is advised to review the estimation procedures of dispersivity values (see Chapter 7) before attempting to design a modeling grid.

EXAMPLE 4-4

For a solute transport numerical model, α_L is estimated to be 20 m. What would be the allowable maximum grid spacing in order to obtain numerically stable results?

SOLUTION

From Eq. (4-152),

$$\frac{\Delta l}{20 \ m} \leq 2$$

Hence $\Delta l \leq 40$ m. Therefore, the horizontal grid spacing should not exceed 40 m.

Practical Considerations for the Design of Grids

Applications of modeling studies, especially during the past three decades, have generated some practical guidelines regarding grid design, which are presented below (Trescott et al., 1976; Mercer and Faust, 1981).

One of the critical steps in developing a ground water model is designing the grid. The finer the grid, the more accurate the solution. Therefore, fine grids should be used where accurate solutions are desired, and coarse grids should be used where accurate solutions are not important. The following general guidelines should be followed for the design of grids:

1. Locate wells near the physical location of the pumping (or injection) well or the center of the well field.
2. Locate boundaries as accurately as possible. For distant boundaries the grid may be expanded, but avoid large grid spacings next to small ones. Follow the rule of thumb mentioned for adjacent grid sizes mentioned above.
3. Nodes should be located closer together in areas where there are spatial changes in hydraulic conductivity or hydraulic head.
4. Align axes of grid with the major directions of anisotropy.

4.3.13 DISCRETIZATION OF TIME

4.3.13.1 Input Parameters for Time in the MODFLOW Family Codes

In the MODFLOW family codes, the total simulation time is divided into stress periods. A *stress period* is defined as a time segment during which all external stresses (pumping rate of a well, recharge rate, etc.) are constant. Each stress period is divided into time steps. The length of each stress period is specified explicitly by the user. Within a stress period, the time steps form a geometric series in which the parameters of the series, the number of elements, and the multiplier are specified by the user (Figure 4-32). The MODFLOW program uses those parameters along with the length of the stress period to calculate the length of each time step. Using the time input data, the MODFLOW code calculates the first time step from

$$Delt(1) = \frac{PERLEN(1-TSMULT)}{1-TSMULT^{NSTP}} \qquad (4\text{-}153)$$

where $PERLEN(T)$ is the length of the stress period, $TSMULT$ (unitless) is the time step multiplier, and $NSTP$ is the number of time steps in stress period. New time steps are calculated from

$$Delt(m + 1) = (TSMULT)[Delt(m)] \qquad (4\text{-}154)$$

As indicated above, $Delt(1)$ is not an input value, and is internally calculated using the required input data for $PERLEN$, $TSMULT$, and $NSTP$. The value of $Delt(1)$ is critically important in

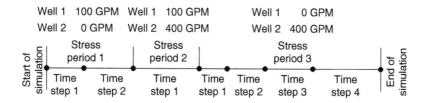

FIGURE 4-32 Division of simulation time into stress periods and time steps. (Adapted from McDonald, M.G. and Harbaugh, A.W., *Open File Report 83–875*, U.S. Geological Survey, National Center, Reston, Virginia, 1984.)

obtaining numerically stable results from the output of MODFLOW. For example, under pumping conditions from highly permeable aquifers, the values of *Delt* (1) may be required to be in orders of seconds depending on the type of problem. Therefore, the user should be aware of the values of *Delt* (1) used in the model. For *TSMULT*, 1.1 generally provides acceptable results. For more accurate results, 1.05 could be used. These are all shown in the following examples. Since the value of *NSTP* is an input for MODFLOW, it needs to be calculated from the given values of *PERLEN*, *TSMULT*, and *DELT* (1). From Eq. (4-153), *NSTP* can be solved as

$$NSTP = \frac{\log\left[\dfrac{DELT(1) - PERLEN(1 - TSMULT)}{DELT(1)}\right]}{\log(TSMULT)} \tag{4-155}$$

EXAMPLE 4-5

For a ground water flow model simulation, *PERLEN* = 150 days and *TSMULT* = 1.1. Determine (a) the value of *NSTP* for *DELT* (1) = 1 min and (b) the value of *NSTP* for *DELT* (1) = 10 sec.

SOLUTION

(a) The initial time step is

$$DELT(1) = 1 \text{ min} = (1 \text{ min})\left(\frac{1 \text{ days}}{1440 \text{ min}}\right) = 694.444 \times 10^{-6} \text{ days}$$

Substitution of this as well as the other values into Eq. (4-155) gives

$$NSTP = \frac{\log\left[\dfrac{694.444 \times 10^{-6} \text{ days} - (150 \text{ days})(1 - 1.1)}{694.444 \times 10^{-6} \text{ days}}\right]}{\log(1.1)}$$

$$= 104.71 \cong 105$$

Therefore, the required input data for MODFLOW are *PERLEN* = 150 days, *TSMULT* = 1.1, and *NSTP* = 105.

   (b) The initial time step is

$$DELT(1) = 1 \text{ min} = (10 \text{ sec})\left(\frac{10 \text{ days}}{86400 \text{ sec}}\right) = 115.741 \times 10^{-6} \text{ days}$$

Substitution of this as well as the other values into Eq. (4-155) gives

$$NSTP = \frac{\log\left[\dfrac{115.741 \times 10^{-6} \text{ days} - (150 \text{ days})(1 - 1.1)}{115.741 \times 10^{-6} \text{ days}}\right]}{\log(1.1)}$$

$$= 123.51 \cong 124$$

Therefore, the required input data for MODFLOW are *PERLEN* = 150 days, *TSMULT* = 1.1, and *NSTP* = 124.

EXAMPLE 4-6

For a ground water flow model simulation *PERLEN* = 150 days and *TSMULT* = 1.05. Determine (a) the value of *NSTP* for *DELT*(1) = 1 min, (b) the value of *NSTP* for *DELT*(1) = 10 sec; and (c) compare the results with Example 4-5.

SOLUTION

(a) The initial time step is

$$DELT(1) = 10 \text{ sec} = (1 \text{ min})\left(\frac{1 \text{ days}}{1440 \text{ min}}\right) = 694.444 \times 10^{-6} \text{ days}$$

Substitution of this as well as the other values into Eq. (4-155) gives

$$NSTP = \frac{\log\left[\dfrac{694.444 \times 10^{-6} \text{ days} - (150 \text{ days})(1 - 1.05)}{694.444 \times 10^{-6} \text{ days}}\right]}{\log(1.05)}$$

$$= 190.35 \cong 191$$

Therefore, the required input data for MODFLOW are $PERLEN = 150$ days, $TSMULT = 1.1$, and $NSTP = 105$.

(b) The initial time step is

$$DELT(1) = 1 \text{ min} = (10 \text{ sec})\left(\frac{10 \text{ days}}{86400 \text{ sec}}\right) = 115.741 \times 10^{-6} \text{ days}$$

Substitution of this as well as the other values in Eq. (4-155) gives

$$NSTP = \frac{\log\left[\dfrac{115.741 \times 10^{-6} \text{ days} - (150 \text{ days})(1 - 1.05)}{115.741 \times 10^{-6} \text{ days}}\right]}{\log(1.05)}$$

$$= 227.08 \cong 227$$

Therefore, the required input data for MODFLOW are $PERLEN = 150$ days, $TSMULT = 1.05$, and $NSTP = 228$.

(c) Both in Example 4-5 and this example, the initial time steps are kept the same. The value of $TSMULT$ is reduced to 1.05 from 1.1 in Example 4-5. The corresponding values of $NSTP$ change significantly.

4.3.13.2 Time Step Selection

Selection of appropriate time steps for a given ground water problem is critically important in solving the problem efficiently. One potential scheme is to increase the size of the time step progressively. For most cases faced in practice, the actual pumping rate vs. time curves are irregular. This type of pumping rate can be efficiently simulated by using a variable time step. A model pumping rate vs. time is also shown in Figure 4-33. As shown in Figure 4-33, the time step is reduced at the beginning of each pumping period and is then allowed to increase. Such progressive adjustments to the time step may be either arbitrary or specified by specific system behavior. The key point is to generate numerically stable results. There are some mathematical expressions regarding time step selection. However, this is not an efficient way of selecting time steps. The other method is to conduct trial-and-error runs in selecting time steps. These are all discussed below.

Expressions for Selection of Time Step
Some expressions regarding selection of time step for numerical ground water flow models are presented below (Bear, 1979; Huyakorn and Pinder, 1983).

Analysis of solving partial differential equations numerically showed that a very restrictive relationship between the sizes of Δx, Δy, and Δt must be satisfied in order to determine a converging and stable solution. The stability criterion is (Bear, 1979)

$$\frac{T}{S}\left[\frac{\Delta t}{(\Delta x)^2} + \frac{\Delta t}{(\Delta y)^2}\right] \le \frac{1}{2} \qquad (4\text{-}156)$$

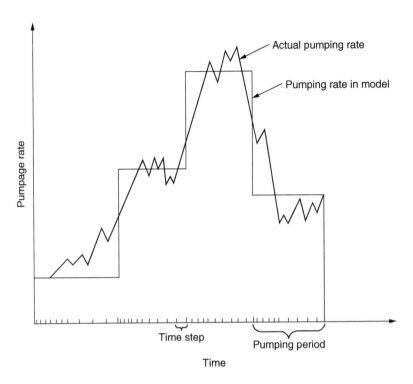

FIGURE 4-33 Idealization of variable pumping rates. (Adapted from Prickett, T.A. and Lonnquist, C.G., *Illinois State Water Surv. Bull.*, 55, 56, 1971.)

where $T(L^2/T)$ is the transmissivity, S (unitless) the storage coefficient, $\Delta t(T)$ the time step, and $\Delta x(L)$ and $\Delta y(L)$ the grid spacings in the x and y directions, respectively.

Based on a stability analysis for one-dimensional cases, Huyakorn and Pinder (1983) determined the following equation for the stability constraint:

$$\Delta t \leq S \frac{(\Delta x)^2}{3T} \tag{4-157}$$

in which the definitions of the parameters are the same as in Eq. (4-156).

Trial-and-Error Runs for Selection of Time Steps
Although Eqs. (4-156) and (4-157) are relatively simple, their usage is not practical, and it is not guaranteed that the results based on their resulting Δt values generate numerically stable results. A relatively speedy and efficient methodology is to conduct runs, say with the MODFLOW code, by varying the time parameters identified in Section 4.3.13.1 and comparing the corresponding head contours or drawdown contours, or, head vs. time or drawdown vs. time curves, at some critical locations in the flow domain. As can be seen from Eqs. (4-153) and (4-154), smaller values of *TSMULT* and higher values of *NSTP* correspond to more accurate results. As shown in Examples 4-5 and 4-6, the other influential quantity is *Delt*(1), the first time step. For example, the parameters in the aforementioned examples can be used to make comparative evaluations for a numerical ground water flow model. It is suggested to use first *PERLEN* = 150 days, *TSMULT* = 1.1, and *NSTP* = 105. Then, another run should be conducted using *TSMULT* = 1.05 and *NSTP* = 191. Once the outputs are available, comparisons should be made by generating the curves mentioned above. If they are close for all practical purposes, *TSMULT* = 1.1 and *NSTP* = 105 should be used for all subsequent runs. If differences are large, additional trial runs should be conducted by changing the parameters, including those of *Delt*(1), as appropriate. The same rule can be applied for time step selection for numerical solute transport models, presented in Section 4.4.5.

4.3.14 THE CONDUCTANCE COMPONENTS OF THE FINITE-DIFFERENCE EQUATIONS IN THE MODFLOW FAMILY CODES

The MODFLOW code is composed of several packages. One of its packages is the Block-Centered Flow (BCF) package, which computes the conductance components of the finite-difference equation. The conductance components determine flow between adjacent cells. To make the required calculations, it is assumed that a node is located at the center of each model cell (Figure 4-22), hence the name "Block-Centered Flow" is given to the package.

In Section 4.3.10, the equation of flow for each cell in the model was developed as Eq. (4-138). The *CV*, *CR*, and *CC* coefficients are conductances between nodes and are also called *branch conductances*. The HCOF and RHS coefficients, as defined by Eqs. (4-139) and (4-140), respectively, are composed of external source terms and storage terms. Besides calculating the conductances and storage terms, the BCF package calculates the flow-correction terms that are added to HCOF and RHS to compensate for the excess vertical flow that the flow equation calculates when part of the lower aquifer becomes unsaturated. A discussion on how all these calculations are made is presented in the following sections.

4.3.14.1 Basic Conductance Equations

Definition of Conductance
Conductance is a combination of several parameters in Darcy's law (see Section 4.3.1.4). As shown in Section 4.3.1.4, Darcy's law defines one-dimensional flow, and for one-dimensional flow in a prism of porous material shown in Figure 4-34, Eq. (4-17) can be written as

$$Q = KA\frac{h_2-h_1}{L} \tag{4-158}$$

where $Q(L^3/T)$ is the flow rate, $K(L/T)$ the hydraulic conductivity in the direction of flow, $A(L^2)$ the cross-sectional area perpendicular to the flow, $h_2 - h_1(L)$ the head difference across the prism parallel to flow (L), and $L(L)$ the length of the flow path. Equation (4-158) can be expressed as

$$Q = C(h_2 - h_1) \tag{4-159}$$

where

$$C = \frac{KA}{L} \tag{4-160}$$

which is called *conductance*. Another form of the conductance definition for horizontal flow in a prism is

$$C = \frac{TW}{L} \tag{4-161}$$

where $T(L^2/T)$ is transmissivity which is K times thickness of the prism in the direction of flow defined by Eq. (4-24) and $W(L)$ the width of the prism (see Figure 4-34).

According to the definition of conductance given above, conductance is defined for a particular prism of material and for a particular direction. As a result, for a prism of porous material, conductance in the three principal directions may be different.

Equivalent Conductance
If a prism of porous material consists of two or more subprisms in series, and the conductance of each subprism is known, a conductance representing the entire prism can be calculated. The

$$Q = \frac{KA\,(h_2 - h_1)}{L}$$

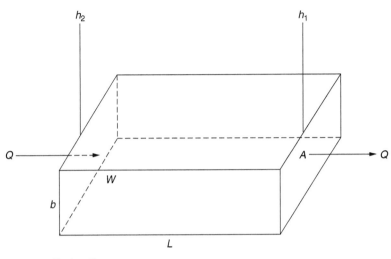

Explanation

K is hydraulic conductivity

h_2 the head at the left end of the prism

h_1 the head at the right end of the prism

Q the flow rate from the left end to the right end

L the length of the flow path

A the cross-sectional area perpendicular to the direction of flow

w the width of the flow path

FIGURE 4-34   Application of Darcy's law to a prism of porous material. (Adapted from McDonald, M.G. and Harbaugh, A.W., *Open File Report 83–875*, U.S. Geological Survey, National Center, Reston, Virginia, 1984.)

equivalent conductance is the rate of flow in the prism divided by the head change across the prism (Figure 4-35). From Eq. (4-159), for points A and B in Figure 4-35, one can write

$$C = \frac{Q}{h_A - h_B} \tag{4-162}$$

The assumption of continuity of head across each section in the series gives

$$\sum_{i=1}^{n} \Delta h_i = h_A - h_B \tag{4-163}$$

From Eq. (4-162), the head drop Δh_i for the ith subprism in Figure 4-35, can be written as

$$\Delta h_i = \frac{q_i}{C_i} \tag{4-164}$$

where $q_i(L^3/T)$ is the flow rate passing through the ith subprism and C_i (L^2/T) is the corresponding conductance. Substitution of Eq. (4-164) into Eq. (4-163) gives

$$\sum_{i=1}^{n} \frac{q_i}{C_i} = h_A - h_B \tag{4-165}$$

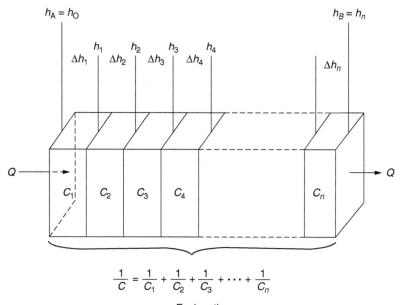

$$\frac{1}{C} = \frac{1}{C_1} + \frac{1}{C_2} + \frac{1}{C_3} + \cdots + \frac{1}{C_n}$$

Explanation

Q is the flow rate

C_n the conductance of prism m

h_n the head at the right side of prism m

Δh_n the head change across prism m

C the conductance of the entire prism

FIGURE 4-35 Conductance calculation through several porous medium prisms in series. (Adapted from McDonald, M.G. and Harbaugh, A.W., *Open File Report 83–875*, U.S. Geological Survey, National Center, Reston, Virginia, 1984.)

Since the flow is one-dimensional and the conservation of mass is satisfied, $q_1 = q_2 = \cdots = q_i = Q$. Therefore, Eq. (4-165) can be written as

$$Q\sum_{i=1}^{n} \frac{1}{C_i} = h_A - h_B \qquad (4\text{-}166)$$

or

$$\frac{h_A - h_B}{Q} = \sum_{i=1}^{n} \frac{1}{C_i} \qquad (4\text{-}167)$$

Using Eq. (4-162) in Eq. (4-167) gives

$$\frac{1}{C_i} = \sum_{i=1}^{n} \frac{1}{C_i} \qquad (4\text{-}168)$$

When there are only two sections, the equivalent conductance takes the form

$$\frac{1}{C} = \frac{1}{C_1} + \frac{1}{C_2} \qquad (4\text{-}169)$$

or

$$C = \frac{C_1 C_2}{C_1 + C_2} \qquad (4\text{-}170)$$

4.3.14.2 Horizontal Conductance

As mentioned previously, *CV*, *CR*, and *CC* coefficients are conductances between nodes and are called as *branch conductances* rather than simply the conductances within cells. Horizontal conductance terms *CR* and *CC* must be calculated between nodes that are adjacent horizontally. The *CR* terms are oriented along rows and they specify conductance between two nodes in the same row. Similarly, the *CC* terms specify conductance between two nodes in the same column. To refer to conductance between two nodes, the subscript notation "1/2" is used. For example, $CR_{i,j+1/2,k}$ represents the conductance between nodes (i, j, k) and $(i, j, + 1, k)$.

Figure 4-36 shows two cells along a row and the parameters used to calculate conductance between nodes in the cells. It is assumed that (1) nodes are in the center of the cells and (2) transmissivity is uniform over a cell. Therefore, the conductance between the nodes is the equivalent conductance of two half cells in series whose conductances are C_1 and C_2. Using Eq. (4-170), one has

$$CR_{i,j+1/2,k} = \frac{C_1 C_2}{C_1 + C_2} \tag{4-171}$$

Using Eq. (4-161) for the conductance terms in Eq. (4-171) with Figure 4-36, one can write

$$CR_{i,j+\frac{1}{2},k} = \frac{\left(\dfrac{TR_{i,j,k}DELC_i}{\frac{1}{2}DELR_j}\right)\left(\dfrac{TR_{i,j+1,k}DELC_i}{\frac{1}{2}DELR_{j+1}}\right)}{\left(\dfrac{TR_{i,j,k}DELC_i}{\frac{1}{2}DELR_j}\right) + \left(\dfrac{TR_{i,j+1,k}DELC_i}{\frac{1}{2}DELR_{j+1}}\right)} \tag{4-172}$$

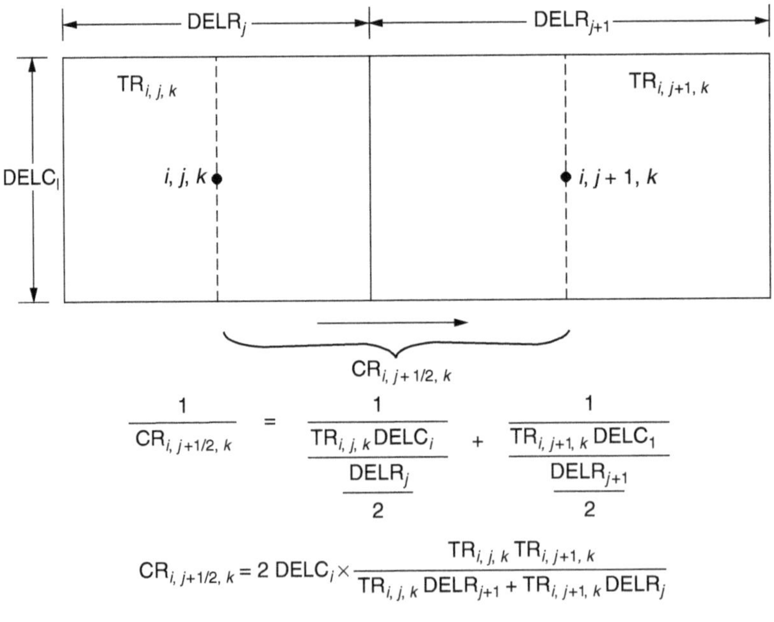

$$\frac{1}{CR_{i,j+1/2,k}} = \frac{1}{\dfrac{TR_{i,j,k}DELC_i}{DELR_j}} + \frac{1}{\dfrac{TR_{i,j+1,k}DELC_1}{DELR_{j+1}}}$$

$$CR_{i,j+1/2,k} = 2\ DELC_i \times \frac{TR_{i,j,k}\ TR_{i,j+1,k}}{TR_{i,j,k}\ DELR_{j+1} + TR_{i,j+1,k}\ DELR_j}$$

Explanation

$TR_{i,j,k}$ is the transmissivity in the row direction in cell i, j, k

$CR_{i,j+1/2,k}$ the conductance in the row direction between nodes i, j, k and $i, j+1, k$

FIGURE 4-36 Conductance calculation between nodes using transmissivity and dimensions of cells. (Adapted from McDonald, M.G. and Harbaugh, A.W., *Open File Report 83–875*, U.S. Geological Survey, National Center, Reston, Virginia, 1984.)

where $TR(L^2/T)$ is the transmissivity in the row direction, $DELR$ (L) the grid width along a row, and $DELC$(L) the grid width along a column. Simplification of Eq. (4-172) gives

$$CR_{i,j+1/2,k} = 2DELC_i \frac{TR_{i,j,k}TR_{i,j+1,k}}{TR_{i,j,k}DELR_{j+1} + TR_{i,j+1,k}DELR_j} \tag{4-173}$$

The same process can be applied for the determination of $CC_{i+1/2,j,k}$, and the result is

$$CC_{i+1/2,j,k} = 2DELR_j \frac{TC_{i,j,k}TC_{i+1,j,k}}{TC_{i,j,k}DELC_{i+1} + TC_{i+1,j,k}DELC_i} \tag{4-174}$$

where $TC(L^2/T)$ is the transmissivity in the column direction. Whenever transmissivity of both cells is zero, the conductance between the nodes in the cells is set equal to zero.

In a confined model, layer horizontal conductance will be constant for the simulation. If a layer is potentially unconfined, new values of horizontal conductance must be calculated as the head fluctuates. This is carried out at the start of each iteration. First, transmissivity is calculated from the hydraulic conductivity and saturated thickness values; and then conductance is calculated from transmissivity and cell dimensions.

4.3.14.3 Vertical Conductance

The calculation of vertical conductance is conceptually similar to the calculation of horizontal conductance. The finite-difference flow equation requires the conductance between two vertically adjacent nodes. $CV_{i,j,k+1/2}$ is the conductance between nodes i, j, k and $i, j, k + 1$ in layers k and $k + 1$. By applying Eq. (4-160) between two vertically adjacent model nodes (Figure 4-37), one has

$$CV_{i,j,k+1/2} = \frac{(KV_{i,j,k+1/2})(DELR_j)(DELC_i)}{DELV_{i,j,k+1/2}} \tag{4-175}$$

where $KV_{i,j,k+1/2}$(L/T) is the hydraulic conductivity between nodes (i, j, k) and $(i, j, k + 1)$ and $DELV$ $i, j, k + 1/2$ (L) the distance between nodes (i, j, k) and $(i, j, k + 1)$.

Vertical Hydraulic Conductivity and Vertical Grid Spacing Parameter (Vcont)
In the MODFLOW program, rather than specifying both vertical hydraulic conductivity and vertical grid spacing, a single term called "Vcont" is specified. Vcont between nodes (i, j, k) and $(i, j, k + 1)$ is given by

$$Vcont_{i,j,k+1/2} = \frac{KV_{i,j,k+1/2}}{DELV_{i,j,k+1/2}} \tag{4-176}$$

The MODFLOW program requires that Vcont between nodes be entered as input data rather than calculating it in the program.

Potentially, several methods can be used to calculate Vcont manually depending on the way that the aquifer is discretized vertically. The general case is that the vertical hydraulic conductivities of two adjacent layers are not the same. The conductance of two half cells in series can be calculated from Eqs. (4-160) and (4-170) as

$$CV_{i,j,k+1/2} = \frac{\left(\dfrac{KV_{i,j,k}DELR_jDELC_i}{\frac{1}{2}DELV_{i,j,k}}\right)\left(\dfrac{KV_{i,j,k+1}DELR_jDELC_i}{\frac{1}{2}DELV_{i,j,k+1}}\right)}{\left(\dfrac{KV_{i,j,k}DELR_jDELC_i}{\frac{1}{2}DELV_{i,j,k}}\right) + \left(\dfrac{KV_{i,j,k+1}DELR_jDELC_i}{\frac{1}{2}DELV_{i,j,k+1}}\right)} \tag{4-177}$$

Equating the right-hand sides of Eq. (4-175) and Eq. (4-177) and rearranging yields

$$Vcont_{i,j,k+1/2} = \frac{KV_{i,j,k+1/2}}{DELV_{i,j,k+1/2}} = \frac{2}{\dfrac{DELV_{i,j,k}}{KV_{i,j,k}} + \dfrac{DELV_{i,j,k+1}}{KV_{i,j,k+1}}} \tag{4-178}$$

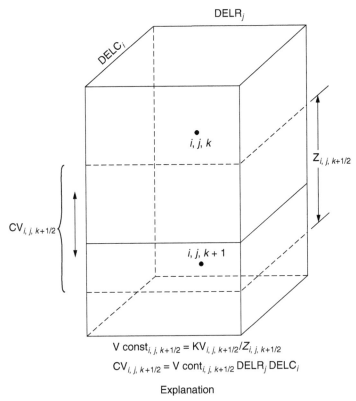

$$V \, \mathrm{const}_{i,\,j,\,k+1/2} = KV_{i,\,j,\,k+1/2}/Z_{i,\,j,\,k+1/2}$$

$$CV_{i,\,j,\,k+1/2} = V \, \mathrm{cont}_{i,\,j,\,k+1/2} \, DELR_j \, DELC_i$$

Explanation

$KV_{i,\,j,\,k+1/2}$ is the hydraulic conductivity between nodes $(1, j, k)$ and $(i, j, k + 1)$

$Z_{i,\,j,\,k+1/2}$ the distance between nodes (i, j, k) and $(i, j, k + 1)$

FIGURE 4-37 Vertical conductance calculation between adjacent cells. (From McDonald, M.G. and Harbaugh, A.W., *Open File Report 83–875*, U.S. Geological Survey, National Center, Reston, Virginia, 1984. with permission.)

4.3.15 Implementation of the Boundary Conditions

The general form of the boundary conditions used in fluid mechanics is given in Section 4.3.11. The implementation of these boundary conditions, which were shaped during the second half of the 20th century, has special importance in ground water modeling application areas. In this section, these boundary conditions will be revisited from the point of view of the practical application of ground water modeling. The details are presented in the following sections.

4.3.15.1 Constant Head Boundaries

The general form of a constant head boundary condition and its physical meaning are described in Section 4.3.11.1. A cell at which the head is constant all the time is called a *constant head cell* (Figure 4-38). Constant head values can be assigned not only on the boundaries of the grid system but also on the boundaries of surface cells at the water table. Some critical points regarding their assignment are presented in Section 4.3.11.1. Examples of constant head cells are shown in Figure 4-39.

The bottom of rivers, canals, lakes, and other surface water bodies are generally permeable. These permeable boundaries may be treated as surfaces of specified head, if the body of the surface water is large enough in volume such that its level is uniform and independent of changes in ground water flow. However, the uniform head on a boundary of this type changes with time as a result of seasonal variations in the surface water level. Streams, rivers, and canals may form boundaries with

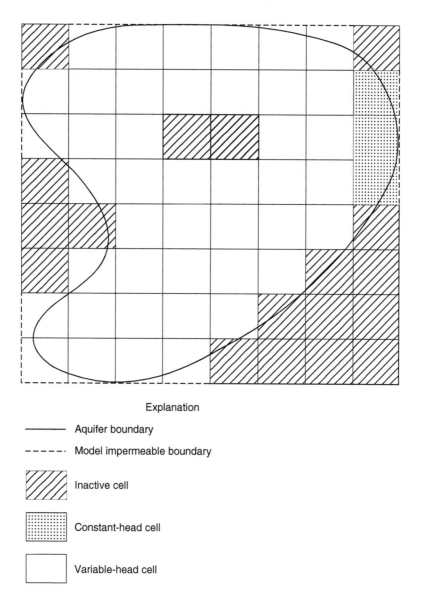

Explanation

—————— Aquifer boundary

- - - - - - Model impermeable boundary

Inactive cell

Constant-head cell

Variable-head cell

FIGURE 4-38 Discretized aquifer showing boundaries and constant head cells. (Adapted from McDonald, M.G. and Harbaugh, A.W., *Open File Report 83–875*, U.S. Geological Survey, National Center, Reston, Virginia, 1984.)

nonuniform distributions of head that may either be constant or variable with time. On the other hand, a small stream might be affected by a nearby withdrawal of ground water from a well if that withdrawal occurred at a rate of the same order of magnitude as the flow in the stream. Under these conditions, the boundary condition would not be independent of ground water flow; it would be a head-dependent flux boundary (see Section 4.3.15.2).

4.3.15.2 Constant Flux Boundaries

The general form of a constant flux boundary condition and its physical meaning are described in Section 4.3.11.1. This type of boundary condition has some special features at wells, recharge areas, and impermeable boundaries, and are discussed below.

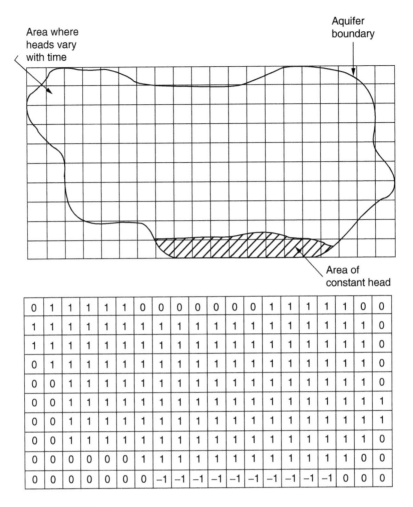

0	1	1	1	1	1	0	0	0	0	0	0	0	1	1	1	1	1	0	0
1	1	1	1	1	1	1	1	1	1	1	1	1	1	1	1	1	1	1	0
1	1	1	1	1	1	1	1	1	1	1	1	1	1	1	1	1	1	1	0
0	1	1	1	1	1	1	1	1	1	1	1	1	1	1	1	1	1	1	0
0	0	1	1	1	1	1	1	1	1	1	1	1	1	1	1	1	1	1	0
0	0	1	1	1	1	1	1	1	1	1	1	1	1	1	1	1	1	1	1
0	0	1	1	1	1	1	1	1	1	1	1	1	1	1	1	1	1	1	1
0	0	1	1	1	1	1	1	1	1	1	1	1	1	1	1	1	1	1	0
0	0	0	0	0	0	1	1	1	1	1	1	1	1	1	1	1	1	0	0
0	0	0	0	0	0	0	-1	-1	-1	-1	-1	-1	-1	-1	-1	-1	0	0	0

IBOUND Codes
<0 Constant head
= 0 No flow
>0 Variable head

FIGURE 4-39   Example of the boundary array (IBOUND) for a single layer. (Adapted from McDonald, M.G. and Harbaugh, A.W., *Open File Report 83–875*, U.S. Geological Survey, National Center, Reston, Virginia, 1984.)

Boundary Conditions at Wells

Terms for Flux Conditions

There are several terms to describe the flux conditions at wells. If water is removed from an aquifer through a well, the well is called an extraction well. If water is added to an aquifer through a well, the well is called an injection well. The term pumpage is also used for wells. Pumpage is the removal or addition of water from or to an aquifer through the use of a well. If water is pumped out of the aquifer, the pumpage is a ground water sink. If water is injected (recharging well), the pumpage is a source of ground water (Mercer et al., 1982).

Units for Well Fluxes

The units of pumpage rate are L 3/T or, more specifically, m³/day, ft³/day, gpm (gallons per minute), L/sec, and so on. According to the MODFLOW convention, if the length unit is "foot" and time unit is "day," the units of pumpage rate must be ft^3/day (cubic foot per day). If they are "meter" and "day," the units of pumpage must be m^3/day (cubic meter per day).

Implementation of the Well Fluxes by the Well Package of MODFLOW
The boundary condition at a well under discharging and recharging conditions is termed specified flux. The mathematical foundations of the specified flux boundary are given in Section 4.3.11.1. The boundary conditions at wells are treated as constant or variable specified flux, depending on which best describes the actual physical conditions.

In MODFLOW, wells are handled by its Well Package. Both recharging and discharging wells can be simulated. A recharging well can be viewed as a source of water that is not affected by the head in the aquifer. A discharging well is also a recharging well with a negative recharge rate according to MODFLOW sign convention. A list containing the location and rate for each well is maintained. The list contains a record for each well based on the layers of the model. A record includes (1) row number, (2) column number, (3) layer number at which the well cell is located, and (4) the rate of the well (recharge or discharge). At each iteration, for each variable head cell (i, j, k) containing a well, the well rate is added to the accumulator in which RHS$_{i,j,k}$ in Eq. (4-138) is formulated.

The top and bottom elevations of screen intervals in a ground water flow model are critically important factors for vertical grid discretization (see Section 4.3.12). This is due to the fact that the pumpage can only be assigned to a single cell. As shown in Figure 4-21, modeling layers are based on cell heights in the vertical direction. If a well is screened only in a single modeling layer, the pumpage rate must be assigned to the corresponding cell in that layer; or if a well is screened in more than one modeling layer, the pumpage rate must be assigned proportionally to the corresponding cells in those modeling layers. This is shown below with an example.

Other Applications of the Well Package
The Well Package of MODFLOW (or other equivalent models) has other potential applications. For example, localized recharge and evaporation can also be simulated by wells.

EXAMPLE 4-7

A ground water flow model is designed to have five modeling layers. The thickness of layer 1 is 2.0 m, layer 2 is 3.0 m, layer 3 is 3.0 m, layer 4 is 2.0 m, and layer 5 is 4.0 m. The discharge rate is $Q = 135$ m^3/day. Under these conditions, (a) if the well is screened at layer 3 ($I = 2, J = 4, K = 3$), what is the value to be assigned to the model: and (b) if the same flow rate is assigned at layers 3 and 4, determine the flow rates to be assigned to the model.

SOLUTION

(a) $Q = 135$ m^3/day is to be assigned at the cell having ($I = 2, J = 4$, and $K = 3$) because the well is only screened at layer 3 ($K = 3$).

(b) The total thickness of layers 3 and 4 is 5.0 m. Therefore, the flow rate per unit length is $q = 135$ m^3/day/5 m $= 27$ m^3/day/m. For layer 3, $Q_3 = (27$ m^3/day/m)(3 m) $= 81$ m^3/day; for layer 4, $Q_4 = (27$ m^3/day/m)(2 m) $= 54$ m^3/day. Notice that $Q_3 + Q_4 = 81$ m^3/day $+ 54$ m^3/day $= 135$ m^3/day. Therefore, $Q_3 = 81$ m^3/day is to be assigned at the cell having $I = 2, J = 4$, and $K = 3$ and $Q_4 = 54$ m^3/day is to be assigned at the cell having $I = 2, J = 4$, and $K = 4$.

Areally Distributed Recharge Boundaries
Definition and Units of Recharge
Recharge is the amount of water that enters an aquifer system. Precipitation is the main source of recharge. Aquifer systems having unconfined uppermost formations receive recharge, which is generally a fraction of the total precipitation as a result of rains and snows. Precipitation occurs evenly over a large area. Therefore, it is called *areally distributed recharge*. It is expressed in terms of flow rate per unit area. Therefore, the units of the recharge rate I are $[I] = L^3/T/L^2 \equiv L/T$. For recharge

rate, units such as cm/sec, m/day, in./hour, ft/day and in./year are used. Accumulated water at the ground surface causes artificial recharge depending on the hydraulic characteristics of the unsaturated zone.

Implementation of the Recharge by the Recharge Package of MODFLOW
The volumetric rate of flow into a cell is the recharge rate multiplied by the horizontal area of the cell. Therefore, for MODFLOW, mathematically,

$$QRCH_{i,j,k} = I_{i,j,k}(DELR_j)(DELC_i)$$
(4-179)

where $I_{i,j,k}$(L/T) is the recharge rate (or infiltration rate), $DELR_j$(L) the dimension of the *j*th column along the row direction, and $DELC_i$(L) the dimension of the *i*th column along the column direction. $QRCH_{i,j,k}$(L^3/T) is the volumetric recharge rate on the rectangular area whose dimensions are $DELR_j$ and $DELC_i$. As a result, recharge boundaries are specified flux boundaries, as discussed in Section 4.3.11. One should bear in mind that the recharge is independent of the head in the cell.

Factors Affecting Recharge
The amount of water that recharges to an unconfined aquifer is determined by a combination of three factors (Fetter, Mercer et al., 1982; 1980): (1) the amount of precipitation that is not lost by evapotranspiration and runoff and is thus available for recharge, (2) the vertical hydraulic conductivity of surficial deposits and other strata in the recharge area of the aquifer, which determines the volume of recharged water capable of moving downward to the aquifer, and (3) the water-moving ability of the aquifer and hydraulic head gradient, which determines how much water can move away from the recharge area. If the water table of an unconfined aquifer is close to the soil surface under natural conditions (unpaved conditions, etc.) the effects of the unsaturated zone are generally negligible. However, if the thickness of the unsaturated zone (the distance between the ground surface and water table) is significant, the unsaturated zone may play a significant role on the recharge depending on its geologic, lithologic, and hydraulic characteristics. The water-moving ability of the aquifer depends on its horizontal and vertical hydraulic conductivities.

Quantification of Recharge
Ground water recharge is one of the most difficult input data to quantify accurately. A good standard technique is currently unavailable. Some data examples are given below (Mercer et al., 1982).

Most often, a percentage of precipitation is used to estimate ground water recharge. The percentage selected is often quite subjective and may be based on estimates of evapotranspiration. For example, the annual rainfall in the Pasco Basin, Washington, is 18.5 cm/year, and Arnett et al. (1980) estimated recharge to be 20% of the rainfall or 3.7 cm/year. Recharge is highly variable both in space and time. As an example, Table 4-7 gives recharge rates in Illinois. It is obvious that in arid or semiarid climates where precipitation is much less than the precipitation in Illinois, recharge will be much less than the data in Table 4-7. Recharge rates can be estimated through a ground water flow model calibration process (see Section 4.5).

Options for Recharge in MODFLOW
In the MODFLOW code all features for recharge are handled by its Recharge Package. The layer to which the recharge is applied can be specified using one of the three options: (1) for option 1, recharge only affects the uppermost layer; (2) for option 2, recharge at each horizontal location affects the layer specified in an indicator array (IRCH) specified by the user; and (3) for option 3, recharge affects the uppermost active cell in each vertical column. These are all shown in Figure 4-40. If option 1 is specified, the appropriate cell is in the top layer of the grid (layer 1). If option 2 is specified, the appropriate cell is the layer specified by the user in the indicator array (IRCH). If option 3 is specified, the appropriate cell is the uppermost active cell at the horizontal location which is not below a constant head cell. If the uppermost active cell is below a constant head cell, recharge is not applied to any cell because this recharge is assumed to be intercepted by the boundary.

TABLE 4-7
Summary of Recharge Rates for Aquifers in Illinois

Location	Recharge Rate		Lithology of Deposits Above Aquifer or Permeable Zones	Aquifer
	m/sec	in./year		
Northeastern Illinois	0.020×10^{-9}	0.025	Maquoketa Formation, largely Shale	Cambrian–Ordovician
DeKalb and Kendall Counties	0.304×10^{-9}	0.378	Glacial drift and units of Cambrian–Ordovician aquifer	Cambrian–Ordovician
DuPage County	1.083×10^{-9}	1.345	Glacial drift, largely till and shaly dolomite	Dolomite of Silurian age
DuPage County	2.334×10^{-9}	2.899	Glacial drift, largely till and Dolomite	Dolomite of Silurian age
DuPage County	2.301×10^{-9}	2.858	Glacial drift, largely till and Dolomite	Dolomite of Silurian age
DuPage County	2.673×10^{-9}	3.320	Glacial drift, largely till and Dolomite	Dolomite of Silurian age
LaGrange, Cook County	2.723×10^{-9}	3.382	Glacial drift, largely till and Dolomite	Dolomite of Silurian age
Chicago Heights, Cook County	3.806×10^{-9}	4.727	Glacial drift, largely till and Dolomite	Dolomite of Silurian age
Libertiville, Lake County	0.880×10^{-9}	1.093	Glacial drift, largely till and shaly dolomite	Dolomite of Silurian age
Woodstock, McHenry County	2.148×10^{-9}	2.668	Glacial drift, largely till	Glacial sand and gravel
Near Joliet, Will County	3.383×10^{-9}	4.202	Glacial drift, largely silt and sand	Glacial sand and gravel
Champaign—Urbana	1.945×10^{-9}	2.416	Glacial drift, largely till	Glacial sand and gravel
Havana region, Mason County, Tazewell County	4.364×10^{-9}	5.420	Glacial drift, largely till	Glacial sand and gravel
Havana region, Mason County, Tazewell County	4.720×10^{-9}	5.962	Glacial drift, largely till	Glacial sand and gravel
Havana region, Mason County, Tazewell County	8.458×10^{-9}	10.505	Glacial drift, largely sand and gravel	Glacial sand and gravel
Havana region, Mason County, Tazewell County	8.221×10^{-9}	10.210	Glacial drift, largely sand and gravel	Glacial sand and gravel
East St. Louis, Madison, and St. Clair Counties	5.870×10^{-9}	7.291	Glacial drift and alluvium, largely sand and gravel	Glacial sand and gravel
East St. Louis, Madison, and St. Clair Counties	5.802×10^{-9}	7.206	Glacial drift and alluvium, largely sand and gravel	Glacial sand and gravel
East St. Louis, Madison, and St. Clair Counties	5.058×10^{-9}	6.282	Glacial drift and alluvium, largely sand and gravel	Glacial sand and gravel
East St. Louis, Madison, and St. Clair Counties	6.259×10^{-9}	7.774	Glacial drift and alluvium, largely sand and gravel	Glacial sand and gravel
Panther Creek basin, Woodford, Livingston, and McLean Counties	6.428×10^{-9}	7.984	Glacial drift, largely till	Glacial drift

Adapted from Walton, W.C., *Ground water Resource Evaluation*, McGraw-Hill, New York, 1970, p. 644. As presented in Mercer et al., U.S. Nuclear Regulatory Commission, NUREG/CR-3066, Washington, DC, 1982.

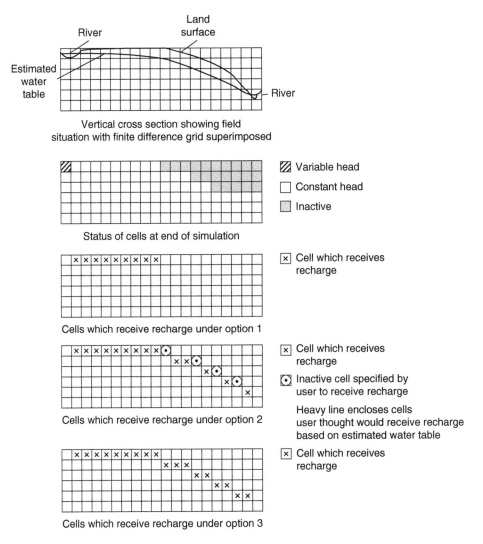

Vertical cross section showing field
situation with finite difference grid superimposed

Status of cells at end of simulation

Cells which receive recharge under option 1

Cells which receive recharge under option 2

Cells which receive recharge under option 3

FIGURE 4-40 Hypothetical problem showing which cells receive recharge under the three options available in the Recharge Package. (Adapted from McDonald, M.G. and Harbaugh, A.W., *Open File Report 83–875*, U.S. Geological Survey, National Center, Reston, Virginia, 1984.)

No-Flow Boundaries

No-flow boundaries are special forms of the specified flux boundaries as described in Section 4.3.11.1. The mathematical expression of a no-flow boundary is given by Eq. (4-146). In numerical models, the implementation of no-flow boundaries is simple. If a model boundary is set free, it virtually means that the boundary is a no-flow boundary. For example, as mentioned previously, a constant head boundary is such that the head is kept constant. If this constancy condition is lifted, the boundary is a no-flow boundary and the MODFLOW and similar codes treat it like that.

4.3.15.3 Head-Dependent Flux Boundaries

In the MODFLOW and family models, a head-dependent flux boundary consists of a source of water outside the modeled area that supplies water to a cell in the modeled area at a rate proportional to the head difference between the source and the cell. In the MODFLOW program, this type of boundary is included in the General-Head Boundary (GHB) package. From Eq. (4-159), the flux into or out of a boundary cell can be expressed as

$$Q_{i,j,k,m} = C_m(HB_m - h_{i,j,k})$$ (4-180)

where $Q_{i,j,k,m}$(L^3/T) is the flux into or out of the boundary cell, C_m(L^2/T) the boundary conductance defined by Eq. (4-160), HB_m(L) the boundary head, and $h_{i,j,k}$(L) the head in the cell computed by the model. The expression for conductance is given by Eq. (4-160).

The source of water could be a gaining river, in which case the constant of proportionality is the conductance of the riverbed (Figure 4-41a). The source could be a buried drain (Figure 4-41b), so that the C_m constant is a function of the material around the drain and the size and spacing of openings in the skin of the drain (Figure 4-41b). The source could also be the head in the aquifer outside the simulated area, in which case the constant of proportionality is the hydraulic conductance of the material between the known head and boundary of the simulated area (Figure 4-41c).

Although the GHB package can be used to simulate the situations in Figure 4-41, it should be done with great care. This package deals with a single linear relationship (Figure 4-42). MODFLOW input for each boundary consists of the location of the boundary cell (layer, row, and column), the boundary head, and the constant of proportionality.

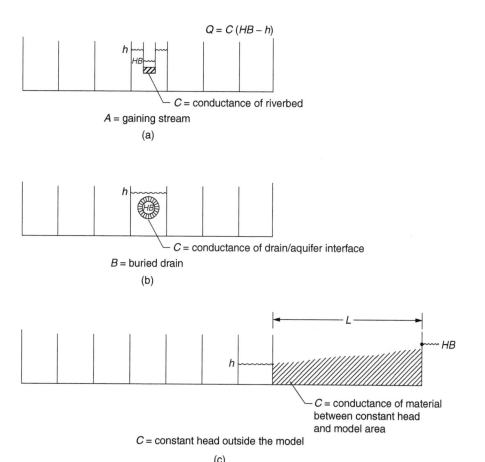

FIGURE 4-41 Head-dependent flux boundaries: (a) A gaining stream; (b) A buried drain; and (c) Outside water source. (Adapted from McDonald, M.G. and Harbaugh, A.W., *Open File Report 83–875*, U.S. Geological Survey, National Center, Reston, Virginia, 1984.)

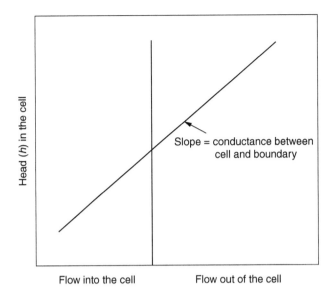

FIGURE 4-42   Flow from a head-dependent boundary as function of head in the aquifer. (Adapted from McDonald, M.G. and Harbaugh, A.W., *Open File Report 83–875*, U.S. Geological Survey, National Center, Reston, Virginia, 1984.)

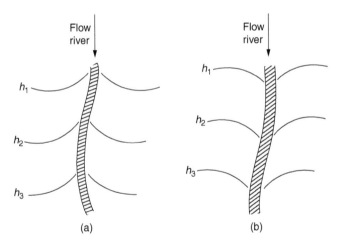

FIGURE 4-43   Water level contours around a river in an unconfined aquifer: (a) Flow from the aquifer to the river; and (b) Flow from the river to the aquifer. (Adapted from McDonald, M.G. and Harbaugh, A.W., *Open File Report 83–875*, U.S. Geological Survey, National Center, Reston, Virginia, 1984.)

4.3.15.4 River Boundaries

Rivers may contribute inflow to the aquifer or outflow from the aquifer depending on the head gradient between the river and aquifer. The effect of leakage through the riverbed on the shape of the water table is shown in Figure 4-43. In the MODFLOW program, rivers are handled by the River (RIV) package and a river is divided into reaches. Each reach is contained completely in a single cell but small reaches are ignored (Figure 4-44). Figure 4-45a shows a cross-section of a grid cell containing a river reach. Figure 4-45b is an idealized form of Figure 4-45a and the riverbed is

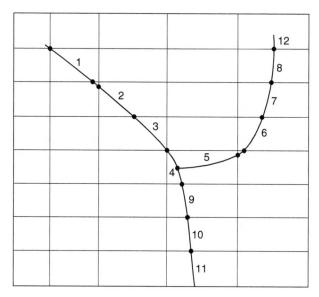

FIGURE 4-44 Discretization of a river into reaches. (From McDonald, M.G. and Harbaugh, A.W., *Open File Report 83–875*, U.S. Geological Survey, National Center, Reston, Virginia, 1984. With permission.)

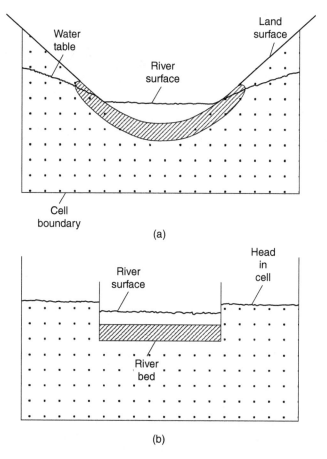

FIGURE 4-45 Cross-section of an aquifer containing a river: (a) A natural cross-section; and (b) Idealized cross-section. (Adapted from McDonald, M.G. and Harbaugh, A.W., *Open File Report 83–875*, U.S. Geological Survey, National Center, Reston, Virginia, 1984.)

represented by a rectilinear prism of homogeneous porous material. The concepts, hydraulics, and modeling aspects regarding river–aquifer interrelationships are presented below (Bear, 1979; McDonald and Harbaugh, 1984).

River–Aquifer Interrelationships
Rivers are generally underlain by unconfined aquifers; in some special cases they may be underlain by confined aquifers as well. Potentially, a river may contribute flow to an underlying aquifer; or the underlying aquifer may contribute flow to the river. Most of the rivers having low water levels receive flow from the underlying aquifers. An aquifer satisfying this condition generally has higher water level elevation than the river water level. Such rivers are called *effluent rivers* or *gaining rivers* (Figure 4-46a). On the other hand, if the water level in a river is higher than the water level

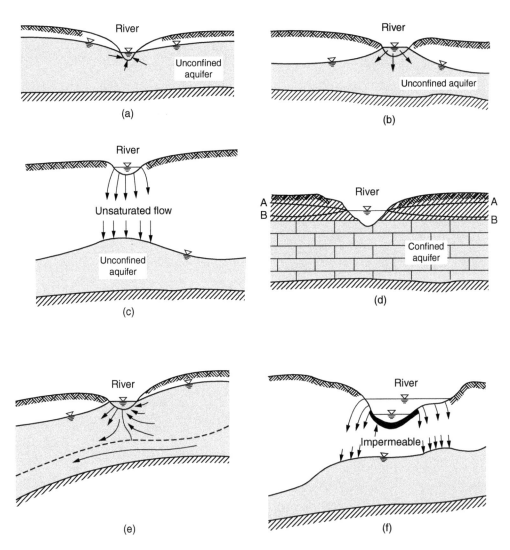

FIGURE 4-46 River–aquifer interrelationships: (a) Flow from an unconfined aquifer to a river; (b) Flow from a river to an unconfined aquifer having shallow water table; (c) Flow from an unconfined aquifer having deep water table; (d) Outflow from the aquifer whose head level is AA or outflow from the river to the aquifer whose head level is BB; (e) River having inflow from the aquifer at one segment and outflow at the other segment; and (f) River having impermeable bed at its lower part. (Adapted from Bear, J., *Hydraulics of Ground water*, McGraw-Hill, New York, 1979, 569 pp.)

in an underlying aquifer, the river is called an *influent river* or *losing river* (Figures 4-46b and c). If a river is located in an impermeable formation and has a direct contact with an underlying confined aquifer, the river may have either outflow or inflow, depending on the levels of the aquifer that are above or below the water level in the river (Figure 4-46d). As shown in Figure 4-46e, a river can have outflow along one cross-sectional segment and inflow along the other segment. The other possibility is that the lower part of the river is covered by an impermeable formation and its upper part is covered by a more permeable formation that permits outflow from the river (Figure 4-46f).

Leakage through a Reach of a Riverbed
In MODFLOW, leakage through a riverbed (Figure 4-47) is approximated by

$$QRIV = KLW\frac{(HRIV-HAQ)}{M} \tag{4-181}$$

where $QRIV(L^3/T)$ is the leakage through the reach of the riverbed, $K(L/T)$ the hydraulic conductivity of the riverbed, $L(L)$ the width of the river, $M(L)$ the thickness of the riverbed, $HAQ(L)$ the head on the aquifer side of the riverbed; and $HRIV(L)$ the head on the river side of the riverbed. Equation (4-181) can be rewritten in terms of the conductance of the reach of the riverbed as

$$QRIV = CRIV(HRIV - HAQ) \tag{4-182}$$

where

$$CRIV = \frac{KLW}{M} \tag{4-183}$$

The head on the river side of the riverbed is the river stage, which is the water level elevation in the river. The head on the aquifer side of the riverbed is slightly complex. Figure 4-48a shows a case in which the porous material adjacent to the riverbed is fully saturated and the head on the aquifer side of the riverbed (HAQ) is equal to the head in the cell. Therefore, Eq. (4-181) can be written as

$$QRIV = CRIV(HRIV - H) \tag{4-184}$$

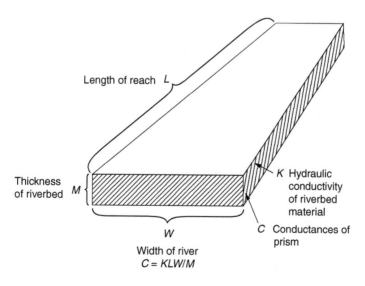

Length of reach L

Thickness of riverbed M

W
Width of river
$C = KLW/M$

K Hydraulic conductivity of riverbed material

C Conductances of prism

FIGURE 4-47 A riverbed with idealized prism of porous material. (Adapted from McDonald, M.G. and Harbaugh, A.W., *Open File Report 83–875*, U.S. Geological Survey, National Center, Reston, Virginia, 1984.)

where $H(L)$ is the head in the cell. If the material adjacent to the riverbed is not saturated (Figure 4-48b), the head on the aquifer side of the riverbed is equal to the elevation of the bottom of the river bed (RBOT). In this case, Eq. (4-182) can be written as

$$QRIV = CRIV(HRIV - RBOT) \qquad (4-185)$$

The Relationship between River Leakage and Head in the Cell
The relationship between river leakage and head in the cell is shown in Figure 4-49. The simple model of river leakage represented by the graph in Figure 4-49 is based on the assumption that leakage from the river is independent of the location of the river within the cell. Although, there may be two river reaches in one cell, they are both assumed to be in the same cell. The MODFLOW model has also assumed that there is always enough water in the river to supply the aquifer. The modeler should compare leakage rates with river discharge rates and ensure that they are in agreement.

Required Data for MODFLOW
MODFLOW input in its RIV package consists of one record for each river reach, which specifies the cell containing the reach (layer, row, and column) and the three parameters needed to calculate seepage, which are river stage, riverbed conductance, and riverbed bottom elevation. The riverbed conductance is calculated by Eq. (4-183).

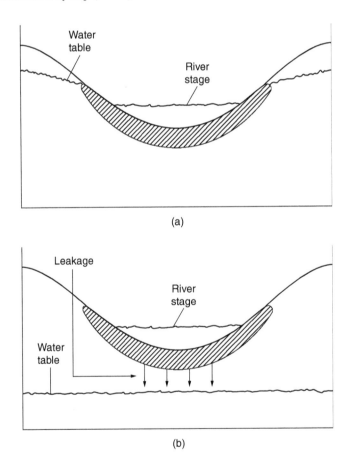

FIGURE 4-48 Cross-section of an aquifer showing the relation between head on the aquifer side of the riverbed and head in the cell: (a) Head on the aquifer side of the river is equal to head in the cell; and (b) Head on the aquifer side of the river is equal to elevation of bottom of river bed. (Adapted from McDonald, M.G. and Harbaugh, A.W., *Open File Report 83–875*, U.S. Geological Survey, National Center, Reston, Virginia, 1984.)

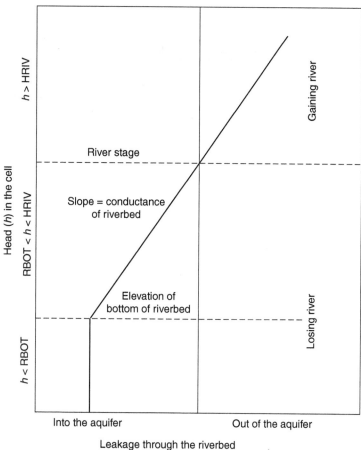

FIGURE 4-49 Leakage through a riverbed into an aquifer as a function of head in the aquifer. (Adapted from McDonald, M.G. and Harbaugh, A.W., *Open File Report 83–875*, U.S. Geological Survey, National Center, Reston, Virginia, 1984.)

4.3.15.5 Drain Boundries

MODFLOW Drain Boundary Approach

In the MODFLOW family codes, drain-type boundary conditions are handled by the Drain (DRN) package. The cross section of a drain in a cell is shown in Figure 4-50. The flow rate into a drain in the saturated zone of an aquifer is approximated in the model by the equation

$$QD_{i,j,k} = CD_{i,j,k}(h_{i,j,k} - d_{i,j,k})$$ (4-186)

where $QD_{i,j,k}(L^3/T)$ is the flow rate into the drain, $d_{i,j,k}(L)$ the head in the drain; $h_{i,j,k}$ (L) the head in the aquifer near the drain, and $QD_{i,j,k}(L^2/T)$ the conductance of the interface between the drain and the aquifer material. Potentially, the conductance depends on (1) the size and frequency of openings in a drain tile; (2) chemical precipitation around a tile; (3) difference in permeability between the aquifer material and the backfill around a tile; and (4) a low permeability bed in an open drain or the converging area of flow as the drain is approached. The head in the drain is assumed to be the

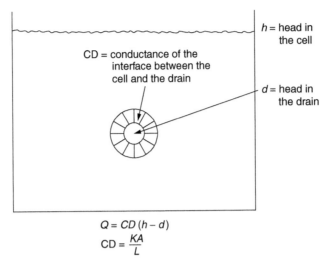

$$Q = CD\,(h - d)$$
$$CD = \frac{KA}{L}$$

A = cross-sectional area perpendicular to the direction of flow

L = length of the flow path

FIGURE 4-50   Cross-section of a drain in an aquifer. (Adapted from McDonald, M.G. and Harbaugh, A.W., *Open File Report 83–875*, U.S. Geological Survey, National Center, Reston, Virginia, 1984.)

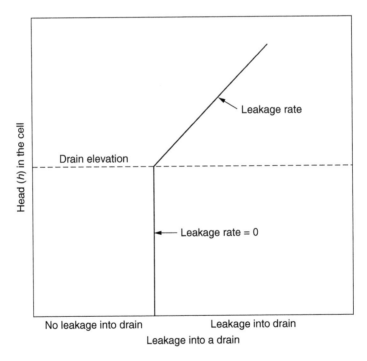

FIGURE 4-51   Leakage into a drain as a function of head in the aquifer. (Adapted from McDonald, M.G. and Harbaugh, A.W., *Open File Report 83–875*, U.S. Geological Survey, National Center, Reston, Virginia, 1984.)

elevation of the drain. As a result, the flow rate into the drain is assumed to be proportional to the head above the drain (Figure 4-51). Equation (4-186) only holds when the head in the aquifer is greater than the head in the drain. When the elevation of the drain is greater than the head in the aquifer, the flow rate into the drain, $QD_{i,j,k}$, is equal to zero.

Required Data for MODFLOW
The required data for MODFLOW for each drain cell are the cell containing the drain (layer, row, and column), drain elevation, and drain conductance.

Application to Springs
A spring is a small area from which ground water comes out of an aquifer onto the ground surface. The discharge rate of a spring depends on the hydraulic conductivity of the aquifer and the hydraulic gradient of the water table (in the case of an unconfined aquifer) or piezometric head surface (in the case of a confined aquifer).

Springs may have different forms, as shown in Figure 4-52. For example, a *depression spring* occurs when a high water table intersects the ground surface. A *perched spring* (Figure 4-52) occurs when a confining layer, which underlies an unconfined aquifer, intersects the ground surface. As

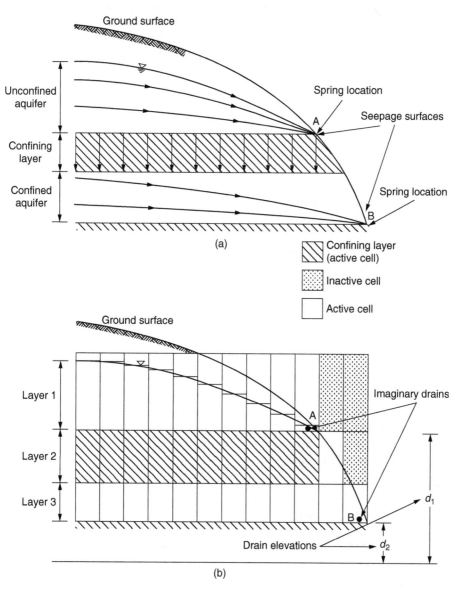

FIGURE 4-52 A simulation approach of springs using the drain boundary of MODFLOW: (a) Conceptual and (b) modeled.

shown in Figure 4-52, a confined aquifer can be drained in the form of a spring as well. In Figure 4-52, the spring is located at point A in the unconfined aquifer and point B in the confined aquifer. The flow rate of a spring basically depends on (1) the difference between the eleations of the water table (the piezometric head elevations for a confined aquifer case) in the vicinity of the spring; (2) the elevation of the spring location (points A and B in Figure 4-52 for the unconfined aquifer and confined aquifer, respectively); and (3) the permeability of the aquifer materials around the spring. If the aquifer is less permeable, flow may not occur at the spring due to the fact that water evaporates as soon as it comes out, depending on the season and the climatic conditions. Sometimes, the spring location may be just wet without any visible flow.

Under the framework described above, springs may be simulated using the DRN package of MODFLOW. The critically important parameter is the drain conductance. As mentioned above, this parameter mainly depends on the permeability of the aquifer materials around the spring. If measured data are available for a flowing spring, the conductance can be determined by comparing the model-predicted flow rates with the measured ones. The difference can be minimized by several trial runs. If the water is lost through evaporation around the spring, model-predicted flow rate comparisons should be made with the available pan evaporation rates at the same site.

4.3.15.6 Evapotranspiration Boundaries

Definitions
The definitions of evaporation, transpiration, and evapotranspiration are given below (Bear, 1979; Mercer et al., 1982).

Evaporation
Evaporation is the transfer of water from the liquid phase to the vapor phase. In the case of ground water, evaporation is the loss of water from the soil surface into the atmosphere. If vegetative cover does not exist, the open soil surface is subject to radiation and wind effects, and soil water evaporates directly from the soil surface. There are three main requirements for evaporation: (1) a continual supply of heat to meet the latent heat requirements; (2) a vapor pressure gradient existing between the soil surface and the atmosphere; and (3) a continual supply of water through the soil layers. The first two conditions determine the evaporative demand and are controlled by micrometeorological factors such as air temperature, humidity, wind velocity, radiation, and crop cover. The third condition, which determines the rate of water supply to the evaporative soil surface, is controlled by soil water content, pressure potential, and relative permeability of the soil. Therefore, the actual evaporation rate is determined by evaporative demand and soil hydraulic properties.

Three stages in the evaporation of water from soils have been characterized (Figure 4-53). In the initial periods of evaporation from a moist soil, the rate of supply is sufficient to meet the evaporation demand, and evaporation proceeds at a constant rate. During the second stage, the falling-rate stage, the evaporation rate falls below the environmental demand. During this stage, the evaporation rate is controlled by the soil's hydraulic properties. During the third stage of evaporation, known as the vapor diffusion stage, the soil is dry enough for liquid water transport to cease, and water losses are due only to vapor transport.

Transpiration
Transpiration is the loss of vapor through plants in response to a vapor pressure deficit between leaves and the atmosphere. To meet this condition, plants must extract water from the root zone. When the soil surface is covered completely by a crop canopy, evaporation losses are negligible, and transpiration is the principal process by which water is lost from the root zone. The environmental factors controlling transpiration are the same as for evaporation, which are given above.

Evapotranspiration
Combined effects due to evaporation and transpiration are commonly referred to as evapotranspiration. In areas where the water table is near the land surface, water can be discharged from the

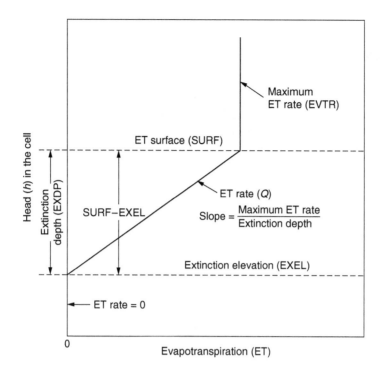

FIGURE 4-53 Schematic representation of MODFLOW evapotranspiration approach. (Adapted from McDonald, M.G. and Harbaugh, A.W., *Open File Report 83–875*, U.S. Geological Survey, National Center, Reston, Virginia, 1984.)

ground water system to the atmosphere by the process of evapotranspiration. This includes both evaporation from the capillary fringe above the water table and transpiration from plants whose roots penetrate to the water table or the capillary fringe.

Representation of Evapotranspiration in a Ground Water Flow Model
A common way to represent evapotranspiration in a ground water flow model is to treat it as a linear function of depth. If the head is at or above land surface, a maximum evapotranspiration rate is specified. The evapotranspiration rate decreases linearly with depth d_{et} (the depth below land surface at which evapotranspiration ceases). Therefore, as the head falls below the land surface, evapotranspiration decreases, and when the head coincides with d_{et}, evapotranspiration ceases.

Application of the approach presented above requires that two values be assigned. The first one is the maximum evapotranspiration rate and the second one is the depth below land surface where evapotranspiration ceases. One should bear in mind that both of these quantities are difficult to estimate. However, the maximum evapotranspiration rate may be estimated from measured water loss. The SI unit of evapotranspiration rate is meter per second (m/sec), but it is generally calculated on annual basis. Typical values of mean annual water loss in meters per year (m/year) are presented in Table 4-8.

The depth at which evapotranspiration ceases is closely related to the depth of plant roots. Most of the water used for evapotranspiration is extracted from the upper soil layers in which plant roots are concentrated. According to Bouwer (1978), for irrigated crops, as a rule of thumb 40% of evapotranspiration is derived from the top one fourth of the root zone, 30% from the second one fourth, 20% from the third one fourth, and 10% from the bottom one fourth of the root zone. Depths of root zones vary from about 0.5 m for shallow-rooted crops like certain grasses and vegetables, to 1 to 2 m for most field crops, and several meters for small to medium trees. Roots of some trees may go down to 10 m or more.

TABLE 4-8
Observed Water Loss

Valley and State	Area		Length of Record	Mean Annual Water
	Acres	Hectares	(years)	Loss (m/year)
Mesilla, New Mexico	109,000	44,112	13	0.8636
Pacas, New Mexico	37,850	15,176	18	0.8966
Sangamon River, Illinois	1,640,000	663,699	16	0.7417
Green River, Kentucky	5,000,000	2,023,473	6	0.7976
Tallapoosa River, Alabama-Georgia	1,060,000	428,976	13	0.8382
Mad River, Ohio	307,000	124,241	13	0.6553
Skunk River, Iowa	1,890,000	764,873	16	0.6858
North Fork of White River, Missouri	755,000	305,544	14	0.7874
North Platte,	462,000	186,969	9	0.6045
Wyoming – Nebraska				
Black River, Wisconsin	494,000	199,919	13	0.5639
Cypress Creek, Texas	545,000	220,558	13	0.9195
Wagon Wheel Gap,Colorado	222	90	14	0.3962
Michigan-Illinois River, Colorado	43,000	17,402	3	0.4572
West River, Vermont	—	—	8	0.5641
Lake Cochituate,	—	—	30	0.5893
Massachusetts				
Swift River, Massachusetts	—	—	15	0.5867

Adapted from Mercer et al., U.S. Nuclear Regulatory Commission, NUREG/CR-3066, Washington DC, 1982; Eagleson, P.S., *Dynamic Hydrology*, McGraw-Hill, New York, 1970; Original Reference: Hamon, W.R., *Proc. Am. Soc. Civil Eng.*, *J. Hydraul. Div.* 87, 107–120, 1961.

MODFLOW Evapotranspiration Modeling Approach
Expressions for Evapotranspiration

In the MODFLOW program all features for evapotranspiration are handled by its Evapotranspiration (ET) package. Evaportranspiration is the mechanism by which water is removed from the liquid phase to the vapor phase. The ET package simulates the effect of evapotranspiration where the source of water is the saturated porous medium, and it deals primarily with water removed by the roots of plants.

The evapotranspiration rate determined by the ET package of MODFLOW depends on the position of the aquifer head relative to two given evapotranspiration reference elevations. They are: (1) ET surface elevation; and (2) ET extinction elevation (Figure 4-53). For a given model cell, the ET rate decreases to zero as the aquifer head declines to the extinction elevation and is set to zero when the aquifer head drops below this elevation. The ET rate increases to a user-controlled maximum limit as the aquifer head rises above the extinction elevation to the given ET surface elevation. The ET rate is assumed to be proportional to the saturated thickness above the given ET extinction elevation. In MODFLOW, the ET rate is expressed in terms of flow into the aquifer as

$$Q = \begin{cases} 0 & \text{when } h < EXEL & (4\text{-}187) \\ \\ EVTR \, \dfrac{h-(EXEL)}{EXDP} & \text{when } SURF \geq h \geq EXEL & (4\text{-}188) \\ \\ EVTR & \text{when } h > SURF & (4\text{-}189) \end{cases}$$

where $Q(L^3/T)$ is the ET rate, $h(L)$ the head in the aquifer, $EXEL(L)$ the extinction elevation, $SURF$ (L) the ET surface elevation, $EXDP$ (L) the extinction depth ($SURF - EXEL$), and $EVTR$ (L^3/T) the maximum ET rate. From Figure 4-53, the term $EXEL$ can be expressed as

$$EXEL = SURF - EXDP \qquad (4\text{-}190)$$

Therefore, Eqs. (4-187) and (4-188), respectively, take the forms

$$Q = \begin{cases} 0 & \text{when} \quad h < SURF - EXDP & (4\text{-}191) \\ EVTR\dfrac{h-(SURF-EXDP)}{EXDP} & \text{when} \quad SURF \geq h \geq SURF - EXDP & (4\text{-}192) \end{cases}$$

Data Required for Evapotranspiration
The data required to calculate the corresponding terms in the finite-difference equations are the three variables: (1) maximum ET rate ($EVTR$); (2) ET surface elevation ($SURF$); and (3) extinction depth ($EXDP$). Therefore, for each horizontal cell, the user must specify the values of these variables. Unlike the Recharge package of MODFLOW (Section 4.3.15.2), in the ET package of MODFLOW, the maximum ET rate (units L^3/T) is required for each horizontal cell location where evapotranspiration occurs. The values in Table 4-8 represent the maximum ET rates in units of meters per year (m/year). By multiplying them with the horizontal area of a cell, the flow rate Q in the dimensions of L^3/T can be determined.

Options for Evapotranspiration
There are two options (Figure 4-54) for indicating from which layer ET is abstracted at a given horizontal cell location. Under option 1, ET is taken from grid layer 1. Under option 2, ET is taken from the layer specified by the user in a special array called IEVT.

4.3.16 SOLUTION TECHNIQUES FOR FINITE-DIFFERENCE EQUATIONS FOR FLOW

Derivation of the finite-difference equation for three-dimensional ground water flow is presented in Section 4.3.10, and Eq. (4-138) is its final form. The matrix-form equations for each cell node are generated from Eq. (4-138). There are a number of methods for solving the matrix-form equations of ground water flow of the MODFLOW family codes. These methods are *direct methods* and *iterative methods*. Iterative methods include: (1) strongly implicit procedure; (2) slice-successive over-relaxation procedure; and (3) preconditioned conjugate-gradient 2(PCG2) procedure. In the following sections, the matrix-form equations, direct methods, and iterative methods are discussed (Mercer and Faust, 1981; Huyakorn and Pinder, 1983; McDonald and Harbaugh, 1984).

4.3.16.1 Matrix-form Equations

The finite-difference numerical approximation, as presented in Section 4.3.10, leads to an algebraic equation for each cell node. These are combined to form a matrix equation, which is a set of N number of equations having N unknowns. Here, N is the number of nodes. The general form of these equations, written in matrix form, is

$$[A]\mathbf{h} = \mathbf{q} \qquad (4\text{-}193)$$

where $[A]$ is a square matrix, and $\mathbf{h}$ and $\mathbf{q}$ are vectors. The $[A]$ matrix contains coefficients related to grid spacing and aquifer properties, such as horizontal and vertical hydraulic conductivities. The $\mathbf{h}$ vector contains the dependent variables to be determined; for example, head values at each node. The $\mathbf{q}$ vector contains all known information, such as extraction and injection rates, and boundary conditions. Further details regarding Eq. (4-193) can be found in McDonald and Harbaugh (1984).

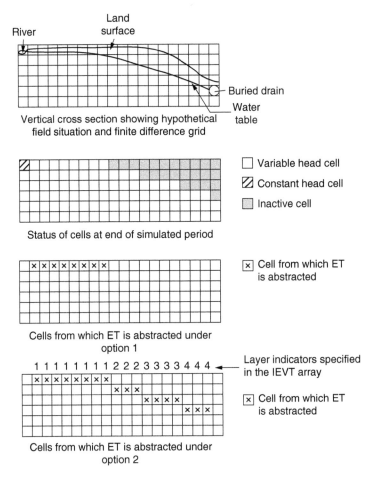

FIGURE 4-54 An example showing from which cells evapotranspiration will be abstracted under the two options available in the ET Package of MODFLOW. (Adapted from McDonald, M.G. and Harbaugh, A.W., *Open File Report 83–875*, U.S. Geological Survey, National Center, Reston, Virginia, 1984.)

Solving the matrix equation is a mathematical problem. However, the hydrologist must be aware of some of its important aspects. This is due to the fact that the matrix solution is the most time-consuming and expensive part of the solution process.

4.3.16.2 Direct Methods

In *direct methods*, a sequence of operations is performed only once, which provides an exact solution to set of equations with round-off errors. These errors depend on the solution algorithm that is used in the solution process. In general, direct methods are based on (1) solution by determinants; (2) solution by successive elimination of the unknowns; and (3) solution by matrix inversion. According to Narasimhan and Witherspoon (1977), perhaps the most widely used direct approach for transient solutions is that of successive elimination and backsubstitution, which includes the Gaussian elimination method (Scarborough, 1966) and the Cholesky decomposition method (Weaver, 1967). The advantages of direct methods are as follows: (1) sequence of operations performed only once; (2) no initial estimates required; (3) no iteration parameters required; and (4) no tolerance required. However, direct methods have two main disadvantages: (1) for large systems, storage requirements and computation time can be potential problems; and (2) direct methods bear round-off errors. The round-off errors are due to the fact that they accumulate for certain types of matrices as a result of many arithmetic operations.

4.3.16.3   Iterative Methods

Iterative methods are attractive because they avoid the need for storing large matrices. Numerous iterative schemes have been developed during the past half century (Mercer and Faust, 1981). Among these schemes are: successive overrelaxation (Varga, 1962), alternating direction implicit procedure (Douglas and Rachford, 1956), iterative alternating direction implicit procedure (Wachpress and Habetler, 1960), and the strongly implicit procedure (Stone, 1968).

Iterative methods start with an initial estimate for the solution. Therefore, the efficiency of the method is dependent on this initial guess. This situation makes the iterative approach less desirable for solving steady-state cases (Narasimhan and Witherspoon, 1977). In the iterative methods, relaxation and acceleration factors are used in order to speed up the solution process. The definition of best values of these factors is often problem-dependent. In addition, iterative solution techniques require that an error tolerance be specified to stop the iterative process. The error tolerance is case-dependent too.

In the MODFLOW code, the solution techniques are based on the iterative methods. These methods are discussed in the following sections.

Strongly Implicit Procedure
The strongly implicit procedure (SIP) is one of the methods used in the MODFLOW code. This is a method for iteratively solving a large system of simultaneous linear equations and was developed by Weinstein et al. (1969). This method is handled in MODFLOW by the SIP package.

Although iteration parameters are important parts of the SIP solution process, how to select the parameters is not well understood. The number of parameters, their values, and the order of cycling through them must be chosen. Weinstein et al. (1969) used the same set of parameters for all three planes and the number of parameters were 4 to 10. Numerous methods of cycling the parameters have been used. Trescott et al. (1976) tested different parameter ordering in two-dimensional problems and found that repeated cycling through the parameters in the order of smallest to largest worked well for a variety of problems. This method of cycling is used in MODFLOW.

In the SIP Package of MODFLOW, "seed" (WSEED) is the most critically important parameter. The choice of this parameter is not straightforward and, accordingly, the user is given the choice of specifying the seed or permitting the model to calculate it. Weinstein et al. (1969) suggested that a trial-and-error method be used to improve the choice of the seed. Thus, provisions have been made to permit the user to specify the seed. By observing the rate of convergence for several different seeds, an optimal seed can be selected.

Slice-Successive Overrelaxation Procedure
Slice-successive overrelaxation (SSOR) is a technique for iteratively solving a system of linear equations. The SSOR package, in particular, solves the finite-difference equations associated with the cells in the finite-difference grid. Instead of solving the entire system of unknowns at the same time, the equations are formulated for a two-dimensional slice (Figure 4-55) under the assumption that the heads in the adjacent two slices are known. Thus, the SSOR package reduces the number of equations by simultaneously solving only those equations representing cells in a single slice. The resulting system of equations is solved by a Gaussian elimination procedure. The slice will usually contain a relatively small number of nodes because in most cases the number of model layers is small. One iteration is complete when all of the slices are processed. After a large number of iterations, the solution converges.

Preconditioned Conjugate-Gradient 2 (PCG2) Procedure
This procedure was not included in the original MODFLOW program, which was released in 1984. It was added to the MODFLOW program by Hill (1990). The PCG2 procedure uses the preconditioned conjugate-gradient method to solve the equations produced by the model for hydraulic head. Using this procedure, linear or nonlinear flow conditions may be simulated. PCG2 includes two preconditioning options: (1) modified incomplete Cholesky preconditioning, which is efficient on scalar computers; and (2) polynomial preconditioning, which requires less computer storage and,

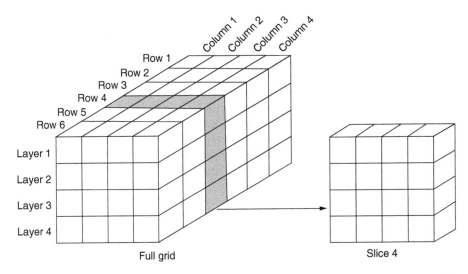

FIGURE 4-55 Conceptual approach of the slice-successive overrelaxation (SSOR) procedure: consider a single vertical slice at a time to reduce the number of equations that must be solved simultaneously. (Adapted from McDonald, M.G. and Harbaugh, A.W., *Open File Report 83–875*, U.S. Geological Survey, National Center, Reston, Virginia, 1984.)

with modifications that depend on the computer used, is most efficient on vector computers. Convergence of the solver is determined using both head change and residual criteria. Nonlinear problems are solved using Picard iterations.

4.4   FINITE-DIFFERENCE MODELING APPROACH FOR SOLUTE TRANSPORT

Modeling the transport of a solute dissolved in ground water involves the solution of coupled partial differential equations for ground water flow and solute transport. The three-dimensional movement of ground water through porous media, assuming constant density and isothermal conditions, is described by Eq. (4-107). Equation (4-107) describes ground water flow under transient conditions in a heterogeneous and anisotropic medium under the condition that the principal axes of the hydraulic conductivity tensor are aligned with the coordinate directions, in which case, the hydraulic conductivity tensor takes the form given by Eq. (4-96).

In the following sections, the finite-difference modeling approach for solute transport is presented especially for MODFLOW family three-dimensional solute transport codes. These codes are FTWORK (Faust et al., 1990), MT3D (Zheng, 1990), and MODFLOWT (Duffield, 1996). In the following sections, the foundations of the three-dimensional solute transport modeling approach will be presented, especially through the MODFLOWT code.

4.4.1   GOVERNING EQUATIONS

4.4.1.1   Governing Differential Equation

From Eqs. (4-47) and (4-96), the Darcy velocity components can be expressed as

$$q_{xx} = - K_{xx}\frac{\partial h}{\partial x} \tag{4-194a}$$

$$q_{yy} = - K_{yy}\frac{\partial h}{\partial y} \tag{4-194b}$$

$$q_{zz} = - K_{zz}\frac{\partial h}{\partial z} \tag{4-194c}$$

The ground water velocities, from Eqs. (2-35) and (4-194), can be expressed as

$$v_x = \frac{q_{xx}}{\varphi_e} = -\frac{1}{\varphi_e}K_{xx}\frac{\partial h}{\partial x} \qquad (4\text{-}195a)$$

$$v_y = \frac{q_{yy}}{\varphi_e} = -\frac{1}{\varphi_e}K_{yy}\frac{\partial h}{\partial y} \qquad (4\text{-}195b)$$

$$v_z = \frac{q_{zz}}{\varphi_e} = -\frac{1}{\varphi_e}K_{zz}\frac{\partial h}{\partial z} \qquad (4\text{-}195c)$$

By substituting Eqs. (4-195) into Eq. (2-47), and by including sink and source terms, one has

$$\frac{\partial}{\partial x}\left(CK_{xx}\frac{\partial h}{\partial x}\right) + \frac{\partial}{\partial y}\left(CK_{yy}\frac{\partial h}{\partial y}\right) + \frac{\partial}{\partial z}\left(CK_{zz}\frac{\partial h}{\partial z}\right) + \frac{\partial}{\partial x}\left(\varphi_e D_{xx}\frac{\partial C}{\partial x} + \varphi_e D_{xy}\frac{\partial C}{\partial y} + \varphi_e D_{xz}\frac{\partial C}{\partial z}\right)$$

$$+ \frac{\partial}{\partial y}\left(\varphi_e D_{yx}\frac{\partial C}{\partial x} + \varphi_e D_{yy}\frac{\partial C}{\partial y} + \varphi_e D_{yz}\frac{\partial C}{\partial z}\right) + \frac{\partial}{\partial z}\left(\varphi_e D_{zx}\frac{\partial C}{\partial x} + \varphi_e D_{zy}\frac{\partial C}{\partial y} + \varphi_e D_{zz}\frac{\partial C}{\partial z}\right)$$

$$= \varphi_e R_d\frac{\partial C}{\partial t} - C'W^+ - CW^- + vR_d\varphi_e C \qquad (4\text{-}196)$$

Using tensorial notation, Eq. (4-196) takes the form

$$\frac{\partial}{\partial x_i}\left(CK_{ij}\frac{\partial h}{\partial x_j}\right) + \frac{\partial}{\partial x_i}\left(\varphi_e D_{ij}\frac{\partial C}{\partial x_j}\right) = \varphi_e R_d\frac{\partial C}{\partial t} - C'W^- - CW^+ + vR_d\varphi_e C \qquad (4\text{-}197)$$

where $C(M/L^3)$ is the concentration of the solute in ground water, φ_e the effective porosity; D_{ij} (L^2/T) are values of the hydrodynamic dispersion tensor, R_d the retardation factor, $C'(M/L^3)$ the concentration of a source of water, $W^+(T^{-1})$ the volumetric flux per unit volume of a source of water, $W^-(T^{-1})$ the volumetric flux per unit volume of a sink of water; and $v(T^{-1})$ the first-order decay or degradation constant. In general, in the mathematical model, φ_e, D_{ij}, R_d, and v are functions of space such that $\varphi_e = \varphi_e(x, y, z)$, $D_{ij} = D_{ij}(x, y, z)$, $R_d = R_d(x, y, z)$; and $v = v(x, y, z)$.

The ground water velocity components, as given by Eqs. (4-195), can be written in tensorial notation as follows:

$$v_i = -\frac{1}{\varphi_e}K_{ii}\frac{\partial h}{\partial x_i} \qquad (4\text{-}198)$$

which are defined along the principal coordinate directions. For this case, the hydraulic conductivity tensor is given by Eq. (4-96). The magnitude of the ground water velocity is given by Eq. (2-138).

4.4.1.2 Hydrodynamic Dispersion Coefficients

The mechanical dispersion coefficients are given by Eqs. (2-139)–(2-144). These expressions were altered slightly by Burnett and Frind (1987) by introducing the vertical transverse dispersivity, α_{TV}. Therefore, Eqs. (2-139)–(2-144) along with Eq. (2-137), take the following forms:

$$D_{xx} = \alpha_L\frac{v_x^2}{v} + \alpha_{TH}\frac{v_y^2}{v} + \alpha_{TV}\frac{v_z^2}{v} + D^* \qquad (4\text{-}199)$$

$$D_{yy} = \alpha_L\frac{v_y^2}{v} + \alpha_{TH}\frac{v_x^2}{v} + \alpha_{TV}\frac{v_z^2}{v} + D^* \qquad (4\text{-}200)$$

$$D_{zz} = \alpha_L\frac{v_z^2}{v} + \alpha_{TH}\frac{v_x^2}{v} + \alpha_{TV}\frac{v_y^2}{v} + D^* \qquad (4\text{-}201)$$

$$D_{xy} = D_{yx} = (\alpha_L - \alpha_{TH})\frac{v_x v_y}{v} \qquad (4\text{-}202)$$

$$D_{xz} = D_{zx} = (\alpha_L - \alpha_{TV})\frac{v_x v_z}{v} \tag{4-203}$$

$$D_{yz} = D_{zy} = (\alpha_L - \alpha_{TV})\frac{v_y v_z}{v} \tag{4-204}$$

where α_L(L) is the longitudinal dispersivity, α_{TH}(L) the transverse horizontal dispersivity, α_{TV} (L) the transverse vertical dispersivity, and D^*(L^2/T) the effective molecular diffusion coefficient. For detailed information about these parameters, the reader should see Section 2.11.

4.4.1.3 Expression for the Retardation Factor

The expression for the retardation factor R_d, which is used in most numerical solute transport models, including MODFLOWT, under linear sorption isotherm conditions, is given by Eq. (2-62).

4.4.1.4 Expression for the Degradation Constant

Degradation refers to both radioactive decay and biological degradation. The general foundations of this phenomena are given in Section 2.6.2. In most solute transport codes including MODFLOWT, Eq. (2-75) is used to simulate the effects of radioactive decay and biological degradation.

4.4.2 FINITE-DIFFERENCE EQUATION

Details regarding the derivation of the finite-difference ground water flow equation are presented in Section 4.3.8. Equation (4-136) is the main expression used in MODFLOW for the numerical solution of the ground water flow equation. The finite-difference approximation of the partial differential equation for solute transport, Eq. (4-197), consists of convective and dispersive terms as follows (Duffield, 1996):

$$[C_{i,j,k}^m(1 - \sigma_{i,j+1/2,k}) + C_{i,j+1,k}^m\sigma_{i,j+1/2,k}]CR_{i,j+1/2,k}(h_{i,j+1,k}^m - h_{i,j,k}^m) -$$

$$[C_{i,j-1,k}^m(1 - \sigma_{i,j-1/2,k}) + C_{i,j,k}^m\sigma_{i,j-1/2,k}]CR_{i,j-1/2,k}(h_{i,j,k}^m - h_{i,j-1,k}^m) +$$

$$[C_{i,j,k}^m(1 - \sigma_{i,j+1/2,k}) + C_{i+1,j,k}^m\sigma_{i+1/2,j,k}]CC_{i+1/2,j,k}(h_{i+1,j,k}^m - h_{i,j,k}^m) -$$

$$[C_{i-1,j,k}^m(1 - \sigma_{i-1/2,j,k}) + C_{i,j,k}^m\sigma_{i-1/2,j,k}]CC_{i-1/2,j,k}(h_{i,j,k}^m - h_{i-1,j,k}^m) +$$

$$[C_{i,j,k}^m(1 - \sigma_{i,j,k+1/2}) + C_{i,j,k+1}^m\sigma_{i,j,k+1/2}]CV_{i,j,k+1/2}(h_{i,j,k+1}^m - h_{i,j,k}^m) -$$

$$[C_{i,j,k-1}^m(1 - \sigma_{i,j,k-1/2}) + C_{i,j,k}^m\sigma_{i,j,k-1/2}]CV_{i,j,k-1/2}(h_{i,j,k}^m - h_{i,j,k-1}^m) +$$

$$DRR_{i,j+1/2,k}(C_{i,j+1,k}^m - C_{i,j,k}^m) - DRR_{i,j-1/2,k}(C_{i,j,k}^m - C_{i,j-1,k}^m) +$$

$$DRC_{i,j+1/2,k}[C_{i+1,j+1,k}^m(1 - \beta_{r_{j+1/2}}) + C_{i+1,j,k}^m{}^m\beta_{r_{j+1/2}} - C_{i-1,j+1,k}^m(1 - \beta_{r_{j+1/2}}) - C_{i-1,j,k}^m\beta_{r_{j+1/2}}] +$$

$$DRC_{i,j-1/2,k}[C_{i+1,j-1,k}^m\beta_{r_{j-1/2}} + C_{i+1,j,k}^m(1 - \beta_{r_{j-1/2}}) - C_{i-1,j-1,k}^m\beta_{r_{j-1/2}} - C_{i-1,j,k}^m(1 - \beta_{r_{j+1/2}})] +$$

$$DRV_{i,j+1/2,k}[C_{i,j+1,k+1}^m(1 - \beta_{r_{j+1/2}}) + C_{i,j,k+1}^m\beta_{r_{j+1/2}} - C_{i,j+1,k-1}^m(1 - \beta_{r_{j+1/2}}) - C_{i,j,k-1}^m\beta_{r_{j+1/2}}] +$$

$$DRV_{i,j-1/2,k}[C_{i,j-1,k+1}^m\beta_{r_{j-1/2}} + C_{i,j,k+1}{}^m(1 - \beta_{r_{j-1/2}}) - C_{i,j-1,k-1}^m\beta_{r_{j-1/2}} - C_{i,j,k-1}^m(1 - \beta_{r_{j-1/2}})] +$$

$$DCC_{i+1/2,j,k}(C_{i+1,j,k}^m - C_{i,j,k}^m) - DCC_{i-1/2,j,k}(C_{i,j,k}^m - C_{i-1,j,k}^m) + \tag{4-205}$$

$$DCR_{i+1/2,j,k}[C_{i+1,j+1,k}^m(1 - \beta_{c_{i+1/2}}) + C_{i,j+1,k}^m\beta_{c_{i+1/2}} - C_{i+1,j-1,k}^m(1 - \beta_{c_{i+1/2}}) - C_{i,j-1,k}^m\beta_{r_{i+1/2}}] +$$

$$DCR_{i-1/2,j,k}[C_{i-1,j+1,k}^m\beta_{c_{i-1/2}} + C_{i,j+1,k}^m(1 - \beta_{c_{i-1/2}}) - C_{i-1,j-1,k}^m\beta_{c_{i-1/2}} - C_{i,j-1,k}^m(1 - \beta_{c_{i-1/2}})] +$$

$$DCR_{i+1/2,j,k}[C_{i+1,j,k+1}^m(1 - \beta_{c_{j+1/2}}) + C_{i,j,k+1}^m\beta_{c_{i+1/2}} - C_{i+1,j,k-1}^m(1 - \beta_{c_{j+1/2}}) - C_{i,j,k-1}^m\beta_{c_{i+1/2}}] +$$

$$DCR_{i-1/2,j,k}[C_{i-1,j,k+1}^m\beta_{c_{i-1/2}} + C_{i,j,k+1}^m(1 - \beta_{c_{i-1/2}}) - C_{i-1,j,k-1}^m\beta_{c_{i-1/2}} - C_{i,j,k-1}^m(1 - \beta_{c_{i-1/2}})] +$$

$$DVV_{i,j,k+1/2}(C_{i,j,k+1}^m - C_{i,j,k}^m) - DVV_{i,j,k-1/2}(C_{i,j,k}^m - C_{i,j,k-1}^m) +$$

$$DVR_{i,j,k+1/2}[C_{i,j+1,k+1}^m(1 - \beta_{v_{i,j+1,k+1/2}}) + C_{i,j+1,k}^m\beta_{v_{i,j+1,k+1/2}} - C_{i,j-1,k+1}^m(1 - \beta_{r_{i,j-1,k+1/2}}) - C_{i,j-1,k}^m\beta_{r_{i,j-1,k+1/2}}]$$

$$DVR_{i,j,k-1/2}[C_{i,j+1,k-1}^m\beta_{v_{i,j+1,k-1/2}} + C_{i,j+1,k}^m(1 - \beta_{v_{i,j+1,k-1/2}}) - C_{i,j-1,k-1}^m\beta_{v_{i,j-1,k-1/2}} - C_{i,j-1,k}^m(1 - \beta_{r_{i,j-1,k+1/2}})]$$

$$DVR_{i,j,k+1/2}[C_{i+1,j,k+1}^m(1 - \beta_{v_{i+1,j,k+1/2}}) + C_{i+1,j,k}^m\beta_{v_{i+1,j,k+1/2}} - C_{i-1,j,k+1}^m(1 - \beta_{v_{i-1,j,k+1/2}}) - C_{i-1,j,k}^m\beta_{v_{i,j-1,k+1/2}}]$$

$$DVR_{i,j,k-1/2}[C_{i,j+1,k-1}^m\beta_{v_{i+1,j,k-1/2}} + C_{i,j+1,k}^m(1 - \beta_{v_{i+1,j,k-1/2}}) - C_{i-1,j,k-1}^m\beta_{v_{i-1,j,k-1/2}} - C_{i-1,j,k}^m(1 - \beta_{v_{i-1,j,k-1/2}})]$$

$$= \Delta r_j \Delta c_i \Delta v_{i,j,k}[\varphi_{e_{i,j,k}}R_{d_{i,j,k}}\frac{(C_{i,j,k}^m - C_{i,j,k}^{m-1})}{(t_m - t_{m-1})} - C_{i,j,k}'W_{i,j,k}^+ - C_{i,j,k}^mW_{i,j,k}' + \lambda_{i,j,k}R_{d,i,j,k}\varphi_{e_{i,j,k}}C_{i,j,k}^m$$

where *DRR*, *DRC*, *DRV*, *DCC*, *DCR*, *DCV*, *DVV*, *DVR*, and *DVC* (L^3/T) are the dispersive flux coefficients evaluated at cell interfaces, σ is the weighting factor, and β_r, β_c, β_v are interpolation factors.

Equation (4-205) approximates the time derivative in the finite-difference equation for solute transport using a backward-in-time formulation. MODFLOWT also provides a central-in-time approximation of the time derivative. The following equation expresses the relationship between the backward and central-in-time approximations used for the time derivative:

$$C_{i,j,k}^{m+1} = (1 - \theta)C_{i,j,k}^m + \theta C_{i,j,k}^{m+1} \tag{4-206}$$

where θ is a time-weighting factor that assumes values of 0.5 and 1, which correspond to central and backward-in-time approximations, respectively.

The coefficients *DRR*, *DRC*, *DRV*, *DCC*, *DCR*, *DCV*, *DVV*, *DVR*, and *DVC* are computed at cell interfaces using components of the hydrodynamic dispersion tensor and dimensions of the finite-difference grid system. For example, the hydrodynamic dispersion coefficient D_{xx} evaluated at the $(i, j+1/2, k)$ interface is given by the following expression:

$$DRR_{i,j+1/2,k} = D_{xx_{i,j+1/2,k}}\frac{2\Delta c_i \Delta v_{i,j,k}}{\Delta r_j + r_{j+1}} \tag{4-207}$$

Cross-product components are computed for the $(i, j + 1/2, k)$ interface as follows:

$$DRC_{i,j+1/2,k} = D_{xy_{i,j+1/2,k}}\frac{\Delta c_i \Delta v_{i,j,k}}{0.5\Delta c_{i-1} + \Delta c_i + 0.5\Delta c_{i+1}} \tag{4-208}$$

$$DRV_{i,j+1/2,k} = D_{xz_{i,j+1/2,k}}\frac{\Delta c_i \Delta v_{i,j,k}}{0.5\Delta v_{i,j,k-1} + \Delta v_{i,j,k} + 0.5\Delta v_{i,j,k+1}} \tag{4-209}$$

The remaining six coefficients appearing in Eq. (4-205) are computed similarly.

The interpolation factors, β_r, β_c, and β_v, are used to interpolate velocities, dispersivities, and effective molecular diffusion coefficients at cell interfaces. The formulas for computing the interpolation factors are as follows:

$$\beta_{r_{j+1/2}} = \frac{\Delta r_{j+1}}{\Delta r_{j+1} + \Delta r_j} \tag{4-210}$$

$$\beta_{c_{i+1/2}} = \frac{\Delta c_{i+1}}{\Delta c_{i+1} + \Delta c_i} \tag{4-211}$$

$$\beta_{v_{i,j,k+1/2}} = \frac{\Delta v_{i,j,k+1}}{\Delta v_{i,j,k+1} + \Delta v_{i,j,k}} \tag{4-212}$$

The computation of Darcy velocities perpendicular to cell interfaces is straightforward. For example, the velocity along the rows perpendicular to the $(i,j + 1/2,k)$ interface is calculated as

$$VR_{i,j+1/2,k} = CR_{i,j+1/2,k}(h_{i,j+1,k}^m - h_{i,j,k}^m) \tag{4-213}$$

The interpolation factors are used in the computation of Darcy velocity components parallel to cell interfaces. For example, the velocity component along the column parallel to the $(i,j + 1/2,k)$ interface is interpolated as follows:

$$VC_{i,j+1/2,k} = \frac{1}{2}[\beta_{r_{j+1/2}}(VC_{i-1/2,j+1,k}+VC_{i+1/2,j+1,k})+(1-\beta_{r_{j+1/2}})(VC_{i-1/2,j,k}+VC_{i+1/2,j,k})] \tag{4-214}$$

The procedures for computing velocities perpendicular and parallel to the cell interfaces are used to compute the components of the hydrodynamic dispersion tensor with Eqs. (4-199)–(4-204).

The interpolation factors given by Eqs. (4-210)–(4-212) are also used in interpolating dispersivities as well as the product of effective porosity and effective molecular diffusion coefficients at cell interfaces. For example,

$$\alpha_{L_{i,j+1/2,k}} = \alpha_{L_{i,j,k}}(1 - \beta_{r_{j+1/2}}) + \alpha_{L_{i,j+1,k}}\beta_{r_{j+1/2}} \tag{4-215}$$

$$(\varphi_e D^*)_{i,j+1/2,k} = (\varphi_e D^*)_{i,j,k}(1 - \beta_{r_{j+1/2}}) + (\varphi_e D^*)_{i,j+1,k}\beta_{r_{j+1/2}} \tag{4-216}$$

The velocity-weighting factors, σ, allow upstream or central-difference spatial weighting of the convective terms in Eq. (4-205). For central-in-space weighting, the velocity-weighting factors are equal to the interpolation factors given by Eqs. (4-210)–(4-212). For upstream or backward difference spatial weighting, σ is determined by upstream values of hydraulic head in adjacent grid blocks. For example, the value of σ along the rows at the $(i,j + 1/2,k)$ interface is computed as

$$\sigma_{i,j+1/2,k} = \begin{cases} 1 & \text{if } h_{i,j+1,k} \geq h_{i,j,k} \\ 0 & \text{if } h_{i,j+1,k} < h_{i,j,k} \end{cases} \tag{4-217}$$

Upstream weighting factors for the remaining block interfaces are computed in a similar way.

4.4.3 INITIAL AND BOUNDARY CONDITIONS

The initial and boundary conditions for solute transport modeling are presented in Section 2.10. Some additional points for finite-difference numerical solute transport modeling are presented below by referring to MODFLOWT.

For each cell in the model grid, MODFLOWT provides options for simulating constant concentration, inactive, and variable concentration conditions. Each type of condition is specified for every cell at the start of a simulation. At constant concentration cells, the concentration remains fixed at a specified value throughout a simulation. No concentrations are simulated for inactive cells that represent the finite-difference grid outside the modeling area. All other cells are variable concentration cells, where the model computes values of concentration for every time step in the modeling process.

4.4.4 DISCRETIZATION OF SPACE

The general principles of discretization of space for solute transport are the same as for ground water flow modeling, presented in Section 4.3.12.

4.4.5 DISCRETIZATION OF TIME

The general principles of discretization of time for solute transport are the same as for ground water flow modeling, presented in Section 4.3.13. Some additional points regarding the time step for solute transport modeling are presented below.

For solute transport simulations, the appropriate time step should be met. The criterion for time step is the Courant number Cr, which is defined as

$$Cr = \frac{v\Delta t}{\Delta l} \qquad (4\text{-}218)$$

where v is the retarded interstitial velocity between any two blocks, Δt the time step length, and Δl the maximum distance between two adjacent grid block centers. The retarded ground water velocity is defined by

$$v = \frac{q}{\varphi_e R_d} \qquad (4\text{-}219)$$

where q is the Darcy velocity.

For stable solutions, the grid spacing and time step should generally be selected such that $Cr \leq 1$ (Huyakorn and Pinder, 1983). This means that the solute should not move any further than one grid block per time step. If large grid blocks exist where solute is not expected to travel, this criterion need only be met where the solute plume is significant.

4.4.6 SOLUTION TECHNIQUES FOR FINITE-DIFFERENCE EQUATIONS FOR SOLUTE TRANSPORT

The general principles of the solution techniques for the finite-difference equations are given in Section 4.3.16. Here, the two solution methods used in MODFLOWT will be briefly described.

4.4.6.1 Slice-Successive Overrelaxation (SSOR) Procedure

As described in Section 4.3.16.3, the slice-successive overrelaxation procedure (SSOR) solves large systems of equations by means of iteration. The SSOR package included in MODFLOW solves systems of linear equations having a symmetric coefficient matrix, the type of coefficient matrix generated by the finite-difference equations for ground water flow. For the solution of the finite-difference equations for solute transport, MODFLOWT extends the SSOR package implemented in MODFLOW to include procedures for solving systems of linear equations with a non-symmetric coefficient matrix.

The SSOR package for solute transport divides the finite-difference grid into vertical slices that are oriented along the rows of the grid system. The SSOR procedure solves for concentrations within each slice by the Gauss–Doolittle procedure, a direct method for banded, nonsymmetric coefficient matrices (Weaver, 1967). When solving each slice, SSOR substitutes the most recently computed values of concentration as known quantities for terms in adjacent slices. The values of concentration change computed for each slice are multiplied by a relaxation factor, ω, that dampens or accelerates the rate of convergence by SSOR. Accelerate convergence occurs when ω values are between 1 and 2, dampen convergence occurs when $\omega < 1$, and no acceleration occurs when $\omega = 1$.

Adjustment of the relaxation parameter for convergence is often required to achieve a satisfactory rate of convergence. A trial-and-error procedure may be applied to determine an optimal value of ω for a given problem. MODFLOWT includes an option for the solute transport SSOR solver that computes the optimal value of relaxation factor, ω_{opt}, as follows:

$$\omega_{opt} = \frac{2}{1+(1+\rho)^{1/2}} \qquad (4\text{-}220)$$

where ρ is the spectral radius. By performing a number of iterations with $\omega = 1$, an estimate of the spectral radius is obtained as follows:

$$\rho = \frac{\Delta C_{i,j,k}{}^{n}}{\Delta C_{i,j,k}{}^{n-1}} \qquad (4\text{-}221)$$

where ΔC is the change in concentration for an iteration, and n and $n-1$ indicate iteration level. When the value of ω converges to an acceptable tolerance, the SSOR iterations continue with the value of ω computed from Eq. (4-220).

4.4.6.2 Orthomin Procedure

The ORTHOMIN (OMN) package is an iterative solution technique for the solution of large sets of simultaneous linear equations that have nonsymmetric coefficient matrices. The method consists of an approximate factorization method for the coefficient matrix followed by an orthogonalization–minimization technique to accelerate convergence. The ORTHOMIN procedure is similar to conjugate gradient methods (see Section 4.3.16.3) for solving systems of equations having a symmetric coefficient matrix developed by Hill (1990). The ORTHOMIN package of MODFLOWT also solves the system of equations arising from the finite-difference approximation of the ground-water flow equation, which has a symmetric coefficient matrix.

4.5 APPLICATION OF NUMERICAL FLOW AND SOLUTE TRANSPORT MODELS

4.5.1 INTRODUCTION

The numerical models used in ground water flow and solute transport studies are general computer programs, and they can be applied to a variety of hydrogeologic conditions. As shown in Sections 4.2 and 4.3, these programs are based on the state-of-the-art numerical solution techniques for the governing partial differential equations of ground water flow and solute transport. Usage of these models requires an understanding of the physical problem and data based on field investigations. In these programs, program input data and output results are all quantitative. However, the appropriate application of these models is partly a subjective procedure. This is mainly due to the fact that understanding of the physical system and its associated available field data will always fall short of what is desired. As the data and information become less, the subjectivity becomes greater. If hydrogeologic data were available for each block of the modeling domain and the boundary conditions along each segment of the modeling domain were known, the subjectivity associated with the modeling application would either be reduced significantly, or there would not be any subjectivity at all. The existence of this ideal case is unlikely for most sites. The modeling efforts and validity level of the modeling application for a given site significantly depends on these factors.

The term "subjectivity" can best be explained by an example. Consider that a modeling study is to be conducted for a hydrogeologic site for which some field data and information are available. The goals of the modeling study are established and more than one separate competent group is selected to conduct modeling studies for the same site. Each group will first develop a conceptual model (see Section 4.5.2.1) based on the available qualitative and quantitative site information. Then, the modeling domain, grid spacings, number of layers, boundary conditions, etc., will be specified. Model calibration and application will follow. It is obvious that each group will fill the data gaps using some assumptions. Under these conditions, one cannot expect that the results of each group will be the same. In other words, the results will bear some subjectivity. Regarding this situation, Mercer and Faust (1981) state:

> The effective application of numerical models to field problems in ground water hydrology is ironically qualitative procedure. The hydrologist must first decide whether a numerical model is necessary for project objectives. (...). Because available data are never as comprehensive as desired, he (the modeler) will probably have to fill data gaps with estimated, interpolated or extrapolated values. Although running the computer program is fairly straightforward, interpreting or analyzing the output can be very difficult.

A similar view by van der Heijde (1987) is as follows:

> The science of ground water flow and solute transport is not yet an exact field of knowledge. Although the physical processes involved obey known mathematical and physical principles, exact aquifer and

contaminant characteristics are hard to obtain and often make even plume definition a difficult task. However, where these characteristics have been reasonably established, ground water models may provide a viable, if not the only, method to predict contaminant transport, to locate areas of potential environmental risk, and to assess possible remedial/corrective actions.

4.5.2 GROUND WATER FLOW MODEL APPLICATION

The general practice for ground water flow modeling application is summarized in the chart in Figure 4-56. In the following sections, some details for ground water flow model applications will be presented based on Figure 4-56.

4.5.2.1 Model Conceptualization and Dimensionality

Model conceptualization is an important step regarding the establishment of the foundational features for the model of a given site. Model dimensionality is heavily dependent on the conceptual model. Once the requirements of these two major steps are met, the appropriate computer codes (or programs) can be selected. These steps are discussed below.

Conceptual Model Development
Once the available site-specific data of a site are compiled, the development of a conceptual model can be initiated. The purpose of a conceptual model is to present the hydrogeologic setting of the site in a simplified form without seriously altering the existing ground water flow and contaminant transport mechanisms at the site. The reason for this conceptualization is twofold: first, the conceptual model condenses large quantities of field data into a concise format that provides a clearer understanding of the fundamental processes occurring beneath the site; and second, the conceptual model is used as an intermediate step from which the actual conditions present beneath the site can be characterized within the constraints of the numerical model. This simplified characterization is then used to construct the ground water flow model.

Development of a conceptual model depends on (see Figure 4-56) (1) objectives of modeling study; (2) surface water bodies; (3) existing extraction, injection, and monitoring wells; (4) extent of contaminant plume(s); (5) type of remedial alternatives (capping, cutoff walls, etc.); (6) numerical stability requirements; (7) bedrock outcrop areas; (8) expected influence zones of extraction and injection wells; and (9) assumptions made for simplifications.

For a given aquifers–aquitards system, the flow in the aquifers and aquitards must be analyzed very carefully. This analysis can potentially simplify the vertical discretization of the numerical model. The general principles for vertical discretization are included in Section 4.3.12.2. As mentioned in Section 4.3.12.2, when the hydraulic conductivities of two aquifers are two or more orders of magnitude greater than that of the aquitard between them, the flow is vertical in the aquitard and horizontal in the aquifers. This situation can simplify the model such that no-flow boundary conditions can be assigned to the boundaries of aquitards.

Selection of Model Dimensionality
A two-dimensional model in the horizontal plane has only one layer; a three-dimensional model has a minimum of two layers. As shown in Figure 4-56, the number of layers depends on (1) objectives of the modeling study; (2) site hydraulics and hydrology; (3) site geology and hydrogeology; (4) screening intervals of the wells; (5) available modeling codes; (6) available data; (7) complexity of the site conditions; (8) type of remedial actions being considered; (9) required level of detail; (10) available funds; and (11) available time period.

Selection of Computer Code(s)
There are a variety of computer codes available publicly. Proprietary codes are available as well. As shown in Figure 4-56, code selection for a ground water flow model depends on (1) status of water density (constant or variable); (2) verification status of the candidate codes; (3) numerical algorithms

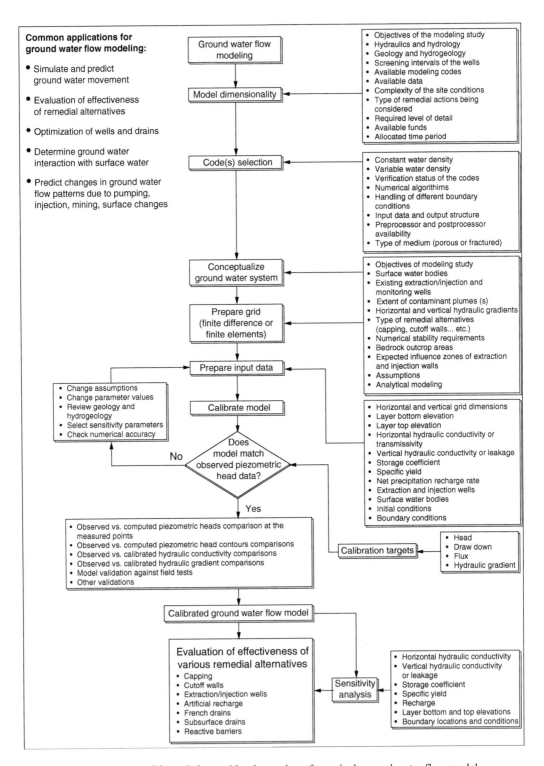

FIGURE 4-56   The steps of formulating and implementing of a typical ground water flow model.

used in the model; (4) handling of different boundary conditions; (5) input data and output structure; (6) preprocessor and postprocessor availability; and (7) type of medium (porous or fractured).

4.5.2.2 Model Construction

The general principles of discretization of a numerical model in space and in time are presented in Sections 4.3.12 and 4.3.13, respectively. Discretization in space includes vertical grid discretization (Section 4.3.12.2) and horizontal grid discretization (Section 4.3.12.3). Once the grid preparation process is complete, input data preparation can be started.

As shown in Figure 4-56, the required data for a ground water flow model are: (1) horizontal grid dimensions; (2) vertical grid dimensions; (3) layer bottom elevations; (4) layer top elevations; (5) horizontal hydraulic conductivities or transmissivities; (6) vertical hydraulic conductivities or leakages; (7) storage coefficients; (8) specific yields; (9) net precipitation recharge rates; (10) extraction and injection wells (locations and rates); (11) initial conditions; and (12) boundary conditions.

In order to prepare the data listed above, the potential sources are as follows: (1) hydrogeologic maps showing areal extent, boundaries, and boundary conditions of all aquifers and aquitards at the site; (2) topographic maps showing all surface water bodies; (3) water table, bedrock, and saturated thickness maps; (4) thickness maps for confined aquifers and confining layers (aquitards) and their boundaries; and (5) maps showing potential precipitation recharge areas.

There are publicly available softwares that allow modelers to interactively design horizontal and vertical model grids and enter the input data mentioned above.

4.5.2.3 Model Calibration

From a ground water flow model, the head variation in different formations of an aquifer system can be determined on the basis of specified hydrogeologic characteristics, boundaries, and stresses. Since ground water velocities are determined from the head distribution, and both convective transport and hydrodynamic dispersion are dependent of the ground water velocity, the accuracy of the solute transport model is heavily dependent on the accuracy of the ground water flow model. Therefore, the success of the ground water flow model calibration for a site is an important foundational success for solute transport modeling efforts for the same site. Calibration of a ground water flow model can be started once the model construction is complete. Definitions for calibration, calibration targets, and evaluation of calibrated model results will be presented (Mercer and Faust, 1981) in the following paragraphs.

Definitions for Calibration
Ground water flow model calibration is the process of adjusting selected model parameters within an expected range until the differences between model-predicted heads and field-observed heads are within the selected criteria for performance. Matching the model head results with field-observed head values is called *ground water flow model calibration*. The calibration process starts with initial estimates for the aquifer's hydrogeologic parameters, and this constitutes the first step in a trial-and-error procedure known as calibration or *history matching*. The model calibration procedure is used to refine initial estimates of aquifer hydrogeologic parameters until the selected criteria are satisfied. The calibration process for a ground water flow model is shown at the center of Figure 4-56. If the modeler finds it difficult to satisfy the criteria, he or she should (1) change the assumptions; (2) change parameter values; (3) review the site geology and hydrogeology; (4) select sensitivity parameters; and (5) check numerical accuracy. Historically, model calibration efforts started by trial and error in the 1960s and still this procedure is widely being used despite the fact that since the early 1990s some programs such as MODFLOWP (see, e.g., Hill, 1992) have been developed for hydrogeologic parameter estimation, but this kind of programs are still in the development stage. More observed data generally make it more time consuming, and sometimes more difficult, even impossible, to calibrate a ground water flow model.

Calibration Targets

A *calibration target* is an observed value that a modeler will attempt to match with model-computed values. Before starting to calibrate a ground water flow model, calibration targets must be specified (Figure 4-56). Calibration targets basically depend on the available observed data and model objectives. For some relatively simple aquifer cases, calibration may not be required.

As shown in Figure 4-56, in general, for a ground water flow model, calibration targets include heads, drawdowns, fluxes, and hydraulic gradients. Head targets are usually ground water levels observed in monitoring wells or piezometers. Drawdowns may be more convenient when trying to match the results of a pumping test. Flux targets are often base flow observations in springs and surface streams. Hydraulic gradients can be derived from the observed head pairs.

In general, a given site, limited observed values exist for hydraulic conductivities, layer bottom elevations, layer top elevations, and boundary conditions. The model further requires net precipitation recharge rate, whereas only total precipitation data are available. Therefore, in order to calibrate a ground water flow model under steady-state conditions, the following quantities are subject to modification: (1) horizontal hydraulic conductivity; (2) vertical hydraulic conductivity; and (3) net precipitation recharge rate. In some cases, formation thicknesses and boundary conditions may have to be modified during the calibration process. If this is the case, as shown in Figure 4-56, the following steps may be followed: (1) change the assumptions; (2) change parameter values; (3) review the site geology and hydrogeology; (4) select sensitivity parameters; and (5) check numerical accuracy.

Sometimes, pumping test data may be available under unsteady-state (transient) conditions. If this is the case, the storage coefficient and specific yield values become part of the parameters. Fortunately, drawdown values are not too sensitive to the ranges of the values of storage coefficient.

Evaluation of Calibrated Model Results

Once a satisfactory match (or a calibrated model) is achieved based on the established calibration targets, the calibrated model results must be evaluated. Regarding a satisfactory match, Mercer and Faust (1981) state:

> No hard and fast rules exist to indicate when a satisfactory match is obtained. The number of "runs" required to produce a satisfactory match depends on the objectives of the analysis, the complexity of the flow system and length of observed history, as well as the patience of the hydrologist. Of course, confidence in any predictive results must be based on (1) a thorough understanding of model limitations; (2) the accuracy of the match with observed historical behavior; and (3) knowledge of data reliability and aquifer characteristics.

Regarding the same topic, Domenico and Schwartz (1990) state:

> Common practice is to set calibration criteria in advance of the calibration exercise. There are no hard and fast rules for what constitutes a good calibration except that errors should be small relative to the total hydraulic head.

A *residual* is the difference between the observed or target value and the model-computed value. During the calibration process, it is attempted to minimize these differences. A common practice is to plot a best-fit line through all data points, as shown in Figure 4-57, and to compare it with the *theoretical ideal line*, which has a geometric slope of 1 and intercept of 0. An ideal calibration occurs when the computed values are all exactly equal to the observed values, which results in a graph where all the points fall on a straight line with a geometric slope of 1 and an intercept of 0. This plot is useful in visualizing the quality of model calibration. For a perfect calibration, the points would lie along a straight line at a 45° angle as shown in Figure 4-57. The degree of scatter around the line determines overall calibration quality; the less scatter, the better quality. Besides this, the residuals for heads and drawdowns are evaluated statistically, and the goodness of a calibrated model is measured generally with the average value of the residuals. Statistical criteria

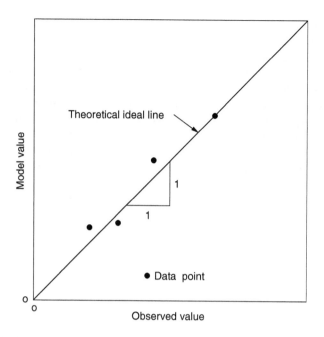

FIGURE 4-57   Evaluation of calibrated data.

may be based on (1) the mean error; (2) the mean absolute error; and (3) the root-mean-squared error (Anderson and Woessner, 1992). These measures of the difference between observed and simulated hydraulic head data are given as

$$\text{Mean error} = \frac{1}{n}\sum_{i=1}^{n}(h_\text{m} - h_\text{s})_i \qquad (4\text{-}222)$$

$$\text{Mean absolute error} = \frac{1}{n}\sum_{i=1}^{n}\left|(h_\text{m} - h_\text{s})_i\right| \qquad (4\text{-}223)$$

$$\text{Root-mean-squared error} = \left[\frac{1}{n}\sum_{i=1}^{n}(h_\text{m} - h_\text{s})_i^2\right]^{1/2} \qquad (4\text{-}224)$$

where n is the number of points where comparisons are made, h_m is the observed hydraulic head at some point i, and h_s is the computed hydraulic head at the same point. Of the three estimates, Anderson and Woessner (1992) point to the root-mean-squared error as the best quantitative measure if the errors are normally distributed. The mean error is not preferred because large positive and negative errors can cancel each other out. A small error estimate may thus hide a poor model calibration. Other requirements are sometimes applied in addition to error estimates, such as quantitatively correct flow directions and flow gradients.

The above-mentioned evaluation process based on Figure 4-57 is necessary, but is not enough. As shown in Figure 4-56, the following evaluations must also be made for a ground water flow model calibration: (1) observed vs. computed piezometric head contours comparison; (2) observed vs. calibrated hydraulic conductivity comparisons; and (3) observed vs. calibrated hydraulic gradient comparisons.

4.5.2.4   Application of A Calibrated Model

Once a calibrated model is achieved for a given site, the calibrated model can be used as a predictive tool to evaluate different cases under steady- and unsteady-state flow conditions. Some potential application areas are shown in Figure 4-56, which include (1) capping; (2) cutoff walls; (3) extraction and injection wells; (4) artificial recharge; (5) french drains; (6) subsurface drains; (7) reactive barriers; and others. In addition to these site-specific applications, models are also used to examine general problems. Hypothetical aquifer problems may be designed to study various types of flow behavior, such as ground water and surface water interactions or flow around a deep radioactive repository. Also, the feasibility of certain proposed mechanisms for observed behavior can be tested. By changing the parameters, it is possible to determine what effect they may have on the overall process. Confidence in predictive results must be based on (1) a clear understanding of model limitations; (2) the accuracy of the match with observed historical behavior; and (3) data reliability knowledge about aquifer characteristics (Mercer and Faust, 1981).

4.5.2.5 Sensitivity Analysis

As shown in Figure 4-56, *sensitivity analysis* is an integral part of the overall modeling process. Sensitivity analysis becomes important as the uncertainty in the key input parameters increases. Sensitivity analysis is generally conducted by varying (1) horizontal hydraulic conductivity; (2) vertical hydraulic conductivity; (3) storage coefficient; (4) specific yield; (5) recharge; (6) layer thickness; and (7) boundary locations and their conditions. Model sensitivity can be expressed as the relative rate of change of selected output caused by unit change in the input. If the change in the input causes a large change in the output, the model is sensitive to that input. Sensitivity analysis parameters should be selected based on the available data and purpose of the modeling study. Of the parameters listed above, hydraulic conductivity is generally the most sensitive parameter for most modeling studies. For low permeable aquifer materials, the recharge rate is generally the most sensitive parameter. Once the sensitivity parameters are selected, their upper and lower limiting values should be selected. Generally, sensitivity analysis is conducted only by varying an input parameter in a run and analyzing the output. Depending on the situation, sometimes more than one input parameter can be changed. But the common practice is to change only one input parameter.

4.5.3 GROUND WATER SOLUTE TRANSPORT MODEL APPLICATION

The general practice for ground water solute transport modeling application is summarized in the chart in Figure 4-58. In the following sections, some details for ground water solute transport model application will be presented based on Figure 4-58.

4.5.3.1 Model Conceptualization and Dimensionality

As mentioned in Section 4.5.2.1, model conceptualization is an important step in the establishment of foundational features for the model of a given site. The ground water model of a site is the foundation of its solute transport model. Therefore, the same conceptual model is a prerequisite for its solute transport modeling efforts as well. Under this framework, in the following sections, some details for ground water solute transport application will be presented based on Figure 4-58.

Conceptual Model Development
In general, the conceptual model developed for the ground water flow modeling portion of the project is used for its solute transport modeling portion as well. However, there is only one significant addition, which is the incorporation of solute sources in the conceptual model for the ground water flow model. All of the assumptions and rationales applied to the ground water flow model are applicable to the conceptual model of the solute transport model.

Selection of Model Dimensionality
The dimensionality of a solute transport model is generally the same as its ground water flow model. In other words, the number of rows, the number of columns, and the number of layers are the same.

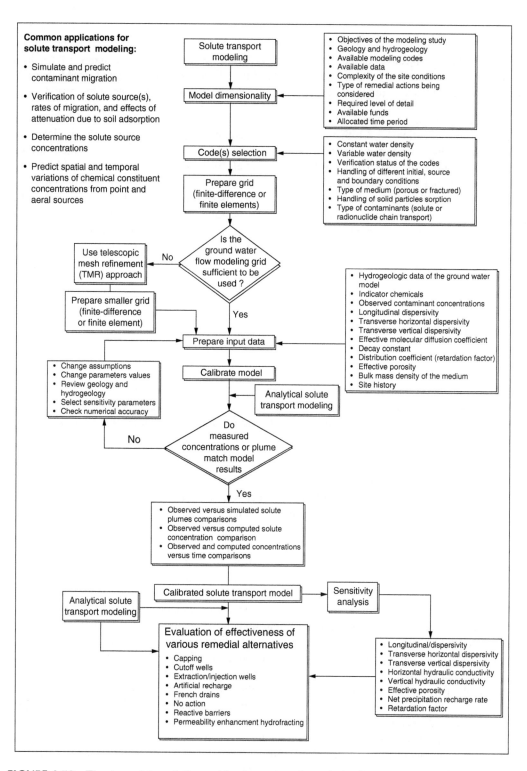

Common applications for solute transport modeling:

- Simulate and predict contaminant migration

- Verification of solute source(s), rates of migration, and effects of attenuation due to soil adsorption

- Determine the solute source concentrations

- Predict spatial and temporal variations of chemical constituent concentrations from point and aeral sources

Solute transport modeling

- Objectives of the modeling study
- Geology and hydrogeology
- Available modeling codes
- Available data
- Complexity of the site conditions
- Type of remedial actions being considered
- Required level of detail
- Available funds
- Allocated time period

Model dimensionality

- Constant water density
- Variable water density
- Verification status of the codes
- Handling of different initial, source and boundary conditions
- Type of medium (porous or fractured)
- Handling of solid particles sorption
- Type of contaminants (solute or radionuclide chain transport)

Code(s) selection

Prepare grid (finite-difference or finite elements)

Is the ground water flow modeling grid sufficient to be used ?

No → Use telescopic mesh refinement (TMR) approach

Prepare smaller grid (finite-difference or finite element)

Yes

Prepare input data

- Hydrogeologic data of the ground water model
- Indicator chemicals
- Observed contaminant concentrations
- Longitudinal dispersivity
- Transverse horizontal dispersivity
- Transverse vertical dispersivity
- Effective molecular diffusion coefficient
- Decay constant
- Distribution coefficient (retardation factor)
- Effective porosity
- Bulk mass density of the medium
- Site history

Calibrate model

Analytical solute transport modeling

- Change assumptions
- Change parameters values
- Review geology and hydrogeology
- Select sensitivity parameters
- Check numerical accuracy

Do measured concentrations or plume match model results

No

Yes

- Observed versus simulated solute plumes comparisons
- Observed versus computed solute concentration comparison
- Observed and computed concentrations versus time comparisons

Analytical solute transport modeling

Calibrated solute transport model → Sensitivity analysis

Evaluation of effectiveness of various remedial alternatives
- Capping
- Cutoff wells
- Extraction/injection wells
- Artificial recharge
- French drains
- No action
- Reactive barriers
- Permeability enhancment hydrofracting

- Longitudinal/dispersivity
- Transverse horizontal dispersivity
- Transverse vertical dispersivity
- Horizontal hydraulic conductivity
- Vertical hydraulic conductivity
- Effective porosity
- Net precipitation recharge rate
- Retardation factor

FIGURE 4-58   The steps of formulating and implementing of a typical solute transport model.

However, in a solute transport model, a lesser number of layers, rows, and columns than the flow model may also be used. As shown in Figure 4-58, the dimensionality of a solute transport model depends on (1) the objectives of the modeling study; (2) site geology and hydrogeology; (3) available modeling codes; (4) available data; (5) complexity of the site conditions; (6) type of remedial actions being considered; (7) required level of detail; (8) available fund; and (9) allocated time period.

Selection of Computer Codes

As mentioned in Section 4.5.2.1, there are a variety of computer codes available publicly. Generally, a numerical solute transport code has a flow modeling portion as well. Typical examples are the FTWORK (Faust et al., 1990) and SWIFT (Reeves et al., 1986) family of codes. Codes like MT3D (Zheng, 1990) and MODFLOWT (Duffield, 1996) are based on the ground water flow modeling code of MODFLOW (McDonald and Harbaugh, 1984). As shown in Figure 4-58, code selection for a ground water solute transport depends on (1) status of water density; (2) verification status of the candidate codes; (3) numerical algorithms; (4) handling of initial, source, and boundary conditions; (5) type of medium (porous or fractured); (6) handling of solid particles sorption; and (7) type of contaminants (solute or radio nuclide chain transport).

4.5.3.2 Model Construction

The principles of model construction for ground water flow (Section 4.5.2.2) are also applicable to solute transport modeling. As mentioned above, the same model grid system for ground water flow can also be used for solute transport. As shown in Figure 4-58, the question to be answered for the ground water flow model grid is as follows: is the ground water flow modeling grid sufficient to be used for solute transport? In some cases, the ground water flow modeling grid cannot be used for solute transport due to the large size of the modeling area. As mentioned in Section 4.3.12.3, in this situation, it may not be possible to satisfy the criteria for the Peclet number (Eq. [4-150]) criteria. If the existing ground water flow model grid cannot be used, an alternative approach is to use the Telescopic Mesh Refinement approach, which is discussed below.

Telescopic Mesh Refinement Approach

The Telescopic Mesh Refinement (TMR) approach was proposed by Ward et al. (1987). This approach provides a means of accurately incorporating regional controlling factors into smaller model domains and also increasing grid resolution in areas of critical importance. Two primary problems associated with the application of solute transport models to hydrogeologic systems are: (1) the model domain is many times larger than that of the contaminated site for correct representation of the ground water flow system; and (2) data point density decreases drastically outside the area proximate to the contaminated site. The application of a three-dimensional solute transport model to a regional flow system can be prohibitive because of computer costs, particularly if proper grid spacing is used to meet numerical constraints and ensure accuracy. In addition, the application may be technically unjustified because of sparse data distribution. A more appropriate modeling approach for this type of application is a structured grid refinement method such as the TMR approach (Figure 4-59). The interlacing of grids is a common procedure that allows investigators to focus on the pertinent area of interest for a particular problem. Examples of areal refinement are typically two-dimensional, with values of hydraulic heads from a regional flow simulation graphically interpolated and applied as prescribed boundary conditions on a local flow model. Grid interlacing has also been used to abstract values of hydraulic head from vertically averaged areal models and to refine simulations in vertical cross-sections. The local grid, defined from the three-dimensional streamlines, is then used for solute transport analysis.

Required Data for Solute Transport Model

As shown in Figure 4-58, the required data for a solute transport model are: (1) the grid and data for the corresponding ground water flow model; (2) indicator chemicals; (3) observed contaminant

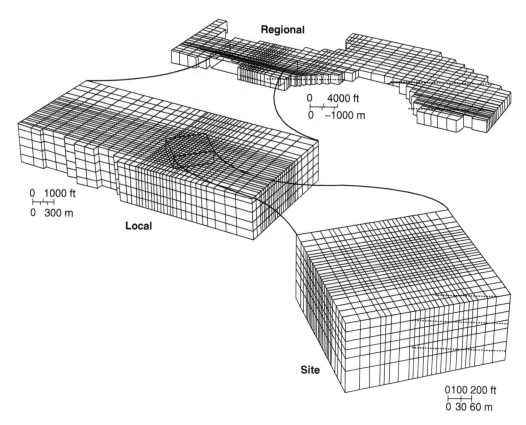

FIGURE 4-59   Conceptual diagram of the telescopic mesh refinement modeling approach. (Adapted from Ward, D.S. et al., *Water Resour. Res.*, 23, 603–617, 1987.)

concentrations; (4) longitudinal and transverse (horizontal and vertical) dispersivities; (5) effective molecular diffusion coefficient; (6) distribution coefficient or retardation factor; (7) effective porosity; (8) bulk density of the medium; and (9) site history.

4.5.3.3   Model Calibration

In general, it is more difficult to calibrate a solute transport model of an aquifer than to calibrate a ground water flow model. Since ground water velocities are determined from the head distribution and since both convective and dispersive transport and hydrodynamic dispersion are functions of the ground water velocity (see Eqs. [4-199]–[4-204]), a reliable calibrated solute transport model can be developed only after a calibrated ground water flow model is developed.

The calibration process of a solute transport model is shown schematically in Figure 4-58. If the flow domain can be assumed to be uniform (usually under no pumping conditions), calibration for solute transport can be conducted using one-, two-, or three-dimensional analytical solute transport models, which are given in detail in Chapter 3. During the calibration process, first observed vs. computed solute plume comparisons should be made at selected elapsed times from the beginning of release. For this purpose, isoconcentration curves should be prepared from the observed concentration data as well as computed results. An example of the comparison of isoconcentration curves is shown in Figure 4-60. Also, observed and computed concentrations vs. time curves at selected points in the solute transport domain can be made. However, one suggestion is to start from observed and computed plume comparisons, which are shown schematically in Figure 4-60. In this example, the observed plume boundary in Figure 4-60, say for trichloroethylene (TCE), is defined

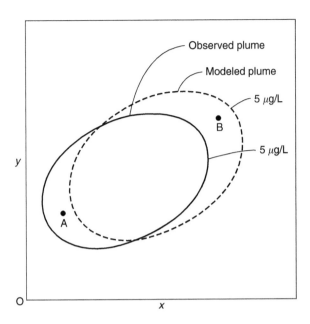

FIGURE 4-60   Observed and computed plume isoconcentration curves comparison.

as 5 μg/L [ppb (parts per billion)], which is the detection limit of TCE. The ground water flow field may be in any form, which is the result of natural flow in the aquifer, extraction wells, and injection wells, and other stress-producing means. As shown in Figure 4-60, this kind of comparison can be made in horizontal planes (x–y) as well as in vertical planes (x–z and y–z). As shown in the Figure 4-60, as a result of a continuous-type or slug-type (or instantaneous-type) solute source (s), there may be discrepancies between the observed and computed plumes. These discrepancies can be minimized, but in most modeling studies they cannot be eliminated. Once the plumes are matched spatially to an acceptable level, the adequacy of the match can be determined by the following procedures: (1) comparison of the observed and computed plume sizes; (2) comparison of the internal concentrations of the observed and computed plumes; and (3) comparison of the maximum concentrations of the observed and computed plumes. In most cases, the maximum concentrations may differ and their places of occurrence may be different as well. Observed vs. computed solute concentration comparisons at different points in the transport domain can be made as shown in Figure 4-57. In order to carry out this type of evaluation, special care must be taken. For example, point A in Figure 4-60 has an observed concentration above the detection limit (above 5 μg/L); but in the computed plume it has a concentration below the detection limit (below 5 μg/L). At point B, the opposite situation occurs. Here, the observed concentration is less than 5 μg/L, and the computed concentration is above 5 μg/L. It is clear that this kind of situation causes undesirable results in developing observed vs. computed isoconcentration curves as well as observed vs. computed concentration comparisons. In conclusion, it can be said that in real-world modeling studies, the comparison of observed and computed isoconcentration curves (or visual comparison) is a widely viable approach.

4.5.3.4   Application of A Calibrated Solute Transport Model

Evaluation of Different Remedial Alternatives
Once a calibrated solute transport model is achieved for a given site, the calibrated model can be used as a predictive tool to evaluate different cases. Some potential application areas are shown in Figure 4-58, which include (1) capping, (2) cutoff walls, (3) extraction and injection wells,

(4) artificial recharge, (5) french drains, (6) subsurface drains, (7) reactive barriers, and others. A calibrated solute transport model plays or can play a role in evaluating reliability, technical feasibility and effectiveness, cost, operation and maintenance, and other aspects of waste disposal facility designs and of alternative remedial actions. With a calibrated solute transport model, the rates and movement of hazardous wastes from sanitary landfills and other contaminated areas can be determined.

Potential Contaminant Source Identification
Solute transport models are also potentially useful in detecting pollutant sources for the contamination of a localized area, such as a production well field. For this purpose, solute transport models have been used to determine potential sources of contamination when a contaminant plume is detected and no sources have been positively identified.

Optimizing Ground Water Monitoring Systems
Solute transport models are effective tools for optimizing ground water monitoring systems. Hazardous waste regulations require the monitoring of well networks to delineate the rate and extent of contamination migration, and specify the number and placement of monitoring wells relative to ambient hydraulic conditions. The goal of this type of investigation is to bring a facility into regulatory compliance and to minimize the number of monitoring wells that must be sampled to provide the required ground water quality data.

Risk Assessment
Solute transport models can also provide important information in risk assessment studies. Estimates of contaminant loadings in different receptors exposed to contaminated ground water are used to calculate the risk level to human health and the environment.

4.5.3.5 Sensitivity Analysis

The general principles of sensitivity analysis are discussed in Section 4.5.2.5. As shown in Figure 4-58, typically, sensitivity analysis can be conducted for a solute transport model by varying the following parameters: (1) longitudinal dispersivity; (2) transverse horizontal dispersivity; (3) transverse vertical dispersivity; (4) horizontal hydraulic conductivity; (5) vertical hydraulic conductivity; (6) recharge; (7) source location and size; and others.

4.6 MODEL TESTING

A major concern in applying ground water flow and solute transport models is the accuracy of modeling evaluations and predictions. This accuracy of model simulation results depends on a number of factors, including how the model is applied, and whether the particular model or computer code was properly verified and validated. This process of *verification* and *validation* is often referred to as *model testing*. These are all discussed in the following paragraphs (van der Heijde et al., 1985; Tsang, 1991).

4.6.1 MODEL VERIFICATION

The objective of the verification process is (1) to check the accuracy of the computational algorithms used to solve the governing equations, and (2) to make sure that the computer code is fully operational. One method by which a computer code can be checked for the accuracy of its theoretical principles and computational algorithms is by executing selected problems for which analytical (closed-form) solutions exist. In this method, a model of a problem is developed for simulation by the computer code. To the extent possible, input data to the analytical and numerical simulations are made identical. Comparisons of the results are a reliable means of verifying the accuracy of the computational method of the computer code. In Chapter 3, a number of verification problems for one-, two-, and three-dimensional solute transport cases are presented by comparing the results of numerical and analytical models. During the past three decades, similar verification studies were carried out and included in the manuals of some well-known numerical computer codes (Sykes et al., 1983; Huyakorn et al., 1984; Ward et al., 1984; Gupta et al., 1984; Faust et al., 1990; and Duffield, 1996).

Verifying complex numerical computer codes with analytical solutions has limitations. The limitations exist because analytical solutions are only available for simplified conditions. To overcome these limitations, the results of the computer code that is being verified are compared with the results of other widely accepted and well-verified codes. Such comparisons assist in verifying the capabilities of the computer code to simulate conditions such as heterogeneity and hydrogeologic stratification, partially penetrating wells, and differing boundary conditions.

4.6.2 MODEL VALIDATION

The objectives of model validation are to determine how well the theoretical foundations of the model describe the actual system behavior, and to compare field data with model-simulated results. In the latter case, a site-specific model is developed that includes site hydrogeologic and geochemical properties, and other existing features such as extraction wells. Computer results of the model, for example, heads, concentrations, etc., are compared with field data. A good correlation of observed data and computer results is indicative of acceptable code performance.

4.7 MODEL MISUSE

The field of ground water modeling is relatively new as compared with other similar areas such as surface water modeling, heat conduction modeling, and others. However, in ground water modeling studies, the generation of a realistic conceptual model and data compilation are much more difficult to accomplish than in the other areas. There are a variety of ways to misuse models and some of them are presented below (Prickett, 1979; Mercer and Faust, 1981; Anderson, 1983).

4.7.1 INVESTIGATOR'S (MODELER) BACKGROUND

The success of a ground water modeling study is heavily dependent on the scientific knowledge of the modeler (s) who are involved in the modeling study. To emphasize the importance of this, Anderson (1983, p. 667) states:

> Model use in general must be combined with educated scientific judgement based on field observations. Faust et al. (1981) give a list of seven questions that should be considered before every modeling exercise. One of the most important questions is, "Does the investigator have sufficient background in both hydrology and modeling?" Such background will help ensure that the model user is aware of the limitations of any given model as well as the geologic complexities of the field. Hence, the key to intelligent model use is the education of model users. Institutionalized "black-boxing" of models without this education could be hazardous.

The conceptual model of a ground water modeling study must be evaluated carefully before starting the modeling study, even if the model is developed by a qualified modeler. If the conceptual model of a site does not match the site-specific features, the results of the model become meaningless. Regarding these points, Mercer and Faust (1981, p. 6) state:

> There are a variety of ways to misuse models (Prickett, 1979). Three common and related ones are: overkill, inappropriate prediction, and misinterpretation. The temptation to apply the most sophisticated computational tool to a problem is difficult to resist. A question that often arises is, under what circumstances should the simulation be three-dimensional as opposed to two- or even one-dimensional. Inclusion of flow in the third (nearly vertical) direction is often recommended only if aquifer thickness is "large" in relation to areal extent or if pronounced heterogeneity exists in the vertical direction (for

example, high stratification). Another type of overkill is that grid sizes are used which are finer (smaller) than necessary considering available information about aquifer properties, resulting in additional work and expense.

4.7.2 PROCEDURAL POINTS

There are some procedures that a modeler should be aware of. For this purpose, Mercer and Faust (1981, p. 6) state:

> In some applications, complex models are used too early in the study. For example, for a hazardous waste problem, one generally should not begin with a solute transport model. Rather, the first step is to be sure the ground water hydrology (velocity in particular) can be characterized satisfactorily, and therefore one begins by modeling ground water flow alone. Once this is done to satisfaction, then solute transport can be included. One must assess the complexities of the problem, the amount of data that is available, and the objectives of the analysis, and then determine the best approach for the particular problem. A general rule might be to start with the simplest model and data until the desired estimation of aquifer performance is obtained.

4.7.3 INSUFFICIENT DATA FOR CALIBRATION

Data for modeling studies are generally limited for all hydrogeologic systems. The level of data generally guides the level of the modeling study for a given site. There are certain difficulties that face the modeling studies conducted for sites with limited data. For this purpose Domenico and Schwartz (1990, p. 145) state:

> The calibration-verification process does not lead to a unique description of a hydrogeologic system. For poorly known systems, a very large number of different models can be developed without knowing which, if any, is correct. Stated another way, different model developers, given the same hydrogeologic data, will probably develop different conceptualizations of the same system, each of which can be calibrated and verified. An example of the difficulties of calibrating a ground water flow model is discussed by Freyberg (1988). Different groups using the same set of synthetic data developed very different predictions concerning system behavior. Sources of variability in the calibration and the resulting prediction were related to (1) the lack of hydrogeologic data, (2) the use of different measures of success in calibration for each group, and (3) differing strategies for calibration (for example, changing local transmissivity values around wells versus changing values over large areas).

4.7.4 BLIND FAITH IN MODEL RESULTS

Proper application of a ground water model for a given site requires an understanding of the site-specific hydrogeologic system and without this, failure of the study is very likely. For this purpose, Mercer and Faust (1981, p. 7) state:

> Perhaps the worst possible misuse of a model is blind faith in model results. Calculations that contradict normal hydrologic intuition almost always are the result of some data entry mistake. A "bug" in the computer program, or misapplication of the model to a problem for which it was not designed. Proper application of a ground water model requires an understanding of the specific aquifer. Without this conceptual understanding the whole exercise may become a meaningless waste of time and money.

4.8 MODELING LIMITATIONS AND SOURCES OF ERROR

All numerical models are based on a set of simplifying assumptions that establish certain limitations in their application to real-world problems. Understanding the limitations and possible sources of error of numerical models is important in order to avoid model misuse. Some important sources of error are given below (Mercer and Faust, 1981).

4.8.1 CONCEPTUAL ERRORS

As mentioned in Section 4.5.2.1, conceptual model development is an important step in the establishment of the foundational features for the model of a given site. For example, a two-dimensional model should be applied with care to a three-dimensional case that involves a series of aquifers connected by aquitards. If the conceptual model of a given site is not realistic, the modeling results may be questionable or completely wrong. Errors of this type are considered *conceptual errors*.

4.8.2 TRUNCATION ERRORS

Numerical modeling practice is the replacement of the model's differential equations by a set of algebraic equations; the solutions of these equations represent an approximation to the exact solution of the original differential equations. Naturally, this method produces *truncation errors*, which means that the exact solutions of the algebraic equations differ from the solutions of the original differential equations. The level of truncation error may be estimated with different computer runs on a model by changing the horizontal and vertical grid spacings, and time steps. Significant sensitivity of computed results to changes in grid spacing and time step indicates a significant level of truncation error. Smaller grid spacing and time steps should be used in order to minimize the truncation errors.

4.8.3 DATA ERRORS

Aquifer hydrogeologic data (hydraulic conductivities, formation thicknesses, etc.) are seldom known accurately or completely, and this situation causes *data errors*. Error caused by erroneous aquifer description data is difficult to assess because the true aquifer description is never known. Aquifer tests and geological studies generally provide valuable information regarding aquifer hydrogeologic parameters. However, much of these data and information may be very local in extent and should be evaluated carefully when used in a model of a relatively large area. As discussed in Sections 4.5.2.3 and 4.5.3.3, the final parameters characterizing the aquifer are usually determined by making the best agreement between observed and computed aquifer behavior during some historical period.

4.9 CASE STUDY FOR GROUND WATER FLOW AND SOLUTE TRANSPORT MODELING

4.9.1 INTRODUCTION

As mentioned previously, numerical models used in ground water flow and solute transport studies are general computer programs that can be applied to a variety of hydrogeologic conditions. In this section, a comprehensive case study regarding ground water flow and solute transport will be presented based on a published work in the literature. The purpose of this section is not to discuss all details associated with the modeling components of this study. Instead, the purpose of this example is to give the reader a broad view regarding the potential successes and limitations of the state-of-the-art models. As mentioned in Section 4.5.3.3, the calibration of solute transport models is more difficult than the calibration of ground water flow models. For some flow fields, an analytical solute transport model can be timewise and costwise more efficient than a numerical solute transport model. One must bear in mind that analytical solute transport models are mostly based on the assumption of a uniform flow field. As long as this assumption can be made, analytical solute transport models can be used.

4.9.2 NUMERICAL GROUND WATER FLOW AND SOLUTE TRANSPORT MODELING EXAMPLE

In Example 4-8, a flow and solute transport modeling example is presented based on Konikow (1977).

EXAMPLE 4-8

This example is based on Konikow (1977), which presents a flow and solute transport modeling application to a chemical pollution problem at the Rocky Mountain Arsenal, near Denver, Colorado. Liquid waste by-products from manufacturing of chemicals for warfare and pesticides were disposed into unlined ponds from 1943 to 1956. The wastes contained chloride concentrations of several thousands of mg/L. In 1954, severe crop damage occurred in fields irrigated with ground water along the South Platte River. This situation prompted the construction of an asphalt-lined evaporation pond. The purpose of this study was to demonstrate the application of a numerical solute transport model to a complex field problem involving contaminant movement in an alluvial aquifer.

SOLUTION

Description of Study Area
The location of the study area is shown in Figure 4-61 and some details regarding its description are presented below.

History of Contamination
The Rocky Mountain Arsenal has been operating since 1942, primarily manufacturing and processing chemical warfare products and pesticides. These operations have produced liquid wastes that contain complex organic and inorganic chemicals, including a characteristically high chloride concentration that apparently ranged up to 5000 mg/L. The liquid wastes were disposed into several unlined ponds (Figure 4-62), resulting in the contamination of the underlying alluvial aquifer. On the basis of available records, it is assumed that contamination first occurred at the beginning of 1943. From 1943 to 1956, primary disposal was into pond A. Alternate and overflow discharges were collected in ponds B, C, D, and E.

Much of the area to the north of the Arsenal is irrigated with surface water and ground water pumped from irrigation wells. Damage to crops irrigated with shallow ground water was observed in 1951, 1952, and 1953. Severe crop damage was reported during 1954, a year when the annual precipitation was about one half of the normal amount, and ground water use was heavier than normal.

Since 1954, several investigations have been carried out to determine the cause of the damage and how to prevent further damages. Additional studies showed that an area of contaminated ground water of several square miles existed north and northwest of the disposal ponds. These data clearly indicated that the liquid wastes had seeped out of the unlined disposal ponds, infiltrated the underlying aquifer, and migrated downgradient toward the South Platte river. To prevent additional contaminant migration towards the aquifer, a 100-acre (0.405 km^2) evaporation pond (Reservoir F in Figure 4-62) with an asphalt lining to hold all subsequent liquid wastes was constructed in 1956.

In 1973 and 1974 there were new claims that crop and livestock damages had been allegedly caused by ground water contamination at the Arsenal. Data collected in 1975 showed the presence of diisopropylmethylphosphonate (DIMP), a nerve-gas by-product of about 0.57 ppb, in a well located 8 miles (12.9 km) downgradient from the disposal ponds and 1 mile (1.6 km) upgradient from the two municipal water supply wells of the City of Brighton. A DIMP concentration of 48 ppm, which is nearly 100,000 times higher, was measured in a ground water sample collected near the disposal ponds. Other contaminants detected in wells or springs in the area included dicyclopentadiene (DCPD), endrin, aldrin, and dieldrin.

Contaminant Pattern
Since 1955, more than 100 observation wells and test holes have been constructed to monitor changes in water levels and water quality in the alluvial aquifer. The areal extent of contamination has been calculated on the basis of chloride concentrations in wells. The concentrations ranged from normal background concentrations of about 40 to 150 mg/L to about 5000 mg/L in ground water near pond A. Data collected during 1955 and 1956 indicated that one main plume of contaminated ground water extended beyond the northwestern boundary of the Arsenal and that a small secondary plume extended

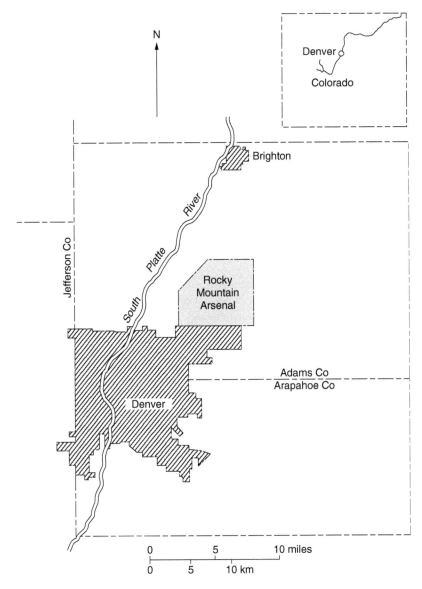

FIGURE 4-61 Location of study area for Example 4-8. (Adapted from Konikow, L.F., U.S. Geological Survey Water-Supply Paper 2044, 1977, pp. 43.)

beyond the northern boundary (Figure 4-63). However, the ground water velocity distribution computed from the water table contour map available at that time could not, in detail, account for the observed pattern of migration from the sources of contamination.

Site Hydrogeology

The records gathered from approximately 200 observation wells, test holes, irrigation wells, and domestic wells were analyzed to estimate the hydrogeologic parameters of the alluvial aquifer. A general ground water contour map for the 1955 to 1971 period is presented in Figure 4-64, which shows that the alluvial aquifer was absent or unsaturated most of the time. These areas, where alluvial aquifer was absent, played the role of posing internal barriers that significantly affected the ground water flow pattern within the aquifer. The isoconcentration contour shown in the Figure 4-63

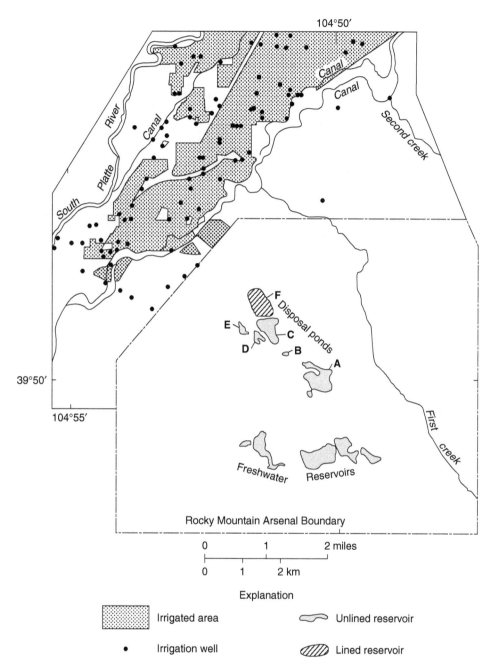

FIGURE 4-62 Major hydrologic features for Example 4-8 (Letters indicate disposal-pond designations assigned by the U.S. Army. (Adapted from Konikow, L.F., U.S. Geological Survey Water-Supply Paper 2044, 1977, pp. 43.)

clearly indicates that the migration of dissolved chloride in the aquifer was also significantly constrained by the aquifer boundaries.

The general direction of ground water flow is from regions of higher water table elevation to those of lower water table elevation, and is approximately perpendicular to the head contours. Deviations from the general flow directions inferred from the head contours may occur in some areas as a result of local variations of aquifer hydrogeologic properties, recharge, and discharge. The nonorthogonality at

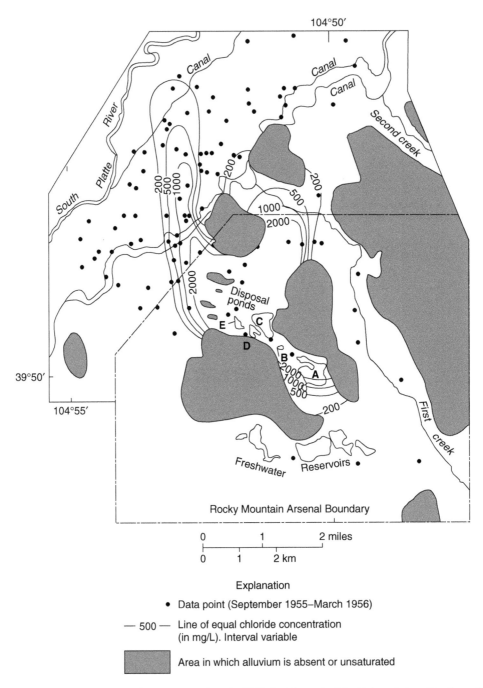

FIGURE 4-63 Observed chloride concentrations in 1956 for Example 4-8. (Adapted from Konikow, L.F., U.S. Geological Survey Water-Supply Paper 2044, 1977, pp. 43.)

locations between head contours and aquifer boundaries indicates that the approximate limit of the alluvial aquifer does not consistently represent a no-flow boundary. At some locations there may be significant flow across this line. Such a condition can occur in areas where the bedrock has relatively high porosity and hydraulic conductivity, or where recharge from irrigation, unlined canals, or other sources is concentrated. Since the bedrock beneath the aquifer is generally much lower than that of the aquifer, ground water flow through the bedrock was assumed to be negligible for the purpose of this investigation.

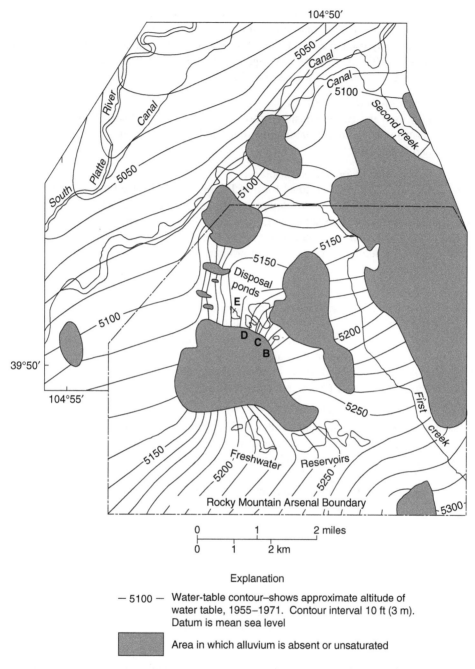

FIGURE 4-64 General ground water table elevation contours in the alluvial aquifer in and adjacent to the Rocky Mountain Arsenal in 1955–1971 for Example 4-8. (Adapted from Konikow, L.F., U.S. Geological Survey Water-Supply Paper 2044, 1977, pp. 43.)

Ground Water Flow and Solute Transport Models Construction

Finite-Difference Grid

The limits of the modeling area were selected to include the entire area with chloride concentrations higher than 200 mg/L and the areas the contaminants would reach in the anticipated time frame. The modeling area was 34 square miles (88 km²) and was subdivided into a finite-difference grid of uniformly spaced squares (Figure 4-65). The grid contained 25 columns (i) and 38 rows (j). Due to the irregular boundaries and discontinuities of the alluvial aquifer, only 516 of the total 950 nodes in the grid were actually used to compute heads (or water-table elevations) in the aquifer. The grid spacing was 1000 ft (305 m) along the rows and column directions. The block-centered discretization convention (see Section 4.3.9) was used in the model grid.

Numerical Solution Method For Flow
Konikow (1977) used the two-dimensional form of Eq. (4-107) and solved the finite-difference equation numerically using an alternative-direction implicit procedure described by Pinder (1970) and Prickett and Lonnquist (1971).

Numerical Solution Method for Solute Transport
Konikow (1977) used a two-dimensional form of Eq. (4-197) and solved it using the method presented by Garder et al. (1964). The development and application of this technique in ground water problems has been presented by Pinder and Cooper (1970), Reddell and Sunada (1970), Bredehoeft and Pinder (1973), Konikow and Bredehoeft (1974), Robertson (1974), Robson (1974), and Zheng (1990). The method actually solves a system of ordinary differential equations that is equivalent to the two-dimensional form of Eq. (4-197).

Boundary Conditions
No-flow and constant head boundaries used in this model are indicated in Figure 4-65 (see Sections 4.3.11 and 4.3.15 for this type of boundary conditions). Konikow (1977) states that the constant head boundaries were specified where it was believed that either underflow into or out of the modeled area or recharge was sufficient to maintain a nearly constant water-table altitude at that point in the aquifer. Altitudes assigned to the constant head cells were determined by superposing the finite-difference grid (Figure 4-65) on the water-table contour map as given in Figure 4-64.

Flow and Solute Transport Parameters
The transmissivity of the alluvial aquifer in this study area ranged from 0 to over 20,000 ft^2/day (over 1,800 m^2/day) and the saturated thickness was generally less than 60 ft (18 m). The highest transmissivities, greatest saturated thicknesses, and lowest hydraulic gradients generally occurred near the South Platte River in the northwestern part of the modeled area. The finite-difference grid was superimposed on maps of transmissivity and saturated thickness and corresponding values were determined for each node of the grid. The storage coefficient was not required because only steady-state ground water flow was considered. Values of effective porosity and dispersivity values were determined by using a trial-and-error adjustment within a range of values determined for similar aquifers in other areas.

Initial Chloride Concentrations
No data were available to describe the chloride concentrations in the aquifer when the Arsenal began its operations. More recent measurements indicated that the normal background concentration may have been as low as 40 mg/L. Therefore, an initial chloride concentration of 40 mg/L was assumed to have existed uniformly throughout the aquifer in 1942.

Aquifer Stresses
No direct measurements of long-term aquifer stresses were available. As a result, these factors were primarily estimated by using a mass balance analysis of the observed flow field. The areas that had probably been irrigated during most of the period from 1943 to 1972 were mapped from aerial photographs. These irrigated areas are shown in Figure 4-62. In the model, irrigation was assumed to occur at 111 nodes of the finite-difference grid, which represents an area of 1.11×10^8 ft^2 (1.03×10^7 m^2). The net rate of recharge from irrigation and precipitation on irrigated areas was estimated through a trial-and-error analysis, which is discussed later. Based on some previous studies for the area, the gross recharge to the aquifer in irrigated parts of the study area was assumed to equal 1.9 ft/year (0.58 m/year). In the study area, irrigation water was derived both from surface water, diverted through canals and ditches, and from ground water, pumped from irrigation wells. The difference

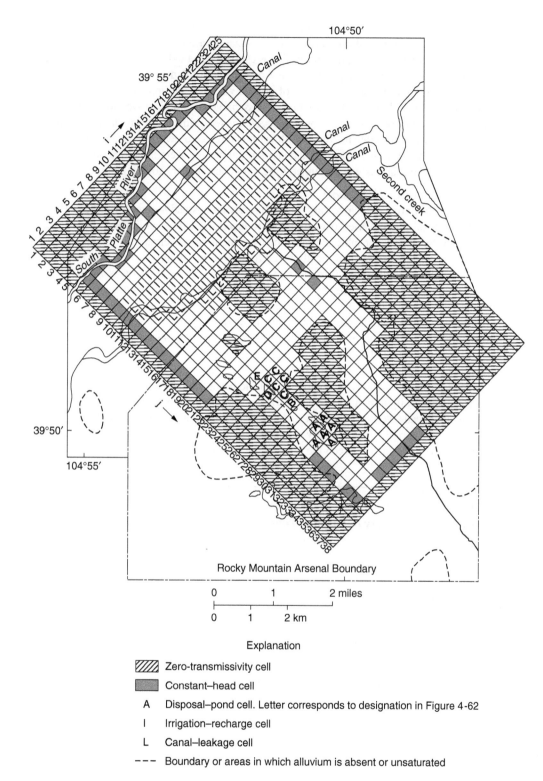

FIGURE 4-65 Finite-difference grid used to model the study area for Example 4-8. (Adapted from Konikow, L.F., U.S. Geological Survey Water-Supply Paper 2044, 1977, pp. 43.)

between the gross recharge and the net recharge, 0.35 ft/year (0.11 m/year), was assumed to equal the total ground water withdrawal rate through wells. Again, based on the previous studies for the area, the average sustained pumping rate per well was estimated to be 0.02 ft^3/sec (5.7×10^{-4} m^3/sec). Leakage from the unlined disposal ponds at the Arsenal represented both a significant source of recharge to the aquifer and the primary source of ground water contamination in the area. Since no records were available to describe the variations in discharge of liquid wastes to the five unlined ponds, the general history of their operation was reconstructed primarily from an analysis of aerial photographs, which were available in 20 sets with varying degrees of average during 1948 and 1971. The summary in Table 4-9 shows that four characteristic subperiods were identified during which the leakage rates and concentrations were assumed to remain constant for modeling purposes.

Calibration of Ground Water Flow Model
Insufficient field data were available to calibrate accurately an unsteady-state flow model. However, the use of the disposal ponds varied over time and induced the only significant transient changes noted in the area. Several water-table elevation measurements in observation wells at the Arsenal showed that the water-table fluctuated locally by up to 20 ft (6 m), mainly in response to filling and emptying of the unlined ponds. Therefore, the hydraulic history of the aquifer was approximated by simulating four separate steady-state flow periods, based on the generalized history of disposal pond operations shown in Table 4-9.

Transmissivity values and boundary conditions in the model were adjusted between successive simulations with an objective of minimizing the differences between observed and computed water table elevations in the irrigated area.

The first period simulated was 1968 to 1972, when it was assumed that pond C was full and that all other unlined ponds were empty. A comparison of the heads computed for 1968 to 1972 with the observed water-table configuration for 1955 to 1971 shows good agreement in most of the modeled area. The computed heads were within 2.5 ft (0.75 m) of the observed heads at more than 84% of the nodes. The other three periods were similarly simulated. As mentioned in Section 4.5.2.3, a residual (or standard error) is the difference between the observed or target value and the model-computed value.

TABLE 4-9
Generalized History of Disposal Pond Operations at the Rocky Mountain Arsenal, 1943–1972 for Example 4-8

Years	Average Use	Computed Leakage (ft^3/sec)	Assumed Chloride Concentration (mg/L)
1943–1956	A — full	0.16	4000
	B, D, E — full	0.18	3000
	C — half full	0.54	3000
1957–1960	A — empty	0.00	N.A.
	B, D, E — empty	0.00	N.A.
	C — full	1.08	1000
1961–1967	A — empty	0.00	N.A.
	B, D, E — one-third full	0.06	500
	C — one-third full	0.36	500
1968–1972	A — empty	0.00	N.A.
	B, D, E — empty	0.00	N.A.
	C — full	1.08	150

N.A., not applicable.

Adapted from Konikow, L.F., Modeling Chloride Movement in the Alluvial Aquifer at the Rocky Mountain Arsenal, Colorado, U.S. Geological Survey Water-Supply Paper 2044, 1977, pp. 43.

The measures of the difference between observed and simulated hydraulic head data are given by Eqs. (4-222)–(4-224). These equations give the mean error for head. Figure 4-66 shows the simulation test number vs. the mean error for head. Figure 4-66 also shows that the mean error for head generally decreased as successive simulation tests were made. After about seven tests, additional adjustments produced only small improvements in the fit between the observed and computed water-table elevations.

A final estimate of the net recharge rate in irrigated areas was made using the set of parameters developed for the final test of Figure 4-66. Figure 4-67 shows the mean of the differences between

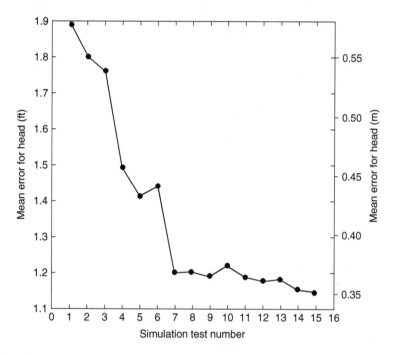

FIGURE 4-66 Change in mean error for head for Example 4-8. (Adapted from Konikow, L.F., U.S. Geological Survey Water-Supply Paper 2044, 1977, pp. 43.)

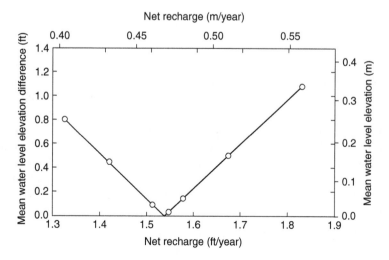

FIGURE 4-67 Relation between the assumed rate of net recharge in irrigated areas and the mean water level elevation difference. (Adapted from Konikow, L.F., U.S. Geological Survey Water-Supply Paper 2044, 1977, pp. 43.)

observed and computed heads at all nodes in the irrigated area minimized (equal to zero) when a net recharge rate of approximately 1.54 ft/year (0.47 m/year) is assumed. Also, irrigation recharge was assumed to have a chloride concentration of 100 mg/L.

The recharge rate due to leakage from the unlined canals was similarly estimated to be approximately 2.37 ft/year (0.72 m/year). The standard error of this estimate was about 1.3 ft (0.40 m). Canal leakage was assumed to contain a chloride concentration of 40 mg/L.

Calibration of Solute Transport Model

The solute transport model for the Rocky Mountain Arsenal area was designed to compute changes in the chloride concentration in the alluvial aquifer during 1943 to 1972. Hydraulic heads and fluxes computed by the ground water flow model were used as input for the solute transport model. A different velocity field was computed for each steady-state flow period outlined in Table 4-9. Comparisons with the observed chloride concentrations at different years and chloride concentrations computed for the corresponding years are discussed in the following paragraphs.

Comparison with the 1956 Observed Chloride Concentrations

The solute transport model was calibrated mainly on the basis of the chloride concentration pattern that was observed in 1956. The chloride concentration contours in 1956 are shown in Figure 4-63. The computed concentrations were most sensitive to variations in the value of effective porosity, φ_e, and least sensitive to the transverse dispersivity, α_T. A comparison of observed and computed chloride concentration patterns indicated that an effective porosity $\varphi_e = 0.30$ and transverse dispersivity of $\alpha_T = 100$ ft (30 m) were best. After appropriate concentrations were assigned to all sources, and an initial background concentration of 40 mg/L was assigned to all nodes in the aquifer to represent conditions at the end of 1942, the solute transport model was run for a 14-year simulation period (1943–1956). The model computed a chloride concentration pattern (Figure 4-68) that agreed closely with the observed pattern (Figure 4-63). The small difference in the directions of the axes of the main plumes between the observed and computed concentrations is probably due mainly to errors in the computed flow field, rather than to errors in the solute transport model.

Comparison with the 1961 Observed Chloride Concentrations

Since 1956, all disposal has been into the asphalt-lined Reservoir F, which eliminated the major source of contamination. However, this could not eliminate the contamination problem because large volumes of contaminants were already present in the aquifer. In January 1961 sufficient data were again available to generate observed chloride concentration contours (Figure 4-69). The computed chloride concentration contours at the start of 1961 are shown in Figure 4-70.

Comparison with the 1969 Observed Chloride Concentrations

Available data suggest that recharge of the aquifer was relatively low from 1961 to about 1968. Data collected in early 1969 indicated the occurrence of a further significant decrease in the overall size of the affected area (Figure 4-71). Apparently, as the contaminated water continued to migrate downgradient, its chloride concentration was diminished by dispersion and dilution. As can be seen from Figure 4-71, chloride concentrations higher than 1 000 mg/L were now limited to only a few isolated areas. The chloride concentration pattern computed for the end of 1968, or start of 1969, using the chloride concentrations computed for the end of 1960 as initial conditions, is presented in Figure 4-72. As can be seen from Figure 4-71 and Figure 4-72, for the most part, the solute transport model reproduced during this period, from 1961 to 1969, compares fairly well with the observed data.

Comparison with the 1972 Observed Chloride Concentrations

From about 1968 or 1969 through 1974, pond C was apparently again maintained full most of the time by the diversion of water from the freshwater reservoirs to the south. Available data showed

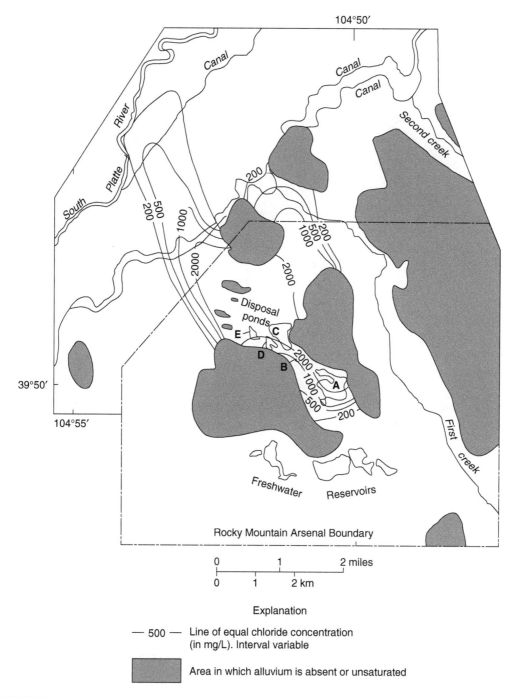

FIGURE 4-68 Computed chloride concentration contours in 1956. (Adapted from Konikow, L.F., U.S. Geological Survey Water-Supply Paper 2044, 1977, pp. 43.)

that by 1972 the areal extent of contamination (Figure 4-73) at concentrations above 1000 mg/L was limited to just small parts of the main zone of contamination. Since both were areas of relatively low hydraulic conductivity it appears that low flow velocities retarded the movement of the

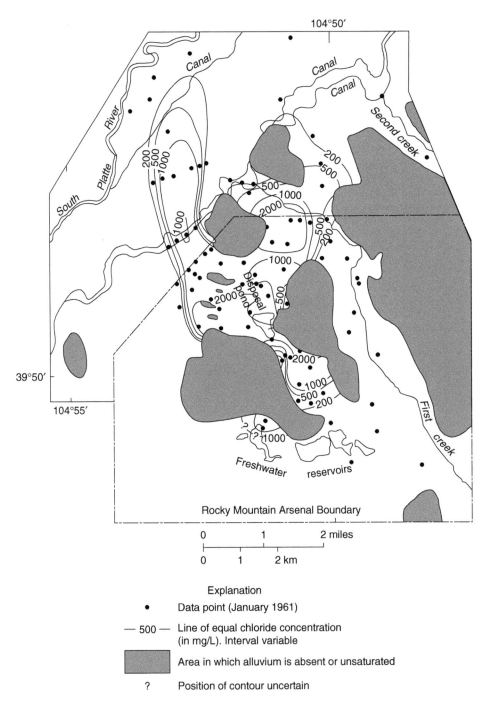

FIGURE 4-69 Observed chloride concentration contours in January 1961. (Adapted from Konikow, L.F., U.S. Geological Survey Water-Supply Paper 2044, 1977, pp. 43.)

contaminated ground water out or through these two areas. Chloride concentrations were almost at normal background levels in the middle of the affected area. This largely reflected the infiltration of fresh water from pond C. The pattern of contamination computed for the start of 1972, by using the chloride concentrations computed for the end of 1968 (Figure 4-72) as initial conditions, is

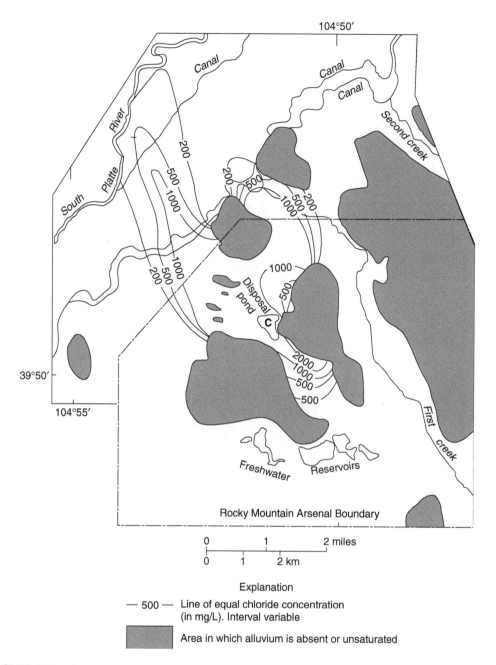

FIGURE 4-70   Computed chloride concentration contours at the start of 1961. (Adapted from Konikow, L.F., U.S. Geological Survey Water-Supply Paper 2044, 1977, pp. 43.)

presented in Figure 4-74. The computed chloride concentration pattern agrees fairly well with the observed concentration pattern (Figure 4-75).

Potential Problems with the Solute Transport Model Calibration
After a 30-year period of simulation, the model has identified (1) the two areas where high chloride concentrations were still present; (2) the reduction in size and strength of the main plume; and (3) the significant reduction in chloride concentration in the middle of the contaminated zone.

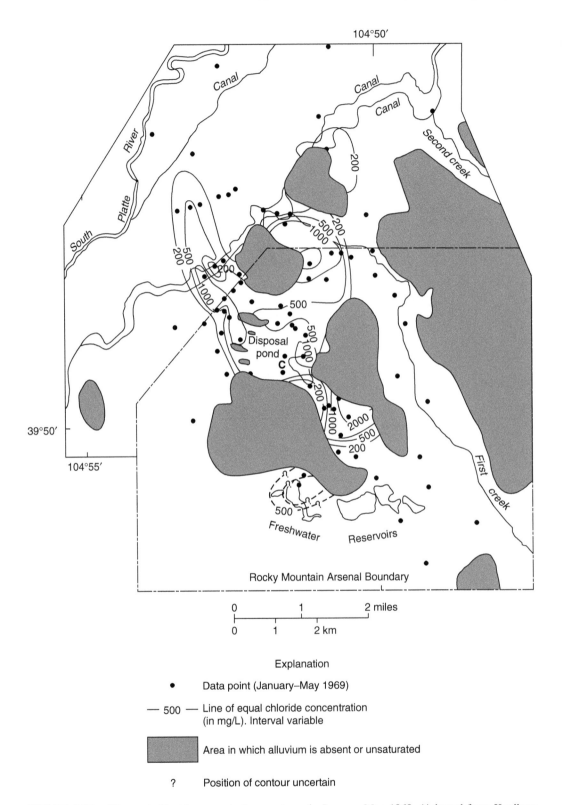

FIGURE 4-71 Observed chloride concentration contours in January–May 1969. (Adapted from Konikow, L.F., U.S. Geological Survey Water-Supply Paper 2044, 1977, pp. 43.)

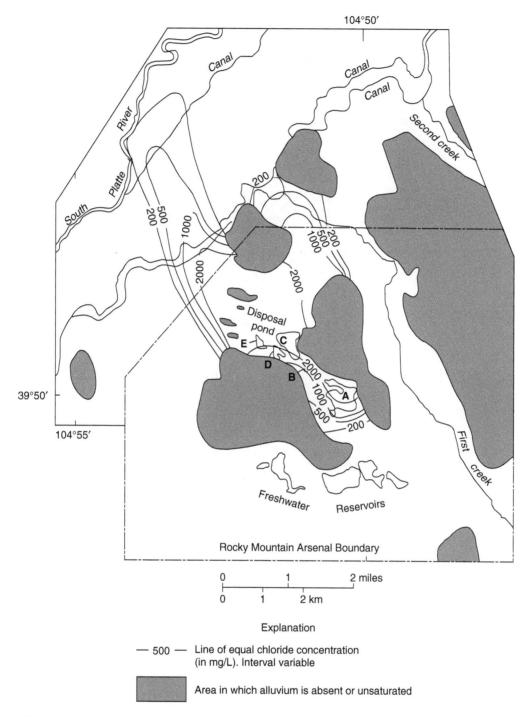

FIGURE 4-72 Computed chloride concentration contours at the start of 1969. (Adapted from Konikow, L.F., U.S. Geological Survey Water-Supply Paper 2044, 1977, pp. 43.)

The model computes the velocity of ground water flow at each node of the model grid, but these data could not be independently verified with field data. The computed velocities ranged from < 1.0 ft/day (0.3 m/day) to > 20 ft/day (6.1 m/day).

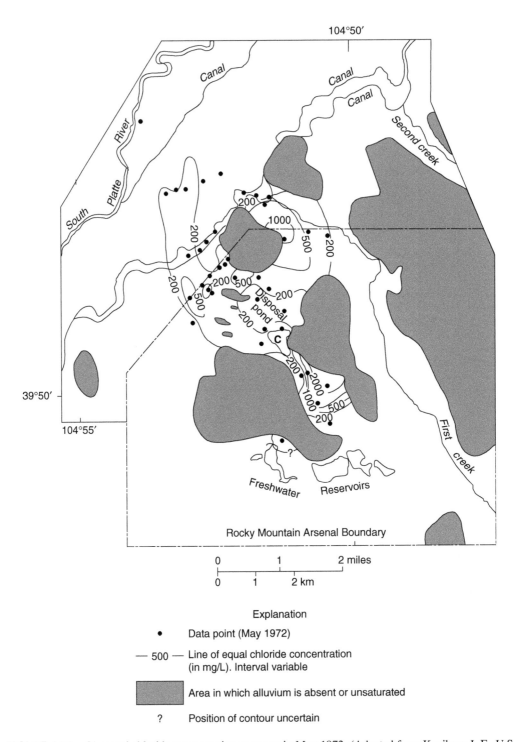

FIGURE 4-73   Observed chloride concentration contours in May 1972. (Adapted from Konikow, L.F., U.S. Geological Survey Water-Supply Paper 2044, 1977, pp. 43.)

Mass balance calculations were performed during the calibration process to check the numerical accuracy of the model simulations. Errors in the mass balance were always less than 1% for the flow model, but averaged about 14% for the solute transport model. The error for solute transport

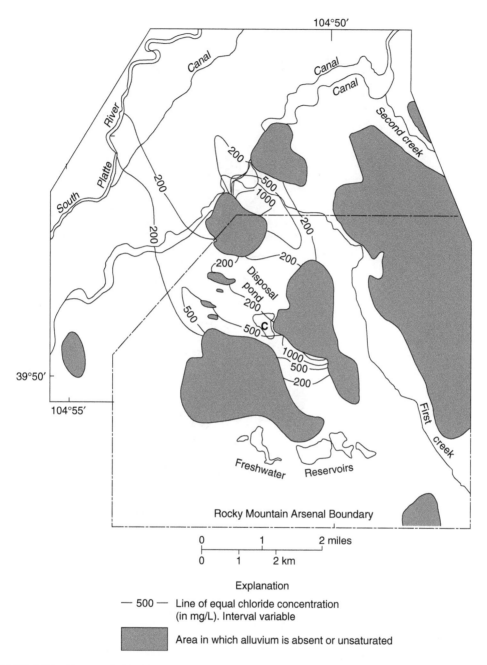

FIGURE 4-74 Computed chloride concentration contours at the start of 1972. (Adapted from Konikow, L.F., U.S. Geological Survey Water-Supply Paper 2044, 1977, pp. 43.)

is somewhat higher than desirable and indicates the need for further refinements in the numerical solution procedure to solve the solute transport equation.

Predictive Capability of the Calibrated Solute Transport Model
Once the model has been adequately calibrated, it can be used to predict or analyze the effects of either future or past changes in stresses or boundary conditions. The Rocky Mountain Arsenal model, which was calibrated over a 30-year historical period, appeared to be reliable enough to be

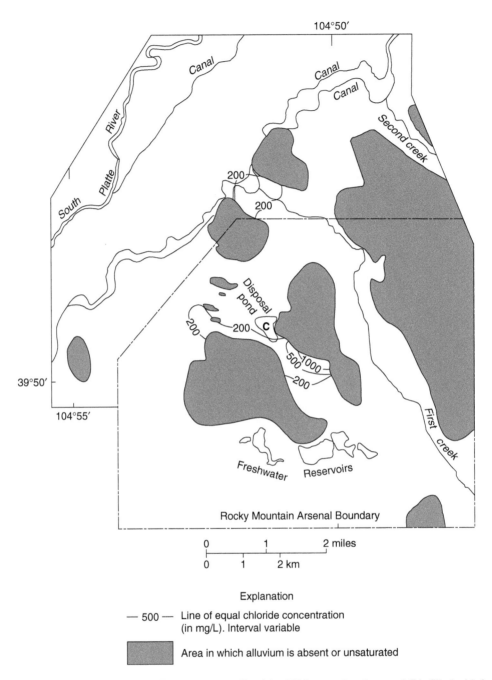

FIGURE 4-75 Chloride concentration contours predicted for 1980, assuming that pond C is filled with fresh water during 1972–1980. (Adapted from Konikow, L.F., U.S. Geological Survey Water-Supply Paper 2044, 1977, pp. 43.)

used for limited predictive purposes. The predictive capability of this model can help to (1) isolate the effects of previous measures; (2) evaluate the causes of present or recurring problems; and (3) predict future concentrations under a variety of assumptions.

Figure 4-76 presents chloride concentrations predicted for 1980, assuming that recharge from pond C is minimal during the period of 1961 to 1980. Additional predictions can be found in Konikow (1977).

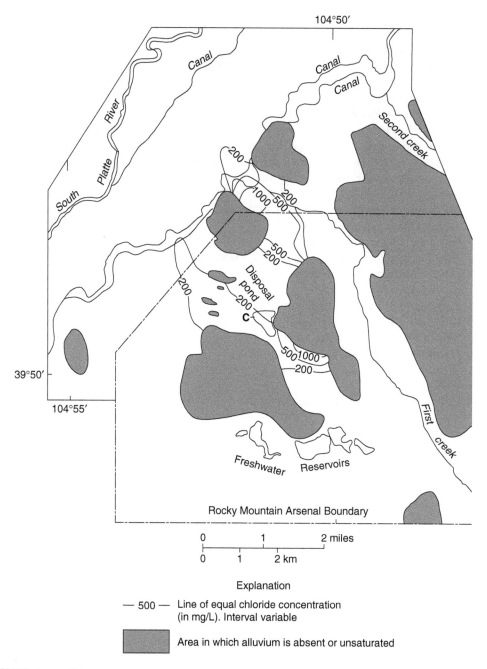

FIGURE 4-76 Chloride concentration contours predicted for 1980, assuming that recharge from pond C is minimal during 1972–1980. (Adapted from Konikow, L.F., U.S. Geological Survey Water-Supply Paper 2044, 1977, pp. 43.)

PROBLEMS

4.1 What are the main numerical modeling techniques? Make a comparative evaluation.

4.2 Why is the finite-difference method for ground water problems so popular?

4-3 Define hydraulic head and name its components?

4.4 Define effective porosity. How can its values be compared with the total porosity?

4.5 Why is the velocity head term $[v^2/(2g)]$ not included in the ground water hydraulic head equation $(h = p/\gamma + z)$?

4.6 Based on experiments, the simple form of Darcy's law is given by Eq. (4-19). Answer the following: (a) Give the physical meaning of Darcy's law? (b) what does the negative sign in Eq. (4-19) imply? and (c) what is statistical macroscopic equivalent of Darcy's law?

4.7 The hydraulic conductivities of an aquifer in the Cartesian coordinates system are given as $K_x = 28$ m/day, $K_y = 20$ m/day, and $K_z = 2$ m/day. Based on these given values, answer the following: (a) what kind of aquifer is this? (b) under what conditions this aquifer can be called an homogeneous and isotropic aquifer? and (c) under what conditions this aquifer can be called a transversely anisotropic aquifer?

4.8 For a sandy aquifer the following values are given: $\gamma = 9.807$ kN/m³, $E_s = 4.78 \times 10^7$ N/m², $\varphi = 0.38$, and $E_w = 2.05 \times 10^9$ N/m² (for water at 5°C). Using these values, (a) determine the specific storage value of the aquifer and (b) compare the specific storage values of the aquifer and water.

4.9 Based on pumping test for an aquifer the following hydraulic conductivity values were determined: $K_{xx} = 14.5$ m/day, $K_{yy} = 17.2$ m/day, and $K_{zz} = 10.0$ m/day. Based on these values (a) determine the principal hydraulic conductivities and (b) determine the angle between the x- and ξ-axis.

4.10 Equation (4-107) is the fundamental differential equation for the MODFLOW family codes. What kind of aquifer does this equation represent? Provide an explanation.

4.11 Explain the finite-difference block- and point-centered formulation. Which formulation is included in the MODFLOW family of codes?

4.12 What are the differences between the "constant head boundary" and "specified head boundary"?

4.13 Define "water divide" and give examples.

5 Analytical and Numerical Modeling Approaches for Solute Travel Times and Path Lines in Aquifers

5.1 INTRODUCTION

As mentioned in the previous chapters, solute transport in ground water occurs mainly through convection (or advection) and dispersion effects. The main focus of the convection–dispersion solute transport theory is to determine the spatial and temporal distribution of solute concentration in a given flow field. Path lines of solute particles as well as their arrival times at receptors (wells, rivers, canals, ditches, etc.) are also important in ground water practices. In the ground water literature, this subject is known as *particle tracking*. Solute particles are tracked under flowing ground water conditions. Studies regarding particle tracking have intensified considerably since the 1970s. Convection models cannot be used to compute solute concentrations in ground water because they do not account for the effect of mixing by dispersion. However, convection models represent a valuable intermediate step between ground water flow models and convection–dispersion solute transport models. In other words, these studies are based mainly on convective ground water flow conditions and dispersion effects are neglected. Therefore, with this approach, only the path lines and arrival times of solute particles can be determined under differing flow configurations. This approach has been applied to both saturated and unsaturated zones (Jury, 1975a, 1975b; Kirkham and Affleck, 1977; Batu and Gardner, 1978). Later, it was generalized by applying it to homogeneous and anisotropic aquifers along with the finite-difference method (Pollock, 1988, 1989). In this chapter, the theoretical as well as practical applications of the approach for particle tracking will be included with practical examples.

Determination of the path lines of solute particles under the convection–dispersion solute transport theory is a relatively new subject in the ground water literature, and one should expect that the results of this approach are closer to the real situation than those of the convection modeling approach. Since the late 1980s, very few studies have appeared in the literature regarding the introduction of this theory to determine the path lines of solute particles under both convective and dispersive solute transport conditions (Batu, 1987, 1988). Based on these studies, in this chapter, both the theoretical foundations as well as practical applications of this approach will be presented.

5.2 PATH LINES AND TRAVEL TIMES OF SOLUTE UNDER CONVECTIVE TRANSPORT CONDITIONS

5.2.1 KINEMATICS OF A SOLUTE PARTICLE

Conceptually, particle tracking is relatively simple. A particle is tracked through a flow field explicitly by first computing the velocity components in the x-, y-, and z-coordinate directions of the particle at its current position. A new location of the same particle is determined by multiplying those velocity components by a finite time step to obtain the incremental changes of the x, y, and z coordinates of the particle over that time interval. As this process is repeated, a series of (x, y, z) and time t coordinates are generated that trace the path of a particle through the flow field as a function of time.

If the velocity of a ground water flow field is known, the path lines of solute particles can be determined spatially and temporally. In order to conduct particle tracking, the ground water velocities (Eq. [2-35]) in the three-dimensional Cartesian coordinates system must be known. Under unsteady-state flow conditions, the expressions of these velocities are

$$v_x(x, y, z, t) = \frac{q_x(x, y, z, t)}{\varphi_e(x, y, z)} \tag{5-1}$$

$$v_y(x, y, z, t) = \frac{q_y(x, y, z, t)}{\varphi_e(x, y, z)} \tag{5-2}$$

$$v_z(x, y, z, t) = \frac{q_z(x, y, z, t)}{\varphi_e(x, y, z)} \tag{5-3}$$

where $v_x(x, y, z, t)$, $v_y(x, y, z, t)$, and $v_z(x, y, z, t)$ are the ground water velocity components in the x-, y-, and z-coordinate directions, respectively, at a point in the flow field having the (x, y, z) coordinates at time t; $\varphi_e(x, y, z)$ is the effective porosity at the same point. In the same equations, $q_x(x, y, z, t)$, $q_y(x, y, z, t)$, and $q_z(x, y, z, t)$ are the Darcy velocity components in the x-, y-, and z-coordinate directions, respectively, at the same (x, y, z) point. Under steady-state flow conditions, Eqs. (5-1), (5-2), and (5-3), respectively, take the following forms:

$$v_x(x, y, z) = \frac{q_x(x, y, z)}{\varphi_e(x, y, z)} \tag{5-4}$$

$$v_y(x, y, z) = \frac{q_y(x, y, z)}{\varphi_e(x, y, z)} \tag{5-5}$$

$$v_z(x, y, z) = \frac{q_z(x, y, z)}{\varphi_e(x, y, z)} \tag{5-6}$$

The velocity vector, $\mathbf{v}(x, y, z)$, is at a tangent to the path line of a solid particle and its expression is

$$\mathbf{v}(x, y, z) = v_x(x, y, z)\mathbf{i} + v_y(x, y, z)\mathbf{j} + v_z(x, y, z)\mathbf{k} \tag{5-7}$$

where $\mathbf{i}$, $\mathbf{j}$, and $\mathbf{k}$ are the unit vectors in the x, y, and z directions, respectively. The magnitude of the velocity vector is

$$|\mathbf{v}(x, y, z)| \equiv v_s = [v_x^2(x, y, z) + v_y^2(x, y, z) + v_z^2(x, y, z)]^{1/2} \tag{5-8}$$

Similarly, the Darcy velocity vector, $\mathbf{q}(x, y, z)$, can be expressed as

$$\mathbf{q}(x, y, z) = q_x(x, y, z)\mathbf{i} + q_y(x, y, z)\mathbf{j} + q_z(x, y, z)\mathbf{k} \tag{5-9}$$

and its magnitude is

$$|\mathbf{q}(x, y, z)| \equiv q_s = [q_x^2(x, y, z) + q_y^2(x, y, z) + q_z^2(x, y, z)]^{1/2} \tag{5-10}$$

From Eq. (4-19) one can write

$$q_s = - KI \tag{5-11}$$

The magnitude of the ground water velocity is

$$v_s = \frac{q_s}{\varphi_e} = -\frac{1}{\varphi_e} KI \tag{5-12}$$

In this book, only steady-state ground water velocity fields will be considered. If the ground water velocities can be expressed analytically, expressions for particle tracking can be developed. These kinds of expressions are mostly available for homogeneous and isotropic media as well as relatively simple boundary conditions. If the aquifer parameters vary spatially, and boundary conditions are relatively complex, flow fields for ground water velocity can be determined by numerical methods. In this book, both analytical and finite-difference numerical methods will be covered.

5.2.2   ANALYTICAL METHODS FOR THE DETERMINATION OF SOLUTE PATH LINES AND TRAVEL TIMES

Analytical solutions for solute path lines and travel times can only be determined for relatively simple cases. First, the travel time expression for the most simple case, which is based on the Darcy equation, will be presented. Then, travel times to a well in a confined aquifer and travel times beneath a dam will be presented.

5.2.2.1   Travel Time Determination Using Darcy's Equation

Travel time determination based on the application of Darcy's equation is useful for a quick assessment of contamination migration. In the following paragraphs, the associated equations are presented (see, e.g., U.S. EPA, 1985).

The distance traveled by an object at a constant velocity is

$$\Delta l = v_s \Delta t \tag{5-13}$$

where Δl is the distance traveled and Δt is the time of travel. If the transport of a nonreactive and nondispersive solute or contaminant is considered, the velocity v_s in Eq. (5-13) becomes equal to the ground water velocity. In Eq. (5-13), the ground water velocity is used because solutes only travel through the water-filled portion of soil pores. The travel time, (Δt) can now be solved from Eq. (5-13) to give

$$\Delta t = \frac{\Delta l}{v_s} \tag{5-14}$$

The ground water velocity is given by Eq. (5-12) and its combination with Eq. (5-14) gives the travel time for saturated flow conditions:

$$\Delta t = -\frac{\varphi_e \Delta l}{KI} \tag{5-15}$$

where Δt (T) is the travel time, Δl (L) the travel distance, φ_e (unitless) the effective porosity, K (L/T) the hydraulic conductivity, and I (L/L) is the hydraulic gradient. The estimated values of porosity are given in Table 4-1. The definition of effective porosity is given by Eq. (4-3); effective porosity of a porous medium is smaller than its porosity.

It must be remembered that travel times computed from Eqs. (5-14) and (5-15) are for nonreactive and nondispersive conservative solutes moving at a constant velocity. Retardation by sorption and attenuation by other solute–soil interactions may substantially decrease the velocity of solute movement and increase the travel time. Conversely, dispersive processes can either substantially increase or decrease the solute velocity, as a portion of the solute molecules moves and changes the travel time.

In many real-world situations, the ground water flow velocity may vary in both direction and magnitude in an aquifer. Variable velocity and variable soil characteristics can easily be incorporated into the calculation of solute travel time and its schematic representation is shown in Figure 5-1. Equation (5-15) is applied over each subregion and then all travel times are summed. Therefore, the total travel time T_t is

$$T_t = \sum_{i=1}^{n} \Delta t_i = - \sum_{i=1}^{n} \left(\frac{\varphi_e \Delta l}{KI} \right)_i = - \sum_{i=1}^{n} \frac{(\varphi_i)(\Delta l)_i}{K_i I_i} \qquad (5\text{-}16)$$

where the subscript i refers to the ith subregion, $(\varphi_e)_i$ is the effective porosity, $(\Delta l)_i$ is the travel distance, K_i is the hydraulic conductivity, and I_i the hydraulic gradient of the ith subregion. The parameters φ_e, Δl, K, and I can be different for each subregion i.

EXAMPLE 5-1

A pond shown in Figure 5-2 releases several chemicals in the underlying unconfined aquifer, which discharges to a nearby river. Determine the travel time of chloride from the edge of the pond in the river by assuming that the thickness of the unsaturated zone is relatively thin. The aquifer is mainly composed of fine sands and its parameters are given in Table 5-1. Using these data, determine (1) the average hydraulic parameters; (2) the average ground water velocity; and (3) the travel time.

SOLUTION

(1) Average hydraulic conductivities. *Arithmetic average hydraulic conductivity:*

$$K_{ar\,av} = \frac{K_1 + K_2 + K_3}{n} = \frac{18.84 \text{ m/day} + 66.70 \text{ m/day} + 38.71 \text{ m/day}}{3} = 41.42 \text{ m/day}$$

Geometric average hydraulic conductivity:

$$K_{geo\,av} = (K_1 K_2 K_3)^{1/3} = [(18.84 \text{ m/day})(66.70 \text{ m/day})(38.71 \text{ m/day})]^{1/3} = 36.50 \text{ m/day}$$

Since the K values are relatively scattered, the geometric average value is a better representative value.

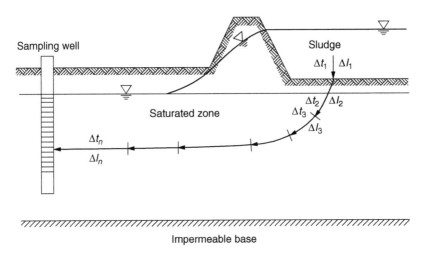

FIGURE 5-1 Schematic representation for travel time calculation when the ground water velocity varies. (Adapted from U.S. Environmental Protection Agency, *Water Quality Assessment: A Screening Procedure for Toxic and Conventional Pollutants in Surface and Ground Water – Part II*, EPA/600/6-85/002b, Environmental Research Laboratory, Athens, Georgia, September, 1985.)

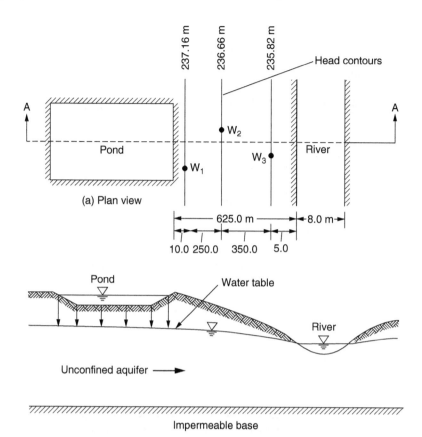

FIGURE 5-2　The geometry of Example 5-1.

TABLE 5-1
Aquifer Hydraulic Parameters and Observed Heads for Example 5-1

Well No.	Hydraulic Head (m MSL[a])	Horizontal Hydraulic Conductivity (cm/sec)	(m/day)	Porosity (φ)
W_1	237.16	2.18×10^{-2}	18.84	0.40
W_2	236.66	7.72×10^{-2}	66.70	0.40
W_3	235.82	4.48×10^{-2}	38.71	0.40

[a] MSL = mean sea level.

Hydraulic gradients:

$$I_{W_1-W_2} = \frac{237.16 \text{ m} - 236.66 \text{ m}}{250.00 \text{ m}} = 0.0020 \text{ m/m}$$

$$I_{W_2-W_3} = \frac{236.66 \text{ m} - 235.82 \text{ m}}{350.00 \text{ m}} = 0.0024 \text{ m/m}$$

$$I_{W_1-W_3} = \frac{237.16 \text{ m} - 235.82 \text{ m}}{600.00 \text{ m}} = 0.0022 \text{ m/m}$$

Average hydraulic gradient:

$$I_{av} = \frac{0.0020 \text{ m/m} + 0.0024 \text{ m/m} + 0.0022 \text{ m/m}}{3} = 0.00213 \text{ m/m}$$

Effective porosity selected value: $\varphi_e = 0.25$.
  (2) From Eq. (5-12), the ground water velocity is

$$v_s = \frac{(36.50 \text{ m/day})(0.00213 \text{ m/m})}{0.25} = 0.31 \text{ m/day}$$

  (3) From Eq. (5-14) the travel time is

$$\Delta t = \frac{625.00 \text{ m}}{0.31 \text{ m/day}} = 2016.13 \text{ days} = 5.52 \text{ years}$$

5.2.2.2 Travel Time Determination Using Analytical Solutions

If an analytical solution exists for a ground water problem, expressions for travel time can be developed using some general equations based on the kinematics of ground water flow. In the following paragraphs, first the equations are presented; then, travel time equations are derived for some cases.

General Equations for Travel Time
In some special cases, analytical expressions can be derived if the ground water velocity components are given with some analytical expressions. From Eqs. (5-4), (5-5), and (5-6) respectively, one can write

$$\frac{dx}{dt} = v_x(x, y, z) = \frac{q_x(x, y, z)}{\varphi_e(x, y, z)} \tag{5-17}$$

$$\frac{dy}{dt} = v_y(x, y, z) = \frac{q_y(x, y, z)}{\varphi_e(x, y, z)} \tag{5-18}$$

$$\frac{dz}{dt} = v_z(x, y, z) = \frac{q_z(x, y, z)}{\varphi_e(x, y, z)} \tag{5-19}$$

Solving Eqs. (5-17), (5-18), and (5-19) for dt, after integration, they, respectively, give

$$T_t = \int_0^{T_t} dt = \int_{x_0}^x \frac{\varphi_e(x, y, z)}{q_x(x, y, z)} \, dx \tag{5-20}$$

$$T_t = \int_0^{T_t} dt = \int_{y_0}^y \frac{\varphi_e(x, y, z)}{q_y(x, y, z)} \, dy \tag{5-21}$$

$$T_t = \int_0^{T_t} dt = \int_{z_0}^z \frac{\varphi_e(x, y, z)}{q_z(x, y, z)} \, dz \tag{5-22}$$

where T_t is the travel time for a solute particle to move from a point (x_0, y_0, z_0) to another point (x, y, z) in the x–y–z Cartesian coordinates system. It must be pointed out that the three equations for T_t must give the same values.

  Based on the general approach presented above, in the following paragraphs, two cases of practical importance will be presented. The first one is related to travel times to a well in a confined aquifer. The second one is related to the determination of the minimum travel time under a dam.

Travel Times to a Well in a Confined Aquifer
In the following paragraphs, several cases will be presented for solute travel times to a well based on Kirkham and Affleck (1977). Although the derivations that follow apply to a confined aquifer, the derivations will also apply to an unconfined aquifer if the steady-state drawdown of the well is small relative to the aquifer's thickness. In these derivations, it is assumed that the dissolved solute moves in piston flow along horizontal stream lines and that the aquifer's characteristics are unchanged by the solute. The assumption of piston flow along horizontal stream lines is a more restrictive assumption than may be needed. For example, if travel times are large, then small differences in density will make the solute sink to the bottom of the aquifer or rise to the top very slowly. However, the travel times would be approximately the same because the horizontal component of flow of the dissolved solute will be very nearly that of the water, despite any up or down movement of the solute (Kirham and Affleck, 1977).

A Single Well Case with One Hydraulic Conductivity
In Figure 5-3, a planar view of a confined aquifer having a fully penetrating well is shown, where R is the zone of influence radius under steady-state extraction conditions. The flow rate equation was first developed by Thiem (1906) and its derivational details can be found in many books see, (e.g., Batu, 1998, pp. 113–116).

The continuity equation is

$$- Q = 2\pi r b q_r \tag{5-23}$$

where $Q(L^3/T)$ is the extraction rate, $r(L)$ the radial coordinate, $b(L)$ the aquifer's thickness, and $q_r(L/T)$ the radial Darcy velocity and is given by

$$q_r = - K_r \frac{\partial h}{\partial r} \tag{5-24}$$

According to the assumed sign convention, $-Q$ corresponds to withdrawal and $+Q$ corresponds to injection. Substitution of the q_r expression into Eq. (5-23) with $K_r \equiv K_h$ gives

$$K_h \frac{\partial h}{\partial r} = \frac{Q}{2\pi r b} \tag{5-25}$$

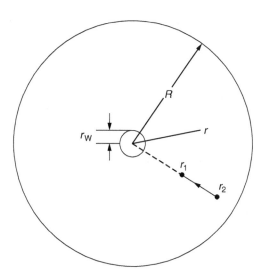

FIGURE 5-3 Solute movement from r_2 to r_1 in a well in a confined aquifer. (Adapted from Kirkham, D. and Affleck, S.B., *Ground Water*, 15, 231–242, 1977.)

The boundary conditions are

$$
h = \begin{cases} h_{\mathrm{w}} & \text{for } r = r_{\mathrm{w}} \quad (5\text{-}26) \\[2mm] H & \text{for } r = R \quad (5\text{-}27) \end{cases}
$$

where h_{w} is the piezometric head at the well boundary, r_{w} the well radius, H the piezometric head before pumping, and R the well influence radius at which the drawdown is zero.

The solution of Eqs. (5-25), (5-26), and (5-27) for an extraction well is (see, e.g., Batu, 1998, p. 115, Eq. [3-8])

$$
Q_{\mathrm{ext}} = \frac{2\pi K_h b (H - h_{\mathrm{w}})}{\ln(R/r_{\mathrm{w}})} \tag{5-28}
$$

For an injection well, Eq. (5-23) becomes

$$
+ Q = 2\pi r b q_r \tag{5-29}
$$

Therefore, combination of Eq. (5-29) with Eq. (5-24) gives

$$
K_h \frac{\partial h}{\partial r} = - \frac{Q}{2\pi r b} \tag{5-30}
$$

Similarly, the solution of Eqs. (5-26), (5-27), and (5-30) for an injection well is

$$
Q_{\mathrm{inj}} = - \frac{2\pi K_h b (H - h_{\mathrm{w}})}{\ln(R/r_{\mathrm{w}})} \tag{5-31}
$$

For the extraction well case, by combining Eqs. (5-23) and (5-28) one has

$$
q_r = \frac{K_h}{r} \frac{(H - h_{\mathrm{w}})}{\ln(R/r_{\mathrm{w}})} \tag{5-32}
$$

For simplicity, the minus sign on the right-hand side of Eq. (5-32) is omitted. In other words, the absolute value is considered; this form represents both extraction and injection well cases.

From Eq. (5-12), the radial ground water velocity v_r (L/T) is

$$
v_r = \frac{q_r}{\varphi_e} \tag{5-33}
$$

which is the solute velocity toward the well. This velocity may also be expressed as

$$
v_r = - \frac{dr}{dt} \tag{5-34}
$$

Here, the minus sign is used because v_r is directed toward the well instead of the outward positive direction. From Eqs. (5-32)–(5-34) one can write

$$
dt = - \frac{dr}{v_r} = - \varphi_e \frac{dr}{q_r} = - \frac{\varphi_e}{K_h} \frac{\ln(R/r_{\mathrm{w}})}{(H - h_{\mathrm{w}})} r\, dr \tag{5-35}
$$

Referring to Figure 5-3, let t_2 be the clock time when solute is at r_2 and t_1 a later clock time when solute is at r_1. Then, after integration and substitution of limits, Eq. (5-35) gives

$$
t_1 - t_2 = \frac{\varphi_e \ln(R/r_{\mathrm{w}})(r_2^2 - r_1^2)}{2 K_h (H - h_{\mathrm{w}})} \tag{5-36}
$$

Equation (5-36) gives the time interval $(t_1 - t_2)$ for a solute to move from r_2 to r_1 in Figure 5-3. Of special interest are the conditions $t_2 = 0$, $t_1 = t$, $r_1 = r_w$, and $r_2 = r$, which when substituted in Eq. (5-36) give

$$t = \frac{\varphi_e \ln(R/r_w)}{K_h(H - h_w)} \frac{(r^2 - r_w^2)}{2} \tag{5-37}$$

where t is the time interval required for a solute initially at distance r from the well center in the aquifer to reach the well. In terms of Q, using Eq. (5-28), Eq. (5-37) takes the form

$$t = \frac{2\pi\varphi_e b}{Q} \frac{(r^2 - r_w^2)}{2} \tag{5-38}$$

Equation (5-36) gives the time it takes for solute to move from a point r_2 to a point r_1 along a radial stream line as shown in Figure 5-3. When the point r_1 becomes a point r_w on the well and r_2 is r, the time of solute travel along the radial stream line is given either by Eq. (5-37) or Eq. (5-38).

Kirkham and Affleck (1977) performed a laboratory test for Eq. (5-37) with a sand tank sector model of $30°$; $r_w = 2.5$ cm, $R = 48.5$ cm, $b = 2$ cm, and $H - h_w = 24.5$ cm. The experimental results, shown in Figure 5-4, reveal that the expected straight line with slope $m = 0.0128$ min/cm^2. The theoretical value of m from Eq. (5-37) is

$$m = \frac{\varphi_e}{K_h} \frac{\ln(R/r_w)}{(H - h_w)} \tag{5-39}$$

After making appropriate substitutions in Eq. (5-39) for R, r_w, and $(H - h_w)$, and the value $m = 0.0128$ min/cm^2, one finds that Eq. (5-39) gives

$$\frac{\varphi_e}{K_h} = \frac{m(H - h_w)}{\ln(R/r_w)} = \frac{(0.0128 \text{ min/cm}^2)(24.5 \text{ cm})}{\ln(48.5 \text{ cm}/2.5 \text{ cm})} = 0.1058 \text{ min/cm}$$

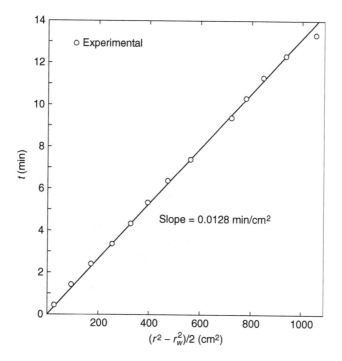

FIGURE 5-4 Experimental test of solute movement to a well. (Adapted from Kirkham, D. and Affleck, S.B., *Ground Water*, 15, 231–242, 1977.)

Independent determinations on the sand used in the model gave $\varphi_e = 0.40$ and $K_h = 3.9$ cm/min. For $\varphi_e/K_h = 0.1026$ min/cm, which checks the value 0.1058 min/cm given above.

EXAMPLE 5-2

This example is adapted from Kirkham and Affleck (1977). For an aquifer and a well, the following parameters are given: $\varphi_e = 0.25$, $K_h = 45$ m/day, $H - h_w = 3$ m, $R = 180$ m, and $r_w = 0.1$ m. Using these values, (1) determine the travel time of a solute to the well starting at $r = 18$ m; (2) determine the travel time of a solute to the well starting at $r = 9$ m; and (3) using Eq. (5-38), show that t varies directly with r^2 when r_w is small relative to r.

SOLUTION

(1) at $r_a = 18$ m, Eq. (5-37) gives

$$t_a = \frac{(0.25)\ln(180 \text{ m}/0.1 \text{ m})}{(45 \text{ m/day})(3 \text{ m})} \frac{[(18 \text{ m})^2 - (0.1 \text{ m})^2]}{2} = 2.2486 \text{ days}$$

(2) at $r_b = 9$ m, Eq. (5-37) gives

$$t_b = \frac{(0.25)\ln(180 \text{ m}/0.1 \text{ m})}{(45 \text{ m/day})(3 \text{ m})} \frac{[(9 \text{ m})^2 - (0.1 \text{ m})^2]}{2} = 0.5621 \text{ days}$$

(3) the t values in (1) and (2) are in the ratio

$$\frac{(2.2486 \text{ days})}{(0.5621 \text{ days})} = 4.0004 \cong 4$$

From Eq. (5-38) when r_w is small

$$t_a = \frac{2\pi\varphi_e b}{Q} \frac{r_a^2}{2}, \qquad t_b = \frac{2\pi\varphi_e b}{Q} \frac{r_b^2}{2}$$

Their ratio

$$\frac{t_a}{t_b} = \frac{r_a^2}{r_b^2} = \frac{(18 \text{ m})^2}{(9 \text{ m})^2} = 4$$

which agrees with the above value. This shows that t varies directly with r^2 when r_w is small relative to r, the distance of the source of solute from the well center.

A Single Well Case with Two Hydraulic Conductivities
In Figure 5-5, a planar view of a confined aquifer with two different hydraulic conductivity zones and a fully penetrating well is shown. For this case, some new notation is needed. In Figure 5-5, K_{h_I}, $K_{h_{II}}$, φ_{e_I}, and $\varphi_{e_{II}}$ are the horizontal hydraulic conductivities and effective porosities of two annular flow regions. For these regions, $r_w < r < r_c$ and $r_c < r < R$. The hydraulic heads at r_w, r_c, and R are h_w, h_c, and H, respectively. Equations for this case are derived to show effects on travel times when a small region around the well has a highly different value of hydraulic conductivity compared with the hydraulic conductivity elsewhere. The low hydraulic conductivity in region I may be the result of clogging of the aquifer's pores. It is considered that solutes are introduced only in region II.

The total time it takes for a solute to move from any point r in region II to the well will be the sum of the time $t_{r\to r_c}$ it takes to move to the interface and the time $t_{r_c\to r_w}$ it takes to move the interface to the well. Thus

$$t = t_{r\to r_c} + t_{r_c\to r_w} \tag{5-40}$$

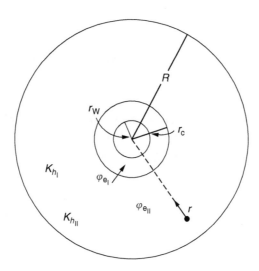

FIGURE 5-5 Solute movement in a confined aquifer with two horizontal hydraulic conductivities K_{h_I} and $K_{h_{II}}$, and effective porosities φ_{e_I} and $\varphi_{e_{II}}$. (Adapted from Kirkham, D. and Affleck, S.B., *Ground Water*, 15, 231–242, 1977.)

From Eq. (5-37) and comparison of Figures 5-3 and 5-5, the time $t_{r \to r_c}$ may be expressed as

$$t_{r \to r_c} = \frac{\varphi_{e_{II}}}{K_{h_{II}}} \frac{\ln(R/r_w)}{(H - h_c)} \frac{(r^2 - r_c^2)}{2} \tag{5-41}$$

and $t_{r_c \to r_w}$ may be expressed as

$$t_{r_c \to r_w} = \frac{\varphi_{e_I}}{K_{h_I}} \frac{\ln(R/r_w)}{(h_c - h_w)} \frac{(r_c^2 - r_w^2)}{2} \tag{5-42}$$

In Eqs. (5-41) and (5-42), it is assumed that all values on the right-hand sides, except h_c, are known.

To get h_c, let Q_I and Q_{II} be the equations of water per unit time that follow through regions I and II. By the continuity principle, Q_I and Q_{II} must be equal. From Eq. (5-28), Q_I and Q_{II} can, respectively, be written as

$$Q_I = \frac{2\pi K_{h_I} b(h_c - h_w)}{\ln(r_c/r_w)} \tag{5-43}$$

and

$$Q_{II} = \frac{2\pi K_{h_{II}} b(H - h_c)}{\ln(R/r_c)} \tag{5-44}$$

Making equal Q_I and Q_{II}, as given by Eqs. (5-43) and (5-44), respectively, and solving the result for h_c gives

$$h_c = \frac{h_w K_{h_I} \ln(R/r_c) + H K_{h_{II}} \ln(r_c/r_w)}{K_{h_I} \ln(R/r_c) + K_{h_{II}} \ln(r_c/r_w)} \tag{5-45}$$

Use of Eq. (5-45) in Eq. (5-41) gives

$$t_{r \to r_c} = \frac{\varphi_{e_{II}}}{K_{h_I} K_{h_{II}}} \frac{[K_{h_I} \ln(R/r_c) + K_{h_{II}} \ln(r_c/r_w)]}{(H - h_w)} \frac{(r^2 - r_c^2)}{2} \tag{5-46}$$

and in Eq. (5-42) gives

$$t_{r_c \to r_w} = \frac{\varphi_{e_I}}{K_{h_I} K_{h_{II}}} \frac{[K_{h_I} \ln(R/r_c) + K_{h_{II}} \ln(r_c/r_w)]}{(H - h_w)} \frac{(r_c^2 - r_w^2)}{2} \tag{5-47}$$

and use of Eqs. (5-46) and (5-47) in Eq. (5-40) gives

$$t = \frac{K_{h_\mathrm{I}}\ln(R/r_\mathrm{c}) + K_{h_\mathrm{II}}\ln(r_\mathrm{c}/r_\mathrm{w})}{2K_{h_\mathrm{I}}K_{h_\mathrm{II}}(H - h_\mathrm{w})}[r^2\varphi_{e_\mathrm{II}} + r_\mathrm{c}^2(\varphi_{e_\mathrm{I}} - \varphi_{e_\mathrm{II}}) - r_\mathrm{w}^2\varphi_{e_\mathrm{I}}] \qquad (5\text{-}48)$$

As a special case for $\varphi_{e_\mathrm{I}} = \varphi_{e_\mathrm{II}}$ and $K_{h_\mathrm{I}} = K_{h_\mathrm{II}} = K_h$, Eq. (5-48) reduces to Eq. (5-37) in view of $\ln (R/r_\mathrm{c}) + \ln (r_\mathrm{c}/r_\mathrm{w}) = \ln (R/r_\mathrm{w})$.

An approximate form of Eq. (5-48) can be developed. If in Eq. (5-48) the last two terms in the brackets are small compared with the first, the travel time t is given approximately by the expression

$$t \cong \frac{\varphi_{e_\mathrm{II}}r^2[K_{h_\mathrm{I}}/K_{h_\mathrm{II}}\ln(R/r_\mathrm{c}) + \ln(r_\mathrm{c}/r_\mathrm{w})]}{2K_{h_\mathrm{I}}(H - h_\mathrm{w})} \qquad (5\text{-}49)$$

Equation (5-49) should only be used as an approximation of travel time when the $K_{h_\mathrm{I}}/K_{h_\mathrm{II}}$ ratio is less than 0.01. If in Eq. (5-49) the expression $K_{h_\mathrm{I}}/K_{h_\mathrm{II}}\ln (r/r_\mathrm{c})$ is very small compared with $\ln(R/r_\mathrm{c})$, the travel time of Eq. (5-49) becomes approximately

$$t = \frac{\varphi_{e_\mathrm{II}}}{K_{h_\mathrm{I}}}\frac{\ln(r_\mathrm{c}/r_\mathrm{w})}{(H - h_\mathrm{w})}\frac{r^2}{2} \qquad (5\text{-}50)$$

Eqation (5-48) gives the travel time of dissolved solute to a single well when the annular region around the well is divided by radius r_c into two regions of different horizontal hydraulic conductivities, K_{h_I} and K_{h_II} and different effective porosities, φ_{e_I} and φ_{e_II}, where region I is in contact with the well and region II extends from r_c to the outer boundary R. Region I in many old wells becomes clogged with fines and chemical precipitates. This results in a lowered effective porosity φ_{e_I} and a highly reduced K_{h_I}.

EXAMPLE 5-3

Consider the well parameters in Example 5-2. If clogging occurred out to a radius $r_\mathrm{c} = 1$ m such that K_{h_I} is reduced from 45 to 0.50 m/day and effective porosity is reduced from $\varphi_{e_\mathrm{II}} = 0.25$ to $\varphi_{e_\mathrm{I}} = 0.20$, (1) calculate the travel time of a solute from $r = 18$ m using Eq. (5-48) and (2) calculate the travel time using the approximate equation as given by Eq. (5-49), and compare the results.

SOLUTION

(1) Substitution of the values $r_\mathrm{c} = 1$ m, $K_{h_\mathrm{I}} = 0.50$ m/day, $K_{h_\mathrm{II}} = 45$ m/day, $\varphi_{e_\mathrm{I}} = 0.20$, $\varphi_{e_\mathrm{II}} = 0.25$, $r = 18$ m, $R = 180$ m, $H - h_\mathrm{w} = 3$ m, and $r_\mathrm{w} = 0.1$ m, into Eq. (5-48) gives

$$t = \frac{(0.50 \text{ m/day})\ln(180 \text{ m/1 m}) + (45 \text{ m/day})\ln(1 \text{ m/0.1 m})}{2(0.50 \text{ m/day})(45 \text{ m/day})(3 \text{ m})}$$

$$\times [(18 \text{ m})^2(0.25) + (1 \text{ m})^2(0.20 - 0.25) - (0.1 \text{ m})^2(0.20)] = 63.69 \text{ days}$$

Notice that in Example 5-3, for the same distance the travel time is 2.2486 days. This is due to the fact that K_{h_I} is much less than K_{h_II}.

(2) Using the same values in the approximate equation as given by Eq. (5-49)

$$t = \frac{(0.25)(18 \text{ m})^2[(0.50 \text{ m/day})/(45 \text{ m/day}) \ln(180 \text{ m/1 m}) + \ln(1 \text{ m/0.1 m})]}{2(0.50 \text{ m/day})(3 \text{ m})} = 63.73 \text{ days}$$

which compares closely with the value determined in (1). The ratio of hydraulic conductivities is

$$\frac{K_{h_\mathrm{I}}}{K_{h_\mathrm{II}}} = \frac{0.50 \text{ m/day}}{45 \text{ m/day}} = 0.011$$

Minimum Travel Time under a Dam

In this case, travel times of solutes will be analyzed beneath a dam. For this analysis, problem statement and assumptions, Darcy velocity beneath the dam, travel time expression, and examples will be presented.

Problem Statement and Assumptions

The cross-section of a dam and ground water system in a semiinfinite aquifer is shown in Figure 5-6 with the representative streamlines and equipotential lines. The goal is to find the minimum travel time from the reservoir side toward the downgradient direction. Particles follow the streamlines shown in Figure 5-6. The shorter the total length of a stream line, the shorter the travel time is. In Figure 5-6, the streamline BOC is the shortest among all the other ones. Therefore, the minimum travel time occurs on this streamline. The analysis is based on the following assumptions: (1) the dam itself is impermeable; (2) the aquifer beneath the dam is homogeneous and isotropic, which means that the hydraulic conductivities in all directions and locations are the same ($K_x = K_y = K_z = K$, where K_x, K_y, and K_z are the hydraulic conductivities in the x, y, and z directions, respectively); (3) the upper boundary of the aquifer coincides with the lower boundary of the dam; and (4) the ground water flow is under steady-state conditions.

Darcy Velocity beneath the Dam

The variation of the Darcy velocity, $q_x(x)$ (L/T), with the x coordinate along BOC in Figure 5-6 is given as (Polubarinova-Kochina, 1962, p. 57; Harr, 1962, p. 90)

$$q_x(x) = \frac{KH}{\pi(b^2 - x^2)^{1/2}} \tag{5-51}$$

where K(L/T) is the hydraulic conductivity, H(L) the hydraulic head difference between the reservoir and downgradient sides of the dam, b(L) the half-width of the bottom of the dam, and x is the horizontal coordinate. Since BOC in Figure 5-6 is a straight line, the vertical Darcy velocity, q_y, is zero.

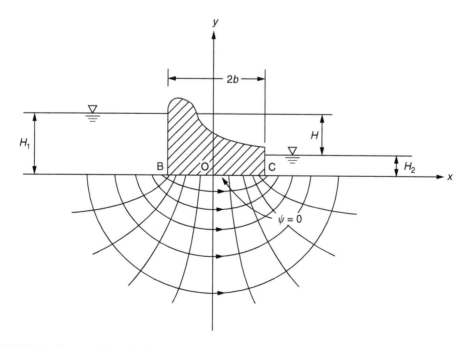

FIGURE 5-6   Cross-section of a dam and ground water system with streamlines and equipotential lines.

Travel Time Expression

The expression for the total travel time, T_t of a solute from point B to C along the BOC line is

$$T_t = \frac{\pi^2 b^2 \varphi_e}{2KH} \tag{5-52}$$

where φ_e is the effective porosity and the other parameters are defined previously. Derivation of Eq. (5-52) is presented in Example 5-4.

EXAMPLE 5-4

Derive Eq. (5-52).

SOLUTION

From Eq. (5-17), the ground water velocity can be expressed as

$$v_x(x) = \frac{dx}{dt} = \frac{q_x(x)}{\varphi_e} \tag{E5-4-1}$$

where $x(L)$ is the horizontal coordinate, t the time, $q_x(x)$ (L/T) the Darcy velocity in the x direction, and φ_e the effective porosity. From Eq. (E5-4-1)

$$dt = \frac{dx}{q_x(x)} \varphi_e \tag{E5-4-2}$$

After integration, Eq. (E5-4-2) gives

$$\int_0^{T_t} dt = T_t = \varphi_e \int_{-b}^{b} \frac{dx}{q_x(x)} = 2\varphi_e \int_0^{b} \frac{dx}{q_x(x)} \tag{E5-4-3}$$

where T_t is the travel time of a solute from point B to point C along the line BOC in Figure 5-6. Using Eq. (5-51) in Eq. (E5-4-3) results in the following expression:

$$T_t = \frac{2\pi\varphi_e}{KH} \int_0^{b} (b^2 - x^2)^{1/2} \, dx \tag{E5-4-4}$$

After evaluation of the integral in Eq. (E5-4-4), one has the following expression (see, e.g., Spiegel, 1968, Eq. [14.244], p. 70)

$$T_t = \frac{2\pi\varphi_e}{KH} \left| \frac{x(b^2 - x^2)^{1/2}}{2} + \frac{b^2}{2} \sin^{-1}\left(\frac{x}{b}\right) \right|_0^b \tag{E5-4-5}$$

or

$$T_t = \frac{2\pi\varphi_e}{KH} \left[\frac{b^2}{2} \sin^{-1}(1) - \frac{b^2}{2} \sin^{-1}(0) \right] \tag{E5-4-6}$$

Using the equalities

$$\sin^{-1}(1) = \frac{\pi}{2}, \qquad \sin^{-1}(0) = 0 \tag{E5-4-7}$$

in Eq. (E5-4-6), it takes exactly the same form as Eq. (5-52).

EXAMPLE 5-5

For the dam and ground water system shown in Figure 5-6, (1) determine time vs. travel time curve for $H = 2$ m, $2b = 2$ m, and $\varphi_e = 0.25$ and discuss the results; and (2) determine travel time vs. H for $K = 12$ m/day and $\varphi_e = 0.25$, and discuss the results.

SOLUTION

(1) Substituting the values in Eq. (5-52) gives

$$T_t = \frac{\pi^2(1 \text{ m})^2(0.25)}{2K(2 \text{ m})}$$

or

$$T_t \text{ (days)} = \frac{0.617}{K(\text{m/day})}$$

The variation of T_t vs. K is shown in Figure 5-7. Figure 5-7 shows that the travel time of a solute strongly depends on the hydraulic conductivity of the aquifer materials. The curve approaches the coordinate axes asymptotically, which means that for relatively low hydraulic conductivities, the travel times become large, and for relatively large hydraulic conductivities, the travel times become small. For example, for $K = 1$ m/day, the travel time is about 0.5 days; and for $K = 20$ m/day, the travel time is about 0.02 days.

(2) Substitution of the values into Eq. (5-52) gives

$$T_t = \frac{\pi^2(1 \text{ m})(0.25)}{2(12.0 \text{ m/day})H}$$

or

$$T_t \text{ (days)} = \frac{0.103}{H(\text{m})}$$

The variation of T_t vs. H is shown in Figure 5-8. As in (1), Figure 5-8 shows that the travel time of a solute strongly depends on the hydraulic head difference. Similarly, the curve approaches to the x- and y-coordinate axes, asymptotically which means that for relatively low hydraulic head differences the travel times become large and for relatively large hydraulic head differences, the travel times become small. Comparison of Figure 5-7 with Figure 5-8 indicates that the dependence of the travel time on the head difference is much more pronounced than dependence on the hydraulic conductivity. For example, for $H = 1$ m, the travel time is about 0.10 days; and for $H = 5$ m, the travel time is about 0.02 days.

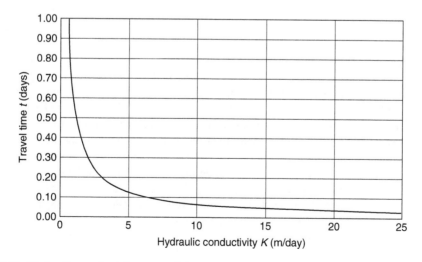

FIGURE 5-7 Hydraulic conductivity vs.travel time for Example 5-5.

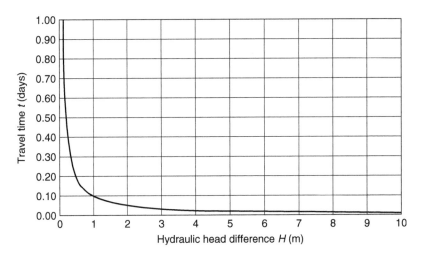

FIGURE 5-8   Hydraulic head difference vs. travel time for Example 5-5.

5.2.3   NUMERICAL METHODS FOR SOLUTE PATH LINES DETERMINATION

5.2.3.1   Introduction

The use of particle-tracking techniques to generate solute path lines and travel time information from the results of numerical ground water flow models can potentially be very helpful in analyzing complex two- and three-dimensional ground water flow and solute transport systems. As mentioned in Section 5.1, convection models cannot be used to compute solute concentrations in ground water because they do not account for the effects of mixing by dispersion. However, these models represent a valuable intermediate step between ground water flow models and convection–dispersion solute transport models.

For analytical solutions, the ground water velocity field is known directly from the solution at every point in the flow field. For numerical models, the ground water flow model first needs to be constructed to determine the hydraulic head distribution in the flow field; then, the ground water velocity is generated based on the hydraulic head distribution. For finite-difference models, it is necessary to establish an interpolation scheme such that the ground water velocity field can be determined at every point in the flow field by using the intercell flow rates from the finite-difference model. A variety of interpolation schemes are possible: (1) step function interpolation; (2) simple linear interpolation; and (3) multilinear interpolation. Step function interpolation assumes that the velocity components are constant between any two nodes, and that they change their values abrubtly at the nodes. Simple linear interpolation assumes that the principal velocity components vary linearly within a grid cell only with respect to their own coordinate system. That is, the velocity component in the x-coordinate direction within a cell varies linearly in the same coordinate direction, but is independent of y and z coordinates within the cell. Multilinear interpolation is similar to simple linear interpolation, except that each of the three principal components of velocity is assumed to be the linear function of all coordinate directions. Although these three interpolation schemes differ slightly from one another, each is capable of producing a reasonable representation of the velocity field for well-discretized ground water flow fields (Pollock, 1988, 1989).

In the following sections, the theory and practical applications of particle tracking based on the finite-difference method will be presented based mainly on Pollock (1988, 1989). This presentation will be made using the definitions and approaches of the U.S. Geological Survey modular three-dimensional ground water flow modeling code MODFLOW (McDonald and Harbaugh, 1984) and threedimensional particle tracking code MODPATH (Pollock, 1988). The MODPATH code has been designed such that it uses the output of MODFLOW for a given model.

5.2.3.2 Theory of Particle Tracking Based on the Finite-Difference Method

Conservation of Mass Equation
As mentioned above, particle tracking will be covered under steady-state ground water flow conditions. From Eq. (4-107) the partial differential equation describing conservation of mass in a steady-state, three-dimensional ground water flow system can be expressed as

$$\frac{\partial}{\partial x}\left(K_{xx}\frac{\partial h}{\partial x}\right) + \frac{\partial}{\partial y}\left(K_{yy}\frac{\partial h}{\partial y}\right) + \frac{\partial}{\partial z}\left(K_{zz}\frac{\partial h}{\partial z}\right) = W \tag{5-53}$$

where x, y, and z (units are L) are the principal coordinate axes of the system. In other words, they are assumed to be aligned along the principal hydraulic conductivities K_{xx}, K_{yy}, and K_{zz} (units L/T), respectively (see Section 4.3.3 for principal hydraulic conductivities). The remaining six components (K_{xy}, K_{xz}, K_{yx}, K_{yz}, K_{zx}, and K_{zy}) in the hydraulic conductivity tensor (Eq. [4-92]) all become equal to zero. Therefore, the corresponding hydraulic conductivity tensor for Eq. (5-53) is the tensor given by Eq. (4-96). In Eq. (5-53), h is the hydraulic head (unit L) and is defined by Eq. (4-1); W (unit 1/T) is the volumetric flux per unit volume and represents sources /or sinks of water. In general, K_{xx}, K_{yy}, and K_{zz} may be functions of space as

$$K_{xx} \equiv K_x(x, y, z), \qquad K_{yy} \equiv K_{yy}(x, y, z), \qquad K_{zz} \equiv K_z(x, y, z) \tag{5-54}$$

and h and W may be functions of space as

$$h \equiv h(x, y, z), \qquad W \equiv W(x, y, z) \tag{5-55}$$

Using Eqs. (4-195a), (4-195b), and (4-195c) in Eq. (5-53), one can write

$$\frac{\partial(\varphi_e v_x)}{\partial x} + \frac{\partial(\varphi_e v_y)}{\partial y} + \frac{\partial(\varphi_e v_z)}{\partial z} = W \tag{5-56}$$

where v_x, v_y, and v_z are the principal components of the average linear ground water velocity vector, and φ_e is the effective porosity. Equation (5-56) expresses conservation of mass for an infinitesimally small volume of aquifer. The finite-difference approximation of Eq. (5-56) can be visualized as a mass balance equation for a finite-sized cell aquifer that accounts for water flowing in and out of the cell, and represents source or sink water within the cell. Figure 5-9 shows a finite-sized cell of aquifer and the components of inflow and outflow across its six faces.

The six faces in Figure 5-9 are referred to as x_1, x_2, y_1, y_2, z_1, and z_2. Face x_1 is the face perpendicular to the x-coordinate direction at $x = x_1$. Similar definitions are valid for the other five faces. The average linear ground water velocity component across each face in the cell (i, j, k) is obtained by dividing the volume flow rate across the face by the cross-sectional area of the face and the effective porosity of the cell. Therefore, the ground water velocities are

$$v_{x_1} = \frac{Q_{x_1}}{\varphi_e \Delta y \Delta z}, \qquad v_{x_2} = \frac{Q_{x_2}}{\varphi_e \Delta y \Delta z} \tag{5-57}$$

$$v_{y_1} = \frac{Q_{y_1}}{\varphi_e \Delta x \Delta z}, \qquad v_{y_2} = \frac{Q_{y_2}}{\varphi_e \Delta x \Delta z} \tag{5-58}$$

$$v_{z_1} = \frac{Q_{z_1}}{\varphi_e \Delta x \Delta y}, \qquad v_{z_2} = \frac{Q_{z_2}}{\varphi_e \Delta x \Delta y} \tag{5-59}$$

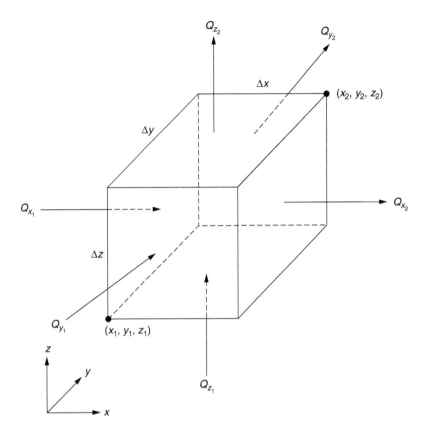

FIGURE 5-9 Orientation of finite-difference cells and definition of cell face flows. (Adapted from Pollock, D.W., *Ground Water*, 26, 743–750, 1988.)

where Q is a volume flow rate across a cell face, and Δx, Δy, and Δz are the dimensions of the cell in the respective coordinate directions. If flow to internal sources or sinks within the cell is specified as Q_s and assuming that the effective porosity (φ_e) is constant within the cell, the finite-difference form of Eq. (5-56) can be written as

$$\frac{\varphi_e(v_{x_2} - v_{x_1})}{\Delta x} + \frac{\varphi_e(v_{y_2} - v_{y_1})}{\Delta y} + \frac{\varphi_e(v_{z_2} - v_{z_1})}{\Delta z} = \frac{Q_s}{\Delta x \Delta y \Delta z} \tag{5-60}$$

The left-hand side of Eq. (5-60) represents the net volume rate of outflow per unit volume of the cell, and the right-hand side represents the net volume rate of production per unit volume due to internal sources or sinks. Equation (5-60) results in a set of algebraic equations expressed in terms of heads at nodes located at the cell centers. The solution of that set of algebraic equations yields the values of head at the node points. Once the head solution has been obtained, the intercell flow rates can be determined from the Darcy equation by using the values of head at the node points. The MODFLOW code (McDonald and Harbaugh, 1984) solves for head and calculates intercell flow rates, and these are input for MODPATH (Pollock, 1989).

Computation Procedure of Path Lines under Unsteady-State Flow Conditions by MODPATH
Mass Balance within a Cell
In order to determine path lines, a method must be established to compute values of the ground water velocity components at every point in the flow field based on the intercell flow rates from the finite-difference ground water flow model. The algorithm described by Pollock (1988, 1989) uses

simple linear interpolation to determine the ground water velocity components at every point within a cell. Using simple linear interpolation, the velocity components can be expressed as

$$v_x = \frac{v_{x_2} - v_{x_1}}{\Delta x}(x - x_1) + v_{x_1} \tag{5-61}$$

$$v_y = \frac{v_{y_2} - v_{y_1}}{\Delta y}(y - y_1) + v_{y_1} \tag{5-62}$$

$$v_z = \frac{v_{z_2} - v_{z_1}}{\Delta z}(z - z_1) + v_{z_1} \tag{5-63}$$

or

$$v_x = A_x(x - x_1) + v_{x_1} \tag{5-64}$$

$$v_y = A_y(y - y_1) + v_{y_1} \tag{5-65}$$

$$v_z = A_z(z - z_1) + v_{z_1} \tag{5-66}$$

where

$$A_x = \frac{v_{x_2} - v_{x_1}}{\Delta x} \tag{5-67}$$

$$A_y = \frac{v_{y_2} - v_{y_1}}{\Delta y} \tag{5-68}$$

$$A_z = \frac{v_{z_2} - v_{z_1}}{\Delta z} \tag{5-69}$$

Linear interpolation generates a continuous velocity field within each individual cell that identically satisfies the conservation of mass equation, (5-56), everywhere within the cell. This situation can be shown as follows: substituting Eqs. (5-64)–(5-66) into Eq. (5-56) and assuming that the effective porosity φ_e is constant within a cell, one can write

$$\varphi_e A_x + \varphi_e A_y + \varphi_e A_z = W \tag{5-70}$$

and using Eqs. (5-67)–(5-69) in Eq. (5-70)

$$\frac{\varphi_e(v_{x_2} - v_{x_1})}{\Delta x} + \frac{\varphi_e(v_{y_2} - v_{y_1})}{\Delta y} + \frac{\varphi_e(v_{z_2} - v_{z_1})}{\Delta z} = W \tag{5-71}$$

The left-hand side of Eq. (5-71) is exactly the same as the left-hand side of Eq. (5-60). Therefore, linear interpolation of the six cell face velocity components results in a velocity field within the cell that automatically satisfies Eq. (5-56) at every point inside the cell on the assumption that internal sources or sinks are considered to be uniformly distributed within the cell.

The Movement of a Particle in a Finite-Difference Cell
Consider the movement of a particle, *p*, through a three-dimensional finite-difference cell. The rate of change in v_x as it moves through the cell is given by

$$\left(\frac{dv_x}{dt}\right)_p = \left(\frac{dv_x}{dx}\right)\left(\frac{dx}{dt}\right)_p \tag{5-72}$$

For notation simplification, the subscript p is used to indicate that a term is evaluated at the location of the particle whose coordinates are (x_p, y_p, z_p). For example, the term $(dv_x/dt)_p$ is the time rate of change of v_x evaluated at the location of the particle. In Eq. (5-72), the term $(dx/dt)_p$ is the time rate of change of the particle at the x-coordinate location. By definition

$$v_{x_p} = \left(\frac{dx}{dt}\right)_p \tag{5-73}$$

where v_{x_p} is the x component of velocity for the particle. Differentiating Eq. (5-64) with respect to x yields the additional relation

$$\frac{dv_x}{dx} = A_x \tag{5-74}$$

Substituting Eqs. (5-73) and (5-74) into Eq. (5-72) gives

$$\left(\frac{dv_x}{dt}\right)_p = A_x v_{x_p} \tag{5-75}$$

Similar equations can be obtained in the y and z directions as

$$\left(\frac{dv_y}{dt}\right)_p = A_y v_{y_p} \tag{5-76}$$

$$\left(\frac{dv_z}{dt}\right)_p = A_z v_{z_p} \tag{5-77}$$

Equations (5-75), (5-76), and (5-77) can also be written in the form

$$\frac{d(v_{x_p})}{v_{x_p}} = A_x\, dt \tag{5-78}$$

$$\frac{d(v_{y_p})}{v_{y_p}} = A_y\, dt \tag{5-79}$$

$$\frac{d(v_{z_p})}{v_{z_p}} = A_z\, dt \tag{5-80}$$

Expressions for Particle Coordinates
First, the x-coordinate expression will be derived from Eq. (5-78). Equation (5-78) can be written as

$$\frac{dv_{x_p}}{v_{x_p}} = A_x\, dt \tag{5-81}$$

in which v_{x_p} is the x component of the particle p as it moves through the medium. In other words, v_{x_p} is the velocity component in the x direction that one would observe if one were moving with the particle. As a result, v_{x_p} will not generally be constant, even under steady flow conditions, because its value may change as a function of position. Integration of Eq. (5-81) over the time interval t_1 to t_2 gives

$$\int_{v_{x_p}(t_1)}^{v_{x_p}(t_2)} \frac{dv_{x_p}}{v_{x_p}} = A_x \int_{t_1}^{t_2} dt \tag{5-82}$$

or after some manipulations

$$\ln\left[\frac{v_{x_p}(t_2)}{v_{x_p}(t_1)}\right] = A_x(t_2 - t_1) = A_x \Delta t \tag{5-83}$$

where $v_{x_p}(t_1)$ is the x component of the velocity of the particle at time t_1, and $v_{x_p}(t_2)$ is the x component of the velocity of the particle at time t_2. From Eq. (5-83) one can write

$$v_{x_p}(t_2) = v_{x_p}(t_1) \exp(A_x \Delta t) \tag{5-84}$$

From Eq. (5-64), $v_{x_p}(t_2)$ is given by the linear interpolation equation as

$$v_{x_p}(t_2) = A_x[x_p(t_2) - x_1] + v_{x_1} \tag{5-85}$$

$x_p(t_2)$ is the x coordinate of the particle at time t_2. Substituting Eq. (5-85) into Eq. (5-84) gives

$$v_{x_p}(t_1) \exp(A_x \Delta t) = A_x[x_p(t_2) - x_1] + v_{x_1} \tag{5-86}$$

or

$$x_p(t_2) = x_1 + \frac{1}{A_x}[v_{x_p}(t_1) \exp(A_x \Delta t) - v_{x_1}] \tag{5-87}$$

Similar equations can be derived in the y and z directions as

$$y_p(t_2) = y_1 + \frac{1}{A_y}[v_{y_p}(t_1) \exp(A_y \Delta t) - v_{y_1}] \tag{5-88}$$

$$z_p(t_2) = z_1 + \frac{1}{A_z}[v_{z_p}(t_1) \exp(A_z \Delta t) - v_{z_1}] \tag{5-89}$$

The velocity components $v_{x_p}(t_1)$, $v_{y_p}(t_1)$, and $v_{z_p}(t_1)$ are known functions of the coordinates of the particle at time t_1. Therefore, the coordinates of the particle at any future time t_2 can be determined directly from Eqs. (5-87)–(5-89).

Computation Procedure of Path Lines under Steady-State Flow Conditions by MODPATH
For steady-state flow conditions, the direct integration method described above can be reduced to a simple algorithm that allows the exit point of a particle from a cell to be determined directly, given any known starting location within the cell. To illustrate the method, consider the two-dimensional example shown in Figure 5-10. Cell (i, j) is in the x–y plane and contains a particle, P, located at (x_p, y_p) at time t_p. In this example, it is assumed that v_{x_1} and v_{x_2} are greater than zero. This means that water flows into the cell through face x_1 and out of the cell through face x_2. Similarly, it is assumed that v_{y_1} and v_{y_2} are also greater than zero, so that water flows into the cell through face y_1 and out of the cell through face y_2. The first step is to determine the face across which particle P passes as it leaves the cell (i, j). In this example, this is accomplished by noting that the velocity components at the four faces require that particle P leaves the cell through either face x_2 or face y_2. Now, let us consider the x direction first. From Eq. (5-64), v_{x_p} can be calculated at the point (x_p, y_p). Since v_x equals v_{x_2} at face x_2, Eq. (5-83) can be used to determine the time that would be required for particle P to reach face x_2 as

$$\Delta t_x = \frac{1}{A_x} \ln\left(\frac{v_{x_2}}{v_{x_p}}\right) \tag{5-90}$$

A similar calculation can be made to determine the time required for particle P to reach face y_2 as

$$\Delta t_y = \frac{1}{A_y} \ln\left(\frac{v_{y_2}}{v_{y_p}}\right) \tag{5-91}$$

In Eqs. (5-90) and (5-91), v_{x_p} and v_{y_p} are the x and y components of the velocity of the particle P at (x_p, y_p). If Δt_x is less than Δt_y, particle P will leave the cell across face x_2 and enter cell $(i, j + 1)$. Conversely, if Δt_y is less than Δt_x, particle P will leave the cell across face y_2 and enter cell $(i - 1, j)$.

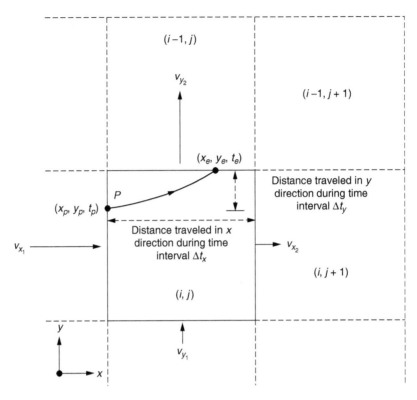

FIGURE 5-10   Schematic representation of a particle path through a two-dimensional cell. (Adapted from Pollock, D.W., *Ground Water*, 26, 743–750, 1988.)

A third possibility is that Δt_x and Δt_y are equal, in which case particle P would leave through the corner cell $(i-1, j+1)$. The particle trajectory shown in Figure 5-10 corresponds to a situation where Δt_y is less than Δt_x. The length of time required for particle P to travel from point (x_p, y_p) to a boundary face of cell (i, j) is taken to be the smaller of Δt_x and Δt_y, and this is denoted by Δt_e. The value of Δt_e is then used in Eqs. (5-87), (5-88), and (5-89) to determine the exit coordinates (x_e, y_e) for particle P as it leaves cell (i, j) as

$$x_e = x_1 + \frac{1}{A_x}[v_{x_p}(t_p) \exp(A_x \Delta t_e) - v_{x_1}] \qquad (5\text{-}92)$$

$$y_e = y_1 + \frac{1}{A_y}[v_{y_p}(t_p) \exp(A_y \Delta t_e) - v_{y_1}] \qquad (5\text{-}93)$$

The time at which particle P leaves the cell is given by

$$t_e = t_p + \Delta t_e \qquad (5\text{-}94)$$

In MODPATH, this sequence of calculations is repeated cell by cell until the particle reaches a discharge point. This approach can be generalized to three-dimensional cases by performing all of the calculations for the z direction in addition to the x and y directions. A flowchart outlining the algorithm for a steady-state flow system is shown in Figure 5-11. The accuracy of path line computations for steady-state flow fields is not affected by time step size because path lines are integrated analytically with respect to time. The ultimate discharge point and total time of travel for a given particle will be identical regardless of the number of discrete time steps taken.

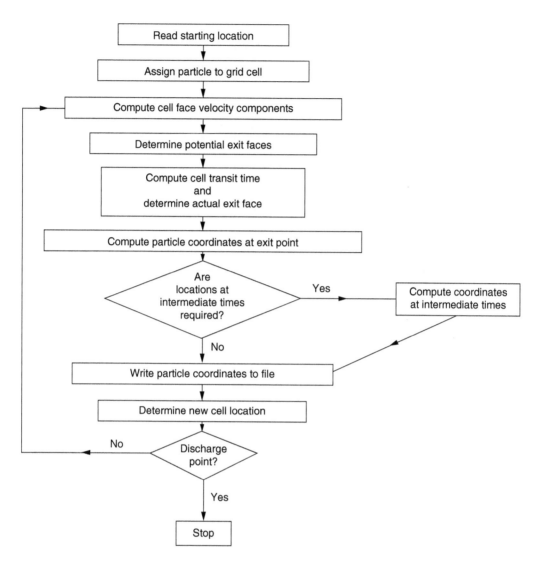

FIGURE 5-11 Flow chart for particle tracking algorithm for steadystate ground water flow systems. (Adapted from Pollock, D.W., *Ground Water*, 26, 743–750, 1988.)

A Special Case Regarding the Velocity Components

There may be a case where all of the velocity components at the cell faces are nonzero and in the positive x or positive y directions. These conditions may not always exist. Figure 5-12 shows all other possible situations that can occur in any of the three coordinate directions. Figure 5-12a shows the case where v_{x_1} and v_{x_2} are in the opposite directions and flow is into the cell through both faces x_1 and x_2. It is obvious the this case that once a particle enters the cell, it cannot leave the cell in the x direction. When this algorithm is implemented, a check is made in MODPATH if this condition exists for a given coordinate direction. If so, a flag is set to indicate that the particle cannot leave the cell across either of the faces in that direction. When this situation prevails in all the three coordinate directions, it indicates that a strong sink is present within the cell and no outflow can occur. Figure 5-12b shows a second alternative in which v_{x_1} and v_{x_2} are in opposite directions and flow is out of the cell through faces x_1 and x_2. This condition implies that a local flow divide exists for the x direction somewhere within the cell. For this situation, the potential exit face in the x direction is

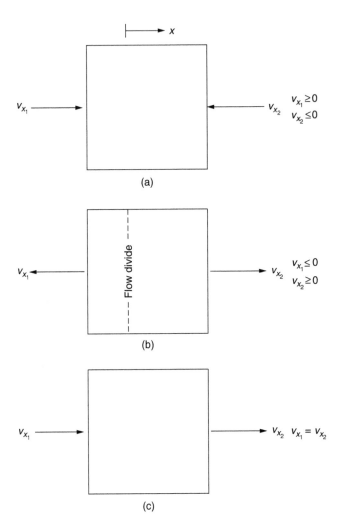

FIGURE 5-12 Possible orientations of cell face velocity components in determining potential exit faces. (Adapted from Pollock, D.W., *Ground Water*, 26, 743–750, 1988.)

determined by checking the sign of v_{x_p}. If v_{x_p} is less than zero, the particle has the potential to leave the cell across face x_1. On the other hand, if v_{x_p} is greater than zero, the particle has the potential to leave the cell in the x direction only through face x_2. After the appropriate potential exit face has been determined for a coordinate direction, the transit time for that direction can be determined as outlined previously. Finally, the case where v_{x_1} is nonzero and equal to v_{x_2} (Figure 5-12c) must be considered a special case because the quotient ln (1)/0 that results makes the formulation indeterminate and cannot be computed. When this is the case, Eq. (5-90) is bypassed and the transit time in the x direction is determined from the simple relations:

$$\Delta t_x = \frac{(x_2 - x_p)}{v_{x_1}}, \quad v_{x_1} > 0 \tag{5-95a}$$

or

$$\Delta t_x = \frac{(x_1 - x_p)}{v_{x_1}}, \quad v_{x_1} < 0 \tag{5-95b}$$

Modifications for Special Cases

The basic algorithm described in the preceding section has been adapted in the computer program MODPATH (Pollock, 1989) to deal with three special cases: (1) grids with nonrectangular vertical discretization; (2) water table layers; and (3) quasi-three-dimensional representation of confining layers. These are described below.

Nonrectangular Vertical Discretization

The development presented previously is based on the assumption that the flow domain is discretized into a three-dimensional rectangular grid of horizontal rectangular cells. However, in practice, potentially three-dimensional finite-difference models use a rectangular grid in the horizontal plane and a deformed grid in the vertical direction to allow grid cells to conform to stratigraphic units with variable thickness, and are not perfectly horizontal. The particle tracking algorithm presented previously for MODPATH can be used to determine path lines for deformed grids. Figure 5-13a shows a hydrogeologic system having cells that conform the stratigraphy. Figure 5-13b and Figure 5-13c show how deformed finite-difference cells are represented in the particle tracking algorithm.

The MODPATH code uses local coordinate, z_L, and its definition is

$$z_L = \frac{z - z_1}{z_2 - z_1} \tag{5-96}$$

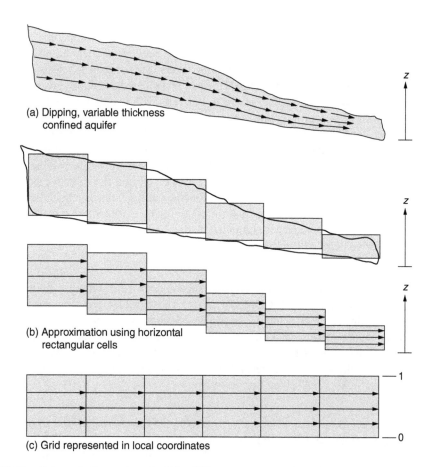

(a) Dipping, variable thickness confined aquifer

(b) Approximation using horizontal rectangular cells

(c) Grid represented in local coordinates

FIGURE 5-13 Schematic illustration of a finite-difference representation of an inclined aquifer with variable thickness. (Adapted from Pollock, D.W., *Ground Water*, 26, 743–750, 1988.)

where z_1 and z_2 are the elevations of the bottom and top of the cell, respectively. According to Eq. (5-96), the local z coordinate equals 0 at the bottom of the cell and 1 at the top of the cell. When a particle is transferred laterally from one cell to another, its local coordinate remains the same. If a particle leaves a cell at a position halfway between the top and bottom of the cell, it is assumed that it enters the neighboring cell halfway between the top and bottom of the cell, regardless of how the thickness or absolute elevation of the layer changes from one cell to the next. This procedure is illustrated schematically in Figure 5-13 for the case of lateral flow in a confined aquifer of variable thickness and dip. When all the layers are constant in thickness and are horizontal, this approach reduces exactly to the algorithm presented above for true rectangular grids.

Water Table Layers
For water table layers, the saturated thickness of cells changes areally with the slope of the water table and the bottom elevation of cells within the layers. The top elevation of a cell in a water table layer is set equal to the head in the cell. Therefore, water table layers vary in thickness even for true three-dimensional rectangular grids. The MODPATH particle tracking algorithm treats water table layers in the same way as the variable thickness stratigraphic layers described above.

Quasi-Three-Dimensional Representation of Confining Layers
A ground water system generally includes high-permeable formations (aquifers) and less-permeable formations (confining layers or aquitards). As mentioned in Section 4.3.12.2, the assumption made is that the flow is vertical through the confining layers and horizontal in the aquifers. The validity of this assumption has been investigated and the details are given in Section 4.3.12.2. In MODFLOW and MODPATH, confining layers are often not simulated as active layers in finite-difference models. Instead, their effect on vertical flow between aquifers is implicitly accounted for by determining the effective vertical hydraulic conductance between aquifers based on the vertical hydraulic conductivity and thickness of the confining layers. This approach is referred as a quasi-three-dimensional representation. In MODPATH, each unsimulated confining layer is assumed to be part of the active model layer directly above it. For cells with underlying confining layers, the local z coordinate within the confining layer varies linearly from -1 at the base of the confining layer to 0 at the top of the confining layer (Figure 5-14). It is assumed that one-dimensional, steady-state, vertical flow exists throughout the confining layer. This means that the average vertical linear velocity is constant throughout the confining layer and its magnitude equals the volumetric flow rate between adjacent model layers divided by the area of the cell and the effective porosity of the confining layer. When a particle reaches a top or bottom face of a cell that is a boundary of a confining layer, the particle is moved vertically across the confining layer into the next active model layer. Travel time across the confining layer is computed by dividing the thickness of the confining layer by the average linear velocity within the confining layer.

Boundary Conditions and Discharge Points
The intercell flow rates for all active cells are computed and stored as output from the block centered flow (BCF) package of the MODFLOW code. Those values are then used as input in MODPATH to determine cell face velocity components. Special consideration is necessary for cells that incorporate boundary conditions. As mentioned in Section 4.3.11 of Chapter 4, the MODFLOW code accounts for three types of boundary conditions: (1) specified head at a node; (2) specified flow rate to or from a cell; and (3) head-dependent flow rate to or from a cell. The treatment of these boundary conditions by MODPATH are discussed below.

Specified Head Cells
Specified head cells act as net sources or sinks of water to the flow system. For the purpose of path line computation of MODPATH, specified head cells are treated as active cells that contain a net source or sink. Path lines, computed through an isolated specified head cell surrounded entirely by variable head cells, are consistent with the existence of a uniformly distributed source or sink of water within the cell. When a number of specified head cells are adjacent to one another, path lines

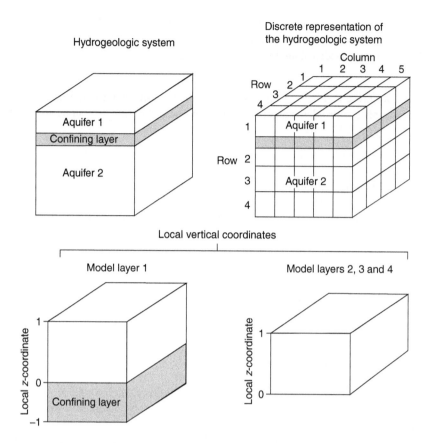

FIGURE 5-14 Definition of local vertical coordinates for the cases of true three-dimensional and quasi-three-dimensional systems. (Adapted from Pollock, D.W., *Ground Water*, 26, 743–750, 1988.)

computed through those cells will usually be misleading and inappropriate due to the fact that the MODFLOW code does not compute rates of flow between adjacent specified head cells. Therefore, path lines in the vicinity of specified head cells should always be examined carefully to make sure that the specified head cells are not exerting an unrealistic artificial constraint on direction or travel time.

Specified Flow-Rate Cells

Finite-difference flow equations represent volumetric water balances for individual grid cells. At the scale of an individual grid cell, the finite-difference representation of the ground water flow equation contains no information about the spatial distribution of specific components of flow into or out of the cell. For example, the finite-difference flow equations cannot distinguish between the case where water flows into a cell across a boundary face and the case where water is injected into the cell at the same rate by a well with no water flowing across the boundary face. For this reason, specified flux boundaries are accounted for in the MODFLOW code by placing wells in cells that contain flux boundaries. In the preceding example, the head distributions resulting from the solution of the finite-difference equations would be identical. In contrast, path lines computed for these two cases depend strongly on the distribution of flow within the cell.

Head-Dependent Flow-Rate Cells

The MODFLOW code includes options to simulate rivers, drains, general boundary fluxes, and evapotranspiration as head-dependent fluxes. Unlike specified fluxes, head-dependent fluxes are part of the solution to the flow problem; therefore, they must be stored as budget output from the MODFLOW code. Those values are then entered as data to MODPATH and either assigned to a specific cell face or treated as distributed sources or sinks.

Examples
Two examples are presented here to illustrate the application of the particle-tracking technique. In these examples, only the problems and results are presented. Intermediate steps are not presented.

EXAMPLE 5-6

This example is taken from Pollock (1988). Water is injected at a constant rate of 160,000 ft ^{3}/day into a well located in a confined aquifer with a thickness of 100 ft and hydraulic conductivity of 10 ft/day, and effective porosity of 0.30. The hydraulic head at a radial distance of 4000 ft from the well is held constant at 100 ft. Schematic description of the problem is shown in Figure 5-15. Determine the path lines and arrival locations of particles released from the well.

SOLUTION

For this problem, the velocity at any point in the flow field is one-dimensional in the radial direction, and an exact analytical solution describing the position of a particle as a function of time can be obtained (see Section 5.2.2.2). The symmetry of the problem requires that only one fourth of the circular flow field be considered. Figure 5-16 shows the finite-difference grid and boundary conditions for flow through one fourth of the system as shown in Figure 5-15. The center of the well is located in the lower left corner of Figure 5-16, and an injection rate of 40,000 ft³/day is assigned to the cell (row = 40, column = 1). The finite-difference solution was obtained from the MODFLOW code (McDonald and Harbaugh, 1984).

Ten particles were placed at a radial distance of 150 ft from the center of the well. The points in Figure 5-17 show the positions of the particles after 2500, 5000, and 7500 days. The dashed curves represent the predicted position of the particles at these points in time based on the analytical

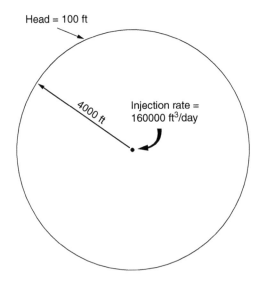

FIGURE 5-15 Schematic description of Example 5-6. (Adapted from Pollock, D.W., *Ground Water*, 26, 743–750, 1988.)

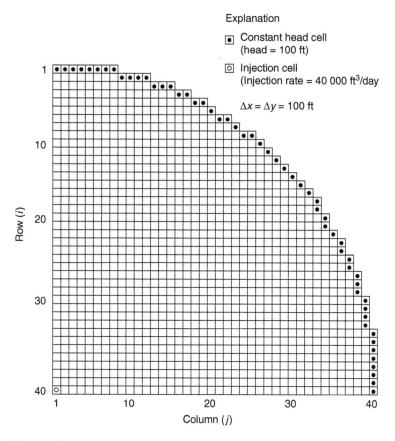

FIGURE 5-16 Finite-difference grid and boundary conditions for Example 5-6. (Adapted from Pollock, D.W., *Ground Water*, 26, 743–750, 1988.)

solution. The numerically generated particle path lines are shown as solid lines. The numerical results agree well with the analytically determined particle positions.

EXAMPLE 5-7

This example is taken from Pollock (1988), which is illustrated schematically in Figure 5-18. On the left-hand side there is an impermeable retaining wall having 50 -ft thickness of fine sand, which is overlain by a 5 ft of water. On the right-hand side, there is seepage face at an elevation of 25 ft above the base of the aquifer. Determine the path lines of water particles.

SOLUTION

The head difference between the reservoir and the seepage face causes water to flow downward from the reservoir, laterally under the wall, and discharge upward to the seepage face. Figure 5-19 shows the finite-difference grid and boundary conditions used to simulate constant heads at 55 ft at the base of the reservoir and 25 ft at the top of the cells (row=6, column=12) through (row=6, column=21) in row=7. Figure 5-20 shows path lines for five particles. Figure 5-20 also shows the equipotential lines generated from the hydraulic heads output of MODFLOW. It can be observed from Figure 5-20 that a good orthogonal relation exists between the path lines and the equipotential lines.

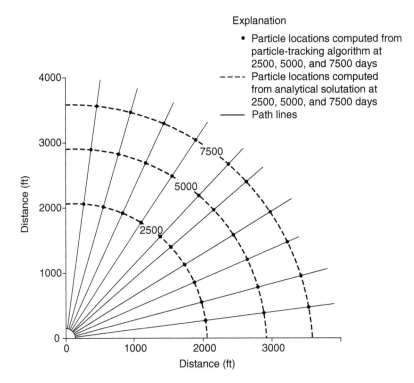

FIGURE 5-17 Comparison of numerically and analytically determined particle locations for Example 5-6. (Adapted from Pollock, D.W., *Ground Water*, 26, 743–750, 1988.)

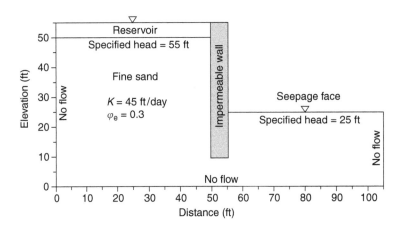

FIGURE 5-18 Schematic description of Example 5-7. (Adapted from Pollock, D.W., *Ground Water*, 26, 743–750, 1988.)

5.3 PATH LINES OF SOLUTE PARTICLES UNDER CONVECTIVE–DISPERSIVE TRANSPORT CONDITIONS

5.3.1 INTRODUCTION

In Section 5.2, the theory and application for path lines of solute particles under only convective (or advective) transport conditions are presented. This approach neglects the dispersive transport effects and is based on the stream function concept. The stream function concept and techniques have been

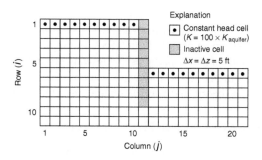

FIGURE 5-19 Finite-difference grid and boundary conditions for Example 5-7. (Adapted from Pollock, D.W., *Ground Water*, 26, 743–750, 1988.)

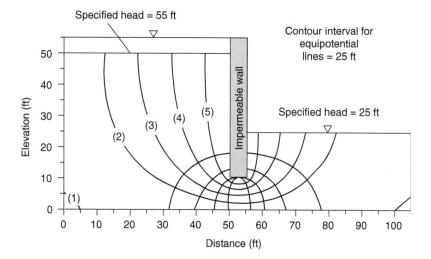

FIGURE 5-20 Path lines and equipotential lines for Example 5-7. (Adapted from Pollock, D.W., *Ground Water*, 26, 743–750, 1988.)

widely used in different branches of hydrodynamics for the analysis of fluid motion both in surface and subsurface flow systems. A fluid motion may be visualized as a flow pattern consisting of streamlines that are tangential to the instantaneous velocity vectors. On the basis of this definition, numerous publications have been presented for the analysis of potential stream functions in the 19th and 20th centuries. Solutions given in those publications have been widely used for the design of various hydraulic structures and water resources engineering area. These solutions are variable tools, but they cannot account for dispersion effects during transport of solutes in porous media. Solutes move under the effects of convection, the fluctuations in the solute velocity field (or dispersion), and molecular diffusion. Because of dispersion, the path lines of solute particles may not match the ground water flow streamlines even in uniform ground water flow-field cases. Solute plumes become larger and larger as the distance from the source increases. Some zones in the flow field remain clean. The transition zone or the line between solute and no-solute transport portions of the flow field is very important in controlling the solute plume. If one could predict the path lines of solute transport in a ground water flow field, the transition zones or lines could be specified. In other words, with this methodology, the zone or zones of solute can be determined.

Introduction of the stream function concept for the hydrodynamic dispersion in porous media is fairly new as compared with the stream function concept for flow. This approach was introduced in the literature toward the end of the 1980s (Batu, 1987, 1988). To the author's knowledge, the application of the stream function concept to hydrodynamic dispersion was not introduced in the literature previously. As an introductory approach, Batu (1987, 1988) presented a stream function concept for the analysis of the path lines of solute particles in a flowing ground water flow field under steady-state flow and transport conditions. In the following sections, the hydrodynamic dispersion stream function (HDSF) concepts, its theory, and application are presented based on Batu (1987, 1988).

5.3.2 Stream Function Concept for Hydrodynamic Dispersion

It is well known from classical hydrodynamics that stream function is a concept based on the continuity principle and the properties of the streamline provide a mathematical means of plotting and interpreting fluid fields. Here, the application of the stream function concept to the analysis of solute transport in porous media will be presented. As shown in Figure 5-21, consider a uniform ground water velocity field through which a steady-state solute transport from constant solute sources occurs. This means that at any point in the flow field, the concentration is not a function of time. It is assumed that solutes are released from continuous sources in the flow field. The solute particles are under the effects of convection and dispersion. As shown in Figure 5-21, the velocity vector of a solute particle $\mathbf{v}^*$ is tangent to the stream line B. In Figure 5-21, v_x^* and v_y^* are the velocity components of the solute particles. For a uniform ground water flow field, the convective–dispersive flux components in the two-dimensional coordinate system are (Eqs. [2-41a] and [2-41c])

$$F_x = \varphi_e U C - \varphi_e D_x \frac{\partial C}{\partial x} \tag{5-97}$$

and

$$F_z = - \varphi_e D_z \frac{\partial C}{\partial z} \tag{5-98}$$

From Eqs. (5-97) and (5-98), the flux components in the x–z plane may be written as

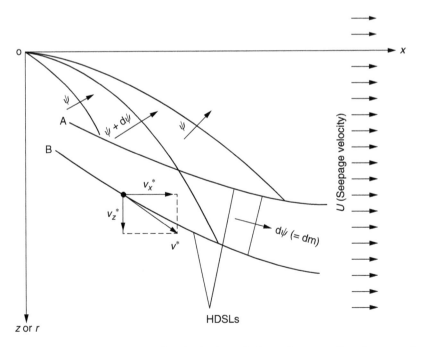

FIGURE 5-21 Schematic representation of the hydrodynamic dispersion streamlines in a ground water flow and solute transport field. (Adapted from Batu, V., *Water Resour. Res.*, 23, 1175–1184, 1987.)

$$F_x = v_x^* C \tag{5-99}$$

and

$$F_z = v_z^* C \tag{5-100}$$

In these equations, F_x and F_z are called *instantaneous mass flux components* (Bear, 1972, p. 101). From Eqs. (5-97)–(5-100) the velocity components of solute particles can be written as

$$v_x^* = \varphi_e U - \frac{\varphi_e D_x}{C} \frac{\partial C}{\partial x} \tag{5-101}$$

$$v_z^* = -\frac{\varphi_e D_z}{C} \frac{\partial C}{\partial z} \tag{5-102}$$

Equations (5-101) and (5-102) show that the velocity components of solute particles are functions of concentration and the water and solute transport parameters of the medium.

A streamline for hydrodynamic dispersion will be called *hydrodynamic dispersion stream line* (HDSL). As shown in Figure 5-21, consider the streamlines A and B in the solute transport field under a uniform ground water flow case. As given in the classical definition of a streamline, it is defined that no solute particles cross streamline A, and the mass rate ψ across all lines OA is the same. Accordingly, ψ is a constant of the streamline, and, if ψ can be found as a function of x and z, the streamline can be plotted. Similarly, the mass rate between O and a closely adjacent stream line B will be $\psi + d\psi$, and the mass rate between the streamlines A and B (i.e., in the stream tube) will be $d\psi$. It is known that for fluid flow, a partial differential equation exists that satisfies the stream function. In the following sections, the partial differential equations satisfying the stream function for hydrodynamic dispersion in the Cartesian and axially symmetric cylindrical coordinates are derived.

5.3.3 PARTIAL DIFFERENTIAL EQUATIONS FOR THE STREAM FUNCTION FOR HYDRODYNAMIC DISPERSION

In the following sections, the partial differential equations for the stream function for hydrodynamic dispersion will be first derived. Then, special cases in the Cartesian and in the cylindrical coordinates under uniform ground water flow case will be presented. Also, the properties of the hydrodynamic dispersion stream function, the proof of the nonorthogonality of the hydrodynamic dispersion streamlines and isoconcentration curves will be included.

5.3.3.1 General Case

In the x–z two-dimensional Cartesian coordinate system, from Eqs. (2-31a) and (2-31c), the convective–dispersive flux components can be written as

$$F_x = \varphi_e v_x C - \varphi_e D_{xx} \frac{\partial C}{\partial x} - \varphi_e D_{xz} \frac{\partial C}{\partial z} \tag{5-103}$$

$$F_z = \varphi_e v_z C - \varphi_e D_{zx} \frac{\partial C}{\partial x} - \varphi_e D_{zz} \frac{\partial C}{\partial z} \tag{5-104}$$

where D_{xx}, D_{xz}, D_{zx}, and D_{zz} are the components of the two-dimensional form of the dispersion tensor D_{ij}, as given by Eg. (2-32); and v_x and v_z are the magnitudes of the ground water velocities in the x- and z-coordinate directions, respectively. Equations (5-103) and (5-104) indicate that the convective–dispersive fluxes occur from higher concentrations toward lower concentrations.

Under steady-state solute transport conditions, from Eq. (2-46), the conservation of mass equation for two-dimensional x–z coordinate system takes the form

$$\frac{\partial F_x}{\partial x} + \frac{\partial F_z}{\partial z} = 0 \qquad (5\text{-}105)$$

Equation (5-105) is satisfied by the hydrodynamic dispersion stream function (HDSF) ψ defined as

$$F_x = \varphi_e \frac{\partial \psi}{\partial z} \qquad (5\text{-}106)$$

and

$$F_z = -\varphi_e \frac{\partial \psi}{\partial x} \qquad (5\text{-}107)$$

By combining the expressions for the convective–dispersive flux components from Eqs. (5-103), (5-104), (5-106), and (5-107), an analog of the Cauchy–Riemann conditions can be obtained:

$$v_x C - D_{xx}\frac{\partial C}{\partial x} - D_{xz}\frac{\partial C}{\partial z} = \frac{\partial \psi}{\partial z} \qquad (5\text{-}108)$$

$$v_z C - D_{zx}\frac{\partial C}{\partial x} - D_{zz}\frac{\partial C}{\partial z} = -\frac{\partial \psi}{\partial x} \qquad (5\text{-}109)$$

By differentiating with respect to z and x and multiplying both sides by D_{zz} and D_{xx}, respectively, the following equations can be obtained:

$$D_{zz}v_x\frac{\partial C}{\partial z} + D_{zz}\frac{\partial v_x}{\partial z}C - D_{xx}D_{zz}\frac{\partial}{\partial z}\left(\frac{\partial C}{\partial x}\right) - D_{zz}\frac{\partial C}{\partial x}\left(\frac{\partial D_{xx}}{\partial x}\right)$$
$$- D_{zz}D_{xz}\frac{\partial^2 C}{\partial z^2} - D_{zz}\frac{\partial C}{\partial z}\frac{\partial D_{xz}}{\partial z} = D_{zz}\frac{\partial^2 \psi}{\partial z^2} \qquad (5\text{-}110)$$

and

$$D_{xx}v_z\frac{\partial C}{\partial x} + D_{xx}\frac{\partial v_z}{\partial x}C - D_{xx}D_{zx}\frac{\partial^2 C}{\partial x^2} - D_{xx}\frac{\partial C}{\partial x}\frac{\partial D_{zx}}{\partial x}$$
$$- D_{xx}D_{zz}\frac{\partial}{\partial x}\left(\frac{\partial C}{\partial z}\right) - D_{xx}\frac{\partial C}{\partial z}\left(\frac{\partial D_{zz}}{\partial x}\right) = -D_{xx}\frac{\partial^2 \psi}{\partial x^2} \qquad (5\text{-}111)$$

Subtracting Eq. (5-111) from Eq. (5-110) gives

$$D_{xx}\frac{\partial^2 \psi}{\partial x^2} + D_{zz}\frac{\partial^2 \psi}{\partial z^2} = -D_{xx}v_z\frac{\partial C}{\partial x} + D_{zz}v_x\frac{\partial C}{\partial z} - D_{xx}\frac{\partial v_z}{\partial x}C + D_{zz}\frac{\partial v_x}{\partial z}C + D_{xx}\frac{\partial C}{\partial z}\left(\frac{\partial D_{zz}}{\partial x}\right)$$
$$- D_{zz}\frac{\partial C}{\partial x}\left(\frac{\partial D_{xx}}{\partial x}\right) + D_{xx}D_{zx}\frac{\partial^2 C}{\partial x^2} - D_{zz}D_{xz}\frac{\partial^2 C}{\partial z^2} + D_{xx}\frac{\partial C}{\partial x}\frac{\partial D_{zx}}{\partial x} - D_{zz}\frac{\partial C}{\partial z}\frac{\partial D_{xz}}{\partial z} \qquad (5\text{-}112)$$

Equation (5-112) is the general partial differential equation, which includes the stream and concentration functions.

5.3.3.2 Special Case in the Cartesian Coordinates: Uniform Ground Water Flow Field

For a uniform ground water flow field, i.e.,

$$v_x = U, \qquad v_z = 0 \qquad (5\text{-}113)$$

and

$$D_{xx} = D_x, \qquad D_{zz} = D_z, \qquad D_{xz} = D_{zx} = 0 \qquad (5\text{-}114)$$

the convective–dispersive flux components given by Eqs. (5-103) and (5-104) take, respectively, the forms of Eqs. (5-97) and (5-98). Also, with Eqs. (5-113) and (5-114), Eq. (5-112) takes the form

$$D_x \frac{\partial^2 \psi}{\partial x^2} + D_z \frac{\partial^2 \psi}{\partial z^2} = D_z U \frac{\partial C}{\partial z} \qquad (5\text{-}115)$$

Similarly, using Eqs. (5-113) and (5-114), Eqs. (5-108) and (5-109), respectively, become

$$UC - D_x \frac{\partial C}{\partial x} = \frac{\partial \psi}{\partial z} \qquad (5\text{-}116)$$

and

$$D_z \frac{\partial C}{\partial z} = \frac{\partial \psi}{\partial x} \qquad (5\text{-}117)$$

Introducing Eq. (5-117) into Eq. (5-115), the following equation can be obtained:

$$D_x \frac{\partial^2 \psi}{\partial x^2} + D_z \frac{\partial^2 \psi}{\partial z^2} = U \frac{\partial \psi}{\partial x} \qquad (5\text{-}118)$$

The corresponding equation for concentration distribution is the two-dimensional steady-state form of Eq. (2-49):

$$D_x \frac{\partial^2 C}{\partial x^2} + D_z \frac{\partial^2 C}{\partial z^2} = U \frac{\partial C}{\partial x} \qquad (5\text{-}119)$$

From Eqs. (2-154) and (2-156) the dispersion coefficients can be expressed as

$$D_x = \alpha_L U + D^* \qquad (5\text{-}120)$$

$$D_z = \alpha_T U + D^* \qquad (5\text{-}121)$$

where α_L(L) is the longitudinal dispersivity, α_T(L) the transverse dispersivity, and D^* (L²/T) the effective molecular diffusion coefficient. For granular materials, D^* can potentially be neglected depending on the hydraulic gradient. If D^* is negligible, Eqs. (5-120) and (5-121), respectively, take the forms

$$D_x = \alpha_L U \qquad (5\text{-}122)$$

$$D_z = \alpha_T U \qquad (5\text{-}123)$$

With Eqs. (5-122) and (5-123), Eqs. (5-118) and (5-119), respectively, take the forms

$$\alpha_L \frac{\partial^2 \psi}{\partial x^2} + \alpha_T \frac{\partial^2 \psi}{\partial z^2} = \frac{\partial \psi}{\partial x} \qquad (5\text{-}124)$$

and

$$\alpha_L \frac{\partial^2 C}{\partial x^2} + \alpha_T \frac{\partial^2 C}{\partial z^2} = \frac{\partial C}{\partial x} \qquad (5\text{-}125)$$

It is interesting to see that the hydrodynamic dispersion stream function ψ and the concentration function C satisfy the same partial differential equation for steady-state solute transport in a uniform ground water flow field.

5.3.3.3 Special Case in the Cylindrical Coordinates: Uniform Ground Water Flow Field

In Figure 5-21, consider steady axially symmetric solute transport in a homogeneous and isotropic medium with the convective term coinciding with the axis of symmetry (x-axis). The z-axis in

Figure 5-21 corresponds to the radial coordinate r. The complete conservation of mass expression in the axially symmetric coordinates is

$$\frac{1}{r}\frac{\partial}{\partial}r(rF_r) + \frac{\partial}{\partial x}(F_c) = 0 \qquad (5\text{-}126)$$

in which F_r [M/(L²T)] is the radial dispersive flux component and F_c [M/(L²T)] is the convective–dispersive flux component in the direction of ground water velocity. The flux components in the r and x directions can be written, respectively, as

$$F_r = -\varphi_e D_r \frac{\partial C}{\partial r} \qquad (5\text{-}127)$$

and

$$F_c = \varphi_e U C - \varphi_e D_c \frac{\partial C}{\partial x} \qquad (5\text{-}128)$$

where D_c(L²/T) is the dispersion coefficient in the convection direction (equivalent to the longitudinal dispersion coefficient) and D_r(L²/T) is the dispersion coefficient in the radial direction (equivalent to the transverse dispersion coefficient). Substituting Eqs. (5-127) and (5-128) into Eq. (5-126), and after some manipulations

$$D_r\frac{\partial^2 C}{\partial r^2} + \frac{1}{r}D_r\frac{\partial C}{\partial r} + D_c\frac{\partial^2 C}{\partial x^2} = U\frac{\partial C}{\partial x} \qquad (5\text{-}129)$$

Equation (5-126) is satisfied by the Stokes' stream function defined as (Stokes, 1842)

$$F_r = -\frac{1}{2\pi r}\varphi_e\frac{\partial\psi}{\partial x} \qquad (5\text{-}130)$$

and

$$F_c = \frac{1}{2\pi r}\varphi_e\frac{\partial\psi}{\partial r} \qquad (5\text{-}131)$$

From Eqs. (5-127), (5-128), (5-130), and (5-131) one can write

$$D_r\frac{\partial C}{\partial r} = \frac{1}{2\pi r}\frac{\partial\psi}{\partial x} \qquad (5\text{-}132)$$

and

$$UC - D_c\frac{\partial C}{\partial x} = \frac{1}{2\pi r}\frac{\partial\psi}{\partial r} \qquad (5\text{-}133)$$

Now, the term C in Eqs. (5-132) and (5-133) will be eliminated. Cross-differentiating Eqs. (5-132) and (5-133) gives

$$D_r\frac{\partial}{\partial x}\left(\frac{\partial C}{\partial r}\right) = \frac{\partial}{\partial x}\left(\frac{1}{2\pi r}\frac{\partial\psi}{\partial x}\right) \qquad (5\text{-}134)$$

and

$$\frac{\partial}{\partial r}(UC) - \frac{\partial}{\partial r}\left(D_c\frac{\partial C}{\partial x}\right) = \frac{\partial}{\partial r}\left(\frac{1}{2\pi r}\frac{\partial\psi}{\partial r}\right) \qquad (5\text{-}135)$$

or, Eqs. (5-134) and (5-135), respectively, take the forms

$$D_r\frac{\partial}{\partial x}\left(\frac{\partial C}{\partial r}\right) = \frac{1}{2\pi r}\frac{\partial^2\psi}{\partial x^2} \qquad (5\text{-}136)$$

and

$$U\frac{\partial C}{\partial r} - D_c\frac{\partial}{\partial r}\left(\frac{\partial C}{\partial x}\right) = \frac{1}{2\pi r}\frac{\partial^2\psi}{\partial x^2} - \frac{1}{2\pi r}\frac{1}{r}\frac{\partial\psi}{\partial r} \qquad (5\text{-}137)$$

Multiplying both sides of Eqs. (5-136) by D_c, and both sides of Eq. (5-137) by D_r, respectively, gives

$$D_c D_r\frac{\partial}{\partial x}\left(\frac{\partial C}{\partial r}\right) = \frac{1}{2\pi r}D_c\frac{\partial^2\psi}{\partial x^2} \qquad (5\text{-}138)$$

and

$$D_r U\frac{\partial C}{\partial r} - D_r D_c\frac{\partial}{\partial r}\left(\frac{\partial C}{\partial x}\right) = \frac{1}{2\pi r}D_r\frac{\partial^2\psi}{\partial r^2} - \frac{1}{2\pi r}D_r\frac{1}{r}\frac{\partial\psi}{\partial r} \qquad (5\text{-}139)$$

Summing both sides of Eqs. (5-138) and (5-139) gives

$$D_r U\frac{\partial C}{\partial r} = \frac{1}{2\pi r}D_c\frac{\partial^2\psi}{\partial x^2} + \frac{1}{2\pi r}D_r\frac{\partial^2\psi}{\partial r^2} - \frac{1}{2\pi r}D_r\frac{1}{r}\frac{\partial\psi}{\partial r} \qquad (5\text{-}140)$$

Using Eq. (5-132) in Eq. (5-140) gives

$$D_r\frac{\partial^2\psi}{\partial r^2} - D_r\frac{1}{r}\frac{\partial\psi}{\partial r} + D_c\frac{\partial^2\psi}{\partial x^2} = U\frac{\partial\psi}{\partial x} \qquad (5\text{-}141)$$

Neglecting the effective molecular diffusion coefficient, the dispersion coefficients take the forms

$$D_c = \alpha_L U \qquad (5\text{-}142)$$

and

$$D_r = \alpha_T U \qquad (5\text{-}143)$$

Substituting Eqs. (5-142) and (5-143) into Eq. (5-141) one has

$$\alpha_T\frac{\partial^2\psi}{\partial r^2} - \alpha_T\frac{1}{r}\frac{\partial\psi}{\partial r} + \alpha_L\frac{\partial^2\psi}{\partial x^2} = \frac{\partial\psi}{\partial z} \qquad (5\text{-}144)$$

5.3.3.4 Properties of the Hydrodynamic Dispersion Stream Function

From the equations given in Section 5.3.3.3, the dimensions of concentration, C, and flux components F_x and F_z, in the three-dimensional sense are $[C] = M/L^3$, $[F_x] = M/(L^2T)$, and $[F_z] = M/(L^2T)$. From Eqs. (5-106) and (5-107), the dimensions of the HDSF are $[\psi] = M/(LT)$. Physically, it can easily be interpreted that ψ is the mass rate in the x–z plane for a unit thickness in the y-coordinate direction. In Figure 5-21, consider the streamlines A and B. As shown in Figure 5-21, $\psi_B - \psi_A = d\psi$ is the mass rate dm passing through the stream tube formed by the streamlines A and B. In other words, $d\psi$ or dm is the mass per unit time passing through this stream tube whose thickness perpendicular to the x–z plane is a unit.

Now, consider that the HDSF is given for two-dimensional solute transport as

$$\psi = \psi(x, z) \qquad (5\text{-}145)$$

From Eq. (5-145), for any HDSL in the solute-transport field one can write

$$d\psi = \frac{\partial\psi}{\partial x}\,dx + \frac{\partial\psi}{\partial z}\,dz = 0 \qquad (5\text{-}146)$$

$$\frac{\partial z}{\partial x} = -\frac{\partial \psi/\partial x}{\partial \psi/\partial z} \tag{5-147}$$

which is the geometric slope of the line tangent to the hydrodynamic dispersion streamline (HDSL). On the other hand, from Eqs. (5-99), (5-100), (5-106), and (5-107) one gets

$$\frac{v_z^*}{v_x^*} = \frac{F_z}{F_x} = -\frac{\partial \psi/\partial x}{\partial \psi/\partial z} \tag{5-148}$$

With Eq. (5-147), Eq. (5-148) means that the resultant velocity

$$v^* = (v_x^{*2} + v_z^{*2})^{1/2} \tag{5-149}$$

is tangential to the HDSL. In Example 5-8, it is shown that the HDSLs are not orthogonal to the isoconcentration curves.

In classical hydrodynamics, the stream function satisfies the Laplace differential equation, which is equivalent only on the left-hand sides of Eqs. (5-118) and (5-141). The terms on the right-hand sides of these equations correspond to the convection effects. The aforementioned HDSF concept is limited to two-dimensional steady-state solute transport cases. Although the HDSF concept is similar to the concept of the stream function of classical hydrodynamics, its physical meaning and interpretation is somewhat different. In classical hydrodynamics, the difference between the values of two adjacent stream lines gives the flow rate passing through the stream tube formed by the stream lines. Consequently, its dimensions are L²/T. As mentioned above, the dimensions of the HDSF are M/(LT). Consequently, the difference between the values of two HDSLs gives the mass of solute passing through the stream tube per unit time.

EXAMPLE 5-8

Show that the HDSLs are not orthogonal to the isoconcentration curves.

SOLUTION

It is assumed that the HDSF is given by Eq. (5-145) and the concentration distribution is given by

$$C = C(x, z) \tag{E5-8-1}$$

The total differential of C can be written as

$$dC = \frac{\partial C}{\partial x}dx + \frac{\partial C}{\partial z}dz \tag{E5-8-2}$$

For an isoconcentration line, $dC = 0$, and Eq. (E5-8-2) takes the form

$$\frac{\partial C}{\partial x}dx + \frac{\partial C}{\partial z}dz = 0 \tag{E5-8-3}$$

or

$$\frac{dz}{dx}\bigg|_{C=\text{constant}} = -\frac{\partial C/\partial x}{\partial C/\partial z} \tag{E5-8-4}$$

From Eq. (5-146), $d\psi = 0$, or

$$\frac{dz}{dx}\bigg|_{\psi=\text{constant}} = -\frac{\partial \psi/\partial x}{\partial \psi/\partial z} \tag{E5-8-5}$$

Using Eqs. (5-116) and (5-117), Eq. (E5-8-5) takes the form

$$\left.\frac{dz}{dx}\right|_{\psi\,=\,constant} = -\frac{D_z\partial C/\partial z}{UC - D_x\,\partial C/\partial x}\tag{E5-8-6}$$

The sets ψ = constant and C = constant are orthogonal if

$$\left(\left.\frac{dz}{dx}\right|_{\psi\,=\,constant}\right)\left(\left.\frac{dz}{dx}\right|_{C\,=\,constant}\right) = -1\tag{E5-8-7}$$

Introducing Eqs. (E5-8-4) and (E5-8-6) in Eq. (E5-8-7) gives

$$\left(-\frac{D_z\partial C/\partial z}{UC-D_x\partial C/\partial x}\right)\left(-\frac{\partial C/\partial x}{\partial C/\partial z}\right) = \left(\frac{D_z\partial C/\partial x}{UC-D_x\partial C/\partial x}\right) \neq -1\tag{E5-8-8}$$

Therefore, the lines ψ = constant and C = constant are not orthogonal unless $D_x = D_z$ and $U = 0$.

5.3.4 SOME ANALYTICAL SOLUTIONS FOR THE HYDRODYNAMIC DISPERSION STREAM FUNCTION

5.3.4.1 Potential Solution Methods

The general two-dimensional partial differential equation for steady-state solute transport in a two-dimensional Cartesian coordinate system including the HDSF and concentration function in an isotropic porous medium is given by Eq. (5-112). Because of its nature, this equation cannot be solved analytically with appropriate boundary conditions. However, this equation can be solved numerically with appropriate boundary conditions either with the finite-difference, finite element, or other numerical methods. To solve this equation numerically, first the concentration distribution, $C(x, z)$, must be obtained from the two-dimensional form of Eq. (2-47). To achieve this, the ground water velocity components v_x (x, z) and v_z (x, z) must be known. Values of these velocities can be determined from the corresponding ground water flow model with appropriate boundary conditions. The components of the dispersion coefficient tensor, D_{xx}, D_{zz}, D_{xz}, and D_{zx}, are all functions of the ground water velocity components. Consequently, ψ is the only unknown in Eq. (5-112) and can be determined numerically using the boundary conditions. A similar equation to Eq. (5-112) can be derived for the axially symmetric solute-transport case. The same numerical procedures, as explained above, can be applied to solve the equation.

Equations (5-118) and (5-144) are the partial differential equations satisfying the HDSF for a uniform ground water flow field in Cartesian and axially symmetric coordinate systems, respectively. As can be seen from Eqs. (5-118) and (5-119), the concentration function $C(x, z)$ and the HDSF $\psi(x, z)$ satisfy exactly the same partial differential equation in the Cartesian coordinate system. For the axially symmetric case, the corresponding equations (5-129) and (5-141) are equivalent with the exception that the second terms on the left-hand sides have opposite signs.

In the following sections, two analytical solutions for the HDSF will be presented (Batu, 1987, 1988, 2002). The first solution is for a strip source in a semi-infinite medium. The second solution is for a strip source in a finite medium.

5.3.4.2 Two-Dimensional Analytical Solutions in a Semi-infinite Porous Medium

Problem Statement and Assumptions
The problem under consideration is a steady-state two-dimensional solute transport from multiple-strip sources in a semiinfinite unidirectional ground water velocity field (Figure 5-22). The medium is semiinfinite through which the x and z coordinates take values in the ranges $(0, +\infty)$ and $(-\infty, +\infty)$,

respectively. The widths of the sources between $z = 0$ and $+ \infty$ are $L_1, L_2, ..., L_{p-1}, L_p$. For the values between $z = 0$ and $-\infty$, the widths are $L'_1, L'_2, ..., L'_{p-1}$, and L'_p. The corresponding source concentrations are all functions of the z coordinate. Consequently, as is shown in Figure 5-22, the convective–dispersive flux components will be the functions of the z coordinate as well. These fluxes are shown by arrows under a curve in Figure 5-22.

Governing Differential Equations and Convective–Dispersive Flux Components
The governing differential equation for the HDSF is given by Eq. (5-118). The corresponding two-dimensional partial differential equation for steady-state concentration distribution is given by Eq. (5-119). The convective–dispersive flux components in terms of HDSF(ψ) are given by Eqs. (5-106) and (5-107). The same flux components in terms of the concentration C are given by Eqs. (2-41a) and (2-41c).

Boundary Conditions
Boundary Conditions for Hydrodynamic Dispersion Stream Function
Using Eq. (5-106), the boundary conditions at the sources can be written as

$$F_x(0, z) = \varphi_e \frac{\partial \psi(0, z)}{\partial z} \tag{5-150}$$

From Figure 5-22, for the sources between $z = 0$ and $z = + \infty$

$$F_x(0, z) = \begin{cases} \varphi_e U C_{M_1}(z), & 0 < z < H_1 & (5\text{-}151) \\ \varphi_e U C_{M_{p-1}}(z), & H_{p-2} < z < H_{p-1} & (5\text{-}152) \\ \varphi_e U C_{M_p}(z), & H_{p-1} < z < H_p & (5\text{-}153) \end{cases}$$

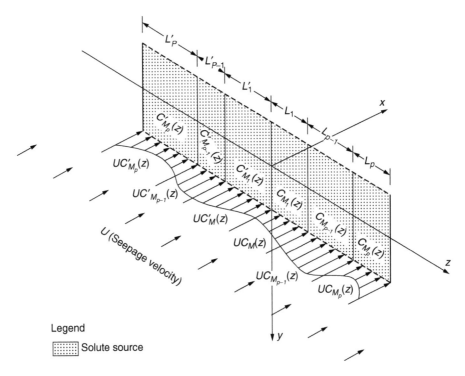

Legend

▦ Solute source

FIGURE 5-22 Schematic representation of strip sources in a semiinfinite porous medium having unidirectional flow field. (Adapted from Batu, V., *Water Resour. Res.*, 23, 1175–1184, 1987.)

For the sources between $z = 0$ and $z = -\infty$, the boundary conditions are

$$F_x(0, z) = \begin{cases} \varphi_e U C'_{M_1}(z), & -H'_1 < z < 0 & \text{(5-154)} \\ \varphi_e U C'_{M_{p-1}}(z), & -H'_{p-1} < z < H'_{p-2} & \text{(5-155)} \\ \varphi_e U C'_{M_p}(z), & -H'_p < z < H'_{p-1} & \text{(5-156)} \end{cases}$$

in which

$$H_1 = L_1 \tag{5-157}$$

$$H_{p-1} = L_1 + L_2 + \cdots + L_{p-1} \tag{5-158}$$

$$H_p = L_1 + L_2 + \cdots + L_{p-1} + L_p \tag{5-159}$$

$$H'_1 = L'_1 \tag{5-160}$$

$$H'_{p-1} = L'_1 + L'_2 + \cdots + L'_{p-1} \tag{5-161}$$

$$H'_p = L'_1 + L'_2 + \cdots + L'_{p-1} + L'_p \tag{5-162}$$

In the rest of the parts of the z-axis ($x = 0$), where the sources do not exist, from Eq. (5-106) the boundary condition is

$$F_x(0, z) = \varphi_e \frac{\partial \psi(0, z)}{\partial z} = 0, \quad z > H_p, \ z < -H'_p \tag{5-163}$$

Equations (5-151)–(5-156) are the third-type (or flux-type) boundary conditions. From Eq. (5-107) and Eq. (2-41c)

$$F_z(x, z) = -\varphi_e D_z \frac{\partial C}{\partial z} = -\varphi_e \frac{\partial \psi}{\partial x} \tag{5-164}$$

With the assumption that for large values of x the concentration C is uniform and no longer changes with z, i.e.,

$$\frac{\partial C(x, z)}{\partial z} = 0 \quad \text{for } x = \infty \tag{5-165}$$

Equation (5-164) takes the form

$$F_z(x, z) = -\varphi_e \frac{\partial \psi}{\partial x} = 0 \quad \text{for } x = \infty \tag{5-166}$$

Boundary Conditions for Concentration
The boundary conditions given by Eqs. (5-151)–(5-156) are also valid for concentration $C(x, z)$. Additional boundary conditions are as follows:

$$\lim_{r \to \infty} C(x, z) = 0, \quad x > 0, \ |z| \geq 0 \tag{5-167}$$

$$\lim_{r \to \infty} \frac{\partial C(x, z)}{\partial x} = 0, \quad x > 0, \ |z| \geq 0 \tag{5-168}$$

$$\lim_{r \to \infty} \frac{\partial C(x, z)}{\partial z} = 0, \quad x > 0, \ |z| \geq 0 \tag{5-169}$$

where

$$r = (x^2 + z^2)^{1/2} \tag{5-170}$$

Solution for Hydrodynamic Dispersion Stream Function for Multiple Sources
General Solution
The general solution of Eq. (5-118) is given by Eq. (E3-7-3) as

$$\psi(x, z) = X_\psi(x) Z_\psi(z)$$

$$= \left[A_1 \exp\left(\frac{Ux}{2D_x} - qx \right) + B_1 \exp\left(\frac{Ux}{2D_x} + qx \right) \right] [A_2 \sin(\lambda z) + B_2 \cos(\lambda z)] \tag{5-171}$$

where

$$q = \left(\frac{D_z}{D_x} \lambda^2 + \frac{U^2}{4D_x^2} \right)^{1/2} \tag{5-172}$$

and A_1, B_1, A_2, and B_2 are constants; and $X_{\psi(x)}$ and $Z_{\psi(z)}$ are the particular integrals of Eq. (5-118). To satisfy Eq. (5-166), B_1 must vanish. Further, since the equations are linear, the sum of any number of particular integrals is also an integral, and thus Eq. (5-171) becomes (Carslaw and Jaeger, 1959, p. 53)

$$\psi(x, z) = \int_0^\infty \exp\left(\frac{Ux}{2D_x} - qx \right) [A(\lambda) \cos(\lambda z) + B(\lambda) \sin(\lambda z)] \, d\lambda \tag{5-173}$$

in which $A(\lambda) = A_1 B_2$ and $B(\lambda) = A_1 A_2$.

Usage of Boundary Condition at x = 0
Using the boundary conditions at $x = 0$, which are given by Eqs. (5-150)–(5-166), one gets

$$\psi(x, z) = \frac{1}{\pi} \int_0^\infty \exp\left(\frac{Ux}{2D_x} - qx \right) \left(\frac{1}{\lambda} \int_{-\infty}^\infty \frac{F_x(0, p)}{\varphi_e} \sin[\lambda(z - p)] \, dp \right) d\lambda \tag{5-174}$$

Equation (5-174) is the solution for any number of sources shown in Figure 5-22. Derivation of Eq. (5-174) is shown in Example 5-9. Equation (5-173) can also be expressed as

$$\psi(x, z) = \frac{1}{\pi} \int_0^\infty \exp\left(\frac{Ux}{2D_x} - qx \right) I(\lambda) \, d\lambda \tag{5-175}$$

where

$$I(\lambda) = \frac{1}{\lambda} \int_{-\infty}^\infty \frac{F_x(0, p)}{\varphi_e} \sin[\lambda(z-p)] \, dp \tag{5-176}$$

EXAMPLE 5-9

Using the boundary conditions at $x = 0$ given by Eqs. (5-150)–(5-166), derive Eq. (5-174).

SOLUTION

From Eq. (5-173)

$$\frac{\partial \psi}{\partial z} = \int_0^\infty \exp\left(\frac{Ux}{2D_x} - qx \right) [- \lambda A(\lambda) \cos(\lambda z) + \lambda B(\lambda) \sin(\lambda z)] \, d\lambda \tag{E5-9-1}$$

With $x = 0$ and using Eqs. (5-150)–(5-156), Eqs. (5-106) and (E5-9-1) give

$$\frac{F_x(0, z)}{\varphi_e} = UC_M(z) = \frac{\partial \psi(0, z)}{\partial z} = \int_0^\infty [\overline{A}(\lambda) \sin(\lambda z) + \overline{B}(\lambda) \cos(\lambda z)] \, d\lambda \tag{E5-9-2}$$

where

$$\overline{A}(\lambda) = -\lambda A(\lambda), \qquad \overline{B}(\lambda) = \lambda B(\lambda) \qquad \text{(E5-9-3)}$$

Equation (E5-9-2) is in the form of *Fourier integral formula*. The general form of the Fourier integral formula is given by (Churchill, 1941, p. 114)

$$f(z) = \frac{1}{\pi} \int_0^\infty \cos(\lambda z) \int_{-\infty}^\infty f(p) \cos(\lambda p) \, dp \, d\lambda + \frac{1}{\pi} \int_0^\infty \sin(\lambda z) \int_{-\infty}^\infty f(p) \sin(\lambda p) \, dp \, d\lambda \qquad \text{(E5-9-4)}$$

Here,

$$f(z) = \frac{F_x(0, z)}{\varphi_e}, \qquad f(p) = \frac{F_x(0, p)}{\varphi_e} \qquad \text{(E5-9-5)}$$

With these, Eq. (E5-9-4) can be expressed as

$$\frac{F_x(0, z)}{\varphi_e} = \int_0^\infty \left[\frac{1}{\pi} \int_{-\infty}^\infty \frac{F_x(0, p)}{\varphi_e} \cos(\lambda p) \, dp \right] \cos(\lambda z) d\lambda + \int_0^\infty \left[\frac{1}{\pi} \int_{-\infty}^\infty \frac{F_x(0, p)}{\varphi_e} \sin(\lambda p) \, dp \right] \sin(\lambda z) d\lambda \qquad \text{(E5-9-6)}$$

From Eqs. (E5-9-2), (E5-9-3), and (E5-9-6) one can write

$$-\lambda A(\lambda) = \frac{1}{\pi} \int_{-\infty}^\infty \frac{F_x(0, p)}{\varphi_e} \sin(\lambda p) \, dp = \overline{A}(\lambda) \qquad \text{(E5-9-7)}$$

$$\lambda B(\lambda) = \frac{1}{\pi} \int_{-\infty}^\infty \frac{F_x(0, p)}{\varphi_e} \cos(\lambda p) \, dp = \overline{B}(\lambda) \qquad \text{(E5-9-8)}$$

or

$$A(\lambda) = -\frac{1}{\pi} \frac{1}{\lambda} \int_{-\infty}^\infty \frac{F_x(0, p)}{\varphi_e} \sin(\lambda p) \, dp \qquad \text{(E5-9-9)}$$

$$B(\lambda) = \frac{1}{\pi} \frac{1}{\lambda} \int_{-\infty}^\infty \frac{F_x(0, p)}{\varphi_e} \cos(\lambda p) \, dp \qquad \text{(E5-9-10)}$$

Substitution of Eqs. (E5-9-9) and (E5-9-10) into Eq. (5-173) gives exactly the same equation as that given by Eq. (5-174).

Solution for Concentration Function for Multiple-Strip Sources
General Solution
The general solution of Eq. (5-119) is given by Eq. (E3-7-3) as

$$C(x, z) = X_c(x) Z_c(x) = \left[C_1 \exp\left(\frac{Ux}{2D_x} - qx \right) + D_1 \exp\left(\frac{Ux}{2D_x} + qx \right) \right] [C_2 \sin(\lambda z) + C_2 \cos(\lambda z)] \qquad \text{(5-177)}$$

where q is given by Eq. (5-173) and C_1, D_1, C_2, and D_2 are constants; and $X_C(x)$ and $Z_C(z)$ are the particular integrals of Eq. (5-119). To satisfy Eqs. (5-167)–(5-169), D_1 must vanish. Further, since the equations are linear, the sum of any number of particular integrals is also an integral, and thus Eq. (5-177) becomes (Carslaw and Jaeger, 1959, p. 53)

$$C(x, z) = \int_0^\infty \exp\left(\frac{Ux}{2D_x} - qx \right) [C(\lambda)\cos(\lambda z) + D(\lambda)\sin(\lambda z)] \, d\lambda \qquad \text{(5-178)}$$

in which $C(\lambda) = C_1 D_2$ and $D(\lambda) = C_1 C_2$.

Usage of Boundary Conditions at x = 0
Using the boundary conditions at $x = 0$, which are given by Eqs. (5-150)–(5-166) one gets

$$C(x, z) = \frac{1}{\pi} \int_0^\infty \frac{\exp\left[\dfrac{Ux}{(2D_x)} - qx \right]}{(U/2 + D_x q)} J(\lambda) \, d\lambda \qquad \text{(5-179)}$$

where

$$J(\lambda) = \int_{-\infty}^{\infty} \frac{F_x(0,\,p)}{\varphi_e} \cos[\lambda(z-p)]\,dp \qquad (5\text{-}180)$$

Equation (5-179) is the solution for any number of sources shown in Figure 5-22. Derivation of Eq. (5-179) is shown in Example 5-10.

EXAMPLE 5-10

Using the boundary conditions at $x = 0$ given by Eqs. (5-151)–(5-156), derive Eq. (5-178).

SOLUTION

From Eq. (5-178)

$$\frac{\partial C}{\partial x} = \int_0^{\infty} \left(\frac{U}{2D_x} - q\right)\exp\left(\frac{Ux}{2D_x} - qx\right)[C(\lambda)\cos(\lambda z) + D(\lambda)\sin(\lambda z)]\,d\lambda \quad (E5\text{-}10\text{-}1)$$

With $x = 0$, (E5-10-1) gives

$$\frac{\partial C(0,\,z)}{\partial x} = \int_0^{\infty} [C(\lambda)\cos(\lambda z) + D(\lambda)\sin(\lambda z)]\,d\lambda \qquad (E5\text{-}10\text{-}2)$$

Similarly, from Eq. (5-178),

$$C(0,\,z) = \int_0^{\infty} [C(\lambda)\cos(\lambda z) + D(\lambda)\sin(\lambda z)]\,d\lambda \qquad (E5\text{-}10\text{-}3)$$

Introduction of Eqs. (E5-10-2) and (E5-10-3) into Eq. (2-41a) gives

$$\frac{F_x(0,\,z)}{\varphi_e} = U\int_0^{\infty} [C(\lambda)\cos(\lambda z) + D(\lambda)\sin(\lambda z)]\,d\lambda$$
$$- D_x\int_0^{\infty}\left(\frac{U}{2D_x} - q\right)[C(\lambda)\cos(\lambda z) + D(\lambda)\sin(\lambda z)]\,d\lambda \qquad (E5\text{-}10\text{-}4)$$

or

$$\frac{F_x(0,\,z)}{\varphi_e} = \int_0^{\infty} [\overline{C}(\lambda)\cos(\lambda z) + \overline{D}(\lambda)\sin(\lambda z)]\,d\lambda \qquad (E5\text{-}10\text{-}5)$$

where

$$\overline{C}(\lambda) = \left(\frac{U}{2} + D_x q\right)C(\lambda) \qquad (E5\text{-}10\text{-}6)$$

and

$$\overline{D}(\lambda) = \left(\frac{U}{2} + D_x q\right)D(\lambda) \qquad (E5\text{-}10\text{-}7)$$

Using the Fourier integral formula, Eq. (E5-9-4), with Eq. (E5-9-5), as used in Example 5-9, one gets exactly the same equation as Eq. (5-179).

Special Solutions for a Single-Strip Source
Using the general solutions given above, special solutions are obtained for a single-strip source whose geometry is shown in Figure 5-23. Figure 5-23 is a special case of Figure 5-22 and $L_1 = L_1' = B$. For this case, the boundary conditions at $x = 0$, as given by Eqs. (5-151)–(5-156), take the forms

$$F_x(0, p) = \varphi_e U C_M \quad -B < z < B \tag{5-181a}$$

$$F_x(0, p) = 0 \quad \text{otherwise} \tag{5-181b}$$

As can be seen from Eqs. (5-176) and (5-180), only $I(\lambda)$ and $J(\lambda)$ need to be evaluated for a single-strip source in order to determine $\psi(x, z)$ and $C(x, z)$ solutions. These are shown below.

Hydrodynamic Dispersion Stream Function Solution
Introduction of Eqs. (5-181a) and (5-181b) into Eq. (5-176) gives

$$I(\lambda) = \frac{1}{\lambda} \int_{-B}^{B} U C_M \sin[\lambda(z-p)] \, dp \tag{5-182}$$

or after evaluation of the integral

$$I(\lambda) = 2U C_M \frac{\sin(\lambda z)\cos(\lambda B)}{\lambda^2} \tag{5-183}$$

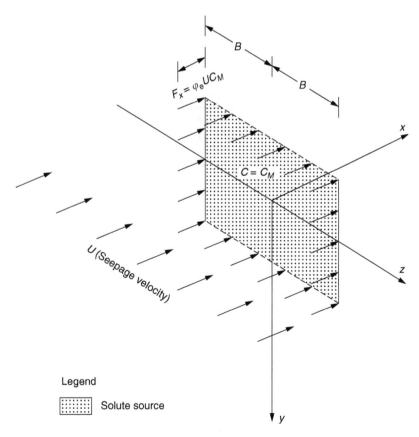

FIGURE 5-23 Geometry of a single strip source. (Adapted from Batu, V., *Water Resour. Res.*, 23, 1175–1184, 1987.)

Substitution of Eq. (5-183) into Eq. (5-175) and using Eq. (5-172) gives

$$\Psi(x, z) = \frac{\psi(x, z)}{UC_M} = \int_0^\infty f_\psi(\lambda) \frac{\sin(\lambda B)}{\lambda} \, d\lambda \tag{5-184}$$

where

$$f_\psi(\lambda) = \frac{2}{\pi} \exp\left[\frac{Ux}{2D_x} - \left(\frac{D_z}{D_x} \lambda^2 + \frac{U^2}{4D_x^2} \right)^{1/2} x \right] \frac{\sin(\lambda z)}{\lambda} \tag{5-185}$$

Because $[\psi] = M/L\,T$, $[U] = L/T$, and $[C_M] = M/L^3$, the dimension of Ψ in Eq. (5-183) is L.

Concentration Function Solution
Introduction of Eqs. (5-181a) and (5-181b) into Eq. (5-180) gives

$$J(\lambda) = \int_{-B}^{B} UC_M \cos[\lambda(z-p)] \, dp \tag{5-186}$$

or after evaluating the integral

$$J(\lambda) = 2UC_M \frac{\sin(\lambda B) \cos(\lambda z)}{\lambda} \tag{5-187}$$

Substitution of Eq. (5-187) into Eq. (5-179), and using Eq. (5-172), gives

$$\frac{C(x, z)}{C_M} = \int_0^\infty f_C(\lambda) \frac{\sin(\lambda B)}{\lambda} \, d\lambda \tag{5-188}$$

where

$$f_C(\lambda) = \frac{2}{\pi} U \exp \frac{\left[\dfrac{Ux}{2D_x} - \left(\dfrac{D_z}{D_x} \lambda^2 + \dfrac{U^2}{4D_x^2} \right)^{1/2} x \right]}{\left[\dfrac{U}{2} + D_x \left(\dfrac{D_z}{D_x} \lambda^2 + \dfrac{U^2}{4D_x^2} \right)^{1/2} \right]} \cos(\lambda z) \tag{5-189}$$

Convective–Dispersive Flux Components — Determination from $\psi\,(x,\,z)$
For $F_x\,(x,\,z)$, from Eqs. (5-106) and (5-184),

$$\frac{F_x(x, z)}{\varphi_e UC_M} = \int_0^\infty f_x(\lambda) \frac{\sin(\lambda B)}{\lambda} \, d\lambda \tag{5-190}$$

where

$$f_x(\lambda) = \frac{2}{\pi} \exp\left[\frac{Ux}{2D_x} - \left(\frac{D_z}{D_x} \lambda^2 + \frac{U^2}{4D_x^2} \right)^{1/2} x \right] \cos(\lambda z) \tag{5-191}$$

As a check, at the origin ($x = z = 0$), Eq. (5-191) takes the form

$$\frac{F_x(0, 0)}{\varphi_e} = \frac{2}{\pi} UC_M \int_0^\infty \frac{\sin(\lambda B)}{\lambda} \, d\lambda \tag{5-192}$$

or (Kreyszig, 1967, p. 719, Eq. [13])

$$\frac{F_x(0, 0)}{\varphi_e} = UC_M \tag{5-193}$$

For $F_z(x, z)$, from Eqs. (5-107) and (5-184),

$$\frac{F_z(x, z)}{\varphi_e U C_M} = \int_0^\infty f_z(\lambda) \frac{\sin(\lambda B)}{\lambda}\, d\lambda \tag{5-194}$$

where

$$f_z(\lambda) = \frac{2}{\pi}\left[\left(\frac{D_z}{D_x}\lambda^2 + \frac{U^2}{4D_x^2}\right)^{1/2} - \frac{U}{2D_x}\right]\exp\left[\frac{Ux}{2D_x} - \left(\frac{D_z}{D_x}\lambda^2 + \frac{U^2}{4D_x^2}\right)^{1/2}x\right]\frac{\sin(\lambda z)}{\lambda} \tag{5-195}$$

Convective–Dispersive Flux Components — Determination from C (x, z)
From Eqs. (2-41a) and (5-188), it can easily be shown that the exact form of Eq. (5-190) for $F_x (x, z)$ can be obtained. This is shown in Example 5-11. Also, from Eqs. (2-41c) and (5-187), it can easily be shown that the exact form of Eq. (5-193) for $F_z (x, z)$ can be obtained. This is shown in Example 5-12.

EXAMPLE 5-11

Derive the expression for $F_x(x, z)$ from Eq. (2-41a) and from the expression of $C (x, z)$ (Eq. [5-188]), and compare the result with the $F_x (x, z)$ expression given Eq. (5-190).

SOLUTION

From Eq. (5-188)

$$\frac{\partial C}{\partial x} = C_M \int_0^\infty \frac{\partial[f_c(\lambda)]}{\partial x}\frac{\sin(\lambda B)}{\lambda}\, d\lambda \tag{E5-11-1}$$

From Eq. (5-189)

$$\frac{\partial[f_c(\lambda)]}{\partial x} = \frac{2}{\pi}U\left[\frac{U}{2D_x} - \left(\frac{D_z}{D_x}\lambda^2 + \frac{U^2}{4D_x^2}\right)^{1/2}\right]\cos(\lambda z)\frac{\left[\frac{Ux}{2D_x} - \left(\frac{D_z}{D_x}\lambda^2 + \frac{U^2}{4D_x^2}\right)^{1/2}x\right]}{\left[\frac{U}{2} + D_x\left(\frac{D_z}{D_x}\lambda^2 + \frac{U^2}{4D_x^2}\right)^{1/2}\right]} \tag{E5-11-2}$$

Substitution of Eqs. (5-188) and (E5-11-1) into Eq. (2-41a) gives

$$F_x = \varphi_e U C_M \int_0^\infty f_C(\lambda)\frac{\sin(\lambda B)}{\lambda}\, d\lambda - \varphi_e D_x C_M \int_0^\infty \frac{\partial[f_c(\lambda)]}{\partial x}\frac{\sin(\lambda B)}{\lambda}\, d\lambda \tag{E5-11-3}$$

Using Eqs. (5-189) and (E5-11-2), Eq. (E5-11-3) takes the form

$$F_x = \varphi_e U C_M \int_0^\infty \frac{2}{\pi}U\frac{\exp\left[\left(\frac{Ux}{2D_x} - \left(\frac{D_z}{D_x}\lambda^2 + \frac{U^2}{4D_x^2}\right)^{1/2}\right)x\right]}{\left[\frac{U}{2} + D_x\left(\frac{D_z}{D_x}\lambda^2 + \frac{U^2}{4D_x^2}\right)\right]^{1/2}}\cos(\lambda z)\frac{\sin(\lambda B)}{\lambda}\, d\lambda \tag{E5-11-4}$$

$$- \varphi_e D_x C_M \int_0^\infty \frac{2}{\pi}U\left[\frac{U}{2D_x} - \left(\frac{D_z}{D_x}\lambda^2 + \frac{U^2}{4D_x^2}\right)^{1/2}\right]\frac{\exp\left[\frac{Ux}{2D_x} - \left(\frac{D_z}{D_x}\lambda^2 + \frac{U^2}{4D_x^2}\right)^{1/2}x\right]}{\left[\frac{U}{2} + D_x\left(\frac{D_z}{D_x}\lambda^2 + \frac{U^2}{4D_x^2}\right)^{1/2}\right]}\cos(\lambda z)\frac{\sin(\lambda B)}{\lambda}\, d\lambda$$

or

$$F_x = \frac{2}{\pi} U C_M \varphi_e \int_0^\infty U \frac{\exp\left[\frac{Ux}{2D_x} - \left(\frac{D_z}{D_x}\lambda^2 + \frac{U^2}{4D_x^2}\right)^{1/2} x\right]}{\left[\frac{U}{2} + D_x\left(\frac{D_z}{D_x}\lambda^2 + \frac{U^2}{4D_x^2}\right)^{1/2}\right]} \cos(\lambda z)\frac{\sin(\lambda B)}{\lambda} d\lambda \quad \text{(E5-11-5)}$$

$$- \frac{2}{\pi} U C_M \varphi_e \int_0^\infty \left[\frac{U}{2} - D_x\left(\frac{D_z}{D_x}\lambda^2 + \frac{U^2}{4D_x^2}\right)^{1/2}\right] \frac{\exp\left[\frac{Ux}{2D_x} - \left(\frac{D_z}{D_x}\lambda^2 + \frac{U^2}{4D_x^2}\right)^{1/2}\right]}{\left[\frac{U}{2} + D_x\left(\frac{D_z}{D_x}\lambda^2 + \frac{U^2}{4D_x^2}\right)^{1/2}\right]} \cos(\lambda z)\frac{\sin(\lambda B)}{\lambda} d\lambda$$

or

$$F_x = \frac{2}{\pi} U C_M \varphi_e \int_0^\infty \frac{\left[U - \frac{U}{2} + D_x\left(\frac{D_z}{D_x}\lambda^2 + \frac{U^2}{4D_x^2}\right)^{1/2}\right]}{\left[\frac{U}{2} + D_x\left(\frac{D_z}{D_x}\lambda^2 + \frac{U^2}{4D_x^2}\right)^{1/2}\right]} \exp\left[\frac{Ux}{2D_x} - \left(\frac{D_z}{D_x}\lambda^2 + \frac{U^2}{4D_x^2}\right)^{1/2} x\right]$$

$$\cos(\lambda z)\frac{\sin(\lambda B)}{\lambda} d\lambda \qquad\qquad\qquad\qquad\qquad \text{(E5-11-6)}$$

or

$$F_x(x, z) = \frac{2}{\pi} U C_M \varphi_e \int_0^\infty \exp\left[\frac{Ux}{2D_x} - \left(\frac{D_z}{D_x}\lambda^2 - \left(\frac{D_z}{D_x}\lambda^2 + \frac{U^2}{4D_x^2}\right)^{1/2}\right) x\right]\cos(\lambda z)\frac{\sin(\lambda z B)}{\lambda} d\lambda \quad \text{(E5-11-7)}$$

and finally it takes the form

$$\frac{F_x(x, z)}{\varphi_e U C_M} = \int_0^\infty f_x(\lambda)\frac{\sin(\lambda B)}{\lambda} d\lambda \qquad\qquad\qquad \text{(E5-11-8)}$$

in which $f_x(\lambda)$ is given by Eq. (5-191). Equation (E5-11-8) is exactly the same as Eq. (5-190).

EXAMPLE 5-12

Derive the expression for $F_z(x, z)$ from Eq. (2-41c) and from the expression of $C(x, z)$ (Eq. [5-188]) and compare the $F_z(x, z)$ expression given by Eq. (5-194).

SOLUTION

From Eq. (5-188)

$$\frac{\partial C}{\partial z} = C_M \int_0^\infty \frac{\partial[f_C(\lambda)]}{\partial z}\frac{\sin(\lambda B)}{\lambda} d\lambda \qquad\qquad \text{(E5-12-1)}$$

From Eq. (5-189)

$$\frac{\partial[f_C(\lambda)]}{\partial z} = -\frac{2}{\pi} U \frac{\exp\left[\frac{Ux}{2D_x} - \left(\frac{D_z}{D_x}\lambda^2 + \frac{U^2}{4D_x^2}\right)^{1/2} x\right]}{\left[\frac{U}{2} + D_x\left(\frac{D_z}{D_x}\lambda^2 + \frac{U^2}{4D_x^2}\right)^{1/2}\right]} \lambda \sin(\lambda z) \qquad \text{(E5-12-2)}$$

Substitution of Eqs. (E5-12-1) and (E5-12-2) into Eq. (2-41c) gives

$$F_z = \frac{2}{\pi} U C_M \varphi_e D_z \int_0^\infty \frac{\exp\left[\frac{Ux}{2D_x} - \left(\frac{D_z}{D_x}\lambda^2 + \frac{U^2}{4D_x^2}\right)^{1/2} x\right]}{\left[\frac{U}{2} + D_x\left(\frac{D_z}{D_x}\lambda^2 + \frac{U^2}{4D_x^2}\right)^{1/2}\right]} \lambda \sin(\lambda z) \frac{\sin(\lambda B)}{\lambda} d\lambda \quad \text{(E5-12-3)}$$

or

$$F_z = \frac{2}{\pi} U C_M \varphi_e D_z \int_0^\infty \frac{\left[\frac{U}{2} - D_x\left(\frac{D_z}{D_x}\lambda^2 + \frac{U^2}{4D_x^2}\right)^{1/2}\right]\exp\left[\frac{Ux}{2D_x} - \left(\frac{D_z}{D_x}\lambda^2 + \frac{U^2}{4D_x^2}\right)^{1/2} x\right]}{\left[\frac{U}{2} - D_x\left(\frac{D_z}{D_x}\lambda^2 + \frac{U^2}{4D_x^2}\right)^{1/2}\right]\left[\frac{U}{2} + D_x\left(\frac{D_z}{D_x}\lambda^2 + \frac{U^2}{4D_x^2}\right)^{1/2}\right]}$$

$$\lambda \sin(\lambda z) \frac{\sin(\lambda B)}{\lambda} d\lambda \quad \text{(E5-12-4)}$$

or after simplification

$$\frac{F_z}{\varphi_e U C_M} = \int_0^\infty f_z(\lambda) \frac{\sin(\lambda B)}{\lambda} d\lambda \quad \text{(E5-12-5)}$$

in which $f_z(\lambda)$ is given by Eq. (5-195). Equation (E5-12-5) is exactly the same as Eq. (5-194).

Evaluation of the Integrals in the Solutions
The integrals in Eqs. (5-184), (5-188), (5-190), and (5-194) are all trigonometric integrals and they are in the form of *Filon integrals* (1928, Eq. [92], p. 38,), for which Filon developed a numerical integration procedure.

The integrals have the form

$$I(x) = \int_a^b \sin(kx)\psi(x)\,dx \quad \text{(5-196)}$$

where $\psi(x)$ is a function with a limited number of turning points in the range of integration and k is a constant that takes large values. This type of integral frequently occurs in mathematical physics and its determination by quadratures is often desirable. In particular, the form

$$I(x) = \int_0^\infty \varphi(x) \frac{\sin(kx)}{x}\,dx \quad \text{(5-197)}$$

has a special importance for evaluation of the integrals in the HDSF application. The evaluation of integrals in the form of Eq. (5-196) or Eq. (5-197) by quadratures when k is large generates difficulties because of the rapid oscillation of the function $\sin(kx)$. Evaluation of these integrals may become very difficult and time-consuming because of the rapid oscillation of the integrands and the singularity at the origin. The ordinary integration methods, such as Simpson's rule, may not be feasible. Filon's numerical integration method (Filon, 1928) is quite convenient to calculate such integrals (Batu, 1979).

Evaluation of the Hydrodynamic Dispersion Stream Function (HDSF)
The HDSF function, as given by Eq. (5-184), is in the form of Eq. (5-197). Using Filon's equation (Filon, 1928, Eq. [20],p 43), Eq. (5-184) can be written as

$$\frac{\psi(x,z)}{UC_M} = \frac{f_\psi(0)}{4}\{\pi[2 - \beta(\xi) - \gamma(\xi)] + 2\beta(\xi)\xi\} + h(\xi)[\eta(\xi)f_\psi'(0) + \beta(\xi)T_{2n} + \gamma(\xi)T_{2n+1}] \quad \text{(5-198)}$$

in which

$$h(\xi) = \frac{\xi}{B} \tag{5-199}$$

$$\eta(\xi) = \frac{1}{\xi} + \frac{\cos(\xi)\sin(\xi)}{\xi^2} - 2\frac{\sin^2(\xi)}{\xi^3} \tag{5-200}$$

$$\beta(\xi) = 2\left[\frac{1+\cos^2(\xi)}{\xi^2} - \frac{\sin(2\xi)}{\xi^3}\right] \tag{5-201}$$

$$\gamma(\xi) = 4\left[\frac{\sin(\xi)}{\xi^3} - \frac{\cos(\xi)}{\xi^2}\right] \tag{5-202}$$

$$T_{2n} = \sum_{n=1}^{\infty}(f_\psi)_{2n}\frac{\sin(2n\xi)}{2nh(\xi)} \tag{5-203}$$

$$T_{2n+1} = \sum_{n=0}^{\infty}(f_\psi)_{2n+1}\frac{\sin[(2n+1)\xi]}{(2n+1)h(\xi)} \tag{5-204}$$

$$(f_\psi)_{2n} = \frac{2}{\pi}\exp\left[\frac{Ux}{2D_x} - \left\langle\frac{D_z}{D_x}[2nh(\xi)]^2 + \frac{U^2}{4D_x^2}\right\rangle^{1/2}x\right]\frac{\sin[2nh(\xi)z]}{2nh(\xi)} \tag{5-205}$$

$$(f_\psi)_{2n+1} = \frac{2}{\pi}\exp\left[\frac{Ux}{2D_x} - \left\langle\frac{D_z}{D_x}[(2n+1)h(\xi)]^2 + \frac{U^2}{4D_x^2}\right\rangle^{1/2}x\right]\frac{\sin[(2n+1)h(\xi)z]}{(2n+1)h(\xi)} \tag{5-206}$$

From Eq. (5-204), for $n = 0$,

$$(T_{2n+1})_{n=0} = [(f_\psi)_{2n+1}]_{n=0}\frac{\sin(\xi)}{h(\xi)} \tag{5-207}$$

where

$$[(f_\psi)_{2n+1}]_{n=0} = \frac{2}{\pi}\left[\frac{Ux}{2D_x} - \left\langle\frac{D_z}{D_x}[h(\xi)]^2 + \frac{U^2}{4D_x^2}\right\rangle^{1/2}\right]\frac{\sin[h(\xi)z]}{h(\xi)} \tag{5-208}$$

Also, from Eq. (5-185), for $\lambda = 0$

$$f_\psi(0) = \frac{2}{\pi}\exp(0)\left[\frac{\sin(\lambda z)}{\lambda}\right]_{\lambda\to 0} = \frac{2}{\pi}(1)(1) = \frac{2}{\pi} \tag{5-209}$$

It can be shown that from Eq. (5-185)

$$f_\psi'(0) = \frac{df_\psi(0)}{d\lambda} = 0 \tag{5-210}$$

Equations (5-198)–(5-210) define the numerical solution of the integral in Eq. (5-184).

Evaluation of $F_x(x,z)$

The $F_x(x, z)$ function, as given by Eq. (5-190), is in the form of Eq. (5-197). Using Filon's equation (Filon, 1928, Eq. [20], p. 43], Eq. (5-190) can be written as

$$\frac{F_x(x, z)}{\varphi_e UC_M} = \frac{f_x(0)}{4}\left\{\pi[2-\beta\{\xi\}-\gamma(\xi)]+2\beta(\xi)\xi\}+h(\xi)[\eta(\xi)f_x'(0)+\beta(\xi)T_{2n}+\gamma(\xi)T_{2n+1}] \tag{5-211}$$

in which $h(\xi)$, $\eta(\xi)$, $\beta(\xi)$, and $\gamma(\xi)$ are given by Eqs. (5-199)–(5-202). Also

$$T_{2n} = \sum_{n=1}^{\infty}(f_x)_{2n}\frac{\sin(2n\xi)}{2nh(\xi)} \tag{5-212}$$

$$T_{2n+1} = \sum_{n=0}^{\infty}(f_x)_{2n+1}\frac{\sin[(2n+1)\xi]}{(2n+1)h(\xi)} \tag{5-213}$$

where

$$(f_x)_{2n} = \frac{2}{\pi}\exp\left[\frac{Ux}{2D_x} - \left\langle\frac{D_z}{D_x}[2nh(\xi)]^2 + \frac{U^2}{4D_x^2}\right\rangle^{1/2}x\right]\cos[2nh(\xi)z] \tag{5-214}$$

$$(f_x)_{2n+1} = \frac{2}{\pi}\exp\left[\frac{Ux}{2D_x} - \left\langle\frac{D_z}{D_x}[(2n+1)h(\xi)]^2 + \frac{U^2}{4D_x^2}\right\rangle^{1/2}x\right]\cos[(2n+1)h(\xi)z] \tag{5-215}$$

From Eq. (5-213), for $n = 0$,

$$(T_{2n+1})_{n=0} = [(f_x)_{2n+1}]_{n=0}\frac{\sin(\xi)}{h(\xi)} \tag{5-216}$$

where, from Eq. (5-215),

$$[(f_x)_{2n+1}]_{n=0} = \frac{2}{\pi}\exp\left[\frac{Ux}{2D_x} - \left\langle\frac{D_z}{D_x}[h(\xi)]^{1/2} + \frac{U^2}{4D_x^2}\right\rangle^{1/2}x\right]\cos[h(\xi)z] \tag{5-217}$$

Also, from Eq. (5-191), for $\lambda = 0$,

$$f_x(0) = \frac{2}{\pi}\exp(0)\cos(0) = \frac{2}{\pi}(1)(1) = \frac{2}{\pi} \tag{5-218}$$

It can be shown that, from Eq. (5-191),

$$f_x'(0) = \frac{df_x(0)}{d\lambda} = 0 \tag{5-219}$$

Equations (5-211)–(5-219) define the numerical solution of the integral in Eq. (5-190).

Evaluation of $F_z(x,z)$
The $F_z(x, z)$ function, as given by Eq. (5-194), is in the form of Eq. (5-197). Using Filon's equation (Filon, 1928, Eq. [20], p. 43), Eq. (5-194) can be written as

$$\frac{F_z(x, z)}{\varphi_e UC_M} = \frac{f_z(0)}{4}\{\pi[2-\beta(\xi)-\gamma(\xi)] + 2\beta(\xi)\xi\} + h(\xi)[\eta(\xi)f_z'(0) + \beta(\xi)T_{2n} + \gamma(\xi)T_{2n+1}] \tag{5-220}$$

in which $h(\xi)$, $\eta(\xi)$, $\beta(\xi)$, and $\gamma(\xi)$ are given by Eqs. (5-199) – (5-202). Also

$$T_{2n} = \sum_{n=1}^{\infty}(f_z)_{2n}\frac{\sin(2n\xi)}{2nh(\xi)} \tag{5-221}$$

$$T_{2n+1} = \sum_{n=0}^{\infty}(f_z)_{2n+1}\frac{\sin[(2n+1)\xi]}{(2n+1)h(\xi)} \tag{5-222}$$

where

$$(f_z)_{2n} = \frac{2}{\pi}\left[\left\langle\frac{D_z}{D_x}[2nh(\xi)]^2 + \frac{U^2}{4D_x^2}\right\rangle^{1/2} - \frac{U}{2D_x}\right]$$
$$\times \exp\left[\frac{Ux}{2D_x} - \left\langle\frac{D_z}{D_x}[2nh(\xi)]^2 + \frac{U^2}{4D_x^2}\right\rangle^{1/2}x\right]\frac{\sin[2nh(\xi)z]}{2nh(\xi)} \tag{5-223}$$

$$(f_z)_{2n+1} = \frac{2}{\pi}\left[\left\langle\frac{D_z}{D_x}[(2n+1)h(\xi)]^2 + \frac{U^2}{4D_x^2}\right\rangle^{1/2} - \frac{U}{2D_x}\right]$$

$$\exp\left[\frac{Ux}{2D_x} - \left\langle\frac{D_z}{D_x}[(2n+1)h(\xi)]^2 + \frac{U^2}{4D_x^2}\right\rangle^{1/2}x\right]\frac{\sin[(2n+1)h(\xi)z]}{(2n+1)h(\xi)} \quad (5\text{-}224)$$

From Eq. (5-222)

$$(T_{2n+1})_{n=0} = [(f_z)_{2n+1}]_{n=0}\frac{\sin(\xi)}{h(\xi)} \quad (5\text{-}225)$$

where, from Eq. (5-224),

$$[(f_z)_{2n+1}]_{n=0} = \frac{2}{\pi}\left[\left\langle\frac{D_z}{D_x}[h(\xi)]^2 + \frac{U^2}{4D_x^2}\right\rangle^{1/2} - \frac{U}{2D_x}\right]$$

$$\times \exp\left[\frac{Ux}{2D_x} - \left\langle\frac{D_z}{D_x}[h(\xi)]^2 + \frac{U^2}{4D_x^2}\right\rangle^{1/2}x\right]\frac{\sin[h(\xi)z]}{h(\xi)} \quad (5\text{-}226)$$

Also, from Eq. (5-195), for $\gamma = 0$

$$f_z(0) = \frac{2}{\pi}(0)\exp(0)\left[\frac{\sin(\lambda z)}{\lambda}\right]_{\lambda\to 0} = \frac{2}{\pi}(0)\,(1)\,(1) = 0 \quad (5\text{-}227)$$

It can be shown that, from Eq. (5-195),

$$(f_z)'(0) = \frac{df_z(0)}{d\lambda} = 0 \quad (5\text{-}228)$$

Equations (5-220)–(5-228) define the numerical solution of the integral in Eq. (5-194).

Evaluation of C (x, z)
The $C(x, z)$ function, as given by Eq. (5-188), is in the form of Eq. (5-197). Using Filon's equation (Filon, 1928, Eq. [20], p. 43), Eq. (5-188) can be written as

$$\frac{C(x, z)}{C_M} = \frac{f_C(0)}{4}\{\pi[2 - \beta(\xi) - \gamma(\xi)] + 2\beta(\xi)\xi\} + h(\xi)[\eta(\xi)f_C'(0) + \beta(\xi)T_{2n} + \gamma(\xi)T_{2n+1}] \quad (5\text{-}229)$$

in which $h(\xi)$, $\eta(\xi)$, $\beta(\xi)$, and $\gamma(\xi)$ are given by Eqs. (5-199)– (5-202). Also

$$T_{2n} = \sum_{n=1}^{\infty}(f_C)_{2n}\frac{\sin(2n\xi)}{2nh(\xi)} \quad (5\text{-}230)$$

$$T_{2n+1} = \sum_{n=0}^{\infty}(f_C)_{2n+1}\frac{\sin[(2n+1)\xi]}{(2n+1)h(\xi)} \quad (5\text{-}231)$$

where

$$(f_C)_{2n} = \frac{2}{\pi}U\frac{\exp\left[\frac{Ux}{2D_x} - \left\langle\frac{D_z}{D_x}[2nh(\xi)]^2 + \frac{U^2}{4D_x^2}\right\rangle^{1/2}x\right]}{\frac{U}{2} + D_x\left\langle\frac{D_z}{D_x}[2nh(\xi)]^2 + \frac{U^2}{4D_x^2}\right\rangle^{1/2}}\cos[2nh(\xi)z] \quad (5\text{-}232)$$

$$(f_C)_{2n+1} = \frac{2}{\pi}U\frac{\exp\left[\frac{Ux}{2D_x} - \left\langle\frac{D_z}{D_x}[(2n+1)h(\xi)]^2 + \frac{U^2}{4D_x^2}\right\rangle^{1/2}x\right]}{\frac{U}{2} + D_x\left\langle\frac{D_z}{D_x}[(2n+1)h(\xi)]^2 + \frac{U^2}{4D_x^2}\right\rangle^{1/2}}\cos[(2n+1)h(\xi)z] \quad (5\text{-}233)$$

From Eq. (5-231), for $n = 0$,

$$(T_{2n+1})_{n=0} = [(f_C)_{2n+1}]_{n=0} \frac{\sin(\xi)}{h(\xi)} \tag{5-234}$$

where, from Eq. (5-233),

$$[(f_C)_{2n+1}]_{n=0} = \frac{2}{\pi} U \frac{\exp\left[\frac{Ux}{2D_x} - \left\langle \frac{D_z}{D_x}[h(\xi)]^2 + \frac{U^2}{4D_x^2} \right\rangle^{1/2} x \right]}{\frac{U}{2} + D_x \left\{ \frac{D_z}{D_x}[h(\xi)]^2 + \frac{U^2}{4D_x^2} \right\}^{1/2}} \cos[h(\xi)z] \tag{5-235}$$

From Eq. (5-189), for $\lambda = 0$,

$$f_C(0) = \frac{2}{\pi} U \frac{\exp(0)}{(U/2 + U/2)} \cos(0) = \frac{2}{\pi} \tag{5-236}$$

It can be shown from Eq. (5-189) that

$$f_C'(0) = \frac{df_C(0)}{d\lambda} = 0 \tag{5-237}$$

Equations (5-229)–(5-237) define the numerical solution of the integral in Eq. (5-188).

To apply the Filon numerical integration method, the values of ξ and the number of terms in the series should be specified. Numerical analysis showed that $\xi = 0.02$ to 0.10 and $N = 500$ to 1000 produce accurate results (Batu, 1987, 1988).

Application of Strip Source Solutions
Hydrodynamic Dispersion Stream Lines (HDSLs): Based on the analytical model presented above, the general theory of the HDSF concept applies to strip sources in a uniform ground water flow field shown in Figure 5-23. The solution for the HDSF is given by Eq. (5-175). In this equation, $I(\lambda)$ represents the source conditions and the geometric distributions of the sources. The term $F_x(0, p)/\varphi_e$ in Eq. (5-176) is given by Eqs. (5-151)–(5-156), which are flux-type (or third-type) boundary conditions. It is interesting to point out that the flux-type boundary conditions are required with regard to Eq. (5-106). In fact, if we intend to use the specified concentration (or first-type) boundary conditions at the sources, the following question must be answered: can we use the concentration-type (or first-type) boundary in the HDSF analysis? The answer is "negative," because as can be seen from the detailed analysis given in the above paragraphs, the general mathematical formulation of the HDSF analysis does not include a direct relationship between the HDSF and concentration function. However, as usual, a concentration-type boundary condition can be used to determine the concentration function and then, using the interrelationships between C and ψ (Eqs. [5-116] and [5-117]), the HDSF can be obtained. This method can only be applied in those cases for which analytical solutions can be developed. An in the HDSF case, $J(\lambda)$ in Eq. (5-179) represents the source boundary conditions and the geometric distribution of the sources. The general solutions for HDSF and concentration function can be applied to any number of strip sources. One may claim that the superposition principle can be applied using the solution for a single-strip source case. If the sizes and source concentrations are all the same for every strip source, the superposition principle can be applied; but for the cases shown in Figure 5-22, the superposition principle cannot be applied.

The final result for the HDSF in Eq. (5-198) is expressed by $\psi(x, z)/(UC_M)$ whose dimension is length (L). Numerical values of Eqs. (5-198), (5-211), (5-220), and (5-229) were determined using a FORTRAN 77 program. In Figure 5-24, Figure 5-25, and Figure 5-26, the HDSLs and normalized concentration profiles are presented for an example with $B = 50$ m (the half-width of the strip), and $\alpha_T = 0.05\alpha_L$, $0.10\alpha_L$, and $0.20\alpha_L$ (transverse dispersivity). As can be seen from these figures, the values of $\psi(x, z)/(UC_M)$ increase as z increases, and become a constant (50 m). The

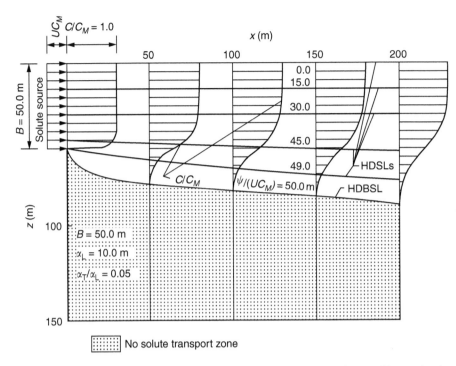

FIGURE 5-24   Hydrodynamic dispersion streamlines and concentration profiles for $B = 50$ m, and $\alpha_T/\alpha_L = 0.05$ ($\alpha_L = 10$ m). (Adapted from Batu, V., *Water Resour. Res.*, 23, 1175–1184, 1987.)

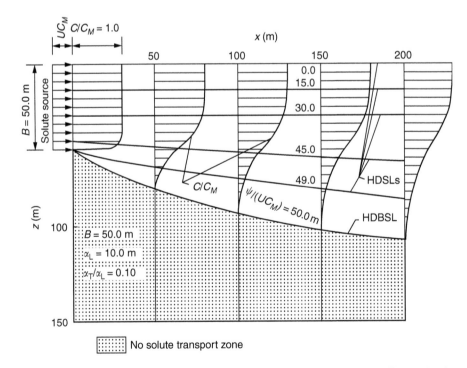

FIGURE 5-25   Hydrodynamic dispersion stream lines and concentration profiles for $B = 50$ m, and $\alpha_T/\alpha_L = 0.10$ ($\alpha_L = 10$ m). (Adapted from Batu, V., *Water Resour. Res.*, 23, 1175–1184, 1987.)

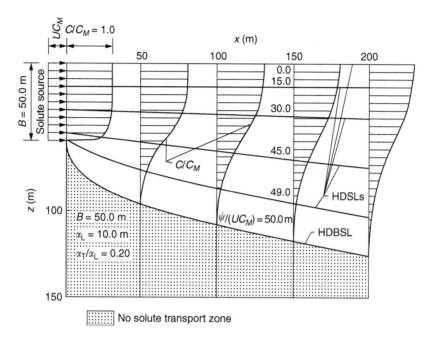

FIGURE 5-26　Hydrodynamic dispersion stream lines and concentration profiles for $B = 50$ m, and $\alpha_T/\alpha_L = 0.20$ ($\alpha_L = 10$ m). (Adapted from Batu, V., *Water Resour. Res.*, 23, 1175–1184, 1987.)

mass rate is $\psi = 50 U C_M$. On the other hand, according to Figure 5-24, the mass influx from the source is $50 U C_M$, which is exactly the same as the above value. In Figures 5-24, 5-25, and 5-26, the normalized concentration profiles C/C_M are shown at $x = 0$, 50, 100, 150, and 200 m. The normalized concentration has a maximum value at $z = 0$ and decreases gradually as the z coordinate increases and becomes zero on the HDSL having the value 50 m. This situation is consistent with the explanation given above. The hydrodynamic dispersion bounding stream line (HDBSL) for this example has the value of 50 m. This value does not depend on the transport parameters α_L and α_T, but depends on the geometry of the sources, B. However, the location of the HDBSL is a function of the ratio of dispersivities. For example, the HDBSL intersects the line $x = 200$ m at $z = 82$ m for $\alpha_L/\alpha_T = 0.05$ (Figure 5-24), $z = 104$ m for $\alpha_L/\alpha_T = 0.10$ (Figure 5-25), and $z = 122$ m for $\alpha_L/\alpha_T = 0.20$ (Figure 5-26). Figures 5-24, 5-25, and 5-26 show that $C/C_M = 1$ at $x = 0$ and $0 < z < B$. As can be seen from the concentration profiles in these figures, the concentrations are all zero on the HDBSLs. This is consistent with the properties of the HDBSL. Beyond the zones of the HDBSLs, the mass rates are all zero because of the constant value of the HDSF.

The analysis presented above can be used for different purposes. Values of α_T are critically important in controlling the lateral extent of plumes. The importance of α_T is shown with examples. Also, practically two important application areas are explained below (Batu, 1988).

Application of Contaminant Plume Analysis to Saturated Zones
Equation (5-184) is used to show the application of the HDSF concept for contaminant plume analysis. Under steady-state ground water flow and steady-state solute transport conditions, a zone exists through which the contaminant particles migrate (Figure 5-27). This zone is bounded by lines that are called HDBSLs. In other words, these stream lines are the boundaries of the convective–dispersive solute transport zone as shown in Figure 5-27. As mentioned previously, the difference of the values of two adjacent streamlines gives the mass rate passing the stream tube formed by the streamlines. This means that this difference is zero everywhere for those streamlines beyond the HDBSLs. The dashed zones beyond the HDBSLs shown in Figure 5-27 correspond to

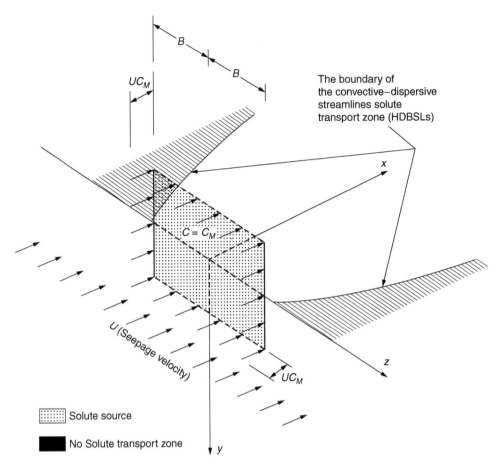

FIGURE 5-27   The geometry of a single-strip source and schematic representation of the HDBSLs. (Adapted from Batu, V., *Ground Water, 26,* 71–77, 1988.)

the no-contaminant transport zones. The HDBSLs are the boundaries of the transport zone along which the concentrations are zero.

Figure 5-28 presents the HDBSLs for $B = 50$ m and $\alpha_L = 10$ m for different α_T values. As can be seen from Figure 5-28, α_T has a significant effect upon the geometrical locations of the HDBSLs. In Figure 5-28, the curves correspond to $\alpha_T = 1, 2, 4,$ and 10 m. The distance between the HDBSLs for $\alpha_T = 1$ and 10 m at $x = 1000$ m is approximately 300 m. The curve for $\alpha_T = 10$ m corresponds to an extreme case because α_T is always less than α_L. Figure 5-28 also shows that $\Psi = \psi$ $(x, z)/(UC_M) = 50$ m for every HDBSL. From Eq. (5-184), for $\Psi = 50$ m, $\Psi = 50UC_M$. This is the total mass passing through the stream tube formed by the HDSL at $z = 0$ and at any HDBSL. The multiplication of the mass flux, UC_M, and the half-width of the source (50 m) gives $50UC_M$, which is exactly the same value given above. Figure 5-29 corresponds to $B = 100$ m. The rest of the values in Figure 5-29 are the same as in Figure 5-28. For this case, the value of HDBSL is 100 m, and this is equal to the mass input from the source. Figure 5-30 presents HDBSLs for $B = 50$ m, $\alpha_L = 1$ m, and $\alpha_T = 0.5$ m. The maximum value of the HDSF is 50 m, and this corresponds to the value of HDBSL. Figure 5-31 corresponds to $B = 50$ m, $\alpha_L = 1$ m, and $\alpha_T = 0.1$ m. Figure 5-30 and Figure 5-31 clearly show that the flux density decreases as the transverse distance increases from the x-coordinate axis. The dashed zones beyond the HDBSLs are the zones through which no contaminant transport occurs. The figures also show that as the transverse dispersivity decreases, the HDSLs tend to be approximately parallel to each other.

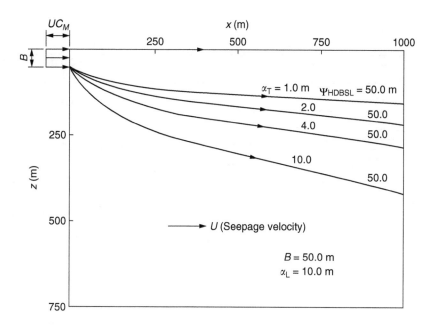

FIGURE 5-28 The HDBSLs for $B = 50$ m, $\alpha_L = 10$ m and different α_T values. (Adapted from Batu V., *Ground Water*, 26, 71–77, 1988.)

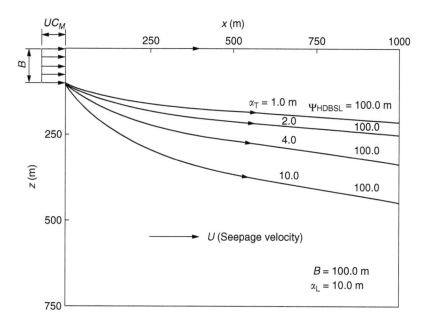

FIGURE 5-29 The HDBSLs for $B = 100$ m, $\alpha_L = 10$ m and different α_T values. (Adapted from Batu V., *Ground Water*, 26, 71–77, 1988.)

Figure 5-32 shows the applicability of the HDSF concept to contaminant migration in a saturated zone. Some important points for the application and assumptions are given as follows (Batu, 1988): (1) the saturated zone is a relatively thin, shallow, homogeneous and isotropic aquifer having steady-state uniform ground water velocity U and a steady contaminant transport; and

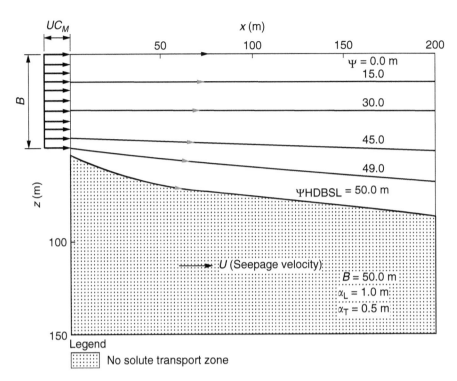

FIGURE 5-30   The HDBSLs for $B = 50$ m, $\alpha_L = 1$ m and $\alpha_T = 0.5$ m. (Adapted from Batu V., *Ground Water*, 26, 71–77, 1988.)

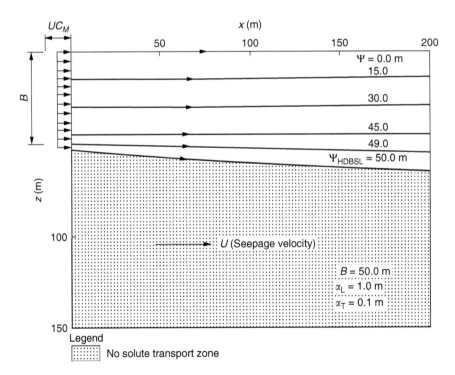

FIGURE 5-31   The HDBSLs for $B = 50$ m, $\alpha_L = 1$ m and $\alpha_T = 0.1$ m. (Adapted from Batu V., *Ground Water*, 26, 71–77, 1988.)

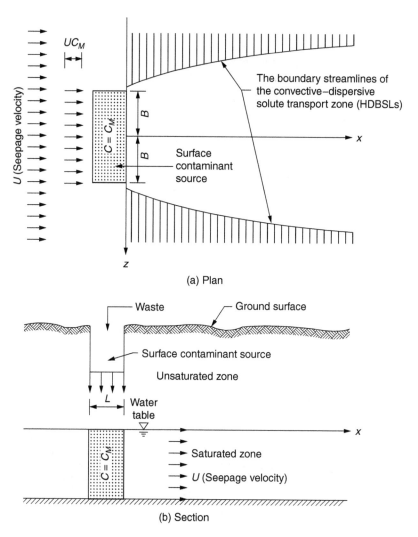

FIGURE 5-32 Schematic representation of the application of the HDSF concept for a saturated aquifer having a contaminant strip source. (Adapted from Batu V., *Ground Water*, 26, 71–77, 1988.)

(2) liquid waste from a surface contaminant source is being discharged to the saturated zone. Landfills, some units of factories, or other industrial complexes may be considered as surface contaminant sources; (3) the length of the surface contaminant source is $2B$ and its length is L; (4) for the sake of simplicity, it is assumed that the entire length of the source ditch reaches to the saturated zone and creates a constant source concentration C_M of a certain solute species in the area beneath the ditch (the dotted zones in Figure 5-32); and (5) the source concentration C_M can also be taken into account as a space-dependent quantity, i.e., $C_M = C_M(z)$ as presented previously.

Application of Contaminant Plume Analysis to Unsaturated Zones
The schematic diagram in Figure 5-33 shows the applicability of the HDSF to contaminant migration in the unsaturated zone. In Section 2.8.1, it is shown that Eq. (5-119) also represents the migration of contaminants in the unsaturated zone under a uniform infiltration rate at the soil surface. Some important points for the application and assumptions are as follows: (1) It is assumed that there is a uniform infiltration rate, v_0, at the soil surface. As shown in Section 2.8.1, the ground water velocity U can be determined using the unsaturated hydraulic conductivity vs. water content relationship for

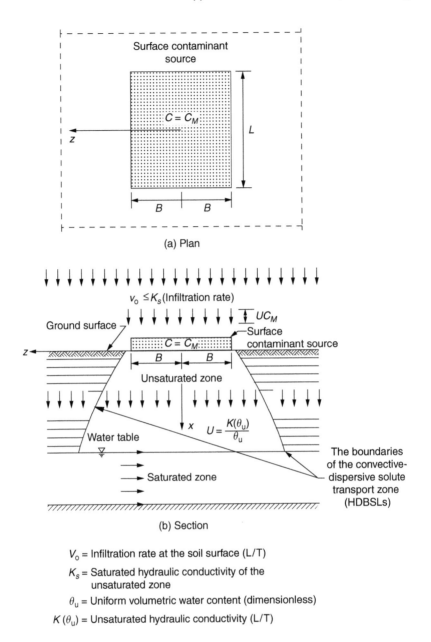

FIGURE 5-33 Schematic representation of the application of the HDSF concept for an unsaturated zone having a contaminant strip source. (Adapted from Batu V., *Ground Water*, 26, 71–77, 1988.)

a given soil. (2) liquid waste from a surface contaminant source is being discharged to the unsaturated zone. Landfills, some units of factories, or other industrial complexes may be considered as surface contaminant sources. (3) the length of the surface contaminant source is 2*B* and its width is L. (4) it is assumed that the waste liquid seeping from the bottom of the source ditch reaches to the top of the unsaturated zone and creates a constant concentration C_M of a certain solute species in the area beneath the ditch (the dotted zones in Figure 5-33). (5) the source concentration C_M can also be taken into account as a space-dependent quantity, as in the saturated zone application case.

The general methodology presented above is applicable to those aquifers in which uniform ground water velocities can be assumed with a reasonable approximation. The results are valid for steady-state water and solute transport cases and the methodology can specify the boundaries of the contaminant zones beyond which there is no contamination. Some aquifers may have approximately uniform flow fields, especially under nonpumping conditions from wells. Due to the effects of pumping and/or injection wells, drains, and other stress-producing effects, the ground water flow field may become highly nonuniform. For these cases, Eq. (5-112) is the governing partial differential equation, which can only be solved numerically.

5.3.4.3 Two-Dimensional Analytical Solutions for Strip Sources in a Finite Porous Medium

Problem Statement and Assumptions
The problem under consideration is steady-state two-dimensional solute transport from multiple-strip sources in a bounded unidirectional ground water flow field as shown in Figure 5-34 (Batu, 2002). The medium goes to infinity in the x-coordinate direction and is finite in the z-coordinate direction. The z coordinate varies between zero and H_p. The widths of the sources between $z = 0$ and $z = H_p$ are $L_1, L_2, ..., L_{p-1}$, and L_p. The corresponding source concentrations are all functions of the z coordinate as well. These fluxes are shown by arrows under a curve in Figure 5-34.

Governing Differential Equations and Convective–Dispersive Flux Components
The governing differential equation for the HDSF is given by Eq. (5-118). The corresponding two-dimensional partial differential equation for steady-state concentration distribution is given by Eq. (5-119). The convective–dispersive flux components in terms of HDSF ψ are given by

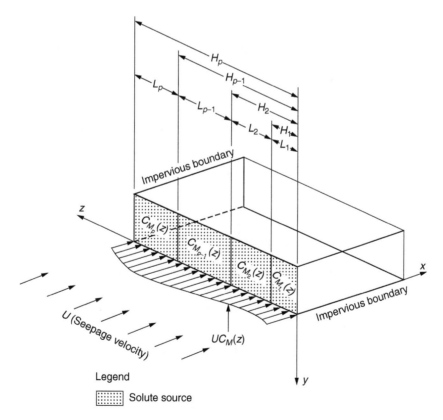

FIGURE 5-34   Schematic representation of strip sources in a finite porous medium having unidirectional flow field. (Adapted from Batu V., Unpublished paper, 2002.)

Eqs. (5-106) and (5-107). The same flux components in terms of the concentration C are given by Eqs. (2-41a) and (2-41c).

Boundary Conditions
Boundary Conditions for Hydrodynamic Dispersion Stream Function
Using Eq. (5-106), the boundary conditions at the sources can be written as

$$F_x(0, z) = \varphi_e \frac{\partial \psi(0, z)}{\partial z} \tag{5-238}$$

and from Figure 5-34 for the sources between $z = 0$ and $z = H_p$,

$$F_x(0, z) = \begin{cases} \varphi_e U C_{M_1}(z), & 0 < z < H_1 \tag{5-239} \\ \varphi_e U C_{M_{p-1}}(z), & H_{p-2} < z < H_{p-1} \tag{5-240} \\ \varphi_e U C_{M_p}(z), & H_{p-1} < z < H_p \tag{5-241} \end{cases}$$

where

$$H_1 = L_1 \tag{5-242}$$

$$H_{p-1} = L_1 + L_2 + \cdots + L_{p-1} \tag{5-243}$$

$$H_p = L_1 + L_2 + \cdots + L_{p-1} + L_p \tag{5-244}$$

Equations (5-239)–(5-241) are the third-type (or flux-type) boundary conditions. From Eqs. (5-107) and (2-41c)

$$F_z(x, z) = - \varphi_e D_z \frac{\partial C}{\partial z} = - \varphi_e \frac{\partial \psi}{\partial x} \tag{5-245}$$

With the assumption that for very large values of x, the concentration C is uniform and no longer changes with z, i.e.,

$$\frac{\partial C(x, z)}{\partial z} = 0, \quad \text{for } x = \infty \tag{5-246}$$

Therefore, Eq. (5-245) takes the form

$$F_z(x, z) = - \varphi_e \frac{\partial \psi}{\partial x} = 0, \quad \text{for } x = \infty \tag{5-247}$$

From Eq. (5-245), the boundary conditions at the impermeable boundaries are

$$\frac{F_z(x, z)}{\varphi_e} = \begin{cases} - \frac{\partial \psi}{\partial x} = 0 & \text{at } z = 0, \ 0 < x < \infty \tag{5-248} \\ - \frac{\partial \psi}{\partial x} = 0 & \text{at } z = H_p, \ 0 < x < \infty \tag{5-249} \end{cases}$$

Boundary Conditions for Concentration Function
The boundary conditions at the sources given by Eqs. (5-238)–(5-241) are also valid for concentration as well. Additional boundary conditions for concentration at the impermeable boundaries are given by Eqs. (3-51)–(3-54).

General Solution for the Hydrodynamic Dispersion Stream Function
General Solution
The general solution of Eq. (5-118) is given by Eq. (5-171).

Usage of Boundary Conditions at Infinity and Impermeable Boundaries
In Example 5-13, it is shown that using the boundary conditions at infinity and at the impermeable boundaries, as given in Eqs. (5-247)–(5-249), the solution can be expressed as

$$\psi(x, z) = \sum_{n=0}^{\infty} A_n \exp\left(\frac{Ux}{2D_x} - q_n x\right) \sin(\lambda_n z) \tag{5-250}$$

where

$$\lambda_n = \frac{n\pi}{H_p}, \quad n = 0, 1, 2, \ldots \tag{5-251}$$

$$q_n = \left(\frac{D_z}{D_x}\lambda_n^2 + \frac{U^2}{4D_x^2}\right)^{1/2} \tag{5-252}$$

and $(A_1 A_2)_n = A_n$.

EXAMPLE 5-13

Using the boundary conditions at infinity and at the impermeable boundaries, as given by Eqs. (5-247)–(5-249), from Eq. (5-171) derive Eq. (5-250).

SOLUTION

From Eq. (5-171), one can write

$$\frac{\partial \psi}{\partial x} = \left[A_1\left(\frac{U}{2D_x}-q\right)\exp\left(\frac{Ux}{2D_x}-qx\right) + B_1\left(\frac{Ux}{2D_x}+q\right)\exp\left(\frac{Ux}{2D_x}+qx\right)\right] \tag{E5-13-1}$$

$$\times [A_2\sin(\lambda z) + B_2\cos(\lambda z)]$$

and

$$\frac{\partial \psi}{\partial z} = \left[A_1\exp\left(\frac{U}{2D_x}-q\right) + B_1\exp\left(\frac{Ux}{2D_x}+q\right)\right][A_2\lambda\cos(\lambda z) - B_2\lambda\sin(\lambda z)] \tag{E5-13-2}$$

To satisfy Eq. (5-247), B_1 must be zero in Eq. (E5-13-1). Therefore, from Eq. (5-171), one can write

$$\psi(x, z) = A_1\exp\left(\frac{Ux}{2D_x} - qx\right)[A_2\sin(\lambda z) + B_2\cos(\lambda z)] \tag{E5-13-3}$$

To satisfy Eq. (5-248), B_2 must be zero in Eq. (E5-13-1). Therefore, Eq. (E5-13-3) takes the form

$$\psi(x, z) = A_1A_2\exp\left(\frac{Ux}{2D_x} - qx\right)\sin(\lambda z) \tag{E5-13-4}$$

Using the boundary condition given by Eq. (5-249), from Eq. (E5-13-1), with $B_1 = B_2 = 0$,

$$A_1A_2\left(\frac{U}{2D_x} - q\right)\exp\left(\frac{Ux}{2D_x} - q\right)\sin(\lambda H_p) = 0 \tag{E5-13-5}$$

In order to have satisfied this condition, the following condition must be satisfied:

$$\sin(\lambda H_p) = 0 = \sin(n\pi), \quad n = 0, 1, 2, \ldots \tag{E5-13-6}$$

Equation (E5-13-6) results in Eq. (5-251). Substitution of Eq. (5-251) into Eq. (E5-13-4) and applying the principle of superposition (since the partial differential equation and boundary conditions are linear), Eq. (5-250) can be obtained.

Usage of Boundary Conditions at x = 0

In Example 5-14 it is shown that using the boundary conditions at $x = 0$, as given in Eqs. (5-239)–(5-241), the solution can be expressed as

$$\psi(x, z) = \frac{z}{H_p}I_0(z) + \frac{2}{H_p}\sum_{n=1}^{\infty}\frac{I_n(z)}{\lambda_n}\exp\left(\frac{Ux}{2D_x} - q_n x\right)\sin(\lambda_n z) \qquad (5\text{-}253)$$

where

$$I_0(z) = \int_0^{H_p}\frac{F_x(0, z)}{\varphi_e}dz \qquad (5\text{-}254)$$

and

$$I_n(z) = \int_0^{H_p}\frac{F_x(0, z)}{\varphi_e}\cos(\lambda_n z)dz \qquad (5\text{-}255)$$

EXAMPLE 5-14

Using the boundary conditions at $x = 0$, as given in Eqs. (5-239)–(5-241), from Eq. (5-250) derive Eq. (5-253).

SOLUTION

From Eq. (5-250)

$$\frac{\partial \psi}{\partial z} = \sum_{n=0}^{\infty}A_n\exp\left(\frac{Ux}{2D_x} - q_n x\right)\cos(\lambda_n z) \qquad (E5\text{-}14\text{-}1)$$

and at $x = 0$

$$\left.\frac{\partial \psi}{\partial z}\right|_{x=0} = \sum_{n=0}^{\infty}A_n\lambda_n\cos(\lambda_n z) \qquad (E5\text{-}14\text{-}2)$$

Substitution of Eq. (E5-14-2) into Eq. (5-238) gives

$$\frac{F_x(0, z)}{\varphi_e} = \sum_{n=0}^{\infty}B_n\cos(\lambda_n z) \qquad (E5\text{-}14\text{-}3)$$

where

$$B_n = A_n\lambda_n = A_n\frac{n\pi}{H_p} \qquad (E5\text{-}14\text{-}4)$$

Consideration of the right-hand side of Eq. (E5-14-3) as a Fourier cosine series for the interval $z = 0$ to $z = H_p$ yields (Churchill, 1941)

$$B_n = \frac{2}{H_p}\int_0^{H_p}\frac{F_x(0, z)}{\varphi_e}\cos(\lambda_n z)\,dz \qquad (E5\text{-}14\text{-}5)$$

From Eqs. (E5-14-4) and (E5-14-5)

$$A_n = \frac{B_n}{\lambda_n} = \frac{2}{H_p}\frac{1}{\lambda_n}\int_0^{H_p}\frac{F_x(0, z)}{\varphi_e}\cos(\lambda_n z)\,dz \qquad (E5\text{-}14\text{-}6)$$

Introduction of Eqs. (E5-14-6) into Eq. (5-250), and by using Eq. (5-251), the first term, A_0 of the cosine series in Eq. (5-250), becomes (bearing in mind that the first term of the cosine series has the coefficient of "1/2")

$$A_0 = \frac{1}{2}\lim_{n\to 0}B_n\left[\frac{\sin(n\pi z/H_p)}{n\pi/H_p}\right] = \frac{1}{2}B_0 z \qquad (E5\text{-}14\text{-}7)$$

From Eq. (E5-14-5), for $n = 0$,

$$B_0 = \frac{2}{H_p} \int_0^{H_p} \frac{F_x(0, z)}{\varphi_e} dz \tag{E5-14-8}$$

Therefore, combining Eqs. (E5-14-7) and (E5-14-8), one has

$$A_0 = \frac{z}{H_p} \left[\int_0^{H_p} \frac{F_x(0, z)}{\varphi_e} dz \right] \tag{E5-14-9}$$

and combining Eqs. (5-250), (E5-14-6), and (E5-14-9) results in Eq. (5-253).

Special Solutions for a Single-Strip Source
Using the general solution, as given by Eq. (5-253), a special solution is obtained for a single-strip source whose geometry is shown in Figure 5-35, which is a special case of Figure 5-34. For this case, the boundary conditions at $x = 0$, as given by Eqs. (5-239)–(5-240), take the form

$$F_x(0, z) = \begin{cases} F_M = \varphi_e U C_M, & D_1 < z < D_1 + 2B & (5\text{-}256) \\ 0, & \text{otherwise} & (5\text{-}257) \end{cases}$$

Hydrodynamic Dispersion Stream Function Solution
As can be seen from Eq. (5-253), only $I_0(z)$ and $I_n(z)$ need to be evaluated for a single strip source in order to determine the $\psi(x,z)$ solution. With Eqs. (5-256) and (5-257), Eqs. (5-254) and (5-255), respectively, take the forms

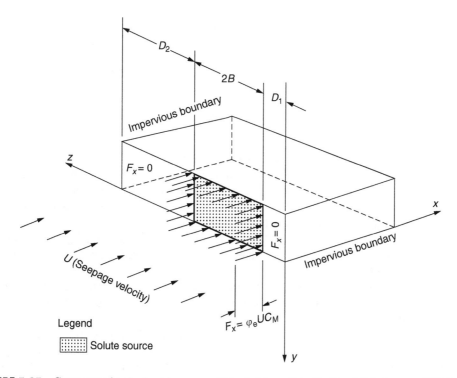

FIGURE 5-35 Geometry of a single-strip source. (Adapted from Batu V., Unpublished paper, 2002.)

$$I_0(z) = \int_{D_1}^{D_1+2B} \frac{F_M}{\varphi_e} \, dz = 2B \frac{F_M}{\varphi_e} \tag{5-258}$$

$$I_n(z) = \int_{D_1}^{D_1+2B} \frac{F_M}{\varphi_e} \cos(\lambda_n z) \, dz = \frac{F_M}{\varphi_e} \frac{1}{\lambda_n} \langle \sin[\lambda_n(D_1 + 2B)] - \sin(\lambda_n d_1) \rangle \tag{5-259}$$

Hydrodynamic Dispersion Stream Function Solution
Substitution of Eqs. (5-258) and (5-259) into Eq. (5-253) gives the solution for HDSF for a single-strip source as

$$\psi(x, z) = \frac{F_M}{\varphi_e} \frac{2B}{D_1 + 2B + D_2} z + \frac{F_M}{\varphi_e} (D_1 + 2B + D_2) \frac{2}{\pi^2} \sum_{n=1}^{\infty} \frac{1}{n^2}$$

$$\times \left\{ \sin[\lambda_n(D_1 + 2B)] - \sin(\lambda_n D_1) \right\} \exp\left(\frac{Ux}{2D_x} - q_n x \right) \sin(\lambda_n z) \tag{5-260}$$

Concentration Solution
The concentration solution for a single-strip source is given by Eq. (3-294).

Convective–Dispersive Flux Components — Determination from $\psi(x, z)$
From Eqs. (5-106) and (5-260), $F_x(x, z)$ can be expressed as

$$F_x(x, z) = \varphi_e \frac{\partial \psi}{\partial z} = \frac{2B}{D_1 + 2B + D_2} F_M + \frac{2}{\pi} F_M \sum_{n=1}^{\infty} \frac{1}{n} \left\{ \sin[\lambda_n(D_1 + 2B)] \right.$$

$$\left. - \sin(\lambda_n D_1) \right\} \exp\left(\frac{Ux}{2D_x} - q_n x \right) \cos(\lambda_n z) \tag{5-261}$$

From Eqs. (5-107) and (5-260), $F_z(x, z)$ can be expressed as

$$F_x(x, z) = -\varphi_e \frac{\partial \psi}{\partial x} = -(D_1 + 2B + D_2) F_M \frac{2}{\pi^2} \sum_{n=1}^{\infty} \frac{1}{n^2} \left\{ \sin[\lambda_n(D_1 + 2B)] \right.$$

$$\left. - \sin(\lambda_n D_1) \right\} \left(\frac{U}{2D_x} - q_n \right) \exp\left(\frac{Ux}{2D_x} - q_n x \right) \sin(\lambda_n z) \tag{5-262}$$

Convective–Dispersive Flux Components — Determination from $C(x, z)$
The expressions for $F_x(x, z)$ can also be determined from $C(x, z)$, which is given by Eq. (3-294). The expression for $F_x(x, z)$ is given in Eq. (3-295) and its form for $v = 0$ corresponds to Eq. (5-261). Equations (3-295) and (5-261) look somewhat different, but their numerical results in Table 5-1 are exactly the same. The values in Table 5-1 were generated for $B = 50$ m, $D_1 = 10,000$ m, $D_2 = 10,000$ m, $\alpha_L = 10$ m, $\alpha_T = 1$ m, $\varphi_e = 0.20$, $C_M = 1$ g/m³, $U = 0.1$ m/day, and $F_M = 0.02$ g/m²/day. In Table 5-1, the convective–dispersive flux components derived from $C(x, z)$ were generated using the ST2B computer program mentioned in Section 3.3.3.2; and the convective–dispersive flux components derived from $\psi(x, z)$ were generated using the computer program HDSF2, which is identified below. As can be seen from Table 5-1, the numerical values of $F_x(x, z)$ derived from $\psi(x, z)$ and $C(x, z)$ are exactly the same.

    Similarly, the expression for $F_z(x, z)$ can also be determined from $C(x, z)$, which is given by Eq. (3-294). The expression for $F_x(x, z)$ is given Eq. (3-296) and its form for $v = 0$ corresponds to Eq. (5-262). Equations (3-296) and (5-262) look somewhat different, but their numerical results as shown in Table 5-1 are exactly the same. The numerical values were generated using the data and computer programs mentioned above.

TABLE 5-1

Comparison of $F_x(x, z)$ and $F_z(x, z)$ Convective–Dispersive Flux Components Determined from $\psi(x, z)$ and $C(x, z)$ Solutions for a Single Strip Source

X(m)	Z(m)	$\psi(x, z)$ Solution		$C(x, z)$ Solution	
		$F_x(x, z)$	$F_z(x, z)$	$F_x(x, z)$	$F_z(x, z)$
100	10050	0.019963	0.000000	0.019963	0.000000
100	10051	0.019962	0.000001	0.019962	0.000001
100	10052	0.019961	0.000002	0.019961	0.000002
100	10053	0.019959	0.000004	0.019959	0.000004
100	10054	0.019955	0.000005	0.019955	0.000005
100	10055	0.019951	0.000006	0.019951	0.000006
100	10056	0.019945	0.000008	0.019945	0.000008
100	10057	0.019938	0.000010	0.019938	0.000010
100	10058	0.019930	0.000012	0.019930	0.000012
100	10059	0.019920	0.000014	0.019920	0.000014
100	10060	0.019908	0.000016	0.019908	0.000016
100	10061	0.019894	0.000019	0.019894	0.000019
100	10062	0.019878	0.000022	0.019878	0.000022
100	10063	0.019859	0.000025	0.019859	0.000025
100	10064	0.019837	0.000029	0.019837	0.000029
100	10065	0.019811	0.000033	0.019811	0.000033

Application of the Analytical Models to Ground Water Systems

The general model and its derivative models under steady-state flow and solute-transport conditions presented above can be applied to solute-transport conditions if the ground water velocity can be assumed to be uniform. These are all discussed in the following paragraphs.

In this section, first a computer program HDSF2, for a single strip source, is described, then the application of this program is presented with examples.

A Computer Program (HDSF2) for HDSF for a Single Strip Source

The author of this book has developed two-dimensional MS–DOS-based FORTRAN program, HDSF2. "HDSF" (Hydrodynamic Dispersion Stream Function and "2" stands for "Two-dimensional"). HDSF2 is a menu-driven program and calculates the values of HDSF and convective-–dispersive flux components under steady-state flow and solute-transport conditions from a strip source in a bounded medium using Eqs. (5-260), (5-261), and (5-262). The HDSF2 program can be run interactively in IBM compatible computers under both DOS and WINDOWS modes. In the series, generally N = 400 − 500 produces reasonably accurate results.

Application to Aquifers

The two-dimensional analytical solute transport model, whose geometry is shown in Figure 5-35, can be used for different purposes. The impervious boundaries around the solute-transport domain increase the potential usage areas of the model. For ground water flow cases, the effects of impervious boundaries such as bedrock outcrops and clay zones surrounding the main ground water flow zone can be taken into account. If the aquifer goes to infinity from both sides of the sources, its effect can be taken into account by assigning large values to D_1 and D_2 as compared with the source width $2B$. The model can also be used along a cross-section of a confined or unconfined aquifer such that the $z = 0$ plane is the bottom of the aquifer, and $z = D_1 + 2B + D_2$ plane is the upper boundary of the aquifer. For a confined aquifer, the $z = D_1 + 2B + D_2$ plane is the bottom of the confining layer and for an unconfined aquifer it is the water table of the unconfined aquifer. For these cases, the source will be infinitely long in the direction of the y-coordinate axis. For a two-dimensional areal model case in the x–z plane, as shown in Figure 5-35, it is assumed that the source extends from the bottom or to the top of the confined aquifer or to the water table of the unconfined aquifer.

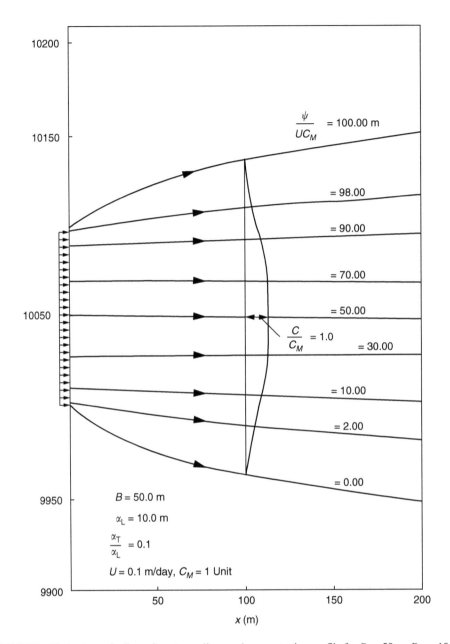

FIGURE 5-36 Hydrodynamic dispersion stream lines and concentration profile for $B = 50$ m, $D_1 = 10,000$ m, $D_2 = 10,000$ m, and $\alpha_T/\alpha_L = 0.10$ ($\alpha_L = 10$ m) for Example 5-13. (Adapted from Batu V., Unpublished paper, 2002.)

EXAMPLE 5-15

For a solute source and aquifer geometry shown in Figure 5-35, the following parameters are given: $C_M = 1$ g/m^3, $B = 50$ m, $D_1 = 10,000$ m, $D_2 = 10,000$ m, $\varphi_e = 0.20$, $K_h = 0.20$ m/day, $I = 0.10$ m/m, $\alpha_L = 10$ m, $\alpha_T = 1$ m, and $D^* = 0$ m^2/day. Using these parameters, (1) calculate the values of U and F_M; (2) using the HDSF2 computer program, determine the HDSLs and discuss the results; and (3) determine the concentration profile at $x = 100$ m using the ST2B program described in Section 3.3.3.2.

SOLUTION

(1) The ground water velocity U, from Eqs. (2-39) and (2-40), is

$$U = \frac{1}{\varphi_e} K_h I = \frac{1}{0.20}(0.20 \text{ m/day}) (0.10 \text{ m/m}) = 0.10 \text{ m/day}$$

From Eq. (5-256),

$$F_M = \varphi_e U C_M = (0.20)(0.10 \text{ m/day}) (1 \text{ g/m}^3) = 0.20 \text{ g/m}^2/\text{day}$$

(2) Using the computer program with the input data, the generated HDSLs are shown in Figure 5-36. In Figure 5-36, $\psi/(UC_M) = 0$ and 100 m are the HDBSLs and beyond these lines solute does not exist. The difference between these HDBSLs is $\psi = (100 \text{ m})UC_M = (100 \text{ m })(0.10 \text{ m/day})(1 \text{ g/m}^3) = 10 \text{ g/m/day}$, which is exactly equal to the mass input at the source given by $(100 \text{ m})UC_M$. Notice that the HDSLs configurations and values in Figure 5-36 compare closely with the ones in Figure 5-25. This is due to the fact that the values of D_1 and D_2 are relatively large.

(3) Using the same data with the ST2B program, the concentration profile at $x = 100$ m is shown in Figure 5-36. Notice that the concentrations on the HDBSLs are all zero, which is consistent with the theory.

EXAMPLE 5-16

For a solute source and aquifer geometry shown in Figure 5-35, the following parameters are given: $C_M = 1$ g/m^3, $B = 0.5$ m, $D_1 = 49$ m, $D_2 = 0$ m, $\varphi_e = 0.20$, $K_h = 0.20$ m/day, $I = 0.10$ m/m, $\alpha_L = 10$ m, $\alpha_T = 1$ m, and $D^* = 0$ m^2/day. These data define a source adjacent to the water table shown

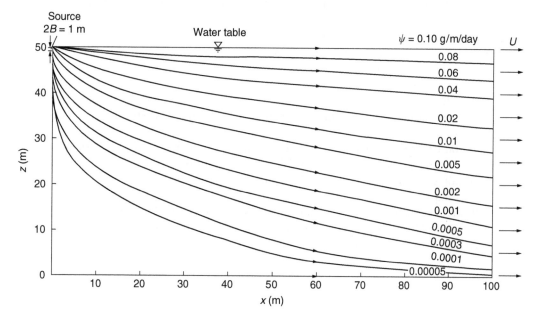

FIGURE 5-37 Hydrodynamic dispersion stream lines for $B = 1$ m, $D_1 = 49$ m, $D_2 = 0$, and $\alpha_T/\alpha_L = 0.05$ ($\alpha_L = 10$ m) for Example 5-14. (Adapted from Batu V., Unpublished paper, 2002.)

in Figure 5-37. Using the given parameters, (1) calculate the values of U and F_M; and (2) using the HDSF2 computer program, determine the HDSLs and discuss the results.

SOLUTION

(1) Using the given data the values of U and F_M are calculated in Example 5-15 as $U = 0.10$ m/day and $F_M = 0.02$ g/m²/day.

(2) The HDSLs generated from the output of the HDSF2 computer program are shown in Figure 5-37. The value of the lowermost HDSL is $\psi = 0.00005$ g/m/day and its value at the water table is $\psi = 0.10$ g/m/day. The lower most HDSL can approximately be considered as the HDBSL because its value is 0.05% of the value of $\psi = 0.10$ g/m/day at the water table. The HDSLs in Figure 5-37 show that after approximately 100 m distance from the source, the HDBSL approaches to the bottom of the aquifer. The HDSLs also show that approximately 90% of solute mass passes through up to approximately 22 m depth from the water table, and only 10% of solute mass passes through the rest of the aquifer thickness of 28 m. The source depth is only 2% of the aquifer thickness. From a practical standpoint, these results can be used in locating and monitoring wells along the aquifer thickness. For this special case, the appropriate location of the monitoring wells is up to 22 m depth from the water table.

PROBLEMS

5.1 What is the main limitation of particle tracking?

5.2 In Eqs. (5-20)–(5-22), assume that φ_e, q_x, q_y, and q_z are independent of the x, y, and z coordinates. Rewrite the same equations under these conditions.

5.3 For an aquifer and a well, the following parameters are given: $\varphi_e = 0.30$, $K_h = 30$ m/day, $H - h_w = 3.5$ m, $R = 200$ m, and $r_w = 0.1$ m. Using these values, (1) determine the travel time of a solute to the well starting at $r = 25$ m; (2) determine the travel time of a solute starting at $r = 12$ m; and (3) using Eq. (5-38), show that t varies directly with r^2 when r_w is small relative to r.

5.4 Derive Eq. (5-38) from Eq. (3-37).

5.5 Consider the well parameters in problem 5.3. Observations showed that clogging occurred out to a radius $r_c = 2$ m. As a result, K_{h_I} is reduced from 30 to 1 m/day and effective porosity is reduced from $\varphi_{e_{II}} = 0.30$ to $\varphi_{e_I} = 0.20$. Using these values, (1) calculate the travel time of a solute from 32 m and using Eq. (5-48) and (2) calculate the travel time using the approximate equation as given by Eq. (5-49) and compare the results.

5.6 The soil under the dam shown in Figure 5-6 has $K = 45$ m/day hydraulic conductivity. The effective porosity is $\varphi_e = 0.25$. With the head difference $H = 8$ m, calculate the travel time of a particle from point B to point C along the line BOC.

5.7 Derive Eq. (5-56).

5.8 What are the main principles of particle tracking of the MODPATH code?

5.9 What is the "hydrodynamic dispersion stream function"? How does it compare with the "stream function" used in fluid mechanics?

6 Stochastic Analysis of Flow and Solute Transport In Heterogeneous Aquifers and Determination of Expressions for Dispersivities Using Stochastic Approaches

6.1 INTRODUCTION

As mentioned in Section 3.2, since the 1950s, a number of analytical and numerical solute transport models were developed by various investigators, using the *deterministic approach*. Since the mid-1970s, *stochastic approaches* started to appear in the literature mainly to evaluate and determine equations for the dispersion coefficients by taking into account the spatial variability of flow and solute transport parameters. Until the 1950s, ground water flow and solute transport parameters such as hydraulic conductivity, effective porosity, and dispersivity have been viewed as well-defined local quantities that can be assigned unique values at each point in the flow and transport domain. Flow and solute transport in aquifers occur in a complex geologic environment whose lithologic, petrophysical, and structural makeups vary in ways that cannot be predicted deterministically in all of their relevant details. These makeups tend to exhibit discrete and continuous variations on a multiplicity of scales, causing flow and transport parameters to act similarly. The flow and solute transport parameters can at best be measured at selected well locations and depth intervals, where their values depend on the scale (support volume) and mode (instrumentation and procedure) of measurement. It is neither practical nor cost-effective to measure the parameters at all points. Therefore, estimating the parameters at points where measurements are not available entails a random error. Also, quite often, the support of measurement is uncertain and the data are corrupted by experimental and interpretive errors. These errors and uncertainties render the flow and solute transport parameters random and the corresponding flow and solute transport equations stochastic (Neuman, 1997).

Stochastic models as predictive tools for flow and solute transport migration, are still at the research level, and the extension of these efforts to real-world practical applications is fairly limited. Regarding this, Carrera (1993) states:

> Stochastic methods have been successful in explaining qualitatively some anomalies of solute transport, but they appear to be far from reaching a stage at which they can be used routinely for solving realistic field problems. On the other hand, when applied with care, deterministic methods have been successfully used in actual problems.

However, the application of stochastic mathematical techniques for the evaluation and determination of dispersion coefficients has made significant strides, especially during the last quarter of the 20th century. The general framework of these approaches is discussed in Section 3.2. However, scientific debate regarding these approaches is still ongoing. For example, Neuman (1997) states:

> Though the uncertain nature of flow and transport parameters is now widely recognized, there does not yet appear to be a consensus about the best way to deal with it mathematically. ...

As mentioned in Section 3.2, according to Freeze et al. (1990), there are three basic stochastic approaches: (1) first-order analysis; (2) perturbation analysis; and (3) monte-Carlo analysis. *First-order analysis* (e.g., Dagan, 1989) is a simple and direct means of propagating uncertainty. This approach can be employed with either analytical or numerical solutions of the governing equations, but is limited to linear or nearly linear systems. Beyond this, this approach requires only the first two moments (the mean and the variance) of the input parameters, and it provides only the first two moments of the predicted output variables. In *perturbation analysis* (e.g., Gelhar, 1993), both the input parameters and the output variables are defined in terms of a mean and a perturbation about the mean. The relationship between input and output uncertainties can be developed using two general techniques. One technique involves developing relationships in the spectral domain using the theory of Fourier–Stieltjes integrals (e.g., Lumley and Panofsky, 1964) and an inverse Fourier transform. This approach is best suited to analytical solutions, an infinite flow domain, and a small coefficient of variation in input uncertainty. The third approach, *Monte-Carlo* analysis, provides the most general approach to uncertainty propagation. The Monte-Carlo approach is not based on stochastic differential equations but rather on deterministic solutions for a number of numerically generated realizations. These realizations are analyzed by statistical methods to estimate means, variances, and possibly, probability density functions. In this chapter, the perturbation analysis technique will mostly be used in evaluating and determining the dispersion coefficients.

6.2 PHYSICAL MECHANISM OF CONVECTIVE–DISPERSIVE SOLUTE TRANSPORT IN HETEROGENEOUS POROUS MEDIA

6.2.1 MECHANICAL DISPERSION COEFFICIENTS BASED ON LABORATORY EXPERIMENTS

As mentioned in Section 2.11, during the past four decades, a number of studies have been conducted to determine universal laws for hydrodynamic dispersion coefficients, using deterministic and stochastic approaches. There are still a lot of controversial aspects regarding dispersion coefficients and research is still being conducted on this subject. The studies initially started with laboratory experiments in the early 1960s. Since the early 1980s, field experiments have been conducted to clarify some of the controversial aspects of dispersion coefficients. There is no doubt that the laboratory studies have played a pivotal role in improving the expressions for dispersion coefficients during the past four decades. As presented in Section 2.11, these expressions in the laboratory column scale along with appropriate models generated values that were quite close to those measured from column tests. However, these comparisons are mostly limited to one-dimensional cases. In the late 1970s, and in the early 1980s field-measured concentration values showed that laboratory-measured dispersion coefficients along with the appropriate models (analytical and numerical) may not be appropriate for most field cases. These studies raised the following questions: (1) Is the dispersive flux under field conditions Fickian? (2) Are the expressions for dispersion coefficients appropriate? (3) Is the dispersion process a deterministic or stochastic phenomenon?

6.2.2 CONVECTIVE–DISPERSIVE SOLUTE TRANSPORT MECHANISM IN NATURAL MATERIALS

The laboratory experiments mentioned in Section 6.2.1 are based on the *homogeneous medium* concept. The concept of homogeneous medium is inappropriate for natural hydrogeologic systems. In particular, the observed degree of spreading of a solute is generally much greater than what can be predicted on the basis of laboratory dispersivity values. This increased spreading is usually attributed to the heterogeneous nature of field environments. Dispersion in granular materials is shown schematically in Figure 6-1. Figure 6-1a shows a simple dilution front and spread of a slug injection in homogeneous materials and the dispersion coefficients mentioned in Figure 6.2.1 are based on this concept. Figure 6-1b shows fingering of tracer front caused by heterogeneities (De Josselin de Jong, 1958). The results of De Josselin de Jong are supported by the results of several studies using two-dimensional laboratory models (Skibitzke and Robertson, 1963; Theis, 1967; and Hillier, 1975). Figure 6-1c shows spreading caused by high permeability lenses (Skibitzke and Robertson, 1963). On the basis of the results shown in Figure 6-1, Skibitzke and Robertson (1963) observed that the presence of small, discrete lenses of high hydraulic conductivity within a sandstone matrix of lower hydraulic conductivity caused much stronger dispersion than that without the lenses.

6.2.3 CONVECTIVE–DIFFUSIVE SOLUTE TRANSPORT MECHANISM IN NATURAL MATERIALS

Figure 6-2 presents a schematic diagram demonstrating the convective–diffusive solute transport mechanism (Gillham and Cherry, 1982b). For the purpose of discussion, the medium is considered

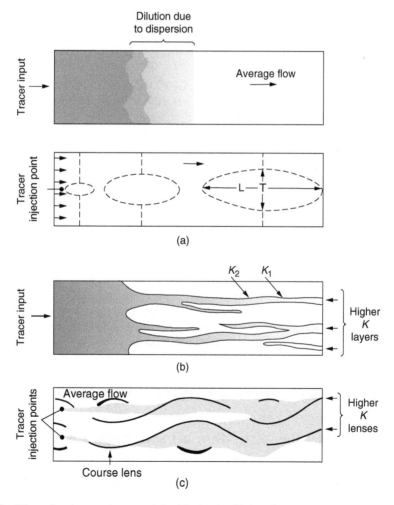

FIGURE 6-1 Dispersion in granular materials: (a) simple dilution front and spread of a slug injection in homogeneous materials; (b) fingering of tracer front caused by heterogeneities; and (c) spreading caused by high permeability lenses. (Adapted from Gilham, R.W. and Cherry, J.A., *Recent Trends in Hydrogeology*, The Geological Society of America, Special Paper 189, Boulder, Colorado, 1982b, pp. 31–62.)

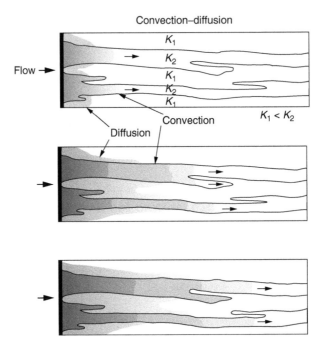

FIGURE 6-2 Schematic diagram showing the convective–diffusive transport mechanism. (Adapted from Gilham, R.W. and Cherry, J.A., *Recent Trends in Hydrogeology*, The Geological Society of America, Special Paper 189, Boulder, Colorado, 1982b, pp. 31–62.)

to have two different hydraulic conductivities, K_1 and K_2 ($K_2 > K_1$). As the solutes are transported, primarily by dispersion in the more permeable heterogeneities (K_2 zones) diffusion causes them to migrate from these heterogeneities into the adjacent heterogeneities of lesser permeability (K_1 zones). The net effect is to reduce the concentrations and solute flux in the permeable zones (K_2 zones) and increase concentrations in the less permeable zones (K_1 zones).With time, the diffusion tends to generate increasing uniformity of concentration distribution in the plume and a concentration front that exhibits increasing spread and a gradual decrease in concentration toward the position of farthest advance. As a result, dispersion at the macroscopic scale and at larger scales within the complex flow system is accomplished primarily by molecular diffusion. The convection–diffusion model predicts more or less continuous plumes showing a marked decrease in concentration with increase in distance from the source. The solutes tend to migrate at velocities lower than those suggested by the heterogeneous-convection model.

6.3   DETERMINATION OF DISPERSIVITIES FROM SOLUTE PLUME SPATIAL MOMENTS

6.3.1   ONE-DIMENSIONAL CONVECTIVE–DIFFUSIVE SOLUTE TRANSPORT AND PLUME SPATIAL MOMENTS

6.3.1.1   One-Dimensional Analytical Solution of a Tracer Slug

Problem Statement and Assumptions
Let us consider a column of porous medium as shown in Figure 3-1. At time equal to zero ($t = 0$), a very thin tracer of marked particles having mass M is injected into the column at $x = 0$. As the

slug moves downstream with the ground water velocity U in the $+x$ direction, it spreads out and occupies an increasing portion of the column. The assumptions are as follows:

1. The solute slug having mass M is instantaneously injected at $x = 0$.
2. The medium goes to infinity along the x-coordinate direction.
3. The ground water flow is steady and has uniform velocity U.
4. The tracer slug is very thin and is perpendicular to the flow direction.

Governing Differential Equation

The governing differential equation is given by Eq. (2-83), and for $R_d = 1$, $D_x = D^*$, and $v = 0$, it takes the following form:

$$\frac{\partial C}{\partial t} = D^* \frac{\partial^2 C}{\partial x^2} - U \frac{\partial C}{\partial x} \qquad (6\text{-}1)$$

in which D^* is the effective molecular diffusion coefficient, and the mechanical dispersion coefficient is assumed to be zero according to Eq. (2-154). By a transformation with

$$\xi = x - Ut, \quad \tau = t \qquad (6\text{-}2)$$

Eq. (6-1) takes the form

$$\frac{\partial C}{\partial \tau} = D^* \frac{\partial^2 C}{\partial \xi^2} \qquad (6\text{-}3)$$

Derivation of Eq. (6-3) is shown in Example 6-1. At $t = 0$, a very thin slug of solute mass is injected into the column at $x = 0$. As the slug moves downstream, the concentration of solute is described by Eq. (6-1). For an observer moving with the average flow, Eq. (6-1) takes the form of Eq. (6-3). It is clear that Eq. (6-1) or Eq. (6-3) is written under the assumption that the convective–dispersive transport phenomenon is Fickian, i. e., it occurs in accordance to Fick's first law for molecular diffusion.

EXAMPLE 6-1

Derive Eq. (6-3) from Eq. (6-1) with the transformation given by Eq. (6-2).

SOLUTION

Using the chain rule of partial differentiation, the expressions in Eq. (6-2) are substituted in Eq. (6-1). The transformation of $\partial C/\partial x$ is given by

$$\frac{\partial C}{\partial x} = \frac{\partial C}{\partial \xi} \frac{\partial \xi}{\partial x} + \frac{\partial C}{\partial \tau} \frac{\partial \tau}{\partial x} = \frac{\partial C}{\partial \xi}(1) + 0 = \frac{\partial C}{\partial \xi} \qquad (\text{E6-1-1})$$

and its second derivative is

$$\frac{\partial^2 C}{\partial x^2} = \frac{\partial^2 C}{\partial \xi^2} \qquad (\text{E6-1-2})$$

The time differentiation follows the same rule and is given by

$$\frac{\partial C}{\partial t} = \frac{\partial C}{\partial \xi} \frac{\partial \xi}{\partial t} + \frac{\partial C}{\partial \tau} \frac{\partial \tau}{\partial t} = \frac{\partial C}{\partial \xi}(-U) + \frac{\partial C}{\partial \tau}(1) = \frac{\partial C}{\partial \tau} - U \frac{\partial C}{\partial \xi} \qquad (\text{E6-1-3})$$

Substitution of Eqs. (E6-1-1), (E6-1-2), and (E6-1-3) into Eq. (6-1) results in Eq. (6-3).

Initial and Boundary Conditions

The initial condition, $C = C(x, 0)$, is in the form of Dirac delta function, $\delta(x)$

$$C(x, 0) = \frac{M}{\varphi_e} \delta(x) \qquad (6\text{-}4)$$

where M is the injected mass, φ_e the effective porosity, and

$$\delta(x) = \begin{cases} \lim_{m \to 0} \delta_m(x) = \dfrac{1}{m} & \text{for} \quad 0 < x < m, \ m > 0 \\[2mm] \lim_{m \to 0} \delta_m(x) = 0 & \text{otherwise} \end{cases} \tag{6-5}$$

The boundary conditions specified for $C(\xi, \tau)$ are

$$M = \int_{-\infty}^{\infty} \varphi_e C(\xi, \tau)\, d\xi \tag{6-6}$$

and

$$\lim_{|\xi| \to 0} C(\xi, \tau) = 0 \tag{6-7}$$

Solution for Concentration Distribution

The solution for the above initial and boundary-value problem is given in Crank (1956) and Bear (1979). The solution is

$$C(x, t) = \frac{M}{\varphi_e (4\pi D^* t)^{1/2}} \exp\left(-\frac{\xi^2}{4D^* t}\right) = \frac{M}{\varphi_e (4\pi D^* t)^{1/2}} \exp\left[-\frac{(x - Ut)^2}{4D^* t}\right] \tag{6-8}$$

and can also be expressed by

$$C(x, t) = \frac{M}{\varphi_e (2\pi)^{1/2} \sigma_x} \exp\left[-\frac{(x - \bar{x})^2}{2\sigma_x^2}\right] \tag{6-9}$$

where

$$\sigma_x = (2D^* t)^{1/2}, \quad \bar{x} = Ut \tag{6-10}$$

According to Eq. (6-2), since $\tau = t$, t is used in Eqs. (6-8)–(6-10).

6.3.1.2 Random Walk: One-Dimensional Form with Only Diffusion

This section presents two rationalizations for the diffusion equation and Fick's first law (see Section 2.3.1), based on the idea that the transport is a result of the random motion of individual solute particles. The developments in this section are adapted mainly from Crank (1956), Fischer et al. (1979), and Bear (1979).

The molecules of a solute in a transport domain are constantly in motion and forever colliding with one another. Any small particle, or molecule, experiences many collisions per second, numbering in billions. There are ~10^{26} molecules in 1 cm^3 of water. The collision rate depends on the solute, the size of the particles, and the density and temperature of the solute. It is obvious that the description of the motion described above must be statistical in nature. As a result, since a large number of particles are involved, the description must be done statistically (Fischer et al., 1979).

The problem is to analyze the statistics of motion of a single molecule (or particle) and generalize them. Although in reality, the transport of a solute is three-dimensional, for the sake of simplicity, only the one-dimensional case is considered. Suppose that the motion of a solute molecule consists of a series of random steps. It is assumed that each step is of equal length, $\Delta\xi$, and takes an interval of time Δt. Since the step is entirely random, the motion may be forward or backward. One asks: how far is the particle likely to get? On average, the answer is nowhere. However after a number of steps, sometimes the particle will have moved forward and sometimes backward. By means of the central limit theorem, it can be shown that in the limit of many steps, the probability of the particle being between $m\Delta\xi$ and $(m + 1)\Delta\xi$ approaches the normal distribution with zero mean and a variance

$$\sigma_x^2 = \frac{t(\Delta\xi)^2}{\Delta t} \tag{6-11}$$

provided the quotient $(\Delta\xi)^2/\Delta t$ approaches a constant value as $\Delta t \to 0$. Designating this quotient by $2D_x$, the probability of the particle being between the point x and $x + dx$ is given by

$$p(x, t)\, dx = \frac{1}{(2\pi)^{1/2}\sigma_x} \exp\left(-\frac{\xi^2}{2\sigma_x^2}\right) dx = \frac{1}{(4\pi D^* t)^{1/2}} \exp\left(-\frac{\xi^2}{4D^* t}\right) dx \qquad (6\text{-}12)$$

Now, if a whole group of particles begins its movement at the origin at time zero, the concentration at station ξ at any later time t will be proportional to the likelihood of any one being in the neighborhood of the station, or in other words

$$C(x, t) = \frac{M}{\varphi_e(4\pi D^* t)^{1/2}} \exp\left(-\frac{\xi^2}{4D^* t}\right) \qquad (6\text{-}13)$$

Comparing Eq. (6-13) with Eq. (6-8), it can be seen that the constant D^* takes the role of the diffusion coefficient, and that the random walk process postulated here leads to the same result as postulating that an initial slug of solute diffuses according to the diffusion equation.

6.3.1.3 Some Statistical Properties of Concentration Distribution

In the stochastic analysis of solute transport, various moments of a concentration distribution are frequently employed. For one-dimensional solute transport cases the defined moments are as follows (e.g., Fischer et al., 1979)

$$\text{zeroth moment} = M_0 = \int_{-\infty}^{\infty} \varphi_e C(x, t)\, dx \qquad (6\text{-}14)$$

$$\text{first moment} = M_1 = \int_{-\infty}^{\infty} \varphi_e x C(x, t)\, dx \qquad (6\text{-}15)$$

$$\text{second moment} = M_2 = \int_{-\infty}^{\infty} \varphi_e x^2 C(x, t)\, dx \qquad (6\text{-}16)$$

$$p\text{th moment} = M_p = \int_{-\infty}^{\infty} \varphi_e x^p C(x, t)\, dx \qquad (6\text{-}17)$$

In general, the moments are functions of time. However, the zeroth moment is simply the mass in the one-dimensional column and is constant. This can be explained as follows: In Eq. (6-14), $\varphi_e\, dx$ gives the differential effective pore volume of the column for a unit cross-section or the volume of water in the same volume under saturated conditions. Multiplying this volume with the solute concentration $C(x, t)$ and integrating it from $-\infty$ to $+\infty$ gives the total mass in the column.

The mean, μ, and spatial variance, σ_x^2, of a concentration distribution can be expressed from the moments as

$$\text{mean} = \mu = \frac{M_1}{M_0} = \frac{\int_{-\infty}^{\infty} \varphi_e x C(x, t)\, dx}{M_0} \qquad (6\text{-}18)$$

and

$$\text{spatial variance} = \sigma_x^2 = \frac{\int_{-\infty}^{\infty} \varphi_e (x - \mu)^2 C(x, t)\, dx}{M_0} = \frac{M_2 - M_0 \mu^2}{M_0} = \frac{M_2}{M_0} - \mu^2 \qquad (6\text{-}19)$$

The various quantities for the normal distribution can be found by carrying out the indicated integrations given above. According to Eq. (6-18), the mean is zero ($\mu = 0$). With $M_0 = M = 1$ and by carrying out the integration in Eq. (6-19) with $C(x, t)$ given by Eq. (6-13), one gets

$$\sigma_x^2 = 2D^* t \qquad (6\text{-}20)$$

Derivation of Eq. (6-20) is shown in Example 6-2.

EXAMPLE 6-2

Using Eq. (6-13) in Eq. (6-19), derive Eq. (6-20).

SOLUTION

Since $M_0 = M = 1$ and $\mu = 0$, Eq. (6-19) takes the form

$$\sigma_x^2 = M_2 = \frac{1}{(4\pi D^* t)^{1/2}} \int_{-\infty}^{\infty} x^2 \exp\left(-\frac{x^2}{4D^* t}\right) dx \tag{E6-2-1}$$

The integral in Eq. (E6-2-1) is evaluated using the method in Spiegel (1975, Problem 3.41, p. 99). With the transformation

$$\frac{x^2}{4D^* t} = v, \quad dv = \frac{2x}{4D^* t} dx \tag{E6-2-2}$$

Equation (E6-2-1) takes the form

$$\sigma_x^2 = \frac{2}{\pi^{1/2}} 2D^* t \int_0^{\infty} v^{1/2} e^{-v} dv \tag{E6-2-3}$$

By performing the integral in Eq. (E6-2-3) (Spiegel, 1968, Eq. [15.76], p. 98), one gets

$$\sigma_x^2 = \frac{2}{\pi^{1/2}} 2D^* t \, \Gamma\left(\frac{3}{2}\right) \tag{E6-2-4}$$

where Γ is the gamma function. Using the recurrence formula

$$\Gamma(n + 1) = n\Gamma(n) \tag{E6-2-5}$$

Therefore, for $n = \frac{1}{2}$,

$$\Gamma(\tfrac{1}{2} + 1) = \tfrac{1}{2}\Gamma(\tfrac{1}{2}) \tag{E6-2-6}$$

And since $\Gamma\left(\frac{1}{2}\right) = \pi^{1/2}$, Eq. (E6-2-4) takes the form given by Eq. (6-20).

6.3.1.4 Standard Deviation

The standard deviation σ_x (the square root of variance) is used as a measure of spread. Using Eq. (6-20), Eq. (6-13) can also be expressed as

$$C(\sigma_x, t) = \frac{M}{\varphi_e (2\pi)^{1/2} \sigma_x} \exp\left(-\frac{\xi^2}{2\sigma_x^2}\right) \tag{6-21}$$

where

$$\sigma_x = (2D^* t)^{1/2} \tag{6-22}$$

6.3.1.5 Fundamental Relationship Between σ_x^2 and D^*

Equation (6-20) yields a fundamental relationship between the rate of change of σ_x^2 (spreading of solute cloud) and effective molecular diffusion coefficient D^*. Taking the derivative of σ_x^2 with respect to t gives the fundamental relationship as

$$\frac{d\sigma_x^2}{dt} = 2D^* \tag{6-23}$$

For the normal distribution, Eq. (6-23) is obvious, but it is also true for any concentration distribution, provided that it is dispersing in accordance with the diffusion equation in a one-dimensional system of infinite extent, and that the concentration is always zero at $x = \pm \infty$. The one-dimensional diffusion equation is also known as *Fick's second law*, and from Eqs. (2-51) and (2-154), with $U = 0$, it can be expressed as

$$\frac{\partial C}{\partial t} = D^* \frac{\partial^2 C}{\partial x^2}$$

(6-24)

Equation (6-23) can be derived directly from Equation (6-24) and this is shown in Example 6-3. Eq. (6-23) states that the variance of a finite distribution increases at the rate $2D^*$ without depending on its shape.

EXAMPLE 6-3

Using Fick's second law, as given by Eq. (6-24), show that Eq. (6-23) is valid for any concentration distribution.

SOLUTION

Multiplying each side of Eq. (6-24) by x^2 and integrating it from $x = -\infty$ to $x = +\infty$ gives

$$\int_{-\infty}^{\infty} \frac{\partial C}{\partial t} x^2 \, dx = \int_{-\infty}^{\infty} D^* x^2 \frac{\partial^2 C}{\partial x^2} \, dx$$

(E6-3-1)

With Eq. (6-16), its left-hand side can be written as

$$\varphi_e \int_{-\infty}^{\infty} \frac{\partial C}{\partial t} x^2 \, dx = \frac{\partial}{\partial t} \int_{-\infty}^{\infty} \varphi_e x^2 C \, dx = \frac{\partial M_2}{\partial t}$$

(E6-3-2)

The right-hand side of Eq. (E6-3-1), integrating by parts, gives

$$\int_{-\infty}^{\infty} D^* x^2 \frac{\partial^2 C}{\partial x^2} \, dx = D^* \int_{-\infty}^{\infty} x^2 \frac{\partial}{\partial x} \left(\frac{\partial C}{\partial x} \right) dx = D^* x^2 \frac{\partial C}{\partial x} \Big|_{-\infty}^{\infty} - D^* \int_{-\infty}^{\infty} 2x \frac{\partial C}{\partial x} \, dx$$

(E6-3-3)

More precisely, it is necessary that

$$\lim_{x \to \pm\infty} \left(x^2 \frac{\partial C}{\partial x} \right) = 0$$

(E6-3-4)

and that Eq. (E6-3-3), by applying again by parts the integration rule, takes the form

$$\int_{-\infty}^{\infty} D^* x^2 \frac{\partial^2 C}{\partial x^2} \, dx = -2D^* \int_{-\infty}^{\infty} x dC = -2D^* x C \Big|_{-\infty}^{\infty} + 2D^* \int_{-\infty}^{\infty} C \, dx = 2D^* \int_{-\infty}^{\infty} C \, dx$$

(E6-3-5)

Multiplying both sides by φ_e and using Eq. (6-14)

$$\int_{-\infty}^{\infty} \varphi_e D^* x^2 \frac{\partial^2 C}{\partial x^2} \, dx = 2D^* M_0$$

(E6-3-6)

The substitution of Eq. (E6-3-2) and (E6-3-6) into Eq. (E6-3-1) gives

$$\frac{\partial M_2}{\partial t} = 2D^* M_0$$

(E6-3-7)

or

$$\frac{1}{M_0} \frac{\partial M_2}{\partial t} = 2D^*$$

(E6-3-8)

From Eq. (6-19)

$$M_2 = M_0 \sigma_x^2 + M_0 \mu^2$$

(E6-3-9)

Taking the derivatives of both sides with respect to time, and by noticing that M_0 and μ are constants, one has

$$\frac{\partial M_2}{\partial t} = M_0 \frac{\partial \sigma_x^2}{\partial t} \tag{E6-3-10}$$

Substituting this into Eq. (E6-3-8) finally gives Eq. (6-23).

6.3.2 Three-Dimensional Plume Spatial Moments

The spatial moments of concentration distributions provide an approximate means for evaluating the convective and dispersive characteristic of tracer plumes that migrate in complex aquifer systems. In this section, three-dimensional spatial moments will be presented by generalizing the expressions for one-dimensional cases that are presented in Section 6.3.1. The ijkth spatial moment of the concentration distribution in space, $M_{ijk}(t)$, is defined here, after Aris (1956), as

$$M_{ijk}(t) = \int_{-\infty}^{\infty}\int_{-\infty}^{\infty}\int_{-\infty}^{\infty} \varphi_e C(x, y, z, t) x^i y^j z^k \, dx \, dy \, dz \tag{6-25}$$

where $C(x, y, z, t)$ is the concentration, φ_e the effective porosity, and x, y, and z are the spatial coordinates. Equation (6-25) was used extensively during the past two decades for the characterization of tracer plumes in heterogeneous aquifers (e.g., Freyberg, 1986; Sudicky, 1986). The volume integral of Eq. (6-25) is defined over all space. However, the integrand will be nonzero only over regions where the concentration $C(x, y, z, t)$ is nonzero. Therefore, the spatial moment, $M_{ijk}(t)$, is an integrated measure of the concentration field over the extent of the solute plume.

In aquifer plume analysis, the zeroth, first, and second spatial moments ($i, j, k = 0, 1$, or 2, respectively) have special importance because they provide measures of the mass, location, and spread of the tracer plumes. They also arise naturally in the parameterization of many mathematical solute transport models, which will be described later in this chapter. These lower-order moments do not, of course, completely describe plume structure. Higher-order moments can be used to examine, for example, the skew or kurtosis of concentration distributions. Under some circumstances, such information may be essential to the complete characterization and interpretation of solute movement. Unfortunately, it is very difficult to develop reliable estimators for the higher order moments (Freyberg, 1986).

Some details regarding the lower order moments are presented in the following sections.

6.3.2.1 Zeroth Moment

If $i = j = k = 0$, Eq. (6-25) takes the form

$$M_{000} = \int_{-\infty}^{\infty}\int_{-\infty}^{\infty}\int_{-\infty}^{\infty} \varphi_e C(x, y, z, t) x^0 y^0 z^0 \, dx \, dy \, dz = \int_{-\infty}^{\infty}\int_{-\infty}^{\infty}\int_{-\infty}^{\infty} \varphi_e C(x, y, z, t) \, dx \, dy \, dz \tag{6-26}$$

Here, $(\varphi_e \, dx \, dy \, dz)$ is the pore volume of the parallelepiped whose dimensions are dx, dy, and dz. As a result, M_{000} is equal to the mass of the tracer in the solution. For a slug solute input of an ideal, nonreactive tracer, the mass in solution must remain constant over time. For a one-dimensional solute transport, Eq. (6-26) takes the form of Eq. (6-14).

6.3.2.2 First Moments

There are three first moments. For $i = 1$, $j = 0$, and $k = 0$, Eq. (6-25) becomes

$$M_{100} = \int_{-\infty}^{\infty}\int_{-\infty}^{\infty}\int_{-\infty}^{\infty} \varphi_e C(x,y,z,t) x \, dx \, dy \, dz \tag{6-27}$$

For $i = 0$, $j = 1$, and $k = 0$; and $i = 0$, $j = 0$, and $k = 1$; Eq. (6-25), respectively, gives

$$M_{010} = \int_{-\infty}^{\infty}\int_{-\infty}^{\infty}\int_{-\infty}^{\infty} \varphi_e C(x,y,z,t) y \, dx \, dy \, dz \tag{6-28}$$

and

$$M_{001} = \int_{-\infty}^{\infty}\int_{-\infty}^{\infty}\int_{-\infty}^{\infty} \varphi_e C(x,y,z,t) z \, dx \, dy \, dz \qquad (6\text{-}29)$$

The first moments about the origin, normalized by mass in solution, define the coordinates of the center of mass (x_c, y_c, z_c):

$$x_c = \frac{M_{100}}{M_{000}}, \qquad y_c = \frac{M_{010}}{M_{000}}, \qquad z_c = \frac{M_{001}}{M_{000}} \qquad (6\text{-}30)$$

It is conventional to define the components of the mean tracer velocity components v_x, v_y, and v_z according to

$$v_x = \frac{dx_c}{dt}, \qquad v_y = \frac{dy_c}{dt}, \qquad v_z = \frac{dz_z}{dt} \qquad (6\text{-}31)$$

Then, the mean velocity vector $\mathbf{v}$ as the time rate of change of the displacement of the center of mass is

$$\mathbf{v} = \frac{dx_c}{dt}\mathbf{i} + \frac{dy_c}{dt}\mathbf{j} + \frac{dz_z}{dt}\mathbf{k} \qquad (6\text{-}32)$$

Here, $\mathbf{v}$ is the plume-scale average solute velocity whose components are v_x, v_y, and v_z under steady flow conditions and these components are given by Eq. (2-35). Under the assumptions of a steady, homogeneous ground water velocity field, a slug input of solute mass, and transport governed by the classical convective–dispersive model, $\mathbf{v}$ is the ground water velocity as given by Eq. (6-32).

6.3.2.3 Second Moments

For $i = 2$, $j = 0$, and $k = 0$, Eq. (6-25) gives

$$M_{200} = \int_{-\infty}^{\infty}\int_{-\infty}^{\infty}\int_{-\infty}^{\infty} \varphi_e C(x,y,z,t) x^2 \, dx \, dy \, dz \qquad (6\text{-}33)$$

For $i = 0$, $j = 2$, and $k = 0$; and $i = 0$, $j = 0$, and $k = 2$, Eq. (6-25), respectively, gives

$$M_{020} = \int_{-\infty}^{\infty}\int_{-\infty}^{\infty}\int_{-\infty}^{\infty} \varphi_e C(x,y,z,t) y^2 \, dx \, dy \, dz \qquad (6\text{-}34)$$

and

$$M_{002} = \int_{-\infty}^{\infty}\int_{-\infty}^{\infty}\int_{-\infty}^{\infty} \varphi_e C(x,y,z,t) z^2 \, dx \, dy \, dz \qquad (6\text{-}35)$$

For $i = 1$, $j = 1$, and $k = 0$, Eq. (6-25) gives

$$M_{110} = \int_{-\infty}^{\infty}\int_{-\infty}^{\infty}\int_{-\infty}^{\infty} \varphi_e C(x,y,z,t) xy \, dx \, dy \, dz \qquad (6\text{-}36)$$

For $i = 1$, $j = 0$, and $k = 1$, Eq. (6-25) gives

$$M_{101} = \int_{-\infty}^{\infty}\int_{-\infty}^{\infty}\int_{-\infty}^{\infty} \varphi_e C(x,y,z,t) xz \, dx \, dy \, dz \qquad (6\text{-}37)$$

For $i = 0$, $j = 1$, and $k = 1$, Eq. (6-25) gives

$$M_{011} = \int_{-\infty}^{\infty}\int_{-\infty}^{\infty}\int_{-\infty}^{\infty} \varphi_e C(x,y,z,t) yz \, dx \, dy \, dz \qquad (6\text{-}38)$$

The one-dimensional spatial variance σ_x^2 as given by Eq. (6-19) can be extended to two-dimensional spatial variance σ_{xx}^2, (by noticing that $\mu = x_c$) as

$$\sigma_{xx}^2 = \frac{\int_{-\infty}^{\infty}\int_{-\infty}^{\infty}\int_{-\infty}^{\infty}\varphi_e C(x, y, z, t)x^2 \, dx \, dy \, dz}{\int_{-\infty}^{\infty}\int_{-\infty}^{\infty}\int_{-\infty}^{\infty}\varphi_e C(x, y, z, t) \, dx \, dy \, dz} - x_c^2 = \frac{M_{200}}{M_{000}} - x_c^2 \qquad (6\text{-}39)$$

Similarly, the following expressions for the spatial variances about y_c and z_c, respectively, can be written as

$$\sigma_{yy}^2 = \frac{M_{020}}{M_{000}} - y_c^2 \qquad (6\text{-}40)$$

and

$$\sigma_{zz}^2 = \frac{M_{002}}{M_{000}} - z_c^2 \qquad (6\text{-}41)$$

Similarly, the expressions for spatial covariance in the x–y, x–z, and y–z planes, respectively, can be written as

$$\sigma_{xy}^2 = \sigma_{yx}^2 = \frac{M_{110}}{M_{000}} - x_c y_c \qquad (6\text{-}42)$$

$$\sigma_{xz}^2 = \sigma_{zx}^2 = \frac{M_{101}}{M_{000}} - x_c z_c \qquad (6\text{-}43)$$

and

$$\sigma_{yz}^2 = \sigma_{zy}^2 = \frac{M_{011}}{M_{000}} - y_c z_c \qquad (6\text{-}44)$$

Using the method in Example 6-3, Eq. (6-23) can be generalized as

$$D_{ij}^* = \frac{1}{2}\frac{d\sigma_{ij}^2}{dt}, \quad i, j = x, y, z \qquad (6\text{-}45)$$

Using stochastic models of field-scale solute transport, investigators have used expressions analogous to Eq. (6-45) to define apparent dispersion coefficients (Gelhar et al., 1979; Matheron and de Marsily, 1980; Dagan, 1984). In Eq. (6-45), σ_{ij} is defined as the ensemble covariance tensor and, therefore, the apparent dispersion coefficient is also an ensemble property.

6.3.3 MECHANICAL DISPERSION COEFFICIENT (D_m) AND DISPERSIVITY TENSORS FOR GENERAL CASE

It is common to assume that the mechanical dispersion coefficient tensor, $\mathbf{D}_m$, is related to the ground water velocity through a relationship given by the form of Eq. (2-134):

$$\left[\mathbf{D}_m\right] = \left[\mathbf{A}\right]\left[v\right] \qquad (6\text{-}46)$$

where $\mathbf{A}$ is the *dispersivity tensor*. The open form of the $\mathbf{D}_m$ tensor is given by Eq. (2-135). In accordance with Eq. (6-45), the mechanical dispersion coefficient $\mathbf{D}_m$ is defined as (Fischer, 1979; Gelhar et al., 1979; Güven et al., 1984; Freyberg, 1986)

$$\left[\mathbf{D}_m\right] = \frac{1}{2}\left[\frac{d\boldsymbol{\sigma}^2}{dt}\right] \qquad (6\text{-}47)$$

and combining with Eq. (6-46) gives

$$[\mathbf{A}] = \frac{1}{2|\mathbf{v}|}\left[\frac{\mathrm{d}\boldsymbol{\sigma}^2}{\mathrm{d}t}\right] \tag{6-48}$$

in which

$$|\mathbf{v}| = (v_x^2 + v_y^2 + v_z^2)^{1/2} \tag{6-49}$$

is the magnitude of the velocity vector. Using

$$\mathbf{x} = \mathbf{v}t \tag{6-50}$$

Equation (6-48) can also be expressed as (Garabedian et al., 1991)

$$[\mathbf{A}] = \frac{1}{2}\left[\frac{\mathrm{d}\boldsymbol{\sigma}^2}{\mathrm{d}\mathbf{x}}\right] \tag{6-51}$$

Equation (6-48) can also be written as

$$\begin{bmatrix} A_{xx} & A_{xy} & A_{xz} \\ A_{yx} & A_{yy} & A_{yz} \\ A_{zx} & A_{zy} & A_{zz} \end{bmatrix} = \frac{1}{2|\mathbf{v}|}\frac{\mathrm{d}}{\mathrm{d}t}\begin{bmatrix} \sigma_{xx}^2 & \sigma_{xy}^2 & \sigma_{xz}^2 \\ \sigma_{yx}^2 & \sigma_{yy}^2 & \sigma_{yz}^2 \\ \sigma_{zx}^2 & \sigma_{zy}^2 & \sigma_{zz}^2 \end{bmatrix} \tag{6-52}$$

In the above expressions, to distinguish the field-scale dispersivities from laboratory values, the field values are designated by the uppercase **A** (Gelhar and Axness, 1981, 1983) and the dispersivity tensor is often called *macrodispersivity tensor*.

6.3.4   Dispersivity and Mechanical Dispersion Coefficient Tensors for Uniform Ground Water Flow Case

For uniform ground water flow cases Eq. (6-52) can be simplified, using a rotated coordinate system (x', y', z') such that the x' axis is aligned with the mean ground water velocity so that the ground water velocity components will be

$$v_x = U, \quad v_y = 0, \quad v_z = 0 \tag{6-53}$$

Then, the nondiagonal elements in Eq. (6-52) will be all zero, and its new form is

$$\begin{bmatrix} A_{x'x'} & 0 & 0 \\ 0 & A_{y'y'} & 0 \\ 0 & 0 & A_{z'z'} \end{bmatrix} = \frac{1}{2U}\frac{\mathrm{d}}{\mathrm{d}t}\begin{bmatrix} \sigma_{x'x'}^2 & 0 & 0 \\ 0 & \sigma_{y'y'}^2 & 0 \\ 0 & 0 & \sigma_{z'z'}^2 \end{bmatrix} \tag{6-54}$$

From this equation, one can

$$A_{x'x'} = \frac{1}{2U}\frac{\mathrm{d}\sigma_{x'x'}^2}{\mathrm{d}t} \tag{6-55}$$

$$A_{y'y'} = \frac{1}{2U}\frac{\mathrm{d}\sigma_{y'y'}^2}{\mathrm{d}t} \tag{6-56}$$

$$A_{z'z'} = \frac{1}{2U}\frac{\mathrm{d}\sigma_{z'z'}^2}{\mathrm{d}t} \tag{6-57}$$

where $A_{x'x'}$ is the longitudinal dispersivity, $A_{y'y'}$ the transverse horizontal dispersivity, and $A_{z'z'}$ the transverse vertical dispersivity. These are well-adopted terms in the literature, however, the terms macrodispersivity, full aquifer dispersivity, and scale-dependent dispersivity are also used in the literature. Equations (6-55)–(6-57) are valid for the case when the position coordinates of local-scale particles follow a Gaussian distribution.

From Eq. (2-135), for the (x', y', z') coordinate system, the mechanical dispersion coefficient tensor takes the form (the nondiagonal elements are all zero)

$$[D_{m_{i'j'}}] = \begin{bmatrix} D_{m_{x'x'}} & 0 & 0 \\ 0 & D_{m_{y'y'}} & 0 \\ 0 & 0 & D_{m_{z'z'}} \end{bmatrix} \tag{6-58}$$

From Eq. (6-46), for the mechanical dispersion coefficient components in Eq. (6-58), one can write

$$D_{m_{x'x'}} = A_{x'x'}U, \quad D_{m_{y'y'}} = A_{y'y'}U, \quad D_{m_{z'z'}} = A_{z'z'}U \tag{6-59}$$

And using Eqs. (6-55)–(6-57), these equations take the following forms:

$$D_{m_{x'x'}} = \frac{1}{2}\frac{d\sigma_{x'x'}^2}{dt} \tag{6-60}$$

$$D_{m_{y'y'}} = \frac{1}{2}\frac{d\sigma_{y'y'}^2}{dt} \tag{6-61}$$

$$D_{m_{z'z'}} = \frac{1}{2}\frac{d\sigma_{z'z'}^2}{dt} \tag{6-62}$$

6.3.5 IMPORTANT POINTS FOR THE APPLICATION OF SPATIAL MOMENTS

Several points should be noted concerning the application of spatial moments to the analysis of tracer tests (Garabedian et al., 1991):

1. The spatial moments are evaluated at points in time, so the finite-difference form was used for the time derivative in Eqs. (6-48) and (6-52). Therefore, the calculated dispersivity values from tracer test results are average values for the periods of time over which the derivative is approximated.
2. Calculation of dispersivity values by Eqs. (6-55)–(6-57) assume a constant ground water velocity U.
3. The time dependence of dispersivity should not be confused with spatial dependence. In spatial moments analysis, dispersivities are assumed to be constant in space, but are allowed to vary with time.
4. The initial shape of the tracer cloud has no effect on calculations of dispersivity values using spatial moments analysis. The solute cloud can have any arbitrary initial shape (with a finite mass) for the analysis to be valid (Fischer et al., 1979).

6.3.6. STOCHASTIC DESCRIPTION OF HYDRAULIC CONDUCTIVITY VARIATION

Most aquifer materials exhibit complex three-dimensional behavior. Variations in hydraulic conductivity have been shown to have a significant effect on dispersivities (Skibitzke and Robinson, 1964; Zilliox and Muntzer, 1975). Studies by Law (1944) and Freeze (1975) have shown that hydraulic conductivity at the local scale is log-normally distributed. Based on these distributions, during the past quarter century, some stochastic approaches have been developed to analyze ground water flow and solute transport in aquifers. Here, for the purposes of explanation, the mathematical presentation has been restricted to a one-dimensional stochastic process. The methods are equally applicable to three-dimensional domains. In this section, basic concepts and definitions are presented for the description of hydraulic conductivity variation in heterogeneous aquifers.

6.3.6.1 Definitions For Nonuniformity, Heterogeneity, and Anisotropy

In the stochastic sense, definitions of nonuniformity, heterogeneity, and anisotropy are different from those in the deterministic sense. Greenkorn and Kessler (1969) have provided a set of definitions for these concepts (Freeze, 1975). They note that in general, the probability density function for hydraulic conductivity (for example) is a function of location and orientation. This function can be described with five independent variables, of three which are rectangular coordinates for location and two are angular coordinates for orientation. If the probability density function is independent of orientation, the medium is called *isotropic*. If the probability density function is dependent on orientation, the medium is called *anisotropic*. If the medium can be expressed by a finite linear combination of delta functions, the medium is called *uniform*. If the distribution cannot be expressed by a finite linear combination of delta functions, the medium is called *nonuniform*. If the distribution is monomodal, the medium is called *homogeneous*. If the distribution is multimodal, the medium is *heterogeneous*. Figure 6-3 shows examples of frequency distributions of hydraulic conductivity for the four possible combinations of homogeneity and uniformity in an isotropic medium. It is reasonable to assume that a uniform medium does not exist in nature. Therefore, Figures 6-3 c and d represent the real-world possibilities.

6.3.6.2 Description of Heterogeneity by the Stochastic Process Theory

Generally, in aquifers, hydraulic conductivity follows a log-normal probability density function (Freeze, 1975). Consider a one-dimensional case shown in Figure 6-4a with log hydraulic conductivities $f_1, f_2, ..., f_n$ at $x_1, x_2, ..., x_n$ distances from the origin given by

$$f_1(x_1) = \ln[K_1(x_1)], \quad f_2(x_2) = \ln[K_2(x_2)], \quad ... \quad, f_n(x_n) = \ln[K_n(x_n)] \qquad (6\text{-}63)$$

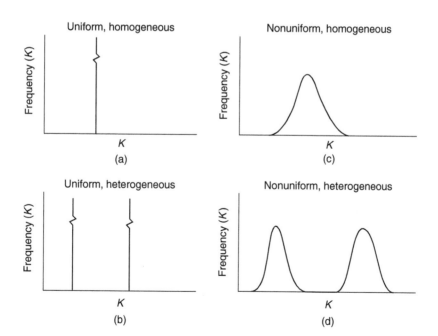

FIGURE 6-3 Greenkorn and Kessler's frequency distribution for hydraulic conductivity K. (Adapted from Freeze, R.A., *Water Resour. Res.*, 11, 725–741, 1975.)

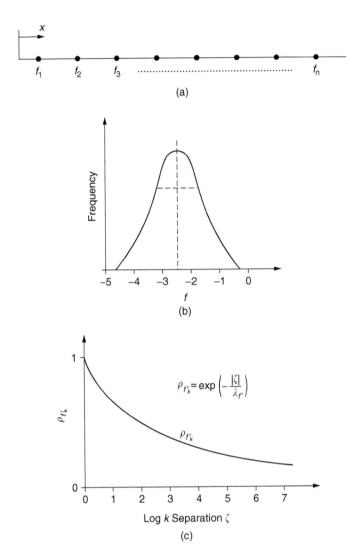

FIGURE 6-4 (a) A one-dimensional sequence of log hydraulic conductivity values; (b) probability density function for f; and (c) autocorrelation function for f. (Adapted from Freeze, R.A., Massmann, J., Smith, L., Sperling, T. and James, B., *Ground Water*, 738–766, 1990.)

or in a combined form

$$f_i(x_i) = \ln[K_i(x_i)], \quad i = 1, 2, \ldots, n \tag{6-64}$$

where $K_i(x_i)$ is the hydraulic conductivity stochastic process and $f_i(x_i)$ is the log hydraulic conductivity, which is assumed to be statistically homogeneous. The parameter $f_i(x_i)$ is normally distributed (Figure 6-4b).

Given a set of measurements of log hydraulic conductivity, the mean, $\bar{f}$, is defined as

$$\bar{f} = \frac{f_1 + f_2 + \cdots + f_n}{n} = \frac{1}{n} \sum_{n=1}^{n} f_i \tag{6-65}$$

By showing $f_i' = f_i - \bar{f}$, the variance is σ_f^2, defined as (Spiegel, 1975, Eq. [15], p. 160)

$$\sigma_f^2 = \frac{(f_1 - \bar{f})^2 + (f_2 - \bar{f})^2 + \cdots + (f_n - \bar{f})^2}{n} = \frac{1}{n} \sum_{n=1}^{n} (f_i - \bar{f})^2 \tag{6-66}$$

The parameter, σ_f^2 is one measure of the heterogeneity of an aquifer. Now, the expression of a common *covariance function* will be given. Some investigators (Freeze et al., 1990) use the term *autocorrelation function* instead of covariance function. The covariance function is defined as

$$R_{f_k'} = \frac{1}{n} \sum_{n=1}^{n} (f_i - \bar{f})(f_{i+1} - \bar{f}) \tag{6-67}$$

where k is the lag. If the measurement f_i is made at position x_i and the measurement f_{i+k} at x_{i+k}, then the separation is defined as

$$\zeta = |x_{i+k} - x_i| \tag{6-68}$$

The covariance function may take a number of forms. One of the most commonly used functions is the exponential decaying covariance function. The expression of this function is (Freeze et al., 1990; Gelhar, 1993)

$$R_{f'}(\zeta) = \sigma_{f'}^2 \exp\left(-\frac{|\zeta|}{\lambda_{f'}}\right) \tag{6-69}$$

where $\sigma_{f'}^2$ is the variance and $\lambda_{f'}$ is the integral scale. Some investigators (Freeze et al., 1990) call it *correlation length*, which is a measure of the distance over which $f_i(x_i)$ is correlated. It is a measure of the distance over which $R_{f'}(\zeta)$ decays to a value of e^{-1}. Values separated by short distances will be highly correlated, and those separated by long distances will only be weakly correlated or not correlated at all. Equation (6-69) is the one-dimensional form of the three-dimensional autocovariance function given by Eq. (6-274). From Eq. (6-67) or (6-69) the *autocorrelation* is defined by (Freeze et al., 1990)

$$\rho_{f_k'} = \frac{R_{f_k'}}{\sigma_{f'}^2} = \exp\left(-\frac{|\zeta|}{\lambda_{f'}}\right) \tag{6-70}$$

The function that displays the drop in correlation with distance is called the *autocorrelation function* (Figure 6-4c).

The geometric mean of f, given by Eq. (6-65), can be expressed as

$$\bar{f} = \ln(K_g), \quad K_g = e^{\bar{f}} \tag{6-71}$$

where K_g is the *geometric mean hydraulic conductivity*. In the above expressions, $\sigma_{f'}^2$ is dimensionless due to the fact that in Eq. (6-66) each term can be written as

$$(f_i - \bar{f}) = \ln[K_i(x_i)] - \ln(K_g) = \ln\left[\frac{K_i(x_i)}{K_g}\right] \tag{6-72}$$

6.3.6.3 Statistical Parameters for Log-Normal Frequency Distributions of Hydraulic Conductivity

As shown in Section 6.3.6.2, if the hydraulic conductivity K is log normally distributed (Figure 6-4b), a new parameter f can be defined based on Eq. (6-63), which is normally distributed as well. As seen from Eqs. (6-64) and (6-65), f is characterized by its mean, $\bar{f}$, and standard deviation, $\sigma_{f'}$. Freeze (1975) summarized the $\sigma_{f'}$ estimates from the various log-normal hydraulic conductivity distributions for various rock and soil types (Table 6-1). From Table 6-1, it appears that the range of values of σ_f lies between 0.5 and 4.0, and the corresponding values of σ_f^2 are 0.25 and 16.0, respectively.

TABLE 6-1

Statistical Parameters for LogNormal Frequency Distributions of Hydraulic Conductivity [$\bar{f}=\ln(K)$, K in cm/sec]

Source	Rock or Soil Type	N^a	$\bar{f}^b$	σ_f^{c}	$\sigma_f^{2,d}$
Bennion and Griffits (1966)	Conglomerate	3,018	−10.57	2.16	4.67
	Sandstone	56,991	−12.39	1.40	1.96
	Marly limestone	7,060	−12.39	1.06	1.12
	Vuggy limestone	17,162	−12.90	1.22	1.49
Law (1944)	Sandstone		−7.74	0.46	0.21
	Sandstone		−8.80	0.92	0.85
	Sandstone		−10.59	0.92	0.85
McMillan (1966)	Sandstone		−12.57	2.30	5.29
	Sandstone		−12.71	1.50	2.25
	Sandstone		−10.59	0.74	0.55
	Sand and gravel		—	1.01	1.02
	Sand and gravel	42	—	1.66	2.76
	Sand and gravel	16	—	1.24	1.54
Willardson and Hurst (1965)	Clay loam	33	−12.46	1.04	1.08
		330	−9.21	1.96	3.84
		287	−10.59	2.14	4.58
	Silty clay	339	−11.42	1.80	3.24
		36	−10.27	3.59	12.89
		352	−6.91	2.14	4.58
	Loamy sand	121	−6.17	1.98	3.92

[a] Number of measurements.
[b] Mean.
[c] Standard deviation.
[d] Variance.
Source: Adapted from Freeze, R.A., *Water Resour. Res.*, 11, 725–741, 1985.

EXAMPLE 6-4

Measured hydraulic conductivity values for a site are given in Table 6-2. The total length of the aquifer portion is 560.0 m and measured value locations are at every 40.0 m along this length. There is a total of 14 measured values along the x-coordinate direction. In the y-coordinate direction, it is assumed that the same measured values are valid. Using the values in Table 6-2, determine the (1) variance and standard deviation of the hydraulic conductivity data, (2) covariance of the hydraulic conductivity data with separations of 40.0 m; (3) value of the autocorrelation; and (4) value of integral scale (λ_f) assuming a one-dimensional exponential form of the covariance function, as given by Eq. (6-69).

SOLUTION

(1) The necessary calculations for variance are given in Table 6-2. The sum of the values in the fourth column of Table 6-2 is

$$\sum_{i=1}^{14} f_i = -56.161$$

and with $n = 14$ and using Eq. (6-65)

$$\bar{f} = \tfrac{1}{14}(-56.161) = -4.012$$

TABLE 6-2
Calculation of Variance for the K Values of Example 6-4

i	x_i (m)	K_i (cm/sec)	$f_i = ln(K_i)$	$f_i - \bar{f}$	$(f_i - \bar{f})^2$
1	40	9.07×10^{-3}	-4.703	-0.691	0.4775
2	80	2.36×10^{-2}	-3.747	0.265	0.0702
3	120	9.63×10^{-3}	-4.643	-0.631	0.3982
4	160	4.94×10^{-3}	-5.310	-1.298	1.6848
5	200	9.91×10^{-3}	-4.614	-0.602	0.3624
6	240	7.09×10^{-3}	-4.949	-0.937	0.8780
7	280	9.03×10^{-3}	-4.707	-0.695	0.4830
8	320	6.19×10^{-2}	-2.782	1.230	1.5129
9	360	1.38×10^{-1}	-1.981	2.031	4.1250
10	400	1.04×10^{-1}	-2.263	1.749	3.0590
11	440	4.76×10^{-3}	-5.348	-1.336	1.7849
12	480	3.85×10^{-3}	-5.560	-1.548	2.3963
13	520	6.91×10^{-2}	-2.672	1.340	1.7956
14	560	5.60×10^{-2}	-2.882	1.130	1.2769
Sum			-56.161		20.3047

The sum of the last column of Table 6-2 is

$$\sum_{i=1}^{14} (f_i - \bar{f})^2 = 20.305$$

and using Eq. (6-66), the variance and standard deviation, respectively, are

$$\sigma_f^2 = \tfrac{1}{14} (20.305) = 1.45, \quad \sigma_{f'} = 1.20$$

(2) The necessary calculations for covariance at separations of 40 m ($K = 1$) are given in Table 6-3. The sum of the last column values is

$$\sum_{i=1}^{13} (f_i - \bar{f})(f_{i+1} - \bar{f}) = 12.448 - 5.616 = 6.832$$

and from Eq. (6-67)

$$R_{f_1'} = \frac{6.832}{13} = 0.526$$

(3) From Eq. (6-70) the value of autocorrelation is

$$\rho_{f_1'} = \frac{0.526}{1.45} = 0.362$$

(4) Using this value and the separation $\zeta = 40$ m in Eq. (6-70)

$$0.362 = \exp\left(-\frac{40.0 \text{ m}}{\lambda_{f'}}\right)$$

and solving for $\lambda_{f'}$ gives $\lambda_{f'} = 39.41$ m.

TABLE 6-3
Calculation of Covariance for the K Values of Example 6-4

i	x_i(m)	K_i(cm/sec)	$f_i - \bar{f}$	$f_{i+1} - \bar{f}$	$(f - \bar{f})(f_{i+1} - \bar{f})$
1	40	9.07×10^{-3}	-0.691	0.265	-0.183115
2	80	2.36×10^{-2}	0.265	-0.631	-0.167215
3	120	9.63×10^{-3}	-0.631	-1.298	0.819038
4	160	4.94×10^{-3}	-1.298	-0.602	0.781396
5	200	9.91×10^{-3}	-0.602	-0.937	0.564074
6	240	7.09×10^{-3}	-0.937	-0.695	0.651215
7	280	9.03×10^{-3}	-0.695	1.230	-0.854850
8	320	6.19×10^{-2}	1.230	2.031	2.498130
9	360	1.38×10^{-1}	2.031	1.749	3.552219
10	400	1.04×10^{-1}	1.749	-1.336	-2.336664
11	440	4.76×10^{-3}	-1.336	-1.548	2.068128
12	480	3.85×10^{-3}	-1.548	1.340	-2.074320
13	520	6.91×10^{-2}	1.340	1.130	1.514200
14	560	5.60×10^{-2}	1.130		
Sum					6.832036

6.3.7 ILLUSTRATIVE EXAMPLES FOR THE DETERMINATION OF DISPERSIVITIES FROM NATURAL GRADIENT EXPERIMENTS USING THE PLUME SPATIAL MOMENTS METHOD

During the past three decades, a number of field experiments were conducted to determine dispersivity values. Some of these experiments are well reported in the ground water literature and they are known as the Borden experiment (Mackay et al., 1986; Freyberg, 1986; Sudicky, 1986); the Twin Lake experiment (Killey and Moltyaner, 1988; Moltyaner and Killey, 1988a, 1988b; Moltyaner and Wills, 1991) and the Cape Cod experiment (LeBlanc et al., 1991; Garebedian et al., 1991; Hess et al., 1992). In the following sections, based on these experiments, some examples regarding the determination of dispersivities using the solute plume spatial methods will be presented.

6.3.7.1 Borden Experiment

Description of the Aquifer
The aquifer is a shallow water table aquifer at Canadian Forces Base, Borden, Ontario. The tracer site is located in a sand quarry that is about 350 m downgradient from an abandoned landfill. The aquifer primarily comprises horizontal, discontinuous lenses of medium-grained, and silty fine-grained sand. The zone of fluctuation of the water table at the tracer site varies seasonally within 1.0 m interval below the ground surface. Figure 6-5 shows a schematic vertical section of the aquifer at the site. The aquifer extends about 9 m beneath the nearly horizontal quarry floor and is underlain by a thick silt clay deposit. In the quarry area, the landfill leachate plume is confined to the bottom 2 to 3 m of the aquifer. As shown in Figure 6-5, the experiment was carried out in the upper uncontaminated portion of the aquifer (Sudicky, 1986; Mackay et al., 1986).

Hydraulic Heads of the Aquifer
Hydraulic heads at the tracer site were monitored periodically using an array of piezometers located at various depths in the aquifer. Contours of the water table position measured on August 10, 1984; and November 15, 1984 are shown in Figure 6-6. Figure 6-6 shows that the hydraulic head field at the experiment site is relatively smooth and uniform, although the aquifer is known to be heterogeneous. The range and annual mean of the horizontal hydraulic gradients are given in Table 6-4.

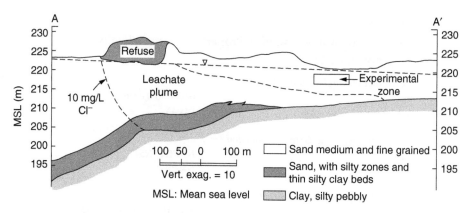

FIGURE 6-5 Vertical geometry of the aquifer for the Borden experiment. Rectangle illustrates the vertical zone in which the experiment was conducted, which is above the landfill leachate plume for the Borden experiment. (Adapted from Mackay, D.M., Freyberg, D.L. and Roberts, P.V., *Water Resour. Res.*, 22, 2017–2029, 1986.)

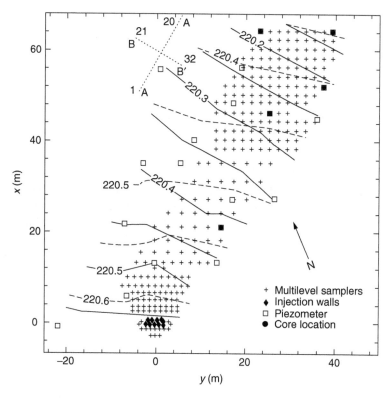

FIGURE 6-6 Hydraulic head distribution at the Borden site. (Adapted from Sudicky, E. A., *Water Resour. Res.*, 22, 2069–2082, 1986.)

Horizontal Hydraulic Conductivity

The hydraulic conductivity distribution in the aquifer at the experimental site has been studied using several techniques. A total of 26 conventional slug tests were conducted and analyzed using the method of Hvorslev (1951). As can be seen from Table 6-4, the resulting estimates of hydraulic conductivity varied from 2.0×10^{-7} to 5.0×10^{-5} m/sec, with a geometric mean value of approximately 7.2×10^{-5} m/sec (Sudicky, 1986; Freyberg, 1986).

TABLE 6-4
Estimates of Borden Aquifer Properties in the Vicinity of the Experimental Site

Horizontal Hydraulic Gradient

Range	0.0035 to 0.0056
Annual mean	0.0043

Horizontal Hydraulic Gradient Direction

Range	N40°E to N54°E
Annual mean	N4E°E

Horizontal Hydraulic Conductivity

Range	5.0×10^{-7} to 2.0×10^{-4} m/sec
Geometric mean	7.2×10^{-5} m/sec
Standard deviation of natural log (σ_f)[a]	0.62
Horizontal correlation scale $(\lambda_1 = \lambda_2)$[b]	2.8 m
Vertical correlation scale (λ_3)[b]	0.12 m

Effective Porosity

Mean	0.33
Standard deviation	0.05

[a] Defined by Eq. (6-66).

[b] Defined in the exponential autocovariance as given by Eq. (6-274).

Adapted from Freyberg, D.L., *Water Resour. Res.*, 22, 2031–2046, 1986.

Spatial Moment Estimation Methodology
Estimation of spatial moments and their derived parameters requires the approximation of the volume integral in Eq. (6-25) using discrete data (Freyberg, 1986). The relatively small amount of vertical displacement and very light amount of vertical spreading observed during the experiment (see Figure 6-8) led Freyberg (1986) to develop an estimation methodology for the zeroth-order, first-, and second-order moments for which the exponent k in Eq. (6-25) is zero. In this case, Eq. (6-25) can be written as

$$M_{ijk}(t) = \int_{-\infty}^{\infty}\int_{-\infty}^{\infty}\int_{-\infty}^{\infty} \varphi_e C_z(x,y,t) x^i y^j \, dx \, dy \qquad (6\text{-}73)$$

with C_z defined as

$$C_z(x,y,t) = \int_{-\infty}^{\infty} C(x, y, z, t) \, dz \qquad (6\text{-}74)$$

This approximation is based on the fact that the spatial resolution of the data is much greater in the vertical direction than in the horizontal. Freyberg used the trapezoidal quadrature in integrating Eq. (6-74) over z. The limits of the integration were set to uniform elevations above and below the plume.

Solute Plume Migration
Figure 6-7 presents vertically averaged isoconcentration contours for chloride at 1, 85, 462, and 647 days after injection. Initially, the plume is nearly rectangular in plain view, and with time, progressively becomes more ellipsoidal. The chloride plume appears to move at an approximately constant velocity. Figure 6-8 presents approximate concentration distributions of the chloride plume after 1 and 462 days from the time of injection. The vertical scale of Figure 6-8 is exaggerated by a factor

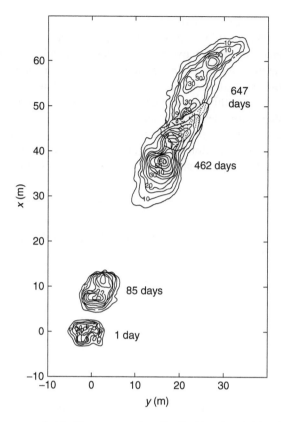

FIGURE 6-7   Vertically averaged chloride concentration distribution after 1, 85, 462, and 647 days of injection for the Borden experiment. (Adapted from Mackay, D.M., Freyberg, D.L. and Roberts, P.V., *Water Resour. Res.*, 22, 2017–2029, 1986.)

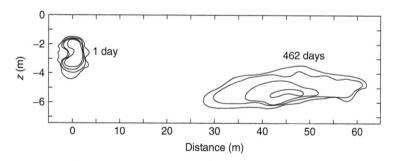

FIGURE 6-8   Approximate concentration distributions of chloride after 1 and 462 days from injection along vertical cross-section AA′ in Figure 6-6 (vertical exaggeration = 3). Contours depicted for the 1- and 462-day plumes are 10, 100, 300, and 600 mg/L and 10, 30, 100, and 300 mg/L, respectively, for the Borden experiment. (Adapted from Mackay, D.M., Freyberg, D.L. and Roberts, P.V., *Water Resour. Res.,* 22, 2017–2029, 1986.)

of 3 for clarity. As can be seen from Figure 6-8, spreading in the horizontal direction is evident and the vertical spreading is very small (Mackay et al., 1986).

Estimation of Mass in Solution

In the estimation process, the effective porosity was treated as spatially uniform with a mean value of 0.33, which is included in Table 6-4 (Freyberg, 1986). Porosity cancels out all parameter estimates except the mass in solution.

As shown in Section 6.3 (Eq. (6-26)), the zeroth spatial moment of concentration distribution measures the mass of a constituent in solution. All evidence in the data indicates that the two tracers, chloride and bromide, are nonreactive with the Borden aquifer materials (Freyberg, 1986). Therefore, except for the mass removed during sampling, one should expect that the mass in solution in the aquifer is constant with time. Freyberg (1986) states that although nearly 10,000 samples have been collected from within and around the chloride and bromide plumes, less than 1% of the total mass of chloride or bromide ions has been removed. This means that one should not expect that the sampling process itself has not resulted in a measurable loss of mass over time.

The estimated mass of solution for bromide and chloride are given in Table 6-5. Figure 6-9 shows the estimated mass of bromide ions in solution as a function of time since injection. As noted

TABLE 6-5
Estimates of Mass in Solution, Location of Center of Mass, and Spatial Covariances for the Bromide and Chloride Plumes

Date	Elapsed Time (days)	Mass in Solution (kg)	Center of Mass[a](m)			Spatial Covariances[b](m²)			s[c](m)
			x_c	y_c	z_c	$\sigma^2_{x'x'}$	$\sigma^2_{y'y'}$	$\sigma^2_{x'y'}$	
Bromide									
Aug. 24, 1982	1	2.02	0.3	−0.2	2.78	1.8	2.0	0.4	0.36
Sept. 1, 1982	9	3.25	0.6	0.4	2.95	2.3	2.8	1.0	0.72
Sept. 8, 1982	16	2.81	1.7	0.9	3.02	1.9	2.8	0.9	1.92
Sept. 21, 1982	29	4.29	2.7	1.0	3.34	2.4	2.5	0.9	2.88
Oct. 5, 1982	43	3.76	4.1	1.5	3.36	3.6	2.5	0.8	4.37
Oct. 25, 1982	63	3.50	5.8	2.2	3.54	4.9	2.7	1.1	6.20
Nov. 16, 1982	85	3.66	7.7	2.9	3.80	5.2	2.6	0.3	8.23
May 9, 1983	259	4.49	22.9	11.3	4.80	16.7	4.6	2.8	25.45
Sept. 8, 1983	381	4.08	31.9	15.1	5.17	20.8	5.0	3.2	35.29
Oct. 26, 1983	429	3.97	36.2	17.2	5.30	22.0	7.3	1.9	40.08
Nov. 28, 1983	462	3.36	38.4	17.2	5.32	30.1	6.6	2.8	42.08
May 31, 1984	647	3.21	53.8	24.6	5.49	50.1	6.9	2.4	59.16
June 26, 1985	1038	2.55	77.8	38.1	5.31	83.3	18.8	13.6	86.63
Chloride									
Aug. 24, 1982	1	6.7	0.2	0.1	2.78	2.1	2.4	0.5	0.22
Sept. 1, 1982	9	9.2	0.7	0.4	3.02	1.7	2.4	0.7	0.81
Sept. 8, 1982	16	9.2	1.6	0.7	3.06	2.3	2.8	0.8	1.75
Sept. 21, 1982	29	11.5	2.9	0.9	3.27	2.5	2.6	0.9	3.04
Oct. 5, 1982	43	11.3	4.1	1.6	3.34	4.4	2.7	1.2	4.40
Oct. 25, 1982	63	9.0	5.7	2.0	3.50	4.4	2.4	1.1	6.04
Nov. 16, 1982	85	11.2	7.7	3.2	3.75	5.7	3.3	0.8	8.34
May 9, 1983	259	11.5	22.7	11.6	4.52	17.8	4.4	3.7	25.49
Sept. 8, 1983	381	9.6	32.3	15.3	5.18	20.6	4.4	3.9	35.74
Oct. 26, 1983	429	9.2	35.9	17.2	5.25	24.3	6.0	3.2	39.81
Nov. 28, 1983	462	8.2	38.2	17.4	5.33	27.8	5.5	2.1	41.98
May 31, 1984	647	9.1	53.1	23.9	5.55	51.5	5.5	3.0	58.23

[a] Given in the field coordinate system (x–y coordinates in Figure 6-6).

[b] Given in the rotated coordinates (x'–y' coordinates in Figure 6-6; x' is parallel to linear horizontal trajectory, y' is perpendicular to linear horizontal trajectory).

[c] $s = (x_c^2 + y_c^2)^{1/2}$

Adapted from Freyberg, D.L., *Water Resour. Res.*, 22, 2031–2046, 1986.

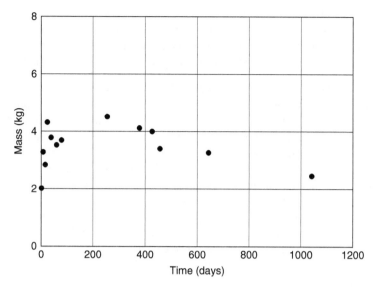

FIGURE 6-9   Estimated mass in solution of bromide ion as a function of time since injection for the Borden experiment.

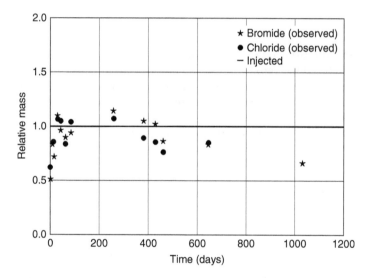

FIGURE 6-10   Estimated masses in solution of chloride and bromide, normalized by injected mass (10.7 kg for chloride, 3.87 kg for bromide) for the Borden experiment.

above, the mass in solution is expected to be effectively constant over the duration of the experiment. The mass estimates shown in Figure 6-9 generally confirm this expectation. However, the data shown in Figure 6-9 are not sufficient to identify any trends in the estimation of variance. The average observed mass is 0.41 kg less than the estimated mass injected. Potential contributing factors regarding this deficiency include (1) bias in point concentrations introduced by the sampling and analytical procedures; (2) consistent over- or underestimation by the quadrature procedure used to approximate Eq. (6-25); (3) consistent errors introduced by the assumption of a spatially uniform porosity field; and (4) a biased estimate of the mean effective porosity.

Figure 6-10 compares the mass estimates for chloride and bromide when each is normalized by the appropriate estimated injected mass. The injected masses of chloride and bromide are 10.7 kg and 3.87 kg, respectively. Paired estimates are shown for all sampling sessions except for the last, because only bromide data are available for the last session. The average relative mass and coefficient of variation for chloride are 0.90 and 0.16, while the same parameters for bromide are 0.89 and 0.20. The bias and variability for chloride are slightly greater than that for chloride (Freyberg, 1986).

Center of Mass and Estimation of its Horizontal Displacement
The estimated coordinates of the centers of mass of the bromide and chloride plumes for each synoptic sessions are given in Table 6-5 (Freyberg, 1986). As mentioned previously, only data for bromide are available for the last sampling session. Using the values in Table 6-5 (x_c and y_c coordinates), the horizontal trajectory of the centers of mass of the two plumes are presented in Figure 6-11. As seen from Figure 6-11, individual locations for the two tracers are almost the same, which indicates that a small estimation variability is introduced by sampling and measurement noise. Further, the data of Figure 6-11 suggest that the tracer plumes followed a remarkably linear horizontal trajectory over the duration of the experiment. The hydraulic gradient at the site is known to undergo small seasonal changes in direction (Mackay et al., 1986). However, the effects of this temporal variability are apparently small and undetectable in the center of mass data. The data points fit excellently to a line shown in Figure 6-11. The best-fit horizontal trajectory is directed 25.5° clockwise from the x-axis, or N47.5° E. This direction is in the center of the range observed for the direction of the hydraulic gradient and is within 2° of the estimated temporal mean direction (see Table 6-5). This suggests that horizontal anisotropy in the mean hydraulic conductivity field is very small to negligible (Freyberg, 1986). Sudicky (1986) reached the same conclusion with an analysis of laboratory-measured hydraulic conductivities from 1279 core samples taken adjacent to the monitored zone.

Estimation of Vertical Displacement of the Mass
Freyberg (1986) estimated the vertical coordinate of the center of mass for each sampling session and estimated each tracer in a manner exactly analogous to that described for horizontal moments. The only difference is that in this case, C_z of Eq. (6-74) is defined as (Freyberg, 1986)

$$C_z(x, y, t) = \int_{-\infty}^{\infty} C(x, y, z, t)z \, dz \qquad (6-75)$$

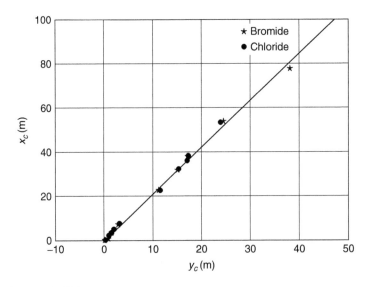

FIGURE 6-11   Horizontal trajectories of the centers of mass of the chloride and bromide plumes for the Borden experiment.

Freyberg (1986) found the vertical center of mass coordinates estimated in this manner to be very insensitive to the smoothing induced by the horizontal quadrature technique. Figure 6-12 presents the vertical locations of the centers of mass of both the bromide and chloride plumes as functions of horizontal displacement. Horizontal displacement distances, given in Table 6-5 for the two tracers, were determined by orthogonal projection of the individual data points onto the linear trajectory shown in Figure 6-11. Figure 6-13 shows the average value of the estimated vertical coordinates for the two tracers as a function of elapsed time. From this analysis, some key conclusions are as follows (Freyberg, 1986): (1) The total vertical displacement over the duration of the experiment is small, about 2.7 m over 1038 days. (2) As a component of the total, the vertical component is negligible. (3) Figure 6-12 and Figure 6-13 reveal that in contrast to the horizontal

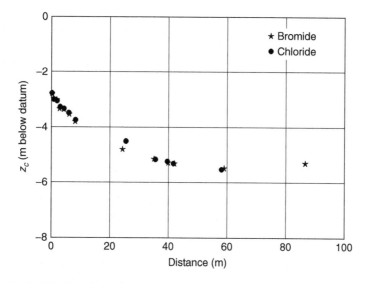

FIGURE 6-12   Vertical displacement of the centers of mass of the chloride and bromide plumes as a function of distance for the Borden experiment.

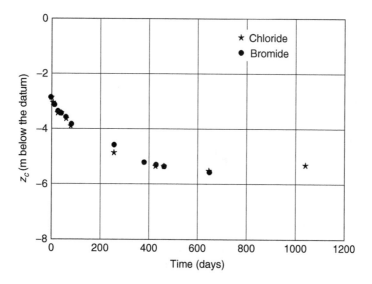

FIGURE 6-13   Average vertical coordinate of the centers of mass of the chloride and bromide plumes as a function of time for the Borden experiment.

trajectory, the vertical trajectory is not uniform over time and space. All of the vertical displacement occurs over the first 650 days of transport with the center of mass actually rising slightly over the last 390 days. and (4) The vertical trajectory is generally concave upward, indicating that the downward vertical velocity decreases with time. Such differences are probably within the estimation error for the vertical coordinate.

The above analysis shows that the reasons for the nature and magnitude of the observed vertical movement shown in Figure 6-12 and Figure 6-13 are not fully understood. Freyberg (1986) states that a number of mechanisms are jointly responsible and they are as follows: (1) a small vertical component in the regional velocity field; (2) the density contrast between the plume and the native ground water; (3) local infiltration and recharge; and (4) interaction with the underlying denser landfill plume at later times.

Solute Velocity
As noted previously, the contribution of the vertical component to the total displacement is negligible. Therefore, attention was focused on the horizontal displacement. Using the values in the second and tenth columns in Table 6-5, the horizontal displacement along the linear trajectory (shown in Figure 6-11) against travel time for the two tracers is shown in Figure 6-14. Not surprisingly, values for the two tracers are practically indistinguishable (Freyberg, 1986). The equation of the fitted line to the data points in Figure 6-14 is $s(m) = 0.091t$ (days). Therefore, the magnitude of the mean velocity is $U = 0.091$ m/day (Freyberg, 1986). This value is reasonably consistent with predictions using the data on hydraulic gradient, horizontal hydraulic conductivity, and effective porosity in Table 6-4.

Estimation of Spatial Covariances and Dispersivities
Spatial covariances for the bromide and chloride plumes are given in Table 6-5 (Freyberg, 1986). The spatial covariances are given in a rotated coordinate system (x', y'). The x'-coordinate direction (the AA' direction in Figure 6-6) is parallel to the linear horizontal trajectory, which coincides with the estimated mean velocity vector, and the y' (the BB' direction in Figure 6-6) coordinate direction is perpendicular to the trajectory. In other words, x' and y' correspond to the longitudinal and transverse directions of traditional convection–dispersion terminology. Freyberg (1986) estimated the covariances using the field coordinate system shown in Figure 6-6 and then rotated 25.5°

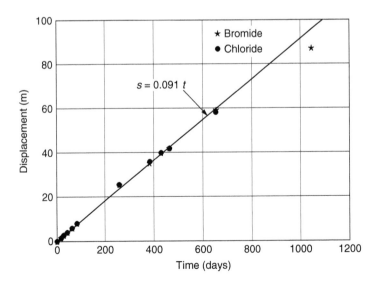

FIGURE 6-14 Horizontal displacement of the centers of mass of the chloride and bromide plumes along the linear trajectory of Figure 6-11 as a function of time since injection for the Borden experiment.

clockwise using the standard Cartesian tensor rotational transformation (Jeffreys, 1974). The estimated values of $\sigma_{x'y'}^2$ should theoretically be zero according to Eq. (6-54) (Moltyaner and Wills, 1991).

As described in Section 6.4, the spatial covariance structure of a concentration distribution provides a measure of the spread of the distribution about its center of mass. Changes in the covariance structure over time or space reflect the changes in the concentration distribution as a result of heterogeneity in the ground water velocity field. In the analysis of the Borden experiment data, Freyberg (1986) focused on the horizontal components of the spatial covariance tensor expressed by Eq. (6-54). The experimental data revealed that the vertical thickness of the tracer plumes remained unchanged over the duration of the experiment (see Figure 6-8). In general, temporal variation in estimates of the vertical components of the spatial covariance tensor cannot be distinguished owing to sampling noise effects. Therefore, only the horizontal components are examined. As shown in Figure 6-7, the initial pulse of the tracer was somewhat rectangular in shape, with its long side parallel to the y-axis field. With time, the plume evolved into an elongated shape, with its major axis roughly aligned with the direction of the mean motion of the plume. The data in Table 6-5 demonstrate this behavior. As can be seen from Table 6-5, for the first four sampling sessions the estimated $y'y'$ components of the covariance tensor are larger than the $x'x'$ components, which reflect that the plume has greater dimension in the transverse direction than in the longitudinal direction. From the fifth sampling session, the $x'x'$ covariance tensor term is larger than the $y'y'$ covariance tensor term and the difference between the two terms increases with time as the plume becomes more and more elongated in the longitudinal direction (Freyberg, 1986).

Using the values of $\sigma_{x'x'}^2$, $\sigma_{y'y'}^2$, and $\sigma_{x'y'}^2$ vs. time in Table 6-5 for chloride and bromide, the variation of the covariance tensor components with time are shown in Figures 6-15–6-17. The scatter between the data for the two tracers is greater than that observed for the locations of the centers of mass (see Figure 6-14). This is expected due to the fact that the sensitivity of the estimation methodology increases with the order of the moment (Freyberg, 1986).

Freyberg (1986) used Eq. (6-45) to describe the experimental plume movement with constant mean ground water velocity and apparent dispersivities. Under those conditions, Eq. (6-45) is valid with constant dispersion coefficient tensor. A linear relationship is expected between the magnitudes of the components of the covariance tensor and time. As a result, as shown by Figures 6-15–6-17, the ordinary least-squares linear fits to all the data points except the last. The last sampling session is excluded because the mean velocity field did not remain homogeneous over the last year of the experiment.

Using the data for the components of the covariance tensor in Table 6-5, the determination of the dispersivity tensor components is shown in Example 6-5.

EXAMPLE 6-5

This example is based on the data presented in Freyberg (1986). The solute velocity is given as $U = 0.091$ m/day (see the previous paragraphs). Using the components of the covariance tensor in Table 6-5 ($\sigma_{x'x'}^2$, $\sigma_{y'y'}^2$, and $\sigma_{x'y'}^2$), determine the components of the dispersivity tensor ($A_{x'x'}^2$, $A_{y'y'}^2$, and $A_{x'y'}^2$) based on Eq. (6-54) using the linear curve-fitting method. Discuss the value of $A_{x'y'}^2$.

SOLUTION

Using the values of $\sigma_{x'x'}^2$, $\sigma_{y'y'}^2$, and $\sigma_{x'y'}^2$ vs. time in Table 6-5, the variation of the covariance tensor components with time are shown in Figures 6-15–6-17. Dispersivities can be determined from the finite-difference forms of Eqs. (6-55)–(6-57). For this purpose, the $\Delta\sigma_{x'y'}^2$ and Δt values are shown in Figures 6-15–6-17. Using the values from Figure 6-15 in the finite-difference form of Eq. (6-55) with $U = 0.091$ m/day gives

$$A_{x'x'} = \frac{1}{2U} \frac{\Delta\sigma_{x'x'}^2}{\Delta t} = \frac{1}{2(0.091 \text{ m/day})} \frac{11.36 \text{ m}^2}{200 \text{ days}} = 0.31 \text{ m}$$

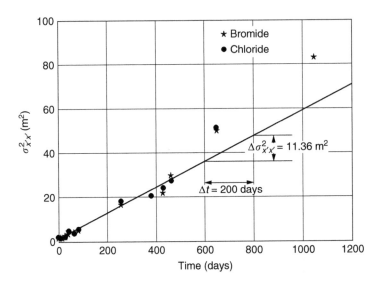

FIGURE 6-15 Longitudinal variance ($\sigma_{x'x'}^2$) vs. time for chloride and bromide for the experiment data.

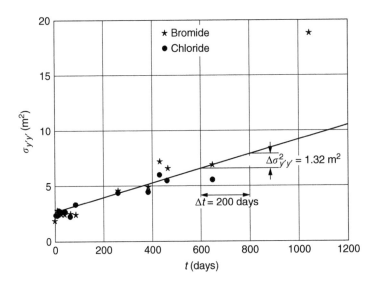

FIGURE 6-16 Transverse variance ($\sigma_{y'y'}^2$) vs. time for chloride and bromide for the Borden experiment data.

Similarly, using the values from Figure 6-16 in the finite-difference form of Eq. (6-56) gives

$$A_{y'y'} = \frac{1}{2U}\frac{\Delta\sigma_{y'y'}^2}{\Delta t} = \frac{1}{2(0.091\ \text{m/day})}\frac{1.32\ \text{m}^2}{200\ \text{days}} = 0.036\ \text{m}$$

Using the values in Figure 6-17 in the finite-difference form for $A_{x'y'}$ gives

$$A_{x'y'} = \frac{1}{2U}\frac{\Delta\sigma_{x'y'}^2}{\Delta t} = \frac{1}{2(0.091\ \text{m/day})}\frac{0.76\ \text{m}^2}{200\ \text{days}} = 0.021\ \text{m}$$

The above results show that $A_{x'y'}$ is almost 60% of $A_{x'y'}$, which is a significant value. As mentioned above, $A_{x'y'}$ is expected to be relatively close to zero according to Eq. (6-54).

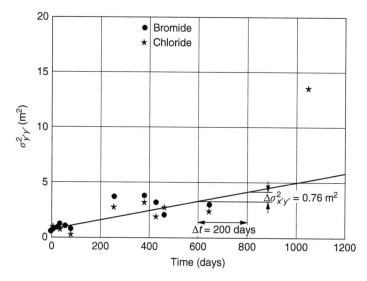

FIGURE 6-17 Covariance ($\sigma^2_{x'x'}$) vs. time for chloride and bromide for the Borden experiment data.

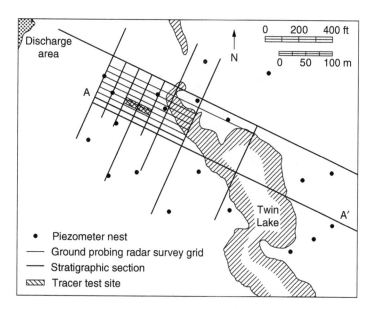

FIGURE 6-18 The Twin Lake experiment area, with locations of site selection boreholes and ground probing radar survey lines. (Adapted from Killey, R.W.D. and Moltyaner, G. L., *Water Resour. Res.*, 24, 1585–1612, 1988.)

6.3.7.2 Twin Lake Experiment

Description of the Aquifer

The aquifer is located at the Chalk River Nuclear Laboratories (CRNL) area in the Ottawa River valley, 180 km northwest of Ottawa, Canada. Figure 6-18 shows the locations of major features, boreholes, where continuous soil cores were collected, and the ground probing radar survey lines. Figure 6-19 shows the stratigraphy of the site along section AA′, based on combined borehole and radar data. Borehole samples showed a consistent sequence of aeolian sands, containing a unit of interstratified very fine sand and minor silt overlying fluvially deposited sands (Killey and Moltyaner, 1988).

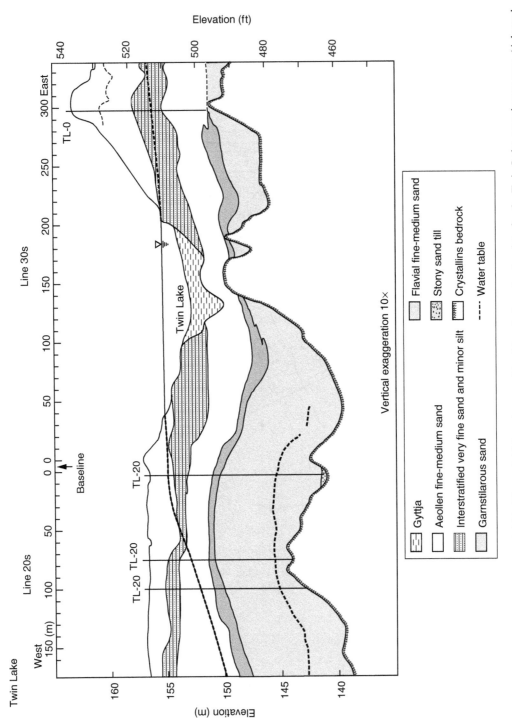

FIGURE 6-19 Stratigraphy along AA' in Figure 6-18 based on combined borehole and radar data for the Twin Lake experiment area. (Adapted from Killey, R.W.D. and Moltyaner, G.L., *Water Resour. Res.*, 24, 1585–1612, 1988.)

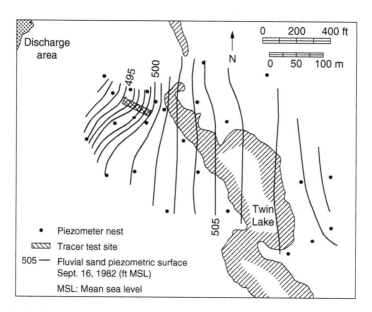

FIGURE 6-20   Hydraulic head distributions in the fluvial sands of the Twin Lake experiment area. (Adapted from Killey, R.W.D. and Moltyaner, G.L., *Water Resour. Res.*, 24, 1585–1612, 1988.)

Hydraulic Heads of the Aquifer

Twin Lake has one seasonal surface inlet and no surface outlet; the lake drains by recharging the underlying sands. Hydraulic head measurements show that most of the aquifer recharge occurs during the spring melt, with an occasional minor recharge event in November to December. Figure 6-20 shows hydraulic head contours in the fluvial sand portion of the aquifer and Figure 6-21 presents hydraulic head distribution in vertical section throughout the site selected for tracer testing. The water table mound created by recharge from Twin Lake is clearly evident in Figure 6-21 (Killey and Moltyaner, 1988).

Horizontal and Vertical Hydraulic Conductivities

The following four methods were used to collect hydraulic conductivity data: (1) permeameter tests of undisturbed sections of thin-wall sediment cores, which provided measurements of vertical hydraulic conductivity, K_v; (2) borehole dilution measurements and hydraulic gradient data, providing, in this case, horizontal hydraulic conductivity, K_h; (3) slug tests (Hvorslev, 1951), combined with permeameter K_v to measure K_h; and (4) estimates of bulk-permeability-based grain size distributions. All hydraulic conductivities are members of lognormal populations at the 20% significance level. Results are summarized in Table 6-6, which lists the logarithmic means and standard deviations for each data set (Killey and Moltyaner, 1988).

Radioiodine Plume Characterization

Radioiodine has been used extensively in field experiments for relatively short transit times. The plume movement and spreading is illustrated by means of two-dimensional contours of reconstructed and interpreted depth-averaged concentration values. Two-dimensional radioiodine vertically averaged isoconcentration contours at different times are shown in Figure 6-23. The concentration contours reveal distinct variability in the tracer concentration caused by the nonuniform ground water velocity field, which resulted in significant tailing of the plume. Starting from $t = 21.65$ days, the tail becomes longer partially due to low tracer concentration measured during the last phase of the monitoring program caused by decay of radioiodine (Moltyaner and Wills, 1991). Figure 6-24 show the positions of the center of mass.

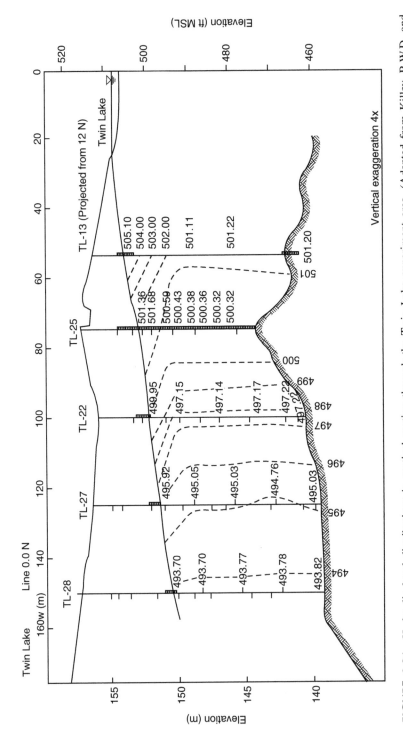

FIGURE 6-21 Hydraulic head distributions in vertical section through the Twin Lake experiment area. (Adapted from Killey, R.W.D. and Moltyaner, G.L., *Water Resour. Res.*, 24, 1585–1612, 1988.)

TABLE 6-6
Log Hydraulic Conductivities of the Twin Lake Experiment Site

Stratigraphic Unit	Permeameter		Borehole Dilution		Slug Test		Grain-Size Estimate	
	N^a	f_v^b	N^a	f_h^b	N^a	f_h^b	N^a	f^b
Aeolian fine-medium sand	35	-5.80 ± 0.74					52	-4.54 ± 0.23
Fluvial fine-medium sand	53	-5.25 ± 0.44	22	-4.28 ± 0.39	8	-5.02 ± 0.58	91	-4.38 ± 0.16
Fluvial medium sand	12	-4.91 ± 0.39					25	-4.19 ± 0.16

[a] Number of analysis.

[b] $f_v = \ln(K_v)$, $f_h = \ln(K_h)$, and $f = \ln(K)$, K, K_v and K_h are in cm/sec.

Adapted from Killey, R.W.D and Moltyaner, G.L., *Water Resour. Res.*, 24, 1585–1612, 1988.

Spatial Moments Analysis

Plume-Scale Statistical Characteristics
Moltyaner and Wills (1991) used Eq. (6-25) without the effective porosity, φ_e term. The authors state that there is no need to include effective porosity in the determination of spatial moments because the concentration field $C(x, y, z, t)$ was recorded using a through-the-wall measurement device. Interpreting normalized $C(x, y, z, t)$ as the probability density function for position coordinates (x, y, z) of local-scale particles, Moltyaner and Wills (1991) used Eq. (6-25) for estimating first moments of the coordinates of local-scale particles and elements of the covariance tensor. In general, the moments are a function of time, but at the zeroth moment, M_{000}, which is given by Eq. (6-26), the mass of the tracer remains invariant over time. As can be seen from Eq. (6-30), the center of mass of the plume (x_c, y_c, z_c) is defined by the normalized first moments M_{100}/M_{000}, M_{010}/M_{000}, and M_{001}/M_{000}.

Moltyaner and Wills (1991) assumed that the tracer is moved at the rate defined by Eq. (6-32). The authors also evaluated the appropriateness of this assumption for the Twin Lake aquifer. Second moments about the center of the tracer plume from the spatial covariance tensor with elements σ_{xx}^2, σ_{yy}^2, σ_{zz}^2, σ_{xy}^2 (or σ_{yx}^2), σ_{xz}^2 (or σ_{zx}^2), and σ_{yz}^2 (or σ_{zy}^2) are defined by Eqs. (6-39), (6-40), (6-41), (6-42), (6-43), and (6-44). The elements of the covariance tensor characterize the spread of the local-scale particles with the center of the plume. As mentioned in Section 6.3.4, it is common to assume that for the case when the x'-axis is aligned with the mean direction of flow, the plume-scale velocity has components given by Eq. (6.53), which are $v_x = U$, $v_y = 0$, and $v_z = 0$. Therefore, the nondiagonal elements of the covariance tensor, given by Eq. (6-52), are zeros and the dispersivities $A_{x'x'}$, $A_{y'y'}$, and $A_{z'z'}$ can be expressed by Eqs. (6-55)–(6-57), respectively. Since the effective molecular diffusion coefficient is neglected, the plume-scale mechanical dispersion coefficients are defined by Eq. (6-59).

Estimation Procedure of Moments and Transport Parameters
In order to estimate the spatial moments and transport parameters on the basis of reconstruction data, Moltyaner and Wills (1991) performed the integration in Eq. (6-25) (without φ_e) using Simpson's approximation. The estimated transport parameters are given in Tables 6-7–6-10. The position of the center of mass of the plume in Table 6-7 is given in field coordinates. The values of the other parameters in Tables 6-8–6-10 are given with respect to the coordinate system, with the positive x'-axis aligned along the mean flow direction.

TABLE 6-7
Estimates of Zero and First Spatial Moments for the Twin Lake Experiment

Time (days)	Vertically Averaged Data				Cross-Sectional Data			Three-Dimensional Data				Areal Data		
	Mass (counts/min)	x_c (m)	y_c (m)	s (m)	Mass (counts/min)	x_c (m)	z_c (m ASL)	Mass (counts/min)	x_c (m)	y_c (m)	z_c (m ASL)	Mass (counts/min)	x_c (m)	y_c (m)
1.74	10,023	1.74	0.19	1.75	5,564	1.78	146.92	12,729	1.78	0.12	147.16	1,421	1.97	0.21
4.44	13,411	4.55	0.48	4.58	6,881	4.82	146.22	15,332	4.79	0.52	146.59	1,975	4.34	0.57
8.69	11,840	9.78	1.33	9.87	5,420	10.06	145.91	12,173	10.14	1.44	146.00	2,267	9.64	1.33
13.39	8,866	15.81	2.09	15.95	5,195	15.23	145.29	10,302	16.02	1.97	145.43	1,763	15.38	1.94
17.06	11,019	20.37	3.08	20.60	3,877	20.56	144.69	10,699	19.77	2.77	144.61	2,208	19.48	2.92
21.65	10,161	24.53	3.54	24.78	3,776	24.03	144.72	11,886	24.73	3.37	144.23	2,343	24.47	3.63
25.16	12,734	29.11	5.14	29.56	4,253	29.05	144.59	14,602	27.94	4.66	144.09	2,726	29.06	5.36
30.37	12,150	33.00	5.74	33.50	6,530	32.24	143.93	14,728	32.16	5.40	143.26	2,993	34.19	6.60
34.38	11,734	37.91	6.73	38.50	6,684	37.53	143.47	12,975	38.52	6.67	142.61	2,817	39.62	7.46

From Mottyaner, G. L. and Wills, C. A., *Water Resour. Res.*, 27, 2007–2026, 1991. With permission.

TABLE 6-8
Estimates of Second Spatial Moments for the Twin Lake Experiment

Time	Vertically Averaged Data			Cross-Sectional Data			Three-Dimensional Data						Areal Data		
(days)	σ^2_{xx} (m²)	σ^2_{yy} (m²)	σ^2_{xy} (m²)	σ^2_{xx} (m²)	σ^2_{zz} (m²)	σ^2_{xz} (m²)	σ^2_{xx} (m²)	σ^2_{yy} (m²)	σ^2_{zz} (m²)	σ^2_{xy} (m²)	σ^2_{xz} (m²)	σ^2_{yz} (m²)	σ^2_{xx} (m²)	σ^2_{yy} (m²)	σ^2_{xy} (m²)
1.74	0.31	0.30	-0.047	0.29	1.63	0.056	0.39	0.41	2.54	-0.099	0.14	-0.45	0.26	0.25	0.20
4.44	1.63	0.45	-0.015	1.91	1.83	0.86	1.18	0.56	1.81	-0.12	0.77	-0.27	0.58	0.43	1.06
8.69	4.80	0.56	-0.16	4.32	1.88	2.07	4.34	0.61	1.71	-0.41	1.74	-0.30	1.23	0.50	0.027
13.39	4.72	0.62	-0.26	7.30	1.75	2.88	5.75	0.85	1.67	-0.47	2.04	-0.23	1.03	0.40	-0.029
17.06	4.69	0.99	-0.51	3.43	1.55	1.12	6.59	1.06	1.60	-1.13	1.82	-0.44	1.30	0.70	0.047
21.65	6.22	0.69[a]	-0.35	7.73	1.04	1.53	7.43	0.75[a]	1.40	-0.56	1.32	-0.25	1.37	0.45[a]	0.14
25.16	12.41	1.28	-0.49	10.50	0.93	0.96	18.02	1.28	1.27	-0.35	1.52	-0.25	2.06	1.01	0.50
30.37	22.99	0.46[a]	-0.18	24.06	0.80	0.73	32.54	0.43[a]	0.93	-0.36	1.52	0.014	4.79	0.57[a]	1.02
34.38	15.19	0.41[a]	0.0033	20.63	0.91	0.88	17.23	0.43[a]	1.24	-0.068	0.87	-0.056	4.62	0.23[a]	-0.38

[a] Data missing in the y direction.

From Moltyaner, G. L. and Wills, C.A., *Water Resour. Res.*, 27, 2007–2026, 1991. With permission.

TABLE 6-9
Velocity Estimates for the Twin Lake Experiment

Time	Vertically Averaged Data		Cross-Sectional Data		Three-Dimensional Data			Areal data		Fitted v_x
	v_x	v_y	v_x	v_z	v_x	v_y	v_z	v_x	v_y	
1.74	1.02	-0.13	1.06	0.54	0.99	-0.00058	0.73	1.14	0.080	1.17
4.44	1.03	-0.0052	1.06	0.088	1.08	0.030	0.22	0.98	0.047	1.06
8.69	1.13	0.0075	1.13	0.037	1.18	0.029	0.088	1.12	0.046	1.18
13.39	1.19	0.028	1.13	-0.0037	1.20	0.058	0.052	1.16	0.060	1.17
17.06	1.21	0.012	1.21	-0.023	1.17	0.043	0.0089	1.15	0.034	1.17
21.65	1.14	0.022	1.14	-0.0092	1.15	0.047	0.0081	1.14	0.035	1.15
25.16	1.18	-0.011	1.19	-0.0033	1.13	0.012	0.0070	1.17	-0.0052	1.19
30.37	1.10	-0.0063	1.06	-0.019	1.08	0.013	-0.011	1.15	-0.014	1.19
34.38	1.12	-0.0090	1.11	-0.023	1.14	0.0082	-0.015	1.17	-0.0096	1.20

Time is given in days and v is in m/day.

From Moltyaner, G. L. and Wills, C. A., *Water Resour. Res.*, 27, 2007–2026, 1991. With permission.

TABLE 6-10
Estimates for Mechanical Dispersion Coefficients for the Twin Lake Experiment

Time	Vertically Averaged Data		Cross-Sectional Data		Three-Dimensional Data			Areal Data		Fitted	
	$D_{m_{x'x'}}$	$D_{m_{y'y'}}$	$D_{m_{x'x'}}$	$D_{m_{y'y'}}$	$D_{m_{x'x'}}$	$D_{m_{y'y'}}$	$D_{m_{z'z'}}$	$D_{m_{x'x'}}$	$D_{m_{y'y'}}$	$D_{m_{x'x'}}$	$D_{m_{y'y'}}$
1.74	0.089	0.086	0.083	0.47	0.11	0.12	0.73	0.057	0.072	0.053	0.12
4.44	0.18	0.051	0.22	0.21	0.13	0.063	0.20	0.065	0.048	0.032	0.037
8.69	0.28	0.032	0.25	0.11	0.25	0.035	0.098	0.071	0.029	0.023	0.0010
13.39	0.18	0.023	0.27	0.065	0.21	0.032	0.062	0.038	0.015	0.014	0.014
17.06	0.14	0.029	0.10	0.045	0.19	0.031	0.047	0.038	0.021	0.010	0.026
21.65	0.14	0.016[a]	0.18	0.024	0.17	0.017[a]	0.032	0.032	0.010[a]	0.011	0.0011
25.16	0.25	0.025	0.21	0.018	0.37	0.026	0.026	0.041	0.020	0.017	0.0006
30.37	0.38	0.0076[a]	0.40	0.013	0.54	0.0071[a]	0.015	0.079	0.008[a]	0.018	0.0017
34.38	0.22	0.0060[a]	0.30	0.013	0.25	0.0063[a]	0.018	0.067	0.003[a]	0.017	0.0016

Time is given in days and $D_{m_{x'x'}}$, $D_{m_{y'y'}}$ and $D_{m_{z'z'}}$ are in m²/day.

[a] Data missing in the y direction.

From Moltyaner, G.L. and Wills, C.A., *Water Resour. Res.*, 27, 2007–2026, 1991. With permission.

The Mass of Tracer (Zeroth Moment)
Table 6-7 summarizes information on the mass of the tracer in the system (zeroth moment) for different times from the beginning of injection. Figure 6-22 shows the estimated mass (in counts per minute) of the tracer as a function of distance for vertically averaged data in Table 6-7. The total mass of the tracer should remain constant over the period of the experiment. As can be seen from Table 6-7 and Figure 6-22, the calculated mass fluctuates with regard to the mean estimate. The variability in mass is probably due to (1) the error associated with the use of the regression analysis for the rotation, (2) the time-inversion procedure, (3) the effect of the interpolation routine, (4) the integration procedure, and (5) sampling errors rather than any true effect within the plume (Moltyaner and Wills, 1991).

The Coordinates of the Center of Mass (Normalized First Moments)
Table 6-7 gives the estimated coordinates of the center of mass of the tracer in the field coordinate system (normalized first moment) for different times from the beginning of injection based on Eq. (6-30). Figure 6-23 graphically presents the results in three planes of the field coordinate system. Data shown in Figure 6-23a (two-dimensional data along the mean direction of flow) suggest that position coordinates of the center of mass of the plume calculated along the mean direction flow follow a trajectory parallel to the bedrock surface. Figure 6-23b shows a regression line drawn through the (x, z) coordinates of the center of mass data points. Figure 6-24c confirms that the mean direction of flow determined as a locus of borehole coordinates of the arrival of the first tracer approximates reasonably well the trajectory of the plume center of mass within the error associated with the spacing between boreholes in the transverse-to-flow direction. Figure 6-23d illustrates a linear trajectory of the center of mass in the y–z plane (Moltyaner and Wills, 1991).

Second Moments
The second moments are presented in Table 6-8, which were calculated for vertically averaged concentration data represented in Figure 6-24, (for data collected along the cross-section of the mean direction of flow). The covariances given in Table 6-8 show more significant variability as compared with the variability of the first moments owing to the high sensitivity of the moments analysis to the order of moments and the error introduced by the estimation methodology. According to Eq. (6-54), the estimated values of $\sigma_{x'y'}^2$, $\sigma_{y'z'}^2$, and $\sigma_{x'z'}^2$ should theoretically be zero. In fact, $\sigma_{x'y'}^2$ and $\sigma_{y'z'}^2$ are negligibly small when compared with $\sigma_{x'x'}^2$ and $\sigma_{x'y'}^2$ (see Table 6-8). The small nonzero values result from using a straight line for the mean direction of flow where as Figure 6-24 shows slight deviations in the position of the plume center mass from a straight line. The values of $\sigma_{x'z'}^2$ indicate

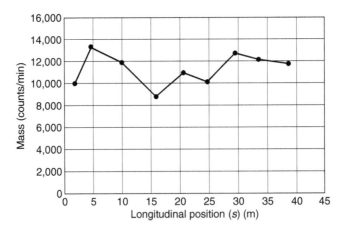

FIGURE 6-22   Mass vs. distance s for vertically averaged data in Table 6-7 for the Twin Lake experiment.

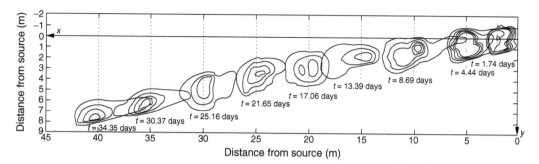

FIGURE 6-23 Concentration contours of transformed and interpolated vertically-averaged data (outer contour is 2% concentration; each subsequent contour shows a 2% increase) for the Twin Lake experiment. (Adapted from Moltyaner, G.L. and Wills, C.A., *Water Resour. Res.*, 27, 2007–2026, 1991.)

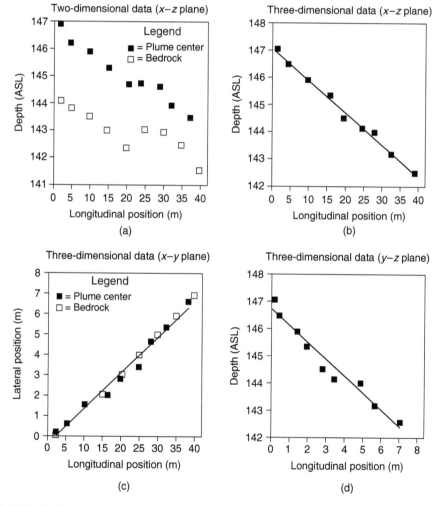

FIGURE 6-24 Positions for the center of mass of the Twin Lake experiment plume: (a) from data along the mean direction of flow; (b) from the three-dimensional x–z plane data; (c) from three-dimensional x–y data; and (d) from the three-dimensional y–z data. (Adapted from Moltyaner, G.L. and Wills, C.A., *Water Resour. Res.*, 27, 2007–2026, 1991.)

the existence of the additional spreading caused by the significant advance of the plume in the horizontal direction in the upper part of the tracer-occupied zone as compared with the bulk of the tracer (Moltyaner and Wills, 1991).

Estimation of Plume Velocity
Table 6-9 presents estimates of the plume velocity from the sources based on moment calculations using Eqs. (6-30) and (6-31). The values of v_y' and v_z' are negligibly small compared with v_x', but are not zero because of the approximation used for the mean direction of flow. The last column of Table 6-9 shows velocity estimates based on fitting a two-dimensional analytical solution [Eq. (3-476)] to the convection–dispersion equation to data observed at the elevation of the center of the plume (Moltyaner and Wills, 1991).

Estimation of Mechanical Dispersion Coefficients
Table 6-10 presents the mechanical dispersion coefficients using moment-based estimates and Eqs. (6-60)–(6-62). In the Borden experiment case (Freyberg, 1986), which was presented previously, $\sigma_{x'x'}^2$ vs. time and $\sigma_{y'y'}^2$ vs. time curves were drawn and their gradients were determined graphically (Figure 6-15 and Figure 6-16). Here, Moltyaner and Wills (1991) used a slightly different approach, which is based on the assumption that the aforementioned curves are linear and pass through the origin. The application is shown below in Example 6-6. The following observations can be made from the values in Table 6-10: (1) The values of $D_{m_{z'z'}}$ show a decrease in spread in the vertical direction with time. This is due to the boundary effect of the bedrock surface, which is not taken into account in the spatial moment calculation. (2) The initial high values of $D_{m_{y'y'}}$ are probably an artifact of the injection procedure. (3) The values indicated by asterisks should probably be ignored because the southern edge of the plume was not sampled in these cases (Moltyaner and Wills, 1991).

EXAMPLE 6-6

This example is based on the data presented in Moltyaner and Wills (1991). Using the components of the covariance tensor in Table 6-8 ($\sigma_{x'x'}^2$, $\sigma_{y'y'}^2$, and $\sigma_{z'z'}^2$) for three-dimensional data, determine the components of the mechanical dispersion coefficient tensor ($D_{m_{x'x'}}$, $D_{m_{y'y'}}$, and $D_{m_{z'z'}}$) based on Eq. (6-58) for (1) $t = 1.74$ days and (2) $t = 8.69$ days.

SOLUTION

(1) From Table 6-8, the pertinent data for $t = 1.74$ days are

$$\sigma_{x'x'}^2 = 0.39 \text{ m}^2, \qquad \sigma_{y'y'}^2 = 0.41 \text{ m}^2, \qquad \sigma_{z'z'}^2 = 2.54 \text{ m}^2$$

Using these values in the finite-difference forms of Eqs. (6-60)–(6-62) with $\Delta t = 1.74$ days, the following values can be obtained:

$$D_{m_{x'x'}} = \frac{1}{2}\frac{\Delta\sigma_{x'x'}^2}{\Delta t} = \frac{1}{2}\frac{0.39 \text{ m}^2}{1.74 \text{ days}} = 0.11 \text{ m}^2/\text{days}$$

$$D_{m_{y'y'}} = \frac{1}{2}\frac{\Delta\sigma_{y'y'}^2}{\Delta t} = \frac{1}{2}\frac{0.41 \text{ m}^2}{1.74 \text{ days}} = 0.12 \text{ m}^2/\text{days}$$

$$D_{m_{z'z'}} = \frac{1}{2}\frac{\Delta\sigma_{z'z'}^2}{\Delta t} = \frac{1}{2}\frac{2.54 \text{ m}^2}{1.74 \text{ days}} = 0.73 \text{ m}^2/\text{days}$$

The values are exactly the same as the corresponding values to 1.74 days in Table 6-10.
(2) From Table 6-8, the pertinent data for $t = 8.69$ days are

$$\sigma_{x'x'}^2 = 4.34 \text{ m}^2, \qquad \sigma_{y'y'}^2 = 0.61 \text{ m}^2, \qquad \sigma_{z'z'}^2 = 1.71 \text{ m}^2$$

Similarly, using these values in the finite-difference forms of Eqs. (6-60)–(6-62) with $\Delta t = 8.69$ days, the following values can be obtained:

$$D_{m_{x'x'}} = \frac{1}{2}\frac{4.34 \text{ m}^2}{8.69 \text{ days}} = 0.25 \text{ m}^2/\text{days}$$

$$D_{m_{y'y'}} = \frac{1}{2}\frac{0.61 \text{ m}^2}{8.69 \text{ days}} = 0.035 \text{ m}^2/\text{days}$$

$$D_{m_{z'z'}} = \frac{1}{2}\frac{1.71 \text{ m}^2}{8.69 \text{ days}} = 0.098 \text{ m}^2/\text{days}$$

The values are exactly the same as the corresponding values to 8.69 days in Table 6-10.

Evaluation of the Observed Plumes with an Analytical Model
The last two columns of Table 6-10 list the mechanical dispersion coefficients determined from fitting a two-dimensional analytical solution of the convection–dispersion equation to data observed at the elevation of the center of mass of the plume. Comparison of the last four columns in Table 6-10 shows that estimates of the longitudinal dispersion coefficient determined from the moments analysis are, on average, larger by a factor of 2 than those determined using the fitting procedure. The three-dimensional dispersion coefficients are listed in Table 6-10 for predictive purposes. Moltyaner and Wills (1991) state that the predicted results fit the observed data poorly.

Moltyaner and Wills (1991) attempted to use solute transport parameters, estimated using vertically averaged data in Table 6-10 since simulating the three-dimensional data proved unsuccessful. The authors used Eq. (3-476), which is a two-dimensional analytical solution for an instantaneous rectangular source (see Section 3.3.4.4 for further details of this solution). Plots determined from this solution superimposed on the breakthrough curves of vertically averaged data along the crosssection of the mean direction of flow predict the behavior of the centroid of the curve reasonably well but not spreading (Figure 6-25). Moltyaner and Wills state that the spatial contours of the observed and simulated data show that the transverse dispersion is significantly overestimated in the spatial moments analysis method. On the other hand, the authors used local-scale solute transport parameters of 0.01 m for longitudinal dispersivity, α_L, 0.001 m for transverse dispersivity, α_T, and 1.137 m/day for ground water velocity, U, in the two-dimensional analytical solute transport model given by Eq. (3-477), and compared its results with the observed data extracted from near the plume's estimated center of mass. As shown in Figure 6-26, the comparison is reasonably well.

6.3.7.3 Cape Cod Experiment

Site Description and Aquifer Characteristics
The Cape Cod experiment was conducted in an abandoned gravel pit on Western Cape Cod near Otis Air Force Base in Massachusetts (Figure 6-27). The tracer test site is located about midway between the Otis Air Base sewage disposal site and Ashumet Pond, a kettle-hole pond in the outwash plain. The Cape Cod area consists of a large sand and gravel outwash plain that was deposited during the retreat of the continental ice sheets from southern New England about 12,000 years ago. The unconsolidated deposits form an unconfined aquifer, which is the main source of drinking water of the area. The characteristics of the aquifer are given in the following paragraphs (LeBlanc et al., 1991).

The aquifer at the Cape Cod experiment site is composed of about 100 m of unconsolidated sediments that overlie a relatively impermeable, crystalline bedrock. The upper 30 m of the aquifer consists of a permeable, stratified, sand and gravel outwash. Beneath the outwash, the sediments consist of a fine-grained sand and silt (Figure 6-28). The median grain size of the outwash is about 0.5 mm, and the outwash generally contains less than 1% silt and clay.

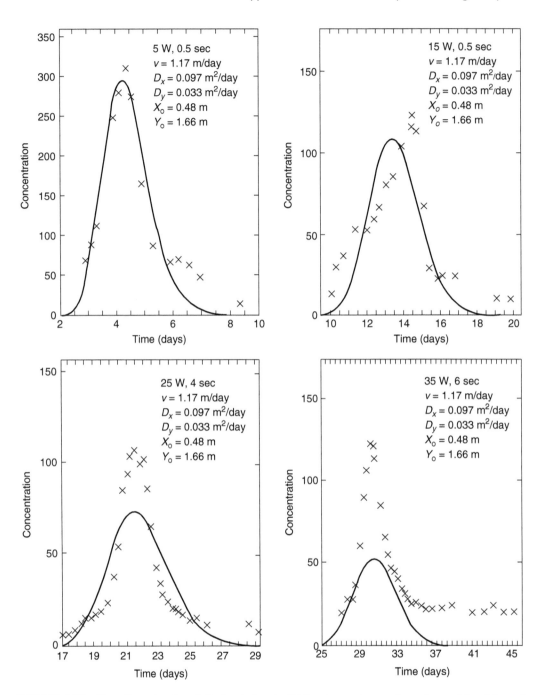

FIGURE 6-25 Two-dimensional analytical solution (solid lines) superimposed on observed breakthrough curves of vertically averaged data (crosses) for the Twin Lake experiment. (Adapted from Moltyaner, G.L. and Wills, C.A., *Water Resour. Res.*, 27, 2007–2026, 1991.) Note: $X_0 = X$ and $Y_0 = Y$ in Eq. (3-476).

The aquifer is highly permeable and its estimated horizontal hydraulic conductivity (K_h) is 110 m/day (0.13 cm/sec), which was determined from an aquifer pumping test conducted in the outwash about 2.2 km south of the site. Based on the results of this test, the ratio of horizontal hydraulic conductivity, K_h, to vertical hydraulic conductivity, K_v, varied between 2 and 5. The effective porosity

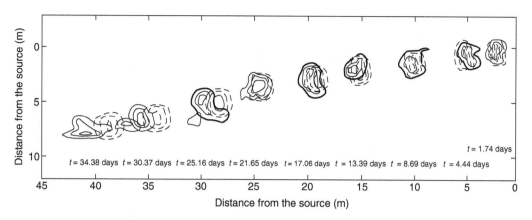

FIGURE 6-26   Predicted concentration contours (dashed lines) superimposed on actual data (solid lines) for the Twin Lake experiment (outer contour is 10% concentration contour; each subsequent shows a 20% increase for $t \leq 8.69$ days and a 10% increase for $t \geq 3.39$ days). (Adapted from Moltyaner, G.L. and Wills, C.A., *Water Resour. Res.*, 27, 2007–2026, 1991.)

φ_e, of the outwash deposits was estimated from several small-scale tracer tests and spatial moments analysis of the test to be about 0.39.

The water table at the test site is generally 3 to 7 m below land surface and slopes to the south at about 0.15 m per 100 m (Figure 6-29). The source of water to the aquifer is recharged from precipitation. In Cape Cod, about 45% of the total annual precipitation, or about 50 cm/year, recharges the ground water system. Most recharge occurs during late fall and winter. The ground water ultimately discharges to the ocean, which is about 10 km south of the study area. The water table altitude typically fluctuates about 1 m annually because of seasonal variations in precipitation and recharge. During the first 17 months of the tracer test (July 1985 to December 1986), however, the water table fluctuated only about 0.3 m because precipitation was less than normal in the fall and winter when recharge is greatest. During this period, the magnitude and direction of the horizontal hydraulic gradient were relatively constant; the magnitude varied from 0.0014 to 0.0018, and the direction varied from 164° to 173° east of the magnetic north. Between December 1986 and May 1987, a period with above-normal precipitation, the water table rose about 1.2 m, and the hydraulic gradient increased to 0.0020 and shifted to 156° east of magnetic north. The direction and magnitude of the hydraulic gradient are influenced by the Ashumet Pond. As water levels rise in the aquifer, the gradient tends to steepen and shift eastward toward the pond. Vertical hydraulic gradients are too small to measure in clusters of monitoring wells at the tracer test site; the vertical gradients must be smaller than 0.3 cm (accuracy of the water level measurements) per 25 m (vertical span of the well clusters). Therefore, ground water flow is nearly horizontal in the aquifer (Le Blanc et al., 1991). Based on the average estimates of horizontal hydraulic conductivity ($K_h = 110$ m/day), hydraulic gradient ($I = 0.0015$ m/m, and effective porosity ($\varphi_e = 0.39$), from Eqs. (2-39) and (2-40), the average ground water velocity U from Darcy's equation is

$$U = \frac{1}{\varphi_e} KI = \frac{1}{0.39}(110 \text{ m/day})(0.0015 \text{ m/m}) = 0.42 \text{ m/day}$$

Design and Operation of the Tracer Test

The tracer test began in July 1985 with the injection of 7.6 m³ of tracer solution into the aquifer. Migration of the tracer cloud was then monitored by the collection of water samples from an array of 656 multilevel samplers. The test was designed and conducted so that the migration of the tracer cloud could be monitored in three dimensions as it traveled about 280 m through the aquifer. Additional information about the test is given in the following paragraphs (LeBlanc et al., 1991).

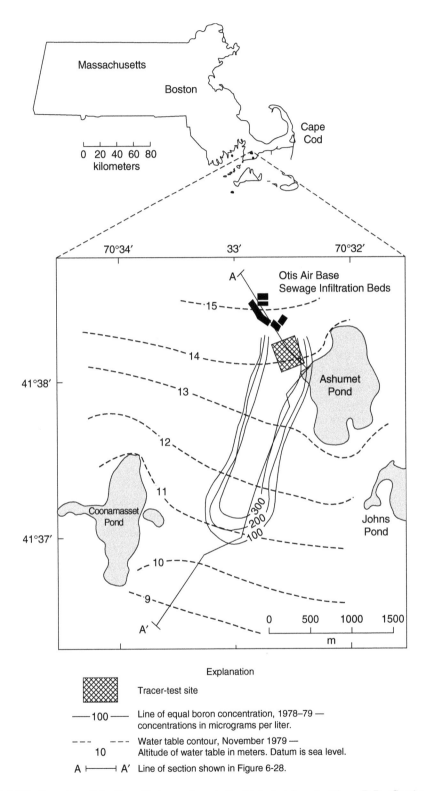

FIGURE 6-27 Location of the Cape Cod experiment site. (Adapted from LeBlanc, D.R., Garabedian, S.P., Hess, K.M., Gelhar, L.W., Quadri, R.D., Stollenwerk, K.G. and Wood, W.W., *Water Resour. Res.,* 27, 895–910, 1991.)

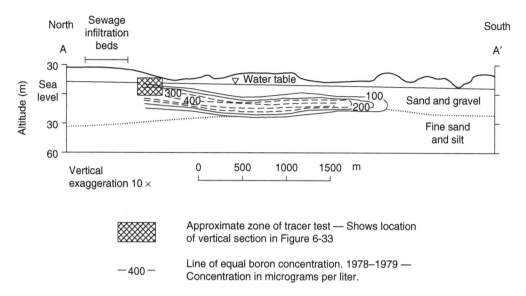

FIGURE 6-28   Vertical location of sewage plume, geohydrologic units, and zone of tracer test of the Cape Cod experiment site (line of section is shown in Figure 6-30). (Adapted from LeBlanc, D.R., Garabedian, S.P., Hess, K.M., Gelhar, L.W., Quadri, R.D., Stollenwerk, K.G. and Wood, W.W., *Water Resour. Res.,* 27, 895–910, 1991.)

Tracer Characteristics

The tracers used in the Cape Cod test were selected to meet the following criteria:

1. They were nontoxic at low concentrations and could be used safely in the aquifer;
2. Natural concentrations of the tracers were low so that the tracers introduced for the test could be followed for a reasonable distance and still be detected (Table 6-11);
3. Both nonreactive and reactive tracers were needed to meet the multiple objectives of the test.

Bromide (Br^-) was expected to be nonreactive and its behavior was similar to that of chloride, but chloride was present in relatively high concentrations in the coastal aquifer of the test site and was not suitable for use in the test. Background bromide concentrations in the aquifer were generally less than 0.1 mg/L and averaged about 0.05 mg/L. As shown in Table 6-11, an injection concentration of 640 mg/L bromide was used. Therefore, the tracer could be detected after as much as a 10,000-fold dilution by mixing with the ambient ground water. This bromide concentration also kept the density difference between the tracer solution and the ambient ground water less than one part in a thousand.

The reactive tracers were lithium (Li^+), molybdate(MoO_4^{2-}), and fluoride (F^-). Fluoride was abandoned as a tracer early in the test because fluoride concentrations were rapidly attenuated by sorption and, as a result, the fluoride cloud became masked by the significant amounts of fluoride already present in the sewage-contaminated ground water.

The tracers were added as salts LiBr, LiF, and Li_2MoO_4 to 7.6 m^3 of water that had been obtained from a shallow well screened in the uncontaminated zone above the sewage plume. The final solution, which was stored in two insulated tanks to maintain the water temperature near the ambient ground water temperature, contained about 890 mg/L dissolved solids.

Injection of the Tracers

The tracers were injected into three 5.08-cm-diameter wells during a 17-h period beginning July 18, 1985 and ending July 19, 1985. Each injection well had a 1.2-m-long, slotted polyvinyl chloride (PVC) screen set at an altitude of 11.9 to 13.1 m or about 1.2 to 2.4 m below the water table.

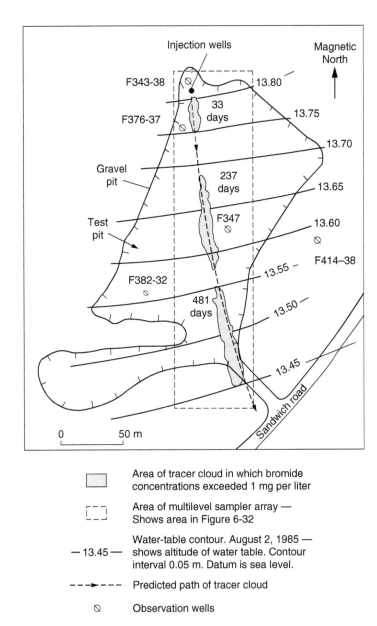

FIGURE 6-29 Cape Cod tracer test site in abandoned gravel pit, showing water table, location of selected monitoring sites, and predicted and observed path of bromide tracer plume. (Adapted from LeBlanc, D.R., Garabedian, S.P., Hess, K.M., Gelhar, L.W., Quadri, R.D., Stollenwerk, K.G. and Wood, W.W., *Water Resour. Res.,* 27, 895–910, 1991.)

Collection and Analysis of Water Samples

The distribution of the tracers in the aquifer was monitored by the collection of water samples from the array of multilevel samplers (MLS). The sampling array consisted of 656 MLS arranged in 71 rows and covered an area 12 m to 22 m wide and 282 m long (Figure 6-30). The vertical spacing between sampling ports is generally constant for a given MLS and varies with horizontal location in the array, from 25.4 cm near the injection wells to 76.2 cm near Sandwich Road, shown in Figure 6-29. This variation is illustrated by the representative section through the array shown in Figure 6-31. Water samples were collected from subsets of the MLS array at about monthly intervals, beginning 13 days after injection, to obtain snapshot views of the three-dimensional distributions of

TABLE 6-11
Characteristics of Tracers for the Cape Cod Experiment

Tracer	Injected Mass(g)	Injected Concentration(mg/L)	Background Concentration(mg/L)
Bromide (Br$^-$)	4900	640	< 0.10
Lithium (Li$^+$)	590	78	< 0.01
Molybdate (MoO$_4^{2-}$ as Mo)	610	80	< 0.02
Fluoride (F$^-$)	380	50	< 0.20

From LeBlanc, D.R., Garabedian, S.P., Hess, K.M., Gelhar, L.W., Quadri, R.D., Stollenwerk, K.G. and Wood, W.W., *Water Resour. Res.,* 27, 895–910, 1991.

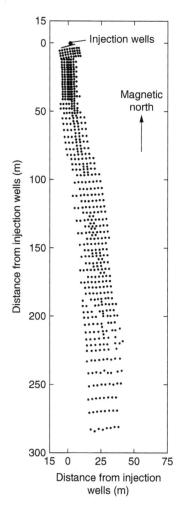

FIGURE 6-30 Location of injection wells and multilevel samplers of Cape Cod tracer test site. (Adapted from LeBlanc, D.R., Garabedian, S.P., Hess, K.M., Gelhar, L.W., Quadri, R.D., Stollenwerk, K.G. and Wood, W.W., *Water Resour. Res.,* 27, 895–910, 1991.)

tracer concentrations. A total of 19 rounds of sampling were completed between July 1985 and June 1988 (Table 6-12). As many as 13,000 water samples were collected from 40 to 290 MLS in a round.

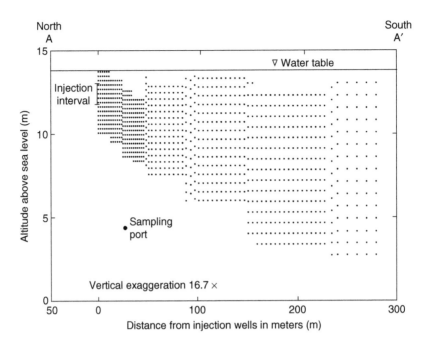

FIGURE 6-31   Vertical location of screened interval of injection wells and ports on multilevel samplers along a representative longitudinal section through the sampler array for the Cape Cod tracer test site. (Adapted from LeBlanc, D.R., Garabedian, S.P., Hess, K.M., Gelhar, L.W., Quadri, R.D., Stollenwerk, K.G. and Wood, W.W., *Water Resour. Res.*, 27, 895–910, 1991.)

Observed Migration of the Bromide Tracer Plume
The migration of the tracer plume was tracked during the test by preparing maps and cross-sections of concentration data. The focus here is on the migration of bromide, the nonreactive tracer (LeBlanc et al., 1991).

Horizontal Migration
In Figure 6-32, the horizontal migration of the bromide plume is illustrated after 33, 237, and 461 days from the time of injection. For each round shown, the maps were prepared by contouring values of the maximum bromide concentration at each MLS location, regardless of the depth at which the maximum occurred. Thus, these maps delineate the maximum areal extent of the tracer plume. It is important to note that the maps of maximum concentration (Figure 6-32) do not necessarily reflect the areal distribution of mass, as would maps of vertically averaged concentrations, nor do they represent horizontal slices through the tracer plume. During the test, the bromide plume moved in a southerly direction along a path predicted from the water-table gradient (Figure 6-29). As will be mentioned later, the average rate of migration of the bromide plume was 0.42 m/day (Garabedian et al., 1991). This observed rate matches the ground water flow velocity estimated earlier by Darcy's equation. As can be seen from Figure 6-32, the longitudinal spreading of the bromide plume in the direction flow is significant. The plume spread much less in the direction transverse to flow. At 461 days, the bromide plume, as delineated by concentrations greater than 1 mg/L, was more than 80 m long but only 14 m wide.

The lithium plume followed a trajectory that was similar to that of the bromide plume. However, its average rate of migration was significantly retarded. As can be seen from Figure 6-32, at 461 days, the lithium plume, which is delineated by concentrations greater than 0.1 mg/L, has spread almost 100 m, but its zone of maximum concentration had traveled only 90 m. Similarly to lithium, the molybdate plume, which is delineated by concentrations greater than 0.1 mg/L, had traveled only about half as far as the bromide plume after 461 days, but it had spread about 150 m (Figure 6-32).

TABLE 6-12
Summary of Sampling Rounds of the Cape Cod Experiment

Nominal Date	Days Since Injection	Number of MLS Sampled	Number of Samples Analyzed			Maximum Observed Concentration (mg/L)			Location of Center of Mass[a] of Bromide Cloud (m)		
			Br⁻	Li⁺	MoO₄²⁻	Br⁻	Li⁺	MoO₄²⁻	x	y	z
July 31, 1985	13	40	597	596	492	576	50.0	72.0	0.8	27.4	12.3
Aug. 20, 1985	33	122	1700	1818	1403	429	30.0	50.3	2.7	−16.9	11.7
Sept. 11, 1985	55	149	2175	1800	1625	311	17.0	37.0	3.0	−25.9	11.1
Oct. 9, 1985	83	176	2264	2379	2188	124	12.0	26.0	5.6	−38.9	10.6
Nov. 6, 1985	111	174	1892	2054	1962	132	7.7	13.2	8.1	−50.9	10.3
Dec. 4, 1985	139	164	1916	1524	1734	76.6	6.2	12.8	10.2	−64.7	10.4
Jan. 8, 1986	174	160	1649	1447	1615	76.6	4.8	6.75	11.1	−77.5	9.6
Feb. 6, 1986	203	125	1592	1863	1314	61.5	3.0	8.29	11.4	−88.8	9.4
March 12, 1986	237	219	2147	1711	1874	65.2	2.3	6.77	11.9	−100.1	9.3
April 17, 1986	273	221	1923	1711	1767	46.9	1.0	4.45	13.3	−114.3	9.2
May 29, 1986	315	241	2270	1772	1960	62.3	1.0	1.84	16.7	−133.9	9.4
July 2, 1986	349	254	2247	1291	1455	64.5	0.80	1.95	19.0	−147.5	9.4
Aug. 6, 1986	384	288	2091	1885	2382	47.3	0.49	1.67	21.9	−162.4	9.2
Sept. 17, 1986	426	197	2002	994	1323	50.6	0.35	2.14	25.4	−180.4	8.9
Oct. 22, 1986	461	286	1656	2029	2706	39.0	0.33	1.76	28.6	−196.3	8.7
Dec. 11, 1986	511	290	1654	2244	2738	42.5	0.35	1.31	32.4	−214.0	8.1
May 6, 1987	657	283	0	2409	3160	b	0.30	0.73	b	b	b
Aug. 12, 1987	755	287	0	3210	4247	b	0.17	0.42	b	b	b
June 22, 1988	1070	142	0	0	2092	b	b	0.08	b	b	b

[a] Location of center of mass calculated from first spatial moment of bromide distribution. The x (positive east) and y (positive north) coordinates are relative to magnetic north-oriented grid with origin at the central injection well. The z coordinate is altitude above sea level.

[b] data were not collected for this sampling date.

From LeBlanc, D.R., Garabedian, S.P., K.M., Hess, Gelhar, L.W., Quadri, R.D., Stollenwerk, K.G. and Word,W.W., *Water Resour. Res., 27,* 895–910, 1991. With permission.

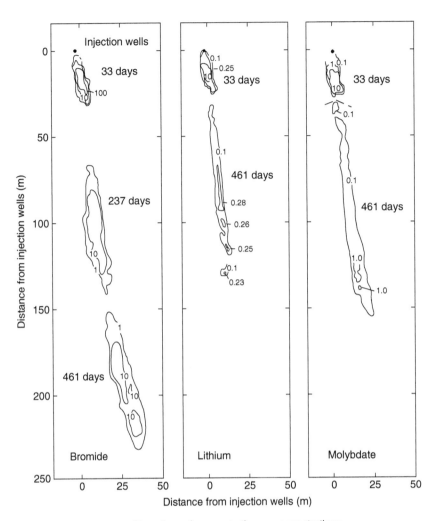

Line of equal concentration — concentrations
— 10 — of bromide, lithium, and molybdate (as Mo) in
milligrams per liter, interval varies.

FIGURE 6-32 Areal distribution of maximum concentrations of bromide at 33, 237, and 461 days after injection for the Cape Cod tracer test. (location of maps shown in Figure 6-29) (Adapted from LeBlanc, D.R., Garabedian, S.P., Hess, K.M., Gelhar, L.W., Quadri, R.D., Stollenwerk, K.G. and Wood, W.W., *Water Resour. Res.*, 27, 895–910, 1991.)

Vertical Migration

The vertical migration of bromide is shown by the longitudinal sections of the bromide distribution at 33, 237, and 461 days in Figure 6-33 and Figure 6-34. An analysis of the spatial moments of the bromide plume (Garabedian et al., 1991), which will be presented later, showed that the center of mass moved downward (Figure 6-35 and Table 6-12) about 3.2 m during the first 237 days of the test (during the same period, the plume moved about 100 m horizontally). Between 237 and 384 days, the bromide plume moved mostly horizontally. After 384 days, the center of mass again began to move downward. Two processes that contribute to the downward movement observed during the test are (1) vertical components of flow associated with areal recharge and (2) sinking of the denser tracer plume into the native ground water. Both processes were probably important during the first 237 days after injection, when about 75% of the total vertical movement observed during the test occurred.

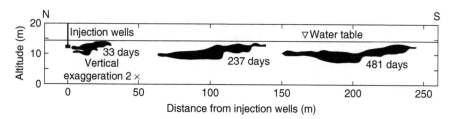

FIGURE 6-33 Vertical location of bromide tracer plume at 33, 237, and 461 days after injection for the Cape Cod tracer test. Plume locations defined by zones in which bromide concentration exceeded 1 mg/L. Line of section approximately along AA' in Figure 6-31. (Adapted from LeBlanc, D.R., Garabedian, S.P., Hess, K.M., Gelhar, L.W., Quadri, R.D., Stollenwerk, K.G. and Wood, W.W., *Water Resour. Res.,* 27, 895–910, 1991.)

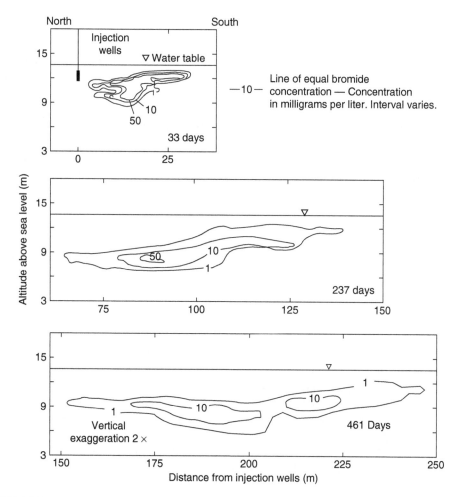

FIGURE 6-34 Vertical distribution of bromide in tracer plume at 33, 237, and 461 days after injection of the Cape Cod tracer test site. Lines of section approximately along AA' in Figure 6-31. (Adapted from LeBlanc, D.R., Garabedian, S.P., Hess, K.M., Gelhar, L.W., Quadri, R.D., Stollenwerk, K.G. and Wood, W.W., *Water Resour. Res.*, 27, 895–910, 1991.)

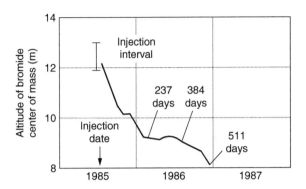

FIGURE 6-35   Vertical position of center of mass of the bromide plume between 1985 and 1987 for the Cape Cod tracer test site. (Adapted from LeBlanc, D.R., Garabedian, S.P., Hess, K.M., Gelhar, L.W., Quadri, R.D., Stollenwerk, K.G. and Wood, W.W., *Water Resour. Res.*, 27, 895–910, 1991.)

Spatial Moments Analysis
Spatial Moment Estimation Methodology
Estimates of the spatial moments were calculated for each sampling round using a numerical integration of the solute distribution (Garabedian et al., 1991). The general form of the spatial moment equation is given by Eq. (6-25). However, Garabedian and others used a slightly modified version of Eq. (6-25) as

$$M_{ijk} = \int_{-\infty}^{\infty}\int_{-\infty}^{\infty}\int_{-\infty}^{\infty} \varphi_e C(x, y, z, t)(x - ux_c)^i(y - uy_c)^j(z - uz_c)^k \, dx \, dy \, dz \qquad (6\text{-}76)$$

where i, j, and k are exponents with values 1, 2 and u is a flag for the central moment, with value 0 or 1. If $u = 0$, Eq. (6-76) becomes exactly the same Eq. (6-25). The mean displacement in the x direction is evaluated as using $u = 0$

$$x_c = \frac{M_{100}}{M_{000}} \qquad (6\text{-}77)$$

which is exactly the same as the first equation of Eq. (6-30). The second central moment in the x direction was evaluated by Garabedian and others as

$$\sigma^2_{xx} = \frac{M_{200}}{M_{000}} \qquad (6\text{-}78)$$

which is the same as Eq. (6-39) for $x_c = 0$.

As the solute concentration was sampled at points rather than continuously in space, Garabedian et al. (1991) used an interpolation approach as part of the numerical integration procedure. As a first step in the procedure, the authors performed integration over the vertical position at each sampler location because the horizontal positions are the same for each set of concentrations in an MLS. The authors assumed that the concentration varies linearly between each pair of adjacent sampling points in the vertical position. The next step was to integrate over the horizontal plane. They assumed a linear interpolation between three sampler locations (i.e., a triangular subregion) of the vertically integrated values and substituted in Eq. (6-76):

$$M_{ijk} = \sum_{p=1}^{N} \iint_{\Omega_p} \varphi_e(A + Bx + Cy)^k(x - ux_c)^i(y - uy_c) \, dx \, dy \qquad (6\text{-}79)$$

where N is the number of triangular subregions in the tracer test domain, A, B, and C are the coefficients defining the vertically integrated concentration distribution, and k is the exponent of the vertical spatial integration. Equation. (6-79) can be solved using a set of equations from finite element analysis. Each of the expressions within the parentheses in Eq. (6-79) is a linear function over the triangular subdomain and can be represented as a sum of linear basis functions. For example,

$$A + Bx + Cy = \sum_{L=1}^{3} I_L \varphi_L(x, y) \tag{6-80}$$

where I_L is the vertically integrated value at point L, and $\varphi_L(x,y)$ is the linear basis function, assuming the value 1 at L and 0 at the other points. A counterclockwise numbering index of the triangle points is assumed for these calculations. Spatial terms are also represented using linear basis functions:

$$x - ux_c = \sum_{L=1}^{3} (x - ux_c) \varphi_L(x, y) \tag{6-81}$$

and

$$y - uy_c = \sum_{L=1}^{3} (y - uy_c) \varphi_L(x, y) \tag{6-82}$$

Substitution of Eqs. (6-80)–(6-82) into Eq. (6-79) gives

$$M_{ijk} = \sum_{p=1}^{N} \iint_{\Omega_p} \varphi_e \left(\sum_{L=1}^{3} I_L \varphi_L \right)_k \left(\sum_{L=1}^{3} (x - ux_c)_L \varphi_L \right)^i \left(\sum_{L=1}^{3} (y - uy_c)_L \varphi_L \right)^j dx \, dy \tag{6-83}$$

A useful formula for the integration of linear basis functions over a triangular area is (Zienkiewicz, 1977)

$$\iint_{\Omega_p} \varphi_1{}^i \varphi_2{}^j \varphi_3{}^k \, dx \, dy = \frac{2A_t i! j! k!}{(i + j + k + 2)!} \tag{6-84}$$

where A_t is the area of the triangular subdomain. Equation (6-83) is solved by expanding the polynomials and sequentially applying Eq. (6-84), then summing the N triangular subregions until the total integral is calculated. Calculation of the second moments for the solute distribution generates a symmetric tensor of solute variance terms, which is given by the matrix on the right-hand side of Eq. (6-52).

Estimation of Mass in Solution
The calculated mass for the 16 bromide sampling rounds are given in Table 6-13 (Garabedian et al., 1991). The values in Table 6-13 indicate that the calculated mass varied from 85 to 105% of the injected mass and their graphical representation is shown in Figure 6-36. The differences are probably due to measurement errors in the bromide concentrations, interpolation of the point values of concentration implicit in the numerical integrations, and the extrapolated data where the concentration distribution was truncated.

The effective porosity ($\varphi_e = 0.39$) used in the mass calculations was obtained by fitting the average calculated mass to the known injected mass. This value is supported by values determined from small-scale tracer tests conducted near the large-scale site (Garabedian et al., 1988) and by porosity measurements reported for sand and gravel (Morris and Johnson, 1967; Perlmutter and Lieber, 1970). It is necessary to point out that effective porosity affects only the mass calculations because it is assumed to be constant in space and time. Therefore, it is canceled from the higher moments calculations (see Eqs. [6-39]–[6-44]).

Center of Mass and Horizontal Velocity
The position of the center of mass for each data is listed in Table 6-13 and the cumulative travel distance with time is shown in Figure 6-37 (Garabedian et al., 1991). The horizontal displacement of the center of mass for bromide followed a nearly constant velocity of 0.42 m/day. The trend displayed by the data in Figure 6-37 is highly linear with a correlation coefficient of 0.997, which

TABLE 6-13
Moment values for Bromide of the Cape Cod Experiment (Length Units in Meters, Angles in Degrees East From South)

Date	Days from Injection	Mass (g)	Center of Mass			Travel Distance	Path Directions		Principal Components of Variance Tensor			Orientation of Cloud Axis	
			x_c^a	y_c^b	z_c^c	s	Θ_i^d	Θ_T^e	$\sigma_{x'x'}^2$	$\sigma_{y'y'}^2$	$\sigma_{z'z'}^2$	Θ_{yx}^f	Θ_{yz}^g
July 1985	13	4440	0.8	−7.4	12.3	7.4	5.9	5.9	1.5	6.5	0.37	18.5	8.0
Aug. 1985	33	5150	2.7	−16.9	11.7	17.1	11.3	8.9	1.8	20.2	0.46	12.8	5.5
Sept. 1985	55	4940	3.0	−25.9	11.1	26.1	2.3	6.6	1.9	34.8	0.50	8.9	2.7
Oct. 1985	83	4990	5.6	−38.9	10.6	39.4	11.4	8.2	2.5	52.4	0.72	9.5	3.5
Nov. 1985	111	4910	8.1	−50.9	10.3	51.7	11.7	9.1	3.1	85.6	0.73	12.5	3.3
Dec. 1985	139	4160	10.2	−64.7	10.4	65.6	8.5	9.0	3.4	118	0.74	11.8	2.4
Jan. 1986	174	4320	11.1	−77.5	9.6	78.7	4.4	8.2	4.3	134	1.03	11.1	0.5
Feb. 1986	203	4850	11.4	−88.8	9.4	90.0	1.5	7.3	3.9	162	1.02	9.3	1.8
March 1986	237	4900	11.9	−100.1	9.3	101.3	2.0	6.8	5.2	189	1.06	7.6	1.4
April 1986	273	4470	13.3	−114.3	9.2	115.3	6.1	6.6	5.2	196	1.05	7.7	1.3
May 1986	315	4810	16.7	−133.9	9.4	135.3	10.0	7.1	6.0	241	1.37	10.4	2.1
July 1986	349	4920	19.0	−147.5	9.4	149.0	9.7	7.3	5.9	227	0.87	11.2	0.8
July 1986[h]	349	5010	19.0	−147.1	9.3	148.7	9.9	7.4	5.8	242	0.89	11.7	0.4
Aug. 1986	384	4930	21.9	−162.4	9.2	164.2	10.8	7.7	7.0	327	0.82	12.9	0.4
Sept. 1986	426	4940	25.4	−180.4	8.9	182.6	11.0	8.0	7.7	334	0.81	14.7	1.1
Oct. 1986	461	5000	28.6	−196.3	8.7	198.8	11.4	8.3	7.1	372	0.98	15.1	1.2
Dec. 1986	511	4950	32.4	−214.0	8.1	216.9	12.3	8.6	10.5	405	1.33	17.6	1.1

$(s = \sum_{i=1}^{n}[(x_{c_i} - x_{c_{i-1}})^2 + (y_{c_i} - y_{c_{i-1}})^2 + (z_{c_i} - z_{c_{i-1}})^2]^{1/2}_c)$ where $x_0 = y_0 = 0.0$, $z_0 = 12.5$.

a Positive is east from magnetic north. Origin is the middle injection well.

b Positive is magnetic north.

c Elevation above sea level.

d $\Theta_i = \tan^{-1}(\Delta x_c / \Delta y_c)$

e $\Theta_T = \tan^{-1}(x_c / y_c)$

f $\Theta_{yx'} = \tan^{-1}(E_{yx}/E_{xx})$, where the eigenvectors are given by the 3×3 matrix whose rows are $[E_{iz}\ E_{ix}\ E_{iy}]$ where $i = x, y, z$, for rows 1, 2, and 3, respectively, and the principal components are $\sigma_{x'x'}^2, \sigma_{y'y'}^2, \sigma_{z'z'}^2$.

g $\Theta_{yz'} = \tan^{-1}(E_{yz}/E_{zz})$, with the eigenvectors and principal components as for $\theta_{yz'}$.

h Moments based on autofluorescence data set.

Source: From Garabedian, S.P., LeBlanc, D.R., Gelhar, L.W. and Celia, M.A., *Water Resour. Res.*, 27, 911–924, 1991. With permission.

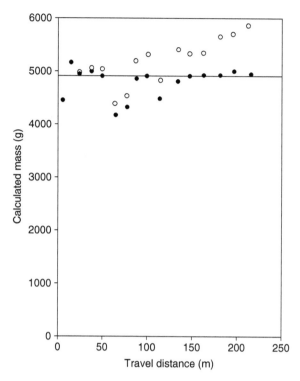

FIGURE 6-36 Calculated bromide mass for each sampling date using corrected (solid circles) and uncorrected (open circles) data for the Cape Cod tracer test. The solid line represents the injected mass (4900 g). (From Garabedian, S.P., LeBlanc, D.R., Gelhar, L.W. and Celia, M.A., *Water Resour. Res.*, 27, 911–924, 1991.)

supports the assumption that a constant velocity over the duration of the test was used in the calculations of dispersivities. This rate of mean movement agrees with the previously estimated velocity of 0.42 m/day, which is calculated using the horizontal hydraulic conductivity, effective porosity, and hydraulic gradient at the site.

The movement of the horizontal location of the center of mass is shown in Figure 6-38. The trajectory closely matches the path predicted from the water-table configuration of August 1985. The strong influence of the water table on the movement of the bromide cloud is also illustrated in Figure 6-39. Figure 6-39 shows the change over time of the direction of displacement of the center of mass of the bromide cloud and the direction of the hydraulic gradient estimated using water level observations at the three wells shown in Figure 6-38. From July 1985 to May 1986, there is a close correspondence between changes in the cloud convective direction and changes in the gradient direction. However, the two directions differ by a consistent amount. The water-table gradient direction averaged about 12° east of magnetic south from July 1985 to May 1986, whereas the direction of the bromide center of mass movement averaged about 7° east of south during the same period. Garabedian et al. (1991) state that it is unlikely that this difference is due to measurement error because sets of gradient directions calculated from other wells in the tracer test area show similar consistent differences between the hydraulic gradient and the bromide cloud displacement. Therefore, the 5° difference between the two directions may reflect a small horizontal anisotropy in the horizontal hydraulic conductivity.

Vertical Movement
The vertical movement of the bromide center of mass is shown in Figure 6-40. The initial rapid downward movement of the bromide plume is the result of density-driven sinking and vertical

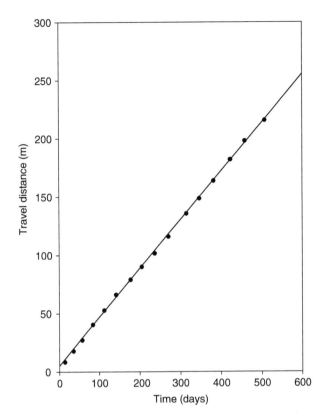

FIGURE 6-37   Distance traveled by the bromide center of mass with time of the Cape Cod tracer test site. The solid line represents the least-squares regression. (Adapted from Garabedian, S.P., LeBlanc, D.R., Gelhar, L.W. and Celia, M.A., *Water Resour. Res.*, 27, 911–924, 1991.)

ground water movement caused by accretion of recharge. Since dilution of the solute concentrations reduced the density contrast over time, the later vertical movement of the tracer plume is entirely due to recharge. This later vertical movement of the plume was at a downward angle below the horizontal equal to about one half the angle at the beginning of the test (Garabedian et al., 1991).

Determination of Variance and Components of the Dispersivity Tensor
In Figure 6-38, the origin of coordinates is the middle injection well. Positive x and x' are east from the magnetic north,: positive y and y' are aligned with the magnetic north. Using the principal components of the variance tensor, the components of the dispersivity tensor can be determined using Eq. (6-51). This is shown in Example 6-7 below.

EXAMPLE 6-7

This example is based on the data presented in Garabedian et al. (1991). Using the components of the covariance tensor in Table 6-13 ($\sigma_{x'x'}^2$, $\sigma_{y'y'}^2$, and $\sigma_{z'z'}^2$), determine (1) the value of longitudinal macrodispersivity (A_{11}) and discuss the result; (2) the value of transverse horizontal macrodispersivity (A_{22}) and discuss the result; (3) the value of transverse vertical macrodispersivity (A_{33}) based on Eq. (6-51) and discuss the result; and (4) compare the results with the ones determined from the Gelhar and Axness theory.

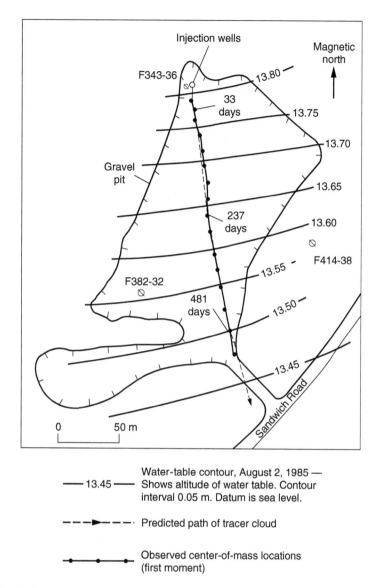

FIGURE 6-38   Trajectory of the bromide center of mass with time for the Cape Cod tracer test site. (Adapted from Garabedian, S.P., LeBlanc, D.R., Gelhar, L.W. and Celia, M.A., *Water Resour. Res.*, 27, 911–924, 1991.)

SOLUTION

(1) In Figure 6-41, the origin is in the middle of the injection well. Positive x and x' are east from the magnetic north, Positive y and y' are aligned with the magnetic north. Using the corresponding data from Table 6-13 (columns 7 and 11), the change in longitudinal variance ($\sigma_{y'y'}^2$) with travel distance of the center of mass is shown in Figure 6-41. The solid line in Figure 6-41 represents the least-squares regression line (using the columns 7 and 11in Table 6-13)

$$\sigma_{y'y'}^2 = -14.3 + 1.91s \qquad\qquad\text{(E6-7-1)}$$

where s is the travel distance. The strong correlation (0.994) between the longitudinal variance ($\sigma_{y'y'}^2$) and travel distance, s, indicates that this trend is linear. By substituting Eq. (E6-7-1) into

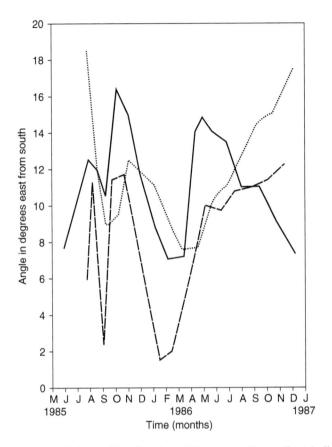

FIGURE 6-39 Comparison with time of the direction of the water-table gradient (solid line), direction of movement of the bromide center of mass (dashed line, Θ_i' Table 6-13), and orientation of the bromide plume dotted line, Θ_{yx} ' Table 6-13) of the Cape Cod tracer test site. (Adapted from Garabedian, S.P., LeBlanc, D.R., Gelhar, L.W. and Celia, M.A., *Water Resour. Res.*, 27, 911–924, 1991.)

Eq. (6-51), the longitudinal macrodispersivity (A_{11}) can be written as

$$A_{11} = \frac{1}{2}\frac{d\sigma^2_{y'y'}}{ds} = \frac{1}{2}\frac{d}{ds}(-14.3 + 1.91s) = \frac{1}{2}(1.91\ m) = 0.955\ m \qquad (E6\text{-}7\text{-}2)$$

Therefore, the resultant longitudinal macrodispersivity is 0.96 m. The linear trend at the scale of this test supports the theoretical result that longitudinal dispersivity in aquifers can reach a constant value (e.g., Gelhar et al., 1979).

A closer inspection shows a nonlinear trend in the longitudinal variance ($\sigma^2_{y'y'}$) during the first 26 m of distance traveled by the center of mass (Figure 6-42). Figure 6-42 is an expansion of the early period of change in longitudinal variance shown in Figure 6-41. The negative intercept for the overall linear trend (Figure 6-41) indicates that there must have been an initial slow increase in variance. This early time nonlinear change in longitudinal variance generates an increase in the apparent longitudinal dispersivity when it is calculated from date to date.

(2) Using the corresponding data from Table 6-13 (columns 7 and 10), the change in the transverse horizontal variance ($\sigma^2_{x'x'}$) with travel distance of the center of mass is shown in Figure 6-43 (the correlation coefficient is 0.974). The solid line in Figure 6-43 represents the least-squares regression line (using the columns 7 and 10 in Table 6-13)

$$\sigma^2_{x'x'} = 1.08 + 0.0364s \qquad (E6\text{-}7\text{-}3)$$

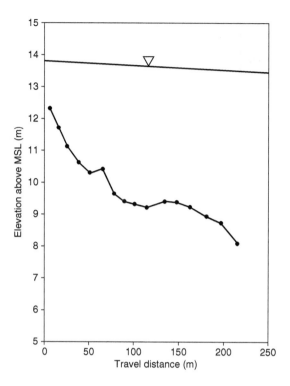

FIGURE 6-40 Elevation of bromide center of mass (solid circles) with travel distance. (Adapted from Garabedian, S.P., LeBlanc, D.R., Gelhar, L.W. and Celia, M.A., *Water Resour. Res.*, 27, 911–924, 1991.)

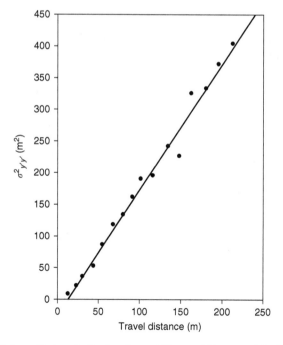

FIGURE 6-41 Change in bromide longitudinal variance with travel distance of the Cape Cod Tracer test. The solid line represents the least squares regression. (Adapted from Garabedian, S.P., LeBlanc, D.R., Gelhar, L.W. and Celia, M.A., *Water Resour. Res.*, 27, 911–924, 1991.)

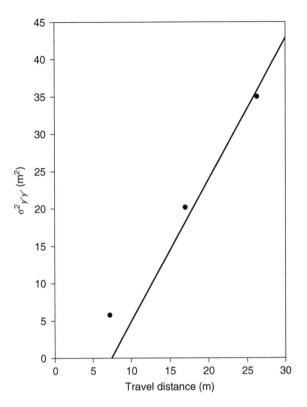

FIGURE 6-42 Early time change in bromide longitudinal variance with travel distance for the Cape Cod tracer test distance. (Adapted from Garabedian, S.P., LeBlanc, D.R., Gelhar, L.W. and Celia, M.A., *Water Resour. Res.*, 27, 911–924, 1991.)

where s is the travel distance. By substituting of Eq. (E6-7-2) into Eq. (6-51), the transverse horizontal macrodispersivity (A_{22}) can be written as

$$A_{22} = \frac{1}{2}\frac{d\sigma^2_{x'x'}}{ds} = \frac{1}{2}\frac{d}{ds}(1.08 + 0.0364s) = \frac{1}{2}(0.0364\ m) = 0.0182\ m \qquad \text{(E6-7-4)}$$

Therefore, the resultant transverse horizontal macrodispersivity is 0.0182 m. This value is much smaller than the longitudinal macrodispersivity; the ratio of $A_{11}/A_{22} = (0.955\ m)/(0.0182\ m) = 52.5$. As can be seen from Figure 6-43, there is more scatter about the linear trend than in the longitudinal case.

(3) Using the corresponding data from Table 6-13 (columns 7 and 12), the change in transverse vertical variance ($\sigma^2_{z'z'}$) with travel distance of the center of mass is shown in Figure 6-44. Although a linear regression line was fitted to these data, the correlation coefficient is relatively low (0.705), and it is not clear that the variances follow a single trend. From the least-squares regression, one gets

$$\sigma^2_{z'z'} = 0.559 + 0.003s \qquad \text{(E6-7-5)}$$

where s is the travel distance. By substituting Eq. (E6-7-5) into Eq. (6-51), the transverse vertical macrodispersivity (A_{33}) can be written as

$$A_{33} = \frac{1}{2}\frac{d\sigma^2_{z'z'}}{ds} = \frac{1}{2}\frac{d}{ds}(0.559 + 0.003s) = \frac{1}{2}(0.003\ m) = 0.0015\ m = 1.5\ mm \qquad \text{(E6-7-6)}$$

Therefore, the resultant transverse vertical macrodispersivity is 1.5 mm.

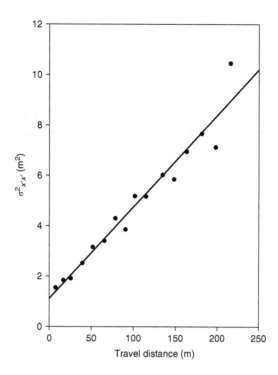

FIGURE 6-43   Change in bromide tranverse horizontal variance with travel distance for the Cape Cod tracer test. The solid line represents the least squares regression. (Adapted from Garabedian, S.P., LeBlanc, D.R., Gelhar, L.W. and Celia, M.A., *Water Resour. Res.*, 27, 911–924, 1991.)

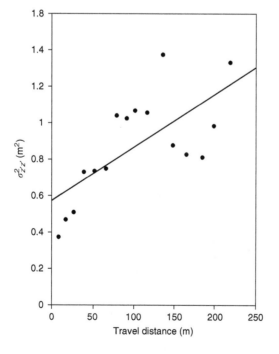

FIGURE 6-44   Change in bromide transverse vertical variance with travel distance of the Cape Cod Tracer test. The solid line represents the least squares regression. (Adapted from Garabedian, S.P., LeBlanc, D.R., Gelhar, L.W. and Celia, M.A., *Water Res. Res.*, 27, 911–924, 1991.)

(4) The results based on the Gelhar and Axness theory are presented in Example 6-14 (Section 6.4.6.11). From Example 6-14 the results are $A_{11} = 0.664$ m, $A_{22} = 5.04 \times 10^{-8}$ m, and $A_{33} = 7.35 \times 10^{-8}$ m. The results show that the A_{22} values compare well; however, the other ones differ significantly.

6.4 DETERMINATION OF EXPRESSIONS FOR DISPERSIVITIES FROM PERTURBATION ANALYSIS USING THE FOURIER–STIELTJES INTEGRAL REPRESENTATION

6.4.1 DEFINITION OF KEY CONCEPT AND QUANTITIES

In order to describe better the recent theories and their applications for the evaluation and determination of dispersion coefficients with stochastic modeling approaches, it is necessary to have a clear understanding of some key definitions regarding stochastic approaches. These are presented in the following sections.

6.4.1.1 Realization and Ensemble

Consider the function $X(t)$, which varies with argument t in an irregular fashion. This is in a general sense specifically it may be the variation of concentration in a ground water flow field with time, $C(t)$, or the variation of the water level in an observation well, $H(t)$ (Figure 6-45). In a mathematical sense, the stochastic process $X(t)$ is defined as being a random variable for any assigned time t. Consider the water level variation in an observation well. If there is only one observation of the time variation for $H(t)$, $H(t)$ or $X(t)$, it is called a *realization* (Figure 6-45a) of the stochastic process. If there is more than one realization, the collection is called the an *ensemble* (Figure 6-45b). Figure 6-45 illustrates the concept of the realization and ensemble.

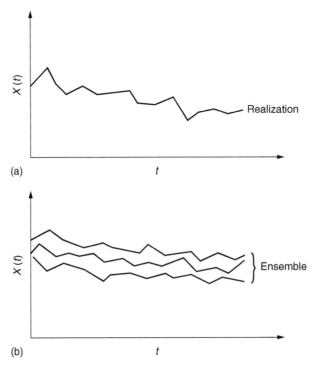

FIGURE 6-45 Schematic description of realization and ensemble: (a) A stochastic process realization and (b) The ensemble with the embedded realization.

6.4.1.2 Expected Value

An important concept in probability and statistics is that of the *expected value*, *mathematical expectation*, or briefly, *expectation* of a random variable. For a discrete random variable X having the possible values $x_1, x_2, ..., x_n$ the expectation of X is defined as (e.g., Spiegel, 1975)

$$E[X] = x_1 P(X = x_1) + x_2 P(X = x_2) + \cdots + x_n P(X = x_n) = \sum_{j=1}^{n} x_j P(X = x_j) \qquad (6\text{-}85)$$

where P stands for "probability." Equivalently, if $P(X = x_j) = f(x_j)$, Eq. (6-85) takes the form

$$E[X] = x_1 f(x_1) + x_2 f(x_2) + \cdots + x_n f(x_n) = \sum_{j=1}^{n} x_j f(x_j) = \sum x f(x) \qquad (6\text{-}86)$$

As a special case of Eq. (6-86) for which the probabilities are all equal, one can write

$$E[X] = \frac{x_1 + x_2 + \cdots + x_n}{n} \qquad (6\text{-}87)$$

which is called the *arithmetic mean* or simply the *mean* of $x_1, x_2, ..., x_n$. For a continuous variable X having probability density function $f(x)$, the expected value of X is defined as

$$\mu = E[X] = \int_{-\infty}^{\infty} x f(x)\, dx \qquad (6\text{-}88)$$

where the analogy with Eq. (6-86) is to be noted.
The simple definition of the expected value based on Eq. (6-88) can be extended to a random variable $X(t)$ having a *probability density function $f(x,t)$* as

$$\mu(t) = E[X(t)] = \int_{-\infty}^{\infty} x f(x, t)\, dx \qquad (6\text{-}89)$$

and the *variance* is defined as

$$\sigma^2(t) = E\langle [X(t)] - \mu(t)]^2 \rangle = \int_{-\infty}^{\infty} [x - \mu(t)]^2 f(x, t)\, dx \qquad (6\text{-}90)$$

In general, both the mean and variance of a stochastic process may vary with time. Also, a stochastic process actually involves a large number of random variables associated with each time point in a continuous time domain. This explicitly means that a complete probabilistic description of a stochastic process must take into account the interrelationship between the random variables corresponding to the stochastic process at different times. Therefore, the probability density function for x at any number of points in time would be required for a complete probabilistic description. Such a description is simply not practical, because for ground water applications, enough data are not available to determine the underlying joint distributions. Therefore, a simpler approach based on moments of first and second order is used to characterize the stochastic process.

6.4.1.3 Variance of Joint Distributions and Covariance

The equations given in Section 6.4.1.2 for one variable can be extended to two or more variables. Thus, if X and Y are two continuous random variables having joint density function $F(x, y)$, the means or expectations of X and Y are (e.g., Spiegel, 1975)

$$\mu_X = E[X] = \int_{-\infty}^{\infty} \int_{-\infty}^{\infty} x f(x, y)\, dx\, dy \qquad (6\text{-}91)$$

and

$$\mu_Y = E[Y] = \int_{-\infty}^{\infty} \int_{-\infty}^{\infty} y f(x, y)\, dx\, dy \qquad (6\text{-}92)$$

The variances are

$$\sigma_X^2 = E[(X - \mu_X)^2] = \int_{-\infty}^{\infty} \int_{-\infty}^{\infty} (x - \mu_X^2) f(x, y) \, dx \, dy \tag{6-93}$$

and

$$\sigma_Y^2 = E[(Y - \mu_Y)^2] = \int_{-\infty}^{\infty} \int_{-\infty}^{\infty} (y - \mu_Y^2) f(x, y) \, dx \, dy \tag{6-94}$$

Another quantity that arises in the case of two variables X and Y is the *covariance*, defined as

$$\sigma_{XY} = \text{cov}[X, Y] = E[(X - \mu_x)(Y - \mu_y)] \tag{6-95}$$

or in terms of the joint density function

$$\sigma_{XY} = \text{cov}[X, Y] = \int_{-\infty}^{\infty} \int_{-\infty}^{\infty} (x - \mu_X)(y - \mu_Y) f(x,y) \, dx \, dy \tag{6-96}$$

If $X = X(t_1)$ and $Y = X(t_2)$, Eq. (6-95) takes the form

$$\text{cov}[X(t_1), X(t_2)] = E\langle [X(t_1) - \mu(t_1)][X(t_2) - \mu(t_2)] \rangle = R(t_1, t_2) \tag{6-97}$$

which means that the covariance function R will be dependent on t_1 and t_2. Note that when $t_1 = t_2$, Eq. (6-97) turns to be the variance as given by Eq. (6-90), which may be a function of time as well.

6.4.1.4 Stationary Process and a Common Covariance Function

If the probabilistic description of a process is independent of time origin, that process is called a *stationary process*. This means that joint probability function distributions will be invariant under shifts of the time origin. Since only first- and second-moment properties are involved in ground water studies, it is sufficient to consider only the *second-order stationary process*, which has the following properties (Gelhar, 1993):

$$E[X(t)] = \mu = \text{constant} \tag{6-98}$$

and

$$R(t_1, t_2) = R(t_1 - t_2) = R(\tau), \quad \tau = t_1 - t_2 \quad \text{for any } t_1, t_2 \tag{6-99}$$

If $t_1 = t_2$ ($\tau = 0$), Eq. (6-99) gives the variance of the process and it will be independent of time. It should be noted that the condition that the mean be constant in Eq. (6-98) is easily relaxed. If the mean is known, one can simply subtract that known mean from X and apply the second-order stationary condition to the resulting zero-mean process. In all of the subsequent analyses, the term "stationary" will mean second-order stationary (Gelhar, 1993).

Figure 6-46 shows a single realization of a stationary process. According to Eqs. (6-97) and (6-99), the covariance function can be evaluated by taking the average of the lagged product of the departure of X from its constant mean, and carrying out this process for all possible time lags. If there is only a single realization, the covariance can be estimated from the time average of the lagged product, again carried out for all possible τ. This time-averaging approach is not sensible for a nonstationary process for which the variance may change with time. Physically, the covariance function indicates the degree of correlation between the process at adjacent points in time. Note from the definition of the covariance function given by Eq. (6-97) that the covariance is an even function of the time lag, i.e., (Gelhar, 1993)

$$R(\tau) = R(-\tau) \tag{6-100}$$

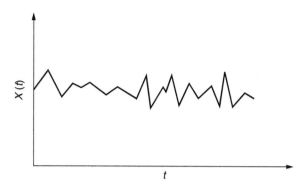

FIGURE 6-46   A realization of a stationary stochastic process.

Usually, the expression for a common *covariance function* will be given. Some investigators (Freeze et al., 1990) use the term *autocorrelation function* instead of covariance function. The covariance function may take a number of forms. One of the most commonly used functions is the exponential decaying covariance function. The expression of this function is (Freeze et al., 1990; Gelhar, 1993)

$$R(\tau) = \sigma^2 \exp\left(-\frac{|\tau|}{\lambda}\right) \tag{6-101}$$

where σ^2 is the variance and λ is the *integral scale*. Some investigators (Freeze et al., 1990) call it *correlation length*, which is a measure of the distance over which $X(t)$ is correlated. Equation (6-101) is shown graphically in Figure 6-47, and λ is a measure of the distance over which $R(\tau)$ decays to a value of e^{-1}. Values separated by short distances will be highly correlated, and those separated by long distances will only be weakly correlated or not correlated at all. Equation (6-101) is the one-dimensional form of the three-dimensional autocovariance function. Application of Eq. (6-101) is shown in Section 6.3.6.2.

6.4.2   ONE-DIMENSIONAL STATIONARY RANDOM PROCESS REPRESENTATION

Let us consider, for the moment, that the quantity under question varies over time, and is given by $X(t)$. Then at each individual time point t, $X(t)$ is a *random variable*, and the complete form of $X(t)$, as t varies over all possible values, is called a *random process*. In essence, a random process is simply a *random function*, but some authors prefer this more descriptive term. All three terms, random process, stochastic process, and random function, may be regarded as completely synonymous (Priestley, 1981).

   If a process is stationary, it is virtually always possible to describe the process in terms of a kind of Fourier representation, which is analogous to Fourier series and Fourier integrals for deterministic functions. Yet random functions are neither periodic nor integrable, so neither Fourier series nor Fourier integrals may be used in the ordinary sense. However, under a set of remarkably weak assumptions, a Fourier representation does exist. The corresponding representation for stationary random processes is called a *Fourier–Stieltjes integral* (Lumley and Panofsky, 1964; Gelhar, 1993). Representation of stationary random processes are presented in the following sections.

6.4.2.1   Spectral Representation Theorem

Consider a zero-mean stationary process represented by $X(t)$, which is a positive, definite, and measurable function from $-\infty$ to $+\infty$. Then, there exists a monotone increasing, real-valued bounded stochastic function $Z(\omega)$ such that

$$X(t) = \int_{-\infty}^{\infty} e^{i\omega t} dZ(\omega) \tag{6-102}$$

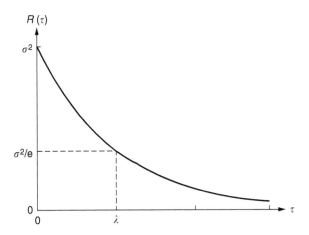

FIGURE 6-47 The negative exponential covariance function from Eq. (6-101).

where $i = (-1)^{1/2}$. Equation (6-102) is the *spectral representation theorem*. If $Z(\omega)$ is nondecreasing and bounded and $X(t)$ is defined as above, then $X(t)$ is called the *stochastic Fourier–Stieltjes integral* of $Z(\omega)$, and is both continuous and positive-definite. $X(t)$ is also called the *spectral representation* (Lumley and Panofsky, 1964; Gelhar, 1993). The differences $dZ(\omega)$ may be thought of as the complex amplitudes of the Fourier modes of frequency ω. In Eq. (6-102), $Z(\omega)$ is a stochastic process having the properties that

$$E[dZ(\omega)] = 0 \qquad (6\text{-}103)$$

and

$$E[dZ(\omega_1)\, dZ^*(\omega_2)] = 0, \quad \omega_1 \neq \omega_2 \qquad (6\text{-}104)$$

in which dZ^* is the complex conjugate and ω is the frequency. Equation (6-103) states that $X(t)$ has zero mean; Equation (6-104) indicates that the increments of Z at two different frequencies are uncorrelated, which means that the process Z has orthogonal increments. When $\omega_1 = \omega_2 = \omega$, then

$$E[dZ(\omega)\, dZ^*(\omega)] = d\Phi(\omega) = S(\omega)\, d\omega, \quad \omega_1 = \omega_2 = \omega \qquad (6\text{-}105)$$

where $\Phi(\omega)$ is the *integrated spectrum* and $S(\omega)$ the *spectral density function*. The integrated spectrum $\Phi(\omega)$ is usually differentiable; however, it contains discrete components at distinct frequencies. One should note that $X(t)$ in Eq. (6-102) is real, and that the term $e^{i\omega t}$ is complex; therefore, Z must be complex as well. A number of rigorous mathematical proofs of the spectral representation theorem, as given by Eq. (6-102), have been developed, and several of these are summarized by Priestley (1981).

6.4.2.2 Covariance Function for A Zero-Mean Stochastic Process $X(t)$

From Eqs. (6-97) and (6-102), the covariance function can be written as

$$R(\tau) = E[X(t + \tau)X^*(t)] = E\left[\int_{-\infty}^{\infty} e^{i\omega(t+\tau)}\, dZ(\omega) \int_{-\infty}^{\infty} e^{-i\omega' t}\, dZ^*(\omega')\right]$$

$$= \int_{-\infty}^{\infty}\int_{-\infty}^{\infty} e^{i[(\omega-\omega')t+\omega\tau]} E[\, dZ(\omega)\, dZ^*(\omega')] \qquad (6\text{-}106)$$

If $X(t)$ is a stationary process, according to Eq. (6-99), its covariance function must be independent of t. Therefore, the term t in the second line of Eq. (6-106) drops out and using Eq. (6-105),

Eq. (6-106) takes the form

$$R(\tau) = \int_{-\infty}^{\infty} S(\omega) e^{i\omega\tau} \, d\omega \qquad (6\text{-}107)$$

which shows that the covariance function can be written as the inverse Fourier transform of the spectral density function $S(\omega)$ (e.g., Spiegel, 1974, Eq. [7] p. 81). The corresponding inverse Fourier transform of $R(\tau)$ is then (e.g., Spiegel, 1974, Eq. [8] p. 81,)

$$S(\omega) = \frac{1}{2\pi} \int_{-\infty}^{\infty} e^{-i\omega\tau} R(\tau) \, d\tau \qquad (6\text{-}108)$$

Equations (6-107) and (6-108) are the classical results for stationary stochastic processes.

6.4.2.3   Important Features of the Covariance and Spectral Density Function Relationships

There are several important features of the covariance and spectral density function relationships and are as follows (Gelhar, 1993):

1. The spectral density function $S(\omega)$ must be nonnegative at all frequencies because, according to Eq. (6-105), it represents a squared amplitude.
2. With $\tau = 0$, Eq. (6-107) takes the form

$$R(0) = \sigma^2 = \int_{-\infty}^{\infty} S(\omega) \, d\omega \qquad (6\text{-}109)$$

which means that the spectral density function represents a distribution of variance over the frequency.
3. The spectral density function is an even function of ω, i.e.,

$$S(\omega) = S(-\omega) \qquad (6\text{-}110)$$

As a result, it is customary to show only the positive arguments of $R(\tau)$ or $S(\omega)$ in graphical representations.
4. A sufficient condition for the existence of the spectral density function is that $R(\tau)$ be absolutely integrable, i.e.,

$$\int_{-\infty}^{\infty} |R(\tau)| \, d\tau < \infty \qquad (6\text{-}111)$$

which is the usual condition for the existence of the Fourier transform in Eq. (6-108).
5. In the spectral representation theorem given by Eq. (6-102), $X(t)$ is a process that varies with time. When referring to a record of the process, what is meant is a graph of its values plotted against time as the variable. However, the variation of some quantity (hydraulic conductivity, concentration, and other quantities) over a region of space can also be studied. This phenomenon is usually referred to as *spatial processes*.

6.4.3   THREE-DIMENSIONAL STATIONARY RANDOM PROCESS REPRESENTATION

In Section 6.4.2.1, the spectral representation theorem for a zero-mean stationary process, $X(t)$, is presented for one-dimensional stochastic processes. If there is a one-dimensional variation, this approach can be applied directly to a spatial process with only a change in notation. For example, the hydraulic characteristics of aquifer materials generally vary significantly from location to location. Therefore, a more realistic description of spatial variability in aquifers requires a more fully three-dimensional representation (Gelhar, 1993).

Let us consider a multidimensional stochastic process, $Y(\mathbf{x})$, where $\mathbf{x}$ is a vector with components x_i ($i = 1, 2, ..., n$, where n can be 1, 2, or 3 coordinates in space). Y may be scalar, a vector, or a tensor quantity in general. Assuming that $Y(\mathbf{x})$ is a zero-mean random field, its covariance function, from Eq. (6-97), can be written as

$$R_{yy}(\mathbf{x}, \mathbf{s}) = E[Y(\mathbf{x} + \mathbf{s})Y(\mathbf{x})] \tag{6-112}$$

where $\mathbf{x}$ is the coordinates vector (its components are x_1, x_2, and x_3 or x, y, and z) and $\mathbf{s}$ is the separation vector (its components are s_1, s_2, and s_3). The covariance function is generally a function of both $\mathbf{x}$ and $\mathbf{s}$. However, if the variability of the phenomenon is independent of the location in space, then the covariance function will only depend on the separation vector between the locations. This phenomenon is called *statistical homogeneity*, which is directly analogous to the stationarity of time-dependent stochastic processes as described by Eq. (6-99). Statistically nonhomogeneous cases are presented in Section 6.4.6.5.

The spectral representation of a three-dimensional process similar to that of a one-dimensional process as given by Eq. (6-102) can be expressed as

$$Y(\mathbf{x}) = \int_{-\infty}^{\infty} e^{i\mathbf{k}.\mathbf{x}} \, dZ_y(\mathbf{k}) \tag{6-113}$$

where

$$\mathbf{k} = (k_1, k_2, ..., k_n) \tag{6-114}$$

is the wave number vector and $Z_y(\mathbf{k})$ is a random function with orthogonal increments. The spectral representation theorem, as presented in Section 6.4.2.1, will also be applied here. In Eq. (6-113), $Z_y(\mathbf{k})$ is a stochastic process having the properties

$$E[dZ_y(\mathbf{k})] = 0 \tag{6-115}$$

and

$$E[dZ_y(\mathbf{k}) \, dZ_y^*(\mathbf{k}')] = 0, \quad \mathbf{k} \neq \mathbf{k}' \tag{6-116}$$

in which dZ_y^* is the complex conjugate and $\mathbf{k}$ is the wave number. Equation (6-116) states that $Y(\mathbf{x})$ has zero mean; Eq. (6-116) indicates that the increments of Z_y at two different frequencies are uncorrelated, which means that the process Z_y has orthogonal increments. When $\mathbf{k} = \mathbf{k}'$, then

$$E[dZ_y(\mathbf{k}) \, dZ_y^*(\mathbf{k})] = d\Phi(\mathbf{k}) = S_{yy}(\mathbf{k}) \, d\mathbf{k} \tag{6-117}$$

where $(\mathbf{k})$ is the integrated spectrum and $S(\mathbf{k})$ is the spectral density function. As in the case for Eqs. (6-107) and (6-108), the corresponding Fourier transform components are

$$R_{yy}(\mathbf{s}) = \int_{-\infty}^{\infty} e^{i\mathbf{k}.\mathbf{s}} S_{yy}(\mathbf{k}) \, d\mathbf{k} \tag{6-118}$$

and

$$S_{yy}(\mathbf{k}) = \frac{1}{(2\pi)^n} \int_{-\infty}^{\infty} e^{-i\mathbf{k}.\mathbf{s}} R_{yy}(\mathbf{s}) \, d\mathbf{s} \tag{6-119}$$

The integrals in Eqs. (6-118) and (6-119) are n-dimensional over separation or wave number space. If $n = 1$, Eqs. (6-118) and (6-119) reduce to Eqs. (6-107) and (6-108), respectively, with $s_1 = \tau$ and $\mathbf{k}_1 = \omega$.

6.4.4 STOCHASTIC ANALYSIS OF ONE-DIMENSIONAL STEADY-STATE FLOW

6.4.4.1 Perturbed Forms of the Flow Equation

From Eq. (4-43), the equation for one-dimensional steady-state saturated flow in the x direction through a homogeneous porous medium is

$$q = -K(x)\frac{dh}{dx} \tag{6-120}$$

where h is the hydraulic head, $K(x)$ is the hydraulic conductivity at any point x, and q the specific discharge, which is a constant. Equation. (6-120) can also be expressed as

$$\frac{dh}{dx} = -qW(x) \tag{6-121}$$

where $W(x)$ is the *hydraulic resistivity* (Bear, 1972, p. 140) and is defined as

$$W(x) = \frac{1}{K(x)} \tag{6-122}$$

The form given by Eq. (6-121) has the advantage of treating large fluctuations in hydraulic conductivity more exactly. Therefore, a solution for the problem of one-dimensional flow in the form of Eq. (6-121) is called "the exact solution" and will serve as a basis of comparison for the appropriate solutions presented in the following paragraphs (Bakr, 1976; Bakr et al., 1978; Gutjahr et al., 1978).

Combining Eq. (6-120) with the continuity equation (e.g., Eq. [4-90] under steady-state conditions)

$$\frac{dq}{dx} = 0 \tag{6-123}$$

gives the differential form of the flow equation:

$$\frac{d}{dx}\left[K(x)\frac{dh}{dx}\right] = 0 \tag{6-124}$$

Expressing the Variables in Terms of Mean and Perturbation
Expressing the variables in Eq. (6-121) in terms of a mean and perturbation gives

$$h(x) = \overline{h}(x) + h'(x), \qquad E[h'(x)] = 0 \tag{6-125}$$

and

$$W(x) = \overline{W}(x) + W'(x), \qquad E[W'(x)] = 0 \tag{6-126}$$

In these equations, the bar indicates the mean of these quantities and the prime indicates perturbation or fluctuation about the mean; E indicates the expected value. The expected values are defined according to Eq. (6-87) and the arithmetic mean of the functions are zero. Taking the expected value of Eq. (6-121) gives

$$E\left[\frac{dh(x)}{dx}\right] = -E[qW(x)] \tag{6-127}$$

The substitution of Eqs. (6-125) and (6-126) into Eq. (6-127) gives

$$E\left[\frac{d\overline{h}(x)}{dx}\right] + E\left[\frac{dh'(x)}{dx}\right] = -qE[\overline{W}(x)] - qE[W'(x)] \tag{6-128}$$

The second expression on the left-hand side of Eq. (6-128) is zero because of the second expression of Eq. (6-125). Similarly, the second expression on the right-hand side of Eq. (6-128) is zero because of the second expression of Eq. (6-126). Therefore, Eq. (6-128) takes the form for the mean

$$E\left[\frac{d\bar{h}(x)}{dx}\right] = -qE[\bar{W}(x)] \tag{6-129}$$

or

$$\frac{d\bar{h}(x)}{dx} = -q\bar{W}(x) \tag{6-130}$$

Equation (6-128) can also be expressed without E as

$$\frac{d\bar{h}(x)}{dx} + \frac{dh'(x)}{dx} = -q\bar{W}(x) - qW'(x) \tag{6-131}$$

Subtracting Eq. (6-130) from Eq. (6-131) gives

$$\frac{dh'(x)}{dx} = -qW'(x) \tag{6-132}$$

Equation (6-132) is the differential equation relating fluctuations in hydraulic conductivity and hydraulic head.

Spectral Representation
Similar to Eq. (6-102), consider the case of statistically homogeneous input perturbation represented by Fourier–Stieltjes integrals in wave number k for hydraulic resistivity, as follows

$$W'(x) = \int_{-\infty}^{\infty} e^{ikx}\, dZ'_W(k) \tag{6-133}$$

Similarly, the hydraulic head perturbation is represented by

$$h'(x) = \int_{-\infty}^{\infty} e^{ikx}\, dZ'_h(k) \tag{6-134}$$

The processes discussed in Section 6.4.2.1 are all examples of a general type called *time series*, i.e., the process varries with time. For example, Eq. (6-102) is a representation of time series. However, the variation of some quantity over a region of space may equally be studied. Such phenomena are usually referred to as spatial processes. Therefore, Eqs. (6-133) and (6-134) represent spatial processes.

By substituting Eqs. (6-133) and (6-134) into Eq. (6-132) one has

$$\frac{d}{dx}\int_{-\infty}^{\infty} e^{ikx}\, dZ_{h'}(k) = -q\int_{-\infty}^{\infty} e^{ikx}\, dZ_{w'}(k) \tag{6-135}$$

which reduces to

$$\int_{-\infty}^{\infty}[ik\, dZ_{h'}(k) + q\, dZ_{w'}(k)]e^{ikx} = 0 \tag{6-136}$$

For this expression to be true, the quantity in brackets must be equal to zero identically for every value of x. Hence,

$$ik\, dZ_{h'}(k) + q\, dZ_{w'}(k) = 0 \tag{6-137}$$

from which one obtains the following relationship between the complex Fourier amplitudes of hydraulic head and resistivity:

$$dZ_{h'}(k) = -\frac{q}{ik}\, dZ_{w'}(k) \tag{6-138}$$

According to the spectral representation theorem as presented in Section 6.4.2.1, the complex conjugate ($z^* = x - iy$ is the complex conjugate of $z = x + iy$ by definition; therefore, $zz^* = x^2 + y^2$) is

$$dZ_{h'}^*(k) = \frac{q}{ik} \, dZ_{w'}^*(k) \tag{6-139}$$

Multiplying both sides of Eqs. (6-138) and (6-139) and taking their expected values result in the following expression:

$$E[dZ_{h'}(k) \, dZ_{h'}^*(k)] = -\frac{q^2}{i^2 k^2} E[dZ_{w'}(k) \, dZ_{w'}^*(k)] \tag{6-140}$$

Applying the spectral representation theorem in Section 6.4.2.1 (Eq. [6-105]) and using the equality of $i^2 = -1$, Eq. (6-140) takes the form

$$S_{h'h'}(k) \, dk = \frac{q^2}{k^2} S_{w'w'}(k) \, dk \tag{6-141}$$

or

$$S_{h'h'}(k) = \frac{q^2}{k^2} S_{w'w'}(k) \tag{6-142}$$

Equation (6-142) is singular at the origin ($k = 0$) and will produce an infinite hydraulic head variance unless $S_{h'h'}$ is proportional to k^n ($n > 1$) for small k (Bakr, 1976).

A Form of the Resistivity Autocovariance
A form of the resistivity autocovariance function whose inverse Fourier transform gives rise to a spectrum that is proportional to k^2 is (Erdelyi, 1954, p. 9)

$$R_{w'w'}(\zeta) = \overline{W'^2} \left(1 - \frac{|\xi|}{\lambda} \right) e^{-|\xi|/\lambda} \tag{6-143}$$

where

$$\overline{W'^2} = \sigma_W'^2 \tag{6-144}$$

is the variance, ζ the separation or lag, and λ the correlation length as defined in Section 6.3.6.2. In addition to removing the singularity of Eq. (6-143) at the origin, the form of Eq. (6-143) for $R_{w'w'}$ is a physically realistic way of characterizing correlations between neighboring values of a property in a given medium (Bakr, 1976).

Determination of the Spectrum $S_{h'h'}$
The spectrum $S_{w'w'}$ obtained by taking the inverse Fourier transform defined by Eq. (6-108), gives

$$S_{w'w'} = \frac{1}{2\pi} \int_{-\infty}^{\infty} e^{-ik\zeta} R_{w'w'}(\zeta) \, d\zeta \tag{6-145}$$

Using Eq. (6-143), Eq. (6-145) can also be expressed as

$$S_{w'w'} = \frac{1}{2\pi} \sigma_W'^2 (I_1 + I_2) \tag{6-146}$$

where

$$I_1 = \int_0^{\infty} e^{-ik\zeta}(1 - a\zeta) e^{-a\zeta} \, d\zeta \tag{6-147}$$

and

$$I_2 = \int_{-\infty}^{0} e^{-ik\zeta}(1 + a\zeta) e^{a\zeta} \, d\zeta \tag{6-148}$$

Performing the integrals in Eqs. (6-147) and (6-148) and substituting them into Eq. (6-146) gives

$$S_{w'w'} = \frac{2k^2\lambda^3\sigma_{w'}^2}{\pi(1 + k^2\lambda^2)^2}$$
(6-149)

The spectrum of $S_{h'h'}$ of hydraulic head fluctuations is then found by substituting Eq. (6-149) into Eq. (6-142) as

$$S_{h'h'}(k) = \frac{2q^2\lambda^3\sigma_{w'}^2}{\pi(1 + \lambda^2k^2)^2}$$
(6-150)

Determination of the Autocovariance Function for the Hydraulic Head ($R_{h'h'}$)
From Eq. (6-107)

$$R_{h'h'}(\zeta) = \int_{-\infty}^{\infty} e^{ik\zeta} S_{h'h'}(k)\, dk$$
(6-151)

and using Eq. (6-150) one gets (Bakr, 1976; Bakr et al., 1978)

$$R_{h'h'}(\zeta) = q^2\lambda^2\sigma'^2_w\left(1 + \frac{\zeta}{\lambda}\right)e^{-\zeta/\lambda} \quad \text{if } \zeta > 0$$
(6-152)

According to Eq. (6-109), $\zeta = 0$ in Eq. (6-151) corresponds to the variance

$$R_{h'h'}(0) = \sigma'^2_h = \int_{-\infty}^{\infty} S_{h'h'}(k)\, dk = q^2\lambda^2\sigma_w^{'2}$$
(6-153)

On dividing Eq. (6-152) by Eq. (6-153) one gets the autocorrelation function as

$$\rho_{h'h'}(\zeta) = \frac{R_{h'h'}(\zeta)}{R_{h'h'}(0)} = \left(1 + \frac{\zeta}{\lambda}\right)e^{-\zeta/\lambda}, \quad \zeta > 0$$
(6-154)

Variation of the autocorrelation function, $\rho_{h'h'}(\zeta)$, is shown in Figure 6-48. Note that the neighboring values of the hydraulic head are highly correlated and that the correlation decreases as the separation increases.

6.4.4.2 Linearized Solution in Terms of Hydraulic Conductivity *K*

Expressing the Variables in Terms of Mean and Perturbation
The method of small perturbations is used to solve the one-dimensional flow equation given by Eq. (6-120). If the parameters are expressed in terms of a mean part and a small perturbation, as in

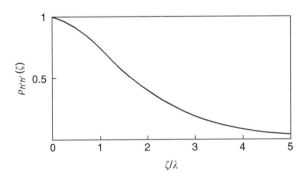

FIGURE 6-48   Autocorrelation function, $\rho_{h'h'}$ (ζ), of hydraulic head fluctuations versus the dimensionless lag ζ/λ. (Adapted from Bakr, A.A., Ph. D. dissertation, New Mexico Institute of Mining and Technology, Socorro, New Mexico, 1976.)

Eqs. (6-125) and (6-126),

$$K(x) = \overline{K} + K'(x), \quad E[K'(x)] = 0 \tag{6-155}$$

and

$$h(x) = \overline{h}(x) + h'(x), \quad E[h'(x)] = 0 \tag{6-156}$$

Introducing these expressions into Eq. (6-120) and taking its expected values

$$E[q] = -E\left[K(x)\frac{dh(x)}{dx}\right] = -E\left[\overline{K}\frac{d\overline{h}(x)}{dx} + \overline{K}\frac{dh'(x)}{dx} + K'(x)\frac{d\overline{h}(x)}{dx} + K'(x)\frac{dh'(x)}{dx}\right] \tag{6-157}$$

Equation (6-157) is linearized by neglecting products of primed quantities (the last term on the right-hand side)

$$E[q] = -\overline{K}\frac{d\overline{h}(x)}{dx} - \overline{K}E\left[\frac{dh'(x)}{dx}\right] - \frac{d\overline{h}(x)}{dx}E[K'(x)] \tag{6-158}$$

or by using the second expressions in Eqs. (6-155) and (6-156)

$$\overline{K}\frac{d\overline{h}}{dx} = -\overline{q} \tag{6-159}$$

Equation (6-158) without E's gives

$$\overline{K}\frac{d\overline{h}}{dx} + K'\frac{d\overline{h}}{dx} + \overline{K}\frac{dh'}{dx} = -\overline{q} \tag{6-160}$$

Subtracting these two equations gives

$$K'\frac{d\overline{h}}{dx} + \overline{K}\frac{dh'}{dx} = 0 \tag{6-161}$$

Using

$$q = -\overline{K}\frac{d\overline{h}}{dx} \tag{6-162}$$

and substituting it in Eq. (6-161), the following expression can be obtained

$$\frac{dh'}{dx} = q\frac{K'}{\overline{K}^2} \tag{6-163}$$

Note that $K'/\overline{K}^2$ corresponds to $-W'$ in the exact solution (cf. with Eq. [6-132]).

Spectral Representation and Determination of the Spectra
Similar to Eq. (6-102), consider the case of statistically homogeneous input perturbation that is represented by Fourier–Stieltjes integrals in wave number k, as follows:

$$K'(x) = \int_{-\infty}^{\infty} e^{ikx}\, dZ_{K'}(k) \tag{6-164}$$

with the head perturbations represented by Eq. (6-134). By substituting Eqs. (6-134) and (6-164) into Eq. (6-163) and following the procedure in Section 6.4.4.1 one has

$$ik\, dZ_{h'}(k) = \frac{q}{\overline{K}^2}\, dZ_{K'}(k) \tag{6-165}$$

According to the spectral representation theorem as presented in Section 6.4.2.1, the complex conjugate is

$$- ik dZ_{h'}^*(k) = \frac{q}{K^2} dZ_{K'}^*(k)$$ (6-166)

Multiplying both sides of Eqs. (6-165) and (6-166) and taking their expected values and applying the spectral representation theorem in Section 6.4.2.1 (Eq. (6-105)) and using the equality of $i^2 = -1$, one gets

$$S_{h'h'}(k) = \frac{q^2}{K^2 k^2} S_{K'K'}(k)$$ (6-167)

Equation (6-167) is singular at the origin and the discussion following Eq. (6-142) applies here also. Consequently, the form of the autocovariance $R_{K'K'}$ of hydraulic conductivity is analogous to Eq. (6-143)

$$R_{K'K'}(\zeta) = \overline{K}'^2 \left(1 - \frac{|\xi|}{\lambda} \right) e^{-|\xi|/\lambda}$$ (6-168)

where

$$\overline{K}'^2 = \sigma_{K'}^2$$ (6-169)

is the variance of K'. Similarly in Eq. (6-149), the spectrum $S_{h'h'}$ can be determined as

$$S_{K'K'} = \frac{2k^2\lambda^3\sigma_{K'}^2}{\pi(1 + k^2\lambda^2)^2}$$ (6-170)

and, the spectrum of $S_{h'h'}$ of hydraulic head fluctuations is then found by substituting Eq. (6-170) into Eq. (6-167) as

$$S_{h'h'}(k) = \frac{2\lambda^3 q^2}{\pi(1 + \lambda^2 k^2)^2} \left(\frac{\sigma_{K'}^2}{\overline{K}^4} \right)$$ (6-171)

Note that the term in the parentheses on the right-hand side of Eq. (6-171) corresponds to $\sigma_{W'}^2$ in Eq. (6-150).

Determination of the Autocovariance Function for the Hydraulic Head ($R_{h'h'}$)
By substituting Eq. (6-171) into Eq. (6-151) one has

$$R_{h'h'}(\zeta) = \int_{-\infty}^{\infty} e^{ik\zeta} \frac{2\lambda^2 q^2}{\pi(1 + \lambda^2 k^2)^2} \frac{\sigma_{K'}^2}{\overline{K}^4} dk$$ (6-172)

and carrying out the integration over wave number space or evaluating Eq. (6-168) at $\zeta = 0$ one has

$$\sigma_h^2 = \frac{\sigma_{K'}^2}{\overline{K}^4} \lambda^2 q^2$$ (6-173)

The results of this analysis will be compared in a later chapter, and the results from both the exact solution and the linearized solution will be developed in the next section.

6.4.4.3 Linearized Solution in Terms of ln(K)

Purpose of the Usage of ln(K) and the Flow Equation
The solution of the differential equations by the method of perturbation requires that the fluctuation of the perturbed parameter be small compared with its mean part. Thus, the results of the perturbed solution to the flow equation in terms of the hydraulic conductivity K may not be appropriate in

many cases since K may vary by several orders of magnitude. In an effort to relax this limitation, the flow equation is solved by perturbing the logarithm of K, since the variability of $\ln(K)$ is considerably smaller than that of K itself (Bakr, 1976; Gelhar, 1993).

One-dimensional steady and saturated water flow through a porous medium is described by Eq. (6-124). From Eq. (6-124) one can write

$$\frac{1}{K}\frac{dK}{dx}\frac{dh}{dx} + \frac{d^2h}{dx^2} = 0 \tag{6-174}$$

or

$$\frac{d^2h}{dx^2} + \frac{d[\ln(K)]}{dx}\frac{dh}{dx} = 0 \tag{6-175}$$

It should be noted that the logarithm of hydraulic conductivity occurs naturally in Eq. (6-175). If $\ln(K)$ is constant, the head distribution is described by the Laplace equation.

Normally Distributed ln(K)

As mentioned in Section 6.3.6.2, in aquifers, hydraulic conductivity generally follows a lognormal probability density function (Freeze, 1975). If $\ln(K)$ is a Gaussian process as shown in Figure 6-4a, it is completely characterized by its mean and covariance. Also, the logarithm of K is less variable than K itself. In the framework of all these reasons, it is relatively easier to work with $\ln(K)$ as given by Eq. (6-63). Here, $\ln(K)$ is regarded as a stochastic process with mean and perturbation. Based on Eq. (6-63), it is convenient to introduce the variable

$$f(x) = \ln[K(x)] \tag{6-176}$$

Expressing the Variables in Terms of Mean and Perturbation

Expressing the variables in terms of a mean and a perturbation

$$h = \bar{h} + h', \quad E[h] = \bar{h}, \quad E[h'] = 0 \tag{6-177}$$

and

$$f = \ln(K) = \bar{f} + f', \quad \bar{f} = E[\ln(K)] = \ln(K_g), \quad E[f'] = 0 \tag{6-178}$$

Substituting these expressions into Eq. (6-175) and taking the expected value of each term results in

$$E\left[\frac{d^2(\bar{h} + h')}{dx^2}\right] + E\left[\frac{d(\bar{f} + f')}{dx}\frac{d(\bar{h} + h')}{dx}\right] = 0 \tag{6-179}$$

$$\frac{d^2\bar{h}}{dx^2} + E\left[\frac{d^2h'}{dx^2}\right] + \frac{d\bar{f}}{dx}\frac{d\bar{h}}{dx} + \frac{d\bar{f}}{dx}E\left[\frac{dh'}{dx}\right] + \frac{d\bar{h}}{dx}E\left[\frac{df'}{dx}\right] + E\left[\frac{df'}{dx}\frac{dh'}{dx}\right] = 0 \tag{6-180}$$

In Eq. (6-180), the second, the fourth, and the fifth terms are zero as a result of $E[h'] = 0$ and $E[f'] = 0$. Therefore, Eq. (6-180) becomes

$$\frac{d^2\bar{h}}{dx^2} + \frac{d\bar{f}}{dx}\frac{d\bar{h}}{dx} + E\left[\frac{df'}{dx}\frac{dh'}{dx}\right] = 0 \tag{6-181}$$

Introducing Eqs. (6-177) and (6-178) into Eq. (6-175), after some manipulations, the following equation can be obtained:

$$\frac{d^2\bar{h}}{dx^2} + \frac{d^2h'}{dx^2} + \frac{d\bar{f}}{dx}\frac{d\bar{h}}{dx} + \frac{d\bar{f}}{dx}\frac{dh'}{dx} + \frac{df'}{dx}\frac{d\bar{h}}{dx} + \frac{df'}{dx}\frac{dh'}{dx} = 0 \tag{6-182}$$

Subtracting Eq. (6-181) from Eq. (6-182) gives

$$\frac{d^2h'}{dx^2} + \frac{d\bar{f}}{dx}\frac{dh'}{dx} + \frac{df'}{dx}\frac{d\bar{h}}{dx} + \frac{df'}{dx}\frac{dh'}{dx} - E\left[\frac{df'}{dx}\frac{dh'}{dx}\right] = 0 \tag{6-183}$$

or

$$\frac{d^2h'}{dx^2} + \frac{d\bar{f}}{dx}\frac{dh'}{dx} + \frac{df'}{dx}\frac{d\bar{h}}{dx} = E\left[\frac{df'}{dx}\frac{dh'}{dx}\right] - \frac{df'}{dx}\frac{dh'}{dx} = 0 \tag{6-184}$$

As can be seen from Eq. (6-184), the terms on the right-hand side of the equal sign involve products of perturbation terms, whereas the terms on the left-hand side involve only terms of first order in perturbations. As the head perturbations are generated by log hydraulic conductivity perturbations, if the ln(K) perturbations are small, one can expect small head perturbations to result. Under this framework, the perturbation equation. (6-184) will be approximated by setting the right-hand side equal to zero. This is based on the fact that the hydraulic conductivity variations must in some sense be small for this log linearization to be a satisfactory approximation (Gelhar, 1993).

Spectral Representation
Equation (6-184) can be written as

$$\frac{d^2h'}{dx^2} + A\frac{dh'}{dx} = I\frac{df'}{dx} \tag{6-185}$$

where

$$A = \frac{d\bar{f}}{dx} \quad I = -\frac{d\bar{h}}{dx} \tag{6-186}$$

Note that here the hydraulic gradient I and the parameter A representing the trend in mean log hydraulic conductivity will in general be functions of x; if $A = 0$, I will be strictly constant.

Similar to Eq. (6-102), consider the case of statistically homogeneous log-hydraulic conductivity, which is represented by Fourier–Stieltjes integrals in wave number k, as follows

$$f'(x) = \int_{-\infty}^{\infty} e^{ikx} \, dZ_{f'}(k) \tag{6-187}$$

Similarly, the hydraulic head perturbation is represented by Eq. (6-134) for a locally stationary statistically homogeneous solution for head. This local homogeneity assumption implies that in Eq. (6-185), A and I are regarded as varying slowly in space. This means that they do not change significantly over a distance corresponding to the correlation scale of h.
Introducing Eqs. (6-134) and (6-187) into Eq. (6-185) gives

$$\frac{d^2}{dx^2}\int_{-\infty}^{\infty} e^{ikx}\,dZ_h' + A\frac{d}{dx}\int_{-\infty}^{\infty} e^{ikx}\,dZ_{h'} = I\frac{d}{dx}\int_{-\infty}^{\infty} e^{ikx}\,dZ_{f'} \tag{6-188}$$

or following the similar procedure in deriving Eq. (6-137) one gets

$$dZ_h' = -\frac{iI}{k - iA}\,dZ_{f'} \tag{6-189}$$

Its complex conjugate (see Section 6.4.4.1 for the definition of "complex conjugate") is

$$dZ_{h'}^* = \frac{iI}{k + iA}\,dZ_{f'}^* \tag{6-190}$$

Multiplying both sides of Eqs. (6-189) and (6-190) and taking their expected values results in the following expression:

$$E[dZ_{h'}(k)dZ_{h'}^*(k)] = -\frac{i^2I^2}{k^2 - i^2A^2}E[dZ_{f'}(k)\,dZ_{f'}^*(k)] \tag{6-191}$$

Applying the spectral representation theorem in Section 6.4.2.1 (Eq. [6-105]) and using the equality of $i^2 = -1$, Eq. (6-191) takes the form

$$S_{h'h'}(k) = \frac{I^2}{k^2 + A^2} S_{f'f'}(k) \tag{6-192}$$

It is necessary to point out that at the origin, the singularity no longer occurs as in Eq. (6-142). However, to be able to compare with the results for the case when there is no trend, the spectrum of $\ln(K)$ will be taken to be in the form of Eq. (6-170), i.e., (Gelhar, 1993, p. 101),

$$S_{f'f'} = \frac{2}{\pi} \frac{\lambda^3 k^2}{(1 + \lambda^2 k^2)^2} \sigma_{f'}^2 \tag{6-193}$$

and

$$R_{f'f'}(\zeta) = f'^2 \left(1 - \frac{|\xi|}{\lambda}\right) e^{-|\xi|/\lambda} \tag{6-194}$$

where

$$f'^2 = E[f'^2] = \sigma_f^2 \tag{6-195}$$

According to Eq. (6-109), $\zeta = 0$ in Eq. (6-151) corresponds to the variance for $A = 0$. Using Eqs. (6-192) and (6-193) in Eq. (6-153) and performing the integration gives

$$R_{h'h'}(0) = \sigma_h'^2 = \int_{-\infty}^{\infty} S_{h'h'}(k)\, dk = \frac{2}{\pi} I^2 \lambda^3 \sigma_{f'}^2 \int_{-\infty}^{\infty} \frac{dk}{(1 + \lambda^2 k^2)^2} = I^2 \lambda^2 \sigma_{f'}^2 \tag{6-196}$$

which is the same as the form of Eq. (6-153).

6.4.4.4 Effective Hydraulic Conductivity for One-Dimensional Case

The concept of an effective hydraulic conductivity of a heterogeneous porous medium has been explored in several studies since the 1960s (Warren and Price, 1961; Bouwer, 1969; Freeze, 1975; Gutjahr et al., 1978). The approach of Gutjahr et al. (1978) allows a direct evaluation of this concept for the steady flow configurations, and this approach will be presented in the following paragraphs. For one-dimensional flow, the exact solution for the mean flow shows that the effective hydraulic conductivity, K_e, is the harmonic mean (Gutjahr et al., 1978)

$$K_e = [E[W]]^{-1} = [E[K^{-1}]]^{-1} \tag{6-197}$$

Gutjahr et al. (1978) state that this result is to be expected from the well-known behavior of flow normal to the layers in a deterministic stratified medium (e.g., Bear, 1972). The authors also add that if $f = \ln(K)$ in Eq. (6-178) is normally distributed, the effective hydraulic conductivity is

$$K_e = K_g \exp\left(-\tfrac{1}{2}\sigma_f^2\right) \tag{6-198}$$

Approximate Analysis of One-Dimensional Flow
From the first and second expressions of Eq. (6-178) one can write

$$K = e^{\bar{f}+f'}, \quad K_g = e^{\bar{f}} \tag{6-199}$$

Substituting the first expression of Eq. (6-199) and the first expression of Eq. (6-125) into the one-dimensional form of Darcy's law, as given by Eq. (6-120), gives

$$q_x = -e^{\bar{f}+f'}\left(\frac{d\bar{h}}{dx} + \frac{dh'}{dx}\right) \tag{6-200}$$

Taking the expected values of both sides

$$E[q_x] = Q = -e^{\bar{f}} E\left[e^{f'}\left(\frac{d\bar{h}}{dx} + \frac{dh'}{dx}\right)\right] \tag{6-201}$$

By opening the exponential term into a Taylor series and making approximations, the following expressions can be obtained:

$$Q = -K_g E\left[\left(1 + f' + \frac{f'^2}{2!} + \frac{f'^3}{3!} + \cdots\right)\left(\frac{d\bar{h}}{dx} + \frac{dh'}{dx}\right)\right]$$

$$\cong -K_g E\left[\left(1 + f' + \frac{f'^2}{2!}\right)\left(\frac{d\bar{h}}{dx} + \frac{dh'}{dx}\right)\right] \tag{6-202}$$

Using Eqs. (6-177), (6-178), and (6-195), Eq. (6-202) can also be expressed as

$$Q = -K_g E\left[\left(1 + f' + \frac{\sigma_f^2}{2}\right)\frac{d\bar{h}}{dx} + \frac{dh'}{dx} + f'\frac{dh'}{dx} + \frac{f'^2}{2}\frac{dh'}{dx}\right]$$

$$= -K_g\left[\frac{d\bar{h}}{dx}\left(1 + \frac{\sigma_f^2}{2}\right) + E\left[f'\frac{dh'}{dx}\right]\right] \tag{6-203}$$

In the above expressions, the hydraulic conductivity has been expressed in terms of the mean and perturbation of $\ln(K)$ from Eqs. (6-177) and (6-178) and the exponential term in the Taylor series. In Eq. (6-203), terms up to second order in perturbations have been retained. The expected value of the product of the perturbations in the head gradient and $\ln(K)$ is a very important term that reflects the relationship between the hydraulic conductivity variation and the head perturbation it generates (Gelhar, 1993).

Spectral Representation
Similar to Eq. (6-102), consider the case of statistically homogeneous input perturbation, which is represented by Fourier–Stieltjes integrals in wave number k as follows:

$$h'(x) = \int_{-\infty}^{\infty} e^{ikx}\, dZ_h'(k) \tag{6-204}$$

and

$$j \equiv \frac{dh'}{dx} = \int_{-\infty}^{\infty} ik e^{ikx}\, dZ_{h'}(k) \tag{6-205}$$

Similarly, $f'(x)$ can be expressed as

$$f'(x) = \int_{-\infty}^{\infty} e^{ikx}\, dZ_{f'}(k) \tag{6-206}$$

The cross-covariance between f' and dh'/dx, if these are parts of a stationary field, according to Eq. (6-102), $Z_{f'}(k)$ and $Z_j(k)$, are stochastic processes having properties such that (Lumley and Panofsky, 1964, pp. 19–23)

$$E[dZ_{f'}(k)] = 0, \quad E[dZ_j(k)] = 0 \tag{6-207}$$

and

$$E[dZ_{f'}(k_1)\, dZ_j^*(k_2)] = 0 \quad k_1 \neq k_2 \tag{6-208}$$

in which dZ_j^* is the complex conjugate and k is the frequency. Equation. (6-207) states that both $f'(x)$ and j have zero mean. Equation (6-208) indicates that the increments of $Z_{f'}$ and Z_j at two different frequencies are uncorrelated. When $k_1 = k_2 = k$, then

$$E[dZ_{f'}(k) \, dZ_j^*(k)] = S_{f'j}(k) \, dk, \quad k_1 = k_2 = k \tag{6-209}$$

where $S_{f'j}(k)$ is the *cross-spectrum* of f' and j, and the asterisk ($*$) denotes the complex conjugate of Z_j. From Eqs. (6-187) and (6-205)

$$E\left[f' \frac{dh'}{dx}\right] = E\left[\left(\int_{-\infty}^{\infty} e^{ikx} dZ_{f'}\right)\left(-\int_{-\infty}^{\infty} ike^{-ikx} dZ_{h'}^*\right)\right]$$
$$= E\left[-ik\int_{-\infty}^{\infty}\int_{-\infty}^{\infty} dZ_{f'} dZ_{h'}^*\right] \tag{6-210}$$

For $A = 0$ in Eq. (6-189)

$$dZ_{h'}^* = \frac{il}{k}dZ_{f'}^* \tag{6-211}$$

Substituting this into Eq. (6-210) gives

$$E\left[f' \frac{dh'}{dx}\right] = E\left[\left(\int_{-\infty}^{\infty} dZ_{f'}(-ik)\int_{-\infty}^{\infty} \frac{il}{k}dZ_{f'}^*\right)\right] \tag{6-212}$$

For $j = f'$, Eq. (6-210) becomes

$$E[dZ_{f'}(k) \, dZ_{f'}^*(k)] = S_{f'f'}(k) \, dk, \quad k_1 = k_2 = k \tag{6-213}$$

The substitution of Eq. (6-213) into Eq. (6-212) gives

$$E\left[f' \frac{dh'}{dx}\right] = Il\int_{-\infty}^{\infty} S_{f'f'}(k) \, dk = I\sigma_{f'}^2 \tag{6-214}$$

The right-hand side of the equation is in accordance with Eq. (6-109).

Therefore, the substitution of Eq. (6-214) and I (the second expression in Eq. [6-186]) into Eq. (6-202) results in

$$Q = -K_g[-I\left(1 + \tfrac{1}{2}\sigma_{f'}^2\right) + I\sigma_{f'}^2] = K_g\left(1 - \tfrac{1}{2}\sigma_{f'}^2\right)I \tag{6-215}$$

which implies that the effective hydraulic conductivity is $K_g (1 - \sigma_{f'}^2/2)$. This approximate result agrees with the exact expression given by Eq. (6-198) to the first order in $\sigma_{f'}^2$. The approximation is evaluated in Example 6-8 given below.

EXAMPLE 6-8

Compare the approximate and exact equations for $\sigma_{f'}^2 = 1$ and 0.5.

SOLUTION

From Eqs. (6-198) and (6-215) one can write

$$Error = \frac{K_g \exp(-1/2\sigma_{f'}^2) - K_g(1 - 1/2\sigma_{f'}^2)}{K_g \exp(-1/2\sigma_{f'}^2)} = \frac{\exp(-1/2\sigma_{f'}^2) - (1 - 1/2\sigma_{f'}^2)}{\exp(-1/2\sigma_{f'}^2)} \tag{E6-8-1}$$

For $\sigma_{f'} = 1$, Eq. (E6-8-1) gives

$$Error = \frac{\exp[-1/2(1)^2)] - [1 - 1/2(1)^2]}{\exp[-1/2(1)^2]} = 0.176$$

For $\sigma_{f'} = 0.5$, Eq. (E6-8-1) gives

$$Error = \frac{\exp[-1/2(0.5)^2)] - [1 - 1/2(0.5)^2]}{\exp[-1/2(0.5)^2]} = 0.0079$$

The above results show that when $\sigma_{f'} = 1$, the approximate result is about 18% below the exact value, but when $\sigma_{f'} = \frac{1}{2}$, the difference is less than 1%.

6.4.5 STOCHASTIC ANALYSIS OF THREE-DIMENSIONAL STEADY-STATE FLOW IN ISOTROPIC POROUS MEDIA

In this section, the effects of three-dimensional spatial variations of hydraulic conductivity to hydraulic head variations are presented. In solving the three-dimensional ground water flow problem, the flow equation is linearized in terms of the natural logarithm of the hydraulic conductivity as was done for the one-dimensional case, which is presented in Section 6.4.4. In the following sections, the three-dimensional perturbed flow equation, effective hydraulic conductivity for three-dimensional cases, and the hydraulic head and log-hydraulic conductivity processes are presented (Bakr, 1976; Bakr et al., 1978; Gutjahr et al., 1978; Gelhar, 1993).

6.4.5.1 Three-Dimensional Perturbed Flow Equation

Consider an unbounded saturated porous medium that has a mean hydraulic gradient I in a particular direction (e.g., the horizontal x-coordinate axis). It is assumed that the medium is isotropic on the local scale. The form of Darcy's law in three dimensions is given by Eq. (4-91) and the mass balance equation under transient conditions is given by Eq. (4-93) Under steady-state conditions, Eq. (4-93) takes the form

$$\frac{\partial^2 h}{\partial x^2} + \frac{\partial^2 h}{\partial y^2} + \frac{\partial^2 h}{\partial z^2} + \frac{\nabla K}{K} \cdot \nabla h = 0 \tag{6-216}$$

in which ∇ is defined by Eq. (4-89). Since

$$\frac{\nabla K}{K} = \nabla \ln(K) \tag{6-217}$$

Equation (6-216), in scalar form, becomes

$$\frac{\partial^2 h}{\partial x^2} + \frac{\partial^2 h}{\partial y^2} + \frac{\partial^2 h}{\partial z^2} + \frac{\partial[\ln(K)]}{\partial x}\frac{\partial h}{\partial x} + \frac{\partial[\ln(K)]}{\partial y}\frac{\partial h}{\partial y} + \frac{\partial[\ln(K)]}{\partial z}\frac{\partial h}{\partial z} = 0 \tag{6-218}$$

In tensorial notation, Eq. (6-218) takes the form

$$\frac{\partial^2 h}{\partial x_i^2} + \frac{\partial[\ln(K)]}{\partial x_i}\frac{\partial h}{\partial x_i} = 0 \tag{6-219}$$

One should note that the logarithm of hydraulic conductivity occurs naturally in the above equations, and the effect of heterogeneity is manifested entirely through gradients in $\ln(K)$. If $\ln(K)$ is constant, the head distribution is described by the Laplace equation.

Using the same approach used in Section 6.4.4.3, the equivalent form of Eq. (6-184) can be written as

$$\frac{\partial^2 h'}{\partial x^2} + \frac{\partial \bar{f}}{\partial x}\frac{\partial h'}{\partial x} + \frac{\partial f'}{\partial x}\frac{\partial \bar{h}}{\partial x} + \frac{\partial^2 h'}{\partial y^2} + \frac{\partial \bar{f}}{\partial y}\frac{\partial h'}{\partial y} + \frac{\partial f'}{\partial y}\frac{\partial \bar{h}}{\partial y} + \frac{\partial^2 h'}{\partial z^2} + \frac{\partial \bar{f}}{\partial z}\frac{\partial h'}{\partial z} + \frac{\partial f'}{\partial z}\frac{\partial \bar{h}}{\partial z}$$

$$= E\left[\frac{\partial f'}{\partial x}\frac{\partial h'}{\partial x}\right] - \frac{\partial f'}{\partial x}\frac{\partial h'}{\partial x} + E\left[\frac{\partial f'}{\partial y}\frac{\partial h'}{\partial y}\right] - \frac{\partial f'}{\partial y}\frac{\partial h'}{\partial y} + E\left[\frac{\partial f'}{\partial z}\frac{\partial h'}{\partial z}\right] - \frac{\partial f'}{\partial z}\frac{\partial h'}{\partial z} \cong 0 \tag{6-220}$$

or in tensorial notation

$$\frac{\partial^2 h'}{\partial x_i^2} + \frac{\partial \bar{f}}{\partial x_i}\frac{\partial h'}{\partial x_i} + \frac{\partial f'}{\partial x_i}\frac{\partial \bar{h}}{\partial x_i} = E\left[\frac{\partial f'}{\partial x_i}\frac{\partial h'}{\partial x_i}\right] - \frac{\partial f'}{\partial x_i}\frac{\partial h'}{\partial x_i} \cong 0 \qquad (6\text{-}221)$$

For two- or three-dimensional flow, the first-order small perturbation approximation of Eq. (6-221) is

$$\frac{\partial^2 h'}{\partial x_i^2} - I_i\frac{\partial f'}{\partial x_i} + A_i\frac{\partial h'}{\partial x_i} = 0 \qquad (6\text{-}222)$$

where

$$I_i = -\frac{\partial \bar{h}}{\partial x_i}, \quad A_i = \frac{\partial \bar{f}}{\partial x_i} \qquad (6\text{-}223)$$

The purpose is to find locally stationary solutions for the head process by treating the parameters I_i and A_i as being locally constant.

6.4.5.2   Spectral Representation

Equation (6-222) can be written as

$$\frac{\partial^2 h'}{\partial x^2} + \frac{\partial^2 h'}{\partial y^2} + \frac{\partial^2 h'}{\partial z^2} - I_x\frac{\partial f'}{\partial x} - I_y\frac{\partial f'}{\partial y} - I_z\frac{\partial f'}{\partial z} + A_x\frac{\partial h'}{\partial x} + A_y\frac{\partial h'}{\partial y} + A_z\frac{\partial h'}{\partial z} = 0 \quad (6\text{-}224)$$

and Eq. (6-223) can be written as

$$I_x = -\frac{\partial \bar{h}}{\partial x}, \quad I_y = -\frac{\partial \bar{h}}{\partial y}, \quad I_z = -\frac{\partial \bar{h}}{\partial z}$$

$$A_x = \frac{\partial \bar{f}}{\partial x}, \quad A_y = \frac{\partial \bar{f}}{\partial y}, \quad A_z = \frac{\partial \bar{f}}{\partial z} \qquad (6\text{-}225)$$

If there is no trend in the mean log hydraulic conductivity, and the mean hydraulic gradient is selected to be in the x-coordinate direction, i.e.,

$$I_x = I, \quad I_y = I_z = 0, \quad A_x = A_y = A_z = 0 \qquad (6\text{-}226)$$

then Eq. (6-224) takes the form

$$\frac{\partial^2 h'}{\partial x^2} + \frac{\partial^2 h'}{\partial y^2} + \frac{\partial^2 h'}{\partial z^2} - I\frac{\partial f'}{\partial x} = 0 \qquad (6\text{-}227)$$

Similar to Eq. (6-102), consider the case of statistically homogeneous input perturbation for h' represented by Fourier-Stieltjes integrals in wave number k, given by

$$h'(\mathbf{x}) = \int_{-\infty}^{\infty} e^{i\mathbf{k}\cdot\mathbf{x}}\, dZ'_h(\mathbf{k}) \qquad (6\text{-}228)$$

Similarly, the hydraulic head perturbation is represented by

$$f'(\mathbf{x}) = \int_{-\infty}^{\infty} e^{i\mathbf{k}\cdot\mathbf{x}}\, dZ'_f(\mathbf{k}) \qquad (6\text{-}229)$$

In Eqs. (6-228) and (6-229), $\mathbf{x} = \{x, y, z\}$ is the position vector and $\mathbf{k} = \{k_x, k_y, k_z\}$ is the wave number vector. The processes discussed in Section 6.4.2.1 are all examples of a general type called *time series*, i.e., the process varies with time. For example, Eq. (6-102) is a representation of time series. However, it may equally be studied the variation of some quantity over a region of space. Such phenomena are usually referred to as *spatial processes*. Therefore, Eqs. (6-228) and (6-229) represent spatial processes.

Substituting Eqs. (6-228) and (6-229) into Eq. (6-227) and with some manipulations one gets

$$\int_{-\infty}^{\infty} [(-k_x^2 - k_y^2 - k_z^2)\, dZ_{h'}(\mathbf{k}) - ilk_x\, dZ_{f'}(\mathbf{k})] e^{i\mathbf{k}\cdot\mathbf{x}} = 0 \tag{6-230}$$

Equation (6-230) holds only if the quantity in brackets is identically equal to zero, i.e.,

$$(-k_x^2 - k_y^2 - k_z^2)\, dZ_{h'}(\mathbf{k}) - ilk_x\, dZ_{f'}(\mathbf{k}) = 0 \tag{6-231}$$

and from this

$$dZ_{h'}(\mathbf{k}) = -\frac{ilk_x}{k_x^2 + k_y^2 + k_z^2}\, dZ_{f'}(\mathbf{k}) \tag{6-232}$$

According to the spectral representation theorem as presented in Section 6.4.2.1, the complex conjugate ($z^* = x - iy$ is the complex conjugate of $z = x + iy$ by definition, therefore, $z^* = x^2 + y^2$) is

$$dZ_{h'}^*(\mathbf{k}) = \frac{ilk_x}{k_x^2 + k_y^2 + k_z^2}\, dZ_{f'}^*(\mathbf{k}) \tag{6-233}$$

Multiplying both sides of Eqs. (6-232) and (6-233) and taking their expected values result in the following expression

$$E[dZ_{h'}(\mathbf{k})dZ_{h'}^*(\mathbf{k})] = \frac{l^2 k_x^2}{k_x^2 + k_y^2 + k_z^2} E[dZ_{f'}(\mathbf{k})\, dZ_{f'}^*(\mathbf{k})] \tag{6-234}$$

Applying the spectral representation theorem in Section 6.4.2.1 (Eq. [6-105]) and using the equality of $i^2 = -1$, Eq. (6-234) takes the form

$$S_{h'h'}(\mathbf{k}) = \frac{l^2 k_x^2}{k^4} S_{f'f'}(\mathbf{k}) \tag{6-235}$$

in which

$$k = |\mathbf{k}| = (k_x^2 + k_y^2 + k_z^2)^{1/2} \tag{6-236}$$

represents the magnitude of the wave number vector $\mathbf{k}$.

6.4.5.3 Effective Hydraulic Conductivity for Three-Dimensional Case

In the three-dimensional case, Eq. (6-203) is still valid. The only problem is that $E\,[f'dh'/dx]$ is different and needs to be evaluated for three-dimensional flow cases. Using the method in Section 6.4.4.4 and from Eq. (6-228) one can write

$$\frac{\partial h'}{\partial x} = \int_{-\infty}^{\infty} ik_x e^{i\mathbf{k}\cdot\mathbf{x}}\, dZ_{h'}(\mathbf{k}) \tag{6-237}$$

From Eq. (6-229) and the complex conjugate of Eq. (6-237)

$$E\left[f'\frac{\partial h'}{\partial x}\right] = E[(\int_{-\infty}^{\infty} e^{i\mathbf{k}\cdot\mathbf{x}}\, dZ_{f'}(\mathbf{k}))(-\int_{-\infty}^{\infty} ik_x e^{-i\mathbf{k}\cdot\mathbf{x}}\, dZ_{h'}^*(\mathbf{k}))] \tag{6-238}$$

Substitution of Eq. (6-232) with the form

$$dZ_{h'}^*(\mathbf{k}) = \frac{ilk_x}{k_x^2 + k_y^2 + k_z^2}\, dZ_{f'}^*(\mathbf{k}) \tag{6-239}$$

into Eq. (6-238) gives

$$E\left[f'\frac{\partial h'}{\partial x}\right] = E\left[\int_{-\infty}^{\infty}\int_{-\infty}^{\infty} dZ_{f'}(\mathbf{k})\left(-ik_x\frac{ilk_x}{k_x^2 + k_y^2 + k_z^2}\,dZ_f^*(\mathbf{k})\right)\right] \tag{6-240}$$

or

$$E\left[f'\frac{\partial h'}{\partial x}\right] = \int_{-\infty}^{\infty}\int_{-\infty}^{\infty}\frac{k_x^2}{k_x^2 + k_y^2 + k_z^2}IE[dZ_{f'}(\mathbf{k})dZ_f^*(\mathbf{k})] = I\int_{-\infty}^{\infty}\int_{-\infty}^{\infty}\int_{-\infty}^{\infty}\frac{k_x^2}{k_x^2 + k_y^2 + k_z^2}S_{f'f'}(\mathbf{k})\,d\mathbf{k} \tag{6-241}$$

If $f'(\mathbf{x})$ is a statistically isotropic process, then Eq. (6-241) can be expressed as (Gutjahr et al., 1978)

$$E\left[f'\frac{\partial h'}{\partial x}\right] = \frac{I}{3}\sum_{i=1}^{3}\int_{-\infty}^{\infty}\int_{-\infty}^{\infty}\int_{-\infty}^{\infty}\frac{k_i^2}{k^2}S_{f'f'}(\mathbf{k})\,d\mathbf{k} = \frac{I}{3}\int_{-\infty}^{\infty}\int_{-\infty}^{\infty}\int_{-\infty}^{\infty}S_{f'f'}(\mathbf{k})\,d\mathbf{k} = I\frac{\sigma_{f'}^2}{3} \tag{6-242}$$

where $\sigma_{f'}^2 = R_{f'f'}(0)$. The substitution of Eq. (6-242) into Eq. (6-203) gives

$$Q = -K_g\left[-I\left(1 + \frac{\sigma_{f'}^2}{2}\right) + I\frac{\sigma_{f'}^2}{3}\right] = IK_g\left(1 + \frac{\sigma_{f'}^2}{6}\right) = IK_e \tag{6-243}$$

where K_e is the *effective hydraulic conductivity* and is defined as

$$K_e = K_g\left(1 + \frac{\sigma_{f'}^2}{6}\right) \tag{6-244}$$

in which K_g is defined by the second expression of Eq. (6-199) and is the geometric mean. Following the same reasoning as above, it can be shown that the equivalent correlation term for a two-dimensional plane flow is (Gutjahr et al., 1978)

$$E\left[f'\frac{dh'}{dx}\right] = I\frac{\sigma_{f'}^2}{2} \tag{6-245}$$

and the mean flow equation from Eq. (6-202) is

$$Q = IK_g\left(1 + \frac{\sigma_{f'}^2}{2} - \frac{\sigma_{f'}^2}{2}\right) = IK_g \tag{6-246}$$

Therefore, in the three-dimensional case, the effective hydraulic conductivity is slightly greater than K_g and in the two-dimensional case it is equal to the geometric mean K_g.

6.4.5.4　The Hydraulic Head and Loghydraulic Conductivity Processes

Darcy Flux Expressions
Equations in Sections 6.4.4.3 and 6.4.4.4 are adopted for the hydraulic head and log-hydraulic conductivity processes. Equation (6-176) presents the hydraulic head in terms of its mean and perturbation. Similarly as in Eq. (6-178), the hydraulic conductivity can be expressed as

$$f(\mathbf{x}) = \ln[K(\mathbf{x})] = \overline{f} + f'(\mathbf{x}) \tag{6-247}$$

where

$$\overline{f} = E[\ln(K)] = \ln(K_g), \quad E[f'(\mathbf{x})] = 0 \tag{6-248}$$

and

$$K = \exp(\overline{f} + f'), \quad K_g = \exp(\overline{f}) \tag{6-249}$$

Using the first expression of Eq. (6-249) in Darcy's equation

$$q_i = -K\frac{\partial h}{\partial x_i} = -e^{\bar{f}+f'}\frac{\partial h}{\partial x_i} = -K_g\exp(f')\frac{\partial h}{\partial x_i} \tag{6-250}$$

As in Eq. (6-202), using the first expression of Eq. (6-177), Eq. (6-250) can be expressed as,

$$q_i = -K_g\left(1 + f' + \frac{f'^2}{2} + \cdots\right)\left(\frac{\partial \bar{h}}{\partial x_i} + \frac{\partial h'}{\partial x_i}\right) \tag{6-251}$$

Assuming

$$\exp(f') \cong 1 + f' \tag{6-252}$$

Equation (6-251) takes the form

$$q_i = -K_g(1+f')\left(\frac{\partial \bar{h}}{\partial x_i} + \frac{\partial h'}{\partial x_i}\right) = -K_g\left(\frac{\partial \bar{h}}{\partial x_i} + \frac{\partial h'}{\partial x_i} + f'\frac{\partial \bar{h}}{\partial x_i} + f'\frac{\partial h'}{\partial x_i}\right) \tag{6-253}$$

Dropping the products of perturbed quantities, $f'\,\partial h'/\partial x_i$, Eq. (6-253) takes the form

$$q_i = -K_g\left(\frac{\partial \bar{h}}{\partial x_i} + \frac{\partial h'}{\partial x_i} + f'\frac{\partial \bar{h}}{\partial x_i}\right) \tag{6-254}$$

Since

$$q_i = \bar{q}_i + q'_i, \quad \bar{q}_i = -K_g\frac{\partial \bar{h}}{\partial x_i} \tag{6-255}$$

the mean-removed form of Eq. (6-254) is

$$q'_i = -K_g\left(f'\frac{\partial \bar{h}}{\partial x_i} + \frac{\partial h'}{\partial x_i}\right) \tag{6-256}$$

Spectral Representations
Making use of Fourier–Stieltjes representations for q'_i, f', and h' similar to those in Section 6.4.5.2, Eq. (6-256) takes the form

$$dZ_{q'_i} = K_g(I_i dZ_{f'} - ik_i dZ_{h'}) \tag{6-257}$$

where

$$I_i = -\frac{\partial \bar{h}}{\partial x_i} \tag{6-258}$$

is the *mean hydraulic gradient* in the x-coordinate direction. The perturbed three-dimensional flow equation is given by Eq. (6-227), and can be expressed as

$$\nabla^2 h' = I_i\frac{\partial f'}{\partial x_i} \tag{6-259}$$

Using Fourier–Stieltjes representations, Eq. (6-259) is expressed by Eq. (6-232), and in tensorial notation, it takes the form

$$dZ_{h'} = -\frac{iI_i k_i}{k^2} dZ_{f'} \tag{6-260}$$

where k is given by Eq. (6-236). Finally, combining Eq. (6-260) and Eq. (6-257) produces a relationship between the complex Fourier amplitudes of specific discharge and log-hydraulic conductivity perturbations:

$$dZ_{q'_i} = K_g\left(I_i - \frac{I_j k_i k_j}{k^2}\right) dZ_{f'} \tag{6-261}$$

Since the coordinate system was chosen such that $\overline{q_2} = \overline{q_3} = 0$, and the medium is isotropic, the average hydraulic gradient in these directions will also be zero. Therefore, Eq. (6-261) takes the form

$$dZ_{q'_i} = K_g I_1 \left(\delta_{i1} - \frac{k_1 k_i}{k^2} \right) dZ_{f'} \tag{6-262}$$

where δ_{ij} is the Kronecker delta ($\delta_{11} = 1$, $\delta_{21} = \delta_{31} = 0$). According to the spectral representation theorem as presented in Section 6.4.2.1, its complex conjugate is

$$dZ_{q'_i}^* = K_g I_1 \left(\delta_{ji} - \frac{k_1 k_j}{k^2} \right) dZ_{f'}^* \tag{6-263}$$

Multiplying both sides of Eqs. (6-262) and (6-263) and taking their expected values gives

$$E[dZ_{q'_i} \, dZ_{q'_i}^*] = K_g^2 I_1^2 \left(\delta_{i1} - \frac{k_1 k_i}{k^2} \right) \left(\delta_{j1} - \frac{k_1 k_j}{k^2} \right) E[dZ_{f'} \, dZ_{f'}^*] \tag{6-264}$$

Applying the spectral representation theorem in Section 6.4.2.1 (Eq. [6-105]), one gets

$$S_{q'_j q'_i} = S_{q'_i q'_j} = K_g^2 I_1^2 \left(\delta_{i1} - \frac{k_1 k_i}{k^2} \right) \left(\delta_{j1} - \frac{k_1 k_j}{k^2} \right) S_{f'f'} \tag{6-265}$$

Usage of Effective Hydraulic Conductivity Expressions
As shown in Section 6.4.5.3, a first-order analysis for isotropic log-hydraulic conductivity field shows that (see Eq. (6-243))

$$q = I_1 K_g \left(1 + \frac{\sigma_{f'}^2}{6} \right) = I_1 K_g \gamma \tag{6-266}$$

where

$$\gamma = 1 + \frac{\sigma_{f'}^2}{6} \tag{6-267}$$

which is valid only when the variance of f' process is small (see Section 6.4.4.4). It is conjectured that for larger variances, the relation (Gelhar, 1993)

$$\gamma = \exp\left(\frac{\sigma_{f'}^2}{6} \right) \tag{6-268}$$

may provide a more suitable estimate of the parameter γ.
Making use of Eq. (6-266) in Eq. (6-265),

$$S_{q'_j q'_i} = q^2 \gamma^{-2} \left(\delta_{i1} - \frac{k_1 k_i}{k^2} \right) \left(\delta_{j1} - \frac{k_1 k_j}{k^2} \right) S_{f'f'} \tag{6-269}$$

which relates the spectral density of flow perturbations to the spectral density of log-hydraulic conductivity perturbations.

6.4.6 ANALYSIS OF MACRODISPERSION IN THREE-DIMENSIONALLY HETEROGENEOUS AQUIFERS UNDER STEADY FLOW CONDITIONS USING STOCHASTIC SOLUTE TRANSPORT THEORY: GELHAR AND AXNESS' FIRST-ORDER ANALYSIS APPROACH

6.4.6.1 Objectives and Literature Review

In this section, a general formulation of Gelhar and Axness (1981, 1983), which incorporates the effects of fully three-dimensional heterogeneity, local dispersive mixing, and the transient development of the macrodispersion process from near-source conditions to possible asymptotic behavior far downstream, will be presented. Before the presentation of this approach, some supplementary background information is given below:

Studies conducted in the 1970s showed that field-observed dispersion coefficients are significantly larger than those measured by laboratory tests on cores (e.g., Fried, 1975; Anderson, 1979). Similar studies conducted in the early 1980s showed that the magnitude of dispersion coefficients measured in the field increases with the scale of the experiment (e.g., Dieulin, 1980; Dieulin et al., 1980). Figure 6-49 presents a summary of observations compiled by Lallemand-Barres and Peaudecerf (1978), which illustrates the wide range of longitudinal dispersivity, α_L, values encountered in the field. In Chapter 7, similar figures with much more data are presented for dispersivity values.

In the late 1960s, 1970s, and the early 1980s, several simplified modeling approaches (see, e.g., Mercado, 1967; Gelhar et al., 1979; Pickens and Grisak, 1981a, 1981b) were used to describe the mixing process in heterogeneous media. Each of those models involved assumptions that were not

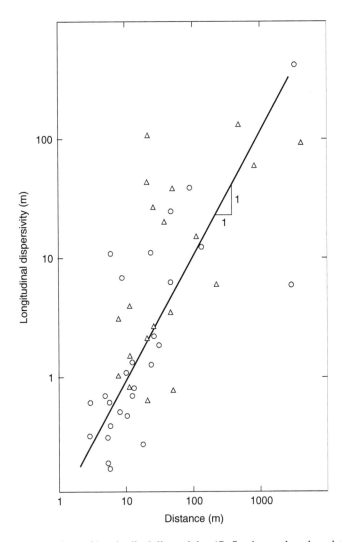

FIGURE 6-49 Field observations of longitudinal dispersivity. (O: Sand, gravel, and sandstone; Δ: Limestone, basalt, granite, and schist) (Adapted from Lallemand-Barres, A. and Peaudecerf, P., *Bulletin de Recherches Géologiques et Minières*, Ze Serie, Section III, No.4, Orleans, France, 1978; Gelhar, L.W and Axness, C.L., Report No. H-8, 1981, 140 pp.)

likely to be valid over a wide range of natural conditions. However, efforts to understand this scale phenomenon of dispersivities (Naff, 1978; Gelhar et al., 1979) led to the proposal of a *macroscopic dispersion coefficient* or *global dispersion coefficient* that was dependent upon the variability of hydraulic parameters (e.g., hydraulic conductivity) at a smaller scale and the average distance over which those parameters were correlated. As mentioned in Section 6.4.4, this average distance called the *correlation scale*, may be multidimensional, reflecting anisotropic heterogeneity within the aquifer (Bakr et al., 1978).

A one-dimensional stochastic analysis of Gelhar et al. (1979, Eq. [20] p. 1389,) for perfectly stratified heterogeneous aquifers with flow parallel to the stratification showed that the macroscopic dispersion coefficient is in the form of the product of the mean flow velocity and a macroscopic dispersivity. This macroscopic dispersivity increases with time or distance of displacement, and for some special forms of the hydraulic conductivity covariance function, it approaches an asymptotic limiting value upon simple statistical properties of the medium. This limiting mechanism is identified herein as the *Taylor mechanism*, after G.I. Taylor (Taylor, 1953), who performed a classical analysis of dispersion in tubes.

The approach of Gelhar and Axness (1981, 1983) is similar to that used by Gelhar et al. (1979). In other words, Gelhar and Axness assumed that variations in hydraulic conductivity and concentration are statistically homogeneous and represent flow and solute transport processes through stochastic equations. The authors applied spectral analysis techniques to the steady mean-removed solute transport equation to derive a Fickian relationship for the dispersive flux. Then they derived equations for macroscopic dispersivities in terms of a general integral expression dependent upon the statistical properties of the medium. These macroscopic dispersivity integrals were evaluated first for the simple isotropic case, and then increasingly complex models were presented for various cases. For each case, mathematical expressions were presented for dispersivities. The analyzed cases are as follows: (1) macroscopic dispersion in statistically isotropic media with local dispersivity isotropic (see Section 6.4.6.9); (2) macroscopic dispersion in statistically isotropic media with local dispersivity anisotropic (see Section 6.4.6.10); and (3) macroscopic dispersion in statistically anisotropic media with arbitrary orientation of stratification (see Section 6.4.6.11). The last case has the following subcases: *Case* I: Anisotropy in the plane of stratification with $\lambda_1 = \lambda_2$ and the mean flow inclined at angle θ with the bedding *Case* II: Horizontal stratification with horizontal and vertical anisotropy *Case* III: Two-dimensional horizontal flow. The statistically isotropic results, which are presented in Sections 6.4.6.9 and 6.4.6.10, are primarily of theoretical interest because most of the earth materials exhibit some degree of stratification, which can be represented in terms of a statistically anisotropic $\ln(K)$ process. The statistically anisotropic case extensively covered examples of some real world cases (Section 6.4.6.11).

6.4.6.2 Scale Classification

The importance of scales of heterogeneity has been emphasized during the past four decades by several investigators (see, e.g., Groult et al., 1966; Alpay, 1972; Bear, 1972). The smallest scale of interest is called the *local scale*, which is defined by using the concept of the representative elementary volume (REV) (Bear, 1972). The REV may be considered as the minimum volume of porous medium for which the porosity or hydraulic conductivity can be represented as properties of a continuum. If this REV is considered as a cube, then the local scale would correspond to the length of one of the sides of the cube. Generally, in aquifers, variations in parameters at the local scale occur. The average distance over which the variations in continuum properties of the aquifer are correlated is called the *aquifer correlation scale*. In their analysis, Gelhar and Axness (1981, 1983) implicitly assumed that the scale of the REV is much smaller than the correlation scale of the aquifer properties. Therefore, the classical continuum flow description is valid at the local scale.

In the stochastic approach, of Gelhar and Axness' the assumption of ergodicity is implicit, i.e, the authors assumed that solute transport in an ensemble of aquifers with the assigned statistical properties approximates a real situation that involves solute transport in a single heterogeneous aquifer. This assumption is reasonable only if the scale of the flow system is large compared with the correlation scale of the aquifer (Lumley and Panofsky, 1964, p. 36). Therefore, the macroscopic dispersivity developed by Gelhar and Axness is meaningful only when the overall scale of the solute transport problem is large compared with the correlation scale of the hydraulic conductivity.

6.4.6.3 Generalization of the Spectral Representation Expressions and First-Order Analysis

In section 6.4.4.4, spectral representation for one-dimensional stochastic processes are presented. As seen in the previous sections, the process of water flow and solute transport is characterized by porosity, hydraulic conductivity, and dispersion coefficients. Of the three parameters, variations of hydraulic conductivity at the local scale play the most important role in the dispersive process. Three-dimensional studies by Warren and Skiba (1964) have shown that the effect of porosity on the macroscopic dispersion process is second-order. Naff (1978) explicitly evaluated the effect of porosity variations and also found that this effect is second-order. In the Gelhar and Axness first-order analysis, porosity is lumped into a bulk dispersion coefficient and its variations are considered insignificant.

The ergodicity assumption in the probabilistic scheme implies that the spatially averaged flow behavior of a single heterogeneous porous medium can be represented by an ensemble of porous media distributed according to a probabilistic distribution. As a result, the various properties attributed to the random porous medium are described by stochastic processes dependent upon the spatial parameter $\mathbf{x}$ (see Eq. [6-113]). The most important tool in the analysis of Gelhar and Axness is commonly referred to as the "representation theorem" in the literature of spectral analysis. Although various forms of the representation theorem exist, here, the form that is presented by Eqs. (6-113)–(6-119) will be used.

As shown in detail in the previous sections, a random function $Y(\mathbf{x})$ can be expressed as

$$Y(\mathbf{x}) = \overline{Y}(\mathbf{x}) + Y'(\mathbf{x}) \tag{6-270}$$

where $Y(\mathbf{x})$ is the mean (first moment) and $Y'(\mathbf{x})$ is zero-mean random fluctuation. Equations (6-125), (6-126), and (6-155) are the special forms of Eq. (6-270) for $h(x), W(x)$, and $K(x)$, respectively. Generally, it is assumed that perturbations are small in order to linearize and drop second-order terms in the analysis. However, this procedure may not always be feasible because variations in some properties, such as hydraulic conductivity, can be large. In order to overcome this difficulty for hydraulic conductivity, the perturbed form of the Darcy equation can be expressed with respect to the logarithm of hydraulic conductivity, which has significantly less variation than the hydraulic conductivity itself (see Eqs. [6-199]–[6-203]). However, as mentioned in Section 6.3.6.2, in field situations, the variance of the logarithm of hydraulic conductivity, by itself, can be large (see Table 6-1). The assumption that the variation of perturbed parameters is small compared with the means of the parameters is usually required for a first-order analysis, which is a process where the terms higher than first-order variations are dropped in order to linearize a differential equation. The rationale behind this approach is that when first-order terms are small, the higher order terms are too small to make an appreciable difference in the solution.

6.4.6.4 Description of the Hydraulic Conductivity Variation

Studies during the past four decades showed that variations in hydraulic conductivity have a significant effect on dispersivity values. Studies by Law (1944) and Freeze (1975) have shown that hydraulic conductivity at the local scale is log-normally distributed (see Figure 6-4). The expressions for log-normal hydraulic conductivities are given by Eqs. (6-64) and (6-176). For three-dimensional cases, Eq. (6-176) can be expressed as

$$f(\mathbf{x}) = \ln[K(\mathbf{x})] \tag{6-271}$$

where $K(\mathbf{x})$ is the hydraulic conductivity stochastic process and $f(\mathbf{x})$ is the log-hydraulic conductivity, and is assumed to be statistically homogeneous. If $\ln[K(\mathbf{x})]$ is a Gaussian process, it is completely characterized by its mean and covariance. Also, the logarithm of $K(\mathbf{x})$ is less than $K(\mathbf{x})$ itself. For all of these reasons, it is therefore advantageous to use the variable $\ln[K(\mathbf{x})]$, which is regarded as a stochastic process. For a three-dimensional case (Eq. [6-178] is the one-dimensional case), the process can be expressed as

$$f(\mathbf{x}) = \overline{f}(\mathbf{x}) + f'(\mathbf{x}), \quad E[f'(\mathbf{x})] = 0 \tag{6-272}$$

where $f(\mathbf{x})$ is a three-dimensional statistically homogeneous random field characterized by its covariance function, similar to that in Eq. (6-67), and its covariance function is

$$R_{f'f'}(\boldsymbol{\zeta}) = E[f'(\mathbf{x} + \boldsymbol{\zeta})f'(\mathbf{x})] \tag{6-273}$$

where $\boldsymbol{\zeta}$ is the separation vector. Generally, one should expect that $R_{f'f'}(\boldsymbol{\zeta})$ depends on both the magnitude and direction of $\boldsymbol{\zeta}$, i.e., the $\ln[K(\mathbf{x})]$ is statistically anisotropic. Such anisotropy would represent the layering and lenticular character in sedimentary rocks (Smith, 1981).

6.4.6.5 Correlation Structure of the Random Hydraulic Conductivity Field

Autocovariance Functions for Hydraulic Conductivity Fields
During the past three decades studies, were conducted an the effect of the correlation structure of the random hydraulic conductivity field (Gelhar, 1976; Bakr et al., 1978). These studies were based on the assumption of the statistical homogeneity of the random medium. For three-dimensional anisotropic flow systems, two forms of autocovariance functions for random hydraulic conductivity fields have been suggested by Naff (1978) and are described below (Gelhar and Axness, 1981, 1983; Gelhar, 1993).

The Exponential Autocovariance
The exponential autocovariance, as a function of the separation vector $\boldsymbol{\zeta} = [\zeta_1, \zeta_1, \zeta_1]$, is

$$R_{f'f'}(\boldsymbol{\zeta}) = \sigma_f^2 \exp\left[-\left(\frac{\zeta_1^2}{\lambda_1^2} + \frac{\zeta_2^2}{\lambda_2^2} + \frac{\zeta_3^2}{\lambda_3^2}\right)^{1/2}\right] \tag{6-274}$$

which has the spectrum

$$S_{f'f'}(\mathbf{k}) = \sigma_f^2 \frac{\lambda_1 \lambda_2 \lambda_3}{[\pi^2(1 + k_1^2\lambda_1^2 + k_2^2\lambda_2^2 + k_3^2\lambda_3^2)^2]} \tag{6-275}$$

where σ_f^2 is the variance of the $f(x)$ process, given by Eq. (6-272), λ_1, λ_2, and λ_3 (the distances to e^{-1} correlation) are the correlation scales in the principal coordinate directions, and k_1, k_2, and k_3, are the components of the of the wave number vector, $\mathbf{k}$.

The Modified Exponential Autocovariance
The modified exponential autocovariance is

$$R_{f'f'}(\boldsymbol{\zeta}) = \sigma_f^2\left(1 - \frac{\zeta_3^2}{l_3^2}s\right)\exp(-s) \tag{6-276}$$

where

$$s^2 = \frac{\zeta_1^2}{l_1^2} + \frac{\zeta_2^2}{l_2^2} + \frac{\zeta_3^2}{l_3^2} \tag{6-277}$$

which has the spectrum

$$S_{f'f'}(\mathbf{k}) = 4\sigma_f^2 \frac{l_1 l_2 l_3^3 k_3^2}{[\pi^2(1 + l_1^2 k_1^2 + l_2^2 k_2^2 + l_3^2 k_3^2)^3]} \tag{6-278}$$

The selection of an autocovariance to represent the log-hydraulic conductivity field depends on the medium involved. For example, Naff (1978) states that Eq. (6-274) is not a suitable autocovariance form for the limiting anisotropic case of a perfectly stratified medium ($\lambda_1 = \lambda_2 \rightarrow \infty$) with the mean flow parallel to the layers, since the resulting head variance is nonzero. This limiting case corresponds to a layered medium with flow parallel to the layers. Therefore, the resulting head field has a constant gradient in the x-coordinate direction with no flow perpendicular to the layers, which means that the head variance must be zero. Equation (6-276) satisfies this condition. Gelhar and Axness (1981, 1983) used both autocovariance and spectrum pairs in their analysis, which is presented in the following sections.

Random Fields That Are Not Statistically Homogeneous
As mentioned earlier, hydraulic characteristics of aquifer materials, such as hydraulic conductivity, generally vary significantly from location to location. The general expressions for three-dimensional statistically homogeneous random field representation are presented in Section 6.4.3. When dealing with random fields that are not statistically homogeneous, the structure of the variability can sometimes be characterized by considering second-moment characteristics of the difference of values of the random fields at two locations. This approach is presented below (Gelhar, 1993).
For a variable hydraulic conductivity field, the variogram is

$$\gamma_{f'f'}(\mathbf{x}, \mathbf{s}) = \tfrac{1}{2}E[[f'(\mathbf{x} + \mathbf{s}) - f'(\mathbf{x})]^2], \quad E[f'(\mathbf{x})] = 0 \tag{6-279}$$

where f' is the mean-removed random hydraulic conductivity field. If the variogram is independent of position $\mathbf{x}$, and dependent only on the separation vector $\mathbf{s}$, Eq. (6-279) can be expressed as

$$\gamma_{f'f'}(\mathbf{x}, \mathbf{s}) = \gamma_{f'f'}(\mathbf{s}) \tag{6-280}$$

The variogram in Eq. (6-279) can be expressed as

$$\gamma_{f'f'}(\mathbf{s}) = \tfrac{1}{2}[E[f'^2(\mathbf{x} + \mathbf{s})] - 2E[f'(\mathbf{x} + \mathbf{s})f(\mathbf{x})] + E[f'^2(\mathbf{x})]] \tag{6-281}$$

If the random field is statistically homogeneous (stationary) (see Section 6.4.1.4 for definition)

$$E[f'^2(\mathbf{x} + \mathbf{s})] = E[f'^2(\mathbf{x})] = \sigma_{f'}^2 \tag{6-282}$$

and

$$E[f'(\mathbf{x} + \mathbf{s})f'(\mathbf{x})] = R_{f'f'}(\mathbf{s}) \tag{6-283}$$

The substitution of Eqs. (6-282) and (6-283) into Eq. (6-281) gives the relationship between the variogram, $\gamma_{f'f'}(s)$, and the covariance, $R_{f'f'}(s)$ as

$$\gamma_{f'f'}(\mathbf{s}) = \tfrac{1}{2}[\sigma_{f'}^2 - 2R_{f'f'}(\mathbf{s}) + \sigma_{f'}^2] = \sigma_{f'}^2 - R_{f'f'}(\mathbf{s}) \tag{6-284}$$

Figure 6-50 graphically shows the relationship between the variogram and the covariance function for a statistically homogeneous process. The variogram approaches to a constant value with an increase in separating , as observed in figure 6-50. This constant value is called the *sill of the variogram*. However, if the random field is not stationary, the variogram may increase indefinitely with the increase in separation

Relationship between the Variogram and Covariance for One-Dimensional Case
Consider a three-dimensional random field represented by Eqs. (6-274) and (6-275). The one-dimensional covariance function can be determined by making the separations $\zeta_2 = \zeta_3 = 0$ in

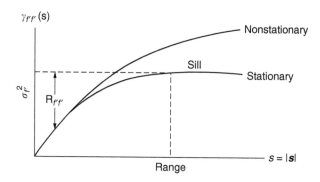

FIGURE 6-50 Variograms for an isotropic process. (Adapted from Gelhar, L.W., *Stochastic Subsurface Hydrology*, Prentice Hall, Inc., Englewood Cliffs, NJ, 1993, 390 pp.)

Eq. (6-274)

$$R_{f'f'}(\zeta_1, 0, 0) = \sigma_{f'}{}^2\exp\left(-\frac{|\zeta_1|}{\lambda_1}\right), \quad \zeta_1 \geq 0 \tag{6-285}$$

The substitution of Eq. (6-285) into Eq. (6-284) gives

$$\gamma_{f'f'}(\zeta_1) = \sigma_{f'}^2\left[1 - \exp\left(-\frac{|\zeta_1|}{\lambda_1}\right)\right] \tag{6-286}$$

Stochastic Characterization of Aquifers

As can be seen from the above expressions, the covariance structure of an aquifer is a function of the variance of the hydraulic conductivity field, $\sigma_{f'}^2$ and the correlation scales λ_1, λ_2, and λ_3. Therefore, in order to apply the results of the stochastic theory to real-world problems, the covariance structure of the subject aquifer must be known. At most aquifers, limited data are available from which the covariance structure of the aquifer can be determined. Gelhar (1993) states that such limitations to determine the covariance structure of an aquifer or statistical parameters are the weakest link in the process of applying stochastic modeling approaches. Characterization of the statistical parameters of an aquifer is still in the development process. Beyond this, it is also equally important to understand how reliably the necessary statistical parameters can be estimated from a given type and amount of data. Some of these issues will be discussed below by presenting an extensive summary of field observations for sites where statistical characteristics have been determined (Gelhar, 1993).

Aquifer Covariance Data from the Literature

In Table 6-14, aquifer covariance data from the literature are summarized (Gelhar, 1993). The same data are graphically presented in Figure 6-51 and Figure 6-52. The data in Table 6-14 involve three general types of observations. The largest scale measurements are generally depth-averaged observations based on aquifer pumping tests and specific capacity measurements from a single well. The depth-average data are based on transmissivity, T, values and they do not include any information about vertical variation. A second category of information is based on observations of infiltration rate on the soil surface and include only information about horizontal variations in hydraulic conductivity. A third category of information involves three-dimensional observations. Figure 6-51 shows that the standard deviation of $f = \ln(K)$, $\sigma_{f'}$, varies over an order of magnitude, but does not show a systematic dependence on scale. Figure 6-52 shows that the correlation scale increases systematically with an increase in the overall scale. Figure 6-53 shows only the data up to an overall scale of 1 km. The horizontal and vertical correlation scales determined at a given site are connected by lines. Typically, the vertical scales are an order of magnitude less than the horizontal scale.

TABLE 6-14
Aquifer Covariance Data from the Literature

Source	Medium	Type[a]	σ_f^b	Correlation Scale λ (m) λ_v^c	λ_v^d	Overall Scale (m) Horizontal	Vertical
Aboufirassi and Marino (1984)	Alluvial-basin aquifer	T	1.22	4,000		30,000	
Bakr (1976)	Sandstone aquifer	A	1.5–2.2		0.3–1.0		100
Binsariti (1980)	Alluvial-basin aquifer	T	1.0	800		20,000	
Byers and Stephen (1983)	Fluvial sand	A	0.9	> 3	0.1	14	5
Delhomme (1979)	Limestone aquifer	T	2.3	6,300		30,000	
Delhomme (personal com.)							
Aquitane	Sand stone aquifer	T	1.4	17,000		50,000	
Durance	Alluvial aquifer	T	0.6	150		5,000	
Kairouan	Alluvial aquifer	T	0.4	1,800		25,000	
Normandy	Limestone aquifer	T	2.3	3,500		40,000	
Nord	Chalk	T	1.7	7,500		80,000	
Devary and Doctor (1982)	Alluvial aquifer	T	0.8	820		5,000	
Gelhar et al. (1983)	Fluvial soil	S	1.0	7.6		760	
Goggin et al. (1988)	Eolian sandstone outcrop	A	0.4	8	3	30	60
Hess (1989)	Glacial outwash sand	A	0.5	5	0.26	20	5
Hoeksema and Kitanidis (1985)	Sandstone aquifer	T	0.6	4.5×10^4		5×10^4	
Hufschmied (1986)	Sand and gravel aquifer	A	1.9	20	0.5	100	20
Loague and Gander (1990)	Prairies soil	S	0.6	8		100	
Luxmoore et al. (1981)	Weathered shale subsoil	S	0.8	< 2		14	
Rehfeldt et al. (1989a, 1989b)	Fluvial sand and gravel aquifer	A	2.1	13	1.5	90	7
Russo and Bressler (1981)	Homra red mediterranean soil	S	0.4–1.1	14–19		100	
Russo (1984)	Gravelly loamy sand soil	S	0.7	500		1,600	
Sisson and Wierenga (1981)	Alluvial silty-clay loam soil	S	0.6	0.1		6	
Smith (1978)	Glacial outwash sand and gravel outcrop	A	0.8	5	0.4	30	30
Sudicky (1986)	Glacial-lacustrine aquifer	A	0.6	3	0.12	20	2
Viera et al. (1981)	Alluvial soil (Yolo)	S	0.9	15		100	

[a] Types of data: T, transmissivity; S, soils; A, three-dimensional aquifer.
[b] Standard deviation of $f = \ln(K)$.
[c] Horizontal correlation scale.
[d] Vertical correlation scale.

Adapted from Gelhar, L.W., *Stochastic Subsurface Hydrology*, Prentice Hall, Inc., Englewood Cliffs, NJey, 1993, 390 pp.

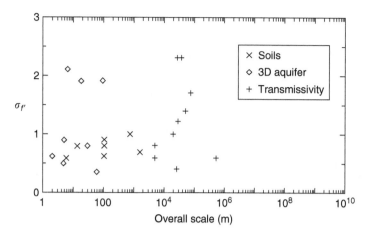

FIGURE 6-51 Loghydraulic conductivity standard deviation versus scale. (Adapted from Gelhar, L.W., Stochastic Subsurface Hydrology, Prentice Hall, Inc., Englewood Cliffs, NJ, 1993, 390 pp.)

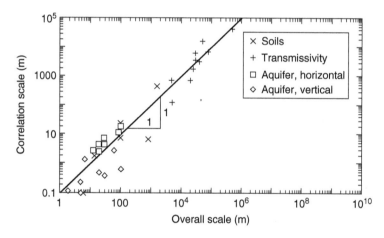

FIGURE 6-52 Correlation scale versus overall scale. (Adapted from Gelhar, L.W., *Stochastic Subsurface Hydrology*, Prentice Hall, Inc., Englewood Cliffs, NJ, 1993, 390 pp.)

Figure 6-54 shows the product of the log-hydraulic conductivity variance and the horizontal correlation scale ($\sigma_f^2 \lambda_h$). This type of plot is interesting because the longitudinal macrodispersivity, A_{11}, is generally proportional to the product of $\sigma_f^2 \lambda_h$ (see Eq. [6-384]). It is also interesting to observe that the increase in this product with scale is similar to the increase in longitudinal macrodispersivity with scale, as observed from field observations (see Figure 6-49).

Aquifer Covariance Data from Field Tests
Table 6-15 presents log-hydraulic conductivity covariance parameters from some field tests (Gelhar, 1993). The Borden site described in section 6.3.7.1, located in Ontario, Canada. It consists of fine to medium sand of glaciolacustrine origin and is the first field site where three-dimensional hydraulic conductivity characterization was conducted. Gelhar (1993) states that the sampling network at the Borden site was probably not adequate to make tight estimates of the parameters. A second site where detailed hydraulic conductivity characterization was conducted is the Cape Cod site in Massachusetts, which is described in Section 6.3.7.3. A third major research site where extensive characterization of the variability of hydraulic conductivity was conducted is the tracer test site near Columbus, Mississippi (Boggs et al., 1990). The aquifer at the Columbus site includes shallow

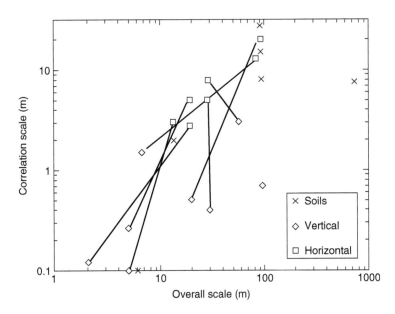

FIGURE 6-53 Correlation scale versus overall scale less than 1 km. (Adapted from Gelhar, L.W., *Stochastic Subsurface Hydrology*, Prentice Hall, Inc., Englewood Cliffs, NJ, 1993, 390 pp.)

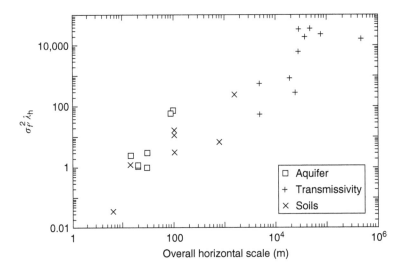

FIGURE 6-54 Product of log-hydraulic conductivity variance and correlation scale versus overall scale. (Adapted from Gelhar, L.W., *Stochastic Subsurface Hydrology*, Prentice Hall, Inc., Englewood Cliffs, NJ, 1993, 390 pp.)

alluvial terrace deposits of sand and gravel, and is distinguished from the Borden and Cape Cod sites by a much higher degree of heterogeneity.

6.4.6.6 Macroscopic Dispersive Fluxes

The Effects of Variations of Porosity and Local Dispersion Coefficients
The effects of variations of porosity (Warren and Skiba, 1964; Naff, 1978) and local dispersion coefficients (Naff, 1978; Gelhar et al., 1979) have been considered and were found to be secondary, relative to the effects of hydraulic conductivity variations. Therefore, in the Gelhar and Axness analysis, these parameters are treated as constants.

TABLE 6-15
Log-Hydraulic Conductivity Covariance Parameters for Field Research Sites

Site	$\sigma_f^{2\,a}$	Correlation Scale λ (m)		Conditions
		λ_h^b	λ_v^c	
Borden site				
Sudicky (1986)	0.38	2.8	0.12	
Woodbury and Sudicky (1991)	0.24	5.1	0.21	Transect A–A
Woodbury and Sudicky (1991)	0.37	8.3	0.34	Transect B–B
Robin et al. (1991)	0.37	1–7.5	0.1–0.3	
Cape Cod site				
Hess (1989)	0.24	5.1	0.26	Best estimates
	0.20–0.32	3.1–8.8	0.16–0.91	$\pm\,2\,\sigma_f$ intervals
Columbus site				
Rehfeldt et al. (1989a, 1989b)	3.5	12		15 wells
	4.6	13	1.6	30 wells, best estimates
	3.3–6.8	7–23	0.75–2.5	30 wells, $\pm\,2\,\sigma_f$ intervals
	3.6	9.5	1.1	30 wells; 2nd order, 2D horizontal trend removed
Boggs et al. (1990)	4.5	12	1.5	48 wells, no trend

[a] Variance.
[b] Horizontal correlation scale.
[c] Vertical correlation scale.

Adapted from Gelhar, L.W., *Stochastic Subsurface Hydrology*, Prentice Hall, Inc., Englewood Cliffs, NJ, 1993, 390 pp.

Macroscopic Dispersive Flux Expressions
The physical and mathematical foundations for convective–dispersive fluxes in the deterministic sense are included in Chapter 2. Based on the generalized form of the convective–dispersive fluxes (Eq. [2-28]), the dispersive flux is defined as

$$\text{Dispersive Flux} = \text{Total Mass Flux} - \text{Mean Convective Flux} \qquad (6\text{-}287)$$

or in terms of expectations (indicated by overbar)

$$E[q_i' C'] \equiv \overline{q'_i C'} = \overline{q_i C} - \overline{q_i}\,\overline{C}, \quad i = x, y, z \qquad (6\text{-}288)$$

where q_i is the specific discharge and C The concentration. The expanded forms of Eq. (6-288) are:

$$F_x' = \overline{q'_x C'} = \overline{q_x C} - \overline{q_i}\,\overline{C} \qquad (6\text{-}289a)$$

$$F_y' = \overline{q'_y C'} = \overline{q_y C} - \overline{q_i}\,\overline{C} \qquad (6\text{-}289b)$$

$$F_z' = \overline{q'_z C'} = \overline{q_z C} - \overline{q_i}\,\overline{C} \qquad (6\text{-}289c)$$

If the dispersive flux is Fickian in nature, similar to Eq. (2-28) of, Eq. (6-288) can be written as

$$F_i' = \overline{q'_i C'} = -q A_{ij} \frac{\partial \overline{C}}{\partial x_j}, \quad i = x, y, z, \quad j = x, y, z \qquad (6\text{-}290)$$

where A_{ij} is the effective dispersivity tensor or macroscopic dispersivity tensor and q is the magnitude of the specific discharge vector. The expanded form of Eq. (6-290) can be written as

$$F_x' = \overline{q_x' C'} = -q\left(A_{xx}\frac{\partial \overline{C}}{\partial x} + A_{xy}\frac{\partial \overline{C}}{\partial y} + A_{xz}\frac{\partial \overline{C}}{\partial z}\right) \qquad (6\text{-}291a)$$

$$F_y' = \overline{q_y' C'} = -q\left(A_{yx}\frac{\partial \overline{C}}{\partial x} + A_{yy}\frac{\partial \overline{C}}{\partial y} + A_{yz}\frac{\partial \overline{C}}{\partial z}\right) \qquad (6\text{-}291b)$$

$$F_z' = \overline{q_z' C'} = -q\left(A_{zx}\frac{\partial \overline{C}}{\partial x} + A_{zy}\frac{\partial \overline{C}}{\partial y} + A_{zz}\frac{\partial \overline{C}}{\partial z}\right) \qquad (6\text{-}291c)$$

Notice that Eqs. (6-291) are of the same nature as the second terms on the right-hand side on Eqs. (2-31) (deterministic-based equations), with the exception that they are written in terms of macroscopic dispersivities. For a uniform flow case, the usual dispersion coefficient tensor, as given by Eq. (2-135), is related to A_{ij} by

$$D_{m_{ij}} = UA_{ij} = \frac{q}{\varphi_e}A_{ij} \qquad (6\text{-}292)$$

where φ_e is the effective porosity.

Equation (6-290) is the resulting generalized form of the Fickian relationship in terms of macrodispersivity tensor A_{ij}, which is related to the traditional dispersion coefficient tensor D_{ij} by Eq. (6-292).

6.4.6.7 Determination of the General Expression for the Macroscopic Dispersion Tensor A_{ij}

In this section, first the stochastic differential equation for steady-state macroscopic dispersion for a three-dimensional porous medium is derived; and then, its general spectral solution is presented (Gelhar and Axness, 1981, 1983).

Stochastic Solute Transport Differential Equation: General Case

Unsteady State Deterministic Solute Transport Differential Equation
The general three-dimensional solute transport differential equation in the deterministic sense is derived in Section 2.5.1 and is given by Eq. (2-47). The tensorial form of Eq. (2-47) is

$$\varphi_e \frac{\partial C}{\partial t} = \frac{\partial}{\partial x_i}\left(E_{ij}\frac{\partial C}{\partial x_i} - Cq_i\right), \quad i = x, y, z, \quad j = x, y, z \qquad (6\text{-}293)$$

where C is the concentration of the transported solute, q_i the specific discharge in the x_i direction, and

$$E_{ij} = \varphi_e D_{ij} \qquad (6\text{-}294)$$

is the local bulk dispersion coefficient tensor, in which φ_e is the effective porosity, and D_{ij} the dispersion coefficient tensor including hydrodynamic dispersion and molecular diffusion. The open forms of Eqs. (6-293) and (6-294) are

$$\varphi_e \frac{\partial C}{\partial t} = \frac{\partial}{\partial x}\left(E_{xx}\frac{\partial C}{\partial x} + E_{xy}\frac{\partial C}{\partial y} + E_{xz}\frac{\partial C}{\partial z} - Cq_x\right)$$

$$+ \frac{\partial}{\partial y}\left(E_{yx}\frac{\partial C}{\partial x} + E_{yy}\frac{\partial C}{\partial y} + E_{yz}\frac{\partial C}{\partial z} - Cq_y\right) \qquad (6\text{-}295)$$

$$+ \frac{\partial}{\partial z}\left(E_{zx}\frac{\partial C}{\partial x} + E_{zy}\frac{\partial C}{\partial y} + E_{zz}\frac{\partial C}{\partial z} - Cq_z\right)$$

and

$$E_{xx} = \varphi_e D_{xx}, \quad E_{xy} = \varphi_e D_{xy}, \quad E_{xz} = \varphi_e D_{xz}$$

$$E_{yx} = \varphi_e D_{yx}, \quad E_{yy} = \varphi_e D_{yy}, \quad E_{yz} = \varphi_e D_{yz} \tag{6-296}$$

$$E_{zx} = \varphi_e D_{zx}, \quad E_{zy} = \varphi_e D_{zy}, \quad E_{zz} = \varphi_e D_{zz}$$

as well as

$$q_x = \varphi_e v_x, \quad q_y = \varphi_e v_y, \quad q_z = \varphi_e v_z \tag{6-297}$$

Equation. (6-295), (6-296), and (6-297) are similar to Eq. (2-47).

Steady-State Deterministic Solute Transport Differential Equation
Under steady-state solute transport conditions, the time derivative in Eq. (6-295) vanishes and, in the absence of sources or sinks, the equation of continuity,

$$\frac{\partial q_i}{\partial x_i} = 0, \quad i = x, y, z \tag{6-298}$$

or its open form,

$$\frac{\partial q_x}{\partial x} + \frac{\partial q_y}{\partial y} + \frac{\partial q_z}{\partial z} = 0 \tag{6-299}$$

is imposed. Therefore, Eq. (6-293) takes the form

$$\frac{\partial}{\partial x_i}(q_i C) = q_i \frac{\partial C}{\partial x_i} + C \frac{\partial q_i}{\partial x_i} = \frac{\partial}{\partial x_i}\left(E_{ij} \frac{\partial C}{\partial x_j}\right), \quad i = x, y, z, \quad j = x, y, z \tag{6-300}$$

and by using Eq. (6-299), it takes the form

$$q_i \frac{\partial C}{\partial x_i} = \frac{\partial}{\partial x_i}\left(E_{ij} \frac{\partial C}{\partial x_j}\right), \quad i = x, y, z, \quad j = x, y, z \tag{6-301}$$

The open form of Eq. (6-301) is

$$q_x \frac{\partial C}{\partial x} + q_y \frac{\partial C}{\partial y} + q_z \frac{\partial C}{\partial z} = \frac{\partial}{\partial x}\left(E_{xx} \frac{\partial C}{\partial x} + E_{xy} \frac{\partial C}{\partial y} + E_{xz} \frac{\partial C}{\partial z}\right)$$

$$+ \frac{\partial}{\partial y}\left(E_{yx} \frac{\partial C}{\partial x} + E_{yy} \frac{\partial C}{\partial y} + E_{yz} \frac{\partial C}{\partial z}\right) + \frac{\partial}{\partial z}\left(E_{zx} \frac{\partial C}{\partial x} + E_{zy} \frac{\partial C}{\partial y} + E_{zz} \frac{\partial C}{\partial z}\right) \tag{6-302}$$

Assumptions and Expressing of the Variables in Terms of a Mean and Perturbation
In addition to the above equations, the following assumptions are made: (1) the local bulk dispersion coefficient, E_{ij}, is constant; (2) "small" random perturbations about the mean occur in concentration, specific discharge, and log-hydraulic conductivity; and (3) these perturbations are stationary (homogeneous) spatial random fields and are not dependent on time. The expressions for concentration, log-hydraulic conductivity, and specific discharge, respectively, are as follows:

$$C(x, y, z) = \overline{C}(x, y, z) + C'(x, y, z) \tag{6-303}$$

$$f(x, y, z) = \overline{f}(x, y, z) + f'(x, y, z) \tag{6-304}$$

and

$$q_i(x, y, z) = \bar{q}_i(x, y, z) + q_i{}'(x, y, z) \tag{6-305}$$

where over bar indicates the mean of these quantities and the prime indicates perturbation.

General Stochastic Differential Equation
By using the perturbed forms of Eqs. (6-303), (6-304), and (6-305) into Eq. (6-301), we get

$$\frac{\partial}{\partial x_i}(\bar{q}_i\bar{C} + \bar{q}_iC' + q_i{}'\bar{C} + q_i{}'C') = E_{ij}\frac{\partial^2(\bar{C} + C')}{\partial x_i\partial x_j} \tag{6-306}$$

Taking the expected values of each term in Eq. (6-306) gives

$$E\left[\frac{\partial}{\partial x_i}(\bar{q}_i\bar{C})\right] = \frac{\partial}{\partial x_i}(\bar{q}_i\bar{C}) \tag{6-307}$$

$$E\left[\frac{\partial}{\partial x_i}(\bar{q}_iC')\right] = \frac{\partial}{\partial x_i}(\bar{q}_iE[C']) = 0 \tag{6-308}$$

$$E\left[\frac{\partial}{\partial x_i}(q_i{}'\bar{C})\right] = \frac{\partial}{\partial x_i}(\bar{C}E[q_i{}']) = 0 \tag{6-309}$$

$$E\left[\frac{\partial}{\partial x_i}(q_i{}'C')\right] = \frac{\partial}{\partial x_i}(E[q_i{}'C')] \equiv \frac{\partial}{\partial x_i}(\overline{q_i{}'C'}) \tag{6-310}$$

and

$$E\left[E_{ij}\frac{\partial^2(\bar{C} + C')}{\partial x_i\partial x_j}\right] = E_{ij}\frac{\partial^2\bar{C}}{\partial x_i\partial x_j} \tag{6-311}$$

By substituting Eqs. (6-307)–(6-311) into Eq. (6-306) one has

$$\frac{\partial}{\partial x_i}(\bar{q}_i\bar{C}) + \frac{\partial}{\partial x_i}(\overline{q_i{}'C_i}) = E_{ij}\frac{\partial^2\bar{C}}{\partial x_i\partial x_j} \tag{6-312}$$

The second term on the left-hand side of Eq. (6-312) reflects the dispersive flux due to random variation in C and q_i. The evaluation of the dispersive flux, $\overline{q_i{}'C'}$, is the main focus of the Gelhar and Axness investigation. In writing Eq. (6-312), the assumption in the item (1) given above regarding the constancy of E_{ij} is implemented.

The stochastic differential equation is determined by subtracting the mean equation, Eq. (6-312), from Eq. (6-306) as

$$\frac{\partial}{\partial x_i}(\bar{q}_iC' + q_i{}'\bar{C} + q_i{}'C' - \overline{q_i{}'C'}) = E_{ij}\frac{\partial^2C'}{\partial x_i\partial x_j} \tag{6-313}$$

Implementing the assumption that primed quantities are small, assumption in the item (2) given above, the second-order term $q'_i C' - \overline{q'_i C'}$ is neglected; from Eq. (6-313) the first-order approximation describing the concentration perturbation becomes

$$\frac{\partial}{\partial x_i}(q_i{}'\bar{C} + \bar{q}_iC') = E_{ij}\frac{\partial^2C'}{\partial x_i\partial x_j} \tag{6-314}$$

which is the general three-dimensional stochastic differential equation. In the x, y, and z coordinates, or in open form, Eq. (6-314), takes the form

$$\frac{\partial}{\partial x}(q_x{}'\overline{C} + \overline{q}_x C') + \frac{\partial}{\partial y}(q_y'\overline{C} + \overline{q}_y C') + \frac{\partial}{\partial z}(q_z'\overline{C} + \overline{q}_z C')$$

$$= E_{xx}\frac{\partial^2 C'}{\partial x^2} + E_{xy}\frac{\partial^2 C'}{\partial x \partial y} + E_{xz}\frac{\partial^2 C'}{\partial x \partial z}$$

$$+ E_{yx}\frac{\partial^2 C'}{\partial y \partial x} + E_{yy}\frac{\partial^2 C'}{\partial y^2} + E_{yz}\frac{\partial^2 C'}{\partial y \partial z} \tag{6-315}$$

$$+ E_{zx}\frac{\partial^2 C'}{\partial z \partial x} + E_{zy}\frac{\partial^2 C'}{\partial z \partial y} + E_{zz}\frac{\partial^2 C'}{\partial z^2}$$

Stochastic Solute Transport Differential Equation: Special Case

For convenience, the x-coordinate axis is aligned in the direction of mean flow so that

$$\overline{q}_x = q \neq 0, \qquad \overline{q}_y = \overline{q}_z = 0 \tag{6-316}$$

From Eqs. (2-152) and (6-294), the local mechanical dispersion coefficient tensor can be expressed as

$$[E_{m_{ij}}] = \varphi_e \begin{bmatrix} \alpha^*_{L}U & 0 & 0 \\ 0 & \alpha^*_{T}U & 0 \\ 0 & 0 & \alpha^*_{T}U \end{bmatrix} = \begin{bmatrix} \alpha^*_{L}q & 0 & 0 \\ 0 & \alpha^*_{T}q & 0 \\ 0 & 0 & \alpha^*_{T}q \end{bmatrix} = \varphi_e[D_{m_{ij}}] \tag{6-317}$$

where α^*_{L} and α^*_{T} are the local longitudinal and transverse dispersivities, respectively. If the specific discharge is very small, an isotropic effective molecular diffusion term could be added to these components. In order to distinguish them from α_L and α_T values in Eq. (2-152), the asterisk (*) is used. Notice that Eq. (6-317) is in the form of Eq. (2-152) and it is the mechanical dispersion coefficient tensor.

Now, expanding the left-hand side of Eq. (6-314) and utilizing Eqs. (6-298) and (6-317), Eq. (6-314) reduces to

$$\frac{\partial q_i'}{\partial x_i}\overline{C} + q_i'\frac{\partial \overline{C}}{\partial x_i} + \frac{\partial \overline{q}_i}{\partial x_i}C' + \overline{q}_i\frac{\partial C'}{\partial x_i} = E_{ij}\frac{\partial^2 C'}{\partial x_i \partial x_j} \tag{6-318}$$

or

$$q_i'\frac{\partial \overline{C}}{\partial x_i} + q\frac{\partial C'}{\partial x_1} = E_{11}\frac{\partial^2 C'}{\partial x_1^2} + E_{22}\frac{\partial^2 C'}{\partial x_2^2} + E_{33}\frac{\partial^2 C'}{\partial x_3^2} \tag{6-319}$$

or by using

$$E_{11} = \varphi_e\alpha^*_{L}U = q\alpha^*_{L}, \qquad E_{22} = \varphi_e\alpha^*_{T}U = q\alpha^*_{T}, \qquad E_{33} = \varphi_e\alpha^*_{T}U = q\alpha^*_{T} \tag{6-320}$$

Equation (6-319) takes the form

$$q_i'\frac{\partial \overline{C}}{\partial x_i} + q\frac{\partial C'}{\partial x_1} = q\left[\alpha^*_{L}\frac{\partial^2 C'}{\partial x_1^2} + \alpha^*_{T}\left(\frac{\partial^2 C'}{\partial x_2^2} + \frac{\partial^2 C'}{\partial x_3^2}\right)\right] \tag{6-321}$$

By using (x, y, z) instead of (x_1, x_2, x_3), Eq. (6-321) takes the form

$$q_x'\frac{\partial \overline{C}}{\partial x} + q_y'\frac{\partial \overline{C}}{\partial y} + q_z'\frac{\partial \overline{C}}{\partial z} + q\frac{\partial C'}{\partial x} = q\left[\alpha^*_{L}\frac{\partial^2 C'}{\partial x^2} + \alpha^*_{T}\left(\frac{\partial^2 C'}{\partial y^2} + \frac{\partial^2 C'}{\partial z^2}\right)\right] \tag{6-322}$$

Equation (6-321) or (6-322) is an approximate stochastic differential equation describing the concentration perturbation C' produced as a result of specific discharge perturbations q_i' (q_x', q_y', q_z').

Spectral Solution and the Macroscopic Dispersion Tensor A_{ij}
As in Eq. (6-102), consider the case of statistically homogeneous input perturbation for C', which is represented by Fourier–Stieltjes integrals in wave number k as follows

$$C'(\mathbf{x}) = \int_{-\infty}^{\infty} e^{i(\mathbf{k}.\mathbf{x})} \, dZ_{C'(x)} \tag{6-323}$$

Similarly, the Darcy flux perturbation is represented by

$$q_i'(\mathbf{x}) = \int_{-\infty}^{\infty} e^{i(\mathbf{k}.\mathbf{x})} \, dZ_{qi'(x)}(\mathbf{k}) \tag{6-324}$$

In Eqs. (6-323) and (6-324), $\mathbf{x} = \{x_1, x_2, x_3\}$ is the position vector and $\mathbf{k} = \{k_1, k_2, k_3\}$ is the wave number vector. The processes discussed in Section 6.4.2.1 are all examples of a general type called *time series*, i.e., the process varyies with time. For example, Eq. (6-102) is the representation of a time series. However, the variation of some quantity over a region of space can be also studied equally. Such phenomena are usually referred to as *spatial processes*. Therefore, Eqs. (6-323) and (6-324) represent spatial processes.
Let

$$G_i = -\frac{\partial \overline{C}}{\partial x_i} \tag{6-325}$$

From Eq. (6-323) we get

$$\frac{\partial C'}{\partial x_1} = ik_1 \int_{-\infty}^{\infty} e^{i(\mathbf{k}.\mathbf{x})} \, dZ_{C'}, \quad \frac{\partial^2 C'}{\partial x_1^2} = -k_1^2 \int_{-\infty}^{\infty} e^{i(\mathbf{k}.\mathbf{x})} \, dZ_{C'} \tag{6-326}$$

and

$$\frac{\partial^2 C'}{\partial x_2^2} = -k_2^2 \int_{-\infty}^{\infty} e^{i(\mathbf{k}.\mathbf{x})} \, dZ_{C'}, \quad \frac{\partial^2 C'}{\partial x_3^2} = -k_3^2 \int_{-\infty}^{\infty} e^{i(\mathbf{k}.\mathbf{x})} \, dZ_{C'} \tag{6-327}$$

The substitution of Eqs. (6-324), (6-325), (6-326), and (6-327) into Eq. (6-321) gives

$$-G_i \int_{-\infty}^{\infty} e^{i(\mathbf{k}.\mathbf{x})} \, dZ_{q_i} + q \int_{-\infty}^{\infty} ik_1 e^{i(\mathbf{k}.\mathbf{x})} \, dZ_{C'}$$

$$= q[-\alpha_L^* k_1^2 \int_{-\infty}^{\infty} e^{i(\mathbf{k}.\mathbf{x})} \, dZ_{C'} + \alpha_T^*(-k_2^2 \int_{-\infty}^{\infty} e^{i(\mathbf{k}.\mathbf{x})} \, dZ_{C'} - k_3^2 \int_{-\infty}^{\infty} e^{i(\mathbf{k}.\mathbf{x})} \, dZ_{C'})] \tag{6-328}$$

By removing the integrals, we get

$$[ik_1 + \alpha_L^* k_1^2 + \alpha_T^*(k_2^2 + k_3^2)] \, q \, dZ_{C'} = G_i \, dZ_{q_i'} = G_j \, dZ_{q_j'} \tag{6-329}$$

since the repeated subscript, representing summation, is arbitrary. Equation. (6-329) can also be expressed as

$$[ik_x + \alpha_L^* k_x^2 + \alpha_T^*(k_y^2 + k_z^2)] \, q \, dZ_{C'} = G_x \, dZ_{q_x'} + G_y \, dZ_{q_y'} + G_z \, dZ_{q_z'} \tag{6-330}$$

Multiplying both sides of Eq. (6-329) by the complex conjugate Fourier amplitude $dZ_{q_j'}^*$ and taking mean values and using the spectral representation theorem (see Section 6.4.2.1), results in a spectral relationship

$$[ik_1 + \alpha_L^* k_1^2 + \alpha_T^* (k_2^2 + k_3^2)]qE[dZ_{C'}\,dZ_{q_i'}^*] = G_j\,E[dZ_{q_j'}\,dZ_{q_i'}^*] \tag{6-331}$$

By applying Eq. (6-105), Eq. (6-331) takes the form

$$[ik_1 + \alpha_L^* k_1^2 + \alpha_T^* (k_2^2 + k_3^2)]qS_{C'q_i'}(\mathbf{k})\,d\mathbf{k} = G_j\,S_{q_j'q_i'}(\mathbf{k})d\mathbf{k}, \quad i,j = 1,2,3 \tag{6-332}$$

or

$$S_{C'q_i'}(\mathbf{k}) = \frac{qG_j\,S_{q_j'q_i'}(\mathbf{k})}{[ik_1 + \alpha_L^* k_1^2 + \alpha_T^* (k_2^2 + k_3^2)]q^2}, \quad i,j = 1,2,3 \tag{6-333}$$

which is the cross-spectrum of C' and q_i'. The cross covariance between C' and q_i', if these are part of a stationary field, can be written as (Lumley and Panofsky, 1964, pp. 19-23)

$$E[C'q_i'] = \overline{C'q_i'} = \int_{-\infty}^{\infty} S_{C'q_i'}(\mathbf{k})\,d\mathbf{k} = qA_{ij}G_j, \quad i,j = 1,2,3 \tag{6-334}$$

where A_{ij} is the macropscopic dispersivity tensor and its expression is

$$A_{ij} = \int_{-\infty}^{\infty} \frac{S_{q_j'q_i'}(\mathbf{k})\,d\mathbf{k}}{[ik_1 + \alpha_L^* k_1^2 + \alpha_T^* (k_2^2 + k_3^2)]q^2} \tag{6-335}$$

Equation (6-334) is a general tensorial expression of the Fickian-type gradient solute transport relationship. This A_{ij} is the same A_{ij} in Eqs. (6-290) and Eq. (6-291).

In order to evaluate the macroscopic dispersivity tensor, A_{ij}, it is necessary to relate the spectrum of the local variation to that of the local loghydraulic conductivity perturbations. The relationship given by Eq. (6-335) depends on the type of medium that is being examined. In the following sections, based on Eq. (6-335), several special cases are presented using $\ln(K)$ structure for hydraulic conductivity. The local hydraulic conductivity is assumed to be isotropic in all cases. The autocovariance functions for these special cases are given in Section 6.4.6.5.

6.4.6.8 Principal Macrodispersivities for Two-Dimensional Solute Transport Cases

In this section, first, the two-dimensional Fickian dispersive fluxes are presented using their three-dimensional forms presented in Section 6.4.6.6. Then, the expressions for the principal macrodispersivities will be derived. Using these expressions, the Mohr's circle for principal macrodispersivities for two-dimensional cases will be presented, as in solid mechanics (e.g., Highdon et al., 1976). For the derivational details, the derivations for the Mohr's circle method for principal hydraulic conductivities in Batu (1998, pp. 84–88) are adapted.

Two-Dimensional Fickian Dispersive Fluxes
For two-dimensional solute transport in the x–y plane , the three-dimensional macroscopic dispersive flux expressions as given by Eqs. (6-291a)–(6-291c) take the following forms:

$$F_x' = \overline{q_x'C'} = -qA_{xx}\frac{\partial \overline{C}}{\partial x} - qA_{xy}\frac{\partial \overline{C}}{\partial y} \tag{6-336a}$$

$$F_y' = \overline{q_y'C'} = -qA_{yx}\frac{\partial \overline{C}}{\partial x} - qA_{yy}\frac{\partial \overline{C}}{\partial y} \tag{6-336b}$$

where

$$q = U\varphi_e \tag{6-337}$$

Principal Macrodispersivities

The two-dimensional Fickian dispersive fluxes in the x–y Cartesian coordinates system are given by Eqs. (6-336a) and (6-336b). Let us now consider the description of this situation with respect to coordinates x^* and y^*, which are obtained from x and y by a rotation through an angle η. The z and z^* axis are assumed to be the same.

The x–y and x^*–y^* coordinates are shown in Figure 6-55. The orientation angle η and the macrodispersivities on the x^* and y^* coordinates are not known. The corresponding expressions regarding these quantities in terms of A_{xx}, A_{yy}, A_{xy}, and A_{yx} derived, in this section.

As can be seen from Figure 6-55, the Fickian dispersive flux vector is the same in x–y and x^*–y^* coordinate systems. Therefore, one can write

$$F_x' = F_{x^*}'\cos(\eta) - F_{y^*}'\sin(\eta) \tag{6-338a}$$

and

$$F_y' = F_{y^*}'\cos(\eta) + F_{x^*}'\sin(\eta) \tag{6-338b}$$

From the geometry in Figure 6-55, the coordinates x^* – y^* are related to x – y coordinates by

$$x = x^*\cos(\eta) - y^*\sin(\eta) \tag{6-339a}$$

and

$$y = y^*\cos(\eta) + x^*\sin(\eta) \tag{6-339b}$$

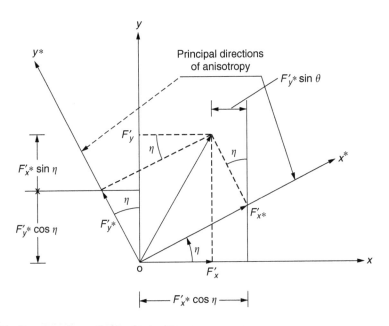

FIGURE 6-55   Rotation of coordinate axes.

On the other hand, from the rules of derivatives and Eqs. (6-338) and (6-339), the following equations can be written:

$$\frac{\partial \overline{C}}{\partial x^*} = \frac{\partial \overline{C}}{\partial x}\frac{\partial x}{\partial x^*} + \frac{\partial \overline{C}}{\partial y}\frac{\partial y}{\partial x^*} = \frac{\partial \overline{C}}{\partial x}\cos(\eta) + \frac{\partial \overline{C}}{\partial y}\sin(\eta) \tag{6-340a}$$

and

$$\frac{\partial \overline{C}}{\partial y^*} = \frac{\partial \overline{C}}{\partial x}\frac{\partial x}{\partial y^*} + \frac{\partial \overline{C}}{\partial y}\frac{\partial y}{\partial y^*} = -\frac{\partial \overline{C}}{\partial x}\sin(\eta) + \frac{\partial \overline{C}}{\partial y}\cos(\eta) \tag{6-340b}$$

From the definition of principal directions of anisotropy, the flux components on the x^* and y^* coordinates in Figure 6-55, respectively, are

$$F_{x^*}' = -qA_{x^*x^*}^*\frac{\partial \overline{C}}{\partial x^*} \tag{6-341a}$$

and

$$F_{y^*}' = -qA_{y^*y^*}^*\frac{\partial \overline{C}}{\partial y^*} \tag{6-341b}$$

where $A_{x^*x^*}^*$ and $A_{y^*y^*}^*$ are the principal macrodispersivities on the x^* and y^* axis, respectively. From Eqs. (6-338), (6-340), and (6-341), the flux components on the x- and y-coordinate directions can be written, respectively, as

$$F_x' = -q[A_{x^*x^*}^*\cos^2(\eta) + A_{y^*y^*}^*\sin^2(\eta)]\frac{\partial \overline{C}}{\partial x} + q(A_{y^*y^*}^* - A_{x^*x^*}^*)\sin(\eta)\cos(\eta)\frac{\partial \overline{C}}{\partial y} \tag{6-342a}$$

$$F_y' = -q[A_{x^*x^*}^*\sin^2(\eta) + A_{y^*y^*}^*\cos^2(\eta)]\frac{\partial \overline{C}}{\partial y} + q(A_{y^*y^*}^* - A_{x^*x^*}^*)\sin(\eta)\cos(\eta)\frac{\partial \overline{C}}{\partial x} \tag{6-342b}$$

From Eqs. (6-336a) and (6-342a), one can write

$$A_{xx} = A_{x^*x^*}^*\cos^2(\eta) + A_{y^*y^*}^*\sin^2(\eta) \tag{6-343a}$$

$$A_{xy} = (-A_{y^*y^*}^* + A_{x^*x^*}^*)\sin(\eta)\cos(\eta) \tag{6-343b}$$

Similarly, from Eqs. (6-336b) and (6-342b), one get

$$A_{yy} = A_{x^*x^*}^*\sin^2(\eta) + A_{y^*y^*}^*\cos^2(\eta) \tag{6-344a}$$

and

$$A_{yx} = (-A_{y^*y^*}^* + A_{x^*x^*}^*)\sin(\eta)\cos(\eta) \tag{6-344b}$$

It is interesting to see from Eqs. (6-343b) and (6-344b) that $A_{xy} = A_{yx}$. Finally, the substitution of the equivalents

$$\sin^2(\eta) = \tfrac{1}{2}[1-\cos(2\eta)], \quad \cos^2(\eta) = \tfrac{1}{2}[1+\cos(2\eta)], \quad \sin(2\eta) = 2\sin(\eta)\cos(\eta) \tag{6-345}$$

into Eqs. (6-343) and (6-344) gives

$$A_{xx} = \tfrac{1}{2}(A^*_{x^*x^*} + A^*_{y^*y^*}) + \tfrac{1}{2}(A^*_{x^*x^*} - A^*_{y^*y^*})\cos(2\eta) \qquad (6\text{-}346)$$

$$A_{yy} = \tfrac{1}{2}(A^*_{x^*x^*} + A^*_{y^*y^*}) - \tfrac{1}{2}(A^*_{x^*x^*} - A^*_{y^*y^*})\cos(2\eta) \qquad (6\text{-}347)$$

$$A_{xy} = A_{yx} = \tfrac{1}{2}(A^*_{x^*x^*} - A^*_{y^*y^*})\sin(2\eta) \qquad (6\text{-}348)$$

Mohr's Circle for Principal Macrodispersivities for Two-Dimensional Case
In the mechanics of materials, Mohr's circle is a pictorial interpretation of the formulas for determining the principal stresses and the maximum shearing stresses at a point in a stressed rigid body. The German engineer Otto Mohr (1835 to 1918) developed a useful graphic interpretation for stress equations. The particular case for plane stress was presented by another German engineer, K. Culmann (1829 to 1881), about 16 years before Mohr's paper was published. This method, commonly known as *Mohr's circle* in solid mechanics, involves the construction of a circle such that the coordinates of each point on the circle represent the normal and shearing stresses on one plane through the stressed point, and the angular position of the radius to the point gives the orientation of the plane. Mohr's circle is also used as a pictorial interpretation of the formulas in determining maximum and minimum second moments of areas and maximum products of inertia (see, e.g., Terzaghi, 1943; Highdon et al., 1976). Mohr's circle is also used to determine principal hydraulic conductivities and their orientation angle (e.g., Verruijt, 1979; Bear, 1972; Batu, 1998).

Mohr's circle approach also provides a convenient way to interpret Eqs. (6-346)–(6-348) to determine the principal macrodispersivities and their orientation angle (Gelhar and Axness, 1981, 1983), and is presented below.

Anisotropy is in the x–y Plane and the z and z* Axis Are the Same
Mohr's circle corresponding to Eqs. (6-346)– (6-348) is shown in Figure 6-56, where A_{xx} and A_{yy} are plotted as horizontal coordinates, and A_{xy} and A_{yx} are plotted as vertical coordinates. The circle is centered on the A_{xx} and A_{yy} axes at a distance $(A_{xx} + A_{yy})/2$ from the A_{yx} and A_{xy} axes and the radius of the circle is given by

$$r = \left[\left(\frac{A_{xx} - A_{yy}}{2}\right)^2 + A_{xy}^2\right]^{1/2} \qquad (6\text{-}349)$$

From the horizontal coordinates of points E and D in Figure 6-56, the expressions for $A^*_{x^*x^*}$ and $A^*_{y^*y^*}$ can be determined as

$$A^*_{x^*x^*} = \tfrac{1}{2}(A_{xx} + A_{yy}) + \left[\tfrac{1}{4}(A_{xx} - A_{yy})^2 + A_{xy}^2\right]^{1/2} \qquad (6\text{-}350)$$

$$A^*_{y^*y^*} = \tfrac{1}{2}(A_{xx} + A_{yy}) - \left[\tfrac{1}{4}(A_{xx} - A_{yy})^2 + A_{xy}^2\right]^{1/2} \qquad (6\text{-}351)$$

The angle 2η from CV to CD is counterclockwise or positive, and from Figure 6-56, we get

$$\tan(2\eta) = \frac{2A_{xy}}{A_{xx} - A_{yy}} \qquad (6\text{-}352)$$

Since $\beta = \eta$ in Figure 6-56, the following equation can also be derived:

$$\tan(\eta) = \tan(\beta) = \frac{A^*_{x^*x^*} - A_{xx}}{A_{xy}} \qquad (6\text{-}353)$$

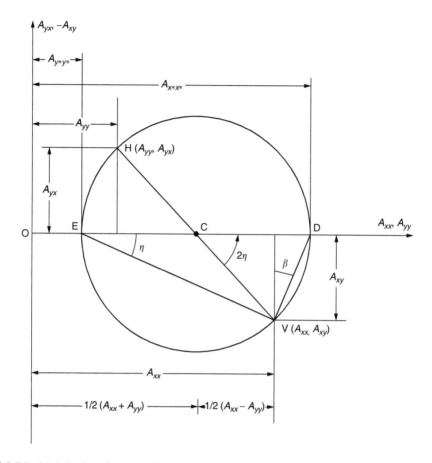

FIGURE 6-56   Mohr's circle for macrodispersivity for two-dimensional solute transport cases in aquifers.

Using the tensorial notation of Gelhar and Axness (1981, 1983), Eqs. (6-350), (6-351), (6-352), and (6-353) can, respectively, be written as

$$A_{11}^* = \tfrac{1}{2}(A_{11} + A_{22}) + \left[\tfrac{1}{4}(A_{11} - A_{22})^2 + A_{12}^2\right]^{1/2} \tag{6-354}$$

$$A_{22}^* = \tfrac{1}{2}(A_{11} + A_{22}) - \left[\tfrac{1}{4}(A_{11} - A_{22})^2 + A_{12}^2\right]^{1/2} \tag{6-355}$$

$$\tan(2\eta) = \frac{2A_{12}}{A_{11} - A_{22}} \tag{6-356}$$

and

$$\tan(\eta) = \tan(\beta) = \frac{A_{11}^* - A_{11}}{A_{12}} \tag{6-357}$$

Anisotropy Is in the x–z Plane and the z and z Axis Are the Same*
Similarly, if the anisotropy is in the x–z plane and the y and y*axis are the same, the following equations, corresponding to Eqs. (6-350), (6-351), (6-352), and (6-353), respectively, can be obtained:

$$A_{x^*x^*}^* = \tfrac{1}{2}(A_{xx} + A_{zz}) + \left[\tfrac{1}{4}(A_{xx} - A_{zz})^2 + A_{xz}^2\right]^{1/2} \tag{6-358}$$

$$A^*_{z^*z^*} = \tfrac{1}{2}(A_{xx} + A_{zz}) - \left[\tfrac{1}{4}(A_{xx} - A_{zz})^2 + A^2_{xz}\right]^{1/2} \tag{6-359}$$

$$\tan(2\eta) = \frac{2A_{xz}}{A_{xx} - A_{zz}} \tag{6-360}$$

$$\tan(\eta) = \tan(\beta) = \frac{A^*_{x^*x^*} - A_{xx}}{A_{xz}} \tag{6-361}$$

Using the tensorial notation of Gelhar and Axness (1981, 1983), Eqs. (6-358), (6-359, (6-360), and (6-361) can, respectively, be written as

$$A^*_{11} = \tfrac{1}{2}(A_{11} + A_{33}) + \left[\tfrac{1}{4}(A_{11} - A_{33})^2 + A^2_{13}\right]^{1/2} \tag{6-362}$$

$$A^*_{33} = \tfrac{1}{2}(A_{11} + A_{33}) - \left[\tfrac{1}{4}(A_{11} - A_{33})^2 + A^2_{13}\right]^{1/2} \tag{6-363}$$

$$\tan(2\eta) = \frac{2A_{13}}{A_{11} - A_{33}} \tag{6-364}$$

$$\tan(\eta) = \tan(\beta) = \frac{A^*_{11} - A_{11}}{A_{13}} \tag{6-365}$$

6.4.6.9 Determination of the Macroscopic Dispersion Tensor A_{ij} in a Statistically Isotropic Aquifer — Local Dispersivity Isotropic

To simplify the initial analysis, the local dispersive process is assumed to be isotropic and the aquifer is assumed to be statistically isotropic as well. The macroscopic dispersivity integral, given by Eq. (6-335), is evaluated by both asymptotic and exact means (Gelhar and Axness, 1981). In this section, first, some pertinent definitions are given. Then, the results of Gelhar and Axness (1981, 1983) are presented.

Statistically Isotropic Media
Definitions
The log-hydraulic conductivity field, as given by Eq. (6-176), is taken to be statistically isotropic, which means that the variation in hydraulic conductivity is independent of direction. This is shown in Figure 6-57 for an unconfined aquifer, and in Figure 6-58 for a confined aquifer. For example, at point 1 in Figure 6-57, the hydraulic conductivity $K_1 (x_1, y_1, z_1)$ is the same in all directions. Similarly, at point 2, the hydraulic conductivity $K_2 (x_2, y_2, z_2)$ is the same in all directions. The same situation is valid at point n, i.e., $K_n (x_n, y_n, z_n)$ is the same in all directions. Mathematically speaking.

$$K_1(x_1, y_1, z_1) \neq K_2(x_2, y_2, z_2) \neq \cdots \neq K_n(x_n, y_n, z_n) \tag{6-366}$$

Although the assumption of statistical isotropy is not realistic for most natural materials, this case is of interest as a conceptually simple situation for which stochastic analysis has been done (e.g., Bakr et al., 1978; Dagan, 1979). From Eq. (6-366)

$$\ln[K_1(x_1, y_1, z_1)] \neq \ln[K_2(x_2, y_2, z_2)] \neq \cdots \neq \ln[K_n(x_n, y_n, z_n)] \tag{6-367}$$

or using Eq. (6-176)

$$f_1(x_1, y_1, z_1) \neq f_2(x_2, y_2, z_2) \neq \cdots \neq f_n(x_n, y_n, z_n) \tag{6-368}$$

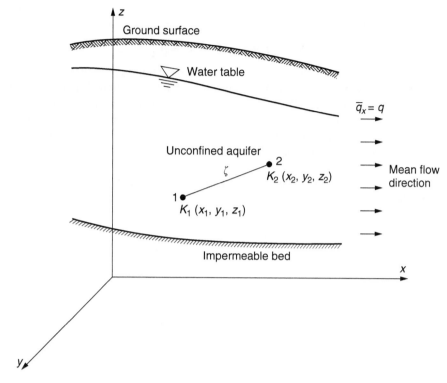

FIGURE 6-57   Flow in a statistically isotropic unconfined aquifer.

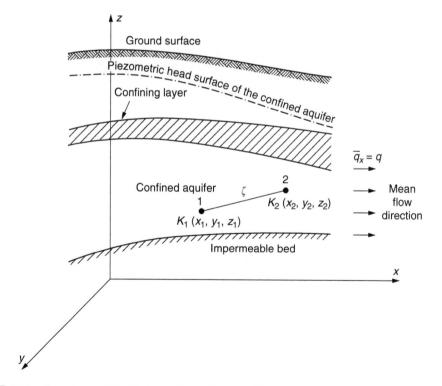

FIGURE 6-58   Flow in a statistically isotropic confined aquifer.

A homogeneous random field is statistically isotropic if the covariance function depends on the distance between the two locations and not on the orientation of the separation vector, (Gelhar, 1993), i.e.,

$$R_{f'f'}(\boldsymbol{\zeta}) = R_{f'f'}(\zeta), \quad \zeta = |\boldsymbol{\zeta}| \tag{6-369}$$

As shown in Figure 6-59 and Eq. (6-369), a homogeneous random field is statistically isotropic if the covariance function depends only on the distance between the locations 1 and 2, and not on the orientation of the separation vector. The orientation of ζ in Figure 6-59 is defined by its $(\zeta_1, \zeta_2, \zeta_3)$ components, and for a statistically isotropic medium, the covariance function is only dependent on ζ and not on $(\zeta_1, \zeta_2, \zeta_3)$. The x- coordinate axis in Figure 6-59 is assumed to be aligned in the direction of mean flow according to Eq. (6-316).

Exponential Autocovariance Function for a Statistically Isotropic Medium
For a statistically isotropic hydraulic conductivity field, $\lambda_1 = \lambda_2 = \lambda_3$, and therefore, Eq. (6-274) takes the form

$$R_{f'f'}(\boldsymbol{\zeta}) = \sigma_{f'}^2 \cdot \exp\left[-\left(\frac{\zeta_1^2 + \zeta_2^2 + \zeta_3^2}{\lambda^2} \right) \right] \tag{6-370}$$

Since

$$|\boldsymbol{\zeta}|^2 = \zeta^2 = \zeta_1^2 + \zeta_2^2 + \zeta_3^2 \tag{6-371}$$

Eq. (6-370) takes the form

$$R_{f'f'}(\boldsymbol{\zeta}) = \sigma_{f'}^2 \exp\left(-\frac{\zeta}{\lambda} \right), \qquad R_{f'f'}(0) = \sigma_{f'}^2 \tag{6-372}$$

and from these

$$\frac{R_{f'f'}(\boldsymbol{\zeta})}{R_{f'f'}(0)} = \frac{R_{f'f'}(\zeta)}{\sigma_{f'f'}^2} = \rho_{f'f'}(\zeta) = \exp\left(-\frac{\zeta}{\lambda} \right) \tag{6-373}$$

which is the same as Eq. (6-69).

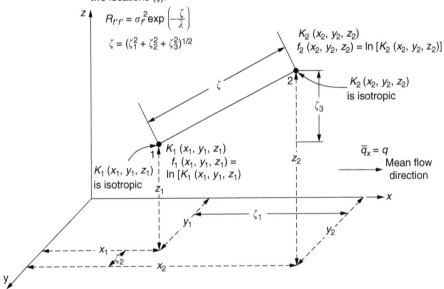

FIGURE 6-59 Statistically isotropic hydraulic conductivity field.

Macroscopic Dispersion Tensor A_{ij} for a Statistically Isotropic Medium
For an isotropic local dispersive process, Eq. (6-317) can be expressed as

$$E_{m_{ij}} = \alpha^* q \delta_{ij} \tag{6-374}$$

where δ_{ij} is the Kronecker delta ($\delta_{ij} = 1$ if $i = j$, zero otherwise) and α^* is the constant local dispersivity. Also, for $\alpha_L^* = \alpha_T^* = \alpha^*$, Eq. (6-335) reduces to

$$A_{ij} = \int_{-\infty}^{\infty} \frac{S_{q'_i q'_i}(\mathbf{k}) \, d\mathbf{k}}{(ik_1 + \alpha^* k^2)q^2} \tag{6-375}$$

where k^2 is given by Eq. (6-236). If the local dispersivity is isotropic ($\alpha_L^* = \alpha_T^*$), as would be the case if local dispersion is dominated by molecular diffusion, Eq. (6-335) takes a relatively simple form, given by Eq. (6-375).

The evaluation of the macroscopic dispersivity integrals of Eq. (6-375) proceeds using Eq. (6-269) in conjunction with the spectral form of Eq. (6-275). One should note that since the medium is assumed to be statistically isotropic, $\lambda_i = \lambda$ for $i = 1, 2, 3$. The spectrum given by Eq. (6-275) is even in k_i, and the integrand is odd for $i \neq j$. In other words, the natural coordinates x_i associated with the mean flow direction are the principal axes of the macroscopic dispersivity tensor in the case of a statistically isotropic log-hydraulic conductivity covariance. In the following section, evaluation of A_{11}, A_{22}, and A_{33} are presented.

Evaluation of A_{11}
Substitution of Eq. (6-269) into Eq. (6-375) gives, for $i = 1, j = 1$,

$$A_{11} = \int_{-\infty}^{\infty} \frac{q^2 \gamma^{-2} (\delta_{11} - k_1^2/k^2)(\delta_{11} - k_1^2/k^2) S_{f'f'}}{(ik_1 + \alpha^* k^2)q^2} \, d\mathbf{k} \tag{6-376}$$

or since $\delta_{11} = 1$

$$A_{11} = \frac{1}{\gamma^2} \int_{-\infty}^{\infty}\int_{-\infty}^{\infty}\int_{-\infty}^{\infty} \frac{(1-k_1^2/k^2)^2(\alpha^* k^2 - ik_1)}{(\alpha^{*2}k^4 + k_1^2)} S_{f'f'}(k_1, k_2, k_3) \, dk_1 \, dk_2 \, dk_3 \tag{6-377}$$

where k is defined by Eq. (6-236). Note that the spectrum given by Eq. (6-274) is even and, therefore, $S_{q'_i q'_i}$ is even in k_i. The integrand of Eq. (6-377) has real and imaginary parts and the imaginary part is zero since the integrand is odd with respect to k_1. Therefore, Eq. (6-377) takes the form

$$A_{11} = \frac{\alpha^*}{\gamma^2} \int_{-\infty}^{\infty}\int_{-\infty}^{\infty}\int_{-\infty}^{\infty} \frac{(1-k_1^2/k^2)k^2}{(\alpha^{*2}k^4 + k_1^2)} S_{f'f'}(k_1, k_2, k_3) \, dk_1 \, dk_2 \, dk_3 \tag{6-378}$$

By letting

$$u_1 = \lambda k_1, \qquad u_2 = \lambda k_2, \qquad u_3 = \lambda k_3, \qquad \varepsilon = \frac{\alpha^*}{\lambda} \tag{6-379}$$

Eq. (6-378) takes the form

$$A_{11} = \frac{1}{\varepsilon\gamma^2} \int_{-\infty}^{\infty}\int_{-\infty}^{\infty}\int_{-\infty}^{\infty} \frac{(1-u_1^2/u^2)u^2}{(u^4 + u_1^2/\varepsilon^2)} \frac{S_{f'f'}(u_1, u_2, u_3)}{\lambda^2} \, du_1 \, du_2 \, du_3 \tag{6-380}$$

where

$$u^2 = u_1^2 + u_2^2 + u_3^2 \tag{6-381}$$

Gelhar and Axness (1981) evaluated the integral in Eq. (6-380) by using both the "asymptotic approach" and the "exact approach," and the results are presented below.

Asymptotic Approach Result for A_{11}

Typical local longitudinal dispersivities are of the order of a centimeter or less, but the correlation scale, λ, is of the order of a meter. Therefore, the parameter ε, defined by the last expression of Eq. (6-379), is small. As a result, useful results can be obtained with the limit $\varepsilon \to 0$. However, if $\varepsilon = 0$ is used in Eq. (6-380), the result becomes indeterminate. For this situation, the main contribution to the integral will be from the region near $u_1 = 0$ (Gelhar, 1993). By letting $v_1 = u_1/\varepsilon$, Eq. (6-381) becomes

$$u^2 = \varepsilon^2 v_1^2 + u_2^2 + u_3^2 \to u_2^2 + u_3^2 \quad \text{as } \varepsilon \to 0 \tag{6-382}$$

Then, Eq. (6-380) takes the following form:

$$A_{11} \cong \frac{1}{\gamma^2} \int_{-\infty}^{\infty}\int_{-\infty}^{\infty} \frac{S_{f'f'}(0, u_2, u_3)}{\lambda^2}\left[\int_{-\infty}^{\infty} \frac{u^2}{u^4 + v_1^2}\, dv_1\right] du_2\, du_3 \tag{6-383}$$

where the integral is evaluated by making the substitution $w = v_1/u^2$. Using Eq. (6-153) for $\sigma_{f'}^2$, Eq. (6-383) takes the form

$$A_{11} = \frac{\sigma_{f'}^2 \lambda}{\pi\gamma^2}\left[\int_{-\infty}^{\infty}\int_{-\infty}^{\infty} \frac{du_2\, du_3}{1 + u_2^2 + u_3^2)}\right] = \sigma_{f'}^2 \frac{\lambda}{\gamma^2} \tag{6-384}$$

where the integral is evaluated by changing to polar coordinates. In Eq. (6-384), γ is given by Eq. (6-267).

Exact Approach Result for A_{11}

The exact solution requires extensive manipulations, which are available in Gelhar and Axness (1981). The authors obtained the exact solution by substituting the spectrum given by Eq. (6-275) into Eq. (6-380), and the final result is (Gelhar and Axness, 1981, Eq. [B18]), p. 99).

$$A_{11} = \frac{\lambda\sigma_{f'}^2}{\gamma^2}\left\{1 - \varepsilon\left(\frac{1}{1+\varepsilon} + \frac{5}{\varepsilon}\right) + \varepsilon^2\left[-3 + 4\ln\left(1 + \frac{1}{\varepsilon}\right)\right]\right.$$

$$\left. + \varepsilon^3\left(3 + \frac{2}{1+\varepsilon}\right) + \varepsilon^4\left[\frac{1}{1+\varepsilon} - 4\ln\left(1 + \frac{1}{\varepsilon}\right)\right]\right\} \tag{6-385}$$

where ε is given by in Eq. (6-379) and ε^{-1} is a Peclet number (see Section 4.3.12.3 for the definition of Peclet number) with the length scale taken to be the integral scale of the log-hydraulic conductivity, λ. It can be observed from Eq. (6-385) that as ε approaches zero, the result for A_{11} becomes exactly the same as in Eq. (6-384), which is the asymptotic result.

Evaluation of $A_{22} = A_{33}$

By subsituting Eq. (6-275) into Eq. (6-269) and using Eq. (6-275), one has

$$A_{ii} = \frac{\sigma_{f'}^2 \alpha^*}{\gamma^2\pi^2} \int_{-\infty}^{\infty}\int_{-\infty}^{\infty}\int_{-\infty}^{\infty} \frac{k_i^2 k_1^2 \lambda^3\, dk_1\, dk_2\, dk_3}{k^2(\alpha^{*2}k^4 + k_1^2)(1 + k^2\lambda^2)^2}, \quad i = 2, 3; \text{ no summation on } i \tag{6-386}$$

where the integrand has been reduced as in Eq. (6-378). Due to symmetry, $A_{22} = A_{33}$. Since the exact solution of Eq. (6-386) is of the order of ε, a first-order asymptotic solution is not given as it would yield zero. The exact solution is obtained by Gelhar and Axness (1981, Eq.[B23], p. 100), by expressing Eq. (6-386) in spherical coordinates, and the final result is

$$A_{22} = A_{33} = \frac{\sigma_{f'}^2 \lambda}{\gamma^2}\left\{\frac{\varepsilon}{3} + \varepsilon^2\left[1 - \ln\left(1 + \frac{1}{\varepsilon}\right)\right] - 2\varepsilon^3 + 2\varepsilon^4\ln\left(1 + \frac{1}{\varepsilon}\right)\right\} \tag{6-387}$$

When $1/\varepsilon$ is large, Eq. (6-387) becomes

$$A_{22} = A_{33} \cong \frac{\sigma_{f'}^2 \lambda}{\gamma^2}\frac{\varepsilon}{3}, \quad \text{when } \frac{\lambda}{\alpha^*} = \frac{1}{\varepsilon} \text{ is large} \tag{6-388}$$

Discussion of the Results

Variation of A_{11} and A_{22} with the Parameters

Some key points are as follows (Gelhar and Axness, 1981; Gelhar, 1993): (1) Logarithmic and rectilinear plots of $\gamma^2 A_{11}/(\lambda \sigma_{f'}^2)$ and $\gamma^2 A_{22}/(\lambda \sigma_{f'}^2)$ as a function of ε are shown in Figure 6-60 and Figure 6-61. Figure 6-62 shows A_{22}/A_{11} as a function of ε. Figure 6-60 shows a comparison of the exact and approximate results, which indicates that the approximate solution is adequate for $\varepsilon < 0.01$. As shown in Figure 6-60, when ε is large, A_{22} is approximately $\lambda \sigma_{f'}^2 \varepsilon/(3\gamma^2)$ and A_{11} is approximately $\lambda \sigma_{f'}^2/\gamma^2$; then, the ratio A_{22}/A_{11} is $\varepsilon/3$, which will be very small under typical conditions. (2) As can be seen from Figure 6-60 and Figure 6-61, the isotropic local dispersivity ($\alpha^* = \alpha_L^* = \alpha_T^*$), which is equivalent to molecular diffusion, has the effect of reducing both the longitudinal and transverse macrodispersivities (A_{11} and A_{22}) (3). The results for small local dispersivity relative to the $\ln(K)$ correlation scale illustrate some important features of the isotropic case. As can

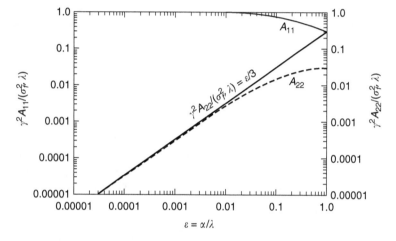

FIGURE 6-60 Logarithmic plot of longitudinal (A_{11}) and transverse (A_{22}) macrodispersivities for a statistically isotropic medium with the correlation scale λ and isotropic local dispersivity α^*. (Adapted from Gelhar, L.W and Axness, C.L., Report No. H-8, 1981, 140 pp.)

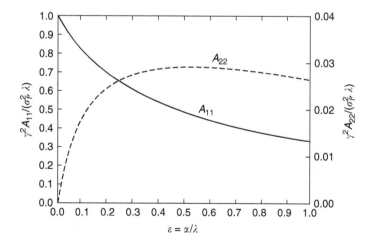

FIGURE 6-61 Linear plot of longitudinal (A_{11}) and transverse (A_{22}) macrodispersivities for a statistically isotropic medium with the correlation scale λ and isotropic local dispersivity α^*. (Adapted from Gelhar, L.W and Axness, C.L., Report No. H-8, 1981, 140 pp.)

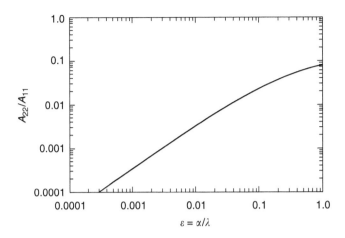

FIGURE 6-62 Ratio of transverse to longitudinal macrodispersivities for statistically isotropic medium. (Adapted from Gelhar, L.W and Axness, C.L., Report No. H-8, 1981, 140 pp.)

be seen from Eq. (6-382), for longitudinal macrodispersivity (A_{11}), the magnitude of A_{11} becomes independent of the local dispersivity, and is determined by the variance of $f = \ln(K)$, i. e., $\sigma_{f'}^2$, and it's correlation scale λ. The dominant influence in the longitudinal macroscopic mixing is that of convective velocity variation due to the variation in hydraulic conductivity. (4) The log-hydraulic conductivity field is taken to be statistically isotropic (see Figure 6-59), which means that the variation of hydraulic conductivity is assumed to be independent of direction. The assumption of statistical isotropy is not realistic for most natural materials. However, the results based on this assumption are conceptually important (e.g., Bakr et al., 1978; Dagan, 1979).

Evaluation of the First-Order Approximation for Flow
As can be seen from Eqs. (6-250)–(6-252), only the first and second terms for f' are taken into account. The rest of the values are neglected. In other words, the exponential term f' is assumed to be expressed by Eq. (6-252). Hence the process is called "first-order approximation," or sometimes "small perturbations assumption." By definition, the variance of f' is

$$\sigma_{f'}^2 = E[f'^2] = E(f - \bar{f})^2] \qquad (6-389)$$

The stochastic theory was developed under the assumption that $\sigma_{f'}^2$ is small. In the case of flow equations, there are indications that the logarithmic linearization is relatively strong. For one-dimensional finite domain flows, Gutjahr and Gelhar (1981) have shown that the head variance found by logarithmic linearization is in reasonable agreement with Monte-Carlo simulations (see Section 6.1 for definition), even for $\sigma_{f'}^2 = 21$. Also, for three-dimensional flow with a statistical $\ln(K)$ field, Dagan (1979) has shown that the effective hydraulic conductivity for the largest $\sigma_{f'}^2 = 21$ was within 1% of the perturbation result using logarithmic linearization. However, one should note that for large $\sigma_{f'}^2$ this result is significantly smaller than the Gelhar and Axness hypothesized generalization (see, Eqs. [6-452]–[6-454], which gives $\gamma = \exp(\sigma_{f'}^2/6)$ for the isotropic case. The exponential generalization of Eqs. (6-452)–(6-454) is exact for flow parallel or normal to perfect layering and, of course, it is possible that the approximations of Dagan (1979) and Gutjahr et al. (1978) introduce similar errors (Gelhar and Axness, 1981). In conclusion, it can be said that the magnitude of A_{ij} for large $\sigma_{f'}^2$ is strongly influenced by γ and, consequently, there is a need for "exact" three-dimensional results in order to evaluate the adequacy of Eqs. (6-452)–(6-454).

Evaluation of the First-Order Approximation for Solute Transport
The adequacy of the first-order approximation of the solute transport equation, as given by Eq. (6-314) or (6-315) for large $\sigma_{f'}^2$ is, a similar to the flow case, uncertain. The comparisons with

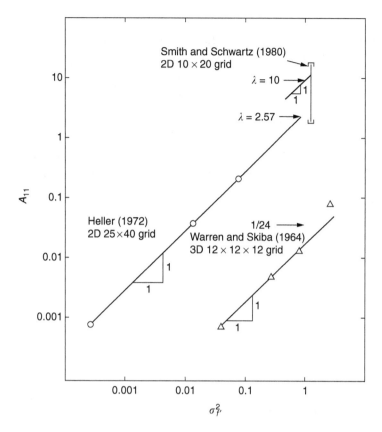

FIGURE 6-63 Comparison of theory ($A_{11} = \sigma_{f'}^2\, \lambda/\gamma^2$ with Monte Carlo simulations. (Adapted from Gelhar, L.W. and Axness, C.L., Report No. H-8, 1981, 140 pp.)

Monte-Carlo simulations in Figure 6-63 indicate that the theory for A_{11} is reasonable for $\sigma_{f'}^2 < 1$, but the results for larger values of $\sigma_{f'}^2$ are not definitive (Gelhar and Axness, 1981). Gelhar and Axness (1981) also state that there is a need for carefully designed Monte-Carlo simulations of demonstrable accuracy in order to evaluate the large $\sigma_{f'}^2$ limitations of the theory.

EXAMPLE 6-9

For a statistically isotropic aquifer, the following values are given: $\sigma_{f'}^2 = 0.78$, $\lambda = \lambda_1 = \lambda_2 = \lambda_3 = 3.5$ m, $\alpha^* = \alpha_L^* = \alpha_T^* = 0.01$ m. Determine the values of longitudinal macrodispersivities (A_{11} and A_{22}).

SOLUTION

The parameters in the A_{11} and A_{22} expressions need to be calculated first. From Eq. (6-267)

$$\gamma = 1 + \frac{0.78}{6} = 1.13$$

γ can also be calculated from Eq. (2-268) as

$$\gamma = \exp\left(\frac{0.78}{6}\right) = 1.14$$

Notice that both equations give close values. From the last expression of Eq. (6-379)

$$\varepsilon = \frac{\alpha^*}{\lambda} = \frac{0.01\ \text{m}}{3.5\ \text{m}} = 2.857 \times 10^{-3}$$

A_{11} can be calculated from Eq. (6-384) or Eq. (6-385) based on the value of ε. Since the value of ε is very small, the terms in the parenthesis of Eq. (6-385) are negligible compared with 1. It can be seen from Figure 6-60 that Eq. (6-384) can be used safely:

$$A_{11} = (0.78)\frac{(3.5 \text{ m})}{(1.13)^2} = 2.14 \text{ m}$$

Similarly, Eq. (6-388) can be used for A_{22} and A_{33} and their values is 2.04×10^{-3}m.

6.4.6.10 Approximate Determination of A_{ij} for Macroscopic Dispersion in Statistically Isotropic Media — Local Dispersivity Anisotropic

Expression for A_{ij}

Gelhar and Axness (1983) determined an approximate solution of the macroscopic dispersivity integral given by Eq. (6-335) by assuming that the local longitudinal dispersivity α_L^* is small compared with the correlation scale $\lambda = \lambda_1 = \lambda_2 = \lambda_3$ in Eq. (6-274). Gelhar and Axness state that this assumption is reasonable because α_L^* is typically of the order of a centimeter or less, while λ is in the order of meters (Bakr, 1976; Smith, 1981). The authors state that the corresponding spectrum, given by Eq. (6-275), is even in k_i so that from Eq. (6-269), the integrand of Eq. (6-335) is odd in k_i when i is not equal to j, and therefore $A_{ij} = 0$ in those cases. Also, when $i = j$, the imaginary term in Eq. (6-335) produces a term odd in k_1 and is therefore zero. The final result is (Gelhar and Axness, 1983, Eq. (29), p. 164):

$$A_{ij} = \int_{-\infty}^{\infty} \frac{(\delta_{i1} - k_1 k_i/k^2)[\alpha_L^* k_1^2 + \alpha_T^*(k_2^2 + k_3^2)]S_{f'f'}(k)}{\gamma^2\{k_1^2 + [\alpha_L^* k_1^2 + \alpha_T^*(k_2^2 + k_3^2)]^2\}}\,d\mathbf{k}, \quad i = 1, 2, 3 \text{ (no summation on } i\text{)} \quad (6\text{-}390)$$

where $S_{f'f'}$ is given by Eq. (6-275).

Evaluation of A_{11}

Gelhar and Axness (1983) determined A_{11} as follows: from Eq. (6-390), the longitudinal macrodispersivity A_{11}, after normalization with

$$u_i = \lambda k_i, \qquad \varepsilon = \frac{\alpha_L^*}{\lambda}, \qquad \kappa = \frac{\alpha_L^*}{\alpha_T^*} \qquad (6\text{-}391)$$

becomes

$$A_{11} = \int_{-\infty}^{\infty}\int_{-\infty}^{\infty}\int_{-\infty}^{\infty} \frac{(1-u_1^2/u^2)^2\varepsilon[u_1^2 + \kappa(u_2^2 + u_3^2)]S_{ff'}}{\gamma^2\lambda^2\{u_1^2 + \varepsilon^2[u_1^2 + \kappa(u_2^2 + u_3^2)]^2\}}\,du_1\,du_2\,du_3 \qquad (6\text{-}392)$$

where

$$u^2 = \lambda^2 k^2 \qquad (6\text{-}393)$$

Owing to the form of the denominator in Eq. (6-392), when $\varepsilon \ll 1$ the main contribution to the integral will come from $u_1 \cong 0$. This can be seen by writing Eq. (6-392) in terms of $v = u_1/\varepsilon$:

$$A_{11} = \int_{-\infty}^{\infty}\int_{-\infty}^{\infty}\int_{-\infty}^{\infty} \frac{(1-\varepsilon^2 v^2/u^2)[\varepsilon^2 v^2 + \kappa(u_2^2 + u_3^2)]S_{ff'}(\varepsilon v, u_2, u_3)}{\gamma^2\lambda^2\{v^2 + [\varepsilon^2 v^2 + \kappa(u_2^2 + u_3^2)]^2\}}\,dv\,du_2\,du_3 \qquad (6\text{-}394)$$

and taking $\varepsilon \to 0$, with

$$a = \kappa(u_2^2 + u_3^2) \qquad (6\text{-}395)$$

$$A_{11} \cong \frac{1}{\gamma^2\lambda^2}\int_{-\infty}^{\infty}\int_{-\infty}^{\infty}\left[\int_{-\infty}^{\infty}\frac{a}{v^2 + a^2}\,dv\right]S_{ff'}(0, u_2, u_3)\,du_2\,du_3$$

$$= \frac{\pi}{\gamma^2}\int_{-\infty}^{\infty}\int_{-\infty}^{\infty}S_{ff'}(0, k_2, k_3)\,dk_2\,dk_3 \qquad (6\text{-}396)$$

Using Eq. (6-275), the last integral is evaluated in polar coordinates to yield

$$A_{11} = \sigma_{f'}^2 \frac{\lambda}{\gamma^2}$$ (6-397)

Evaluation of A_{22}
Gelhar and Axness (1983), from Eq. (6-390), expressed transverse dispersivity A_{22} as

$$A_{22} = \frac{\alpha_L^*}{\gamma^2 \lambda^3} \int_{-\infty}^{\infty} \int_{-\infty}^{\infty} \int_{-\infty}^{\infty} \frac{u_1^2 u_2^2 [u_1^2 + \kappa(u_2^2 + u_3^2)] S_{f'f'}}{u^4 \{u_1^2 + \varepsilon^2 [u_1^2 + \kappa^2(u_2^2 + u_3^2)]^2\}} \, du_1 \, du_2 \, du_3$$ (6-398)

Taking $\varepsilon \to 0$ in the integrand, it takes the form

$$A_{22} \cong \int_{-\infty}^{\infty} \int_{-\infty}^{\infty} \int_{-\infty}^{\infty} \frac{S_{f'f'} u_2^2 [u_1^2 + \kappa(u_2^2 + u_3^2)]}{\lambda^3 u^4} \, du_1 \, du_2 \, du_3$$ (6-399)

By substituting Eq. (6-275) into Eq. (6-399) and expressing it in spherical coordinates, Gelhar and Axness (1983) obtained the following result:

$$A_{22} = \frac{\sigma_{f'}^2 \alpha_T^*}{15\gamma^2} \left(1 + \frac{4\alpha_T^*}{\alpha_L^*} \right)$$ (6-400)

From Eq. (6-390), it is evident that $A_{22} = A_{33}$.

Special Case: Two-Dimensional Flow
It is frequently assumed that ground water flow is two-dimensional, because the horizontal scale of many aquifers is orders of magnitude larger than the vertical scale. Gelhar and Axness (1983) and Gelhar (1993) note that if such a two-dimensional flow field is used in a solute transport analysis, the corresponding macrodispersivities can be found. Gelhar and Axness (1983) applied the asymptotic approach to a two-dimensional flow in the (x_1, x_2) plane by using $S_{f'f'}$ from Eq. (6-275) with $\lambda_1 = \lambda_2 = \lambda$ and $\lambda_3 \to \infty$ in Eq. (6-390), and obtained the following results:

$$A_{11} = \sigma_{f'}^2 \frac{\lambda}{\gamma^2}$$ (6-401)

and

$$A_{22} = \frac{\sigma_{f'}^2 \alpha_L^*}{8\gamma^2} \left(1 + \frac{3\alpha_T^*}{\alpha_L^*} \right)$$ (6-402)

Notice that Eq. (6-401) is exactly the same as Eq. (6-397).

Typical Parameter Values in the A_{11} and A_{22} Expressions
For field conditions, the ratio α_L^*/λ is typically 10^{-2} or smaller and α_T^*/α_L^* is of the order of 10^{-1}. Thus, the ratio A_{22}/A_{11} is of the order of 10^{-3}. The analysis based on an isotropic $\ln(K)$ process predicts extremely small transverse dispersion. The results for the statistically isotropic media indicate that the longitudinal dispersion is convection-controlled, but the transverse dispersion is determined by local dispersion (Gelhar and Axness, 1981).

EXAMPLE 6-10

For a statistically isotropic aquifer, the following values are given: $\sigma_{f'}^2 = 0.78$, $\lambda = \lambda_1 = \lambda_2 = \lambda_3 = 3.5$ m, $\alpha_L^* = 0.01$ m, and $\alpha_T^*/\alpha_L^* = 0.1$. Determine the values of the longitudinal macrodispersivities (A_{11} and A_{22}).

SOLUTION

The parameters in the A_{11} and A_{22} expressions need to be calculated first. From Eq. (6-267)

$$\gamma = 1 + \frac{0.78}{6} = 1.13$$

γ can also be calculated from Eq. (2-268) as

$$\gamma = \exp\left(\frac{0.78}{6}\right) = 1.14$$

Notice that both equations give close values. From Eq. (6-397)

$$A_{11} = (0.78)\frac{(3.5 \text{ m})}{(1.13)^2} = 2.14 \text{ m}$$

and from Eq. (6-400)

$$A_{22} = \frac{(0.78)(0.01 \text{ m})}{15(1.13)^2}[1 + 4(0.1)] = 5.7 \times 10^{-4} \text{ m}$$

6.4.6.11   Determination of A_{ij} for Macroscopic Dispersion in a Statistically Anisotropic Aquifer with Arbitrary Orientation of Stratification

The statistically isotropic results, which are presented in Sections 6.4.6.9 and 6.4.6.10, are primarily of theoretical interest and provide some general insights about macrodispersivities. In real world cases, most natural aquifer materials exhibit some degree of stratification. In these kinds of media, the mean ground water flow is not generally parallel to the stratification direction. Therefore, the medium will behave hydraulically as an anisotropic medium and the direction of the mean specific discharge and the mean hydraulic gradient will not be the same. In general, all three components of the mean hydraulic gradient can be nonzero (Gelhar and Axness, 1981, 1983; Gelhar, 1993).

For this purposes, first the mean flow behavior of the heterogeneous statistically anisotropic medium in terms of an effective hydraulic conductivity tensor is established, and then the flow perturbation spectra in terms of $\ln(K)$ spectrum is expressed. This analysis is a generalization of the isotropic three-dimensional analysis given by Gutjahr et al. (1978), which is presented in Section 6.4.5. Finally, the results of macroscopic dispersion are presented by treating it such that the mean flow is not necessarily aligned in the direction of stratification.

Aquifer Description and General Formulation
Figure 6-64 and Figure 6-65 show the stratification and coordinate axes in an unconfined aquifer and in a confined aquifer, respectively. The (x, y, z) or (x_1, x_2, x_3) coordinates system is aligned in the direction of mean flow. Therefore,

$$\overline{q_1} = q \neq 0, \qquad \overline{q_2} = \overline{q_3} = 0 \tag{6-403}$$

The primed x_i' system (see Figure 6-66) is oriented with the positive x_1' axis in the direction of stratification. The covariance function given by Eq. (6-274) applies in the primed coordinate system. Since the two coordinate systems are not necessarily identical, the mean hydraulic gradient vector may not coincide with the direction of the mean flow. Again, assuming the local anisotropy of the hydraulic conductivity, Eq. (6-259) describes the head perturbation as

$$\nabla^2 h' = I_j \frac{\partial f'}{\partial x_j} \tag{6-404}$$

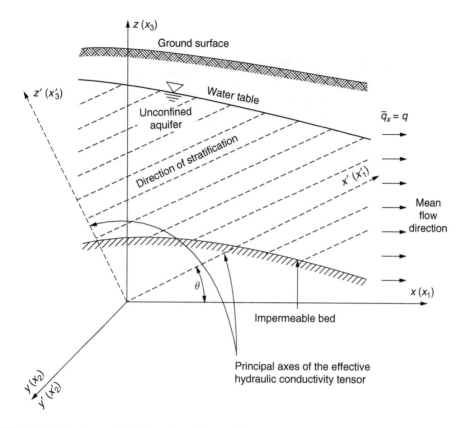

FIGURE 6-64   Flow in a statistically anisotropic unconfined aquifer with arbitrary orientation of stratification.

Making use of Fourier–Stieltjes representations in Eq. (6-404)

$$k^2\, dZ_{h'} = -ik_j I_j\, dZ_{f'} \tag{6-405}$$

or

$$dZ_{h'} = -\frac{ik_j I_j}{k^2}\, dZ_{f'} \tag{6-406}$$

which will be used to evaluate the effective hydraulic conductivity and the flow spectra.

Effective Hydraulic Conductivity Tensor
The Darcy equation expressed by Eq. (6-251) can be written as

$$q_i = K_g\left(1 + f' + \frac{1}{2}f'^2 + \cdots\right)\left(I_i - \frac{\partial h'}{\partial x_i}\right) \tag{6-407}$$

where I_i is given by Eq. (6-258). Taking the expected values of Eq. (6-407) and dropping beyond the second-order terms yield the mean flow expression:

$$
\begin{aligned}
E[q_i] &\cong E\left[K_g\left(1 + f' + \frac{1}{2}f'^2\right)\left(I_i - \frac{\partial h'}{\partial x_i}\right)\right] \\
&= E\left[K_g\left(1 + f' + \frac{1}{2}f'^2\right)I_i - K_g\left(1 + f' + \frac{1}{2}f'^2\right)\frac{\partial h'}{\partial x_i}\right] \\
&= K_g\left\{\left(1 + E[f'] + \frac{1}{2}E[f'^2]\right)I_i - E\left[\frac{\partial h'}{\partial x_i}\right] - E\left[f'\frac{\partial h'}{\partial x_i}\right] - \frac{1}{2}E\left[f'^2\frac{\partial h'}{\partial x_i}\right]\right\}
\end{aligned} \tag{6-408}
$$

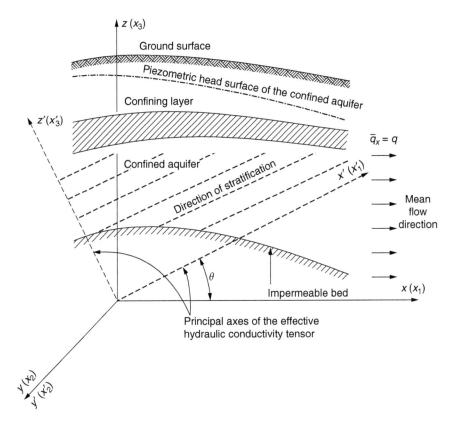

FIGURE 6-65   Flow in a statistically anisotropic confined aquifer with arbitrary orientation of stratification.

Since $E[f'^2] = \sigma_{f'}^2$ according to Eq. (6-195), $E[f'] = 0$ and $E[h'] = 0$, Eq. (408) takes the form

$$\bar{q}_i = K_g\left[\left(1 + \frac{1}{2}\sigma_{f'}^2\right)I_i - \overline{f'\frac{\partial h'}{\partial x_i}}\right] \qquad (6\text{-}409)$$

$$= K_g\left[\left(1 + \frac{1}{2}\sigma_{f'}^2\right)\delta_{ij} - F_{ij}\right]I_j$$

$$= \overline{K_{ij}}\,I_j$$

where

$$\overline{K_{ij}} = K_g\left[\left(1 + \tfrac{1}{2}\sigma_{f'}^2\right)\delta_{ij} - F_{ij}\right] \qquad (6\text{-}410)$$

is the effective hydraulic conductivity tensor and

$$\overline{f'\frac{\partial h'}{\partial x_i}} = F_{ij}I_j \qquad (6\text{-}411)$$

which is the term to be evaluated.

Evaluation of the Term $\overline{f'\partial h'/\partial x_i}$
Using the spectral representation of $h'(x, y, z)$ and $f'(x, y, z)$, which are given by Eqs. (6-228) and (6-229), respectively, and following the method in Section 6.4.5.3, as in Eq. (6-238), in tensorial

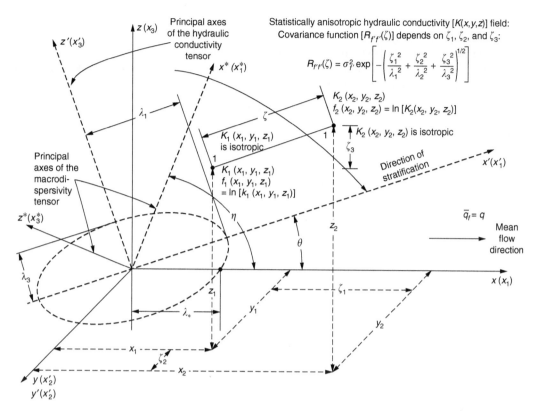

FIGURE 6-66 Coordinate systems for the case $x_2 = x_2'$; the dashed ellipse represents the e^{-1} level of correlation in the covariance function given by Eq. (6-274), the x_1' direction being that of bedding. The x_1' coordinates are the principal axes of the effective hydraulic conductivity tensor and x_i^* coordinates are the principal axes of the macrodispersivity tensor. (Adapted from Gelhar, L.W and Axness, C.L., Report No. H-8, 1981, 140 pp.)

notation, the following equation can be obtained:

$$E\left[f'\frac{\partial h'}{\partial x_i}\right] = \overline{f'\frac{\partial h'}{\partial x_i}} = E\left[\left(\int_{-\infty}^{\infty}e^{i\mathbf{k}\cdot\mathbf{x}}dZ_{f'}(\mathbf{k})\right)\left(-\int_{-\infty}^{\infty}ik_ie^{-i\mathbf{k}\cdot\mathbf{x}}dZ_{h'}^{*}(\mathbf{k})\right)\right] \tag{6-412}$$

$$= -\int_{-\infty}^{\infty}\int_{-\infty}^{\infty}ik_iE[dZ_{f'}(\mathbf{k})\,dZ_{h'}^{*}(\mathbf{k})]$$

From Eq. (6-406), we get

$$dZ_{h'}^{*} = \frac{ik_jI_j}{k^2}\,dZ_{f'}^{*} \tag{6-413}$$

Applying the spectral representation theorem in Section 6.4.2.1 [Eq. (6-105)] and using the equality of $i^2 = -1$, the combination of Eqs. (6-412) and (6-413) gives

$$\overline{f'\frac{\partial h'}{\partial x_i}} = \int_k \frac{1}{k^2}k_ik_jI_jS_{ff'}(\mathbf{k})\,d\mathbf{k} \tag{6-414}$$

where

$$F_{ij} = \int_k \frac{1}{k^2}k_ik_jS_{ff'}(\mathbf{k})\,d\mathbf{k} \tag{6-415}$$

The General Expression for the F_{ij} Integral

Evaluation of the F_{ij} integral is a lengthy process and is given in Gelhar and Axness (1981). Here, some intermediate steps will be presented in order to use the final equations better. The authors transformed the F_{ij} integral in Eq. (6-415) from the (k_1, k_2, k_3) coordinate system to the (k_1', k_2', k_3') coordinate system. From Figure 6-66 the transformation rules

$$k_i' = a_{ij}k_j, \quad k_i = a_{ji}k_j' \tag{6-416}$$

apply, where

$$a_{ij} = \cos(x_i', x_j) = \cos(u_{ij}) \tag{6-417}$$

denotes the direction cosine matrix. Equation. (6-415) may then be transformed to the (k_1', k_2', k_3') system as

$$F_{ij} = \int_k \frac{k_i k_j}{k^2} S_{ff'}(\mathbf{k})\, d\mathbf{k} = \int_{k'} \frac{|J|(a_{mi}k_m')(a_{nj}k_n')S_{ff'}(k_1', k_2', k_3')dk_1'\, dk_2'\, dk_3'}{k'^2} \tag{6-418}$$

where

$$|J| = |a_{ij}| \tag{6-419}$$

is the Jacobian of the transformation and

$$k^2 = k_1^2 + k_2^2 + k_3^2 = k_1'^2 + k_2'^2 + k_3'^2 = k'^2 \tag{6-420}$$

The direction cosines further satisfy the relations (Eisenhart, 1939) for $j = 1, 2, 3$ and j is not equal to k

$$\sum_{i=1}^{3} a_{ji}^2 = 1, \quad \sum_{i=1}^{3} a_{ij}^2 = 1, \quad \sum_{j=1}^{3} a_{ji}a_{ki} = 0, \quad \sum_{i=1}^{3} a_{ij}a_{ik} = 0 \tag{6-421}$$

simplify the direction cosine matrix to

$$a_{ij} = \begin{bmatrix} a_{12} & a_{12} & (1-a_{11}{}^2-a_{11}{}^2)^{1/2} \\ a_{21} & a_{22} & (1-a_{21}{}^2-a_{22}{}^2)^{1/2} \\ (a_{12}a_{23} - a_{12}a_{13}) & (a_{13}a_{21} - a_{23}a_{11}) & (a_{11}a_{22} - a_{21}a_{12}) \end{bmatrix} \tag{6-422}$$

from which it can be seen that $|J| = |a_{ij}| = 1$. Then the integral given by Eq. (6-418) becomes

$$F_{ij} = a_{mi}a_{nk}\int_{k'} \frac{k_m' k_n'}{k'^2} S_{ff'}(\mathbf{k}')\, d\mathbf{k}' = \sigma_{f'}{}^2 a_{mi}a_{nj}g_{mn} \tag{6-423}$$

where

$$g_{mn} = \int_{k'} \frac{k_m' k_n'}{k'^2} S_{f'f'}(\mathbf{k}')\, d\mathbf{k}' \tag{6-424}$$

The Expressions for g_{11}, g_{22}, and g_{33}

Since $S_{f'f'}(\mathbf{k}')$ from Eq. (6-275) is even in k_i' ($i = 1, 2, 3$), $g_{mn} = 0$ for $m \neq n$. For $m = n = 1$, letting

$$u_i = \lambda_i k_i' \quad \text{(no summation)} \tag{6-425}$$

one gets

$$g_{11} = \frac{\lambda_1 \lambda_2 \lambda_3}{\pi^2} \int_{-\infty}^{\infty}\int_{-\infty}^{\infty}\int_{-\infty}^{\infty} \frac{k_1'^2}{k'^2} \frac{dk_1'\, dk_2'\, dk_3'}{(1+ k_1'^2\lambda_1^2 + k_2'^2\lambda_2^2 + k_3'^2\lambda_3^2)^2}$$

$$= \frac{8}{\pi^2} \int_0^{\infty}\int_0^{\infty}\int_0^{\infty} \frac{u_1^2\, du_1\, du_2\, du_3}{(1 + u^2)^2(u_1^2 + \sigma^2 u_2^2 + \rho^2 u_3^2)} \tag{6-426}$$

where

$$\sigma = \frac{\lambda_1}{\lambda_2}, \qquad \rho = \frac{\lambda_1}{\lambda_3} \tag{6-427}$$

The coordinates (x_1', x_2', x_3') are defined so that $\lambda_1 \geq \lambda_2 \geq \lambda_3$ or $\rho \geq \sigma \geq 1$. Changing the spherical coordinates with u_3 as the polar axis and then letting $t = \cos(\varphi)$, Eq. (6-426) becomes

$$g_{11} = \frac{2}{\pi} \int_0^{\pi/2} F(a) \cos^2(\theta) \, d\theta \tag{6-428}$$

where

$$F(a) = \frac{[(1/a + a) \tan^{-1}(1/a) - 1]}{\rho^2 - f^2} \tag{6-429}$$

$$f = [\cos^2(\theta) + \sigma^2 \sin^2(\theta)]^{1/2} \tag{6-430}$$

and

$$a = \left(\frac{f^2}{\rho^2 - f^2} \right)^{1/2} \tag{6-431}$$

Similarly, the g_{22} integral can be reduced to

$$g_{22} = \frac{2}{\pi} \sigma^2 \int_0^{\pi/2} F(a) \sin^2(\theta) \, d\theta \tag{6-432}$$

and the g_{33} integral can be reduced to

$$g_{33} = \frac{2}{\pi} \int_0^{\pi/2} \frac{\rho^2}{\rho^2 - 1} \left[1 - a \tan^{-1}\left(\frac{1}{a}\right) \right] d\theta \tag{6-433}$$

For the general case $\rho > \sigma > 1$, the g_{ij} integrals, in Eqs. (6-428), (6-432), and (6-433) are integrated numerically, and the results are given in Figure 6-67 and Figure 6-68 as a function of the parameters $\rho = \lambda_1/\lambda_3$ and $\sigma = \lambda_1/\lambda_3$.

Special Cases for g_{11}, g_{22}, g_{33}, a_{ij}, and F_{ij} for $\lambda_1 = \lambda_2$ ($\sigma = 1$)
For the special case $\lambda_1 = \lambda_2$ ($\sigma = 1$), Eq. (6-428) is integrated as

$$g_{11}(\rho,1) = \frac{1}{2} \frac{1}{\rho^2 - 1} \left[\frac{\rho^2}{(\rho^2 - 1)^{1/2}} \tan^{-1}(\rho^2 - 1) - 1 \right] \tag{6-434}$$

In this case, it can easily be shown that

$$g_{22}(\rho, 1) = g_{11}(\rho, 1) \tag{6-435}$$

For $\sigma = 1$, Eq. (6-433) becomes

$$g_{33}(\rho,1) = \frac{\rho^2}{\rho^2 - 1} \left[1 - \frac{1}{(\rho^2 - 1)^{1/2}} \tan^{-1}(\rho^2 - 1) \right] \tag{6-436}$$

For the special case $\lambda_1 = \lambda_2$ ($\sigma = 1$), with the mean flow in the (x_1', x_2') plane as shown in Figure 6-66, the a_{ij} matrix given by Eq. (6-422) becomes

$$a_{ij} = \begin{bmatrix} \cos(\theta) & 0 & \sin(\theta) \\ 0 & 1 & 0 \\ -\sin(\theta) & 0 & \cos(\theta) \end{bmatrix} \tag{6-437}$$

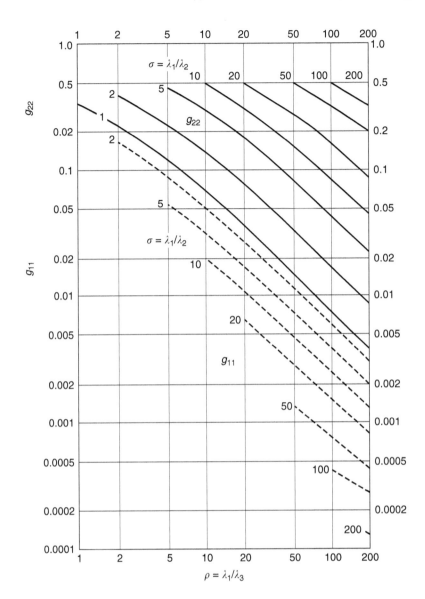

FIGURE 6-67 g_{11} and g_{22} as function of λ_1/λ_3 and λ_1/λ_2. (Adapted from Gelhar, L.W and Axness, C.L., Report No. H-8, 1981, 140 pp.)

In this case, the components of F_{ij} are

$$F_{11} = \sigma_f^2[g_{11} \cos^2(\theta) + g_{33} \sin^2(\theta)] \tag{6-438}$$

$$F_{22} = \sigma_f^2 g_{22} \tag{6-439}$$

$$F_{13} = F_{31} = \sigma_f^2(g_{11} - g_{33}) \cos(\theta) \sin(\theta) \tag{6-440}$$

and

$$F_{33} = \sigma_f^2[g_{11} \sin^2(\theta) + g_{33} \cos^2(\theta)] \tag{6-441}$$

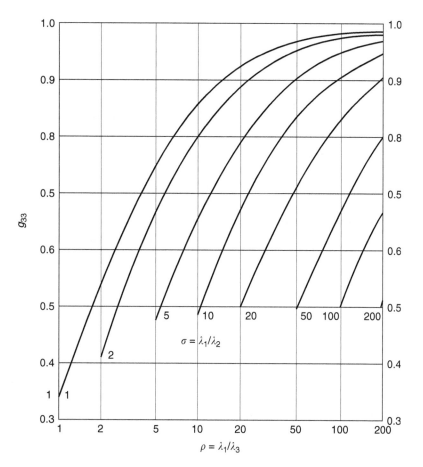

FIGURE 6-68 g_{33} as function of λ_1/λ_3 and λ_1/λ_2. (Adapted from Gelhar, L.W and Axness, C.L., Report No. H-8, 1981, 140pp.)

General Expression for the Effective Hydraulic Conductivity Tensor
By using Eq. (6-423), the effective hydraulic conductivity tensor given by Eq. (6-410) becomes

$$\overline{K}_{ij} = K_g \left[\left(1 + \tfrac{1}{2}\sigma_{f'}^2 \right) \delta_{ij} - \sigma_{f'}^2 a_{mi} a_{nj} g_{mn} \right] \tag{6-442}$$

From the form of g_{mn} in Eq. (6-424), it is evident that $g_{mn} = 0$ if $m \neq n$, since $S_{f'f'} (k')$ is always even in k_i'. Therefore, Eq. (6-442) shows that the effective hydraulic conductivity tensor is a second rank, symmetric tensor.

Effective Hydraulic Conductivity Tensor for a Special Case: Statistical Isotropy in the
$x_1' - x_2'$ *Plane* $(\lambda_1 = \lambda_2)$
In the case of statistical isotropy in the $x_1' - x_2'$ plane $(\lambda_1 = \lambda_2)$ with the x_1- and x_1'-axis separated by angle θ as shown in Figure 6-66, the components of the effective hydraulic conductivity tensor in the unprimed coordinates system are

$$\overline{K}_{11} = K_g \left\{ 1 + \sigma_{f'}^2 \left[\tfrac{1}{2} - g_{11} \cos^2(\theta) - g_{33} \sin^2(\theta) \right] \right\} \tag{6-443}$$

$$\overline{K}_{22} = K_g \left[1 + \sigma_{f'}^2 \left(\tfrac{1}{2} - g_{22} \right) \right] \tag{6-444}$$

540 Applied Flow and Solute Transport Modeling in Aquifers

$$\overline{K}_{13} = \overline{K}_{31} = K_g \sigma_{f'}^2 \sin(\theta) \cos(\theta)(g_{33} - g_{11}) \tag{6-445}$$

$$\overline{K}_{21} = \overline{K}_{12} = \overline{K}_{23} = \overline{K}_{32} = 0 \tag{6-446}$$

and

$$\overline{K}_{33} = K_g \left\{ 1 + \sigma_{f'}^2 \left[\tfrac{1}{2} - g_{33} \cos^2(\theta) - g_{11} \sin^2(\theta) \right] \right\} \tag{6-447}$$

The functional form of g_{11} is given by Eq. (6-434), $g_{11} = g_{22}$ from Eq. (6-435), and g_{33} is given by Eq. (6-436).

There are three special cases of interest and they are given below.

1. *Statistically Isotropic Media* $(\lambda_1 = \lambda_2 = \lambda_3)$: In this case, $\rho = 1$ according to the second expression of Eq. (6-427) and $g_{11} = g_{22} = g_{33} = 1/3$ (Gelhar and Axness, 1981). Then, from Eqs. (6-443)–(6-447)

$$\overline{K}_{ij} = K_g \left(1 + \frac{\sigma_{f'}^2}{6} \right) \delta_{ij} \tag{6-448}$$

which checks with the first-order result for a statistically isotropic media given by Eq. (6-244), originally developed by Gutjahr et al. (1978).

2. *Mean Flow Parallel to Bedding* $(\lambda_1 = \lambda_2 > \lambda_3)$: This situation corresponds to $\theta = 0$ in Figure 6-64 for which the mean flow becomes parallel to bedding. Therefore, from Eq. (6-437), $a_{11} = a_{22} = a_{33} = 1$, and from Eqs. (6-443)–(6-447), the expressions for the effective hydraulic conductivity become

$$\overline{K}_{11} = \overline{K}_{22} = K_g \left[1 + \sigma_{f'}^2 \left(\tfrac{1}{2} - g_{11} \right) \right] \tag{6-449}$$

$$\overline{K}_{33} = K_g \left[1 + \sigma_{f'}^2 \left(\tfrac{1}{2} - g_{33} \right) \right] \tag{6-450}$$

Equations (6-449) and (6-450) can also be expressed as

$$\overline{K}_{ii} = K_g \left[1 + \sigma_{f'}^2 \left(\frac{1}{2} - g_{ii} \right) \right], \quad i = 1, 2, 3 \text{ (no summation on } i) \tag{6-451}$$

It should be noted that there is no sum on the repeated index, which means that it is used to identify the three principal components of the effective hydraulic conductivity tensor.

The principal hydraulic conductivity tensor corresponds to the three orthogonal directions associated with the correlation scales in Eq. (6-274). Note from Eq. (6-434) that as $\rho = \lambda_1/\lambda_3 \rightarrow \infty$, $g_{11} \rightarrow 0$, and g_{33} in Eq. (6-433) approaches 1. As a result, $\overline{K}_{33}$ becomes negative for $\sigma_{f'}^2 > 2$. This impossible situation results from the neglect of higher order terms in Eq. (6-407). For the special case where $\sigma_{f'} = 1$ and $\rho = \infty$, Gutjahr et al. (1978) found an 18% error between the approximate solution and exact solution for one-dimensional flow perpendicular to layering with a log-normally distributed K (see Example 6-8). Therefore, Eqs. (6-449) and (6-450) are good approximations only when $\sigma_{f'} < 1$. Since for some natural aquifer materials $\sigma_{f'} > 1$, Gelhar and Axness (1981, 1983) considered a generalization that properly approaches the limiting case of one-dimensional flow parallel and perpendicular to layers. The authors generalized Eqs. (6-449) and (6-450) by considering the quantities in parentheses as the first two terms of the Taylor series expansion for $\exp(x)$. The generalization of Eqs. (6-449) and (6-450) for $\sigma_{f'} > 1$ becomes

$$\overline{K}_{11} = K_g \exp\left[\sigma_{f'}^2\left(\tfrac{1}{2} - g_{11}\right)\right] \qquad (6\text{-}452)$$

$$\overline{K}_{22} = K_g \exp\left[\sigma_{f'}^2\left(\tfrac{1}{2} - g_{22}\right)\right] \qquad (6\text{-}453)$$

$$\overline{K}_{33} = K_g \exp\left[\sigma_{f'}^2\left(\tfrac{1}{2} - g_{33}\right)\right] \qquad (6\text{-}454)$$

For $\rho \rightarrow \infty$, these equations become

$$\overline{K}_{11} = \overline{K}_{22} = K_g \exp\left(\frac{\sigma_{f'}^2}{2}\right) \qquad (6\text{-}455)$$

$$\overline{K}_{33} = K_g \exp\left(-\frac{\sigma_{f'}^2}{2}\right) \qquad (6\text{-}456)$$

which correspond to the arithmetic and harmonic means of K, respectively, if K is log-normally distributed. The harmonic mean corresponds to Eq. (6-198), as the effective hydraulic conductivity for flow normal to a stratified medium, whereas the arithmetic mean is the effective hydraulic conductivity for flow parallel to perfect layering. Note from Figure 6-67 and Figure 6-68 that for $\theta = 0$ and $\rho > 100$, the anisotropy ratio $\overline{K}_{11}/\overline{K}_{33}$ from Eqs. (6-455) and (6-456) is very close to $\exp(\sigma_{f'}^2)$. Figure 6-69 compares $\overline{K}_{11}/\overline{K}_1$ and $\overline{K}_{33}/\overline{K}_{11}$ from Eqs. (6-443)–(6-447) and Eqs. (6-452)–(6-454) for various aspect ratio ρ as a function of $\sigma_{f'}^2$ with $\theta = 0$. Figure 6-70 gives the anisotropy ratio $\overline{K}_{11}/\overline{K}_{33} = \exp[\sigma_{f'}^2(g_{33} - g_{11})]$ as a function of $\sigma_{f'}^2$ with the aspect ratio $\rho = \lambda_1/\lambda_3$ as a parameter.

3. *Mean Flow Inclined to Bedding ($\theta \neq 0$):* In this case, the generalization to large $\sigma_{f'}^2$, Eqs. (6-452)–(6-454) can be used to determine the principal components of the effective hydraulic conductivity tensor. Then, the tensor transformation rule can be used to determine the component of K_{ij} in the inclined system.

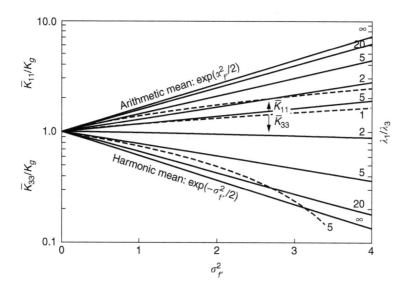

FIGURE 6-69 Effective hydraulic conductivities parallel $\overline{K}_{11}$ and perpendicular $\overline{K}_{33}$ to bedding; the solid lines are the generalization of Eqs. (6-451)–(6-453) and the dashed lines are the first-order results based on Eqs. (6-443) through (6-447). (Adapted from Gelhar, L.W and Axness, C.L., Report No. H-8, 1981, 140 pp.)

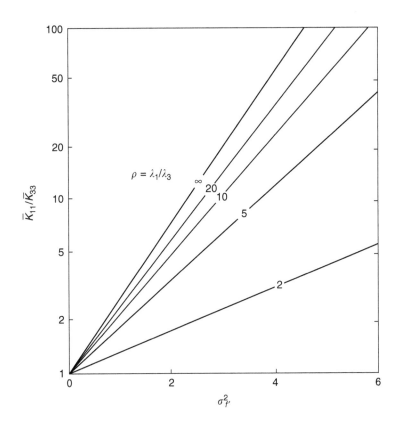

FIGURE 6-70 Anisotropy of effective hydraulic conductivity as a function of the variance of ln (K), $\sigma_{f'}^2$, and aspect ratio λ_1/λ_3 for the case $\lambda_1 = \lambda_2$. (Adapted from Gelhar, L.W and Axness, C.L., Report No. H-8, 1981, 140pp.)

The general results for $\lambda_1 \neq \lambda_2 \neq \lambda_3$, as given by Figure 6-67 and Figure 6-68, can be used to determine K_{ij} for a three-dimensionally anisotropic material. The principal components can be determined from the generalization of Eqs. (6-452)–(6-454), using g_{ij} from Figure 6-67 and Figure 6-68, and the transformation rule can be used to find K_{ij} in the required coordinate system. This more general case will be considered in the analysis of macrodispersivity tensor, which is presented below.

Specific Discharge Spectra
The specific discharge perturbation is found by subtracting Eq. (6-409) from Eq. (6-407) such that

$$q_i' = q_i - \bar{q}_i = K_g \left(1 + f' + \frac{1}{2}f'^2 + \cdots\right)\left(I_i - \frac{\partial h'}{\partial x_i}\right) - K_g \left[\left(1 + \frac{1}{2}\sigma_{f'}^2\right)I_i - \overline{f'\frac{\partial h'}{\partial x_i}}\right] \quad (6\text{-}457)$$

After some manipulation and leaving to first order we get

$$q_i' \cong K_g\left(f'I_i - \frac{\partial h'}{\partial x_i}\right) \quad (6\text{-}458)$$

Using the Fourier–Stieltjes representations for q_i', f', and h', which are given by Eqs. (6-324), (6-229), and (6-228), respectively, in Eq. (6-457)

$$dZ_{q_i'} = K_g I_i \, dZ_{f'} - K_g ik_i\left(-\frac{ik_j I_j}{k^2}\,dZ_{f'}\right) = K_g I_j\left(\delta_{ij} - \frac{k_i k_j}{k^2}\right)dZ_{f'} \quad (6\text{-}459)$$

Its complex conjugate is

$$dZ^*_{q'_i} = K_g I_j \left(\delta_{ij} - \frac{k_i k_j}{k^2} \right) dZ^*_{f'} \tag{6-460}$$

Applying the spectral representation theorem in Section 6.4.2.1 (Eq. [6-105])

$$S_{q'_i q'_j} = S_{q'_j q'_i} = K_g^2 I_m I_n \left(\delta_{im} - \frac{k_i k_m}{k^2} \right)\left(\delta_{jn} - \frac{k_j k_n}{k^2} \right) S_{f'f'} \tag{6-461}$$

This is an important relationship in the analysis of microdispersion, which is presented below.

Formulation and General Solution for Macroscopic Dispersion in Statistically Anisotropic Media with Arbitrary Orientation of Stratification
Gelhar and Axness (1981, 1983) generalized their approaches to statistically anisotropic media in which the mean flow is not necessarily aligned in the direction of stratification. The authors used the results presented above in conjunction with the spectral relationship given by Eq. (6-335) to determine a generalized Fickian transport relationship for this type of medium. The resulting integral expression for the macroscopic dispersivity tensor is then evaluated for the special case where the following assumptions are made: (1) The medium hydraulic conductivity is isotropic at the local scale as assumed previously; (2) local longitudinal dispersivity is small compared with the aquifer correlation scale in the direction of stratification ($\alpha^*_L << \lambda_1$); and (3) the log-hydraulic conductivity autocovariance is assumed to be exponential, as given by Eq. (6-274).

Generalized Fickian Transport Relationship and General Solution
Equation. (6-335) still holds for the general anisotropic medium. Using the specific discharge spectrum, given by Eq. (6-461), in Eq. (6-335), we get

$$A_{ij} = \int_{-\infty}^{\infty} \frac{K_g^2 I_m I_n (\delta_{im} - k_i k_m / k^2)(\delta_{jn} - k_j k_n / k^2) S_{f'f'} \, d\mathbf{k}}{[ik_1 + \alpha^*_L k_1^2 + \alpha^*_T (k_2^2 + k_3^2)]q^2} \tag{6-462}$$

Since

$$S_{q'_i q'_j} = S_{q'_j q'_i} \tag{6-463}$$

Therefore,

$$A_{ij} = \frac{1}{I_1^2 \gamma^2} [I_i I_j P - I_j I_n Q_{in} - I_m I_i Q_{jm} + I_m I_n R_{ijmn}] \tag{6-464}$$

where

$$P = \int_k \frac{S_{f'f'}(\mathbf{k})}{B(\mathbf{k})} \, d\mathbf{k} \tag{6-465}$$

$$Q_{ij} = \int_k \frac{k_i k_j S_{f'f'}(\mathbf{k})}{k^2 B(\mathbf{k})} \, d\mathbf{k} \tag{6-466}$$

$$R_{ijmn} = \int_k \frac{k_i k_j k_m k_n S_{f'f'}(\mathbf{k})}{k^4 B(\mathbf{k})} \, d\mathbf{k} \tag{6-467}$$

$$B(\mathbf{k}) = ik_1 + \alpha^*_L k_1^2 + \alpha^*_T (k_2^2 + k_3^2) \tag{6-468}$$

and

$$\gamma = \frac{q}{K_g I_1} \tag{6-469}$$

Using Eq. (6-464) it can be seen that $A_{ij} = A_{ji}$, and from Eq. (6-467) $R_{ijmn} = R_{jimn}$. These confirm that the macroscopic dispersivity is a second rank symmetric tensor.

Evaluation of the Integrals
The details for the evaluation of the integrals given by Eqs. (6-465)–(6-467) for the anisotropic exponential covariance given by Eq. (6-274) can be found in Gelhar and Axness (1983). The authors evaluated the integrals using an asymptotic approach on the assumption that $\alpha_L^* \ll \lambda_1$. The resultant asymptotic solutions are independent of the local dispersivity. In the following paragraphs, several special cases are presented.

Special Case 1: Isotropy is in the Plane of Stratification with $\lambda_1 = \lambda_2$ and the Mean Flow Inclined at Angle θ with the Bedding
For the special case where the medium is assumed to be statistically isotropic in the x_1'–x_2' plane (see Figure 6-66), $\lambda_1 = \lambda_2$. As shown in Figure 6-66, the mean flow is inclined at an angle θ with the bedding ($I_2 = 0$). Gelhar and Axness (1981) determined the A_{ij} terms in terms of the hydraulic gradient ratio I_3/I_1, $\rho = \lambda_1/\lambda_3$, θ, and the flow correction factor γ given by Eq. (6-469). For γ, the hypothesized relationships for the effective hydraulic conductivities are given by Eqs. (6-452)–(6-454).

Determination of the Ratio I_3/I_1 and γ
The ratio I_3/I_1 is evaluated by noting that

$$\bar{q}_3 = 0 = \bar{K}_{13}I_1 + \bar{K}_{33}I_3 \tag{6-470}$$

and by using the tensor transformation

$$\bar{K}_{ij} = \bar{K}'_{mn} a_{mi} a_{nj} \tag{6-471}$$

between the primed and unprimed systems (see Figure 6-66), I_3/I_1 can be expressed in terms of the angle θ and the principal values of $\bar{K}_{ij}$ as follows:

$$\frac{I_3}{I_1} = -\frac{\bar{K}_{13}}{\bar{K}_{33}} = -\frac{(\bar{K}'_{11} - \bar{K}'_{33}) \sin(\theta) \cos(\theta)}{\bar{K}'_{11} \sin^2(\theta) + \bar{K}'_{33} \cos^2(\theta)} = -\frac{(1-\beta) \sin(\theta) \cos(\theta)}{\sin^2(\theta) + \beta \cos^2(\theta)} \tag{6-472}$$

where

$$\beta = \frac{\bar{K}'_{33}}{\bar{K}'_{11}} = \exp[\sigma_f^2(g_{11} - g_{33})] \tag{6-473}$$

Similarly, using the Darcy equation

$$q = \bar{q}_1 = \bar{K}_{11}I_1 + \bar{K}_{13}I_3 \tag{6-474}$$

the expression for γ in Eq. (4-469) can be expressed as

$$\gamma = \frac{q}{K_g I_1} = \frac{\bar{K}_{11}I_1 + \bar{K}_{13}I_3}{K_g I_1} \tag{6-475}$$

After transformation

$$\gamma = \frac{\bar{K}'_{11}\bar{K}'_{33}}{K_g \bar{K}_{33}} = \frac{\bar{K}'_{11}\bar{K}'_{33}}{K_g[\bar{K}'_{11} \sin^2(\theta) + \bar{K}'_{33} \cos^2(\theta)]} = \frac{\exp[\sigma_f^2(1/2 - g_{33})]}{\sin^2(\theta) + \beta \cos^2(\theta)} \tag{6-476}$$

where the relations given by Eqs. (6-452)–(6-454) are used as hypothesized previously.

Components of the Macrodispersivity Tensor
With isotropy in the plane of stratification, Gelhar and Axness (1981) determined the macrodispersivity tensor A_{ij} from Eq. (6-464) with $I_2 = 0$. The result for A_{11} is:

$$A_{11} = \sigma_{f'}^2 \frac{\lambda_1 \mu}{\gamma^2 \xi} \tag{6-477}$$

where

$$\mu = \frac{\lambda_3}{\lambda_1} \tag{6-478}$$

and

$$\xi = [\sin^2(\theta) + \mu^2 \cos^2(\theta)]^{1/2} \tag{6-479}$$

The general expression for g_{11} is given by Eq. (6-428) and its graphical form is given in Figure 6-67 as a function of $\rho = \lambda_1/\lambda_3 = 1/\mu$. The general expression for g_{33} is given by Eq. (6-433) and its graphical form is given in Figure 6-68 as a function of $\rho = \lambda_1/\lambda_3 = 1/\mu$. The result in Eq. (6-477) shows that the longitudinal macrodispersivity component A_{11} is proportional to $\sigma_f^2 \lambda_1$ in a form similar to that of the isotropic case as given by Eq. (6-384).
The expressions for A_{22}, A_{33}, and A_{13}, respectively are (Gelhar and Axness, 1981)

$$A_{22} = \frac{\sigma_f^2 \lambda_1 \mu (1-\beta)^2 \sin^2(\theta) \cos^2(\theta)}{2[\sin^2(\theta) + \beta \cos^2(\theta)]^2 (1+\xi)^2 \gamma^2} \tag{6-480}$$

$$A_{33} = \frac{\sigma_f^2 \lambda_1 \mu (1-\beta)^2 \sin^2(\theta) \cos^2(\theta)(1+2\xi)}{2[\sin^2(\theta) + \beta \cos^2(\theta)]^2 (1+\xi)^2 \gamma^2} \tag{6-481}$$

and

$$A_{13} = A_{31} = -\frac{\sigma_f^2 \lambda_1 \mu (1-\beta) \sin(\theta) \cos(\theta)}{(1+\xi)[\sin^2(\theta) + \beta \cos^2(\theta)]\gamma^2} \tag{6-482}$$

Note that Eq. (6-482) shows that $A_{13} \neq 0$, which means that the principal axes of the macrodispersivity tensor do not coincide with the direction of the mean flow. The principal axes and principal components of A_{ij} are determined by standard tensor transformation relationships (see Section 6.4.6.8). For example, the angle η between x (x_1) and $x^*(x^*_1)$, the principal axis of A_{ij} (see Figure 6-66) is given by Eq. (6-364).

Illustrative Example for the Principal Components of the Macrodispersivity Tensor
To illustrate the structure of Eqs. (6-477)–(6-482), the dependence of A_{ij} on θ was determined by Gelhar and Axness (1981) for a specified medium using the following parameters:

$$\beta = \frac{\overline{K}'_{33}}{\overline{K}'_{11}} = \frac{1}{5}, \qquad \rho = \frac{\lambda_1}{\lambda_3} = 5, \qquad \sigma_f^2 = 2.58$$

Figure 6-71 shows the principal components of A_{ij}, A^*_{11}, A^*_{22}, and A^*_{33} in normalized form along with the angle η. As shown in Figure 6-71a, η is negative when θ is positive. In other words, the principal axis associated with the maximum principal component of A_{ij} is deflected in a direction opposite to that of the stratification. It can be observed that the transverse dispersivities A^*_{22} and A^*_{33} are of a similar magnitude and attain a maximum of 8% of A^*_{11}. For $\theta = 0°$ and $90°$, $A^*_{22} = A^*_{33} = 0$. This should be expected since local dispersion is neglected in the asymptotic analysis leading Eqs. (6-477), (6-480)–(6-482). These cases with flow parallel or perpendicular to stratification are similar to those analyzed in Sections 6.4.6.8 and 6.4.6.9, including the effect of local dispersion. Typically, for these special cases, the transverse macrodispersivities will be several orders of

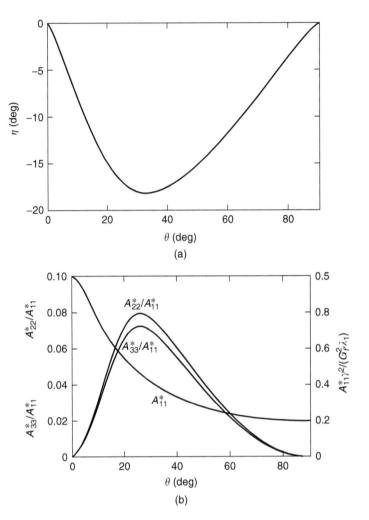

FIGURE 6-71   (a) Principal angle η and (b) Principal components of macrodispersivity tensor for Case I. (Adapted from Gelhar, L.W and Axness, C.L., Report No. H-8, 1981, 140 pp.)

magnitude smaller than the longitudinal dispersivity. When the stratification is at an angle to the mean flow, the transverse dispersivities may range from 1 to 10% of A_{11}^* (see Figure 6-71).

EXAMPLE 6-11

This example is adapted from Gelhar and Axness (1981). The Mt. Simon aquifer is approximately isotropic in the plane of stratification, which means that $\lambda_1 = \lambda_2$. The angle with the bedding, θ, is given as $20°$ (refer to Figure 6-64 for its definition). Representative permeability and porosity variations of the Mt. Simon aquifer are shown in Figure 6-72. With the values given as $\sigma_f^2 = 2.5$, $\lambda_1 = \lambda_2 = 10$ m, and $\lambda_3 = 0.5$ m, determine (1) the values of A_{11}, A_{22}, A_{33}, and A_{13}; (2) the values of A_{11}^*, A_{22}^*, A_{33}^* and η; and (3) the ratio of $\overline{K}_{11}'/K_{33}'$, and demonstrate the results graphically.

SOLUTION

First, the values of ρ, σ, μ, β, ξ, and γ need to be calculated. From Eqs. (6-427) and (6-478),

$$\rho = \frac{10 \text{ m}}{0.5 \text{ m}} = 20, \qquad \sigma = \frac{10 \text{ m}}{10 \text{ m}} = 1, \qquad \mu = \frac{0.05 \text{ m}}{10 \text{ m}} = 0.05$$

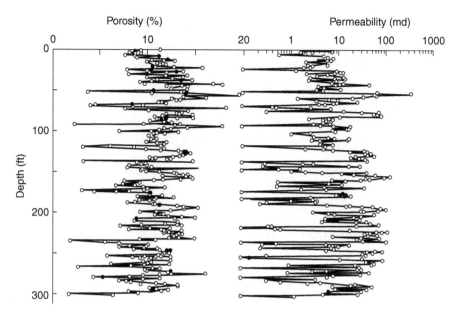

FIGURE 6-72   Permeability and porosity of cores collected at 1-ft intervals from a borehole in the Mt. Simon aquifer in Illinois based on data from Bakr, A.A., Ph.D. dissertation, New Mexico Institute of Mining and Technology, Socorro, NMexico, 1976; Gelhar, L.W., *Stochastic Subsurface Hydrology*, Prentice Hall, Inc., Englewood Cliffs, NJ, 1993, 390 pp.

From Figure 6-67, one get $g_{11} = 0.0375$; and from Figure 6-68, one get $g_{33} = 0.926$. Using the given values, from Eq. (6-473)

$$\beta = \exp[2.5(0.0375 - 0.9260] = 0.108472$$

From Eq. (6-479),

$$\xi^2 = [\sin(20°)]^2 + (0.05)^2 [\cos(20°)]^2 = 0.140470$$
$$\xi = 0.374793$$

From Eq. (6-476),

$$\gamma = \frac{\exp[2.5(0.5 - 0.926)]}{[\sin(20°)]^2 + (0.108472)[\cos(20°)]^2} = 1.620247$$

(1)   From Eq. (6-477)

$$A_{11} = \frac{(2.5)(10 \text{ m})(0.05)}{1.620247^2(0.374793)} = 1.270 \text{ m}$$

From Eq. (6-480),

$$A_{22} = \frac{(2.5)(10 \text{ m})(0.05)(1 - 0.108472)^2[\sin(20°)]^2[\cos(20°)]^2}{2\langle[\sin(20°)]^2 + (0.108472)[\cos(20°)]^2\rangle^2(1 + 0.374793)^2(1.620247)^2} = 0.229 \text{ m}$$

From Eq. (6-481),

$$A_{33} = \frac{(2.5)(10 \text{ m})(0.05)(1 - 0.108472)^2[\sin(20°)]^2[\cos(20°)]^2[1 + 2(0.374793)]}{2\langle[\sin(20°)]^2 + (0.108472)[\cos(20°)]^2\rangle^2(1 + 0.374793)^2(1.620247)^2} = 0.400 \text{ m}$$

and from Eq. (6-482),

$$A_{13} = A_{31} = -\frac{(2.5)(10 \text{ m})(0.05)(1 - 0.108472) \sin(20°) \cos(20°)}{(1 + 0.374793)\langle[\sin(20°)]^2 + (0.108472)[\cos(20°)]^2\rangle(1.620247)^2} = -0.454 \text{ m}$$

(2) Since in the plane of stratification, which is $x - y$ or $x_1 - x_2$, the medium is statistically isotropic, $A_{22}^* = A_{22} = 0.229$ m. Therefore, anisotropy exists only in the x–z plane (or $x_1 - x_3$ plane). As a result, the macrodispersivities in the principal directions in the x–z plane can be calculated from Eq. (6-362) for A_{11}^* and Eq. (6-363) for A_{33}^*. From Eq. (6-362),

$$A_{11}^* = \tfrac{1}{2}(1.270 \text{ m} + 0.400 \text{ m}) + \left[\tfrac{1}{4}(1.270 \text{ m} - 0.400 \text{ m})^2 + (0.454 \text{ m})^2 \right]^{1/2} = 1.464 \text{ m}$$

From Eq. (6-363),

$$A_{33}^* = \tfrac{1}{2}(1.270 \text{ m} + 0.400 \text{ m}) - \left[\tfrac{1}{4}(1.270 \text{ m} - 0.400 \text{ m})^2 + (0.454 \text{ m})^2 \right]^{1/2} = 0.206 \text{ m}$$

From Eq. (6-364),

$$\tan(2\eta) = \frac{2(-0.454 \text{ m})}{1.270 \text{ m} - 0.400 \text{ m}} = -1.044$$

$$\eta = -23.10°$$

(3) From Eq. (6-473),

$$\overline{K}_{11}' = \frac{\overline{K}_{33}'}{\beta} = \frac{\overline{K}_{33}'}{0.108472} = 9.219\overline{K}_{33}'$$

The graphical demonstration is shown in Figure 6-73. The effects of off-diagonal terms illustrated graphically in Figure 6-73, which depicts, in a vertical plane, the configuration of average sediment

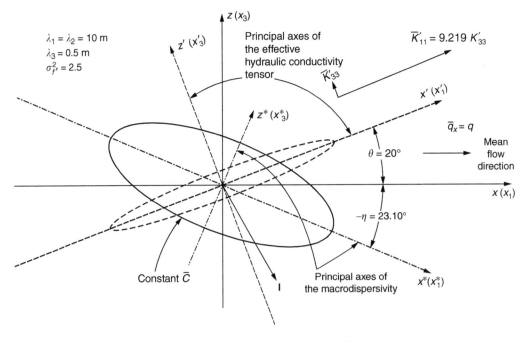

FIGURE 6-73 Graphical demonstration of the results for Example 6-11.

lenses as in the dashed ellipse with major and minor axes λ_1 and λ_3, and the relative orientation of a mean solute plume resulting from an instantaneous point injection of tracer after large downstream displacement, shown as the solid-line ellipse with major and minor axes proportional to $(A_{11}^*)^{1/2}$ and $(A_{33}^*)^{1/2}$. In other words, the dashed –line ellipse represents an e^{-1} correlation line of the covariance of ln (K) and the solid line ellipse indicates a line of constant mean concentration in the plume. The anisotropy calculated from Eq. (6-473) is of the order of 10 and leads to a mean hydraulic gradient with orientation shown. The turning of the plume in the direction of the mean hydraulic gradient reflects the influence of the negative diagonal term, A_{13} . In Figure 6-73, the direction of the mean flow $\bar{q}_x = q$, is shown which is always in the x (x_1) direction and the direction of the mean hydraulic gradient. Although the orientation of the solute plume relative to the sediment lenses at first seems counterintuitive, this feature simply reflects the influence of the transverse hydraulic gradients that occur in Eq. (6-464) (Gelhar and Axness, 1981; Gelhar, 1993).

Special Case 2: Horizontal Stratification with Horizontal and Vertical Anisotropy

A second relatively simple anisotropic configuration is one in which the flow is parallel to horizontal stratification. Both horizontal and vertical anisotropy are depicted in Figure 6-74 for an unconfined aquifer and in Figure 6-75 for a confined aquifer. The plan view of each case is shown in Figure 6-76. As shown in Figure 6-74 and Figure 6-75, this case represents horizontal flow in a horizontally stratified aquifer $(\lambda_3 << \lambda_1)$, with anisotropy in the horizontal plane (x, y) [or

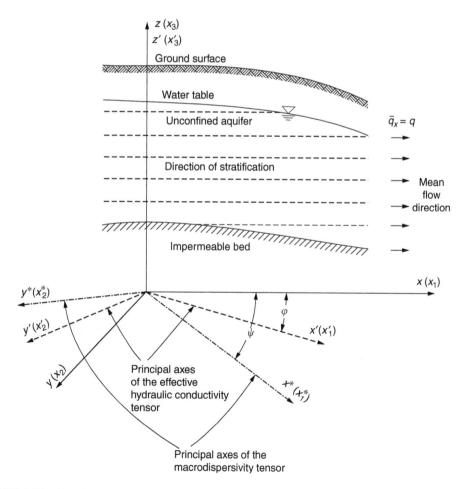

FIGURE 6-74   Flow in a statistically horizontal and vertical anisotropic aquifer with horizontal stratification.

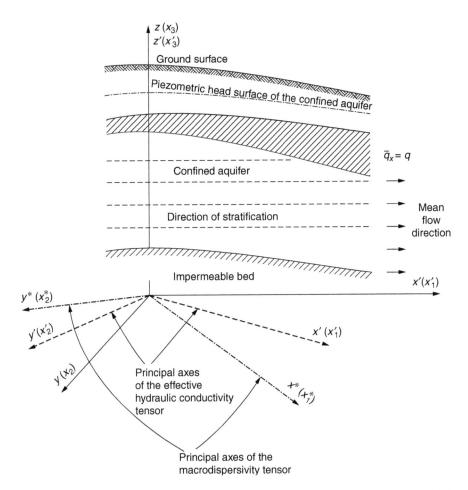

FIGURE 6-75 Flow in a statistically horizontal and vertical anisotropic confined aquifer with horizontal stratification.

(x_1, x_2)] with $\lambda_2 < \lambda_1$. This is the commonly encountered aquifer situation in which the flow is predominantly horizontal in stratified sedimentary deposits. The horizontal anisotropy is associated with a preferred direction in the depositional process. As shown in Figure 6-76, φ is the angle between the direction of the mean flow and that of a maximum correlation in the $\ln(K)$ process (Gelhar, 1993).

Components of the Macrodispersivity Tensor
Gelhar and Axness (1981) evaluated the effects of horizontal anisotropy by considering $\lambda_1 > \lambda_2$ with x_1' (the direction of maximum correlation scale λ_1) at an angle φ with the mean flow direction x_1, as shown in Figure 6-76. Then, the authors evaluated the macroscopic dispersivity tensor for this case and obtained its components. The result for A_{11} is

$$A_{11} = \frac{\sigma_{f}^2 \lambda_1 \lambda_2}{\gamma^2 \lambda_3 \zeta} \tag{6-483}$$

where

$$\gamma = \frac{\exp[\sigma_{f}^2(1/2 - g_{22})]}{\sin^2(\varphi) + \beta \cos^2(\varphi)} \tag{6-484}$$

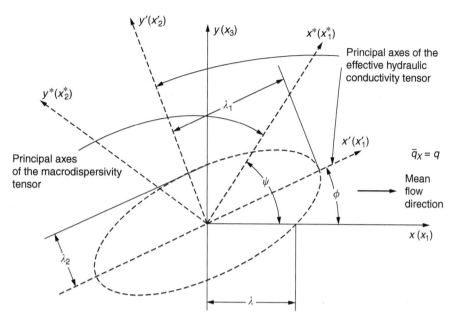

FIGURE 6-76 Coordinate systems for Case 2 ($x_3 = x_3'$). The dashed ellipse represents the e^{-1} level of correlation in the covariance function given by Eq. (6-274); the x_1' direction is the maximum correlation scale in the horizontal; the x_i' coordinates are the principal axes of the effective hydraulic conductivity tensor; and x_i^* coordinates are the principal axes of the macrodispersivity tensor. (Adapted from Gelhar, L.W and Axness, C.L., Report No. H-8, 1981, 140 pp.)

in which β is given by

$$\beta = \frac{\overline{K}'_{22}}{\overline{K}'_{11}} = \exp[\sigma_{f'}^2(g_{11} - g_{22})] \tag{6-485}$$

The expression for ζ is given by

$$\zeta = \left[\left(\frac{\lambda_1}{\lambda_2} \right)^2 \sin^2(\varphi) + \left(\frac{\lambda_2}{\lambda_3} \right)^2 \cos^2(\varphi) \right]^{1/2} \tag{6-486}$$

The expression for the ratio A_{22}/A_{11} is given by

$$\frac{A_{22}}{A_{11}} = \left(\frac{I_2}{I_1} \right)^2 \frac{\zeta(2\zeta + 1)}{2(1 + \zeta)^2} \tag{6-487}$$

The expression for I_2/I_1 is given as

$$\frac{I_2}{I_1} = \frac{(\beta - 1) \sin(\varphi) \cos(\varphi)}{\sin^2(\varphi) + \beta \cos^2(\varphi)} \tag{6-488}$$

The expression for the ratio A_{33}/A_{11} is given by

$$\frac{A_{33}}{A_{11}} = \left(\frac{I_2}{I_1} \right)^2 \frac{\zeta}{2(1 + \zeta)^2} \tag{6-489}$$

The expression for the ratio A_{12}/A_{11} is given by

$$\frac{A_{22}}{A_{11}} = \left(\frac{I_2}{I_1} \right) \frac{\zeta}{1 + \zeta} \tag{6-490}$$

Applied Flow and Solute Transport Modeling in Aquifers

and

$$A_{32} = A_{23} = 0 \qquad (6\text{-}491)$$

In this case, A_{12} is nonzero and the orientation of the principal axes of A_{ij}, x_i^*, is determined from Eq. (6-356) by replacing η by ψ, which is shown in Figure 6-76.

Important Features for Case II

There are some important features for Case II: (1) when $\varphi = 0$ or $90°$, $A_{22} = A_{33} = A_{12} = 0$ because I_2/I_1 is zero. The transverse dispersion cos become very small when the mean flow direction coincides with one of the principal axes of the effective hydraulic conductivity tensor. (2) this case represents a commonly conceptualized aquifer for which the mean flow is horizontal and parallel to bedding. As a result, this case represents the solute transport process in a two-dimensional depth-averaged system with horizontal mean flow.

EXAMPLE 6-12

This example is adapted from Gelhar (1993). For an aquifer with horizontal stratification the following values are given: $\sigma_f^2 = 2.5$, $\lambda_1 = 10$ m, $\lambda_2 = 2$ m, $\lambda_3 = 0.5$ m, and $\varphi = 45°$ (refer see Figure 6-76 for its definition). Using these values, (1) the values of A_{11}, A_{22}, A_{33}, and A_{12}; (2) the values A_{11}^*, A_{22}^*, A_{33}^* and ψ; and (3) the ratio of $\bar{K}_{11}'/\bar{K}_{22}'$, and demonstrate the results graphically.

SOLUTION

First, the values of ρ, σ, μ, β ξ, and γ need to be calculated. From Eqs. (6-427) and (6-478)

$$\rho = \frac{10 \text{ m}}{0.5 \text{ m}} = 20, \qquad \sigma = \frac{10 \text{ m}}{2 \text{ m}} = 5, \qquad \mu = \frac{0.05 \text{ m}}{10 \text{ m}} = 0.05$$

From Figure 6-67, one get $g_{11} = 0.018$ and $g_{22} = 0.180$. By using the given values, from Eq. (6-485)

$$\beta = \exp[2.5(0.018 - 0.180] = 0.666975$$

From Eq. (6-484),

$$\gamma = \frac{\exp[2.5(0.5 - 0.180)]}{[\sin(45°)]^2 + (0.666975)[\cos(45°)]^2} = 2.670168$$

From Eq. (6-486),

$$\zeta = \left[\left(\frac{10 \text{ m}}{0.5 \text{ m}} \right)^2 \sin^2(45°) + \left(\frac{2 \text{ m}}{0.5 \text{ m}} \right)^2 \cos^2(45°) \right]^{1/2} = 14.212670$$

From Eq. (6-488),

$$\frac{I_2}{I_1} = \frac{(0.666975 - 1) \sin(45°) \cos(45°)}{\sin^2(45°) + (0.666975) \cos^2(45°)} = -0.199779$$

(1) From Eq. (6-483),

$$A_{11} = \frac{(2.5)(10 \text{ m})(2 \text{ m})}{(2.670168)^2(0.5 \text{ m})(14.212670)} = 0.987 \text{ m}$$

From Eq. (6-487),

$$\frac{A_{22}}{A_{11}} = (-0.199779)^2 \frac{(14.212670)[2(14.212670) + 1]}{2(1 + 14.212670)^2} = 0.036063$$

and by using the A_{11} value

$$A_{22} = 0.036063 A_{11} = (0.036063)(0.987 \text{ m}) = 0.036 \text{ m}$$

From Eq. (6-489),

$$\frac{A_{33}}{A_{11}} = (-0.199779)^2 \frac{14.212670}{2(1 + 14.212670)^2} = 0.001226$$

and by using the A_{11} value

$$A_{33} = 0.001226 A_{11} = (0.001226)(0.987 \text{ m}) = 0.0012 \text{ m}$$

From Eq. (6-490),

$$\frac{A_{12}}{A_{11}} = (-0.199779) \frac{14.212670}{1 + 14.212670} = -0.186647$$

By sing the A_{11} value

$$A_{12} = -0.186647 A_{11} = -(0.186647)(0.987 \text{ m}) = -0.184 \text{ m}$$

(2) Anisotropy exists in the x–y (or $x_1 - x_2$) plane only. Therefore $A_{33}^* = A_{33} = 0.0012\text{m}$. As a result, the macrodispersivities in the principal directions in the x–y plane can be calculated from Eq. (6-354) for A_{11}^* and Eq. (6-355) for A_{22}^*. From Eq. (6-354),

$$A_{11}^* = \tfrac{1}{2}(0.987 \text{ m} + 0.036 \text{ m}) + \left[\tfrac{1}{4}(0.987 \text{ m} - 0.036 \text{ m})^2 + (-0.184 \text{ m})^2 \right]^{1/2} = 1.021 \text{ m}$$

From Eq. (6-355),

$$A_{22}^* = \tfrac{1}{2}(0.987 \text{ m} + 0.036 \text{ m}) - \left[\tfrac{1}{4}(0.987 \text{ m} - 0.036 \text{ m})^2 + (-0.184 \text{ m})^2 \right]^{1/2} = 0.0016 \text{ m}$$

From Eq. (6-356),

$$\tan(2\psi) = \frac{2(-0.184 \text{ m})}{0.987 \text{ m} - 0.036 \text{ m}} = -0.386961$$

$$\psi = -10.58°$$

(3) From Eq. (6-485),

$$\frac{\overline{K'_{11}}}{K'_{22}} = \frac{1}{\beta} = \exp[\sigma_f^2(g_{22} - g_{11})] = \exp[2.5(0.180 - 0.018)] = 1.500$$

The graphical demonstration is shown in Figre 6-77. The calcualted longitudinal dispersivity (A_{11}) is about 1m (0.987 m) and the transverse horizonatal dispersivity (A_{22}) is about 4 cm (3.6 cm).

However the transverse vertical dispersivity (A_{33}) is only 1.2 mm, which is an order of magnitude lower than the transverse horizontal dispersivity (A_{22}). As can be seen from these values, the vertical transverse dispersivity is a very small quantity and is of the same order as a typical local transverse dispersivity. Therefore, an effect of local dispersivity is likely. This calculation also indicates that the transverse dispersion process is significantly anisotropic. The off-diagonal dispersivity term A_{12} is negative and numerically significant compared to the transverse dispersion terms. The influence of this off-diagonal term is illustrated graphically in Figure 6-77. The fashion of Figure 6-77 is similar to Figure 6-73 (Example 6-11), but, in this case, in a horizontal plane. In Figure 6-77 the dashed-line ellipse corresponds to the orientation of constant ln (K) and the solid-line ellipse corresponds to the mean concentration resulting from a pulse injection source (Gelhar, 1993).

The horizontal anisotropy of the effective hydraulic conductivity is relatively modest. Therefore, the deviation of the mean hydraulic gradient from the direction of the mean flow is small. The plume is again rotated so that its primary principal axis x^*(or x_1^*) nearly coincides with the direction of the mean hydraulic gradient (Gelhar, 1993).

Special Case 3: Two-Dimensional Horizontal Flow

The horizontal stratification, which is presented above, with horizontal and vertical anisotropy involves horizontal mean flow to bedding. This mean condition is similar to the commonly conceptualized two-dimensional depth-averaged flow in aquifers. The results of the two approaches are presented below (Gelhar and Axness, 1981, 1983; Gelhar,1993). The first approach is the two-dimensional depth-averaged approximation, and the second one is the usage of the three-dimensional equations with $\lambda \to \infty$.

Two-Dimensional Depth-Averaged Approximation

For the two-dimensional horizontal flow case, the perturbation flow field is considered to vary only in the horizontal plane (x_1, x_2) in relation to the field situation. Therefore, the flow perturbations represent the depth-averaged flow. The results for this case can be found simply by considering $\lambda_3 >> \lambda_1, \lambda_2$ in the three-dimensional horizontal flow situation, which is presented as Case 2. One should note that by taking $\lambda_3 \to \infty$ in Eq. (6-274), the covariance function becomes exponential in the two horizontal directions. By taking $\lambda_3 \to \infty$ in the three-dimensional expression for A_{ij} given

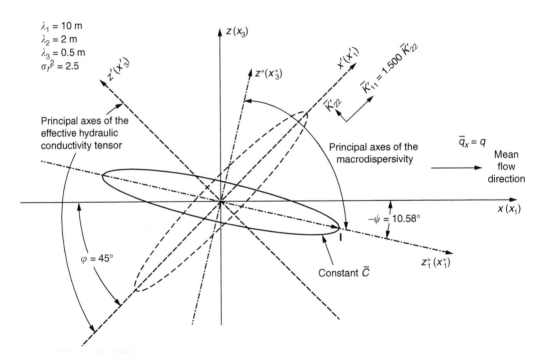

FIGURE 6-77   Graphical demonstration of the results for Example 6-12.

by Eq. (6-464), the following results were determined for the two-dimensional case (Gelhar and Axness, 1981, 1983):

$$A_{11} = \frac{\sigma_{f'}^2 \lambda_1 \lambda_2}{\gamma^2 [\lambda_1^2 \sin^2(\varphi) + \lambda_2^2 \cos^2(\varphi)]^{1/2}} \tag{6-492}$$

and

$$A_{22} = A_{33} = A_{12} = 0 \tag{6-493}$$

where γ is given by Eq. (6-484) with g_{22} given as

$$g_{22} = \frac{\sigma}{1 + \sigma} \tag{6-494}$$

in which $\sigma = \lambda_1/\lambda_2$ according to Eq.(6-427) and β is given by Eq. (6-485).

The two-dimensional longitudinal macrodispersivity expression for A_{11}, as given by Eq. (6-492), is identical to that of the three-dimensional case given by Eq. (6-483), but the factor γ is different. In addition, the interpretation of the variance and correlation scales are different because of the fact that in the two-dimensional case, these parameters represent a depth-averaged flow. This means that the variance is smaller and the correlation scales are larger than the corresponding three-dimensional flow case. However, the transverse dispersivities are significantly different between the two cases. The more realistic three-dimensional description produces significant convectively- controlled transverse dispersion, but in the depth-averaged two-dimensional approximation, the transverse dispersivities are always zero for the convective limit. One should bear in mind that by using the depth-averaged two-dimensional approximation to represent the flow perturbations, important features of the real transverse mixing process are lost. This example demonstrates that the mean concentration behavior by real three-dimensional variations is not correctly represented if the flow field is depth-averaged before calculating the flow perturbations. In other words, the order of averaging makes a difference in this case (Gelhar and Axness, 1981, 1983; Gelhar, 1993).

Some important features regarding the two-dimensional horizontal flow case are as follows: (1) in Eqs. (6-492) and (6-493), the transverse dispersivities are zero even when the hydraulic gradient I_2 is nonzero (I_2/I_1 from Eq. [6-488] also applies in this case). (2) by comparing Eqs. (6-483) through (6-491) and Eqs. (6-492) and (6-493), one can see that the two- and three-dimensional results for the longitudinal dispersivity (A_{11}) are in identical form; they will differ numerically because the factor γ differs in the two cases (g_{ij} changes). (3) for the two- and three-dimensional cases, the transverse dispersivities differ significantly; the more realistic three-dimensional description produces significant convectively-controlled transverse dispersion, whereas for the two-dimensional depth-averaged approximation, the transverse dispersivities are zero for the convective limit.

Usage of the Three-Dimensional Equations by Taking the Limit $\lambda_3 \to \infty$

Gelhar and Axness (1981, 1983) evaluated the transverse macroscopic dispersivity integral asymptotically for the two dimensional case for an isotropic covariance ($\lambda_1 = \lambda_2$) with small finite local dispersivities. The result for A_{22} is

$$A_{22} = \frac{\sigma_f^2 \alpha_L^*}{8\gamma^2}\left(1 + 3\frac{\alpha_T^*}{\alpha_L^*}\right) \tag{6-495}$$

From Eq. (6-477) and Eqs. (6-483) – Eq. (6-493), Gelhar and Axness (1981, 1983) showed that

$$A_{11} = \frac{\sigma_{f'}^2 \lambda_*}{\gamma^2} \tag{6-496}$$

where λ_* is the correlation distance in the direction of the mean flow as depicted in Figures 6-66 and 6-76.

Illustrative Example from the Borden Experiment
The Borden experiment is presented in detail in Section 6.3.7.1. As mentioned in this Section, the aquifer is a shallow water-table aquifer at the Canadian Forces Base, Borden, Ontario. The aquifer is comprised of primarily horizontal, discontinuous lenses of medium-grained and silty fine-grained sand. Figure 6-5 shows a schematic vertical section of the aquifer at the site. All other pertinent information is presented in Section 6.3.7.1. Here, the spatial variability of hydraulic conductivity (Sudicky, 1986) and determination of macrodispersivities based on the Gelhar and Axness theory (Gelhar and Axness, 1981, 1983) will be presented with examples.

Hydraulic Conductivity Measurements
The hydraulic conductivity measurement procedure is described by Sudicky (1986). Measurements were conducted to determine the spatial structure of the hydraulic conductivity field of the Borden tracer test site by extracting 32 cores of the aquifer material along two core lines, one along the direction of ground water flow (A–A$'$) and one transverse to the mean flow direction (B–B$'$), and their locations are shown in Figure 6-6. A total of 20 cores, each spaced at 1m horizontal intervals, were obtained along section A–A$'$. The transverse core line B–B$'$ is comprised of 13 cores, again with 1m horizontal spacings. Core 10 forms the intersection between core lines A–A$'$ and B–B$'$. Each of the cores was approximately 2m in length and was obtained from depths between about 2.5 and 4.5m below the ground surface. This depth interval is approximately coincident with the vertical extent of the tracer zone. Hydraulic conductivity profiles along each of 32 cores were obtained as follows: first, each core was subdivided into 0.05m equal length sub-samples and allowed to dry. As small intervals of 0.05 m were chosen, the material within each segment was relatively homogeneous in texture. Each subsample was then placed into a specially designed falling head permeameter that permitted rapid sample emplacement, testing and removal. Simultaneous use of five permeameters allowed the testing of about 75 core subsamples each day. The packing of the material was based on a field estimate for porosity equal to 0.34. A hydraulic conductivity value was then determined using the standard falling head procedure. A total of 1279 hydraulic conductivity measurements were made in this manner along the two core lines at the Borden tracer test site.

Hydraulic Conductivity Profiles
The hydraulic conductivity data described above were analyzed using statistical methods (Sudicky, 1986). Figure 6-78 compares the measured hydraulic conductivity profiles for cores 15 and 16, which are located along the core line A–A$'$ in Figure 6-6 and are separated by a horizontal distance of 1 m. From the results shown in Figure 6-78, which are typical of the depth variability at the Borden site, it can be seen that the hydraulic conductivity ranges between 6.0×10^{-4} and 2.0×10^{-2} cm/scc, which is a contrast in values of slightly more than a factor 30. There is no evidence for a trend in the hydraulic conductivity with depth.

Contours of Hydraulic Conductivity
Contour plots of the measured hydraulic conductivity field for the two vertical cross-sections along A–A$'$ and B–B$'$ in Figure 6-6 were generated by linear interpolation and are shown in Figure 6-79 and Figure 6-80, respectively (Sudicky, 1986). The negative of the natural logarithm of hydraulic conductivity, $-\ln(K)$, was contoured because the range of $-\ln(K)$ is less than that of K. For contouring purposes, the elevation of each sampling point for hydraulic conductivity shown in Figure 6-79 and Figure 6-80 was assigned a value corresponding to the elevation of the midpoint of each 0.05 m core subsample. A total of 720 measurements of hydraulic conductivity with constant vertical (0.05 m) and horizontal (1.0 m) spacing are contained in results for cross-section A–A$'$ in Figure 6-6. The total number of data points along B–B$'$ in Figure 6-6 is 468, with the horizontal and vertical spacing between individual points being the same as that along A–A$'$ in Figure 6-6.

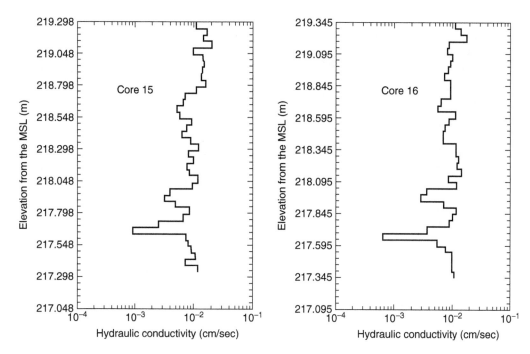

FIGURE 6-78   Comparison of hydraulic conductivity profiles for cores separated by a 1 - m horizontal distance for the Borden experiment. (Adapted from Sudicky, E.A., *Water Resour. Res.*, 22, 2069–2082, 1986.)

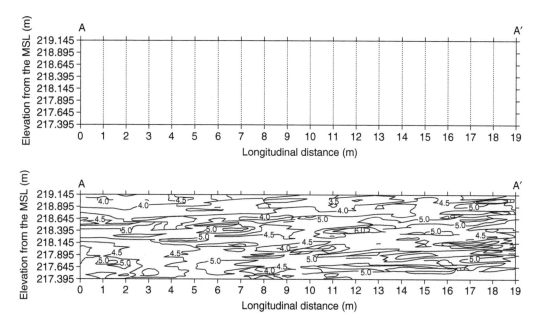

FIGURE 6-79   Location of measurements and distribution of $-\ln(K)$ along A–A′ in Figure 6-6 for the Borden experiment (contour interval = 0.5; vertical exaggeration = 2; and $K < 10^{-3}$ cm/s in stippled zone). (Adapted from Sudicky, E.A., *Water Resour. Res.*, 22, 2069–2082, 1986.)

Ground Water Velocity Field

Given the natural heterogeneity of the Borden aquifer shown in Figure 6-79 and Figure 6-80, it can be expected that the local ground water velocity field will vry irregularly in three-dimensional space

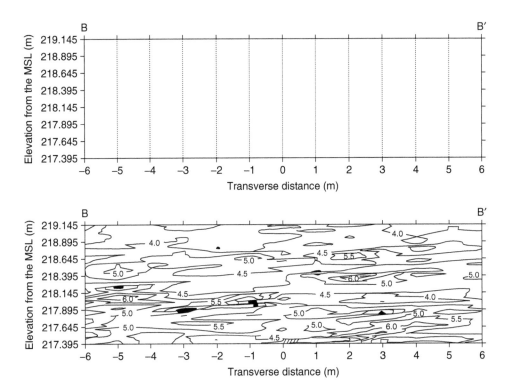

FIGURE 6-80   Location of measurements and distribution of -ln (K) along B-B′ in Figure 6-6 for the Borden experiment (contour interval = 0.5; vertical exaggeration = 2; and $K < 10^{-3}$ cm/s in stippled zone). (Adapted from Sudicky, E.A., *Water Resour. Res.*, 22, 2069–2082, 1986.)

over scales of a few centimeters to a few tens of centimeters in the vertical direction and over distances of the order of a 1 or 2 m in the horizontal plane. It is these scales of velocity fluctuations that are believed to be responsible for the irregularity in the chloride concentration distribution shown in Figure 6-7 (Sudicky, 1986).

Characterization of Hydraulic Conductivity Field
As mentioned above, the Borden aquifer is a heterogeneous aquifer. In fact, Sudicky (1986) calls it a "homogeneously heterogeneous aquifer." A fundamental question is how to deal with this type of material heterogeneity if it is desired to make predictions of solute transport over hundreds of meters, as would be the case for modeling leachate plume originated from the Borden landfill. Although it is possible to map the structure of dispersion-controlled heterogeneities in some detail over distances of a few meters to a few tens of meters, it would be impractical to make similar detailed measurements for the purpose of constructing a detailed deterministic model for the entire aquifer. As mentioned in Section 6.3.6.2, the hydraulic conductivity field ln (K) (Eq. [6-64]) can be considered a as random field. Therefore, the spatial persistence exhibited by Figure 6-79 and Figure 6-80 can be represented by its covariance function, which is defined by Eq. (6-274). If the covariance function can be estimated from measurements within the sample domain, then it is possible to estimate the components of the effective hydraulic conductivity and macrodispersivity tensors applicable to the large-scale problem using a stochastic theory such as that developed by Gelhar and Axness (1981, 1983).

Sudicky (1986) estimated the autocorrelation structure of the Borden aquifer obtained from hydraulic conductivity measurements along the different directions and presented a covariance model for the Borden aquifer. As mentioned in Section 6.3.6.2, it is normally held that hydraulic conductivity follows a log normal probability density function. Frequency histograms were constructed by Sudicky (1986) according to Eq. (6-64) by selecting every fifth hydraulic conductivity value from every second core such that the minimum vertical separation distance between adjacent points corresponds to 0.25 m, and the minimum horizontal separation to 2.0 m. Results obtained from one of the many subsets of hydraulic conductivity values separated by these distances are shown in Figure 6-81. Visual inspection of Figure 6-81 indicates that the log-hydraulic conductivity (f) more closely resembles the *Gaussian distribution*, which is sometimes known as *normal distribution*. The mean is defined by Eq. (6-65) and the variance is defined by Eq. (6-66). Sudicky (1986) determined the mean ($\bar{f}$) and variance of $\bar{f}$ ($\sigma_{f'}^2$), respectively, as $\bar{f} = -4.63$ and $\sigma_{f'}^2 = 0.38$ at 22°C temperature from all 1279 measured values of hydraulic conductivity.

The mean ($\bar{f}$) and variance of $\bar{f}$ ($\sigma_{f'}^2$) determined by Sudicky (1986) from each core are plotted with distance along core lines A-A' and B-B' (see Figure 6-6 for their location) in Figure 6-82 and are compared to the overall mean and variance. Although the mean and variance from each core fluctuate about the overall values, stationarity (a stochastic process is said to be stationary in the strict sense if its statistics are invariant with a shift distance origin or in time) in the mean and variance of ln(K) appears to be a reasonable assumption for the Borden aquifer, at least over the horizontal distances covered by the core lines. The fluctuations in the variance are seen to be larger than that of the mean, with small variances generally occurring over the first 5m along A–A'. Sudicky (1986) states that these smaller variance values are mainly the result of an absence of thin, low hydraulic

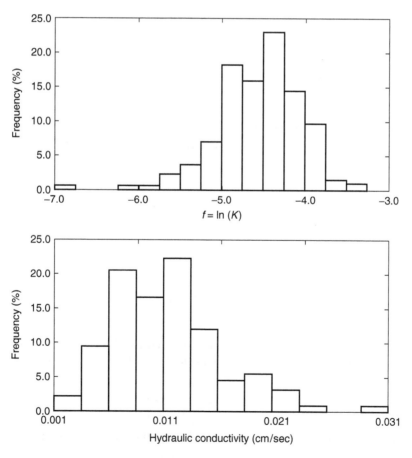

FIGURE 6-81   Frequency histograms for ln (K) and K for the Borden Experiment site. (Adapted from Sudicky, E. A., *Water Resour. Res.*, 22, 2069–2082, 1986.)

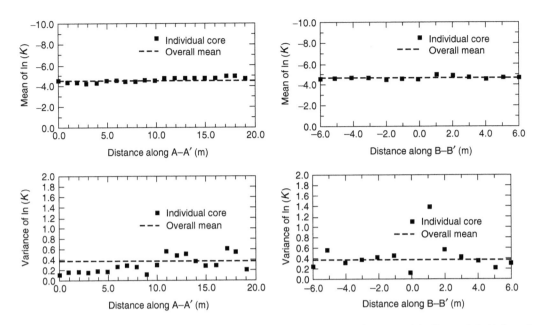

FIGURE 6-82   Mean and variance of ln (K) versus distance along A–A$'$ and B–B$'$ in Figure 6-6. (Adapted from Sudicky, E.A., *Water Resour. Res.*, 22, 2069–2082, 1986.)

conductivity lenses over this distance as can be seen from Figure 6-79. The largest variance value, equal to 1.4, occurs at a distance equal to 1.0 m along B–B$'$ and is the result of measured hydraulic conductivity values in the range of $5.0 \times 10^{-5} - 7.0 \times 10^{-4}$ cm/sec over a thin zone near the bottom of core 27.

Horizontal and Vertical Autocorrelation Functions of f = ln(K)
Sudicky (1986) estimated the horizontal and vertical autocorrelation function of $f = \ln(K)$, ρ_f, from the equally spaced measurements along A–A$'$ and B–B$'$ (see Figure 6-6 for their locations) using (Jenkins and Watts, 1968)

$$\overline{\rho_{f'}} = \frac{\dfrac{1}{n}\sum_{i=0}^{n(\zeta_j)}(f_i - \bar{f})(f_i - \bar{f})_{i + k\Delta\zeta_j}}{\dfrac{1}{n}\sum_{i=0}^{n}(f_i - \bar{f})2}, \quad k = 0, 1,\dots, m \tag{6-497}$$

Alternatively, using Eq. (6-66), one gets

$$\overline{\rho_{f'}} = \frac{1}{\sigma_{f'}^2}\frac{1}{n}\sum_{i=0}^{n(\zeta_j)}(f_i - \bar{f})(f_i - \bar{f})_{i + k\Delta\zeta_j}, \quad k = 0, 1,\dots, m \tag{6-498}$$

which is the same form of Eq. (6-70).
The lag-distance increment $\Delta\varsigma_j$ is 0.05 and 1.0 m in the vertical ($j = 3$) and horizontal ($j = 1, 2$), directions, respectively. The results are shown in Figure 6-83, where they are compared to an exponential model of the form

$$\rho_{f'}(\zeta_j) = \eta \exp\left(-\frac{\zeta_j}{\lambda_j}\right) \tag{6-499}$$

which is similar to Eq. (6-154). The estimate given by Eq. (6-497) is normalized with respect to

$$\eta = \frac{\sigma_{f'}^2}{\sigma_{f'}^{*2}} = \frac{\sigma_{f'}^2}{\sigma_{f'}^2 + \sigma_0^2} \tag{6-500}$$

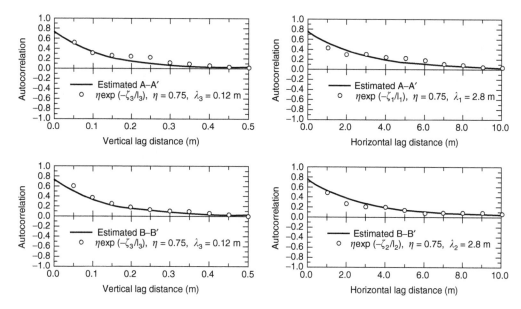

FIGURE 6-83 Horizontal and vertical autocorrelation functions for ln (K) along cross-sections A–A′ and B–B′ in Figure 6-6. (Adapted from Sudicky, E.A., *Water Resour. Res.*, 22, 2069–2082, 1986.)

in which $\sigma_{f'}^2 = \sigma_{f'}^{2*} - \sigma_0^2$ is an estimate of the variance of the variability of $\ln(K)$, and σ_0^2 is the variance due to essentially uncorrelated variations below the scale of measurement (nugget effect) (Sudicky, 1986).

It can be seen from Figure 6-83 that the exponential model with a vertical correlation scale length $\lambda_3 = 0.12$m and $\eta = 0.75$ closely approximates the estimated vertical autocorrelation function for B–B′ and provides a reasonable fit to that estimated from A–A′ (refer Figure 6-6 for the locations of A–A′ and B–B′). In the transverse horizontal direction along B–B′, use of correlation length $\lambda_2 = 2.8$ m and $\eta = 0.75$ provides good agreement with the estimated function, as does the use of $\lambda_1 = \lambda_2 = 2.8$ m and $\lambda_3 = 0.12$. Also, for 1279 measurements of hydraulic conductivity, Sudicky (1986) found that $\sigma_{f'}^{2*} = 0.38$. Therefore, from Eq. (6-500), $\sigma_{f'}^2 = (0.75)(0.38) = 0.285$. It should be recalled that horizontal isotropy in the covariance function is based on the finding that equal horizontal correlation length scales were obtained from the analysis performed along core lines A–A′ and B–B′ (refer Figure 6-6 for their locations). However, it is plausible that possible anisotropy was not detected because the two core lines are perpendicular to each other.

The Covariance Model
Sudicky used the covariance model given by Eq. (6-274), which requires that $\sigma_{f'}^2$, λ_2, λ_2, and λ_3 be known. These values are given above.

EXAMPLE 6-13

This example is adapted from Sudicky (1986) regarding the Border experiment. The statistical parameters of the Borden aquifer are discussed above and they are: $\sigma_{f'}^2 = 0.285$, $\lambda_1 = \lambda_2 = 2.8$ m, and $\lambda_3 = 0.12$ m. The angle with the bedding, θ, is very small such that $\sin(\theta) \approx 0$ and $\cos(\theta) \approx 1$. Using these values, determine(1) the values of A_{11}, A_{22}, A_{33}, and A_{13} and compare the results with the ones determined from the moments method (Example 6-5); and (b) the values A_{11}^* and A_{33}^*.

SOLUTION

First, the values of ρ, σ, μ, β, ξ, and γ need to be calculated. From Eqs. (6-427) and (6-478)

$$\rho = \frac{2.8 \text{ m}}{0.12 \text{ m}} = 23.33, \qquad \sigma = \frac{2.8 \text{ m}}{2.8 \text{ m}} = 1, \qquad \mu = \frac{0.12 \text{ m}}{2.8 \text{ m}} = 0.042857$$

From Figure 6-67, one gets $g_{11} = 0.0324$ and from Figure 6-68 one gets $g_{33} = 0.9385$. By using the given values, from Eq. (6-473)

$$\beta = \exp[0.285(0.0324 - 0.9385)] = 0.772409$$

From Eq. (6-479),

$$\xi^2 = [\sin(0°)]^2 + (0.042857)^2[\cos(0°)]^2 = (0.042857)^2$$
$$\xi = 0.042857$$

From Eq. (6-476),

$$\gamma = \frac{\exp[0.285(0.5 - 0.9385)]}{[\sin(0°)]^2 + (0.772409)[\cos(0°)]^2} = 1.142555$$

(1) From Eq. (6-477),

$$A_{11} = \frac{(0.285)(2.8 \text{ m})(0.042857)}{1.142455)^2(0.042857)} = 0.61 \text{ m}$$

As can be seen from the expressions of A_{22}, A_{33}, and A_{13} given by Eqs. (6-480)–(6-482), they are all the functions of $\sin(\theta)$. Since $\sin(\theta) \approx 0$, $A_{22} \approx 0$, $A_{33} \approx 0$. and $A_{13} \approx 0$. As can be observed from Example 6-5, $A_{x'x'} = 0.31$ m and $A_{y'y'} = 0.021$ m, which correspond, to A_{11} and A_{22}, respectively.

The above results clearly show that there are significant differences among the results determined by the two different methods. The differences can be attributed to (1) the accuracy of the hydrogeologic and chemical data determined under field conditions; (2) errors in determining the statistical aquifer parameters; and (3) the assumptions made in deriving the theoretical equations in determining the components of the macrodispersivity tensor.

(2) Anisotropy exists in the x–z (or $x_1 - x_3$) plane only. Therefore $A^*_{33} = A_{33} = 0$. The macrodispersivities in the principal directions in the x–y plane can be calculated from Eq. (6-354) for A^*_{11} and Eq. (6-355) for A^*_{33}. From Eq. (6-354)

$$A^*_{11} = \frac{1}{2}(0.61 \text{ m} + 0 \text{ m}) + \left[\frac{1}{4}(0.61 \text{ m} - 0 \text{ m})^2 + (0 \text{ m})^2 \right]^{1/2} = A_{11} = 0.61 \text{ m}$$

From Eq. (6-363)

$$A^*_{33} = \frac{1}{2}(0.61 \text{ m} + 0 \text{ m}) - \left[\frac{1}{4}(0.61 \text{ m} - 0 \text{ m})^2 + (0. \text{ m})^2 \right]^{1/2} = A_{33} = 0 \text{ m}$$

EXAMPLE 6-14

This example is adapted from Hess et al. (1992) and pertains to the Cape Cod experiment. The statistical parameters of the Cape Cod aquifer are given as (Hess, 1989; Hess et al., 1992) $\sigma^2_f = 0.24$, $\lambda_1 = \lambda_2 = 3.5$ m, and $\lambda_3 = 0.19$ m. The mean flow is at a slight angle to the bedding ($\theta = 0.5°$), which corresponds to Case 1. Using these values, determine (1) the value of A_{11} and compare it with the one determined from the moments method; (2) the value of A_{22} and compare it with the one determined from the moments method and discuss the results; (3) the value of A_{33} and compare it with the one determined from the moments method and discuss the results; and (4) the value of A_{13}. The results based on the moments method are given in Example 6-7.

SOLUTION

First, the values of ρ, σ, μ, β, ξ, and γ need to be calculated. From Eqs. (6-427) and (6-478),

$$\rho = \frac{3.5 \text{ m}}{0.19 \text{ m}} = 18.42, \qquad \sigma = \frac{3.5 \text{ m}}{3.5 \text{ m}} = 1, \qquad \mu = \frac{0.19 \text{ m}}{3.5 \text{ m}} = 0.0543$$

From Figure 6-67, one get $g_{11} = 0.0377$ and from Figure 6-68 one get $g_{33} = 0.915$. By using the given values, from Eq. (6-473),

$$\beta = \exp[0.24(0.0377 - 0.915] = 0.810136$$

From Eq. (6-479),

$$\xi^2 = [\sin(0.5°)]^2 + (0.0543)^2[\cos(0.5°)]^2 = 0.054361$$
$$\xi = 0.054993$$

From Eq. (6-476),

$$\gamma = \frac{\exp[0.24(0.5 - 0.915)]}{[\sin(0.5°)]^2 + (0.810136)[\cos(0.5°)]^2} = 1.117322$$

(1) From Eq. (6-477),

$$A_{11} = \frac{(0.24)(3.5 \text{ m})(0.0543)}{(1.117322)^2(0.054993)} = 0.664 \text{ m}$$

The calculated value of $A_{11} = 0.664$ m for the asymptotic longitudinal dispersivity based on the Gelhar and Axness theory is in reasonable agreement with the observed value $A_{11} = 0.955$ m from the tracer test analysis using the moments method (Example 6-7). It should be recognized that the two values were determined independently, and the difference between them is more likely due to errors in estimating statistical parameters from field data than a failure of the theoretical approach.

(2) From Eq. (6-480),

$$A_{22} = \frac{(0.24)(3.5 \text{ m})(0.0543)(1 - (810136)^2[\sin(0.5°)]^2[\cos(0.5°)]^2}{2\langle[\sin(0.5°)]^2 + (0.810136)[\cos(0.5°)]^2\rangle^2(1 + 0.054993)^2(1.117322)^2} = 5.54 \times 10^{-8} \text{ m}$$

This value does not agree with the observed value (based on the moments method) $A_{22} = 0.0182$ m Example 6-7) and indicates that the transverse mixing mechanism may not be properly represented by the steady–flow assumption used in the development of the Gelhar and Axness theory. The work by Rehfeldt (1988) and Rehfeldt and Gelhar (1992) account for unsteady flow effects by the incorporation of temporal fluctuations in the mean hydraulic gradient in the stochastic development and shows that these fluctuations significantly increase transverse horizontal dispersivity (see Section 6.4.7). For the Cape Cod site, Rehfeldt and Gelhar (1992) calculated a transverse horizontal dispersivity $A_{22} = 0.8$ cm, a value that is in reasonable agreement with the observed value 1.82 cm given in Example 6-7.

(3) From Eq. (6-481),

$$A_{33} = \frac{(0.24)(3.5 \text{ m})(0.0543)(1 - 0.810136)^2[\sin(0.5°)]^2[\cos(0.5°)]^2[1 + 2(0.054993)]}{2[[\sin(0.5°)]^2 + (0.810136)[\cos(0.5°)]^2]^2(1 + 0.054993)^2(1.117322)^2} = 6.16 \times 10^{-8} \text{ m}$$

As can be seen, A_{33} is essentially zero. It can be seen from the moments method values that the slope of the initial rise in vertical variance was larger and more linear (correlation coefficient 0.705) than the overall trend (Figure 6-44). The transverse vertical dispersivity calculated from this early trend (July 1985 to January 1986) is 4 mm. This large change in vertical variance was, most likely, due to the density-induced sinking of the cloud in the early part of the test (Figure 6-40). The sinking caused an increased vertical "smearing" of the cloud and resulted in an increased vertical spreading and dispersivity of the bromide cloud. The density-induced vertical movement also could have affected the early time change of the longitudinal variance, but additional study is needed before this potential effect can be quantified (Garabedian et al., 1991).

(4) From Eq. (6-482),

$$A_{13} = A_{31} = -\frac{(0.24)(3.5 \text{ m})(0.0543)(1 - 0.810136) \sin(0.5°) \cos(0.5°)}{(1 + 0.233154)\langle[\sin(0.5°)]^2 + (0.810136)[\cos(0.5°)]^2\rangle(1.117322)2} = -7.08 \times 10^{-5} \text{ m}$$

6.4.7 ANALYSIS OF MACRODISPERSION IN THREE-DIMENSIONALLY HETEROGENEOUS AQUIFERS UNDER UNSTEADY FLOW CONDITIONS USING STOCHASTIC SOLUTE TRANSPORT THEORY: THE REHFELDT AND GELHARS' FIRST-ORDER ANALYSIS APPROACH

6.4.7.1 Introduction

Rehfeldt (1988) and Rehfeldt and Gelhar (1992) developed a general theory of the dispersion of solutes in unsteady flow in heterogeneous aquifers, and showed quantitatively by stochastic analysis that unsteady flow leads to increased longitudinal and transverse dispersive fluxes. The authors used the resulting predictive expressions to show the magnitude of the macrodispersivities at several natural gradient experiment sites. For steady flow in a heterogeneous aquifer, Gelhar and Axness (1983), as given in Section 6.4.6, have shown that a small-scale variation in the specific discharge vector, due to the spatial variability of hydraulic conductivity, leads to a field-scale dispersive flux characterized by the macrodispersivity. If the flow in this heterogeneous aquifer is transient, then the small-scale variation in the specific discharge vector is due to spatial and temporal variability. This combination of temporal and spatial variability will produce small-scale fluctuations in the specific discharge vector that differ from variations due to spatial variability alone.

Examination of the Influence of Temporal Variability on Dispersive Mixing
To examine the influence of temporal variability on dispersive mixing, it is necessary to understand how the specific discharge is influenced by hydraulic head variation. The specific discharge is given by Darcy's law, $q = KI$ (Chapter 2, Eq. [2-39]), where K is the hydraulic conductivity and I is the hydraulic gradient. The hydraulic conductivity is variable in space, but not in time. Therefore, the temporal variation in q must be the result of temporal variation in the hydraulic gradient.

Limitations of the Approach and Assumptions
The analysis is based on the small perturbation analysis of Gelhar and Axness (1983), which is presented in Section 6.4.6, and has the same restrictions. The assumptions are as follows: (1) the log-hydraulic conductivity and hydraulic gradient are assumed to be stationary random processes; (2) expressions resulting from stochastic analysis are ensemble averages; and (3) ergodicity, which means that solute transport in an ensemble of aquifers with assigned statistical properties is assumed to approximate the real-field situation of transport in a single heterogeneous aquifer (a realization of the stochastic process), is invoked. The third assumption is reasonable only if the spatial scale of the flow system is large compared with the correlation scale of the aquifer and if the temporal scale of the flow system is large compared with the hydraulic gradient correlation scale. Therefore, the macrodispersivities are meaningful only if the overall scale of the problem is large compared with the correlation scales of hydraulic conductivity and hydraulic gradient.

6.4.7.2 Governing Equations

The governing equations are derived based on the theory of Rehfeldt and Gelhar (1992). The governing equation is given by Eq. (6-293), which can be expressed as

$$\varphi_e \frac{\partial C}{\partial t} = E_{ij} \frac{\partial^2 C}{\partial x_i \partial x_j} - q_i \frac{\partial C}{\partial x_i}, \quad i, j = 1, 2, 3 \tag{6-501}$$

where φ_e is the effective porosity, C the solute concentration, E_{ij} the i, jth component of the local bulk dispersion coefficient tensor (assumed constant) and expressed by Eq. (6-294), and q_i the ith component of the specific discharge vector. The conservation of fluid mass for an incompressible fluid in an elastic aquifer requires

$$\frac{\partial q_i}{\partial x_i} = 0 \tag{6-502}$$

Following the developments in Section 6.4.6.7 (Eqs. [6-303] –[6-314]), the concentration and specific discharge are assumed to be random variables composed of the sum of a mean and a small perturbation

$$C(x, y, z, t) = \overline{C}(x, y, z, t) + C'(x, y, z, t) \tag{6-503}$$

$$q_i(x, y, z, t) = \overline{q}_i(x, y, z, t) + q'_i(x, y, z, t), \quad i = 1, 2, 3 \tag{6-504}$$

where mean quantities are denoted with an overbar and primed variables are the perturbations. By substituting Eqs. (6-503) and (6-504) into Eqs. (6-501) and (6-502) and following the same procedure in Section 6.4.6.7, one gets

$$\varphi_e \frac{\partial \overline{C}}{\partial t} + \frac{\partial}{\partial x_i}(\overline{q}_i\,\overline{C} + \overline{q'_i C'}) = E_{ij} \frac{\partial^2 \overline{C}}{\partial x_i \partial x_j} \tag{6-505}$$

$$\frac{\partial \overline{q}_i}{\partial x_i} = 0 \tag{6-506}$$

where the second term $\overline{q_i C'}$ represents the dispersive flux due to random variation in q_i and C. Equation (6-505) is exactly the same as Eq. (6-312), with the exception of the first term on its left-hand side. By analogy with turbulent diffusion, and because the term occupies the position of the dispersive flux in the classical convection–dispersion equation, $(\overline{q'_i C'})$ is referred to as the mean macrodispersive flux. Subtracting the mean equations, Eqs. (6-505) and (6-506), from Eqs. (6-501) and (6-502), and retaining first-order terms in the perturbations, we get the equation governing the perturbations

$$\varphi_e \frac{\partial C'}{\partial t} + q_{i'} \frac{\partial \overline{C}}{\partial x_i} + q \frac{\partial C'}{\partial x_1} = q \left[\alpha_L^* \frac{\partial^2 C'}{\partial x_1^2} + \alpha_T^* \left(\frac{\partial^2 C'}{\partial x_2^2} + \frac{\partial^2 C'}{\partial x_3^2} \right) \right] \tag{6-507}$$

where the mean flow is aligned along the coordinate axis x_1 (x) so that $\overline{q}_1 = q \neq 0$, and $\overline{q}_2 = \overline{q}_3 = 0$. Equation (6-507) is the same as Eq. (6-322) with exception of the notation and the first term on its left-hand side. The local dispersion tensor E_{ij} is given by Eq. (6-317), as

$$[E_{ij}] = \begin{bmatrix} \alpha_L^* q & 0 & 0 \\ 0 & \alpha_T^* q & 0 \\ 0 & 0 & \alpha_T^* q \end{bmatrix} \tag{6-508}$$

where α_L^* and α_T^* are the local longitudinal and transverse dispersivities, respectively. In fact, E_{ij} is $E_{m_{ij}}$, but for simplicity it will be shown by E_{ij}, which represents only the mechanical dispersion effects (m stands for "mechanical").

6.4.7.3 Spectral Solution

Similarly, as in the case of Section 6.4.6.7 (Eqs. [6-323]–[6-335]), by assuming local statistical homogeneity in space and stationarity in time, the solution of Eq. (6-507) can be developed using space–time Fourier–Stieltjes representations for the perturbed quantities (Lumley and Panofsky, 1964):

$$C'(x, y, z, t) = \int_{-\infty}^{\infty} e^{i(\mathbf{k} \cdot \mathbf{x} + \omega t)} \, dZ_C(\mathbf{k}, \omega) \tag{6-509}$$

$$q_{i'}(x, y, z, t) = \int_{-\infty}^{\infty} e^{i(\mathbf{k} \cdot \mathbf{x} + \omega t)} \, dZ_C(\mathbf{k}, \omega) \tag{6-510}$$

where

$$\mathbf{k} = (k_1, k_2, k_3) \equiv (k_x, k_y, k_z) \tag{6-511}$$

is the wave number vector and

$$\mathbf{x} = (s_1, x_2, x_3) \equiv (x, y, z) \tag{6-512}$$

is the position vector and ω is the frequency. Substituting Eqs. (6-509) and (6-510) into Eq. (6-507), using the unique properties of the spectral representation theorem (Lumley and Panofsky, 1964), and following the same procedure as in Section 6.4.6.7 (Eqs. [6-322]–[6-330]) we get an equation in terms of the complex Fourier amplitudes, dZ:

$$\{i\varphi_e\omega + [ik_1 + \alpha_L^* k_1^2 + \alpha_T^*(k_2^2 + k_3^2)] \, q\} \, dZ_C = G_i \, dZ_{qi} = G_j \, dZ_{q_j} \tag{6-513}$$

where

$$G_i = -\frac{\partial \overline{C}}{\partial x_i} \tag{6-514}$$

or

$$G_x = -\frac{\partial \overline{C}}{\partial x}, \qquad G_y = -\frac{\partial \overline{C}}{\partial y}, \qquad G_z = -\frac{\partial \overline{C}}{\partial z} \tag{6-515}$$

Equation (6-513) is exactly the same as Eq. (6-329), with the exception of the $i\varphi_e\omega$ term. It is assumed that G_i is nearly constant (i.e., the concentration gradient changes slowly) at the scale of the $\ln(K)$ correlation length.

Multiplying both sides of Eq. (6-513) by $dZ_{q_i}^*$, and taking the expected value (the procedure is the same as the one used in Eqs. [6-330] through [6-335]), we get an equation relating the specific discharge spectrum $(S_{q_iq_j})$ to the spectrum of the mean dispersive flux (S_{Cq_i}):

$$\{i\varphi_e\omega + [ik_1 + \alpha_L^* k_1^2 + \alpha_T^*(k_2^2 + k_3^2)] \, q\} \, S_{Cq_i}(\mathbf{k}, \omega) = G_j \, S_{q_iq_j}(\mathbf{k}, \omega) \tag{6-516}$$

Equation (6-516) is the same as Eq. (6-332), with the exception of the first term on the left-hand side.

The mean dispersive flux, $\overline{q_i'C'}$, obtained by the integration of its spectrum, is assumed to be Fickian in nature and, and in like Eq. (6-290), can be represented as

$$\overline{q_i'C'} = -qA_{ij}\frac{\partial \overline{C}}{\partial x_j}, \quad i, j = 1, 2, 3 \tag{6-517}$$

where A_{ij} is the macrodispersivity tensor. From Eq. (6-516),

$$S_{Cq_i}(\mathbf{k}, \omega) = \frac{qG_j S_{q_i q_j}(\mathbf{k}, \omega)}{i\varphi_e \omega q + [ik_1 + \alpha^*_L k_1^2 + \alpha^*_T(k_2^2 + k_3^2)]q^2}, \quad i, j = 1, 2, 3 \tag{6-518}$$

$S_{Cq_i}(\mathbf{k}, \omega)$ is the cross-spectrum of C and q_i. By using the procedure used for Eqs. (6-331)–(6-335), one gets

$$E[C'q_i'] = \overline{C'q_i'} = \int_{-\infty}^{\infty}\int_{-\infty}^{\infty} S_{Cq_i}(\mathbf{k}, \omega) \, d\mathbf{k} \, d\omega = qA_{ij}G_j, \quad i, j = 1, 2, 3 \tag{6-519}$$

where

$$A_{ij} = \int_{-\infty}^{\infty}\int_{-\infty}^{\infty} \frac{S_{q_i q_j}(\mathbf{k}, \omega) \, d\mathbf{k} \, d\omega}{i\varphi_e \omega q + [ik_1 + \alpha^*_L k_1^2 + \alpha^*_T(k_2^2 + k_3^2)]q^2} \tag{6-520}$$

which is the same as Eq. (6-335) with the exception of the first term in its denominator.

6.4.7.4 Development of the Specific Discharge Spectrum and Flow Equation

Field-scale dispersion is caused by perturbations in the specific discharge vector, $\mathbf{q}$, which are induced by hydraulic conductivity variations. The influence of transient flow on the macrodispersive flux can be evaluated by examining the Darcy equation, as given by Eq. (6-250), which relates the specific discharge to hydraulic conductivity and the hydraulic gradient. Since hydraulic conductivity is not temporally variable, temporal perturbations in q_i must be introduced by the hydraulic gradient term. It is clear from Eq. (6-250) that temporal variations in h are uniform in space, and that temporal perturbations in q_i are not caused by them. Therefore, the simple rising and falling of h, but without corresponding changes in the hydraulic gradient, should not influence dispersion at the field scale. For the temporal variability of h to influence q_i, and hence dispersion, the spatial gradient of h must vary with respect to time (Rehfeldt and Gelhar, 1992).

Decomposing the Hydraulic Gradient into the Sum of Three Parts
To utilize the stochastic method, the hydraulic gradient is decomposed into the sum of three parts (Rehfeldt and Gelhar, 1992)

$$-\frac{\partial h}{\partial x_i} = I_i + \iota_i - \frac{\partial h'}{\partial x_i} \tag{6-521}$$

where I_i is the ensemble mean gradient and is given as

$$I_i = -\frac{\partial \overline{h}}{\partial x_i} \tag{6-522}$$

ι_i is a temporally variable but spatially constant gradient at the scale of the $\ln(K)$ correlation length, and $\partial h'/\partial x_i$ is the local scale perturbation of the hydraulic gradient in space and time. Substitution of Eq. (6-522) into Eq. (6-521) gives

$$-\frac{\partial h}{\partial x_i} = -\frac{\partial \overline{h}}{\partial x_i} + \iota_i - \frac{\partial h'}{\partial x_i} \tag{6-523}$$

Alternatively, Eq. (6-523) can be derived from Eq. (6-177) by adding the term ι_i.

    To understand the terms in Eq. (6-523) better, consider a portion of an aquifer, ΔV, that has dimensions similar to the $\ln(K)$ correlation scale. Within ΔV, the hydraulic gradient fluctuates due to some distant fluctuating boundary condition, such as a river or a recharge area. The hydraulic gradient averaged over ΔV and time is I_i. The temporal fluctuations in the hydraulic gradient due to this distance boundary are considered to be uniform over ΔV. This implies that if the direction of the hydraulic gradient is changed by $5°$, then all of ΔV would experience the same rotated gradient. This temporally variable component of the hydraulic gradient that is uniform over ΔV is ι_i. This term serves as the temporal forcing to the flow system and is treated independently of the local perturbations, $\partial h'/\partial x_i$. This can be viewed as the flow system's response to the spatial variability in K and the temporal variability of the boundary conditions manifested in ι_i (Rehfeldt and Gelhar, 1992).

Specific Discharge Perturbations Equation
The specific discharge perturbations are obtained, in a manner similar to Eq. (6-251), as

$$q_i = -K_g \exp(f') \frac{\partial h}{\partial x_i} = K_g (1 + f' + \frac{f'^2}{2} + \cdots)\left(I_i + \iota_i - \frac{\partial h'}{\partial x_i}\right) \qquad (6\text{-}524)$$

where K_g is given by Eq. (6-249). Equation (6-524) is the same as Eq. (6-251) with the exception of the ι_i term. Retaining terms up to the second order, and taking the expected values of the terms of Eq. (6-524) yields,

$$E[q_i] \cong K_g E\left[\left(1 + f' + \frac{f'^2}{2}\right)\left(I_i + \iota_i - \frac{\partial h'}{\partial x_i}\right)\right]$$

$$= K_g E\left[I_i\left(1 + f' + \frac{f'^2}{2}\right) + \iota_i\left(1 + f' + \frac{f'^2}{2}\right) - \left(1 + f' + \frac{f'^2}{2}\right)\frac{\partial h'}{\partial x_i}\right] \qquad (6\text{-}525)$$

As $E[h'] = 0$, $E[f'] = 0$, $E[\iota_i] = 0$, and $E[q_i] = \bar{q}_i$, Eq. (6-525) takes the form

$$\bar{q}_i = K_g\left[I_i\left(1 + \frac{\sigma_f'^2}{2}\right) + \overline{\iota_i f'} - \overline{f'\frac{\partial h'}{\partial x_i}}\right] \qquad (6\text{-}526)$$

which is the same as Eq. (6-201) with the exception of the $\overline{\iota_i f'}$ and the fact that the notation (Eq. [6-202]) is written for the x-coordinate direction, whereas Eq. (6-526) is written in the tensorial notation. Subtracting Eq. (6-526) from Eq. (6-524) we get

$$q_i - \bar{q}_i = q_i' = K_g\left(1 + f' + \frac{f'^2}{2}\right)\left(I_i + \iota_i - \frac{\partial h'}{\partial x_i}\right) - K_g\left[I_i(1 + \sigma_f'^2) + \overline{\iota_i f'} - \overline{f'\frac{\partial h'}{\partial x_i}}\right]$$

$$= K_g(\iota_i - \frac{\partial h'}{\partial x_i} + I_i f') \qquad (6\text{-}527)$$

The local gradient term in Eq. (6-527), $\partial h'/\partial x_i$, is evaluated from the ground water flow equation as explained below.

Flow Equation
Simplified Flow Equation
The equation governing the transient, three-dimensional flow of an incompressible fluid, based on Eq. (6-219), is written in terms of the natural logarithm of K as

$$\frac{\partial^2 h}{\partial x_i^2} + \frac{\partial[\ln(K)]}{\partial x_i}\frac{\partial h}{\partial x_i} = \frac{S_s}{K}\frac{\partial h}{\partial t} \qquad (6\text{-}528)$$

where h is the hydraulic head and S_s the specific storage, and the rest of parameters are as defined previously. Assuming that $S_s \approx 0$, so that the right-hand side of Eq. (6-528) is approximately zero, we get a flow equation consistent with the conservation of fluid mass equation given by Eq. (6-506). The derivation of the transient ground water flow equation can be found in many books on the subject of ground water hydraulics (e.g., Bear, 1979; Batu, 1998). Therefore, Eq. (6-528) takes the form of Eq. (6-219):

$$\frac{\partial^2 h}{\partial x_i^2} + \frac{\partial[\ln(K)]}{\partial x_i}\frac{\partial h}{\partial x_i} = 0 \qquad (6\text{-}529)$$

The expected influence of internal storage (fluid and solid matrix compressibility) will be small. This restriction is appropriate for natural flow variation in most unconfined aquifers because storage changes in that case are primarily due to water-table movement. As noted previously, temporal variation in the hydraulic gradient, not hydraulic head alone, will influence macrodispersion in heterogeneous porous media. Therefore, the $S_s \approx 0$ assumption simplifies the analysis without impacting the results significantly.

Quasi-Steady Flow System

The quasi-steady flow assumption produces a quasi-steady flow system where the transients are compared with a series of steady-state snapshots of zero storativity, which presumes that the statistics of the hydraulic gradient fluctuations in the aquifer and the boundary are the same. By using $f = \ln(K) = \bar{f} + f'$ from Eq. (6-178) in Eq. (6-528) we get

$$\frac{\partial^2 h}{\partial x_i^2} + \left(\frac{\partial \bar{f}}{\partial x_i} + \frac{\partial f'}{\partial x_i} \right) \frac{\partial h}{\partial x_i} = 0 \qquad (6\text{-}530)$$

or using Eq. (6-521)

$$-\frac{\partial}{\partial x_i}\left(I_i + \iota_i - \frac{\partial h'}{\partial x_i} \right) + \left(\frac{\partial \bar{f}}{\partial x_i} + \frac{\partial f'}{\partial x_i} \right)\left(-I_i - \iota_i + \frac{\partial h'}{\partial x_i} \right) = 0 \qquad (6\text{-}531)$$

or

$$-\frac{\partial I_i}{\partial x_i} - \frac{\partial \iota_i}{\partial x_i} + \frac{\partial^2 h'}{\partial x_i^2} + \left(\frac{\partial \bar{f}}{\partial x_i} + \frac{\partial f'}{\partial x_i} \right)\left(-I_i - \iota_i + \frac{\partial h'}{\partial x_i} \right) = 0 \qquad (6\text{-}532)$$

taking the expected value of Eq. (6-532) gives,

$$-E\left[\frac{\partial I_i}{\partial x_i} \right] - E\left[\frac{\partial \iota_i}{\partial x_i} \right] + E\left[\frac{\partial^2 h'}{\partial x_i^2} \right] - E\left[\frac{\partial \bar{f}}{\partial x_i} I_i \right] - E\left[\frac{\partial \bar{f}}{\partial x_i} \iota_i \right]$$

$$+ E\left[\frac{\partial \bar{f}}{\partial x_i} \frac{\partial h'}{\partial x_i} \right] - E\left[\frac{\partial f'}{\partial x_i} I_i \right] - E\left[\frac{\partial f'}{\partial x_i} \iota_i \right] + E\left[\frac{\partial f'}{\partial x_i} \frac{\partial h'}{\partial x_i} \right] = 0 \qquad (6\text{-}533)$$

In Eq. (6-533), the second, third, fifth, sixth, and seventh terms are zero because ι_i is temporally variable but spatially constant, $E[h'] = 0$, and $E[f'] = 0$. Therefore, Eq. (6-533) takes the form

$$-\frac{\partial I_i}{\partial x_i} - \frac{\partial \bar{f}}{\partial x_i} I_i - \frac{\overline{\partial f'}}{\partial x_i} \iota_i + \frac{\overline{\partial f'}}{\partial x_i} \frac{\partial h'}{\partial x_i} = 0 \qquad (6\text{-}534)$$

Subtracting Eq. (6-534) from Eq. (6-532) and retaining first-order terms in the perturbations yields the equation for perturbations in the hydraulic head resulting from hydraulic gradient and hydraulic conductivity perturbations:

$$-\frac{\partial \iota_i}{\partial x_i} + \frac{\partial^2 h'}{\partial x_i^2} - \frac{\partial \bar{f}}{\partial x_i} \iota_i + \frac{\partial \bar{f}}{\partial x_i} \frac{\partial h'}{\partial x_i} - I_i \frac{\partial f'}{\partial x_i} = 0 \qquad (6\text{-}535)$$

If there is no large-scale trend in $\ln(K)$, $\bar{f}$ is constant. By recognizing that ι_i is uniform in space at the scale of the $\ln(K)$ correlation length, Eq. (6-535) can be simplified to

$$\frac{\partial^2 h'}{\partial x_i^2} - I_i \frac{\partial f'}{\partial x_i} = 0 \qquad (6\text{-}536)$$

Spectral Representation for Flow and Flux Equations

Spectral Representation of the Flow Equation

The perturbed quantities in Eqs. (6-527) and (6-536) are represented by Fourier–Stieltjes integrals in wave number and frequency as

$$h'(x, y, z, t) = \int_{-\infty}^{\infty} e^{i(\mathbf{k} \cdot \mathbf{x} + \omega t)} \, dZ_{h'}(\mathbf{k}, \omega) \qquad (6\text{-}537)$$

$$f'(x, y, z, t) = \int_{-\infty}^{\infty} e^{i(\mathbf{k}.\mathbf{x} + \omega t)} \, dZ_{f'}(\mathbf{k}, \omega) \tag{6-538}$$

$$\iota_i(x, y, z, t) = \int_{-\infty}^{\infty} e^{i(\mathbf{k}.\mathbf{x} + \omega t)} \, dZ_{\iota_i}(\mathbf{k}, \omega) \tag{6-539}$$

$$q_i'(x, y, z, t) = \int_{-\infty}^{\infty} e^{i(\mathbf{k}.\mathbf{x} + \omega t)} \, dZ_{q_i'}(\mathbf{k}, \omega) \tag{6-540}$$

Substituting Eqs. (6-537) and (6-538) into Eq. (6-536) and following the same procedure in Section 6.4.6.7 (Eqs. [6-322]–[6-330]) results in an equation in terms of the complex Fourier amplitudes, dZ:

$$-k^2 \, dZ_{h'} - ik_i I_i \, dZ_{f'} = 0 \tag{6-541}$$

where

$$k^2 = k_1^2 + k_2^2 + k_3^2 \equiv k_x^2 + k_y^2 + k_z^2 \tag{6-542}$$

Eq. (6-541) is the same as Eq. (6-231) with the exception that the second term of Eq. (6-541) represents a more general case. More importantly, Eq. (6-541) is exactly the same as Eq. (6-260).

Spectral Representation of the Flux Equation
Substituting Eqs. (6-537)–(6-540) into Eq. (6-527) gives

$$dZ_{q_i'} = K_g(dZ_{\iota_i} - ik_i \, dZ_{h'} + I_i \, dZ_{f'}) \tag{6-543}$$

which is the same as Eq. (6-257), with the exception of dZ_{ι_i}. From Eq. (6-541),

$$dZ_{h'} = -\frac{iI_i k_i}{k^2} \, dZ_{f'} \tag{6-544}$$

substituting this into Eq. (6-543) and using $i^2 = -1$ gives

$$dZ_{q_i'} = K_g\left(dZ_{\iota_i} - \frac{I_j k_i k_j}{k^2} \, dZ_{f'} + I_i \, dZ_{f'}\right) \tag{6-545}$$

which is the same as Eq. (6-261) with the exception of the term dZ_{ι_i}. The complex conjugate of Eq. (6-545) is ($z^* = x - iy$ is the complex conjugate of $z = x + iy$ by definition)

$$dZ^*_{q_i'} = K_g\left(dZ^*_{\iota_i} - \frac{I_j k_i k_j}{k^2} \, dZ^*_{f'} + I_i \, dZ^*_{f'}\right) \tag{6-546}$$

Multiplying both sides of Eqs. (6-545) and (6-546) and taking their expected values, and using the Kronecker δ as in Eq. (6-265), gives

$$S_{q_i' q_j'} = K_g^2\left[S_{\iota_i \iota_j} + \left(\delta_{im} - \frac{k_i k_m}{k^2}\right)\left(\delta_{jn} - \frac{k_j k_n}{k^2}\right)I_n I_m S_{f'f'}\right] \tag{6-547}$$

where the cross-spectrum $S_{\iota_i f'}(\mathbf{k}, \omega) = 0$. The expression for the specific discharge spectrum has been significantly simplified by recognizing the special form of the input spectra. One should bear in mind that f' and ι_i have been represented as space-time random processes even though f' is time-invariant and ι_i is spatially uniform. There is no inconsistency in the above representation, because a constant can be represented as a random variable with a covariance having infinite correlation length and a spectrum with all the power concentrated at zero frequency, ω, or wave number, k, (Rehfeldt and Gelhar, 1992).

The hydraulic conductivity is time invariant; hence, its spectrum can be represented by

$$S_{f'f'}(\mathbf{k},\ \omega) = S_{f'f'}(\mathbf{k})\delta(\omega) \tag{6-548}$$

where $\delta(\omega)$ is the Dirac delta function., similarly the hydraulic gradient fluctuation is assumed to be spatially uniform, and its spectrum is of the form

$$S_{t_i t_i}(\mathbf{k},\ \omega) = S_{t_i t_i}(\omega)\delta(\mathbf{k}) \tag{6-549}$$

The Expression for Macrodispersivity
Substitution of Eq. (6-547) into Eq. (6-520) gives

$$A_{ij} = \int_{-\infty}^{\infty}\int_{-\infty}^{\infty} \frac{K_g^2 S_{t_i t_i}(\mathbf{k},\ \omega)}{i\varphi_e\omega q + [ik_1 + \alpha_L^* k_1^2 + \alpha_T^*(k_2^2 + k_3^2)]q^2}\ \mathbf{dk}\ d\omega$$

$$+ \int_{-\infty}^{\infty}\int_{-\infty}^{\infty} \frac{K_g^2\left(\delta_{im} - \dfrac{k_i k_m}{k^2}\right)\left(\delta_{jn} - \dfrac{k_j k_n}{k^2}\right)I_n I_m S_{f'f'}(\mathbf{k},\ \omega)}{i\varphi_e\omega q + [ik_1 + \alpha_L^* k_1^2 + \alpha_T^*(k_2^2 + k_3^2)]q^2}\ \mathbf{dk}\ d\omega \tag{6-550}$$

Using the special forms of the spectra, Eqs., (6-548)–(6-550) are decomposed into the sum of two components, one incorporating the effect of temporal variability and the other spatial variability:

$$A_{ij} = A_{ij}^{(t)} + A_{ij}^{(s)} \tag{6-551}$$

where $A_{ij}^{(t)}$ is due to temporal variability and $A_{ij}^{(s)}$ is due to spatial variability. From eqs. (6-550) and (6-551), the components are given by

$$A_{ij}^{(t)} = \int_{-\infty}^{\infty} \frac{K_g^2 S_{t_i t_i}(\omega)}{i\varphi_e\omega q}\ d\omega \tag{6-552}$$

and

$$A_{ij}^{(s)} = \int_{-\infty}^{\infty}\int_{-\infty}^{\infty} \frac{K_g^2\left(\delta_{im} - \dfrac{k_i k_m}{k^2}\right)\left(\delta_{jn} - \dfrac{k_j k_n}{k^2}\right)I_n I_m S_{f'f'}}{[ik_1 + \alpha_L^* k_1^2 + \alpha_T^*(k_2^2 + k_3^2)]q^2}\ \mathbf{dk} \tag{6-553}$$

Eq. (6-552) is written by making the wave number zero ($k_1 = k_2 = k_3 = 0$). Equation (6-553) is identical to Eq. (6-462).

Evaluation of the Unsteady Flow Component
The component resulting from unsteady flow is evaluated by rewriting Eq. (6-552) as

$$A_{ij}^{(t)} = \lim_{\varepsilon \to 0} K_g^2 \int_{-\infty}^{\infty} \frac{S_{t_i t_i}(\omega)}{\varepsilon + i\varphi_e\omega q}\ d\omega \tag{6-554}$$

where ε is a small real parameter. Equation. (6-554) can also be written as

$$A_{ij}^{(t)} = \lim_{\varepsilon \to 0} K_g^2 \int_{-\infty}^{\infty} \frac{S_{t_i t_i}(\omega)(\varepsilon - i\varphi\omega q)}{\varepsilon^2 + \varphi^2\omega^2 q^2}\ d\omega \tag{6-555}$$

where ε is a small real parameter. Choosing a form for $S_{t_i t_i}(\omega)$ that is even in ω, such as the exponential, renders the imaginary part in Eq. (6-555) zero upon integration. Eq. (6-555) reduces to

$$A_{ij}^{(t)} = \lim_{\varepsilon \to 0} \frac{K_g^2}{\varphi_e q} \int_{-\infty}^{\infty} \frac{(\varphi_e q/\varepsilon)\, S_{t_i t_i}(\omega)}{\varphi_e^2 \omega^2 q^2/\varepsilon^2 + 1}\ d\omega \tag{6-556}$$

By substituting

$$v = \frac{\varphi_e q}{\varepsilon}\ \omega \tag{6-557}$$

Eq. (6-556) reduces

$$A_{ij}^{(t)} = \lim_{\varepsilon \to 0} \frac{K_g^2}{\varphi_e q} \int_{-\infty}^{\infty} \frac{S_{\iota_i \iota_j}(\varepsilon v/(\varphi_e q))}{v^2 + 1} \, \mathrm{d}v = \frac{K_g^2}{\varphi_e q} S_{\iota_i \iota_j}(0) \int_{-\infty}^{\infty} \frac{\mathrm{d}v}{v^2 + 1} \tag{6-558}$$

or

$$A_{ij}^{(t)} = 2 \frac{K_g^2}{\varphi_e q} S_{\iota_i \iota_j}(0) \int_0^{\infty} \frac{\mathrm{d}v}{v^2 + 1} = \pi \frac{K_g^2}{\varphi_e q} S_{\iota_i \iota_j}(0) \tag{6-559}$$

Using Eq. (6-469), Eq. (6-559) can be written as

$$A_{ij}^{(t)} = \frac{\pi q}{\gamma^2 \varphi_e I_1^2} S_{\iota_i \iota_j}(0) \tag{6-560}$$

6.4.7.5 Hydraulic Gradient Spectrum and Expression $A_{ij}(t)$

From Eq. (6-108), the transform relationship for the spectrum is

$$S_{\iota_i \iota_j}(\omega) = \frac{1}{2\pi} \int_{-\infty}^{\infty} \mathrm{e}^{-i\omega\tau} R_{\iota_i \iota_j}(\tau) \, \mathrm{d}\tau \tag{6-561}$$

Since $\omega = 0$,

$$S_{\iota_i \iota_j}(0) = \frac{1}{2\pi} \int_{-\infty}^{\infty} R_{\iota_i \iota_j}(\tau) \, \mathrm{d}\tau \tag{6-562}$$

Assuming that the covariance function is exponential, as in Eq. (6-285),

$$R_{\iota_i \iota_j}(\tau) = \sigma_{\iota_i \iota_j}^2 \exp\left(-\frac{\tau}{\lambda_{\iota_i \iota_j}}\right) \tag{6-563}$$

Equation. (6-562), after some manipulations, becomes

$$S_{\iota_i \iota_j}(0) = \frac{1}{2\pi} \int_{-\infty}^{\infty} \sigma_{\iota_i \iota_j}^2 \exp\left(-\frac{\tau}{\lambda_{\iota_i \iota_j}}\right) \mathrm{d}\tau = \frac{1}{\pi} \sigma_{\iota_i \iota_j}^2 \lambda_{\iota_i \iota_j} \tag{6-564}$$

where $\sigma_{\iota_i}^2$ is the variance of the ith component of the ι when $i = j$ and the covariance between the ith and jth component when $i \neq j$ and $\lambda_{\iota_i \iota_j}$ is the integral scale. Substituting Eq. (6-564) into Eq. (6-560) gives

$$A_{ij}^{(t)} = \frac{1}{\gamma^2} \frac{q}{\varphi_e} \frac{\sigma_{\iota_i \iota_j}^2}{I_1^2} \lambda_{\iota_i \iota_j} \tag{6-565}$$

which is a general expression for the contribution of unsteady flow to macrodispersion.

6.4.7.6 Examination of Hydraulic Gradient Variability

It is instructive to examine two specific cases of hydraulic gradient variability (Rehfeldt and Gelhar, 1992). In the first case, only the magnitude of the gradient changes over time. In the second case, only the direction varies over time. To simplify the following examples, hydraulic gradient fluctuations are assumed to be in the (x_1, x_2) (x, y) plane only, so that $I_3 = \iota_3 = 0$.

Hydraulic Gradient Magnitude Variation
In this special case, only the magnitude of the gradient varies over time. In addition, it is assumed that the hydraulic conductivity field is statistically anisotropic and that the directions of the mean hydraulic gradient and the mean specific discharge may be different. Figure 6-84 illustrates the general coordinate configuration in two dimensions (its three-dimensional form is illustrated in Figure 6-74 and Figure 6-75, for unconfined aquifer and confined aquifer, respectively). In this case of

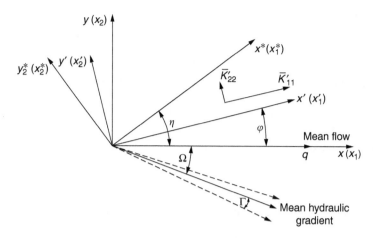

FIGURE 6-84 Coordinate systems for transient flow through heterogeneous porous media. The x_i' and x_i^* coordinates are the principal axes of the bulk hydraulic conductivity tensor and macrodispersivity tensor, respectively. The angles Ω and φ are related through the anisotropic form of Darcy's law by $\tan(\varphi) = \overline{K}_{22}'/\overline{K}_{11}'$ $\tan(\Omega + \varphi)$. The angle Γ is the fluctuation in the direction of the hydraulic gradient about its mean. (Rehfeldt, K. R. and Gelhar, L. W., *Water Resour. Res.*, 28, 2085–2099, 1992.)

gradient magnitude variation, assume $\xi = 0$. In this two-dimensional example, the $x(x_1)$ and $y(x_2)$ components of the hydraulic gradient vector are

$$I_1^* = I^* \cos(-\Omega) = I^* \cos(\Omega) \tag{6-566}$$

$$I_2^* = I^* \sin(-\Omega) = -I^* \sin(\Omega) \tag{6-567}$$

where I^* is the magnitude of the hydraulic gradient and Ω is the angle between the mean flow direction and the mean hydraulic gradient. The magnitude and the components of the hydraulic gradient vector are decomposed into means and perturbations such that

$$I^* = I + \iota, \qquad I_1^* = I_1 + \iota_1, \qquad I_2^* = I_2 + \iota_2 \tag{6-568}$$

where I is the mean hydraulic gradient magnitude, I_1 the mean of the $x(x_1)$ component; I_2 the mean of the $y(x_2)$ component; and ι, ι_1, and ι_2 are the perturbations in the magnitude, $x(x_1)$, and $y(x_2)$ components, respectively. The notation in the expressions in Eq. (6-568) is consistent with the definition of the terms in Eq. (6-521). Substituting the expressions in Eq. (6-568) into Eqs. (6-566) and (6-567) and taking the expected value yields the mean and perturbation of the gradient components, which are given as

$$I_1 = I \cos(\Omega), \qquad I_2 = -I \sin(\Omega) \tag{6-569}$$

$$\iota_1 = \iota \cos(\Omega), \qquad \iota_2 = -\iota \sin(\Omega) \tag{6-570}$$

The perturbed quantities are described by Fourier–Stieltjes integrals (Lumley and Panofsky, 1964) as

$$\iota_1 = \int_{-\infty}^{\infty} e^{i\alpha x} \, dZ_{\iota_1} \tag{6-571}$$

$$\iota_2 = \int_{-\infty}^{\infty} e^{i\alpha x} \, dZ_{\iota_2} \tag{6-572}$$

and

$$\iota = \int_{-\infty}^{\infty} e^{i\omega t} \, dZ_t \tag{6-573}$$

The substitution of Eqs. (6-571) and (6-573) into the first expression in Eq. (6-570) gives

$$dZ_{\iota_1} = \cos(\Omega) \, dZ_c \tag{6-574}$$

and its complex conjugate is ($z^* = x - iy$ is the complex conjugate of $z = x + iy$ by definition)

$$dZ_{\iota_1}^* = \cos(\Omega) \, dZ_t^* \tag{6-575}$$

Multiplying both sides of Eqs. (6-574) and (6-575) and taking the expected values gives

$$E[dZ_{\iota_1} \, dZ_{\iota_1}^*] = \cos^2(\Omega) E[dZ_t \, dZ_t^*] \tag{6-576}$$

Applying the spectral representation theorem in Section 6.4.2.1 [Eq. (6-105)], the spectra can be expressed as

$$S_{\iota_1 \iota_1}(\omega) = \cos^2(\Omega) S_{\iota t}(\omega) \tag{6-577}$$

Similarly, from the second expression of Eq. (6-570) and Eqs. (6-572) and (6-573) we can write

$$S_{\iota_2 \iota_2} = \sin^2(\Omega) S_{\iota t}(\omega) \tag{6-578}$$

From the expressions in Eq. (5-570), and Eqs. (6-571) and (6-572), the cross-spectrum of ι_1 and ι_2 can be expressed as

$$S_{\iota_1 \iota_2}(\omega) = -\sin(\Omega) \cos(\Omega) \, S_{\iota t}(\omega) \tag{6-579}$$

Here, in Eqs. (6-577) and (6-578), $S_{\iota t}(\omega)$ is the spectrum of the gradient magnitude fluctuations. For this case of temporal variation in the magnitude of the hydraulic gradient only, the spectra $S_{\iota t}(\omega)$ are functions of the spectrum of the gradient magnitude fluctuations and the mean angle Ω between **q** and **I** vectors. Nonzero Ω is possible only in the case of bulk anisotropy of hydraulic conductivity. Gelhar and Axness (1983) have shown that (see also Section 6.4.6.6) anisotropy of the local hydraulic conductivity covariance function leads to large-scale anisotropy of bulk hydraulic conductivity. Therefore, the hydraulic gradient spectra, given by Eqs. (6-577)–(6-579), are functions not only of the temporal characteristics of the magnitude, but also of the spatial variability of hydraulic conductivity via large-scale hydraulic anisotropy. The determination of the components of the macrodispersivity tensor $[A_{11}^{(t)}, A_{11}^{(t)}, A_{11}^{(t)}]$ are presented below.

Determination of $A_{11}^{(t)}$
From Eq. (6-560) one can write

$$A_{11}^{(t)} = \frac{1}{\gamma^2} \frac{q\pi}{\varphi_e I_1^2} S_{\iota_1 \iota_1}(0) \tag{6-580}$$

From Eq. (6-577), for $\omega = 0$,

$$S_{\iota_1 \iota_1}(0) = \cos^2(\Omega) S_{\iota t}(0) \tag{6-581}$$

and by combining with Eq. (6-580)

$$A_{11}^{(t)} = \frac{1}{\gamma^2} \frac{q\pi}{\varphi_e I_1^2} \cos^2(\Omega) S_{tt}(0) \tag{6-582}$$

From Eq. (6-564), assuming the gradient magnitude spectrum (S_{tt}) is in the exponential form as in Eq. (6-563), Eq. (6-582) gives

$$A_{11}^{(t)} = \frac{1}{\gamma^2} \frac{q}{\varphi_e I_1^2} \cos^2(\Omega) \sigma_t^2 \lambda_t \tag{6-583}$$

and using the first expression in Eq. (6-569),

$$A_{11}^{(t)} = \frac{1}{\gamma^2} \frac{q}{\varphi_e} \frac{\sigma_t^2}{I^2} \lambda_t \tag{6-584}$$

where σ_t^2 is the variance of gradient magnitude fluctuations and λ_t the time integral scale.

Determination of $A_{22}^{(t)}$

From Eq. (6-560) one can write

$$A_{22}^{(t)} = \frac{1}{\gamma^2} \frac{q\pi}{\varphi_e I_1^2} S_{t_2 t_2}(0) \tag{6-585}$$

From Eq. (6-578), for $\omega = 0$,

$$S_{t_2 t_2}(0) = \sin^2(\Omega) S_{tt}(0) \tag{6-586}$$

and combination with Eq. (6-585) gives

$$A_{22}^{(t)} = \frac{1}{\gamma^2} \frac{q\pi}{\varphi_e I_1^2} \sin^2(\Omega) S_{tt}(0) \tag{6-587}$$

From Eq. (6-564), assuming the gradient magnitude spectrum (S_{tt}) is in the exponential form, as in Eq. (6-563), Eq. (6-587) gives

$$A_{22}^{(t)} = \frac{1}{\gamma^2} \frac{q}{\varphi_e I_1^2} \sin^2(\Omega) \sigma_t^2 \lambda_t \tag{6-588}$$

Using the first expression in Eq. (6-569)

$$A_{22}^{(t)} = \frac{1}{\gamma^2} \frac{q}{\varphi_e} \frac{\sigma_t^2}{I^2} \tan^2(\Omega) \lambda_t \tag{6-589}$$

where σ_t^2 is the variance of gradient magnitude fluctuations and λ_t the time integral scale.

Determination of $A_{12}^{(t)}$

From Eq. (6-560) one can write

$$A_{12}^{(t)} = \frac{1}{\gamma^2} \frac{q\pi}{\varphi_e I_1^2} S_{t_1 t_2}(0) \tag{6-590}$$

From Eq. (6-579), for $\omega = 0$,

$$S_{t_1 t_2}(0) = -\sin(\Omega) \cos(\Omega) S_{tt}(0) \tag{6-591}$$

and combination with Eq. (6-590) gives

$$A_{12}^{(t)} = -\frac{1}{\gamma^2} \frac{q\pi}{\varphi_e I_1^2} \sin(\Omega) \cos(\Omega) S_{tt}(0) \tag{6-592}$$

From Eq. (6-564), assuming the gradient magnitude spectrum (S_{tt}) is in the exponential form, as in Eq. (6-563), Eq. (6-592) gives

$$A_{22}^{(t)} = -\frac{1}{\gamma^2} \frac{q\pi}{\varphi_e I_1^{\,2}} \sin(\Omega)\cos(\Omega)\sigma_t^2\lambda_t \qquad (6\text{-}593)$$

and using the first expression in Eq. (6-569)

$$A_{22}^{(t)} = -\frac{1}{\gamma^2} \frac{q}{\varphi_e} \frac{\sigma_t^2}{I^2} \tan(\Omega)\lambda_t \qquad (6\text{-}594)$$

where σ_t^2 is the variance of gradient magnitude fluctuations and λ_t the time integral scale.

Evaluation of the Macrodispersivity Tensor Components
(1) From Eqs. (6-584), (6-589), and (6-594), it is apparent that $A_{11}^{(t)}$ is larger than the other two components for typical aquifer conditions where the angle between the direction of the hydraulic gradient and the specific discharge is less than 45°; (2) $A_{11}^{(t)}$ is always nonzero, whereas the other two components are zero if the bulk hydraulic conductivity is isotropic or if flow, is aligned with the principal directions of the hydraulic conductivity tensor; (3) When $\Omega \neq 0$, $A_{11}^{(t)} \neq 0$, indicating that the principal component axes of the macrodispersivity tensor due to transient flow will rotate with respect to the direction of mean flow; (4) The negative value of $A_{12}^{(t)}$ implies that the major principal component axis of the macrodispersivity tensor will rotate toward the mean hydraulic gradient; (5) For the steady-state case, Gelhar and Axness (1983) also found that (see also Section 6.4.6.6) the major principal axis of the macrodispersivity tensor, $A_{11}^{(s)}$, rotates towards the mean hydraulic gradient, and (6) From Eq. (6-551) it is known that the behavior of the plume will be governed by the sum of the components of the steady and transient parts of the full dispersivity tensor; therefore, relative to the steady flow results of Gelhar and Axness (1983) (see also Section 6.4.6.6), the combined effect of temporal and spatial variability will be an enhanced longitudinal dispersive flux, and to a lesser extent, a larger transverse-dispersive flux; and in addition, rotation of the plume toward the mean hydraulic gradient will be enhanced.

6.4.7.7 Examination of Hydraulic Gradient Direction Variability

Determination of the Spectrum Expressions
For this special case, the magnitude of the hydraulic gradient will remain constant, but the direction of the gradient will vary in time. Again, for illustration, Rehfeldt and Gelhar (1992) examined only the variation in the horizontal plane (Figure 6-84). To investigate the temporal variation in the direction of the hydraulic gradient, characterized by fluctuations about the mean angle Ω, the components of the hydraulic gradient vector are given by

$$I_1 = I\cos(-\Omega + \Gamma) = I[\cos(\Omega)\cos(\Gamma) + \sin(\Omega)\sin(\Gamma)] \qquad (6\text{-}595)$$

and

$$I_2 = I\sin(-\Omega + \Gamma) = -I[\sin(\Omega)\cos(\Gamma) - \cos(\Omega)\sin(\Gamma)] \qquad (6\text{-}596)$$

where Ω defines the mean direction of the gradient; Γ the variation about the mean; and I the magnitude of the gradient, which is taken to be constant and equal to the mean. Assuming Γ is small (less than 10°), the trigonometric functions of primed quantities are approximated by the first term of a Taylor expansion such that $\cos(\Gamma) \approx 1$ and $\sin(\Gamma) \approx \Gamma$, Eqs. (6-595) and (6-596) take the forms

$$I_1 = I[\cos(\Omega) + \sin(\Omega)\Gamma] \qquad (6\text{-}597)$$

and

$$I_2 = -I[\sin(\Omega) - \cos(\Omega)\Gamma] \qquad (6\text{-}598)$$

and evaluating the mean, by using $E[\Gamma] = 0$,

$$I_1 = I\cos(\Omega), \qquad I_2 = -I\sin(\Omega) \qquad (6\text{-}599)$$

From Eqs. (6-597) and (6-598), the perturbation equations can be written as

$$\iota_1 = I[\Gamma\sin(\Omega)], \qquad \iota_2 = I[\Gamma\cos(\Omega)] \qquad (6\text{-}600)$$

Similarly as in the case of Section 6.4.6.7 (Eqs. [6-323]–[6-335]), by assuming stationarity in time, the perturbed quantities are represented by Fourier–Stieltjes representations (Lumley and Panofsky, 1964) as

$$\iota_1 = \int_{-\infty}^{\infty} e^{i\omega t}\, dZ_{\iota_1} \qquad (6\text{-}601)$$

$$\iota_2 = \int_{-\infty}^{\infty} e^{i\omega t}\, dZ_{\iota_2} \qquad (6\text{-}602)$$

$$\Gamma = \int_{-\infty}^{\infty} e^{i\omega t}\, dZ_{\Gamma} \qquad (6\text{-}603)$$

Substituting of Eqs. (6-601)–(6-603) into the first expression in Eq. (6-600) gives

$$dZ_{\iota_1} = I\sin(\Omega)\, dZ_{\Gamma} \qquad (6\text{-}604)$$

and its complex conjugate ($z^* = x - iy$ is the complex conjugate of $z = x + iy$ by definition) is

$$dZ_{\iota_1}^* = I\sin(\Omega)\, dZ_{\Gamma}^* \qquad (6\text{-}605)$$

Multiplying both sides of Eqs. (6-604) and (6-605) and taking their expected values gives

$$E[dZ_{\iota_1}\, dZ_{\iota_1}^*] = I^2 \sin^2(\Omega) E[dZ_{\Gamma}\, dZ_{\Gamma}^*] \qquad (6\text{-}606)$$

Applying the spectral representation theorem in Section 6.4.2.1 [Eq. (6-105)], the spectrum can be expressed as

$$S_{\iota_1\iota_1}(\omega) = I^2 \sin^2(\Omega) S_{\Gamma\Gamma}(\omega) \qquad (6\text{-}607)$$

Similarly, substituting Eqs. (6-602) and (6-603) into the second expression of Eq. (6-600) and following the same steps as above,

$$S_{\iota_2\iota_2}(\omega) = I^2 \cos^2(\Omega) S_{\Gamma\Gamma}(\omega) \qquad (6\text{-}608)$$

Similarly, from Eqs. (6-600)–(6-603), the cross spectrum of ι_1 and ι_2 is

$$S_{\iota_1\iota_2}(\omega) = I^2 \sin(\Omega)\cos(\Omega)\, S_{\Gamma\Gamma}(\omega) \qquad (6\text{-}609)$$

As was observed for the case where the magnitude of the hydraulic gradient varied over time (see Section 6.4.7.7), the hydraulic gradient spectra are functions of the mean angle Ω between the mean flow direction and the mean hydraulic gradient. Again, properties of the spatial variability of K will influence the spectra in Eqs. (6-607)–(6-609) through the bulk anisotropy of the hydraulic conductivity.

Components of the Macrodispersivity Tensor

Determination of $A_{11}^{(t)}$
Substituting Eq. (6-607) with $\omega = 0$ into Eq. (6-580) gives

$$A_{11}^{(t)} = \frac{1}{\gamma^2} \frac{q\pi}{\varphi_e I_1^2} \sin^2(\Omega) S_{\Gamma\Gamma}(0) \tag{6-610}$$

From Eq. (6-564), assuming the angle magnitude spectrum $(S_{\Gamma\Gamma})$ is in exponential form as in Eq. (6-563), Eq. (6-610) becomes

$$A_{11}^{(t)} = \frac{1}{\gamma^2} \frac{q}{\varphi_e I_1^2} I^2 \sin^2(\Omega) \sigma_\Gamma^2 \lambda_\Gamma \tag{6-611}$$

and using the first expression of Eq. (6-599),

$$A_{11}^{(t)} = \frac{1}{\gamma^2} \frac{q}{\varphi_e} \tan^2(\Omega) \sigma_\Gamma^2 \lambda_\Gamma \tag{6-612}$$

Determination of $A_{22}^{(t)}$
Substituting Eq. (6-608) with $\omega = 0$ into Eq. (6-585) gives

$$A_{22}^{(t)} = \frac{1}{\gamma^2} \frac{q\pi}{\varphi I_1^2} I^2 \cos^2(\Omega) S_{\Gamma\Gamma}(0) \tag{6-613}$$

From Eq. (6-564), assuming the angle magnitude spectrum $(S_{\Gamma\Gamma})$ is in exponential form as in Eq. (6-563), Eq. (6-613) becomes

$$A_{22}^{(t)} = \frac{1}{\gamma^2} \frac{q}{\varphi_e I_1^2} I^2 \cos^2(\Omega) \sigma_\Gamma^2 \lambda_\Gamma \tag{6-614}$$

and using the first expression of Eq. (6-599)

$$A_{22}^{(t)} = \frac{1}{\gamma^2} \frac{q}{\varphi_e} \sigma_\Gamma^2 \lambda_\Gamma \tag{6-615}$$

Determination of $A_{12}^{(t)}$
Substituting of Eq. (6-609) with $\omega = 0$ into Eq. (6-590) gives

$$A_{12}^{(t)} = \frac{1}{\gamma^2} \frac{q\pi}{\varphi_e I_1^2} I^2 \sin(\Omega)\cos(\Omega) S_{\Gamma\Gamma}(0) \tag{6-616}$$

From Eq. (6-564), assuming the angle magnitude spectrum $(S_{\Gamma\Gamma})$ is in exponential form as in (6-563), Eq. (6-616) becomes

$$A_{12}^{(t)} = \frac{1}{\gamma^2} \frac{q\pi}{\varphi_e I_1^2} I^2 \sin(\Omega)\cos(\Omega) \sigma_\Gamma^2 \lambda_\Gamma \tag{6-617}$$

and using the first expression of Eq. (6-599)

$$A_{12}^{(t)} = \frac{1}{\gamma^2} \frac{q}{\varphi_e} \tan(\Omega) \sigma_\Gamma^2 \lambda_\Gamma \tag{6-618}$$

In the above expressions, Ω is the mean hydraulic gradient direction, σ_Γ^2 (rad^2) is the variance of the hydraulic gradient fluctuations, and λ_Γ (time) is the integral scale of the hydraulic gradient direction fluctuations.

Evaluation of the Macrodispersivity Tensor Components

The relationships between $A_{11}^{(t)}$, $A_{22}^{(t)}$, and $A_{12}^{(t)}$ as a function of Ω are given by Eqs. (6-612), (6-615), and (6-618), respectively. Although these equations are similar in form to Eqs. (6-584), (6-589), and (6-594), there are two significant differences between the results of gradient magnitude fluctuation and gradient direction fluctuation: (1) In the case of gradient direction variation, for typical ground water flow situations where $|\Omega| < 45°$, it is the transverse macrodispersivity coefficient, $A_{22}^{(t)}$, is the largest. The transverse term is nonzero even in the case of isotropic bulk hydraulic conductivity; (2) Another important difference between the influence of the gradient magnitude and direction on $A_{ij}^{(t)}$ is the sign of the off-diagonal term. In the case of $\Omega \neq 0$, gradient direction fluctuations, Eq. (6-618) produces a positive $A_{12}^{(t)}$ term, which implies that the transient component of the macrodispersivity tensor may cause the plume to rotate toward the principal component of the bulk hydraulic conductivity tensor; (3) The direction of plume rotation due to gradient direction fluctuations is opposite to the direction of rotation due to spatial variability; therefore, the combined effect of spatial and temporal variability will depend on the relative magnitudes of the transient and steady components of the macrodispersivity tensor.

EXAMPLE 6-15

This example is adapted from Garabedian et al. (1991) and Hess et al. (1992). The ground water velocity of the tested Cape Cod site is $U = 0.42$ m/day. The rest of the parameters are as follows: $\sigma_f^2 = 0.24$, $\lambda_1 = \lambda_2 = 3.5$ m, $\lambda_3 = 0.19$ m, $\theta = 0.5°$, $\lambda_\Gamma = 60$ days, and $\sigma_\Gamma = 0.25(8°) = 2° = 3.49 \times 10^{-2}$ rad. Here, σ_Γ is the standard deviation of the flow direction observed in the tracer test (Garabedian et al., 1991) and λ_Γ is the correlation scale of the gradient changes, which is estimated to be 60 days based on the observed temporal changes in the gradient. Using these data determine (1) the value of $A_{22}^{(t)}$ as a result of hydraulic gradient direction variability and (2) the value of A_{22}, and compare it with the A_{22} value determined from the tracer test.

SOLUTION

(a) Equation (6-615) to be used. In Eq. (6-615), γ is another parameter to be used. For the same parameters, it is calculated as $\gamma = 1.117322$ (see Example 6-14). Substituting these values into Eq. (6-615) and noting that $q/\varphi_e = U$, gives

$$A_{22}^{(t)} = \frac{1}{(1.117322)^2}(0.42 \text{ m/day})(3.49 \times 10^{-2} \text{ rad})^2(60 \text{ days}) = 0.025 \text{ m}$$

This unsteady component is added to the steady component (see Example 6-7), $A_{22}^{(s)} = 5.54 \times 10^{-8}$ m. Therefore, from Eq. (6-551), the total apparent transverse horizontal dispersivity is

$$A_{22} = A_{22}^{(t)} + A_{22}^{(s)} = 0.025 \text{ m} + 5.54 \times 10^{-8} = 0.025 \text{ m}$$

The steady component, $A_{22}^{(s)}$, is essentially zero. This value, $A_{22} = 0.025$ m, compares favorably with the value $A_{22} = 0.0182$ m (see Example 6-7) determined from the tracer test.

PROBLEMS

6.1 Explain the importance of stochastic approaches in ground water hydrology.

6.2 List the basic stochastic approaches and explain them briefly.

6.3 What is the fundamental relationship between the spatial variance (σ_x^2) and the effective molecular diffusion coefficient (D^*)? Explain its physical meaning.

TABLE P6-4-1
Measured Hydraulic Conductivity Values for Problem 6.4

i	x_i(m)	K_i (cm/sec)
1	10	2.56×10^{-3}
2	20	4.72×10^{-3}
3	30	1.75×10^{-3}
4	40	9.65×10^{-4}
5	50	1.35×10^{-1}
6	60	1.15×10^{-1}
7	70	5.45×10^{-3}
8	80	4.27×10^{-3}
9	90	5.52×10^{-3}
10	100	6.55×10^{-3}

6.4 Measured hydraulic conductivity values for a site are given in Table $P6$-4-1. The total length of the aquifer portion is 100 m. Measured value locations are at every 10 m along this length. Using these values determine (1) the variance and standard deviation of the hydraulic conductivity data; (2) the covariance of the hydraulic conductivity data with separations of 10 m; and (3) the value of the autocorrelation. Assuming a one-dimensional exponential form of the covariance function as given by Eq. (6-69), determine the value of the integral scale, ($\lambda_{f'}$).

6.5 Using the components of the covariance tensor in Table 6-8 ($\sigma^2_{x'x'}$, $\sigma^2_{y'y'}$, $\sigma^2_{z'z'}$) for the three-dimensional data, determine the components of the mechanical dispersion coefficient tensor ($D_{m_{z'z'}}, D_{m_{x'x'}}$ and $D_{m_{y'y'}}$) based on Eq. (6-58) for $t = 25.16$ days. Compare the values with the corresponding values in Table 6-10.

6.6 Define the terms "realization" and "ensemble" with an example.

6.7 Describe the "stationary process"?

6.8 Describe the "statistical homogeneity"?

6.9 Describe the "macrodispersivity tensor"? Also describe its special form under uniform ground water flow conditions.

6.10 What is the main assumption of "Gelhar and Axness" first-order analysis approach" for the determination of the dispersivity expressions?

6.11 (1) What is the correlation structure of the random hydraulic conductivity field? and (2) How is this structure selected? Give examples.

6.12 Based on the Gelhar and Axness first-order analysis approach, the values $A_{11} = 0.78$ m, $A_{22} = 0.025$ m, and $A_{12} = -0.095$ m were determined for the macrodispersivity tensor. Determine the values of the principal macrodispersivities and the rotation angle η between the (x, y) coordinates and (x^*, y^*) coordinates.

6.13 For a statistically isotropic aquifer, the following values are given: $\sigma^2_{f'} = 0.72$, $\lambda = \lambda_1 = \lambda_2 = \lambda_3 = 4.4$ m, $\alpha^*_L = 0.005$m, $\alpha^*_L/\alpha^*_T = 0.1$. Determine longitudinal macrodispersivity (A_{11}) and transverse macrodispersivity (A_{22}) values.

6.14 An aquifer is approximately isotropic in the plane of stratification, which means that $\lambda_1 = \lambda_2$. The angle with the bedding (θ) is $18°$. With the values given as $\sigma^2_{f} = 2.85$, $\lambda = \lambda_1 = \lambda_2 = 15$ m, and $\lambda_3 = 0.75$ m, determine (1) the values of A_{11}, A_{22}, A_{33}, and A_{13}; and (2) the values of A^*_{11}, A^*_{22}, and A^*_{33}, and the angle η.

7 Estimation of Dispersivities for Solute Transport in Aquifers

7.1 INTRODUCTION

As presented in detail in Chapter 6, dispersivities can be determined for an aquifer by using the expressions based on the Gelhar and Axness first-order stochastic analysis approach. However, the requirement of this approach is to define the covariance structure of the aquifer in terms of the variance of the hydraulic conductivity field (σ_f^2,) and the correlation scales (λ_1, λ_2, λ_3). For most aquifers, although limited data are available, from these data, the covariance structure of the aquifer can be determined. As mentioned in Chapter 6, characterization of the statistical parameters of an aquifer is still in the development process. In spite of these difficulties, during the past decades, various investigators have developed some empirical equations for the estimation of dispersivities. The purpose of this chapter is to present these estimation procedures.

The prediction of concentration variation of a contaminant plume in ground water by using analytical or numerical models requires that hydrodynamic dispersion coefficients, or simply, dispersion coefficients in the corresponding differential equations, as given by Eqs. (2-79)–(2-83), be known. The effective molecular diffusion coefficient (D^*) in Eqs. (2-27), (2-137), (2-154), (2-155), and (2-156) can be estimated based on the type of constituent and the medium that use the values in Table 2-1. As can be seen from Eqs. (2-139)–(2-144), the mechanical dispersion coefficient, D_m has relatively complex expressions as compared with the effective molecular diffusion coefficient D^*. The aforementioned expressions take simpler forms for uniform flow fields (see Eqs. [2-145]–[2-156]). In these expressions, A_L (longitudinal dispersivity), A_{TH} (transverse horizontal dispersivity), and A_{TV} (transverse vertical dispersivity) are critically important quantities not only on account of their effect on solute transport, but also from the point of view of their estimation procedures. To distinguish the field-scale dispersivities from laboratory values, the field-scale values are designated by "A" (see Gelhar and Axness, 1981, 1983; as presented in Section 6).

In this section, the state-of-the-art estimation procedures for dispersivities will be presented with examples.

7.2 LABORATORY VS. FIELD-SCALE DISPERSIVITIES

7.2.1 LABORATORY AND FIELD SCALES

The expressions for the mechanical dispersion coefficients given in Section 2.11.1 were generated by analyzing laboratory experimental data for concentration. These expressions are also used to predict concentration variations under field conditions. During the past decade, studies by various investigators have shown that dispersivity values based on laboratory experiments are relatively small and often fail to predict the behavior of a solute plume under field conditions. Furthermore, solute transport models under field conditions have demonstrated that field-observed dispersion coefficients can be several orders of magnitude higher than those indicated by laboratory tests with similar materials (e.g., Bredehoeft and Pinder, 1973; Konikow and Bredehoeft, 1974; Fried, 1975). The characteristic length of a solute constituent migration zone (e.g., the column length) under laboratory conditions is generally less than a meter (m), whereas under field conditions, the characteristic length (e.g., the length of the plume or the distance between the source and the toe of the plume) is generally tens of hundreds of meters or even longer. For this reason, the terms *field scale* or *field-scale dispersion* have been introduced in the literature (Gelhar et al., 1985; Gelhar et al., 1992). The term "field" refers to the vertical dimensions of tens to hundreds of meters and horizontal dimensions of tens of meters to kilometers.

7.2.2 Dispersivities under Field Conditions

Solute transport in geologic media is known to be strongly influenced by spatial variations in hydraulic conductivity. Hydraulic conductivity variations can be significant in stratified heterogeneous porous media. Such variations produce fluctuations in the ground water velocity and these variations cause solutes to spread at rates considerably greater than those normally observed in laboratory column experiments. Using a stochastic technique based on spectral analysis, Gelhar et al. (1979) took into account the spatial variations in ground water velocity and theoretically explained the field-scale dispersion effects on dispersivities. In the following year, Matheron and de Marsily (1980) confirmed the conclusions of Gelhar and others'.

The differential equations for hydrodynamic dispersion (e.g., Eq. [2-98]) with dispersion coefficients given by Eqs. (2-154), (2-155), and (2-156) are used for solute transport modeling for natural porous materials under field conditions. Gelhar et al. (1979) theoretically showed that the mechanical dispersion coefficient over large periods of time is in the form of a product of the mean velocity and dispersivity, as indicated in Eqs. (2-154), (2-155), and (2-156). This means that dispersivity may be a function of space, but in most applications it is assumed to be constant over the region of the aquifer that encompasses the entire plume both horizontally and vertically (Gelhar et al., 1992).

7.3 APPROACHES FOR THE DETERMINATION OF DISPERSIVITIES

During the past two decades, a number of theoretical studies based on deterministic and stochastic approaches have been carried out to describe field-scale dispersive mixing and develop methods for the determination of dispersivities. All of these theories present the view that the small-scale heterogeneity or spatial variability of the aquifer is mainly responsible for field-scale dispersion. Although these studies have generated important answers to some key conceptual dispersion phenomena questions regarding scale effects, the estimation of dispersivities from the practitioner's point of view still faces a lot of difficulties on account of the subjectivity of the approach. In other words, estimated dispersivities for a site by one applicant may turn out to be significantly different from the dispersivities estimated by another applicant for the same site. As mentioned above, the existing methods for the determination of dispersivities require a significant amount of data to determine the statistical characteristics of hydraulic conductivity variation for a given site. It is generally rare to find a site that has enough data points for this kind of statistical evaluation. In view of these limitations, there are three different approaches regarding the determination of dispersivities may be considered. These approaches, based on Gelhar et al. (1985, 1992), are presented below.

7.3.1 Approach I: Perfectly Layered Aquifer Assumption

Analysis of field-scale dispersion effects has started in analyzing aquifers based on the assumption that aquifers are perfectly layered (Pickens and Grisak, 1981a, 1981b; Molz et al., 1983, 1986). These investigators suggested measuring the variability in detail and modeling the transport of a solute in each layer with local dispersivities, thus eliminating the need for a field-scale dispersivity. This approach assumes that the aquifer is perfectly stratified, which means that hydraulic conductivity varies only in the direction perpendicular to the layers. Models for this kind of situation have been developed by deterministic methods (e.g., Marle et al., 1967; Dieulin, 1980; Molz et al., 1983) and stochastic methods (Mercado, 1967; Gelhar et al., 1979; Matheron and de Marsily, 1980). Most of these approaches are similar to the original work by Taylor (1953) on dispersion in tubes, but they have doubtful applicability on the field scale (10^2 to 10^4 m) because real aquifers are not perfectly stratified over these dimensions. However, one must emphasize the fact that these studies have contributed significantly to some of the concepts of field-scale dispersion phenomena.

7.3.2 APPROACH II: THE SPATIAL STATISTICS OF THE HETEROGENEOUS VELOCITY FIELD IS ASSUMED TO BE GIVEN

This approach presumes that the statistics of a heterogeneous velocity field is given and determines the dispersion coefficient from the Lagrangian covariance of the velocity field (e.g., Dieulin, 1981; Simmons, 1982a, 1982b; Tang et al., 1982; Winter, 1982). This method is similar to the work of Taylor (1921) on turbulent diffusion, but its mathematical techniques are more complex. The velocity covariance is not a directly measurable quantity, and the requirement that the velocity covariance be known is a key weakness. Moreover, even if the velocity covariance can be measured one potential problem is that it will change in accordance with the flow conditions. This method can be used in a nonpredictive mode by adjusting the velocity covariance function until agreement with an observed plume is accomplished. This approach was attempted by Simmons (1982a) for a field data case.

7.3.3 APPROACH III: THE STATISTICAL DESCRIPTION OF THE SPATIAL VARIABILITY OF HYDRAULIC CONDUCTIVITY IS ASSUMED TO BE GIVEN

This approach is based on stochastic methods with the statistical description of the one-, two-, and three-dimensional spatial variability of hydraulic conductivity. The stochastic methods are based on stochastic differential equations with perturbation methods (Dagan, 1982; Gelhar and Axness, 1983; Neuman et al., 1987) and Monte Carlo simulations (Warren and Skiba, 1964; Heller, 1972; Smith and Schwartz, 1980). These stochastic approaches incorporate the effects of practically unknowable small-scale variations in ground water flow by means of macrodispersivities that are used in a deterministic transport model describing large-scale variations in flow by means of convective terms.

The stochastic differential equations approach (Gelhar, 1993; Gelhar and Axness, 1981, 1983; Dagan, 1984, 1989) depends on perturbed steady flow and solute transport equations to construct the macroscopic dispersive flux and evaluate the resulting macrodispersivity tensor in terms of a three-dimensional, statistically anisotropic input covariance describing the hydraulic conductivity. All of these theories introduce some assumptions about relatively small perturbations, and this constitutes the key weakness of these approaches (see Section 6). In other words, these stochastic theories were developed under the assumption that the variance of hydraulic conductivity distribution ($\sigma_{f'}^2$), is relatively small. If $\sigma_{f'}^2 < 1$, these theories generally produce accurate results. If $\sigma_{f'}^2 > 1$, the predicted values for the macrodispersivity tensor components become approximate, depending on the value of $\sigma_{f'}^2$.

Moreover, in addition to $\sigma_{f'}^2$, these approaches also require that the correlation scales in three different coordinate directions ($\lambda_1, \lambda_2, \lambda_3$) be known. Therefore, another practical limitation of these methods is that they require a significant amount of hydraulic conductivity data in order to determine the statistical characteristics of a given aquifer site.

7.4 ESTIMATION OF DISPERSIVITIES USING RELATIONSHIPS BETWEEN DISPERSIVITIES AND FIELD SCALE

7.4.1 INTRODUCTION

Mathematical solute transport models are recognized to be powerful tools in predicting the concentration variation of contaminant constituents in aquifers. For example, in order, to predict the spatial and temporal variations of a contaminant plume in an aquifer, both the ground water flow and solute transport parameters of the aquifer are needed. Ground water flow parameters, especially hydraulic conductivity, can be determined based on various field tests including pumping and slug tests. These methods are well established in the literature and have been used by practitioners over the past half century (e.g., Batu, 1998). The situation for the determination of dispersivities is relatively more complex, and the available number of methods for their estimation is rather limited and controversial. As mentioned in Section 7.3, the state-of-the-art stochastic theories have generated approaches for the determination of dispersivities based on the statistical characteristics of formations (variance, covariance, etc.), but these approaches are valid only for relatively less heterogeneous formations. Beyond this, determination of the statistical characteristics for a given aquifer requires a tremendous amount of hydraulic conductivity data, and from the vantage point of funding and time, it is difficult, if not impossible, to generate satisfactory data for statistical evaluation.

Under the framework of the problems regarding dispersivities outlined above, currently, the only viable method is to estimate dispersivities by means of empirical approaches, which are generally based on the analysis of field-determined dispersivities. During the past three decades, a number of investigators have conducted field experiments to determine dispersivity values by analyzing observed concentration data by different methods. These data have been compiled by Gelhar et al. (1985, 1992) and are listed in Table 7-1. Gelhar and others have provided some charts and practical guidelines for the estimation of dispersivities. Table 7-1 presents a number of dispersivities for different types of tests and media. The same data were also interpreted and analyzed by Neuman (1990), and the analysis resulted in some empirical formulas. In this section, the data for dispersivities and their estimation methods will be presented with examples.

7.4.2 FIELD DATA ON DISPERSIVITIES

7.4.2.1 Summary of Field Observations

Gelhar et al. (1985) conducted a thorough review of published literature and numerous unpublished sources, and this review yielded field-scale values of dispersivities for various sites across the United States and ten other countries. Dispersivity values were determined by the analysis of field-scale tracer tests and contaminant transport modeling efforts. The literature sources and pertinent data characterizing each reviewed study are summarized in Gelhar et al. (1992) and are presented in Table 7-1. Table 7-1 includes information on 59 different field sites. The information compiled from each study includes site location, description of aquifer material, average aquifer saturated thickness, hydraulic conductivity or transmissivity, effective porosity, mean pore velocity, flow configuration, dimensionality of monitoring network, tracer type and input conditions, length scale of the test problem, reported values of longitudinal and transverse horizontal and transverse vertical dispersivities, and classification of the reliability of the reported data. Blank entries indicate that the information was not provided in the cited documents. Table 7-1 summarizes information for comparison purposes only; more details can be found in the original referenced documents. The aforementioned features potentially affect the determined dispersivity values from the field data. Therefore, in order to interpret these data better both quantitatively and qualitatively, some key features of the tests, such as characteristics of the aquifers, types of subsurface solute transport tests, dimensionality and monitoring, and field dispersivities and scales, are amplified in the following sections.

7.4.2.2 Aquifer Characteristics

Table 7-1, generated by Gelhar et al. (1985, 1992), includes study sites that represent a wide variety of aquifer conditions and settings. Columns 2 to 6 present aquifer characteristics and settings. Information regarding aquifers is related to their material, saturated thickness, hydraulic conductivity or transmissivity, and velocity. The aquifer thickness for each site is the arithmetic average of

TABLE 7-1

Summary of Field Observations for Field-Scale Dispersivities

Reference and Site Name	Aquifer Material	Average Aquifer Thickness, b (m)	Horizontal Hydraulic Conductivity K_h (m/sec) or Transmissivity T (m²/sec)	Effective Porosity, φ_e (%)	Velocity, U (m/day)	Flow Configuration	Monitoring	Tracer and Input[a]	Method of Data Interpretation	Scale of Test s (m)	Dispersivity $A_L/A_{TH}/A_{TV}$[b] (m)	Classification of Reliability of $A_L/A_{TH}/A_{TV}$[c] (I, II, III)
Adams and Gelhar (1992), Columbus, Mississippi	Very heterogeneous sand and gravel	8	10^{-5}–10^{-3} m/sec	35	0.03–0.5	Ambient	Three-dimensional	Br⁻ (pulse)	Spatial moments	200	7.5	II
Ahlstrom et al. (1977), Hanford, Washington	Glaciofluviatile sands and gravels	64	5.7×10^{-4} – 3.0×10^{-2} m/sec			Ambient	Two-dimensional	³H (contamination)	Two-dimensional numerical model	20,000	30.5/18.3	III
Bentley and Walter (1983), WIPP	Fractured dolomite	5.5		18	0.3	Two-well recirculating	Two-dimensional	PFB, SCN (step)	One-dimensional quasi-uniform flow solution [Grove and Beetem, 1971]	23	5.2	III
Bierschenk (1959) and Cole (1972), Hanford, Washington	Glaciofluviatile sands and gravels	64	1.7×10^{-1} m²/sec	10	26 / 31	Ambient	Two-dimensional	fluorescein (pulse)	One-dimensional uniform flow solution	3,500 / 4,000	6 / 460	III / III
Bredehoeft and Pinder (1973), Brunswick, Georgia	Limestone	50	6.5×10^{-7} – 8.6×10^{-7} m²/sec	35		Radial converging	Two-dimensional	Cl⁻ (contamination)	Two-dimensional numerical model	2,000	170/52	III

(Continued)

TABLE 7-1 (Continued)

Reference and Site Name	Aquifer Material	Average Aquifer Thickness, b (m)	Horizontal Hydraulic Conductivity K_h (m/sec) or Transmissivity T (m²/sec)	Effective Porosity, φ_e (%)	Velocity, U (m/day)	Flow Configuration	Monitoring	Tracer and Input[a]	Method of Data Interpretation	Scale of Test s (m)	Dispersivity $A_L/A_{TH}/A_{TV}$[b] (m)	Classification of Reliability of $A_L/A_{TH}/A_{TV}$ (I, II, III)[c]
Claasen and Corders (1975), Amargosa, Nevada	Fractured dolomite and limestone	15	5×10^{-2} – 11×10^{-2} m²/sec	6–60	0.14–3.4	Two-well recirculating	Two-dimensional	³H (pulse)	One-dimensional quasi-uniform flow solution (Grove and Beetem, 1971)	122	15	III
Daniels (1981, 1982), Nevada Test Site	Alluvium derived from tuff	500	1.7×10^{-5} m/sec		0.04	Radial converging	Two-dimensional	³H (contamination)	Radial flow type curve (Sauty, 1980)	91	10–30	III
Dieulin (1981), Le Cellier (Lozere, France)	Fractured granite	20	3×10^{-4} – 9×10^{-4} m/sec	2–8	3	Radial converging	Two-dimensional	Cl⁻, I⁻ (pulse)	Radial flow type curve (Sauty, 1980)	5	0.5	II
Dieulin (1980), Torcy, France	Alluvial deposits	6	3×10^{-4} m/sec		0.5	Ambient	Two-dimensional (resistivity)	Cl⁻ (pulse)	One-dimensional uniform flow solution	15	3	III
Egboka et al. (1983), Borden	Glaciofluvial sand	7–27	10^{-5}–10^{-7} m/sec	38	0.01–0.04	Ambient	Three-dimensional	³H (environmental)	One-dimensional uniform flow solution	600	30–60	III
Fenske (1973), Tatum Salt Dome, Mississippi	Limestone	53	4.7×10^{-6} m/sec	23	1.2	Radial diverging		³H (pulse)	One-dimensional uniform flow solution	91	11.6	III
Freyberg (1986), Borden	Glaciofluvial sand	9	7.2×10^{-5} m/sec	33 (total)	0.09	Ambient	Three-dimensional	Br⁻, Cl⁻ (pulse)	Spatial moments	90	0.43/0.039	I

Reference / Location	Aquifer material		K			Flow configuration	Tracer	Dimensionality	Method / Model			Class
Fried and Ungemach (1971), Rhine aquifer	Sand, gravel, and cobbles	12			9.6	Radial diverging	Cl^- (pulse)		One-dimensional radial flow numerical model	6	11	III
Fried (1975), Rhine aquifer (salt mines) southern Alsace, France	Alluvial; mixture of sand, gravel, and pebbles with clay lenses	125	10^{-3} m/sec			Ambient	Cl^- (contamination)	Three-dimensional	Two-dimensional numerical model	800	15/1	III
Fried (1975), Lyons, France (sanitary landfill)	Alluvial, with sand and gravel and slightly stratified clay lenses	20			5.0	Ambient	EC (contamination)	Two-dimensional	Two-dimensional numerical model	600–1000	12/4	III
Garabedian et al. (1988) Cape Cod, Massachusetts	Medium to coarse sand with some gravel overlying silty sand and till	70	1.3×10^{-3} m/sec	39	0.43	Ambient	Br^- (pulse)	Three-dimensional	Spatial moments	250	0.96/0.018/0.0015	I
Gelhar (1982), Hanford, Washington	Brecciated basalt interflow zone					Two-well without recirculation	^{131}I (pulse)	Two-dimensional	One-dimensional nonuniform flow solution along streamlines (Gelhar, 1982)	17.1	0.60	I
Goblet (1982), site B, France	Fractured granite	50	10^{-5}–10^{-7} m/sec	84		Radial converging	RhWt, SrCl (pulse)	Two-dimensional	One-dimensional uniform flow solution including borehole flushing effects	17	2	III

(Continued)

TABLE 7-1 (Continued)

Reference and Site Name	Aquifer Material	Average Aquifer Thickness, b (m)	Horizontal Hydraulic Conductivity K_h (m/sec) or Transmissivity T (m²/sec)	Effective Porosity, φ_e (%)	Velocity, U (m/day)	Flow Configuration	Monitoring	Tracer and Input[a]	Method of Data Interpretation	Scale of Test s (m)	Dispersivity $A_L/A_{TH}/A_{TV}$[b] (m)	Classification of Reliability of $A_L/A_{TH}/A_{TV}$ (I, II, III)[c]
Grove (1977), NRTS, Idaho	Basaltic lava and sediments	76	1.4×10^{-1} – 1.4×10^{1} m²/sec	10		Ambient	Two-dimensional	Cl⁻ (contamination)	Two-dimensional numerical model	20,000	91/91	III
Grove and Beetem (1971), Eddy County (near Carlsbad), New Mexico	Fractured dolomite	12		12	3.5	Two-well recirculating	Two-dimensional	³H (step)	One-dimensional quasi-uniform flow solution (Grove and Beetem, 1971)	55	38.1	III
Gupta et al. (1975), Sutter Basin, California	Sandstone, shale, sand, and alluvial sediments					Ambient		Cl⁻ (environmental)	Three-dimensional numerical model	50,000	80–200/8–20	III
Halevy and Nir (1962) and Lenda and Zuber (1970), Nahal Oren, Israel	Dolomite	100		3.4	4.0	Radial converging	Two-dimensional	⁶⁰Co (pulse)	One-dimensional uniform flow solution	250	6	II
Harpaz (1965), southern coastal plain, Israel	Sandstone with silt and clay layers	90			14	Radial diverging	Two-dimensional	Cl⁻ (step)	One-dimensional radial flow solution	28	0.1–1.0	II
Helweg and Labadie (1977), Bonsall subbasin, California						Ambient		TDS (contamination)	Two-dimensional numerical model	14,000	30.5/9.1	III
Hoehn (1983), lower	Layered gravel and silty sand	25	9.2×10^{-4} – 6.6×10^{-3} m/sec		3.4 1.8	Ambient	Two-dimensional	uranine (pulse)	One-dimensional	4.4 4.4	0.1 0.01	III

Site/Reference	Aquifer material	Scale (m)	Velocity / Conductivity	Value	Flow condition	Dimensionality	Tracer	Method/Solution	Longitudinal dispersivity	Transverse dispersivity	Class
Glatt Valley, Switzerland				1.2, 8.6, 4.1, 1.7				uniform flow solution for layers	4.4, 10.4, 10.4, 10.4	0.2, 0.3, 0.04, 0.7	III
Hoehn and Santschi (1987), lower Glatt valley, Switzerland	Layered gravel and silty sand	27.5	8.1×10⁻⁵ – 6.6×10⁻³ m/sec	1.5, 3.2, 5.6, 3.9, 3.2	Ambient	Two-dimensional / Two dimensional / Two-dim.	uranine (pulse) / ³H (environmental)	Temporal moments / Temporal moments	4.4, 10.4, 100, 110, 500	1.1, 1.2, 6.7, 10.0, 58.0	II, II, III, III, III
Huyakorn et al. (1986), Mobile, Alabama	Layered medium sand	21.6		0.35	Two-well without recirculation	Two-dimensional	Br⁻ (pulse)	Two-dimensional numerical model	38.3	4.0	I
Iris (1980), Campuget (Gard), France	alluvial deposits	9	3.6×10⁻³ m²/sec	0.05	Radial diverging	Three-dimensional	heat (pulse)	Two-dimensional radial numerical model	40	3/1.5	II
Ivanovitch and Smith (1978), Dorset, England	Fractured chalk		2.2×10⁻³ m/sec (fast pulse)	57.6, 0.5	Radial converging		⁸²Br (pulse)	One-dimensional uniform flow solution	8	3.1	III
Chalk			3.6×10⁻⁴ m/sec (slow pulse)	9.6, 2.3	Radial converging		⁸²Br (pulse)	One-dimensional uniform flow solution	8	1.0	III
Kies (1981), New Mexico State University, Las Cruces	Fluvial sands		9.55×10⁻⁵ m/sec	42 (total)	Ambient	Two-dimensional	NO₃⁻ (pulse)	Two-dimensional uniform flow solution	25	1.6/0.76	III
Klotz et al. (1980), Dormach, Germany	Fluvioglacial gravels	14		20	Radial converging	Two-dimensional	⁸²Br, uranine (pulse)	One-dimensional uniform flow solution	10	5, 1.9	II
Konikow (1976), Rocky Mountain Arsenal	Alluvium			30	Ambient		Cl⁻ (contamination)	Two-dimensional numerical model	13,000	30.5	III

(Continued)

TABLE 7-1 (Continued)

Reference and Site Name	Aquifer Material	Average Aquifer Thickness, b (m)	Horizontal Hydraulic Conductivity K_h (m/sec) or Transmissivity T (m²/sec)	Effective Porosity, φ_e (%)	Velocity, U (m/day)	Flow Configuration	Monitoring	Tracer and Input[a]	Method of Data Interpretation	Scale of Test s (m)	Dispersivity $A_L/A_{TH}/A_{TV}$[b] (m)	Classification of Reliability of $A_L/A_{TH}/A_{TV}$ (I, II, III)[c]
Konikow and Bredehoeft (1974), Arkansas River valley (at La Junta, Colorado)	Alluvium, inhomogeneous clay, silt, sand and gravel	2.5	2.4×10^{-4} – 4.2×10^{-3} m/sec	20		Ambient	Two-dimensional	dissolved solids (contamination)	Two-dimensional numerical model	18,000	30.5/9.1	III
Kreft et al. (1974), Poland	Sand	2.5	3.1×10^{-5} – 1.5×10^{-4} m/sec; 1.2×10^{-4} m²/sec	24	29	Radial converging	Two-dimensional	^{131}I (pulse)	One-dimensional uniform flow solution	5-6	0.18	II
Kreft et al. (1974), Zn–Pb deposits, Poland	Fractured dolomite	57	2.5×10^{-4} – 4.7×10^{-4} m/sec	2.4	7.5 100	Radial converging		^{131}I (pulse)	One-dimensional uniform flow solution	22	44-110	II
	Fractured dolomite	48	2.5×10^{-4} – 4.7×10^{-4} m/sec	2.4	60.1 22.7	Radial converging		^{131}I (pulse)	One-dimensional uniform flow solution	21.3	2.1	II
Kreft et al. (1974), sulfur deposits, Poland	Limestone	7	1.1×10^{-4} m/sec	12.3	10 10.8	Radial converging		^{58}Co (pulse)	One-dimensional uniform flow solution	27	2.7-27	II
	Limestone	7	1.1×10^{-4} m/sec	12.3	8.6	Radial converging		^{58}Co (pulse)	One-dimensional uniform flow solution	41.5	20.8	II
Lau et al. (1957), University of	Sand and gravel with clay lenses	1.5	9×10^{-4} m/s	30	7	Radial diverging		Cl⁻ (step)	One-dimensional radial	19	2-3	I

Reference / Location	Material		Conductivity			Flow	Dimensionality	Tracer	Model		Dispersivity	Class
California, Berkeley									numerical model			
Lee et al. (1980), Perch Lake, Ontario, (lake bed)	sand		3.2×10^{-5} m/sec		0.14	Ambient	Three-dimensional	Cl^- (pulse)	One-dimensional uniform flow solution	≤ 6	0.012	II
Leland and Hillel (1981), Amherst, Massachusetts	Fine sand and glacial till	0.75	$2.4 -$ 3×10^{-5} m/sec	40	0.3-0.6	Ambient	Three-dimensional	Cl^- (pulse)	Two-dimensional uniform flow solution	4	0.05-0.07	III
Mercado (1966), Yavne region, Israel	Sand and sandstone with some silt and clay	80	$2.1 \times 10^{-8} -$ 2.4×10^{-8} m²/sec	23.3	0.84-3.4	Radial diverging/ converging	Three-dimensional	^{60}Co, Cl^- (step)	One-dimensional radial flow solution	≤ 115 (observation wells)	0.5-1.5 (injection phase)	I
Meyer et al. (1981); Koeberg Nuclear Power Station, South Africa	Sand			20	0.12	Ambient	Three-dimensional	^{131}I (pulse)	One-dimensional uniform flow solution for layers	2-8	0.01, 0.03, 0.01, 0.05 for layers; 0.42 for depth average	III
Molinari and Peaudecerf (1977) and Sauty (1977), Bonnaud, France	Sand	3	$8.3 \times 10^{-4} -$ 1.1×10^{-3} m²/sec		2.7	Forced uniform	Two-dimensional	I^-	Two-dimensional uniform flow solution	13	0.79	I
					1.0			^3H		13	1.27	I
					2.4			^{131}I		13	0.72	I
					1.0			^{131}I		26	2.23	I
					2.0			^{131}I (pulse)		33.2	1.94/0.11	I
					2.0			^{131}I (pulse)		32.5	2.73/0.11	I
Moltyaner and Killey (1988a, b), Twin Lake aquifer (Chalk River)	Fluvial sand			40.8 (total)	1.2	Ambient	Three-dimensional		Two-dimensional uniform flow solution	40	0.06– 0.16/.../0.0006– 0.002	II
Naymik and Barcelona (1981), Meredosia, Illinois (Morgan County)	Unconsolidated sand and gravel	27	$2.2 \times 10^{-2} -$ 4.3×10^{-2} m²/sec			Ambient	Two-dimensional	NH_3 (contamination)	Two-dimensional numerical model	16.4	2.13-3.35/0.61– 0.915	III

(Continued)

TABLE 7-1 (Continued)

Reference and Site Name	Aquifer Material	Average Aquifer Thickness, b (m)	Horizontal Hydraulic Conductivity K_h (m/sec) or Transmissivity T (m²/sec)	Effective Porosity, φ_e (%)	Velocity, U (m/day)	Flow Configuration	Monitoring	Tracer and Input[a]	Method of Data Interpretation	Scale of Test s (m)	Dispersivity $A_L/A_{TH}/A_{TV}$[b] (m)	Classification of Reliability of $A_L/A_{TH}/A_{TV}$ (I, II, III)[c]
New Zealand Ministry of Works and Development (1977) Heretaunga aquifer, New Zealand:												
Roys Hill site	Gravel with cobbles	100	0.29 m²/sec	22	150–200	Ambient	Three-dimensional	^{131}I, RhWt, ^{82}Br, Cl⁻, E. Coli (pulse)	Three-dimensional uniform flow solution	54–59	1.4–11.5/0.1–3.3/0.04–0.10	II
Flaxmere site 2	Alluvium (gravels)	120	0.37 m²/sec	22	20–25	Ambient	Three-dimensional	RhWt, ^{82}Br (pulse)	Three-dimensional uniform flow solution	25	0.3–1.5/.../0.06	II
Hastings City rubbish dump	alluvium (gravels)		0.14, 0.35 m²/sec		20	Ambient	Three-dimensional	Cl⁻ (contamination)	Three-dimensional uniform flow solution	290	41/10/0.07	III
Oakes and Edworthy (1977), Clipstone, United Kingdom	Sandstone	44	2.4×10^{-6} – 1.4×10^{-4} m/sec	32–48	5.6, 4.0 9.6	Radial diverging	Two-dimensional	^{82}Br (pulse)	Radial flow numerical model	6	0.16, 0.38	II
										3	0.31	II
					2.4, 3.6	Radial converging	Two-dimensional	Cl⁻, I⁻ (pulse)		6	0.6	II
										3	0.6	II
Papadopulos and Larson (1978),	Medium to fine sand interspersed	21	5×10^{-4} m/sec (horizontal) and 5.1×10^{-5} m/sec	25	0.05	Radial diverging	Two-dimensional	heat (step)	Two-dimensional numerical	57.3	1.5	II

Mobile, Alabama	with clay and silt		(vertical)						model			
Pickens and Grisak (1981a), Chalk River	Sand	8.5	2×10⁻⁵ – 2×10⁻⁴ m/sec	38	0.15	Two-well recirculating	Three-dimensional	^{51}Cr (step)	One-dimensional quasi-uniform flow solution	8	0.5	III
	Sand	8.5	2×10⁻⁵ – 2×10⁻⁴ m/sec	38	0.15	Radial diverging/converging	Three-dimensional	^{131}I (step)	One-dimensional radial flow solution	3	0.03	III
Pinder (1973), Long Island	Glacial outwash	43	7.5×10⁻⁴ m/s	35	0.43	Regional	Three-dimensional	Cr^{+6} (contamination)	Two-dimensional numerical model	1,000	21.3/4.2	III
Rabinovitz and Gross (1972), Roswell Basin, New Mexico	Fractured limestone	61	1.1×10⁻² – 2.9×10⁻¹ m²/s	1	11–21	Regional	Two-dimensional	^{3}H (environmental)	One-dimensional uniform flow solution	32,000	20–23	III
Rajaram and Gelhar (1991), Borden	Glaciofluvial sand	9	7.2×10⁻⁵ m/sec	33 (total)	0.09	Ambient	Three-dimensional	Br⁻, Cl⁻ (pulse)	Spatial moments	90	0.50/0.05/0.0022	I
Roberts et al. (1981), Palo Alto bay lands	Sand, gravel, and silt	2	1.25×10⁻³ m²/sec (lower aquifer); 5.0×10⁻⁴ m²/sec (upper aquifer)	25	15.5	Radial diverging	Two-dimensional	Cl⁻ (step)	One-dimensional uniform flow solution	11	5	III
					12.0					20	2	III
					3.5					40	8	III
					25.6					16	4	III
					7.9					43	11	III
Robertson (1974) and Robertson and Barraclough (1973), NRTS, Idaho	Basaltic lava and sediments	76	1.4×10⁻¹ – 1.4×10¹ m²/sec	10	1.5–8	Regional	Two-dimensional	Cl⁻ (contamination)	Two-dimensional numerical model	20,000	910/1370	III

(Continued)

TABLE 7-1 (Continued)

Reference and Site Name	Aquifer Material	Average Aquifer Thickness, b (m)	Horizontal Hydraulic Conductivity K_h (m/sec) or Transmissivity T (m²/sec)	Effective Porosity, φ_e (%)	Velocity, U (m/day)	Flow Configuration	Monitoring	Tracer and Input[a]	Method of Data Interpretation	Scale of Test s (m)	Dispersivity $A_L/A_{TH}/A_{TV}$[b] (m)	Classification of Reliability of $A_L/A_{TH}/A_{TV}$ (I, II, III)[c]
Robson (1974, 1978), Barstow, California	Alluvial sediments	27	2.1×10^{-4} – 1×10^{-2} m²/sec	40	-	Two-well recirculating	Two-dimensional	Cl⁻ (step)	One-dimensional quasi-uniform flow solution	6.4	15.2	III
				40	3	Regional	Two-dimensional	TDS (contamination)	Two-dimensional numerical model	10,000	61/18	III
Robson (1978), Barstow, California	Alluvial sediments	30.5	5×10^{-4} m/s	40		Regional	Three-dimensional	TDS (contamination)	Two-dimensional numerical model (vertical section)	3,200	61/.../0.2	III
Rousselot et al. (1977), Byles–Saint Vulbas near Lyon, France	Clay, sand, and gravel	12	6.5×10^{-3} – 1.5×10^{-2} m/sec	14	18	Radial converging	Two-dimensional	I⁻ (pulse)	One-dimensional uniform flow solution for layers	9.3	6.9	II
				2.1–18	11.5, 3.8					5.3	0.3, 0.7	III
				1.8–5.9	46.7, 16					10.7	0.46, 1.1	III
				11–24	24					7.1	0.37	II
Sauty (1977), Corbas, France	Sand and gravel	12			125, 100	Radial converging	Two-dimensional	I⁻ (pulse)	One-dimensional uniform flow solution for layers	25	11, 1.25	III
					15.5, 78					50	25, 6.25	III
					6.9					150	12.5	II
Sauty et al (1978), Bonnaud, France	Sand	3	8.3×10^{-4} – 1.1×10^{-3} m²/sec			Radial diverging	Two-dimensional	heat (step)	One-dimensional radial flow	13	1.0	II

Reference	Material		Hydraulic conductivity			Flow	Dimensionality	Tracer	Model			Class
Segol and Pinder (1976), Cutler area, Biscayne Bay aquifer, Florida	Fractured limestone and calcareous sandstone	30.5	0.45×10^{-2} (horizontal) and 0.09×10^{-4} m/sec (vertical)	25	20	Ambient	Three-dimensional	Cl⁻ (environmental)	Two-dimensional numerical model	490	*6.7/.../0.67*	III
Sudicky et al. (1983), Borden	Glaciofluvial sand	7–27	4.8×10^{-5} – 7.6×10^{-5} m/sec	38	0.07–0.25	Ambient	Three-dimensional	Cl⁻ (pulse)	Three-dimensional uniform flow solution	11	0.08/0.03	II
Sykes et al. (1982, 1983), Borden	Sand		5.8 – 7.2×10^{-5} m/sec	35		Ambient	Three-dimensional	Cl⁻ (pulse)	Two-dimensional numerical model	0.75	0.01/0.005	II
										700	*7.6/.../0.31*	III
Sykes et al. (1983), Mobile, Alabama	Sand, silt, and clay	21	5×10^{-4} (horizontal) and 2.5×10^{-5} m/sec (vertical)	25	0.05	Radial diverging	Three-dimensional	heat (step)	Three-dimensional numerical model	57.3	0.76/.../0.15	II
Vaccaro and Bolke (1983), Spokane aquifer, Washington and Idaho	Glaciofluvial sand and gravel	152	9×10^{-5} – 6.5 m²/sec	7–40	0.003–2.8	Ambient		Cl⁻ (contamination)	Two-dimensional numerical model	43,400	91.4/27.4	III
Valocchi et al. (1981), Palo Alto bay lands	Sand, gravel, and silt	2	1.25×10^{-3} m²/sec (lower aquifer); 5.0×10^{-4} m²/sec (upper aquifer)	25	27	Radial diverging		Cl⁻ (step)	Two-dimensional numerical model	16	1.0/0.1	I
Walter (1983), WIPP	Fractured dolomite	7	8.0×10^{-5} m²/sec	0.7 and 11 (along separate paths)	4.7, 2.4	Radial converging	Two-dimensional	MTFMB, PFB, MFB, para-FB (pulse)	One-dimensional uniform flow solution	30	10–15	III
Webster et al. (1970), Savannah River Plant, South Carolina	Crystalline, fractured schist and gneiss	76	3.6×10^{-7} m/sec	1.3		Two-well recirculating	Two-dimensional	⁸⁵Sr	One-dimensional quasi-uniform flow solution	538	134	III
				21.4				⁸⁵Br (pulse)				

(Continued)

TABLE 7-1 (Continued)

Reference and Site Name	Aquifer Material	Average Aquifer Thickness, b (m)	Horizontal Hydraulic Conductivity K_h (m/sec) or Transmissivity T (m²/sec)	Effective Porosity, φ_e (%)	Velocity, U (m/day)	Flow Configuration	Monitoring	Tracer and Input[a]	Method of Data Interpretation	Scale of Test s (m)	Dispersivity $A_L/A_{TH}/A_{TV}$[b] (m)	Classification of Reliability of $A_L/A_{TH}/A_{TV}$ (I, II, III)[c]
Werner et al. (1983), Hydrothermal Test Site, Aefligen, Switzerland	Gravel	20	6×10^{-3} m/sec	17	9.1	Ambient	Three-dimensional	heat (step)	One-dimensional numerical model	700	130–234	III
										37	131	III
										105	208	III
										200	234	III
Wiebenga et al. (1967) and Lenda and Zuber (1970), Burdekin Delta, Australia	Sand and gravel	6.1	5.5×10^{-3} m/s	32	29	Radial converging		^{131}I, ^{3}H (pulse)	One-dimensional uniform flow solution	18.3	0.26	II
Wilson (1971) and Robson (1974), Tucson, Arizona	Unconsolidated gravel, sand, and silt		5.75×10^{-3} m²/sec	38		Two-well without recirculation	Three-dimensional	Cl⁻ (step)	One-dimensional quasi-uniform flow solution	79.2	15.2	III
						Radial diverging	Two-dimensional	Cl⁻ (step)	One-dimensional radial flow solution	4.6	0.55	III

Reference	Material					Type	Tracer	Method			Class
Wood (1981), Aquia Formation, southern Maryland	Sand	1,000	2.9×10^{-4} – 8.7×10^{-4} m^2/sec	35	0.0003–0.0007	Ambient	Na$^+$ (environmental)	One-dimensional uniform flow solution	10^5	5,600–40,000	III
Wood and Ehrlich (1978) and Bassett et al. (1980), Lubbock, Texas	Sand and gravel	17	3.2×10^{-3} – 4.4×10^{-3} m^2/sec	78	Radial converging	Two-dimensional	I$^-$ (pulse)	One-dimensional radial flow solution	1.52	0.015	II

[a] TDS denotes total dissolved solids; EC, electrical conductivity; PFB, pentafluorobenzoate; MTFMB, metatrifluoromethylbenzoate; MFB, metafluorobenzoate; Para-FB, parafluorobenzoate; RhWT, rhodamine-WT dye; and SCN, thiocyanate.

[b] A_L denotes longitudinal dispersivity; A_{TH}, transverse horizontal dispersivity; and A_{TV}, transverse vertical dispersivity. Reported values for A_L, A_{TH}, and A_{TV} are separated by slashes. Absence of slashes means that values were reported for A_L only. A comma or a dash separating entries means that multiple values or a range of values, respectively, were reported for a particular dispersivity component.

[c] For description of classification criteria, see Section 7.4.3.

TABLE 7-2
Number of Longitudinal Dispersivity, A_L, values for Different Types of Tests and Media

Media Type	Tracer Type			
	Artificial	Contamination	Environmental	Total
Porous	68	14	6	88
Fractured	15	1	2	18
Total	83	15	8	106

Source: Adapted from Gelhar et al., *Water Resour. Res.*, 28, 1955–1974, 1992.

the range at that site. Aquifer thicknesses reported ranged from less than 1 m to 1 km. Both aquifer material and thickness were of interest since it was previously reported that field-scale dispersivity was independent of these aquifer characteristics (Lallemand-Barres and Peaudecerf, 1978). The aquifer materials at the various sites were primarily unconsolidated sediments, with the exception of a few occurrences of sandstone and fractured rock. Hydraulic conductivity and transmissivity values show the range reported for the site. Gelhar and others reported that values for effective porosity, φ_e, vary from 0.5% (for fractured media) to 60% (for porous media). When a value was reported as "porosity" in a report for a site, Gelhar et al. (1985, 1992) interpreted the reported porosity value as the effective porosity value used in the convection–dispersion equation (see Section 2.3.2 for more details on total and effective porosities). The reported studies were mostly based on a uniform ground water velocity approach, and ground water velocities, U, were calculated from Eq. (2-40) using the corresponding specific discharge, q, and effective porosity, φ_e, values. The velocities ranged from 0.0003 to 200 m/days.

7.4.2.3 Types of Subsurface Solute Transport Tests

Columns 7 and 10 from the left in Table 7-1 summarize the method used for dispersivity determination at each site. Columns 7 to 9 describe experimental conditions (flow configuration, monitoring, tracer, and input); Column 10 presents methods of data interpretation. Dispersivity values were determined by analyzing data from one of two types of subsurface solute transport events: 1. Large-scale, uncontrolled contamination events (naturally occurring or human-induced); and 2. Controlled field tracer tests. The main differences between these two types of events is that for controlled field tracer tests both the quantity and duration of solute input are known. These are described below.

1. **Large-scale, naturally occurring or human-induced uncontrolled contamination events** If the source history of a contamination event for an aquifer is unknown, and the transport of its contaminants is caused by the ambient flow of ground water by which the solute plumes potentially extend over regional scales that are measured by hundreds of meters, the event is called "uncontrolled contamination event." Examples of naturally occurring events include tritium in ground water from recharge containing atmospheric bomb tritium, seawater intrusion, and mineral dissolution. Examples of human-induced contamination events include leaks and spills to ground water from landfills, industrial plants, storage tanks, surface impoundments, infiltration basins, and others. Values of dispersivities for uncontrolled events are commonly determined by fitting a one-, two-, or three-dimensional solute transport model, which are based on one of the differential equations given by Eqs. (2-79)–(2-83), to historical data. The common practice is that values of dispersivities are altered until the model output matches with an acceptable

approximation to historical spatially and temporally measured solute concentration distributions (Gelhar et al., 1985, 1992).

2. **Controlled Field Tracer Tests** Controlled field tracer tests are conducted under ambient ground water flow conditions, and are also known as "natural gradient tests," or under induced stress conditions including pumping, injection, or recharge. Induced stress conditions include radial, two-well, and forced uniform flow. These are described below (Gelhar et al., 1985, 1992). A discussion of the advantages and disadvantages of different types of tracer tests is presented by Welty and Gelhar (1989).

Radial flow tracer tests: In radial flow tracer tests, a pulse or step input of tracer is injected at an injection (or recharge) well and the tracer concentration vs. time is recorded at an observation well (diverging radial flow test), or the tracer is injected at an observation well and the tracer concentration vs. time is recorded at a distant pumping well (converging radial flow test).

Two-well tests: In a two-well test system, both an injection well and a distant pumping well are operational. The tracer is injected at the injection well and the tracer concentration vs. time is observed at the pumping well. The water that contains the tracer is often recirculated from the pumping well to the injection well.

Forced uniform flow tests: In a forced uniform flow test, a uniform flow field is generated between two lines of equally spaced wells, one line injecting and the other line pumping, with both screened to the full depth of the aquifer.

7.4.2.4 Dimensionality and Monitoring

To obtain dispersivity data, two- or three-dimensional monitoring was conducted at a site. This information is included in column 8 of Table 7-1. Two-dimensional monitoring refers to depth-averaged monitoring, whereas three-dimensional monitoring implies point samples with depth. This information is included in Table 7-1 because vertical mixing in an observation well affects the concentration of tracer in a water sample. When a tracer is not injected over the full aquifer thickness, vertically mixed samples underestimate the tracer concentration, and this results in overestimated longitudinal dispersivity (Meyer et al., 1981; Pickens and Grisak, 1981a). This is due to the fact that the tracer occupies only a portion of the vertical thickness. When a sample from the entire thickness is taken, the true tracer concentration will be diluted in the well with tracer-free water. If an interpretation is made based on the diluted concentration, an overestimated dispersivity will be obtained. For many of the sites listed in Table 7-1, there was no information as to whether point or fully mixed sampling was performed. Gelhar et al. (1992) also points out that from the examination of the cases where three-dimensional measurements of solute concentrations were conducted, it can clearly be observed that vertical mixing of the tracer, as it travels through the aquifer, is often very minimal (LeBlanc, 1982; Sudicky et al., 1983; Freyberg, 1986; Garabedian et al., 1988, 1991).

7.4.2.5 Correlation of Longitudinal Dispersivity with Other Quantities

Gelhar et al. (1985, 1992) determined the correlation between the longitudinal dispersivity α_L, and the scale, s, aquifer horizontal hydraulic conductivity, K_h, and aquifer effective porosity, φ_e, respectively, and the results are presented below.

Longitudinal Dispersivity, A_L, vs. Scale, s
The "scale of test" is presented in Column 11 of Table 7-1. This is the distance traveled by a tracer from the source under ambient conditions, or the distance between injection and observation wells in the case of an induced flow configuration. The values of dispersivities (A_L, A_{TH}, and A_{TV}) at the indicated scales are given in column 12 in Table 7-1. As can be seen from Table 7-1, for all the sites, the longitudinal dispersivity A_L values are available, but transverse dispersivity, A_{TH}, and vertical dispersivity, A_{TV}, values are available, for only a limited number of sites. The reported values for A_L

and A_{TH} are 24 and 9, respectively. In Table 7-1, data from the 59 sites generated 106 A_L values, and their graphical form is presented in Figure 7-1. Gelhar et al. (1992) plotted the arithmetic average in cases where a range was reported either for the scale or dispersivity in Table 7-1. In Figure 7-1, information regarding the medium (fractured medium or porous medium) is also included; open symbols (18 values) are for fractured media and solid symbols (88 values) are for porous media. In almost all cases, the A_{TH} values were found to be 1 to 2 orders of magnitude less than the A_L values, and A_{TV} values 2 to 3 orders of magnitude less than A_L values.

Longitudinal Dispersivity, A_L, vs. Horizontal Hydraulic Conductivity, K_h, and Effective Porosity, φ_e

In Table 7-1, horizontal hydraulic conductivity, K_h, or transmissivity, T, values and qualitative information regarding aquifer materials are also included. Based on these data, longitudinal dispersivity, A_L, vs. horizontal hydraulic conductivity, K_h, values and longitudinal dispersivity, A_L, vs. effective porosity, φ_e, values are shown in Figure 7-2 and Figure 7-3, respectively. These figures clearly indicate that there is no relationship between A_L and these two aquifer properties.

7.4.3 DATA EVALUATION AND METHODS FOR ESTIMATION OF DISPERSIVITIES

Gelhar et al. (1985, 1992) have provided some charts and practical guidelines for the estimation of dispersivities. Here, the word "estimation" is used deliberately because of the fact that these approaches bear empirical features and also because there are controversial views about these approaches. The same data were also interpreted and analyzed by Neuman (1990) by using regression methods, which resulted in the creation of some empirical formulas for estimating dispersivities. In this section, the estimation methods for dispersivities will be presented with examples.

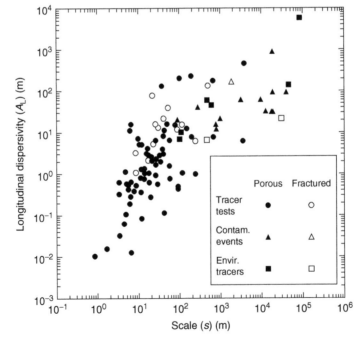

FIGURE 7-1 Longitudinal dispersivity, A_L, vs. scale of observation, s, from 59 field sites characterized by widely differing geologic materials. (Adapted from Gelhar et al., *Water Resour. Res.*, 28, 1955–1974, 1992.)

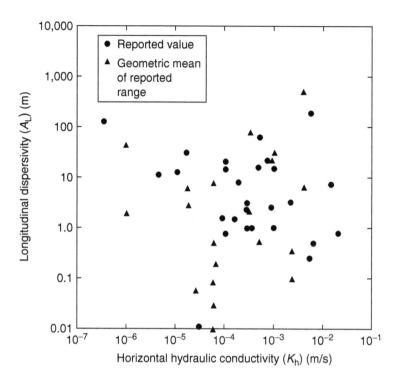

FIGURE 7-2 Longitudinal dispersivity, A_L, vs. horizontal hydraulic conductivity, K_h, data. (Adapted from Gelhar et al., EPRI Report EA-4190, Elect. Power Res. Inst., California, August 1985.)

7.4.3.1 Evaluation of Data for Dispersivities

Gelhar et al. (1985, 1992) evaluated the dispersivity data given in Table 7-1 and Figure 7-1, both quantitatively and qualitatively for the generation of guidelines or "rule of thumb" type approaches for the estimation of dispersivities for a given site. As mentioned previously, there are still controversial views among different investigators regarding the interpretation of the aforementioned data and the deduction of some general approaches from them for the estimation of dispersivities. For example, Gelhar et al. did not provide curve-fitting type approaches to these data, as opposed to the approaches furnished by some other investigators (Neuman, 1990). Results of Gelhar et al. (1985, 1992) are presented in the following paragraphs.

As can be seen from Figure 7-1, the longitudinal dispersivity, A_L, increases with the length of scale, s. And it can be seen from Table 7-1 that field-observed A_L values range from 0.01 to approximately 5500 m at scales of 0.75 m to 100 km. Also, the A_L values for porous and fractured media tend to scatter over a similar range. However, at a smaller scale, fractured media seem to show higher A_L values. Gelhar and others noted a number of problems with the data, listed in Table 7-1 and their interpretation. The typical problems found in the studies reported in Table 7-1 include the following: (1) data analysis not matched with flow configuration; (2) mass input history unknown; (3) nonconservative effects of tracer not accounted for; (4) dimensionality of the monitoring not matched to the dimensionality of the analysis; and (5) assumption of distinct geologic layers in analysis when their actual presence was not documented. Based on these problems, Gelhar and others decided to rate the data listed in Table 7-1 as high (I), medium (II), or low (III) reliability according to the criteria presented below. Table 7-3 lists the criteria used to designate either high- or low-reliability data. No specific criteria were defined for the intermediate classification. The authors' classification of dispersivity is not intended to be a judgment on the quality of the study as a whole, but rather to provide some criteria by which to screen the large number of dispersivity values.

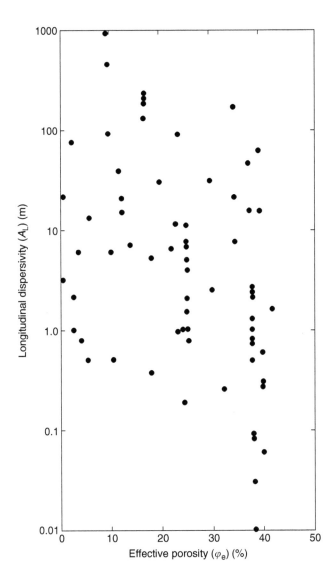

FIGURE 7-3 Longitudinal dispersivity, A_L, vs. porosity, φ_e, data. (Adopted from Gelhar et al., EPRI Report EA-4190, Elect. Power Res. Inst., California, August 1985.)

7.4.3.2 Criteria for High-Reliability Dispersivities Data

For a reported dispersivity value to be classified as high reliability, Gelhar et al. (1985, 1992),mention the following criteria:

1. The tracer test must be either an ambient flow with known input, a diverging radial flow, or a two-well pulse test without recirculation. These three test configurations produce breakthrough curves that are sensitive to dispersion coefficient. Therefore, they generally work well in field applications.
2. The tracer input must be well defined. Both the input concentration and the temporal distribution of the input concentrations must be known from measurements. If this is not the case, the input is another unknown in the solution of the convection–dispersion equation. Therefore, the resulting dispersivity values have less confidence.

TABLE 7-3
Criteria Used to Classify the Reliability of the Reported Dispersivity Values

Classification	Criteria
High reliability	Tracer test was either ambient flow, radial divergent flow, or two-well instantaneous pulse test (without recirculation)
	Tracer input was well defined
	Tracer was conservative
	Spatial dimensionality of the tracer concentration measurements was appropriate
	Analysis of the tracer concentration data was appropriate
Low reliability	Two-well recirculating test with step input was used
	Single-well injection-withdrawal test with tracer monitoring at the single well was used
	Tracer input was not clearly defined
	Tracer breakthrough curve was assumed to be the superposition of breakthrough curves in separate layers
	Measurement of tracer concentration in space was inadequate
	Equation used to obtain dispersivity was not appropriate for the data collected

Source: Adapted from Gelhar et al., EPRI Report EA-4190, Elut. Power. Res. Inst., California, August 1985; *Water. Resour. Res.*, 28, 1955-1974, 1992.

3. The tracer must be conservative. Tracers such as Cl^-, I^-, Br^-, and tritium are considered to be conservative. If a tracer is not conservative (or it is reactive), the governing equations become complicated such that additional parameters must be estimated. Therefore, the resulting dispersivity values have less confidence.
4. The dimensionality of the tracer concentration measurements must be appropriate. In reality, the distribution of a tracer in an aquifer is three-dimensional. If the monitoring of a tracer is three-dimensional, the corresponding dispersivities can be judged to have high reliability. However, if the aquifer tracer had been injected and measured along the full depth of the aquifer, two-dimensional monitoring is acceptable. In all other cases, such as the dimension of the measurement was either not reported or two-dimensional measurements were used instead of three-dimensional measurements, Gelhar and others have judged the derived dispersivity values of being lower reliability.
5. The analysis of the concentration data must be appropriate. Appropriate data interpretation can be done by one of the following methods, as appropriate: 1. breakthrough curve analysis, which is usually applied to uniform ambient flow tests and radial flow tests (e.g., Sauty, 1980); 2. method of spatial moments, applied to uniform ambient flow tests (e.g., Freyberg, 1986; Garabedian et al., 1991); and 3 numerical events applied to contamination events (e.g., Pinder, 1973; Konikow and Bredehoeft, 1974).

Some Major Problems Regarding Dispersivity Determination
There are some major problems or difficulties regarding the determination of dispersivities. These problems are presented below (Gelhar et al., 1985, 1992).

1. *Constant dispersivity assumption*: A common difficulty with the interpretation of concentration data using breakthrough curve matching to determine the longitudinal

dispersivity, A_L, is the assumption that A_L is constant. The field data assembled in the review by Gelhar et al. (1985, 1992) suggest that this assumption is not valid, at least for small-scale tests (tens of meters). At larger scales (hundreds of meters) an asymptotic constant value of A_L is predicted by some theories (Mercado, 1967; Gelhar et al., 1979; Matheron and de Marsily, 1980; Gelhar and Axness, 1983). However, at most sites, the displacement distance after which A_L is constant is not known. Data for which no *a priori* assumptions were made regarding A_L were considered to be highly reliable.

2. *Analysis of three-dimensional plumes with one- or two-dimensional solutions*: A second major problem with many of the analyses reviewed by Gelhar and others was that o one- or two-dimensional solution to the convection–dispersion equation was used when the distribution of the plume under consideration was three-dimensional in nature. If the dimensionality of the solute plume, solute measurements, and the data analyses are consistent, the dispersivities are considered to be highly reliable.

7.4.3.3 Criteria for Low-Reliability Dispersivities Data

If one of the following criteria is met, a reported dispersivity value is classified as low reliability (Gelhar et al., 1985, 1992):

1. The two-well recirculating test with a step input was used. A potential problem with this test configuration is that the breakthrough curve is not strongly influenced by dispersion except in the early stages, when concentrations are low (Welty and Gelhar, 1989). Consequently, the two-well test with a step input is generally insensitive to dispersion. For this reason, tests based on this approach are considered to produce low-reliability dispersivities data.
2. The single-well injection-withdrawal test was used with tracer monitoring at the pumping well. In this case, water is pumped into and out of one well and, therefore, is a small-scale test. Since observations are made at the production well, the dispersion process observed is different from that of unidirectional flow. The problem stems from the fact that macrodispersion near the injection well results from velocity differences, which are caused by the variation of hydraulic conductivity due to layered heterogeneity. As a result, the tracer travels at different velocities in layers as it radiates outward. It will also travel with the same velocity pattern as goes back to the production well. This means that the mixing process is partially reversible and that the dispersivity might be underestimated compared to that for unidirectional flow. Heller (1972) conducted experiments demonstrating the reversibility effect on a laboratory scale.
3. The tracer input was not clearly defined. For example, when a contaminant event or environmental tracer is modeled, the tracer input regarding its quantity and temporal distribution is not well defined. This situation generates another unknown in solving the convection–dispersion equation.
4. The tracer breakthrough curve was assumed to be the superposition of breakthrough curves in separate layers, when the evidence for such layers at the field scale was little or nonexistent. These studies are based on the assumption that the porous medium is perfectly stratified,which, especially at the field scale, may not be a valid assumption. At a small scale, defined as a few meters, the existence of continuous layers may be a reasonable assumption. However, the dispersivity of each layer does not represent the field-scale parameter. One should bear in mind that the field-scale dispersivity is a result of spreading due to different velocities in each layer as a result of the variation of hydraulic conductivity.
5. The measurement of tracer concentration was not adequate. Under ambient flow conditions, the tracer distribution is usually three-dimensional. Under these conditions, if the

measurements are two dimensional, the dispersivity values are classified as low reliability. On the other hand, if the tracer is introduced over the entire saturated thickness of the aquifer, two-dimensional measurements might be adequate.

6. The mathematical model used to determine dispersivity was inappropriate for the data collected. The convection–dispersion equation-based solutions are based on various assumptions regarding flow and solute transport characteristics. Therefore, these assumptions must be consistent with the experimental conditions of the data collected from a field experiment. For example, in a radial flow test, the converging or diverging flow field around the pumping or injection well is nonuniform. Therefore, the application of a one-dimensional uniform velocity flow solution for solute transport will not be appropriate in this case.

7.4.3.4　Results of the Classification for the Longitudinal Dispersivity Data

From the classification process presented in the previous sections, Gelhar et al. (1992) judged that 14 dispersivity values have high reliability in the following sites: Borden, Ontario, Canada; Otis Air Force Base, Cape Cod, Massachusetts; Hanford, Washington; Mobile, Alabama; University of California, Berkeley, California; Yavne region, Israel; Bonnaud, France (six tests); and Palo Alto bay lands, California. A total of 61 values were judged to be of low reliability based on one or more reasons discussed in Section 7.4.3.3, and a total of 31 sites were judged as providing intermediate value of quality data. Figure 7-4 presents the replotted longitudinal dispersivity, A_L, vs. scale length, s, data based on the values in Table 7-1, with symbols denoting the reliability classification; the largest symbols correspond to data that were judged to be of highest reliability.

An Important Feature of the Longitudinal Dispersivity, Data
All data for longitudinal dispersivity, A_L, in Table 7-1 indicates that A_L might increase indefinitely with scale, s. However, based on the data shown in Figure 7-4, this trend cannot be extrapolated

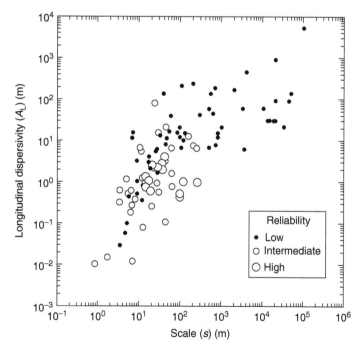

FIGURE 7-4　Longitudinal dispersivity, A_L, vs. scale of observation, s, with data classified by reliability. (Adapted from Gelhar et al., *Water Resour. Res.*, 28, 1955–1974, 1992.)

with confidence to all scale lengths. Also, it is not clear from Figure 7-4 whether A_L increases indefinitely with s, or whether A_L approaches to a constant value for very large s values, as predicted by some theories (Mercado, 1967; Gelhar et al., 1979; Matheron and de Marsily, 1980; Gelhar and Axness, 1983). As shown in Figure 7-4, the values of the scale length s, for high-reliability A_L values is less than 300 m. This potentially means that reliable data at scales larger than 300 m are needed. However, according to Gelhar et al. (1992), the feasibility of conducting controlled tracer tests at these very large scales is open to question.

Reanalysis of Selected Longitudinal Dispersivity, Data
In some cases in Table 7-1, the concentration data were of high reliability but the method of analysis could be improved. Gelhar et al. (1992) reevaluated the data to determine longitudinal dispersivity, A_L, value, which they judged to be of higher reliability. The details of these reevaluations are included in Welty and Gelhar (1989) and their summary is given below.

Corbas, France
The data correspond to a converging radial flow test and are reported in Sauty (1977). An important feature of these data is that the tests were conducted at three different scales in the same aquifer material. A tracer was injected at 25, 50, and 150 m from the pumped well, and Sauty (1977) evaluated the data using uniform flow solutions to the one-dimensional convection–dispersion equation given by Eq. (2-51). Gelhar et al. (1992) state that since Sauty's two-layer-scheme assumption was not supported by geologic evidence, they rated the data at the smaller scales as lower reliability than the data at 150 m. Welty and Gelhar (1989) reevaluated these data using a solution that accounted for nonuniform, convergent radial flow effects without making any assumptions about geologic layers. As seen from Table 7-1, A_L, values reported by Sauty at 25 m are 11 and 1.25 m for the two hypothesized layers, whereas and Welty and Gelhar calculated A_L value 2.4 m without the assumption of layers. At 50 m, Sauty calculated A_L values of 25 and 6.25 m for the two layers; Welty and Gelhar calculated an overall value of 4.6 m for A_L. At a scale of 150 m, Sauty calculated an A_L value of 12.5 m without the assumption of layers; whereas Welty and Gelhar calculated an A_L value of 10.5 m, which is in close agreement with Sauty's value. The calculations of Welty and Gelhar (1989) indicate that A_L increases with scale by accounting for nonuniform flow effects and without the arbitrary assumption of geologic layers.

Savannah River Plant, Georgia
Webster et al. (1970) evaluated concentration data from a two-well recirculating test using the analysis methodology of Grove and Beetem (1971). This analysis methodology assumes that is uniform along stream tubes and sums individual breakthrough curves along the stream tubes to obtain a composite breakthrough curve. Webster and others obtained A_L value 134 m at a scale of 538 m using this method. Gelhar et al. (1992) reevaluated the same data by using the methodology of Gelhar (1982), which accounts for nonuniform flow effects, and obtained a value A_L 47 m. Gelhar et al. (1992) state that they have more confidence in this value because the analysis represents the actual flow configuration more accurately.

Tucson, Arizona
The data reported by Wilson (1971) for a two-well test was later reevaluated by Robson (1974) using a Grove and Beetem-type analysis (Grove and Beetem, 1971). Wilson reported a $A_L = 15.2$ m value at a scale of 79.2 m. Using a nonuniform flow solution based on Gelhar (1982), Gelhar et al. (1992) obtained 1.2 m A_L value, which is an order of magnitude smaller than the value obtained by Robson. Gelhar et al. state that they have more confidence in this value because the analysis reflects the actual situation more accurately.

Columbus, Mississippi
Gelhar et al. (1992) state that a superficial spatial moments interpretation of the natural gradient tracer test at the Columbus site in a heterogeneous aquifer indicated around $A_L = 70$ m by ignoring

the nonuniformity of the flow. Adams and Gelhar (1992) conducted a more refined analysis by including the influence of flow nonuniformity and found $A_L = \sim 7$ m. Gelhar et al. (1992) regarded this refined estimate as having intermediate reliability because of the uncertainty of the mass balance at the Columbus site.

Other Sites

As can be seen from the reanalyzed values, all values of longitudinal dispersivity, A_L, are smaller than the original values. Gelhar et al. (1992) state that they have higher confidence in these values because they are based on more realistic assumptions. In all the cases, Gelhar and others rate the new values to be of intermediate reliability instead of low reliability. The reevaluated data are denoted by solid symbols in Figure 7-5 and connected to their original values by vertical arrows.

Based on the reanalyzed results, Gelhar et al. (1992) suspect that it is most likely that improved analyses would reduce many of the lower-reliability A_L values shown in Figure 7-4. Gelhar and others also state that there are a few cases for which more appropriate observations and data interpretations would probably result in larger A_L values. The authors also state that they suspect that most probably, improved analysis would reduce many of the lower reliability A_L values shown in Figure 7-4. They add that there are a few cases for which more appropriate observations and interpretations would result in larger A_L values. As an example, Gelhar and others state the Twin Lake natural gradient tracer test (Moltyaner and Killey, 1988a, 1988b), for which data interpretation was made by using breakthrough curves at individual boreholes constructed as the average of breakthrough curves in three arbitrarily defined layers. They state that they suspect that this kind of localized observations will produce a significantly low A_L value than that from a spatial moments analysis, which takes into account the overall spreading of the plume. The magnitude of the possible increase in A_L cannot be assessed due to the fact that the sampling network did not completely encompass the plume at the

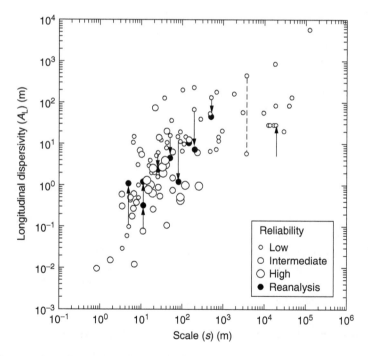

FIGURE 7-5 Longitudinal dispersivity, A_L, vs. scale of observation, s, with data with adjustments resulting from reanalysis. Arrows indicate reported values at tails and corresponding values from reanalyses at heads. Dashed line connects two A_L values determined at the Hanford site. (Adapted from Gelhar et al., *Water Resour. Res.*, 28, 1955–1974, 1992.)

Twin Lake site. The authors mention second example which is the first Borden site natural gradient experiment (Sudicky et al., 1983) (as another example) for which the data were analyzed using an analytical solution with spatially constant dispersivities. In the close proximity to the source where dispersivities actually increasing with displacement, the approach of Sudicky and others tend to underestimate the magnitude of dispersivities. Gelhar et al. (1985) reanalyzed the first Borden experiment based on the method of spatial moments and found that the value of A_L at 11 m was 2 to 4 times that found by Sudicky et al. (1983). The resulting increase in A_L is shown in Figure 7-5, connected to the original point by a vertical line. Gelhar and others regarded this point as having intermediate reliability because of the incomplete plume sampling and plume bifurcation in this test.

Gelhar et al. (1992) state that dispersivities at small displacements will also be underestimated if the analysis is based on breakthrough curves measured in localized samples in individual layers. The authors also add that such effects are probable, for example, in the interpretations for the Perch Lake (Lee et al., 1980) and Lower Glatt Valley (Hoehn, 1983) data. Later interpretations by Hoehn and Santschi (1987) for the Lower Glatt Valley data using temporal moments showed values for A_L an order of magnitude larger and these are connected with the original values by vertical lines as shown in Figure 7-5.

The results presented above for A_L show uncertainty. This uncertainty is further hightlighted by Gelhar et al. (1992) by considering the data for the Hanford site (Bierschenk, 1959; Cole, 1972). Gelhar and others interpreted the same data from breakthrough curves at two different wells at roughly the same distance (around 4000 m) from the injection point and generated A_L values differing by two orders of magnitude (see dashed line in Figure 7-5). This large difference illustrates the difficulty in interpreting point breakthrough curves in heterogeneous aquifers, even at large displacement. Ahlstrom et al. (1977) used 30.5m for A_L for the numerical simulations of the contaminant plume extending 20000 m and this value is denoted by bold arrow in Figure 7-5.

Conclusions
Gelhar et al. (1992) concluds that the results of the above-mentioned reanalysis provide an explicit indication of the uncertainties in the A_L values shown in Figure 7-4, and suggest that for large displacements the low-reliability A_L values are likely to decrease, whereas for small displacements some increases can be expected.

Evaluation of the Longitudinal Dispersivity Values Based on Numerical Simulations
Gelhar et al. (1992) critically evaluated the values of A_L based on numerical simulations of contaminant plumes and their results are as follows:

1. None of the cases for contamination events present any explicit information on the selection of A_L values or the basis on which the values may be considered optimal. Therefore, it is not possible to quantify the uncertainty in A_L values based on contamination event simulations.
2. In order to generate numerically stable results, the values of Peclet number should be inversely dependent on the longitudinal dispersivity, A_L, and this number should not exceed a certain upper limit (e.g., Huyakorn and Pinder, 1983, p. 206). Therefore, because of the possible difficulties associated with large grid Peclet numbers, some of the A_L values based on contaminant plumes are probably biased toward higher values. Such overestimates would mainly occur at larger scales.

7.4.3.5 Results of the Classification for the Transverse Dispersivities Data

As can be seen from Table 7-1, the data on transverse horizontal dispersivity, A_{TH}, are much more limited. In the case of transverse vertical dispersivity, A_{TV}, the data are even more limited. However, these limited data reveal some important features in the applications. The analysis and results of Gelhar et al. (1985, 1992) based on these data are presented below.

Transverse Horizontal Dispersivity, A_{TH}

Using the values in Table 7-1, the transverse horizontal dispersivity, A_{TH}, vs., s, is shown in Figure 7-6. Figure 7-6 shows that there appears to be a trend of α_{TH} increasing with s, but exhibition results from low-reliability data, which are based largely on contaminant event simulations that use two-dimensional depth-averaged approaches. Moreover, in these contamination events, the sources are often ill defined. For example, if the actual source area is larger than that represented in the model, spreading in the transverse directions will be greater, but would be incorrectly attributed to transverse dispersion.

Transverse Vertical Dispersivity, A_{TV}

Using the values in Table 7-1, the transverse vertical dispersivity, A_{TV} vs. scale length, s, is shown in Figure 7-7. As can be seen from Figure 7-7, the data certainly do not imply any significant trend with the overall scale range. It can be noted from Figure 7-7 that there are only two data points of high reliability, which correspond to the Borden (Freyberg, 1986) and Cape Cod (Garabedian et al., 1988, 1991) sites. The estimate of A_{TV} for the Borden site is made from a three-dimensional analysis of Rajaram and Gelhar (1991). The values of A_{TV} are much smaller than the values of A_{TH}, which reflects the roughly horizontal stratification of hydraulic conductivity in permeable sedimentary materials. Gelhar et al. (1992) report that all of the A_{TV} values are less than 1 m and high-reliability values are only a few millimeters, and are of the same order as magnitude as the local transverse dispersivity of sandy materials.

The A_L/A_{TH} and A_L/A_{TV} Ratios

The ratio of longitudinal dispersivity to transverse horizontal dispersivity, A_L/A_{TH}, and the ratio of longitudinal dispersivity to transverse vertical dispersivity, A_L/A_{TV}, is shown in Figure 7-8. This type of presentation is useful because it is a common practice to select constant values for the ratio of

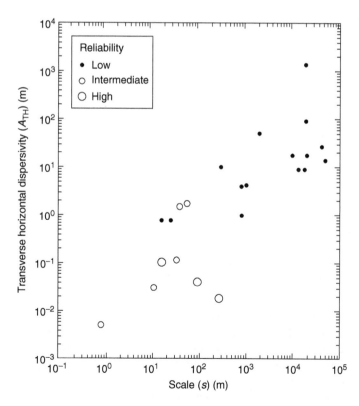

FIGURE 7-6 Transverse horizontal dispersivity, A_{TH}, vs. scale of observation, s, with data classified by reliability. (Adapted from Gelhar et al., *Water Resour. Res.*, 28, 1955–1974, 1992.)

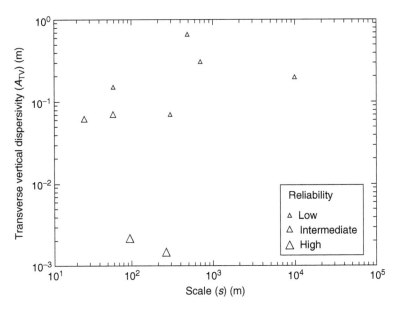

FIGURE 7-7   Transverse vertical dispersivity, A_{TV}, vs. scale of observation, s, with data classified by reliability. (Adapted from Gelhar et al., *Water Resour. Res.*, 28, 1955–1974, 1992.)

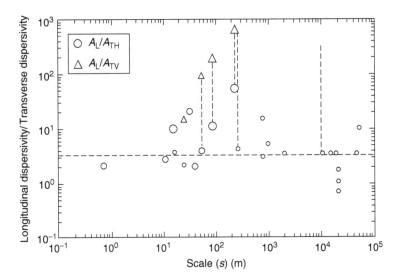

FIGURE 7-8   Longitudinal dispersivity/transverse vertical dispersivity, A_L/A_{TH}, vs. scale of observation, s, with data classified by reliability. (Adapted from Gelhar et al., *Water Resour. Res.*, 28, 1955–1974, 1992.)

longitudinal to transverse dispersivities. The horizontal dashed line in Figure 7-8 conveys that the A_L/A_{TH} ratio is equal to 3, which has been used in numerical simulations. However, there does not appear to be any real justification for using this ratio. Gelhar et al. (1992) also state that they were not aware of any simulation work that systematically demonstrated the appropriateness of this ratio for transverse horizontal dispersivity. In Figure 7-8, two high-reliability points show an order of magnitude A_L/A_{TH} ratio and the vertical dashed lines show three-dimensionally monitored sites for which all three principal components of the dispersivity tensor have been estimated. In all of these cases, the A_{TV}/A_{TH} ratio is around 10^{-1} to 10^{-2}, which emphasizes the small degree of vertical mixing in naturally stratified sediments. This small degree of vertical mixing is important in many applications, such as the design of observation wells to monitor contamination plumes and remediation practices.

7.4.3.6 Interpretations of the Dispersivities Data

The review of field observations of dispersive mixing in aquifers presented in the previous sections, based on Gelhar et al. (1985, 1992), has several overall features. These features are evident from the data for dispersivities that are presented in Figures 7-1, and 7-4–7-7 and Table 7-1. Based on the extensive review of the data for dispersivities, which is presented in the previous sections, Gelhar et al. (1992) emphasize several overall features as follows:

(1) Clear Trend of Systematic Increase in A_L with s
Disregarding reliability, the data in Figures 7-1, 7-4, and 7-5 indicate a clear trend of systematic increase in, A_L, with scale, s.

(2) Porous vs. Fractured Media
In terms of the aquifer type (i.e., porous vs. fractured media), the A_L values at a smaller scale may seem to be higher for fractured media, but in terms of the lower reliability of the fractured media data, this difference is not significant.

(3) Status of High-Reliability Points
When the data on A_L are classified based on reliability, the dependence on the scale of A_L is less clear (see Figure 7-4). As can be seen in Figure 7-4, there are no high-reliability points at scales greater than 300m. The high-reliability points are systematically present in the lower portion of the scattered data. The lack of high-reliability data at scales greater than 300 m reflects the fact that the A_L data beyond these scales are almost exclusively from contamination plume simulations or environmental tracer studies, for which the input of solute is typically ill defined. Carrying out controlled input tracer experiments at these larger scales requires a very long period of time. Hence, such experiments have not been conducted.

(4) A Single Universal Line Does Not Seem Reasonable
Figure 7-4 suggests that some overall trend in the increase of A_L with scale is plausible. But Gelhar et al. (1992) state that it does not seem reasonable to conclude that a single universal line proposed by Neuman (1990), which is presented in Section 7.4.4, can be meaningfully identified by applying standard linear regression to all of the A_L data. Gelhar et al. (1992) state that they would expect a family of curves reflecting different dispersivities in aquifers with different degrees of heterogeneity. They, further state that the presumption of such a universal model ignores the fact that different aquifers will have different degrees of heterogeneity at a given scale.

(5) Explanation Based on the Stochastic Theory
At a given scale, the longitudinal dispersivity, A_L, typically ranges over 2-3 orders of magnitude. This degree of variation can be explained in terms of the established stochastic theory (e.g., Gelhar and Axness, 1983; Dagan, 1984). This theory shows that A_L is proportional to the product of the variance and the correlation scale of the natural logarithm of hydraulic conductivity. The theoretical results for the developing dispersion process (Gelhar et al., 1979; Dagan, 1984; Gelhar, 1987; Naff et al., 1988) show that A_L initially increases linearly with displacement distance and gradually approaches a constant asymptotic value. Gelhar et al. (1992) state that one could visualize the behavior of curve in Figure 7-4 as being the result of the superimposition of several such theoretical curves on different parameters characterizing aquifer heterogeneity.

(6) Uncertainty in Estimating Longitudinal Dispersivity, A_L
The results of the reanalysis of Gelhar et al. (1992) for several of the individual sites, presented in Section 7.4.3.4, serve to illustrate explicitly the uncertainty associated with estimating A_L values.

As mentioned in Section 7.4.3.4, the reanalysis indicated that by using more appropriate models, lower A_L values were determined, except in the case of very small displacements where limited localized sampling could result in underestimates of the bulk spreading and mixing. In cases where A_L estimates were based on numerical simulations of contamination events, the degree of uncertainty is likely to be large.

(7) Strong Cautionary About Linear Regression Representation Through the Dispersivity Data

Gelhar et al. (1992, p.1972) caution users routinely adopting A_L values from Figure 7-4 or from a linear representation of the data as follows :

> ... We feel that the preponderance of evidence favors the use of dispersivity values in the lower half of the range at any given scale. If values in the upper part of the range are adopted, excessively large dilution may be predicted and the environmental consequences misrepresented. ...

(8) Transverse Dispersivities

As mentioned in Section 7.4.3.5, in the case of transverse dispersivities, very low A_{TV} values have been observed at several sites, and it is particularly important to recognize this feature. This means that contamination plumes will potentially show very limited vertical mixing with high concentrations at a given horizon. The recognition of these features is important for the design of monitoring well systems and the restoration of aquifers from contamination.

(9) Typical Ratios for Dispersivities

Gelhar et al. (1992) state that the A_{TH} values of are typically an order of magnitude smaller than A_L values, i.e.,

$$A_{TH} \approx 10^{-1} A_L \tag{7-1}$$

For A_{TV} values, Gelhar et al. (1992) state that they are typically two orders of magnitude smaller than A_L values, i.e.,

$$A_{TV} \approx 10^{-2} A_L \tag{7-2}$$

7.4.4 NEUMAN'S INTERPRETATION OF THE LONGITUDINAL DISPERSIVITY DATA

7.4.4.1 Theoretical Background

Neuman (1990) interpreted the data for dispersivities compiled by Gelhar et al. (1985), presented in Section 7.4.3, based on the quasi-linear theory of Neuman and Zhang (Neuman and Zhang, 1990; Zhang and Neuman, 1990). The Neuman and Zhang's quasi-linear theory (the NZ theory) accounts for nonlinearity caused by the deviation of plume "particles" from their mean trajectory in three-dimensional, statistically homogeneous but anisotropic porous media under an exponential covariance of log-hydraulic conductivity. The authors' study is closely related to earlier work by Dagan (1984, 1987, 1988), who follows Taylor's Lagrangian theory of diffusion by continuous motions and the example of Lundgren and Pointin (1976) to derive an expression for the spatial covariance of ensemble mean concentrations in a statistically random field of mildly fluctuating log-hydraulic conductivities.

Expression for Asymptotic Longitudinal Dispersivity, $A_{\infty L}$

Based on the results of the NZ quasi-linear theory, Neuman (1990, Eq. [7] p. 1751) states that for three-dimensional dispersion, the expression for $A_{\infty L}$, which is the (constant) Fickian asymptote of the longitudinal macrodispersivity A_L as $t \to \infty$, is

$$A_{\infty L} = A_{\infty DL} s \sigma_f^2, \quad A_{\infty L} \leq s \tag{7-3}$$

where $A_{\infty DL}$ is the asymptotic dimensionless longitudinal dispersivity, s is the integral scale along the main ground water velocity (v) direction, and σ_f^2 is the variance of $f = \ln(K)$, K being the hydraulic conductivity. In some strongly anisotropic media (Zhang and Neuman, 1990, Figures 1, 6, and 7), $A_{\infty DL} << 1$, and therefore Eq. (7-3) is valid only when $A_{\infty DL} << s$.

All existing stochastic theories (e.g., Gelhar and Axness, 1983) are nominally restricted to very mildly heterogeneous media in which $\sigma_f^2 << 1$. However, Neuman (1990) states that the quasi-linear theory is believed to be less restricted than the other theories when extended to strongly heterogeneous media in which $\sigma_f^2 \geq 1$. He further states that this is so because the quasi-linear theory deals with non-linearity due to deviation of plume "particles" from their mean trajectory without formally limiting σ_f^2.

7.4.4.2 Interpretation of Scale Effect and Longitudinal Dispersivity Equations

Neuman (1990) turned his attention to over 130 A_L values from laboratory and field tracer studies in porous and fractured media throughout the world [recorded by Lallemand-Bares and Peaudecerf (1978), Pickens and Grisak (1981a), Gelhar et al. (1985), Arya (1986), and Arya et al. (1988)]. The A_L values used by Neuman vary from 1 mm to more than 1 km and correspond to studies conducted on scales ranging from less than 10 cm to 100 km. The A_L values mentioned above were derived from Fickian models which, assume that log-hydraulic conductivity is statistically homogeneous. Neuman (1990, p. 1752) states:

> This and the fact that scale dependence is a non-Fickian phenomenon suggest that the data should be consistent with the recent quasi-linear theory of Neuman and Zhang (1990) and Zhang and Neuman (1990) which accounts for both Fickian and non-Fickian dispersion in homogeneous media.

Like in Figures 7-1 and 7-4, Neuman (1990) generated a double-logarithmic plot of A_L values vs. s. He excluded A_L values obtained by calibrating numerical grid models against solute concentrations of large-scale plumes. Neuman (1993) states that the reason for this exclusion is that such models account deterministically for medium heterogeneity by allowing hydraulic conductivities to vary spatially over the model domain, thereby yielding lower A_L values. Neuman (1990) proposed three interpretations for the remaining 134 A_L values, which are presented below.

Neuman's First Interpretation
Neuman's first interpretation is illustrated in Figure 7-9 and is based on the assumption that Eq. (7-3) is valid if $A_{\infty L} << s$. Neuman excludes the three points shown in Figure 7-9, which correspond to $s \geq 3500$m (Reasons for this exclusion will become clear later), and fits a regression line of the form

$$\log (A_L) = b \log(s) + c \qquad (7\text{-}4)$$

for the remaining 131 points along with the corresponding 95% confidence limit, as shown in Figure 7-9, Neuman's best fit is expressed as

$$(A_L) = 0.175s^{1.46} \qquad (7\text{-}5)$$

with a regression coefficient $R^2 = 0.74$ and 95% confidence intervals [0.0113, 0.0272] around the coefficient 0.0175, and [1.30, 1.61] around the exponent 1.46.

Neuman's Second Interpretation
In his second interpretation, Neuman (1990) considered that many of the replotted tracer studies have been conducted in anisotropic media, and hence, Eq. (7-3) may not be valid unless $A_{\infty L} << s$. To account for this possibility, in Figure 7-10, Neuman fitted two separate regression lines of the form of Eq. (7-4) to 119 data points associated with $s \leq 100$ m (line A) and 16 data points associated with $s \geq 100$ m (line B). One point lies at $s = 100$ m and belongs to both lines. Neuman expressed line A in Figure 7-10 as

$$A_L = 0.0169s^{1.53}, \quad s \leq 100 \text{ m} \qquad (7\text{-}6)$$

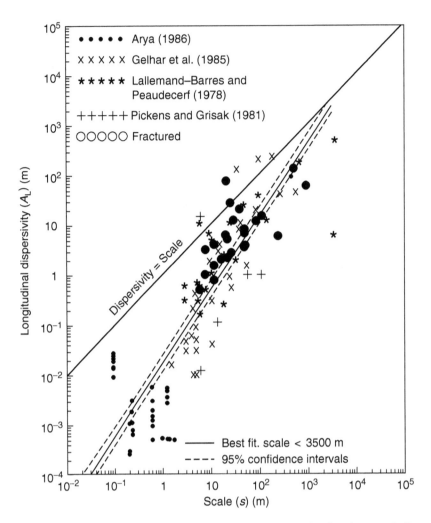

FIGURE 7-9   First interpretation of longitudinal dispersivity, A_L, vs. scale of study, s, excluding numerical model calibration results. (Adapted from Neuman, *Water Resour. Res.*, 29, 1863–1865, 1993.)

with a regression coefficient of $R^2 = 0.71$ and 95% confidence intervals [0.0108, 0.0264] around the coefficient 0.0169, and [1.35, 1.70] around the exponent 1.53. This is similar to Eq. (7-5).

Neuman's Conclusions Based on His Interpretations
On the basis of his two interpretations Neuman (1990, p. 1754) concluded that log-hydraulic conductivities of the geologic media represented in Figure 7-9 and Figure 7-10 constitute a self-similar random field with homogeneous increments. He further states that even though this conclusion was reached by excluding data associated with $s \geq 3500$ m in Figure 7-9 and $s > 100$ m in Figure 7-10, it is nevertheless consistent with these data. Based on his universal semivariogram analysis and by using Eqs. (7-5) and (7-6), Neuman gives the following equation:

$$A_L \cong 0.017s^{1.5}, \quad s \leq 3460 \text{ m} \tag{7-7}$$

Neuman (1990, p. 1755) states that line B in Figure 7-10 can be expressed as

$$A_L \cong 0.32s^{1.83}, \quad s \geq 100 \text{ m} \tag{7-8}$$

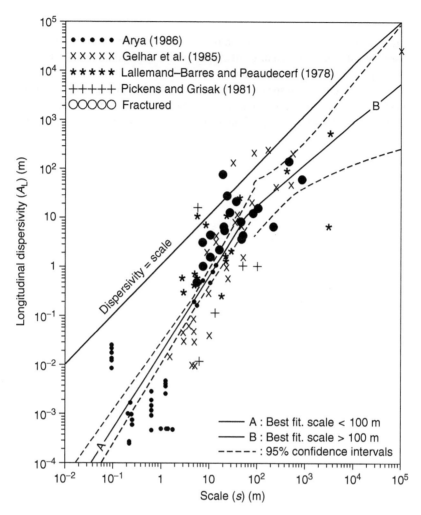

FIGURE 7-10   Second interpretation of longitudinal dispersivity, A_L, vs. scale of study, s, excluding numerical model calibration results. (Adapted from Neuman, *Water Resour. Res.*, 29, 1863–1865, 1993.).

with a regression coefficient $R^2 = 0.44$ and 95% confidence intervals [0.026, 3.95] around the coefficient 0.32, and [0.30, 1.37] around the coefficient 0.83.

7.4.5   USAGE OF THE DISPERSIVITIES ESTIMATION PROCEDURES

The dispersivities estimation procedures presented above should be used with caution. The most important step of this procedure is the selection of the scale length s in the diagrams or equations. A common approach is to use the maximum extent of a plume from the source as the value of s. Figure 7-4 can be used for an initial estimate of longitudinal dispersivity A_L. If $s = 20$ m, by using the high-reliability data points from Figure 7.4, one gets $A_L = 1.0$ m. The estimated values of A_{TH} (from Eq. [7-1]), and A_{TV} (from Eq. [7-2]), respectively are 0.1 m and 0.01 m. By using $s = 20$ m, from Eq. [7-7], one gets $A_L = 1.5$ m. Again, from Eqs. (7-1) and (7-2) one gets $A_{TH} = 0.15$ m and $A_{TV} = 0.015$ m. From a practical standpoint, the A_L values estimated from Figure 7-4 and Eq. (7-7) are approximately the same.

The dispersivity data analysis presented in the previous sections show that there are controversial views regarding a linear regression representation through the data points. These views are further elaborated in Neuman (1993) and Gelhar et al. (1993).

PROBLEMS

7.1   Describe the term "field scale" and compare it with the term "laboratory scale."

7.2   Why are the estimation procedures of dispersivities based on dispersivity vs. scale diagrams (empirical approach) important in practice ?

7.3   Describe the foundations of the empirical approaches to estimate values of dispersivities.

7.4   The maximum extent of a solute plume is $s = 200$ m. Estimate the values of A_L, A_{TH}, and A_{TV} using Figure 7-4 and the equations given by Neuman. Discuss the results.

8 Determination of Solute Sorption Parameters in Aquifers

8.1 INTRODUCTION

Contaminants can potentially enter into aquifers from regional sources or waste spills. Within the ground water zone of aquifers, contaminants migrate under the influence of convection, mechanical dispersion, molecular diffusion, and sorption. All of these components are covered in detail in Chapter 2, and the determination and estimation procedures of the parameters associated with these solute transport components are discussed in the previous chapters, with the exception of sorption parameters. The determination and estimation procedures of sorption parameters are presented in this chapter.

There has been an increase in the number and quantity of organic chemicals since World War II. More than 3,000,000 organic compounds are known to exist and more than 40,000 have been manufactured, according to statistics (Cherry et al., 1984). Many of these chemicals are known to be hazardous or potentially hazardous. During the past decades, many common organic chemicals have been recognized as hazardous and relatively mobile in the water-bearing zones of aquifers. As a result, many publications have appeared in the literature regarding the sorption characteristics of some selected contaminant constituents. Prediction of the migration of contaminants in aquifers requires the quantification of the retardation factor R_d. The parameters to be quantified in the retardation factor R_d equation can be observed in Eq. (2-62); they are bulk density, ρ_b, total porosity, φ, and distribution coefficient K_d. The estimation procedures of bulk density and total porosity are discussed in Chapter 2. If the values of these parameters for a contaminant constituent are known, the retardation factor defined by Eq. (2-60) and Eq. (2-62) can be determined. As a matter of fact, the main goal of this chapter is to determine the retardation factor using these three parameters. Of the parameters in Eq. (2-62), the most difficult one to quantify is the distribution coefficient. Its estimation procedures for some selected organic chemicals, will be included in this chapter.

8.2 THE K_d / K_{oc} APPROACH FOR THE DETERMINATION OF RETARDATION FACTOR

As can be seen from Eq. (2-62) the retardation factor R_d equation is based on the linear sorption isotherm, as a result of which, the distribution coefficient K_d in Eq. (2-62) is constant. This form has wide applications in predicting the migration of contaminants in aquifers. Since K_d is a function of the organic carbon partition coefficient K_{oc}, which is defined later, this approach is called the K_d/K_{oc} approach.

8.2.1 SCHEMATIC ILLUSTRATION OF THE RETARDATION CONCEPT

In this section, the retardation concept is illustrated schematically for laboratory conditions and for a hypothetical field situation (Cherry et al., 1984). To illustrate this concept, Eq. (2-83) for $v = 0$ is expressed as

$$\frac{\partial C}{\partial t} = D'_x \frac{\partial^2 C}{\partial x^2} - U' \frac{\partial C}{\partial x} \tag{8-1}$$

where

$$D'_x = \frac{D_x}{R_d}, \quad U' = \frac{U}{R_d} \qquad (8\text{-}2)$$

and D'_x is the *effective dispersion coefficient* and U' is the *effective solute velocity*. The front of the solute plume is retarded relative to the rate of convection of the front of nonreactive solutes, U. Therefore, the retardation factor equation given by Eq. (2-62) can be expressed as

$$R_d = \frac{U}{U'} = 1 + \frac{\rho_b}{\varphi} K_d \qquad (8\text{-}3)$$

In the absence of dispersion, the front is conceptualized as a plug-displacement front. In the presence of Gaussian dispersion, it represents the 50th percentile concentration level of an advancing slug of contamination (i.e, the middle of the dispersed front) emanating from a continuous source. The retardation concept is illustrated schematically in Figure 8-1 for laboratory conditions and for a hypothetical field situation. In the field example, the ground water velocity is assumed to have little spatial or temporal variability.

As can be seen from Eqs. (8-1) and (8-3), with the K_d/K_{oc} approach, the many analytical and numerical mathematical solutions for the convection–dispersion equation for nonreactive contaminants in saturated homogeneous porous media can be used for predictive purposes. In these models, the velocity of the advancing solute front becomes the effective solute velocity, U/R_d, and the dispersion coefficient becomes the effective dispersion coefficient, D_x/R_d.

8.2.2 CHARACTERISTICS OF THE K_d/K_{oc} APPROACH AND RELATED QUANTITIES

Equation (2-62) is based on the assumption that the adsorption or exchange reactions are always instantaneous, which is called *equilibrium adsorption*. If there is no interaction between the chemical and the solid phase, K_d in Eq. (2-61) becomes 0 and R_d reduces to 1. In some cases, R_d may become < 1, indicating that only a fraction of the liquid phase participates in the solute transport process. This situation occurs when the tracer is subject to anion exclusion or when relatively immobile liquid regions are present. In the case of anion exclusion, $(1-R_d)$ can be viewed as the relative anion exclusion volume and $(-K_d)$ as the specific anion exclusion volume (e.g., expressed in cm^3 water per g of soil) (van Genuchten, 1980).

Theoretically, Eq. (2-60) is only applicable if the reactions are instantaneous or, in practice, if the reactions are fast, relative to the ground water velocity.

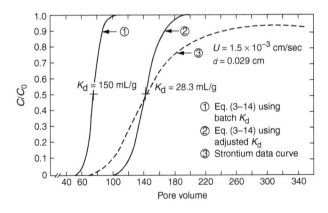

FIGURE 8-1 Experimental and computed column breakthrough curves from strontium. (Adapted from Reynolds, W.D., M.S. Thesis, University of Waterloo, Waterloo, Ontario, 1978, 149 pp; as presented in Cherry et al., *Ground water Contamination*, National Academy Press, Washington, DC, 1984, pp. 46–64.)

8.2.2.1 Relationship between the Distribution Coefficient K_d and Organic Carbon Content f_{oc}

The relationship between the distribution coefficient K_d and organic carbon content f_{oc} is an important step in determining K_d values. Some details regarding this relationship are given below (Anderson and Pankow, 1986).

For a wide variety of organic compounds, it has been shown that the distribution coefficient K_d can be expressed as

$$K_d = K_{oc} f_{oc} \qquad (8\text{-}4)$$

where f_{oc} is the *fraction of organic carbon content* in the soil ($0 \leq f_{oc} \leq 1$) and K_{oc} is the *distribution coefficient* of the compound in pure soil and sediment organic carbon (Karickhoff et al., 1979; Karickhoff, 1981, 1984). If the f_{oc} value of a contaminant constituent and that of soil is measured, the K_d for the compound of interest may be estimated without carrying out sorption isotherm measurements. This is done by using Eq. (8-4) and K_{oc} data obtained in the study of another soil or sediment. When K_{oc} for the compound of interest has been determined accurately and, if the K_d of interest is then determined experimentally, agreement between the predicted and measured K_d values is usually found to be within a factor of 5, and often within a factor of 2 (Mingelgrin and Gerstl, 1983). When f_{oc} is low and large amounts of swelling clays are present, then the application of Eq. (8-4) can lead to larger errors (Hassett et al., 1980; Karickhoff, 1984). Errors or lab-to-lab inconsistencies in the methods used to determine f_{oc} values can also introduce discrepancies. In addition, the study of strongly partitioning compounds such as polychlorinated biphenyls (PCBs) can be complicated by an artifact caused by nonsettling particles (Voice et al., 1983; Gschwend and Wu, 1985). This artifact can lead to the underestimation of large K_d and K_{oc} values.

The approach based on Eq. (8-4) works reasonably well for a wide range of soils, provided that the soil organic carbon content is sufficiently high (e.g., $f_{oc} > 0.001$). In the case of lower organic carbon content cases, sorption of the neutral organics onto the mineral phase can cause significant errors in the estimation of K_d (Chiou et al., 1985). However, the K_{oc} relationship allows estimations of sorption-based retardation factors developed from measured f_{oc} values, rather than conducting more experiments. Values of K_{oc} for many contaminant constituents are unknown. As a result, numerous researchers have developed correlation equations to relate K_{oc} to more commonly available chemical properties, such as solubility or octanol–water partition coefficient.

8.2.2.2 Organic Carbon Partition Coefficient K_{oc}

In Eq. (8-4), K_{oc} is called either *organic carbon partition coefficient* or *partition coefficient*. This parameter is another indicator of the tendency of a chemicals to partition between ground water and the soil. For certain chemicals, K_{oc} is directly related to the distribution coefficient via the fraction of organic carbon content f_{oc} according to Eq. (8-4). Since f_{oc} is a dimensionless quantity, the dimensions of K_{oc} are the same as those of K_d, i.e., $[K_{oc}] = [K_d]$. Therefore, Eqs. (2-64)–(2-67) are valid for K_{oc} as well. From Eq. (8-4)

$$K_{oc} = \frac{K_d}{f_{oc}} \qquad (8\text{-}5)$$

Similar to Eq. (2-67), one can write

$$[K_{oc}] = \frac{\text{mass of solute on the solid phase of oc per unit mass of oc}}{\text{concentration of solute in solution}} \qquad (8\text{-}6)$$

where oc refers to organic carbon. In dimensional form,

$$[K_{oc}] = \frac{M/M}{M/L^3} = \frac{L^3}{M} \qquad (8\text{-}7)$$

Similarly, for K_d, in practice, the following units are used:

$$[K_{oc}] = \frac{\dfrac{\mu g \text{ (sorbed solute mass)}}{g \text{(mass of organic carbon)}}}{\mu g/ml \text{(aqueous concentration)}} = \frac{\dfrac{mg \text{(sorbed solute mass)}}{kg \text{(mass of organic carbon)}}}{mg/L \text{(aqueous concentration)}} = \frac{mL}{g} = \frac{L}{kg} \quad (8\text{-}8)$$

The existence of the K_{oc} parameter plays an important role in determining the fate and transport of chemicals in soils and sediments. The parameter K_{oc} is commonly used in river models, runoff models, and ground water models, where the transport of a specific contaminant constituent is investigated. The values of K_{oc} (in the units of $ML/g \equiv L/kg$) may range from 10^0 to 10^7. Numerous studies have shown that values of K_{oc} determined from Eq. (8-4) for a specific constituent are relatively constant and reasonably independent of the soil or sediment. The spread of values determined from a range of soils and sediments generally results in an uncertainty (coefficient of variation) of 10 to 140% (Lyman et al., 1982).

Certain precautions have to be taken when using K_{oc} values. If the sorption isotherm is nonlinear, for example, Eq. (2-57), the K_{oc} value determined from Eq. (8-4) will not be the same as the one determined from a single data point and Eq. (8-8). Both the values will have discrepancies from the one determined from an isotherm based on several data points (Lyman et al., 1982).

Some earlier investigations regarding soil sorption coefficients are based on the results on a soil–organic matter basis K_{om} rather than K_{oc}. Since the organic carbon content of a soil or sediment can be measured more directly, reported values of K_{oc} is preferred. The ratio of organic matter to organic carbon generally varies from soil to soil, but a value of 1.724 is often assumed when conversion is necessary. Therefore, the relationship between K_{oc} and K_{om} is (Lyman et al., 1982)

$$K_{oc} \cong 1.724 K_{om} \quad (8\text{-}9)$$

8.2.2.3 Octanol–Water Partition Coefficient

The octanol–water partition coefficient, K_{ow} is defined as the ratio of a chemical's concentration in the octanol phase to its concentration in the aqueous phase of a two-phase octanol-water system:

$$K_{ow} = \frac{\text{concentration in octanol phase}}{\text{concentration in aqueous phase}} \quad (8\text{-}10)$$

Therefore, K_{ow} values are dimensionless. Octanol–water partition coefficient indicates the tendency of a chemicals to partition between the ground water and the soil. This parameter is measured using low solute concentrations, where K_{ow} is a very weak function of solute concentration. The values of K_{ow} are usually measured at room temperature (20 or 25°C). The effect of the temperature on K_{ow} is insignificant, usually of the order of 0.001 to 0.01 $\log(K_{ow})$ units per degree, and may be either positive or negative (Leo et al., 1971; Lyman et al., 1982).

Measured values of K_{ow} for organic chemicals have been found to be as low as 10^{-3} and as high as 10^7. Therefore, they have a range of ten orders of magnitude. In terms of $\log(K_{ow})$, this range is from -3 to 7. In practice, it is possible to estimate $\log(K_{ow})$ with an uncertainty of no more than ± 0.1 to $0.2\log(K_{ow})$ units.

Since the 1970s, K_{ow} has become a key parameter in studies regarding the environmental fate of organic chemicals. It has been found that K_{ow} is related to water solubility, soil and sediment sorption coefficients, and bioconcentration factors for aquatic life.

8.2.2.4 Solubility in Water

The solubility of a chemical in water may be defined as the maximum amount of the chemical that will dissolve in pure water at a specified temperature. Solubility in water determines the degree to which the chemical will dissolve in water as well as the maximum possible concentrations of the chemical. High solubility indicates low sorption tendencies. Solubility is one of the most important parameters that affect the fate and transport of organic chemicals in the environment. Chemicals having high solubility are easily and quickly distributed by the hydrologic cycle. Water solubility, as an environmental parameter, is much less important for gases than it is for liquids or solids.

Aqueous concentrations of solubility are usually stated in terms of weight per weight (ppm, ppb, g/kg, etc.) or weight per volume (mg/L, μg/L, mol/L, etc.). Less common units are mole fraction and molal concentration (moles per kg of solvent). At low concentrations, all units are proportional to one another. At high concentrations, this is not the case, and it becomes important to distinguish if the solubility is per volume of pure water or per volume of solution. The solubilities of the most common organic chemicals are in the range of 1 to 100,000 ppm at ambient temperatures.

8.2.3 Evaluation of the K_d/K_{oc} Approach with Experiments

8.2.3.1 Comparison of the Experimental and Computed Column Breakthrough Curves

The usage of the K_d/K_{oc} approach in evaluating the migration of contaminants became common in the 1960s and 1970s. This approach is being used to evaluate the behavior of various nonreactive elements, such as transition metals, heavy metals, metalloids, and trace organic compounds. Although the K_d/K_{oc} approach is undoubtedly convenient, its validity as a means of developing reliable predictions of the behavior of organic contaminants in actual aquifer systems is questionable in many situations (Cherry et al., 1984). It should be emphasized that the key advantage of Eqs. (2-59) and (2-62) is their mathematical convenience and their simple application in laboratory tests, which can be used to determine K_d/K_{oc} values for input in the retardation factor equation. However, this does not imply that the K_d/K_{oc} approach yields predictions of acceptable accuracy for every contaminant constituent when applied to field cases (Gillham and Cherry, 1982b).

Determination of S = f(C) Relationship
The validity of applying Eq. (8-1) to relatively simple hydrogeologic systems can be evaluated by comparing simulated results with the results of laboratory experiments, under the conditions that tracers are passed through columns of saturated porous media under steady flow as well as other experimental constraints. A necessary prerequisite for the comparison is an agreement be obtained between simulated and experimental breakthrough curves for nonreactive species (such as Cl^-, ^3H, and ^{18}O) in the solution that also contains nonreactive species of interest. Another prerequisite is that the reactive species of interest have a partitioning function, determined from batch tests, that can be described by a linear isotherm within the concentration range relevant to the column experiment. The $S = f(C)$ relationship is normally determined in the laboratory by means of batch tests in which a known mass of the porous medium is immersed in a solution representing the leachate of ground water. The solution contains a specified concentration of the contaminant of interest. After the agitation of the liquid–solid mixture over a period of hours or days, the contaminant concentration in the solution is determined and, by difference, the concentration sorbed in the solids is known. When this test is repeated using different concentrations of the contaminant in the solution, the $S = f(C)$ relation, which is known as the *sorption isotherm*, is determined (Gillham and Cherry, 1982b).

Comparison with Experiments
The results described above were attained by Reynolds (1978) in experiments with a column of fine sand of glacial origin. In the experiments, strontium was used as the reactive tracer and Cl^- and ^3H as the nonreactive tracers. In Figure 8-1, the simulated breakthrough curve for strontium is compared with the measured one. The K_d used for the reaction term in the model was obtained from a batch test on the sand. This experiment resulted in a partitioning relation with a high degree of linearity over the entire concentration range of interest. The modeled breakthrough curve was generated using

Eq. (3-14). Figure 8-1 shows that the measured breakthrough curve is displaced to the right of the modeled breakthrough curve. This displacement indicates a retardation that is twice the magnitude of the retardation using the measured K_d. Gillham and Cherry (1982b) point out that the appreciable difference between the breakthrough curves and the modeled curves based on Eq. (8-1), is more important than the differences between the batch and column K_d values. Figure 8-1 presents a breakthrough curve computed using Eq. (8-1) with a K_d value adjusted such that the $C/C_0 = 0.5$ point on the computed curve coincides with the same relative concentration on the measured curve. Inspection of the two curves indicates that although the $C/C_0 = 0.5$ points are matched, the measured breakthrough curve is much more spread out than the computed curve. The measured curve has a pronounced asymmetric shape with an extreme extension of the extreme tailing end.

Normalized Column Breakthrough Curves for Reactive and Nonreactive Solutes
As shown in Figure 8-1, the breakthrough curves based on the convection–dispersion model with linear isotherms are symmetrical or nearly symmetrical, whereas curves obtained from laboratory column experiments are asymmetrical with extended tails. Reynolds (1978) compiled data from several published studies in which tracers with linear isotherms were used to determine breakthrough curves from column experiments. The results were scaled and plotted on the same graph of dimensionless concentration (C/C_0) vs. dimensionless time ($T = Ut/L$) and are shown in Figure 8-2. The graphs in Figure 8-2 are based on the studies of several investigators for a variety of sorbates and sorbents. As shown in Figure 8-2, all the breakthrough curves are asymmetrical and fit within a single narrow band on the dimensionless graph. This situation implies the presence of a common factor that is currently not accounted for in the convection–dispersion formulation (Cherry et al., 1984).

8.2.3.2 Evaluation of the Linear Equilibrium Isotherm Equation with Respect to the Freundlich Isotherm

Comparison of the Linear and Freundlich Isotherm Equations
As mentioned in Section 2.6, Eq. (2-61) or Eq. (2-62) is a special form of Eq. (2-57), the Freundlich isotherm. Eq. (2-61) is the linear adsorption isotherm for $N = 1$. However, there is an error introduced by this approximation. This error was evaluated by Rao and Davidson (1980) and their evaluation is presented below.

The error introduced in the linear adsorption isotherm depends upon the values of C and N in Eq. (2-57). Equations. (2-57) and (2-61) can be written, respectively, as

$$S_{FI} = K_d C^N \tag{8-11}$$

and

$$S_{LI} = K_d C \tag{8-12}$$

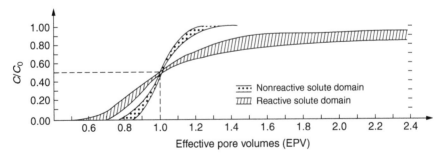

FIGURE 8-2 Normalized column breakthrough curves for reactive and nonreactive solutes. (Adapted from Reynolds, W.D., M.S. Thesis, University of Waterloo, Waterloo, Ontario, 1974, 149 pp; as presented in Cherry et al., *Ground water Contamination*, National Academy Press, Washington, DC, 1984, pp. 46–64.)

Dividing both sides of these equations gives

$$\frac{S_{FI}}{S_{LI}} = C^{N-1} \tag{8-13}$$

where, FI and LI stand for "Freundlich isotherm" and "linear isotherm", respectively. A plot of C^{N-1} vs. C in the range 0.1 to 10 μg/mL and for N values ranging from 0.5 to 1.0 is shown in Figure 8-3. It may be noted that $S_{FI}/S_{LI} = 1.0$ represents a perfect agreement between linear and nonlinear isotherms. Thus, for $C < 1.0$, the amount adsorbed is underpredicted and for $C > 1.0$, the amount adsorbed is overpredicted by assuming linearity. However, for $0.1 < C < 10.0$ and $0.5 < N < 1.0$, as shown in Figure 8-3, the maximum deviation will be a factor of 3, if Eq. (8-1) is used instead of Eq. (8-12). Such errors may be tolerable in many practical applications such as nonpoint source contamination models. However, for large solute concentrations, such as those encountered under pesticide waste disposal sites, the amount absorbed could be easily overestimated by order of magnitude or more (Davidson et al., 1978; Rao and Davidson, 1979).

Comparison of Concentration Profiles Based on Linear and Freundlich Isotherm
Equation. (2-59) is based on linear equilibrium sorption, and this equation as well as its two and three-dimensional versions are widely used in solute transport analysis under laboratory and field conditions, along with analytical solutions that are based on these equations. The same linear equilibrium approach is also used in the differential equations (Eq. [2-79]) for solute transport analysis in homogeneous and anisotropic porous media using numerical modeling techniques. Equation (2-58) is based on the Freundlich isotherm and is more representative of the real situation than Eq. (2-59). Unfortunately, studies regarding the evaluation of the linear equilibrium approach are fairly limited. In this section, the results of Weber et al. (1991) for the comparison of linear and nonlinear Freundlich isotherms are presented for tetrachloroethylene (TTCE).

Weber et al. (1991) selected TTCE as the solute and used a fairly typical aquifer material from the Michigan area called "Wagner soil". The authors collected completely mixed batch reactor

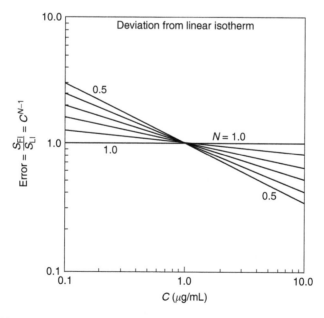

FIGURE 8-3 Graphical representation of the error introduced by the linear sorption isotherm when the sorption is Freundlish isotherm (nonlinear). The numbers on the lines are Freundlish constants N. (Adapted from Rao and Davidson, *Environmental Impact of Nonpoint Source Pollution*, Overcash, M.R., and Davidson, J.M., Eds., Ann Arbor Science Publishers Inc., Ann Arbor, Michigan, 1980, pp. 23–67.)

(CMBR) equilibrium data in laboratories and their results are presented in Figure 8-4. TTCE is a slightly polar chlorinated solvent of high volatility ($\log(K_{ow}) = 2.8$) and the Wagner soil has organic carbon content of 1.2%. Figure 8-4 shows the best fits to the equilibrium data by linear regression

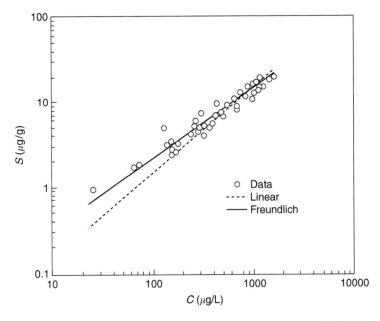

FIGURE 8-4 Comparison of the sorption isotherm data for tetrachloroethylene (TTCE) and Wagner soil with the linear and nonlinear (Freundlish) isotherm models. (Adapted from Weber et al., *Water Res.*, 25, 499-528, 1991.)

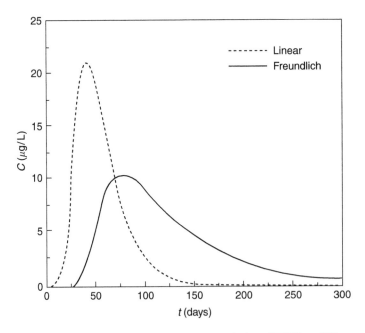

FIGURE 8-5 Time versus concentration curves for tetrachloroethylene (TTCE) and Wagner soil using solute transport models based on linear and nonlinear (Freundlish) sorption isotherm. (Adapted from Weber et al., *Water Res.*, 25, 499-528, 1991.)

with the linear and nonlinear Freundlich partitioning models. Figure 8-4 reveals that the Freundlich model provides a better overall representation of the data, although portions of the data are reasonably well matched by the linear model.

Weber et al. (1991) also compared concentration vs. time profiles for a selected point by using two different forms of the convection–dispersion–reaction equation transport model, incorporating these two alternative isotherm models calibrated with C_i (initial concentration) = 1000 μg/L, U = 1 m/day, D_x = 0.1 m²/day, ρ_b = 2.67 g/cm³, φ_e = 0.40, and time of solute input = 1 day. Comparison of the linear and Freundlich isotherm results is shown in Figure 8-5. It is apparent from this comparison that the use of a linear isotherm relationship to represent the equilibrium sorption behavior of the TTCE with respect to Wagner soil results in a substantially different situation than the use of Freundlich isotherm relationship. The increased retention at low concentrations and the decreased slope of the Freundlich model fit leads to greater asymmetry or tailing of the solute pulse. It is important to note that these effects are significant even though the data are not grossly illfit by the linear model. Weber and others do not present experimental concentration vs. data in Figure 8-5 as in Figure 8-4. Therefore, a quantitative evaluation for the linear and Freundlich isotherms cannot be made. Beyond this, comparisons are only presented for a single contaminant constituent. Therefore, additional research needs to be carried out by using different contaminant constituents and different materials in two- and three-dimensional solute transport cases.

8.3 COMPILED DATA FOR K_{oc} VALUES

8.3.1 DIRECT DATA FOR K_{oc} FOR VARIOUS CHEMICALS

As mentioned in Section 8.2.2.1, for a wide variety of organic compounds, K_d can be expressed by Eq. (8-4). If the values of f_{oc} (the fraction of organic carbon content in the soil) and K_{oc} (the constituent's distribution coefficient) are known, the K_d value can be estimated from Eq. (8-4).

During the past three decades many researchers have published K_{oc} data on numerous contaminant constituents. Mercer et al. (1990) compiled K_{oc} and other parameters for selected chemicals and they are presented in Table 8-1.

TABLE 8-1
Water Solubility, Vapor Pressure, K_{oc}, and K_{ow} for Selected Chemicals

Chemical Name	Water Solubility (mg/L)	Vapor Pressure (mmHg)	Henry's Law Constant (atm m³/mol)	K_{oc} (mL/g)	K_{ow}
Pesticides					
Acrolein (2-Propenal)	2.08E+05	2.69E+02	9.54E−05		8.13E−01
Aldicarb (Temik)	7.80E+03				5.00E+00
Aldrin	1.80E−01	6.00E−06	1.60E−05	9.60E+04	2.00E+05
Captan	5.00E−01	6.00E−05	4.75E−05	6.40E+03	2.24E+02
Carbaryl (Sevin)	4.00E+01	5.00E−03	3.31E−05	2.30E+02	2.29E+02
Carbofuran	4.15E+02	2.00E−05	1.40E−08	2.94E+01	2.07E+02
Carbophenothion (Trithion)				4.66E+04	
Chlordane	5.60E−01	1.00E−05	9.63E−06	1.40E+05	2.09E+03
p-Chloroaniline (4-Chlorobenzenamine)	5.30E+03	2.00E−02	6.40E−07	5.61E+02	6.76E+01
Chlorobenzilate	2.19E+01	1.20E−06	2.34E−08	8.00E+02	3.24E+04
Chlorpyrifos (Dursban)	3.00E−01	1.87E−05	2.87E−05	1.36E+04	6.60E+04
Crotoxyphos (Ciodrin)	1.00E+03	1.40E−05	5.79E−09	7.48E+01	
Cyclophosphamide	1.31E+09			4.20E−02	6.03E−04
					(Continued)

TABLE 8-1 (*Continued*)

Chemical Name	Water Solubility (mg/L)	Vapor Pressure (mmHg)	Henry's Law Constant (atm m^3/mol)	K_{oc} (mL/g)	K_{ow}
DDD	1.00E−01	1.89E−06	7.96E−06	7.70E+05	1.58E+06
DDE	4.00E−02	6.50E−06	6.80E−05	4.40E+06	1.00E+07
DDT	5.00E−03	5.50E−06	5.13E−04	2.43E+05	1.55E+06
Diazonin (Spectracide)	4.00E+01	1.40E−04	1.40E−06	8.50E+01	1.05E+03
1,2-Dibromo-3-chloropropane (DBCP)	1.00E+03	1.00E+00	3.11E−04	9.80E+01	1.95E+02
1,2-Dichloropropane	2.70E+03	4.20E+01	2.31E−03	5.10E+01	1.00E+02
1,3-Dichloropropane (Telone)	2.80E+03	2.50E+01	1.30E−01	4.80E+01	1.00E+02
Dichlorvos	1.00E+04	1.20E−02	3.50E−07		2.50E+01
Dieldrin	1.95E−01	1.78E−07	4.58E−07	1.70E+03	3.16E+03
Dimethoate	2.50E+04	2.50E−02	3.00E−07		5.10E−01
Dinoseb	5.00E+01	5.00E−05	3.16E−07	1.24E+02	1.98E+02
N,N-Diphenylamine	5.76E+01	3.80E−05	1.47E−07	4.70E+02	3.98E+03
Disulfoton	2.50E+01	1.80E−04	2.60E−06	1.60E+03	
α-Endosulfan	1.60E−01	1.00E−05	3.35E−05		3.55E+03
β-Endosulfan	7.00E−02	1.00E−05	7.65E−05		4.17E+03
Endosulfan sulfate	1.60E−01				4.57E+03
Endrin	2.40E−02	2.00E−07	4.17E−06		2.18E+05
Ethion	2.00E+00	1.50E−06	3.79E−07	1.54E+04	
Ethylene oxide	1.00E+06	1.31E+03	7.56E−05	2.20E+00	6.03E−01
Fenitrothion	3.00E+01	6.00E−06	7.30E−08		2.40E+03
Heptachlor	1.80E−01	3.00E−04	8.19E−04	1.20E−04	2.51E+04
Heptachlor epoxide	3.50E−01	3.00E−04	4.39E−04	2.20E+02	5.01E+02
α-Hexachlorocyclohexane	1.63E+00	2.50E−05	5.87E−06	3.80E+03	7.94E+03
β-Hexachlorocyclohexane	2.40E−01	2.80E−07	4.47E−07	3.80E+03	7.94E+03
δ-Hexachlorocyclohexane	3.14E+01	1.70E−05	2.07E−07	6.60E+03	1.26E+04
γ-Hexachlorocyclohexane (Lindane)	7.80E+00	1.60E−04	7.85E−06	1.08E+03	7.94E+03
Isophorone	1.20E+04	3.80E−01	5.75E−06		5.01E+01
Kepone	9.90E−03			5.50E+04	1.00E+02
Leptophos	2.40E+00			9.30E+03	2.02E+06
Malathion	1.45E+02	4.00E−05	1.20E−07	1.80E+03	7.76E+02
Methoxychlor	3.00E−03			8.00E+04	4.75E+04
Methyl parathion	6.00E+01	9.70E−06	5.59E−08	5.10E+03	8.13E+01
Mirex (Dechlorane)	6.00E−01	3.00E−01	3.59E−01	2.40E+07	7.80E+06
Nitralin	6.00E−01	9.30E−09	7.04E−09	9.60E+02	
Parathion	2.40E+01	3.78E−05	6.04E−07	1.07E+04	6.45E+03
Phenylurea (Phenylcarbamide)				7.63E+01	6.61E+00
Phorate (Thimet)	5.00E+01	8.40E−04	8.49E−11	3.26E+03	
Phosmet	2.50E+01	<1.0E−03			6.77E+02
Ronnel (Fenchlorphos)	6.00E+00	8.00E−04	5.64E−05		4.64E+04
Strychnine	1.56E+02				8.51E+01
2,3,7,8-Tetrachlorodibenzo-p-dioxin	2.00E−04	1.70E−06	3.60E−03	3.30E+06	5.25E+06
Toxaphene	5.00E−01	4.00E−01	4.36E−01	9.64E+02	2.00E+03
Trichlorfon (Chlorofos)	1.54E+05	7.80E−06	1.71E−11	6.10E+00	1.95E+02
Herbicides					
Alachlor	2.42E+02			1.90E+02	4.34E+02
Ametryn	1.85E+02			3.88E+02	

(Continued)

TABLE 8-1 (*Continued*)

Chemical Name	Water Solubility (mg/L)	Vapor Pressure (mmHg)	Henry's Law Constant (atm m³/mol)	K_{oc} (mL/g)	K_{ow}
Amitrole (Aminotriazole)	2.80E+05			4.40E+00	8.32E−03
Atrazine	3.30E+01	1.40E−06	2.59E−13	1.63E+02	2.12E+02
Benfluralin (Benefin)	<1.0E+00	3.89E−04		1.07E+04	
Bromocil	8.20E+02			7.20E+01	1.04E+02
Cacodylic acid	8.30E+05			2.40E+00	1.00E+00
Chloramben	7.00E+02	<7.0E−03		2.10E+01	1.30E+01
Chlorpropham	8.80E+01			8.16E+02	1.16E+03
Dalapon (2,2-Dichloropropanoic acid)	5.02E+05				5.70E+00
Diallate	1.40E+01	6.40E−03	1.65E−04	1.90E+03	5.37E+00
Dicamba	4.50E+03	2.00E−05	1.30E−09	2.20E+00	3.00E+00
Dichlobenil (2,6-Dichlorobenzonitrile)	1.80E+01	3.00E−06	3.77E−08	2.24E+02	7.87E+02
2,4-Dichlorophenoxyacetic acid (2,4-D)	6.20E+02	4.00E−01	1.88E−04	1.96E+01	6.46E+02
Dipropetryne	1.60E+01	7.50E−07	1.53E−08	1.18E+03	
Diuron	4.20E+01	<3.1E−06		3.82E+02	6.50E+02
Fenuron	3.85E+03	<1.6E−04		4.22E+01	1.00E+01
Fluometuron	9.00E+01			1.75E+02	2.20E+01
Linuron	7.50E+01	1.50E−05	6.56E−08	8.63E+02	1.54E+02
Methazole (Oxydiazol)	1.50E+00			2.62E+03	
Metobromuron	3.30E+02	3.00E−06	3.10E−09	2.71E+02	
Monuron	2.30E+02	5.00E−07	5.68E−10	1.83E+02	1.33E+02
Neburon	4.80E+00			3.11E+03	
Oxadiazon	7.00E−01	<1.0E−06		3.24E+03	
Paraquat	1.00E+06			1.55E+04	1.00E+00
Phenylmercuric acetate (PMA)	1.67E+03				
Picloram	4.30E+02	<6.2E−07		2.55E+01	2.00E+00
Prometryne	4.80E+01	1.00E−06	6.62E−09	6.14E+02	
Propachlor	5.80E+02			2.65E+02	5.60E+02
Propazine	8.60E+00	1.60E−07	5.63E−09	1.53E+02	7.85E+02
Silvex (Fenoprop)	1.40E+02			2.60E+03	
Simazine	3.50E+00	3.60E−08	2.73E−09	1.38E+02	8.80E+01
Terbacil	7.10E+02			4.12E+01	7.80E+01
2,4,5-Trichlorophenoxyacetic acid	2.38E+02			8.01E+01	4.00E+00
Triclopyr	4.30E+02	1.26E−06	9.89E−10	2.70E+01	3.00E+00
Trifluralin	6.00E−01	2.00E−04	1.47E−04	1.37E+04	2.20E+05
Aliphatic Compounds					
Acetonitrile (Methyl cyanide)	Infinite	7.40E+01	4.00E−06	2.20E+00	4.57E−01
Acrylonitrile (2-Propenenitrile)	7.94E+04	1.00E+02	8.84E−05	8.50E−01	1.78E+00
Bis (2-Chloroethoxy) methane	8.10E+04	<1.0E−01			1.82E+01
Bromodichloromethane (Dichlorobromomethane)	4.40E+03	5.00E+01	2.40E−03	6.10E+01	7.59E+01
Bromomethane (Methyl bromide)	1.30E+04	1.40E+03	1.30E−02		1.26E+01
1,3-Butadiene	7.35E+02	1.84E−03	1.78E−01	1.20E+02	9.77E+01
Chloroethene (Ethyl chloride)	5.74E+03	1.00E+03	6.15E−04	1.70E+01	3.50E+01
Chloroethene (Vinyl chloride)	2.67E+03	2.66E+03	8.19E−02	5.70E+01	2.40E+01
Chloromethane (Methyl chloride)	6.50E+03	4.31E+03	4.40E−02	3.50E+01	9.50E−01

(Continued)

TABLE 8-1 (*Continued*)

Chemical Name	Water Solubility (mg/L)	Vapor Pressure (mmHg)	Henry's Law Constant (atm m³/mol)	K_{oc} (mL/g)	K_{ow}
Cyanogen (Ethanedinitrile)	2.50E+05				
Dibromochloromethane	4.00E+03	1.50E+01	9.90E−04	8.40E+01	1.23E+02
Dichlorodifluoromethane (Freon 12)	2.80E+02	4.87E+03	2.97E+00	5.80E+01	1.45E+02
1,1-Dichloroethane (Ethylidine chloride)	5.50E+03	1.82E+02	4.31E−03	3.00E+01	6.17E+01
1,2-Dichloroethene (Ethylene dichloride)	8.52E+03	6.40E+01	9.78E−04	1.40E+01	3.02E+01
1,1-Dichloroethene (Vinylidine chloride)	2.25E+03	6.00E+02	3.40E−02	6.50E+01	6.92E+01
1,2-Dichloroethene (*cis*)	3.50E+03	2.08E+02	7.58E−03	4.90E+01	5.01E+00
1,2-Dichloroethene (*trans*)	6.30E+03	3.24E+02	6.56E−03	5.90E+01	3.02E+00
Dichloromethane (methylene chloride)	2.00E+04	3.62E+02	2.03E−03	8.80E+00	2.00E+01
Ethylene dibromide (EDB)	4.30E+03	1.17E+01	6.73E−04	4.40E+01	5.75E+01
Hexachlorobutadiene	1.50E−01	2.00E+00	4.57E+00	2.90E+04	6.02E+04
Hexachlorocyclopentadiene	2.10E+00	8.00E−02	1.37E−02	4.80E+03	1.10E+05
Hexachloroethane (Perchloroethane)	5.00E+01	4.00E−01	2.49E−03	2.00E+04	3.98E+04
Iodomethane (Methyl iodide)	1.40E+04	4.00E+02	5.34E−03	2.30E+01	4.90E+01
Isoprene		4.00E+02			
Pentachloroethane (Pentalin)	3.70E+01	3.40E+00	2.44E−02	1.90E+03	7.76E+02
1,1,1,2-Tetrachloroethane	2.90E+03	5.00E+00	3.81E−04	5.40E+01	
1,1,2,2-Tetrachloroethane	2.90E+03	5.00E+00	3.81E−04	1.18E+02	2.45E+02
Tetrachloroethene (PERC)	1.50E+02	1.78E+01	2.59E−02	3.64E+02	3.98E+02
Tetrachloromethane (Carbontetrachloride)	7.57E+02	9.00E+01	2.41E−02	4.39E+02	4.37E+02
Tribromomethane (Bromoform)	3.01E+03	5.00E+00	5.52E−04	1.16E+02	2.51E+02
1,1,1-Trichloroethane (Methylchloroform)	1.50E+03	1.23E+02	1.44E−02	1.52E+02	3.16E+02
1,1,2-Trichloroethane (Vinyltrichloride)	4.50E+03	3.00E+01	1.17E−03	5.60E+01	2.95E+02
Trichloroethene (TCE)	1.10E+03	5.79E+01	9.10E−03	1.26E+02	2.40E+02
Trichlorofluoromethane (Freon 11)	1.10E+03	6.67E+02	1.10E−01	1.59E+02	3.39E+02
Trichloromethane (Chloroform)	8.20E+03	1.51E+02	2.87E−03	4.70E+01	9.33E+01
1,1,2-Trichloro-1,2,2-trifluoroethane	1.00E+01	2.70E+02			1.00E+02

Aromatic Compounds

1,1-Biphenyl (Diphenyl)	7.50E+00	6.00E−02	1.50E−03		7.54E+03
Benzene	1.75E+03	9.52E+01	5.59E−03	8.30E+01	1.32E+02
Bromobenzene (Phenyl bromide)	4.46E+02	4.14E+00	1.92E−03	1.50E+02	9.00E+02
Chlorobenzene	4.66E+02	1.17E+01	3.72E−03	3.30E+02	6.92E+02
4-Chloro-m-cresol (Chlorocresol)	3.85E+03	5.00E−02	2.44E−06	4.90E+02	9.80E+02
2-Chlorophenol (*o*-Chlorophenol)	2.90E+04	1.80E+00	1.05E−05	4.00E+02	1.45E+02
Chlorotoluene (Benzyl chloride)	3.30E+03	1.00E+00	5.06E−05	5.00E+01	4.27E+02
m-Chlorotoluene	4.80E+01	4.60E+00	1.60E−02	1.20E+03	1.90E+03
o-Chlorotoluene	7.20E+01	2.70E+00	6.25E−03	1.60E+03	2.60E+03
p-Chlorotoluene	4.40E+01	4.50E+00	1.70E−02	1.20E+03	2.00E+03
Cresol (Technical) (Methylphenol)	3.10E+04	2.40E−01	1.10E−06	5.00E+02	9.33E+01

(Continued)

TABLE 8-1 (*Continued*)

Chemical Name	Water Solubility (mg/L)	Vapor Pressure (mmHg)	Henry's Law Constant (atm m³/mol)	K_{oc} (mL/g)	K_{ow}
o-Cresol (2-Methylphenol)	2.50E+04	2.43E−01	1.50E−06		8.91E+01
p-Cresol (4-Methylphenol)		1.14E−01			8.51E+01
Dibenzofuran					1.32E+04
1,2-Dichlorobenzene (*o*-Dichlorobenzene)	1.00E+02	1.00E+00	1.93E−03	1.70E+03	3.98E+03
1,3-Dichlorobenzene (*m*-Dichlorobenzene)	1.23E+02	2.28E+00	3.59E−03	1.70E+03	3.98E+03
1,4-Dichlorobenzene (*p*-Dichlorobenzene)	7.90E+01	1.18E+00	2.89E−03	1.70E+03	3.98E+03
2,4-Dichlorophenol	4.60E+03	5.90E−02	2.75E−06	3.80E+02	7.94E+02
Dichlorotoluene (Benzal chloride)	2.50E+00	3.00E−01	2.54E−02	9.90E+03	1.60E+04
Diethylstilbestrol (DES)	9.60E−03			2.80E+01	2.88E+05
2,4-Dimethylphenol (as-*m*-Xylenol)	4.20E+03	6.21E−02	2.38E−06	2.22E+02	2.63E+02
1,3-Dinitrobenzene	4.70E+02			1.50E+02	4.17E+01
4,6-Dinitro-*o*-cresol	2.90E+02	5.00E−02	4.49E−05	2.40E+02	5.01E+02
2,4-Dinitrophenol	5.60E+03	1.49E−05	6.45E−10	1.66E+01	3.16E+01
2,3-Dinitrotoluene	3.10E+03			5.30E+01	1.95E+02
2,4-Dinltrotoluene	2.40E+02	5.10E−03	5.09E−06	4.50E+01	1.00E+02
2,5-Dinitrotoluene	1.32E+03			8.40E+01	1.90E+02
2,6-Dinitrotoluene	1.32E+03	1.80E−02	3.27E−06	9.20E+01	1.00E+02
3,4-Dinitrotoluene	1.08E+03			9.40E+01	1.95E+02
Ethylbenzene (Phenylethane)	1.52E+02	7.00E+00	6.43E−03	1.10E+03	1.41E+03
Hexachlorobenzene (Perchlorobenzene)	6.00E−03	1.09E−05	6.81E−04	3.90E+03	1.70E+05
Hexachlorophene (Dermadex)	4.00E−03			9.10E+04	3.47E+07
Nitrobenzene	1.90E+03	1.50E−01	2.20E−05	3.60E+01	7.08E+01
2-Nitrophenol (*o*-Nitrophenol)	2.10E+03				5.75E+01
4-Nitrophenol (*p*-Nitrophenol)	1.60E+04				8.13E+01
m-Nitrotoluene (Methylnitrobenzene)	4.98E+02				2.92E+02
Pentachlorobenzene	1.35E−01	6.00E−03		1.30E+04	1.55E+05
Pentachloronitrobenzene (Quintozene)	7.11E−02	1.13E−04	6.18E−04	1.90E+04	2.82E+05
Pentachlorophenol	1.40E+01	1.10E−04	2.75E−06	5.30E+04	1.00E+05
Phenol	9.30E+04	3.41E−01	4.54E−07	1.42E+01	2.88E+01
Pyridine	1.00E+06	2.00E+01			4.57E+00
Styrene (Ethenylbenzene)	3.00E+02	4.50E+00	2.05E−03		
1,2,3,4-Tetrachlorobenzene	3.50E+00	4.00E−02		1.80E+04	2.88E+04
1,2,3,5-Tetrachlorobenzene	2.40E+00	7.00E−02		1.78E+04	2.88E+04
1,2,4,5-Tetrachlorobenzene	6.00E+00	5.40E−03		1.60E+03	4.68E+04
2,3,4,6-Tetrachlorophenol	7.00E+00	4.60E−03		9.80E+01	1.26E+04
Toluene (Methylbenzene)	5.35E+02	2.81E+01	6.37E−03	3.00E+02	5.37E+02
1,2,3-Trichlorobenzene	1.20E+01	2.10E−01	4.23E−03	7.40E+03	1.29E+04
1,2,4-Trichlorobenzene	3.00E+01	2.90E−01	2.31E−03	9.20E+03	2.00E+04
1,3,5-Trichlorobenzene	5.80E+00	5.80E−01	2.39E−02	6.20E+03	1.41E+04
2,4,5-Trichlorophenol	1.19E+03	1.00E+00	2.18E−04	8.90E+01	5.25E+03
2,4,6-Trichlorophenol	8.00E+02	1.20E−02	3.90E−06	2.00E+03	7.41E+03
1,2,4-Trimethylbenzene (Pseudocumene)	5.76E+01	2.03E+00	5.57E−03		

(Continued)

TABLE 8-1 (*Continued*)

Chemical Name	Water Solubility (mg/L)	Vapor Pressure (mmHg)	Henry's Law Constant (atm m^3/mol)	K_{oc} (mL/g)	K_{ow}
Xylene (mixed)	1.98E+02	1.00E+01	7.04E−03	2.40E+02	1.83E+03
m-Xylene (1,3-Dimethylbenzene)	1.30E+02	1.00E+01	1.07E−02	9.82E+02	1.82E+03
o-Xylene (1,2-Dimethylbenzene)	1.75E+02	6.60E+00	5.10E−03	8.30E+02	8.91E+02
p-Xylene (1,4-Dimethylbenzene)	1.98E+02	1.00E+01	7.05E−03	8.70E+02	1.41E+03
Polyaromatic Hydrocarbons					
Acenaphthylene	3.93E+00	2.90E−02	1.48E−03	2.50E+03	5.01E+03
Acenapthene	3.42E+00	1.55E−03	9.20E−05	4.60E+03	1.00E+04
Anthracene	4.50E−02	1.95E−04	1.02E−03	1.40E+04	2.82E+04
Benz(*c*)acridine	1.40E+01			1.00E+03	3.63E+04
Benzo(*a*)anthracene	5.70E−03	2.20E−08	1.16E−06	1.38E+06	3.98E+05
Benzo(*a*)pyrene	1.20E−03	5.60E−09	1.55E−06	5.50E+06	1.15E+06
Benzo(*b*)fluoranthene	1.40E−02	5.00E−07	1.19E−05	5.50E+05	1.15E+06
Benzo(*gh i*)perylene	7.00E−04	1.03E−10	5.34E−08	1.60E+06	3.24E+06
Benzo(*k*)fluoranthene	4.30E−03	5.10E−07	3.94E−05	5.50E+05	1.15E+06
2-Chloronapthalene	6.74E+00	1.70E−02	4.27E−04		1.32E+04
Chrysene	1.80E−03	6.30E−09	1.05E−06	2.00E+05	4.07E+05
1,2,7,8-Dibenzopyrene	1.01E−01			1.20E+03	4.17E+06
Dibenz(*a,h*)anthracene	5.00E−04	1.00E−10	7.33E−08	3.30E+06	6.31E+06
7,12-Dimethylbenz(*a*)anthracene	4.40E−03			4.76E+05	8.71E+06
Fluoranthene	2.06E−01	5.00E−06	6.46E−06	3.80E+04	7.94E+04
Fluorene (2,3-Benzidene)	1.69E+00	7.10E−04	6.42E−05	7.30E+03	1.58E+04
Indene					8.32E+02
Indeno(1,2,3-cd)pyrene	5.30E−04	1.00E−10	6.86E−08	1.60E+06	3.16E+06
2-Methylnapthalene	2.54E+01			8.50E+03	1.30E+04
Napthalene (Napthene)	3.17E+01	2.30E−01	1.15E−03	1.30E+03	2.76E+03
1-Napthylamine	2.35E+03	6.50E−05	5.21E−09	6.10E+01	1.17E+02
2-Napthylamine	5.36E+02	2.56E−04	8.23E−08	1.30E+02	1.17E+02
Phenanthrene	1.00E+00	6.80E−04	1.59E−04	1.40E+04	2.88E+04
Pyrene	1.32E−01	2.50E−06	5.04E−06	3.80E+04	7.59E+04
Tetracene (Napthacene)	5.00E−04			6.50E+05	8.00E+05
Amines and Amides					
2-Acetylaminofluorene	6.50E+00			1.60E+03	1.91E+03
Acrylamide (2-Propenamide)	2.05E+06	7.00E−03	3.19E−10		
4-Aminobiphenyl (*p*-Biphenylamine)	8.42E+02	6.00E−05	1.59E−08	1.07E+02	6.03E+02
Aniline (Benzenamine)	3.66E+04	3.00E−01	1.00E−06		7.00E+00
Auramine	2.10E+00			2.90E+03	1.45E+04
Benzidine (*p*-Diaminodiphenyl)	4.00E+02	5.00E−04	3.03E−07	1.05E+01	2.00E+01
2,4-Diaminotoluene (Toluenediamine)	4.77E+04	3.80E−05	1.28E−10	1.20E+01	2.24E+00
3,3-Dichlorobenzidine	4.00E+00	1.00E−05	8.33E−07	1.55E+03	3.16E+03
Diethanolamine	9.54E+05				3.72E−02
Diethylaniline (Benzenamine)	6.70E+02				9.00E+00
Diethylnitrosamine (Nitrosodiethylamine)		5.00E+00			3.02E+00
Dimethylamine	1.00E+06	1.52E+03	9.02E−05	4.35E+02	4.17E−01

(Continued)

TABLE 8-1 (*Continued*)

Chemical Name	Water Solubility (mg/L)	Vapor Pressure (mmHg)	Henry's Law Constant (atm m³/mol)	K_{oc} (mL/g)	K_{ow}
Dimethylaminoazobenzene	1.36E+01	3.30E−07	7.19E−09	1.00E+03	5.25E+03
Dimethylnitrosamine	Infinite	8.10E+00	7.90E−07	1.00E−01	2.09E−01
Diphenylnitrosamine					3.72E+02
Dipropylnitrosamine	9.90E+03	4.00E−01	6.92E−06	1.50E+01	3.16E+01
Methylvinylnitrosamine	7.60E+05	1.23E+01	1.83E−06	2.50E+00	5.89E−01
m-Nitroaniline (3-Nitroaniline)	8.90E+02				2.34E+01
o-Nitroaniline (2-Nitroaniline)	1.47E+04				6.17E+01
p-Nitroaniline (4-Nitroaniline)	7.30E+02				2.45E+01
N-Nitrosodi-n-propylamine					
Thioacetamide (ethanethioamide)	1.63E+05				3.47E−01
o-Toluidine hydrochloride	1.50E+04	1.00E−01	9.39E−07	2.20E+01	1.95E+01
o-Toluidine (2-Aminotoluene)	7.35E+01	<1.0E+00		4.10E+02	7.58E+02
Triethylamine	1.50E+04	7.00E+00	1.30E+05		

Ethers and Alcohols

Allyl alcohol (Propenol)	5.10E+05	2.46E+01	3.69E−06	3.20E+00	6.03E−01
Anisole (Methoxybenzene)	1.52E+03	2.60E+00	2.43E−04	2.00E+01	1.29E+02
Benzyl Alcohol (Benzenemethanol)	8.00E+02	1.10E−01	1.95E−05		1.26E+01
Bis (2-Chloroethyl) ether	1.02E+04	7.10E−01	1.31E−05	1.39E+01	3.16E+01
Bis (2-Chloroisopropyl) ether	1.70E+03	8.50E−01	1.13E−04	6.10E+01	1.26E+02
Bis (Chloromethyl) ether	2.20E+04	3.00E+01	2.06E−04	1.20E+00	2.40E+00
4-Bromophenyl phenyl ether		1.50E−03			1.91E+04
2-Chloroethyl vinyl ether	1.50E+04	2.67E+01	2.50E−04		1.90E+01
Chloromethyl methyl ether					1.00E+00
4-Chlorophenyl phenyl ether	3.30E+00	2.70E−03	2.19E−04		1.20E+04
Diphenylether (Phenyl ether)	2.10E+01	2.13E−02	8.67E−09		1.62E+04
Ethanol	Infinite	7.40E+02	4.48E−05	2.20E+00	4.79E−01

Phthalates

Bis (2-Ethylhexyl) phthalate	2.85E−01	2.00E−07	3.61E−07	5.90E+03	9.50E+03
Butylbenzyl phthalate	4.22E+01				6.31E+04
Di-n-octyl phthalate	3.00E+00				1.58E+09
Dibutyl phthalate	1.30E+01	1.00E−05	2.82E−07	1.70E+05	3.98E+05
Diethyl phthalate	8.96E+02	3.50E−03	1.14E−06	1.42E+02	3.16E+02
Dimethylphthalate	4.32E+03	<1.0E−02			1.30E+02

Ketones and Aldehydes

2-Butanone (Methyl ethyl ketone)	2.68E+05	7.75E+01	2.74E−05	4.50E+00	1.82E+00
2-Hexanone (Methyl butyl ketone)	1.40E+04	3.00E+10	2.82E−05		
4-Methyl-2-Pentanone (Isopropylacetone)	1.70E+04	2.00E+01	1.55E−04		
Acetone (2-Propanone)	Infinite	2.70E+02	2.06E−05	2.20E+00	5.75E−01
Formaldehyde	4.00E+05	1.00E+01	9.87E−07	3.60E+00	1.00E+00
Glycidaldehyde	1.70E+08	1.97E+01	1.10E−08	1.00E−01	2.82E–02
Acrylic acid (2-Propenoic acid)	Infinite	4.00E+00			1.35E+00

Carboxylic Acids and Esters

Azaserine	1.36E+05			6.60E+00	8.32E−02
Benzoic acid	2.70E+03				7.41E+01

(Continued)

TABLE 8-1 (*Continued*)

Chemical Name	Water Solubility (mg/L)	Vapor Pressure (mmHg)	Henry's Law Constant (atm m³/mol)	K_{oc} (mL/g)	K_{ow}
		Herbicides			
Dimethyl sulfate (DMS)	3.24E+05	6.80E−01	3.48E−07	4.10E+00	5.75E−02
Ethyl methanesul fonate (EMS)	3.69E+05	2.06E−01	9.12E−08	3.80E+00	1.62E+00
Formic acid	1.00E+06	4.00E+01			2.88E−01
Lasiocarpine	1.06E+03			7.60E+01	9.77E+00
Methyl methacrylate	2.00E+01	3.70E+01	2.43E−01	8.40E+02	6.17E+00
Vinyl acetate	2.00E+04				
PCBs					
Aroclor 1016	4.20E−01	4.00E−04			2.40E+04
Aroclor 1221	1.50E+01	6.70E−03			1.23E+04
Aroclor 1232	1.45E+00	4.06E−03			1.58E+03
Aroclor 1242	2.40E−01	4.10E−04	5.60E−04		1.29E+04
Aroclor 1248	5.40E−02	4.90E−04	3.50E−03		5.62E+05
Aroclor 1254	1.20E−02	7.70E−05	2.70E−03	4.25E+04	1.07E+06
Aroclor 1260	2.70E−03	4.10E−05	7.10E−03		1.38E+07
Polychlorinated biphenyls (PCBs)	3.10E−02	7.70E−05	1.07E−03	5.30E+05	1.10E+06
Heterpcyclic Compounds					
Dihydrosafrole	1.50E+03			7.80E+01	3.63E+02
1,4-Dioxane (1,4-Diethylene dioxide)	4.31E+05	3.99E+01	1.07E−05	3.50E+00	1.02E+00
Epichlorohydrin	6.00E+04	1.57E+01	3.19E−05	1.00E+01	1.41E+00
Isosafrole	1.09E+03	1.60E−08	3.25E−12	9.30E+01	4.57E+02
N-Nitrosopiperidine	1.90E+06	1.40E−01	1.11E−08	1.50E+00	3.24E−01
N-Nitrosopyrrolidine	7.00E+06	1.10E−01	2.07E−09	8.00E−01	8.71E−02
Safrole	1.50E+03	9.10E−04	1.29E−07	7.80E+01	3.39E+02
Uracil mustard	6.41E+02			1.20E+02	8.13E−02
Hydrazines					
1,2-Diethylhydrazine	2.88E+07			3.00E−01	2.09E−02
1,1-Dimethylhydrazine	1.24E+08	1.57E+02	1.00E−07	2.00E−01	3.80E−03
1,2-Diphenylhydrazine (Hydrazobenzene)	1.84E+03	2.60E−05	3.42E−09	4.18E+02	7.94E+02
Hydrazine	3.41E+08	1.40E+01	1.73E−09	1.00E−01	8.32E−04
Miscellaneous Organic Compounds					
Aziridine (Ethylenimine)	2.66E+06	2.55E+02	5.43E−06	1.30E+00	9.77E−02
Carbon disulfide	2.94E+03	3.60E+02	1.23E−02	5.40E+01	1.00E+02
Diethyl arsine	4.17E+02	3.50E+01	1.48E−02	1.60E+02	9.33E+02
Dimethylcarbamoyl chloride	1.44E+07	1.95E+00	1.92E−08	5.00E−01	4.79E−02
Methylnitrosourea	6.89E+08			1.00E−01	1.54E−04
Mustard gas (bis(2-Chloroethyl)sulfide)	8.00E+02	1.70E−01	4.45E−05	1.10E+02	2.34E+01
Phenobarbital	1.00E+03			9.80E+01	6.46E−01
Propylenimine	9.44E+05	1.41E+02	1.12E−05	2.30E+00	3.31E−01
Tetraethyl lead	8.00E−01	1.50E−01	7.97E−02	4.90E+03	
Thiourea (Thiocarbamide)	1.72E+06			1.60E+00	8.91E−03
Tris-BP (2,3-Dibromo1propanol phosphate)	1.20E+02			3.10E+02	1.32E+04

(Continued)

TABLE 8-1 (*Continued*)

Chemical Name	Water Solubility (mg/L)	Vapor Pressure (mmHg)	Henry's Law Constant (atm m³/mol)	K_{oc} (mL/g)	K_{ow}
Inorganics					
Ammonia	5.30E+05	7.60E+03	3.21E−04	3.10E+00	1.00E+00
Antimony and compounds		1.00E+00			
Arsenic and compounds		0.00E+00			
Beryllium and compounds		0.00E+00			
Cadmium and compounds		0.00E+00			
Chromium III and compounds		0.00E+00			
Chromium VI and compounds		0.00E+00			
Copper and compounds		0.00E+00			
Cyanogen chloride	2.50E+03	1.00E+03	3.24E−02		1.00E+00
Hydrogen cyanide	Infinite	6.20E+02			5.62E−01
Hydrogen sulfide	4.13E+03	1.52E+04	1.65E−01		
Lead and compounds		0.00E+00			
Mercury and compounds (inorganic)	3.00E−02	2.00E−03	1.10E−02		
Nickel and compounds		0.00E+00			
Potassium cyanide	5.00E+05				
Selenium and compounds		0.00E+00			
Silver and compounds		0.00E+00			
Sodium cyanide	8.20E+05				
Thallium chloride	2.90E+03	0.00E+00			
Thallium sulfate	2.00E+02	0.00E+00			
Thallium and compounds		0.00E+00			
Zinc and compounds		0.00E+00			

Source: Adapted from Mercer et al., Basics of Pump and Treat Ground Water Remediation Technology, U.S. Environmental Protection Laboratory, Ada, Oklahoma, 1990

Rao and Davidson (1980) compiled data on sorption isotherm parameters (K_d, N, and K_{oc}) for a number of pesticides and related compounds and they are presented in Table 8-2. The list of compounds included in this table was prepared on the basis of "benchmark" pesticides and covers a broad spectrum of herbicides, insecticides, and fungicides frequently used in agriculture. In Table 8-2, coefficients of variation (CV) are also included. As shown in Table 8-2, the CV values for K_d range from 13.5 to 150.8 % while the CV for K_{oc} are much lower. Rao and Davidson (1980) conclude that this situation supports the view that K_{oc} may be a universal sorption coefficient for each pesticide; the values of N are not as variable as K_d and K_{oc} values; the overall mean and CV for N are, respectively, 0.87 and 14.8 %; therefore, the assumption of linear isotherms required for using K_{oc} values is apparently not valid for many pesticides. Rao and Davidson (1980) also conclude that there seem to be three major reasons for the rather large variability in the K_{oc} values shown in Table 8-2: (1) the relationship between K_d and f_{oc} may be nonlinear. However, multiple regression analysis for a variety of data did not reveal any consistent or significant trend in this regard; (2) although the variability in K_{oc} values from a given reference were generally low (CV = 40 to 60 %), the CV for K_{oc} values pooled from several references was generally higher. This trend suggests that the experimental methods employed by various authors may not have been the same. For example, shaking method, contact time allowed between pesticide and soil, temperature, and the analytic

TABLE 8-2

Summary of Sorption Partition Coefficient Values Compiled from Published Literature for Several Pesticides and Related Organic Compounds

Pesticide	Number of Soils	K_d Mean (%CV[1])	N Mean (%CV[1])	K_{oc} Mean (%CV[1])
Ametryne	32	6.16 (65.1)		388.4 (57.1)
Amiben	12	1.40 (145.2)		189.6 (149.7)
Atrazine	56	3.20 (89.8)	0.86 (9.9)	163.0 (49.1)
Bromacil	2			72.0 (102.1)
Carbofuran	5	1.05 (111.8)	0.96 (6.5)	29.4 (30.0)
Chlorobromuron	5	37.22 (121.2)	0.74 (8.8)	995.6 (55.1)
Chloroneb	1			1652.9 (–)
Chloroxuron	5	234.0 (71.1)		4343.3 (28.8)
Chloropropham	36			816.3 (–)
Chlorthiamid	6	5.57 (39.7)		98.3 (27.5)
Ciodrin	3			74.8 (59.1)
DDT	2			243118.0 (65.0)
Dicamba	5	0.11 (103.9)	1.18 (13.7)	2.2 (73.5)
Dichlobenil	34	3.0 (70.9)		224.4 (77.4)
Dimethyl Amine	5	14.0 (77.3)	1.00 (0.5)	434.9 (19.8)
Dipropetryne	5	13.5 (91.5)	0.86 (3.9)	1180.8 (74.9)
Disulfoton	20	32.3 (91.7)	0.92 (10.3)	1603.0 (144.2)
Diuron	84	8.9 (150.8)	0.83 (28.8)	382.6 (72.4)
Fenuron	10	2.11 (120.8)	0.87 (13.2)	42.2 (84.7)
Lindane	3	20.1 (13.5)		1080.9 (13.0)
Linuron	33	21.2 (100.2)	0.71 (6.6)	862.8 (72.3)
Malathion	20	34.1 (67.1)		1796.9 (65.9)
Methyl Parathion	7	12.7 (67.2)	0.94 (12.9)	5101.5 (113.6)
Methyl urea	5	3.5 (80.1)		58.8 (15.1)
Metobromuron	4	6.7 (62.8)	0.76 (15.7)	271.5 (37.1)
Monolinuron	10	12.3 (83.6)	0.76 (10.1)	284.3 (55.2)
Monuron	18	7.6 (122.5)	0.83 (20.6)	183.5 (60.8)
Neburon	5	166.8 (68.3)		3110.5 (23.5)
p-Chloroaniline	5	16.8 (79.4)	0.70 (2.6)	561.5 (33.6)
Parathion	4	21.9 (63.7)	1.03 (2.9)	10650.3 (74.6)
Phenyl urea	5	4.9 (91.9)		76.3 (12.3)
Picloram	26	0.63 (150.2)	0.85 (13.2)	25.5 (138.5)
Prometone	29	7.2 (147.5)	0.81 (4.4)	524.3 (143.6)
Prometryne	38	10.8 (123.8)	0.82 (5.3)	614.3 (99.1)
Propazine	36	3.1 (135.8)	0.95 (3.6)	153.5 (37.0)
Simazine	147	2.3 (158.5)	0.78 (5.2)	138.4 (12.6)
Telone (cis)	6	54.8 (78.6)		798.1 (44.3)
Telone (trans)	6	93.2 (72.0)		1379.0 (45.4)
Terbacil	4	0.78 (145.0)	0.93 (6.1)	41.2 (42.2)
Thimet	4	8.8 (77.7)	0.97 (4.2)	3255.2 (49.5)
Trithion	4	74.6 (26.6)	0.94 (7.4)	46579.7 (80.2)
2,4-D	9	0.78 (128.6)	1.01 (4.0)	19.6 (72.4)
2,4-D Amine	3	2.0 (112.5)	0.75 (9.0)	109.1 (30.2)
2,4,5-T	4	1.6 (87.3)		80.1 (45.3)

(1) Coefficient of Variation

Source: Adapted from Rao and Davidson, Environmental Impact of Nonpoint Source Pollution, Overcash, M.R. and Davidson, J.M., Eds., Ann Anbor Science Publishers Inco., Ann Arbor, Michigon, 1980, pp. 23–67.

TABLE 8-3
Summary of Octanol-Water Partition Coefficients, K_{ow}, for Pesticides Compiled from Literature

Pesticide	Ref. No.	K_{ow}	log (K_{ow})
A. Insecticides			
Aldicarb	(1)	5.00000E+00	6.98970E−01
Altosid	(1)	1.76000E+02	2.24551E+00
Carbaryl	(1)	6.51000E+02	2.81358E+00
Carbofuran	(1)	2.07000E+02	2.31597E+00
Chlordane	(1)	2.10800E+03	3.32387E+00
Chlorpyrifos	(1)	2.05900E+03	3.31366E+00
Chlorpyrifos	(2)	6.60000E+04	4.81954E+00
Chlorpyrifos	(3)	1.28825E+05	5.11000E+00
Chlorpyrifos, Methyl	(1)	1.97000E+03	3.29447E+00
Chlorpyrifos, Methyl	(3)	2.04170E+04	4.30999E+00
DDD	(1)	1.15000E+05	5.06070E+00
DDE	(1)	7.34450E+04	4.86596E+00
DDE p, p	(2)	4.89779E+05	5.69000E+00
DDT	(2)	3.70000E+05	5.56820E+00
DDT p, p	(2)	1.54882E+06	6.19000E+00
DDVP	(1)	1.95000E+02	2.29003E+00
Dialifor	(3)	4.89780E+04	4.69000E+00
Diazinon	(1)	1.05200E+03	3.02202E+00
Dichlofenthion	(3)	1.38038E+05	5.14000E+00
Dicifol	(1)	3.46100E+03	3.53920E+00
Dieldrin	(1)	4.93000E+03	3.69285E+00
Dinoseb	(1)	1.98000E+02	2.29667E+00
Endrin	(1)	1.61900E+03	3.20925E+00
Ethoxychlor	(1)	1.18000E+03	3.07188E+00
Fenitrothion	(3)	2.29900E+03	3.36154E+00
HCB	(2)	1.66000E+06	6.22011E+00
Heptachlor	(1)	7.36600E+03	3.86723E+00
Leptofos	(1)	4.12200E+03	3.61511E+00
Leptophos	(3)	2.04174E+06	6.31000E+00
Lindane	(1)	6.43000E+02	2.80821E+00
Malathion	(1)	2.30000E+02	2.36173E+00
Malathion	(3)	7.76000E+02	2.88986E+00
Methomyl	(1)	1.20000E+01	1.07918E+00
Methoxychlor	(1)	2.05000E+03	3.31175E+00
Methoxychlor	(4)	1.20000E+05	5.07918E+00
Methyl Parathion	(1)	2.07600E+03	3.31723E+00
Parathion	(3)	6.45500E+03	3.80990E+00
Permethrin	(1)	7.53000E+02	2.87679E+00
Phorate	(1)	8.23000E+02	2.91540E+00
Phosalone	(3)	1.99530E+04	4.30001E+00
Phosmet	(3)	6.76000E+02	2.82995E+00
Propoxur	(1)	2.80000E+01	1.44716E+00
Ronnel	(3)	7.58580E+04	4.88000E+00
Terbufos	(1)	1.67000E+02	2.22272E+00
Toxaphene	(1)	1.69500E+03	3.22917E+00

(Continued)

TABLE 8-3 (*Continued*)

Pesticide	Ref. No.	K_{ow}	$\log (K_{ow})$
B. Herbicides			
Alachlor	(1)	4.34000E+02	2.63749E+00
Atrazine	(5)	2.12000E+02	2.32634E+00
Atrazine	(1)	2.26000E+02	2.35411E+00
Bifenox	(1)	1.74000E+02	2.24055E+00
Bromacil	(1)	1.04000E+02	2.01703E+00
Chloramben	(1)	1.300000E+01	1.11394E+00
Chloropropham	(1)	1.16000E+03	3.06446E+00
Dalapon	(2)	5.70000E+00	7.55875E−01
Dalapon, Na Salt	(2)	1.00000E+00	0.00000E+00
Dicamba	(1)	3.00000E+00	4.77121E−01
Dichlobenil	(1)	7.87000E+02	2.89597E+00
Diuron	(1)	6.50000E+02	2.81291E+00
Monuron	(1)	1.33000E+02	2.12385E+00
Msma	(1)	8.00000E−04	−3.09691E+00
Nitrofen	(1)	1.24500E+03	3.09517E+00
Paraquat .2HCl	(1)	1.00000E+00	0.00000E+00
Picloram	(2)	2.00000E+00	3.01030E−01
Propachlor	(1)	4.10000E+01	1.61278E+00
Propanil	(1)	1.06000E+02	2.02531E+00
Simazine	(1)	8.80000E+01	1.94448E+00
Terbacil	(5)	7.80000E+01	1.89209E+00
Trifluralin	(1)	1.15000E+03	3.06070E+00
2, 4-D	(5)	4.16000E+02	2.61909E+00
2, 4-D	(1)	4.43000E+02	2.64640E+00
2, 4-D	(6)	6.46000E+02	2.81023E+00
2,4, 5-T	(2)	7.00000E+00	8.45098E−01
2, 4, 5-T, Butyl Ester	(2)	6.40000E+04	4.80618E+00
2, 4, 5-T Octyl Ester	(2)	9.09000E+02	2.95856E+00
C. Fungicides			
Benomyl	(1)	2.64000E+02	2.42160E+00
Captan	(1)	3.30000E+01	1.51851E+00
PCP	(1)	1.42900E+04	4.15503E+00

Note: 1 — Metcalf and Lu (1978); 2 — Kenaga (1975); 3 — Chiou and Freed (1977); 4 — Karickhoff and Brown (1978); 5 — Rao and Davidson (1980); 6 — Leo et al. (1971).

Source: Adapted from Rao and Davidson, Environmental Impact of Nonpoint Source Pollution, Overcash, M.R. and Davidson, J.M., Eds., Ann Anbor Science Publishers Inc., Ann Arbor, Michigon, 1980, pp. 23–67.

method used to assay the pesticide contribute to the variability method; and (3) whether the sorption parameter K_d was derived from an entire isotherm or a single concentration.

8.4 REGRESSION EQUATIONS FOR K_{oc} VALUES

8.4.1 RELATIONSHIP BETWEEN THE K_{oc} AND K_{ow}

A linear relationship between $\log (K_{ow})$ and $\log (K_{oc})$ values for 30 pesticides was reported by Briggs (1973). Similarly, Karickhoff and Brown (1978) obtained the following relationships for s-triazines and dinitroanaline herbicides:

$$\log(K_{oc}) = 0.94 \log(K_{ow}) + 0.02 \qquad (8\text{-}14)$$

which is Eq. (8-23) in Table 8-4. For polycylic aromatic hydrocarbons, the authors have given the following equation:

$$\log(K_{oc}) = 1.00\log(K_{ow}) - 0.21 \tag{8-15}$$

which is Eq. (8-22) in Table 8-4.

K_{ow} values for several pesticides reported in the literature are summarized in Table 8-3 (Rao and Davidson, 1980). These K_{ow} values were paired by the authors with the available K_{oc} values (see Table 8-2) and were used to obtain the following regression equation ($r^2 = 0.91$):

$$\log(K_{oc}) = 1.029\log(K_{ow}) - 0.18 \tag{8-16}$$

which is Eq. (8-24) in Table 8-4. A graphical representation of Eqs. (8-14)–(8-16), along with the data for 13 pesticides, is presented in Figure 8-6. As shown in Figure 8-6, an excellent agreement between the regression equations of Karickhoff and Brown (Eqs. [8-14] and [8-16]) is apparent. This is especially important considering the fact that the data used to develop Eq. (8-16) were obtained from several references.

TABLE 8-4
Regression Equations for the Estimation of K_{oc} Based on K_{ow}

Eq. No.	Equation	N[a]	R^2[b]	Chemical Classes Represented	Reference
8-20	$\log(K_{oc}) = 0.544\log(K_{ow}) + 1.377$	45	0.74	Wide variety, mostly pesticides	Kenaga and Goring (1980)
8-21	$\log(K_{oc}) = 0.937\log(K_{ow}) - 1.006$	19	0.95	Aromatics, polynuclear aromatics, triazines, and dinitroaniline herbicides	Brown and Flagg (1981)
8-22	$\log(K_{oc}) = 1.00\log(K_{ow}) - 0.21$	10	1.00	Mostly aromatic or polynuclear aromatics, two chlorinated	Karickhoff et al. (1979)
8-23	$\log(K_{oc}) = 0.94\log(K_{ow}) + 0.02$	9	NA	s-Triazines and dinitroaniline herbicides	Karickhoff and Brown (1978)
8-24	$\log(K_{oc}) = 1.029\log(K_{ow}) - 0.18$	13	0.91	Variety of insecticides, herbicides, and fungicides	Rao and Davidson (1980)
8-25	$\log(K_{oc}) = 0.524\log(K_{ow}) + 0.855$	30	0.84	Substituted phenylureas and alkyl-N-phenylcarbamates	Briggs (1973)

[a] N=number of chemical constituents used to obtain regression equation.

[b] r^2=correlation coefficient for regression equation.

[c] NA=not available.

Source: Adapted from Lyman et al. *Handbook of Chemical Property Estimation Methods*, McGraw-Hill Book Company, New York, 1982.

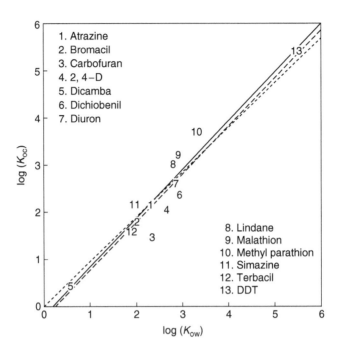

FIGURE 8-6 Relationship between K_{oc} and K_{ow} values for pesticides reported in the literature. The solid line is a plot of Eq. (8-16). The dashed lines are plots of Eqs. (8-14) and (8-15) and represent best fit lines for data reported for s-triazines, dinitroanalines, and polycyclic aromatic hydrocarbons. (Adapted from Rao and Davidson, *Environmental Impact of Nonpoint Source Pollution*, Overcash, M.R., and Davidson, J.M., Eds., Ann Arbor Science Publishers Inc., Ann Arbor, Michigan, 1980, pp. 23–67.)

8.4.2 RELATIONSHIP BETWEEN K_{oc} AND SOLUBILITY S

Direct K_{oc} data for many compounds are unknown. As a result, numerous researchers have developed correlation equations to relate K_{oc} to more commonly available chemical properties, such as solubility S or K_{ow} (Chiou et al., 1982a; Chiou et al., 1982b; 1983; Schwarzenbach and Westall, 1981; Karickhoff, 1981; Kenaga and Goring, 1980; Johnson et al., 1989). Some of the correlations cover a broad range of compounds for which the errors associated with the K_{oc} estimates can be large. However, within a compound class, K_{oc} values derived using these expressions generally provide reasonable estimates for sorption. Based on data obtained from experiments, Kenaga and Goring (1980) derived the following equation:

$$\log(K_{oc}) = -0.55 \log(S) + 3.64 \tag{8-17}$$

where K_{oc} is expressed in mL/g (or L/kg) and S is expressed in mg/L. The graphical representation of Eq. (8-17) is shown in Figure 8-7. With regard to the correlation of K_{ow}, Kenaga and Goring (1980) gave the following equation:

$$\log(K_{oc}) = 0.544 \log(K_{ow}) + 1.377 \tag{8-18}$$

8.5 ESTIMATION OF K_{oc} FROM THE REGRESSION EQUATIONS

The estimation methods for K_{oc} are based on some regression equations developed by various researchers, especially during the past decades. Lyman et al. (1982) compiled most of the regression equations regarding K_{oc} and their compiled results are presented in the following sections.

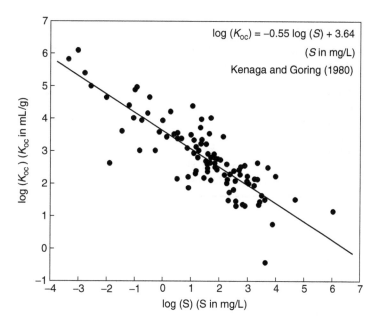

FIGURE 8-7 Correlation of K_{oc} and solubility data. (From Kenaga, E.E. and Goring, C.A.I., ASTM Special Technical Publication 707, ASTM, Washington, DC 1980; as presented in Johnson et al., Seminor Publication, U.S. Environmental Protection Agency, EPA/625/4-89/019, Cincinnati, Ohio, 1989, pp. 41-56.)

8.5.1 FOUNDATIONS OF THE ESTIMATION METHODS

All the methods available in the literature for estimation of K_{oc} are based on empirical equations along with other characteristics of the constituents. Of these characteristics, K_{ow} is one that is widely used. The regression equations obtained from various data sets are usually expressed in log–log form as (Lyman et al., 1982)

$$\log(K_{oc}) + a\, \log(K_{ow}) + b \qquad (8\text{-}19)$$

where a and b are constants. Although K_{ow} has been used most frequently, a few equations of the form expressed by Eq. (8-19) have been reported. Each equation was derived from a different data set representing different chemical constituents and ranges of the parameters involved. Lyman et al. (1982) state that the uncertainty associated with any K_{oc} value estimated by one of these equations is generally less than one order of magnitude, i.e., less than $\pm$ a factor of 10.

8.5.2 AVAILABLE REGRESSION EQUATIONS BASED ON K_{ow}

Table 8-4 presents regression equations for K_{oc} based on K_{ow} values for various chemical constituents (Lyman et al., 1982). Six regression equations, Eqs. (8-20)–(8-25), are listed in Table 8-4 along with some basic information on the data set used to derive each equation. Note that Eqs. (8-22), (8-23), and (8-24) in Table 8-4 are presented as Eqs. (8-15), (8-14), and (8-16), respectively, in Section 8.4.1.

Compounds used by Karickhoff et al. (1979) for Eq. (8-21) (in part) and (8-22) are given in Table 8-5. Compounds used by Brown and Flagg (1981) and Karickhoff and Brown (1978) to determine Eq. (8-21) (in part) and Eq. (8-23) are given in Table 8-6. Compounds used by Rao and Davidson (1980) for Eq. (8-24) are given in Table 8-7. Compounds used by Briggs (1973) for Eq. (8-25) (Lyman et al., 1982) are given in Table 8-8.

Correlations between K_{oc} and K_{ow} for various equations in Table 8-4 are given in Figure 8-8.

TABLE 8-5
Compounds Used by Karickhoff et al. (1979) for Eq. (8-21) (in Part) and Eq. (8-22)

Anthracene	2-Methylnaphthalene
Benzene	Naphthalene
Hexachlorobbiphenyl	Phenanthrene
Metahoxychlor	Pyrene
9-Methylanthracene	Tetracene

Source: Adapted from Lyman et al. *Handbook of Chemical Property Estimation Methods*, McGraw-Hill Book Company, New York, 1982.

TABLE 8-6
Compounds Used by Brown and Flagg (1981) and Karickhoff and Brown (1978) to Determine Eq. (8-21) (in Part) and Eq. (8-23)

Atrazine	Propazine	Trifluralin
Cyanazine	Simazine	Two photodegradation products of trifluralin
Ipazine	Trietazine	

Source: Adapted from Lyman et al. *Handbook of Chemical Property Estimation Methods*, McGraw-Hill Book Company, New York, 1982.

TABLE 8-7
Compounds Used by Rao and Davidson (1980) for Eq. (8-24)

Atrazine	Dicamba	Malathion
Bromacil	Dichlorobenil	Methylparathion
Carbofuran	Diuron	Simazine
2,4-D	Lindane	Terbacil
DDT		

Source: Adapted from Lyman et al. *Handbook of Chemical Property Estimation Methods*, McGraw-Hill Book Company, New York, 1982.

EXAMPLE 8-1

In Table 8-1, for benzene (aromatic compounds), the following parameters are given: $S = 1.75 \times 10^{-3}$ mg/L (water solubility), $K_{oc} = 8.30 \times 10^1$ mL/g, and $K_{ow} = 1.32 \times 10^2$. Based on these given values, (1) evaluate the regression equations to be used for benzene; (2) use Eq. (8-17) to determine the K_{oc} value for benzene; (3) use Eqs. (8-21) and (8-22) in Table 8-1 to determine K_{oc} value for benzene; (4) use Eq. (8-20) to determine the K_{oc} value for benzene; and (5) discuss the results.

TABLE 8-8
Compounds Used by Briggs (1973) Regression Equation[a,b]

X	R$_1$	R$_2$	X	R$_1$	R$_2$
4-Cl	CH$_3$	CH$_3$	3-Cl	H	H
3,4-Cl	CH$_3$	CH$_3$	3,4-Cl	H	H
3-CF$_3$	CH$_3$	CH$_3$	3-Cl, 4-OCH$_3$	H	H
3-Cl, 4-OCH$_3$	CH$_3$	CH$_3$	3-F	H	H
4-Cl	CH$_3$	OCH$_3$	4-F	H	H
3,4-Cl	CH$_3$	OCH$_3$	3-CF$_3$	H	H
4-Br	CH$_3$	OCH$_3$	3-Br	H	H
3-Cl, 4-Br	CH$_3$	OCH$_3$	4-Br	H	H
3-Cl	CH$_3$	H	3-OH	H	H
3,4-Cl	CH$_3$	H	4−SO$_3$	H	H
3-Cl, 4-OCH$_3$	CH$_3$	H	H	H	H

R
CH$_3$
CH$_2$CH$_3$
CH$_2$CH$_2$CH$_3$
CH(CH$_3$)$_2$
CH$_2$CH$_2$CH$_2$CH$_3$

[a] Eq. (8-25).
[b] Although Briggs (1973) states that 30 compounds were used to derive the regression equation, only 28 were listed in the reference cited Briggs (1969) for the original data.
Source: Adapted from Lyman et al. *Handbook of Chemical Property Estimation Methods,*
        McGraw-Hill Book Company, New York, 1982.

SOLUTION

(1) There are several alternatives. First of all, as mentioned in Section 8.4.2, Eq. (8-17) can be used to estimate K_{oc}. According to Table 8-5, Eqs. (8-21) and (8-22) in Table 8-4 cover benzene as well. Therefore, these equations can be used with the known K_{ow} values. On the other hand, Eq. (8-20) covers a variety of chemicals, mostly pesticides. Therefore, this equation can be used as well.

(2) There are several alternatives. First of all, as mentioned in Section 8.4.2, if solubility is known, Eq. (8-17) can be used for estimating K_{oc}. Introducing the S value in Eq. (8-17) gives

$$\log(K_{oc}) = -0.55 \log(1.75 \times 10^3) + 3.64 = 1.856$$
$$K_{oc} = 10^{1.856} = 71.78 \text{ mL/g}$$

which correlates well with $K_{oc} = 83.0$ mL/g, as given in Table 8-1.

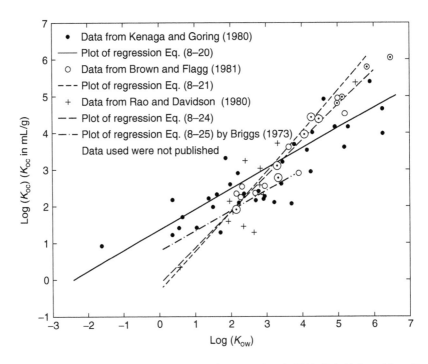

FIGURE 8-8   Correlations between K_{oc} and K_{ow} for various equations in Table 8-4. (Adapted from Lyman et al., *Handbook of Chemical Property Estimation Methods*, McGraw-Hill Book Company, New York, 1982.)

TABLE 8-9
Comparison of the Measured and Estimated Values of K_{oc} for Benzene

Estimation Procedure	Used Measured Value	K_{oc} (mL/g)
Measured	–	83.0[a]
Eq. (8-17)[b]	S[c]	71.78
Eq. (8-21)[d]	K_{ow}^c	95.72
Eq. (8-21)[d]	K_{ow}^c	81.47
Eq. (8-21)[e]	K_{ow}^c	339.63

[a] K_{oc} value is given in Table 8-1.
[b] If solubility is known and other data are not available, Eq. (8-17) is recommended.
[c] The values are given in Table 8-1.
[d] According to Table 8-5, the regression equation covers benzene as well because benzene is in the "aromatic compounds group."
[e] Regression equation covers a wide variety of chemicals, mostly pesticides.

---

(3) According to Table 8-5, Eqs. (8-21) and (8-22) cover benzene as well. Therefore, from Eq. (8-21),

$$\log(K_{oc}) = 0.937 \log(1.32 \times 10^2) - 0.0061.981$$
$$K_{oc} = 10^{1.981} = 95.72 \text{ mL/g}$$

and from Eq. (8-22),

$$\log(K_{oc}) = 1.00 \log(1.32 \times 10^2) - 0.21 = 1.911$$
$$K_{oc} = 10^{1.911} = 81.47 \text{ mL/g}$$

(4) Equation (8-20) in Table 8-4 is valid for a wide variety of chemicals, mostly pesticides. This equation is also given as Eq. (8-18) and the result is

$$\log(K_{oc}) = 0.544 \log(K_{ow}) + 1.377 = 2.531$$
$$K_{oc} = 10^{2.531} = 339.63$$

(5) The results are given in Table 8-9. As can be seen from this table, with the exception of Eq. (8-20) in Table 8-4, the predicted K_{oc} values for benzene are close to the measured value (83.0 mL/g). The results show that Eqs. (8-21) and (8-22) are more representative.

PROBLEMS

8.1 Explain the K_d/K_{oc} approach for the determination of the retardation factor.
8.2 (a) Define "effective dispersion coefficient" and "effective solute velocity"?
8.3 Define "equilibrium adsorption"?
8.4 What is the physical meaning of K_{oc}?
8.5 What is the importance of the octanol–water partition coefficient K_{ow}?

References

Aboufirassi, M., and M.A. Marino, "Cocriking of Aquifer Transmissivities from Field Measurements of Transmissivity and Specific Capacity", *Math. Geol.*, Vol. 16, No.1, pp. 19–35, 1984.

Abramowitz, M. and I. A. Stegun, *Handbook of Mathematical Functions With Formulas, Graphs, and Mathematical Tables*, National Bureau of Standards, United States Government Printing Office, Washington, DC, 1964.

Adams, E. E. and L. W. Gelhar, Field study of dispersion in a heteregeneous aquifer: 2. spatial moments analysis, *Water Resour. Res.*, 28, 3293–3308, 1992.

Ahlstrom, S. W., H. P. Foote, R. C. Arnett, C. P. Cole, and R. J. Serne, Multicomponent Mass Transport Model: theory and numerical implementation (discrete-particle-random-walk-version), Report BNWL-2127, Battelle Pacific Northwest Laboratory, Richland, Washington, 1977.

Alpay, O. A., A practical approach to defining reservoir heterogeneity, *J. Petrol. Technol.*, 24, 841–848, 1972.

Anderson, M. P., "Using models to simulate the movement of contaminants through ground water flow systems", *CRC Crit. Rev. Environ. Control*, Vol. 9, 97–156, 1979.

Anderson, M. P., Ground-water modeling—the emperor has no clothes, Ground Water, 21, 666–669, 1983.

Anderson, M. P. and W. W. Woessner, *Applied Ground water Modeling*, Academic Press, San Diego, California, 1992, 381pp.

Anderson, M. R. and J. F. Pankow, A case study of a chemical spill: polychlorinated biphenyls (PCBs), 3. PCB sorption and retardation in soil underlying the site", *Water Resour. Res.*, 22, 1051–1057, 1986.

Appelt, J., K. Holtzclaw, and P. F. Pratt, Effect of anion exclusion on the movement of chloride through soils, *Soil Sci. Soc. Am. Proc.*, 39, 264–267, 1975.

Aris, R., On the dispersion of a solute in a fluid flowing through a tube, *Proc. R. Soc. London, Ser. A*, 235, 67–78, 1956.

Arnett, R. C., R. G. Baca, J. A. Caggiano, S. M. Price, R. E. Gephart, and S. E. Logan, Preliminary Hydrologic Release Scenarios for a Candidate Repository Site in the Columbia River Basalts, RHO-BWI-ST-12, 1980.

Arya, A., Dispersion and Reservoir Heteregeneity, Ph.D. dissertation, University of Texas, Austin, Texas, 1986.

Arya, A., T. A. Hewett, R. G. Larson and L. W. Lake, Dispersion and reservoir heterogeneity, *SPE Reservoir Eng.*, 3, 139–148, 1988.

Babu, D. K., G. F. Pinder, A. Niemi, and D. P. Ahlfeld, Chemical Transport by Three-Dimensional Ground water Flow, Report B4-WR-3, Department of Civil Engineering, Princeton University, Princeton, New Jersey, 1987.

Baker, L. E., Effects of dispersion and dead-end pore volume in miscible flooding, *Soc. Petrol. Eng. J.*, 17, 219–227, 1977.

Bakr, A. A., Stochastic Analysis of the Effect of Spatial Variations in Hydraulic Conductivity on Ground water Flow, Ph.D. dissertation, New Mexico Institute of Mining and Technology, Socorro, New Mexico, 1976.

Bakr, A. A., L. W. Gelhar, A. L. Gutjahr, and J. R. MacMillan, Stochastic analysis of spatial variability in subsurface flows: 1. comparison of one- and three-dimensional flows, *Water Resour. Res.*, 14, 263–271, 1978.

Baron, F. S., R. K. Rowe, and M. Quigley, Laboratory determination of chloride diffusion coefficient in an intact shale, *Can. Geotechn. J.*, 27, 177–184, 1990.

Bassett, R. L., et al, "Preliminary Data From a Series of Artificial Recharge Experiments at Stanton, Texas", *U. S. Geol. Surv. Open File Rep., 81–0149*, 1980.

Batu, V., Flow net for unsaturated infiltration from strip sources, *J. Irrigation Drainage Div.* Am. Soc. Civil Eng., 105, 233–245, 1979.

Batu, V., Time-dependent, linearized two-dimensional infiltration and evaporation from nonuniform and nonperiodic strip sources, *Water Resour. Res.*, 18, 1725–1733, 1982.

Batu, V., Two-dimensional dispersion from strip sources, *J. Hydraulic Eng.*, Am. Soc. Civil Eng., 109, 827–841, 1983.

Batu, V., A finite element dual mesh method to calculate nodal darcy velocities in nonhomogeneous and anisotropic aquifers, *Water Resour. Res.*, 20, 1705–1717, 1984.

Batu, V., Introduction of the stream function concept to the analysis of hydrodynamic dispersion in porous media, *Water Resour. Res.*, 23, 1175–1184, 1987.

Batu, V., Contaminant plume analysis using the hydrodynamic dispersion stream function (HDSF) concept, *Ground Water*, 26, 71–77, 1988.

Batu, V., A generalized two-dimensional analytical solution for hydrodynamic dispersion in bounded media with the first-type boundary conditions at the source, *Water Resour. Res.*, 25, 1125–1132, 1989.

Batu, V., A generalized two-dimensional analytical solute transport model in bounded media for flux-type finite multiple sources, *Water Resour. Res.*, 29, 2881–2892, 1993.

Batu, V., A generalized three-dimensional analytical solute transport model for multiple rectangular first-type sources, *J. Hydrol.*, 174, 57–82, 1996.

Batu, V., A generalized three-dimensional analytical solute transport model for multiple third-type (flux-type) rectangular sources, Unpublished paper, 1997a.

Batu, V., A generalized two-dimensional analytical solute transport for multiple instantaneous strip sources, Unpublished paper, 1997b.

Batu, V., A generalized three-dimensional analytical solute transport for multiple instantaneous rectangular sources, Unpublished paper, 1997c.

Batu, V., *Aquifer Hydraulics: A Comprehensive Guide to Hydrogeologic Data Analysis*, John Wiley & Sons, Inc., 1998, 727pp. New York.

Batu, V., A generalized two-dimensional analytical model in bounded media for the hydrodynamic dispersion stream stream lines, Unpublished paper, 2002.

Batu, V. and W. R. Gardner, Steady-state solute convection in two dimensions with nonuniform infiltration, *Soil Sci. Soc. Am. J.*, 42, 18–22, 1978.

Batu, V. and M. T. van Genuchten, First- and third-type boundary conditions in two-dimensional solute transport modeling, *Water Resour. Res.*, 26, 339–350, 1990.

Bear, J., The Transition Zone Between Fresh and Salt Waters in Coastal Aquifers, Ph.D. thesis, University of California, Berkeley, California, 1960.

Bear, J., On the tensor form of dispersion in porous media, *J. Geophys. Res.*, 66, 1185–1197, 1961.

Bear, J., *Dynamics of Fluids in Porous Media*, Dover Publications, New York, 1988, Originally published by the American Elsevier Publishing Company, New York, New York, 764 pp., 1972.

Bear, J., *Hydraulics of Ground water*, McGraw-Hill, New York, 569 pp., 1979.

Bear, J. and Y. Bachmat, On the equivalence of areal and volumetric averages in transport phenomena in porous media, *Adv. Water Resources*, 6, 59–62, 1983.

Bennion, D. W., and J. C. Griffiths, A Stochastic model for predicting variations in reservoir rock properties, *Trans. AIME*, Vol. 237, Part 2, 9–16, 1966.

Bentley, H. W. and G. R. Walter, Two-Well Recirculating Tracer Tests at H-2: Waste Isolation Pilot Plant (WIPP), Southwest New Mexico, Draft Paper, Hydro Geochem., Inc., Tucson, Arizona, 1983.

Bierschenk, W. H., Aquifer Characteristics and Ground-Water Movement at Hanford, Report HW-60601, Hanford At. Products Oper., Richland Washington, 1959.

Biggar, J. W. and D. R. Nielsen, Diffusion effects in miscible displacement occuring in saturated and unsaturated Porous materials, *Journal of Geophysical Research*, Vol. 65, 2887–2895, 1960.

Biggar, J. W. and D. R. Nielsen, Miscible displacement: II. behavior of tracers, *Soil Sci. Soc. Am. Proc.*, 26, 125–128, 1962.

Binsariti, A. A., "Statistical Analysis and Stochastic Modeling of the Cortaro Aquifer in Southern Arizona, " *Ph.D. thesis*, Department of Hydrology and Water Resources, University of Arizona, Tucson, Arizona, 1980.

Bittinger, M. W., H. R. Duke, and R. A Longenbaugh, Mathematical simulations for better aquifer management, Pub. No. 72IASH, 509–519, 1967.

Boast, C. W., Modeling the movement of chemicals in soils by water, *Soil Sci.*, 115, 224–230, 1973.

Boggs, J. M., S. C. Young, D. J. Benton, and Y. C. Chung, Hydrogeologic Characterization of the MADE Site, EPRI EN-6915, *Research Project 2485-5*, Interim Report, Electric Power Research Institute, Palo Alto, California, July 1990.

Bouwer, H., Planning and interpreting soil permeability measurements, *J. Irrig. Drainage Div.*, Am. soc. civ. Eng., 95, 391–402, 1969.

Bouwer, H., *Ground water Hydrology*, McGraw-Hill, New York, 1978.

Bredehoeft, J. D. and G. F. Pinder, Mass transport in flowing ground water, *Water Resour. Res.*, 9, 194–210, 1973.

Bresler, E., Simultaneous transport of solutes and water under transient unsaturated flow condition, *Water Resour. Res.*, 9, 975–986, 1973.

Bresler, E., Two-dimensional transport of solutes during nonsteady infiltration from a trickle source, *Soil Sci. Soc. Am. Proc.*, 39, 604–613, 1975.

Brigham, W. E., Mixing equations in short laboratory columns, *Soc. Petrol. Eng. J.*, 14, 91–99, 1974.

Briggs, G. G., Molecular structure of herbicides and their sorption by soils, *Nature*, Vol. 223, p.1288, Great Britain, London, 1969.

Briggs, G. G., A simple relationship between soil sorption of organic chemicals and their octanol/water partition coefficients, *Proceedings of the 7th British Insecticides and Fungicides Conference*, Vol. 11, 1973, pp. 475–478.

Brown, D. S. and E. W. Flagg, Empirical prediction of organic pollutant sorption in natural sediments, *J. Environ. Qual.*, 10, 382–386, 1981.

Bruch, J. C. and R. L. Street, Two-dimensional dispersion, *J. Sanitary Eng. Div.*, Am. Soc. Civ. Eng., 93, 17–39, 1967.

Burnett, R. D. and E. O. Frind, An alternating direction Galerkin technique for simulation of ground water contamination transport in three dimensions, 2. dimensionality effects, *Water Resour. Res.*, 23, 695–705, 1987.

Byers, E. and D. B. Stephens, Statistical and stochastic analyses of hydraulic conductivity and particle size in a fluvial sand, *Soil Sci. Soc. J.*, 47, 1072–1081, 1983.

Carslaw, H. S. and J. C. Jaeger, *Conduction of Heat in Solids*, 2nd ed., Clarendon, Oxford, Great Britain, 510 pp., 1959.

Carrera, J., An overview of uncertainties in modelling ground water solute transport, *J. Contaminant Hydrol.*, 13, 23–48, 1993.

Cherry, J. A., R. W. Gillham, and J. F. Barker, 1984, Contaminants in ground water: chemical processes, in *Ground water Contamination*, National Academy Press, Washington, D.C., 46–64, 1984.

Chiou, C. T., V. H. Freed, D. W. Schmedding, and R. L. Kohnert, Partition coefficients and bioaccumulation of selected organic chemicals, Environ, Sci. Technol., Vol. 11, 475–478, 1977.

Chiou, C. T., L. J. Peters, and V. H. Freed, A physical concept of soil–water equilibria for nonionic organic compounds, *Science*, 206, 831–832, 1979.

Chiou, C. T., D. W. Schmedding, and M. Manes, Partitioning of organic compounds on octanol-water systems, *Env. Sci. Technol.*, 16, 4–10, 1982a.

Chiou, C. T., P. E. Porter, and Q. W. Schmedding, Partition equilibra of nonionic organic compounds between soil organic matter and water, *Env. Sci. Technol.*, 17, 227–231, 1982b.

Chiou, C. T., P. E. Porter, and D. W. Schmedding, Partition equilibria of nonionic organic compounds between soil organic matter and water, *Env. Sci. Tech.*, Vol. 17, 227–231, 1983.

Chiou, C. T., T. D. Shoup, and P. E. Porter, Mechanistic roles of soil humus and minerals in the sorption of nonionic organic compounds from aqueous and organic solutions, *Org. Geochem.*, 8, 9–14, 1985.

Churchill, R. V., *Fourier Series and Boundary Value Problems*, McGraw-Hill, New York, 206 pp., 1941.

Claasen, H. C. and E. H. Cordes, Two-well recirculating tracer test in fractured carbonate rock, Nevada, *Hydrol. Sci. Bull.*, 20, 367–382, 1975.

Cleary, R. W. and M. J. Ungs, Analytical methods for ground water pollution and hydrology, Report 78-WR-15, Water Resources Program, Princeton University, Princeton, New Jersey, 1978.

Cole, J. A., Some interpretations of dispersion measurements in aquifers, in *Ground water Pollution In Europe*, Cole, J. A. Ed., Water Research Association, Reading, England, 1972, pp. 86–95.

Coats, K. H. and B. D. Smith, Dead-end pore volume and dispersion in porous media, *Soc. Petrol. Eng. J.*, 4, 73–84, 1964.

Crank, J., *Mathematics of Diffusion*, Oxford University Press, New York and London, 347 pp., 1956.

Dagan, G., Models of ground water flow in statistically homogeneous porous formations, *Water Resour. Res.*, 15, 47–63, 1979.

Dagan, G., Stochastic modeling of ground water flow by unconditional and conditional probabilities, 2. solute transport, *Water Resour. Res.*, 18, 4, 835–848, 1982.

Dagan, G., Solute transport in heterogeneous porous formations, *J. Fluid Mech.*, 145, 151–177, 1984.

Dagan, G., Theory of solute transport by ground water, *Ann. Rev. Fluid Mech.*, 19, 183–215, 1987.

Dagan, G., Time-dependent macrodispersion for solute transport in anisotropic heterogeneous aquifers, *Water Resour. Res.*, 24, 1491–1500, 1988.

Dagan, G., *Flow and Transport in Porous Formations*, Springer, Berlin, Germany, 465 pp., 1989.

Dagan, G., Stochastic modeling of flow and transport: the broad perspective, *Subsurface Flow and Transport: A Stochastic Approach*, Dagan, G. and Neuman, S. P. Eds., International Hydrology Series, Cambridge University Press, Cambridge, UK, 3–19, 1997.

Daniels, W. R., Ed., Laboratory Field Studies Related to the Radionuclide Migration Project, Progress Report LA-8670-PR, Los Alamos Sci. Lab., Los Alamos, New Mexico, 1981.

Daniels, W. R., Ed., Laboratory Field Studies Related to the Radionuclide Migration Project (Draft), Progress Report LA-9192-PR, Los Alamos Sci. Lab., Los Alamos, New Mexico, 1982.

Darcy, H., Les Fontaines Publiques de la Dijon, Dalmont, Paris, France, 1856 (in French).

Davidson, J. M., L. T. Ou, and P. S. C. Rao, Adsorption, Movement, and Biological Degradation of High Concentrations of Selected Pesticides in Soils, *Proceedings of the 4th Annual Research Symposium Land Disposal of Hazardous Wastes*, EPA-600/9-78-016, 233–244, 1978.

Daus, A. D. and E. O. Frind, An alternating direction Galerkin technique for simulation of contaminant transport in complex ground water systems, *Water Resour. Res.*, 21, 653–664, 1985.

Deans, H. A., A mathematical model for dispersion in the direction of flow in porous media, *J. Petrol. Eng.*, 3, 49–52, 1963.

De Josselin de Jong, G., Longitudinal and transverse diffusion in granular deposits, *Trans. Am. Geophys. Union*, 39, 67–74, 1958.

Delhomme, J. P., Kriging in the hydrosciences, *Adv. Water Res.*, Vol. 1, 251–266, 1978.

Devary, J. L., and P. G. Doctor, Pore velocity estimation uncertainties, *Water Resour. Res.*, 17, 1, 149–161, 1981.

Dieulin, A., Propagation de Pollution dans un Aquifere Alluvial: L'Effet de Parcours, Doctoral Dissertation, L'Université Pierre et Marie Curie-Paris IV and L'Ecole Nationale Superieure des Mines de Paris, Fontainebleau, Cedex, France, 1980 (in French).

Dieulin, A., Lixiviation Situ d'un Gisement d'Uranium en Milieu Granitique, Draft Report No. LHM/RD/81/63, L'Université Pierre et Marie Curie-Paris IV and L'Ecole Nationale Superieure des Mines de Paris, Fontainebleau, Cedex, France, 1981 (in French).

Dieulin, A., B. Beaudoin, and G. de Marsily, Sur le Transfert d'Éléments en Solution Dans Un Aquifere Alluvionnaire Structure, *C.R. Acad. Sci. Paris*, 291, Ser. D, 805, 1980 (in French).

Domenico, P. A. and G. A. Robbins, A new method of contaminant plume analysis, *Ground Water*, 23, 476–485, 1985.

Domenico, P. A., and F. W. Schwartz, *Physical and Chemical Hydrogeology*, John Wiley & Sons, New York, First edition, New York, 824 pp., 1990.

Domenico, P. A. and F. W. Schwartz, *Physical and Chemical Hydrogeology*, John Wiley & Sons, New York, Second edition, 1998, 506 pages.

Douglas, J., Jr., and H. H. Rachford, Jr., On the numerical solution of heat conduction problems in two and three space variables, *Trans. Am. Math. Soc.*, 82, 421–439, 1956.

Duffield, G. M., MODFLOWT: A Modular Three-Dimensional Ground water Flow and Transport Model, *Version 1.1*, Hydrosolve, Inc. and GeoTrans, Inc., Sterling, Virginia, 1996.

Ebach, F. A. and R. R. White, Mixing of fluids flowing through beds of packed solids, *J. Am. Instit. Chem. Eng.*, 4, 161–169, 1958.

Egboka, B. C. E., J. A. Cherry, R.N. Farvolden, and E.O. Frind, Migration of contaminants in ground water at a landfill: a case study, 3, tritium as an indicator of dispersion and recharge, *J. Hydrol.*, 63, 51–80, 1983.

Eisenhart, L. P., *Coordinate Geometry*, Ginn and Company, Boston, Massachusetts, 297 pp., 1939.

Erdelyi, A. (Ed.), *Tables of Integral Transforms, in Bateman Manuscript Project*, Vol. 1, McGraw-Hill, New York, 1954.

Fatt, I., M. Maleki, and R. N. Upadhyay, Detection and estimation of dead-end pore volume in reservoir rock by conventional laboratory tests, *Soc. Petrol. Eng. J.*, 6, 206–212, 1966.

Faust, C. R., L. R. Silka, and J. W. Merces, Computer modeling and ground-water protection, *Ground Water*, Vol. 19, No. 4, 362–365, 1981.

Faust, C. R., P. N. Sims, C. P. Spalding, and P. F. Anderson, FTWORK: Ground water flow and solute transport in three dimensions, version 2.4, *A Finite-Difference Computer Code*, GeoTrans, Inc., Sterling, Virginia, 1990.

Fenske, P. R., Hydrology and Radionuclide Transport, Monitoring Well HT-2m, Tatum Dome, Mississippi, Project Report 25, Technical Report. NVD-1253-6, Center for Water Resources Research, Desert Research Institute, University of Nevada System, Reno, Nevada, 1973.

Ferris, J. G., D. B. Knowles, R. H. Brown, and R. W. Stallman, Theory of Aquifer Tests, U.S. Geological Survey Water Supply Paper 1536-E, 69–174, 1962.

Fetter, C. W., Jr., *Applied Hydrogeology*, Charles E. Merrill and Co., Columbus, Ohio, 488 pp., 1980.

Filon, L. N. G., On a quadrature formula for trigonometric integrals, *Proc. R. Soc. Edinburgh, Sect. A, Math.*, 49, 38–47, 1928.

Fischer, H. B., E. J. List, R. C. Y. Koh, J. Imberger, and N. H. Brooks, *Mixing in Inland and Coastal Waters*, Academic Press, New York, 1979.

Freeze, R. A., A Stochastic – conceptual analysis of one-dimensional ground water flow in nonuniform homogeneous media, *Water Resour. Res.*, 11, 725–741, 1975.

Freeze, R. A. and J. A. Cherry, *Ground water*, Prentice-Hall, Englewood Cliffs, New Jersey, 604 pp., 1979.

Freeze, R. A., J. Massmann, L. Smith, T. Sperling, and B. James, Hydrogeological decision analysis: 1. A framework, *Ground Water*, 28, 738–766, 1990.

Freeze, R. A. and P. A. Witherspoon, Theoretical analysis of regional ground water flow: 1. Analytical and numerical solutions to the mathematical model, *Water Resour. Res.*, 2, 641–656, 1966.

Freyberg, D. L., A natural gradient experiment on solute transport in a sand aquifer, 2. Spatial moments and the advection and dispersion of nonreactive tracers, *Water Resour. Res.*, 22, 2031–2046, 1986.

Freyberg, D. L., An exercise in ground-water model calibration and prediction, *Ground Water*, 26, 350–360, 1988.

Fried, J. J., *Ground water Pollution: Theory, Methodology, Modeling and Practical Rules*, Elsevier Scientific Publishing Company, New York, 330 pp., 1975.

Fried, J. J. and M. A. Combarnous, Dispersion in porous media, in *Advances in Hydroscience*, Ven Te Chow, Ed:, Vol. 7, Academic Press, New York, 169–282, 1971.

Fried, J. J. and P. Ungemach, Determination *in Situ* du Coefficient de Dispersion Longitudinale d'Un Milieu Poreux Naturel, *C. R. Acad. Sci., Ser. 2, 272*, 1327–1329, 1971 (in French).

Garabedian, S. P., L. W. Gelhar, and M. A. Celia, Large-Scale Dispersive Transport in Aquifers: Field Experiments and Reactive Transport Theory, Report 315, Ralph M. Parsons Laboratory for Water Resour. and Hydrodynamics, Massachusetts Institute of Technology, Cambridge, Massachusetts, 290 pp., 1988.

Garabedian, S. P., D. R. LeBlanc, L. W. Gelhar, and M. A. Celia, Large-scale natural gradient tracer test in sand and gravel, cape Cod, Massachusetts, 2. analysis of tracer moments for a nonreactive tracer, *Water Resour. Res.*, 27, 911–924, 1991.

Garder, A. O., D. W. Peaceman, and A. L. Pozzi, Jr., Numerical calculation of multidimensional miscible displacement by the method of characteristics, *Soc. Petrol. Eng. J.*, 1, 26–36, 1964.

Gelhar, L. W., Stochastic analysis of flow in aquifers, Amer. Water Res. Assoc., *Advances in Ground water Hydrology, Symposium Proceedings*, 57–71, 1976.

Gelhar, L. W., Analysis of Two-Well Tracer Tests with a Pulse Input, Report. RHO-BW-CR-131 P, Rockwell International, Richland, Washington, 1982.

Gelhar, L. W., Stochastic Analysis of Solute Transport in Saturated and Unsaturated Porous Media, *NATO ASI Se., Ser. E., 128*, 657–700, 1987.

Gelhar, L. W. *Stochastic Subsurface Hydrology*, Prentice Hall, Inc., Englewood Cliffs, New Jersey, 390 pp., 1993.

Gelhar, L. W. and C. L. Axness, Stochastic Analysis of Macrodispersion in Three-Dimensional Heteregeneous Aquifers, Report No. H-8, Hydrology Research Program, New Mexico Institute of Mining and Technology, Socorro, New Mexico, August, 140 pp., 1981.

Gelhar, L. W. and C. L. Axness, Three-dimensional stochastic analysis of macrodispersion in aquifers, *Water Resour. Res*, 19, 161–180, 1983.

Gelhar, L. W., A. L. Gutjahr, and R. L. Naff, Stochastic analysis of macrodispersion in a stratified aquifer, *Water Resour. Res.*, 15, 1387–1397, 1979.

Gelhar, L. W., A. Mantoglou, C. Welty, and K. R. Rehfeldt, A Review of Field-Scale Physical Solute Transport Processes in Saturated and Unsaturated Porous Media, EPRI Report. EA-4190, Elect. Power Res. Inst., Palo Alto, California, August, 1985.

Gelhar, L. W., C. Welty, and K. R. Rehfeldt, A critical review of data on field-scale dispersion in aquifers, *Water Resour. Res.*, 28, 1955–1974, 1992.

Gelhar, L. W., C. Welty, and K. R. Rehfeldt, Reply, *Water Resour. Res.*, 29, 1867–1869, 1993.

Gelhar, L. W., P. J. Wierenga, K. R. Rehfeldt, C. J. Duffy, M. J. Simonett, T.-C. Yeh, and W. R. Strong, "Irrigation Return Flow Water Quality Monitoring, Modeling and Variability in the Middle Rio Grande Valley, New Mexico," *U.S. Environmental Protection Agency Project Report*, EPA-6001S2-072, 1983.

Gillham, R. W. and J. A. Cherry, Predictability of solute transport in diffusion controlled hydrogeologic regimes, *Proceedings of the Symposium on Low-Level Waste Disposal: Facility Design, Construction and Operating Practices*, 1982a.

Gillham, R. W. and J. A. Cherry, Contaminant migration in saturated unconsolidated geologic deposits, *Recent Trends in Hydrogeology*, The Geological Society of America, Special Paper 189, Boulder, Colorado, 31–62, 1982b.

Gillham, R. W., M. J. L. Robin, D. J. Dytynyshyn, and H. M. Jonston, Diffusion of nonreactive and reactive solutes through fine-grained barrier materials, *Can. Geotechn. J.*, 21, 541–550, 1984.

Gillham, R. W., M. J. L. Robin, and C. J. Ptacek, A device *in situ* determination of geochemical transport parameters 1. Retardation, *Ground Water*, 28, 666–672, 1990.

Glueckauf, E., K. H. Barker, and G. P. Kitt, Theory of chromatography, 8: the separation of lithium isotopes by ion exchange and of neon isotopes by low-temperature adsorption columns," *Trans. Faraday Soc.*, 7, 199–213, 1949.

Goblet, P., Interpretation d'Experiences de Tracage en Milieu Granitique (Site B), Report LHM/RD/82/11, Cent. D'Inf. Geol., Ecole Natl. Super. Des Mines de Paris, Fontainebleau, France, 1982 (in French).

Goggin, M. N., M. A. Chandler, G. Kacurek, and L. W. Lake, Patterns of Permeability in Eolian Deposits: Page Sandstone (Jurassic), Northeastern Arizona, *SPE Formation Evaluation*, 297–306, 1988.

Golden Software, Inc., *SURFER*, Version 4, Golden, Colorado, 1988.

Gröbner, W. and N. Hofreiter, *Integraltafel*, Springer, Vienna, Vol. 1, 1949, Vol. 2, 1950 (in German).

Gschwend, P. M. and S. -C. Wu, On the constancy of sediment–water partition coefficients of hydrophobic organic pollutants, *Environ. Sci. Technol.*, 19, 90–96, 1985.

Golubev, V. S. and A. A. Garibyants, Heterogeneous Processes of Geochemical Migration (Special Research Report), Consultants Bureau (Translated from the Russian), New York, 150 pp., 1971.

Greenkorn, R. A. and D. P. Kessler, Dispersion in heterogeneous nonuniform anisotropic porous media, *Ind. Eng. Chem.*, 61, pp. 14–32, 1969.

Groult, J., L. H. Reiss, and L. Montadert, Reservoir inhomogeneities deduced from outcrop observations and production logging, *J. Petrol. Tech*, 18, 883–891, 1966.

Grove, D. B., The use of Galerkin finite-element methods to solve mass transport equations, Report USGS/WRD/WRI-78/011, U.S. Geological Survey, Denver, Colorado, 1977. (Available as NTIS PB 277-532 from Natl. Tech. Inf. Serv., Springfield, Virginia.)

Grove, D. B. and W. A. Beetem, Porosity and dispersion constant calculations for a fractured carbonate aquifer using the two-well tracer method, *Water Resour. Res.*, 7, 128–134, 1971.

Gupta, S. K., C. R. Cole, F. W. Bond, and A. M. Monti, *Finite-Element Three-Dimensional Ground-Water (FE3DGW) Flow Model: Formulation, Computer Source Listings, and User's Manual*, ONWI–548, Battelle Memorial Institute, Columbus, Ohio, 1984.

Gupta, S. K., K. K. Tanji, and J. N. Luthin, A Three-Dimensional Finite Element Ground Water Model, Report UCAL-WRC-C-152, California Water Resources Center, University of California, Davis, 1975. (Available as *NTIS PB 248–925* from Natl. Tech. Inf. Serv., Springfield, Virginia.)

Gutjahr, A. L. and L. W. Gelhar, Stochastic models of subsurface flow: infinite versus finite domains and stationarity, *Water Resour. Res.*, 17, 337–350, 1981.

Gutjahr, A. L., L. W. Gelhar, A. A. Bakr, and J. R. MacMillan, Stochastic analysis of spatial variability in subsurface flows: 2. evaluation and application, *Water Resour. Res.*, 14, 953–959, 1978.

Guvanasen, V. and R. E. Volker, Experimental investigations of unconfined aquifer pollution from recharge basins, *Water Resour. Res.*, 19, 707–717, 1983.

Güven, O., F. J. Molz, and J. G. Melville, An analysis of dispersion in stratified aquifer, *Water Resour. Res.*, 20, 1337–1354, 1984.

Halevy, E., and A. Nir, Determination of aquifer parameters with the aid of radioactive tracers, *J. Geophys. Res.*, 67, 2403–2409, 1962.

Hamaker, J. W., Decomposition: quantitative aspects, in *Organic Chemicals in the Environment*, Goring, C. A. I. and Hamaker, J. W., Eds., Mercel Dekker, Inc., New York, 1972.

Hamon, W. R., Estimating potential evapo-transpiration, *Proc. Am. Soc. Civil Eng., J. Hydraul. Div.*, 87, 107–120, 1961.

Hantush, M. S., Hydraulics of wells, in *Advances in Hydroscience*, Ven Te Chow, Ed., Academic Press, New York, 281–442, 1964.

Harpaz, Y., Field experiments in recharge and mixing through wells, Underground Water Storage Study Tech. Rep. 17, Publ. 483, Tahal-Water Plann. For Israel, Tel Aviv, 1965.

Harleman, D. R. F. and R. R. Rumer, Longitudinal and lateral dispersion in an isotropic porous medium, *J. Fluid Mech.*, 16, 385–394, 1963.

Harr, M. E., *Ground Water and Seepage*, McGraw-Hill, New York, 315 pp., 1962.

Hashimoto, I., K. B. Deshpande, and H. C. Thomas, Peclet numbers and retardation factors for ion exchange columns, *Ind. Eng.Chem. Fund.*, 3, 213–218, 1964.

Hassett, J. J., J. C. Means, W. L. Banwart, and S. G. Wood, Sorption Properties of Sediments and Energy-Related Pollutants, Rep. EPA-600/3-80-041, U.S. Environmental Protection Agency, Athens, Georgia, 1980.

Heller, J. P., Observations of mixing and diffusion in porous media, *Proceedings, Second Symposium on Fundamentals of Transport Phenomena in Porous Media*, Vol. 1, International Association for Hydraulic Research, University of Guelph, Canada, 1–26, 1972.

Helveg, O. J. and J. W. Labadie, Linked models for managing river basin salt balance, *Water Resour. Res.*, 13, 329–336, 1977.

Hess, K. M., Use of a borehole flowmeter to determine spatial heterogeneity of hydraulic conductivity and macrodispersion in a sand and gravel aquifer, Cape Cod, Massachusetts, in *Proceedings of the Conference on New Field Techniques for Quantifying the Physical and Chemical Properties of Heterogeneous Aquifers*, Molz F. J., Melville, J. G., and Guven, O., Eds., National Water Well Association, Dublin, Ohio, 497–508, 1989.

Hess, K. M., S. H. Wolf, and M. A. Celia, Large-scale natural gradient tracer test in Dand and Gravel, Cape Cod, Massachusetts, 3. hydraulic conductivity variability and calculated macrodispersivities, *Water Resour. Res.*, 2011–2027, 1992.

Higgins, G. H., Evaluation of the ground water contamination hazard from underground nuclear explosions, *J. Geophys. Res.*, 64, 1509–1519, 1959.

Highdon, A., E. H. Ohlsen, W. B. Stiles, J. A. Weese, and W. F. Riley, *Mechanics of Materials*, Wiley, New York, 756 pp., 1976.

Hildebrand, F. B., *Advanced Calculus for Applications*, 2nd ed., Prentice-Hall Inc., Englewood Cliffs, New Jersey, 1976.

Hill, M. C., Preconditioned Conjugate-Gradient 2 (PCG2), A Computer Program for Solving Ground-Water Flow Equations, U.S. Geological Survey, Water-Resources Investigations Report 90-4048, 43 pp., 1990.

Hill, M. C., A Computer Program (MODFLOWP) for Estimating Parameters of a Transient, Three-Dimensional Ground-Water Flow Model Using Nonlinear Regression, U.S. Geological Survey Open-File Report 91–484, 1992.

Hillel, D., *Fundamentals of Soil Physics*, Academic Press, New York, 1980.

Hillier, R., "A Laboratory Investigation of Dispersion and Ground Water Flow Through Heterogeneous Sands", *B.Sc. Thesis*, University of Waterloo, Waterloo, Ontario, Canada, 43 pp., 1975.

Hoehn, E., Geological Interpretation of Local-Scale tracer Observations in a River-Ground Water Infiltration System, Draft Report, Swiss Fed. Inst. Reactor Res. (EIR), Würenlingen, Switzerland, 1983.

Hoehn, E. and P. H. Santschi, Interpretation of tracer displacement during infiltration of river water to groundwater, *Water Resour. Res.*, 23, 633–640, 1987.

Hoeksema, R. J., and P. K. Kitanidis, "Analysis of the spatial structure of properties of selected aquifers", *Water Resour. Res.*, 21, 563–572, 1985.

Hufschmied, P., Estimation of three-dimensional statistically anisotropic hydraulic conductivity field by means of single well pumping tests combined with flowmeter measurements, *Hydrogeologie*, 2, 163–174, 1986.

Hunt, B., Dispersive sources in uniform ground-water flow, *J. Hydraulics Div.*, Am. Soc. Civil Eng., 109, 75–85, 1978.

Huyakorn, P. S., A. G. Kretschek, R. W. Broome, J. W. Mercer, and B. H. Lester, Testing and Validation of Models for Simulating Solute Transport in Ground water: Development, Evaluation, and Comparison of Benchmark Techniques, GWMI 84-13, International Ground Water Modeling Center, Holcomb Research Institute, Butler University, Indianapolis, Indiana, 420 pp., 1984.

Huyakorn, P. S., P. F. Anderson, F. J. Molz, O. Güven, and J. G. Melville, Simulations of two-well tracer tests in stratified aquifers at the Chalk River and he Mobile Sites, *Water Resour. Res.*, 22, 1016–1030, 1986.

Huyakorn, P. S., and G. F. Pinder, *Computational Methods in Subsurface Flow*, Academic Press, New York, 473 pp., 1983.

Huyakorn, P. S., M. J. Ungs, L. A. Mulkey, and E. A. Sudicky, A three-dimensional analytical method for predicting leachate migration, *Ground Water*, 25, 588–598, 1987.

Hvorslev, M. J., Time Lag Soil Permeability in Ground water Operations, Bulletin No. 36, U.S. Army Corps of Engineers, Waterways Experiment Station, Vicksburg, Mississippi, 49 pp., 1951.

International Technology (IT) Corporation, GEOFLOW, *Ground Water Flow and Solute Transport Computer Program, User's Manual*, Monroeville, Pennsylvania, 1986.

Iris, P., Contribution à l'Étude de la Valorisation Energetique des Aquifères Peu Profonds, Thèse de Docteur-Ingenieur, Ecole des Mines de Paris, Fontainebleau, France, 1980 (in French).

Ivanovitch, M., and D. B. Smith, "Determination of aquifer parameters by a two-well pulsed method radioactive tracers," *J. Hydrol.*, 36, 35–45, 1978.

Jacob, C. E., Flow of ground water, in *Engineering Hydraulics*, Ed., Rouse, H., John Wiley & Sons, New York, 321–386, 1950.

Jaeger, J., *Engineering Fluid Mechanics*, Translated from the German by P.O. Wolf, Blackie & Son Ltd., London, Great Britain, 529 pp., 1956.

Jeffreys, H., *Cartesian Tensors*, Cambridge University Press, New York, 1974.

Jenkins, G. M. and D. G. Watts, *Spectral Analysis and Its Applications*, Holden-Day, San Francisco, California, 1968.

Johnson, R. L., C. D. Palmer, and W. Fish, Subsurface Chemical Processes Transport and Fate of Contaminants in the Subsurface, Seminar Publication, U.S. Environmental Protection Agency, EPA/625/4-89/019, Cincinnati, Ohio, 41–56, 1989.

Jury, W. A., Solute travel-time estimates for tile-drained fields: I. Theory, *Soil Sci. Soc. Am. Proc.*, 39, 1020–1024, 1975a.

Jury, W. A., Solute travel-time estimates for tile-drained fields: II. application to experimental studies, *Soil Sci. Soc. Am. Proc.*, 39, 1024–1028, 1975b.

Karickhoff, S. W., Semi-empirical estimation of sorption of hydrophobic pollutants on natural sediments and soils, *Chemosphere*, 10, 833–846, 1981.

Karickhoff, S. W., Organic pollutant sorption in aquatic systems, *J. Hydraulic Eng.*, 110, 707–735, 1984.

Karickhoff, S. W. and D. S. Brown, Paraquat sorption as a function of particle size in natural sediments, *J. Environ. Qual.*, 7, 246–252, 1978.

Karickhoff, S. W., D. S. Brown, and T. A. Scott, Sorption of hydrphobic pollutants on natural sediments, *Water Res.*, 13, 241–248, 1979.

Kemper, W. D. and J. B. Rollins, Osmotic efficiency coefficients across compacted clays, *Soil Sci. Soc. Am. Proc.*, 30, 529–534, 1966.

Kenaga, E. E., "Partitioning and Uptake of Pesticides of Biological Systems", In Environmental Dynamics of Pesticides (Editors: R. Haque and V. H. Freed), Plenum Press, New York, 217–275, 1975.

Kenaga, E. E. and C. A. I. Goring, ASTM Special Technical Publication 707, ASTM, Washington, DC, 1980.

Kies, B., Solute Transport in Unsaturated Field Soil and in Ground water, Ph.D. Dissertation, Department of Agronomy, New Mexico State University, Las Cruces, New Mexico, 1981.

Killey, R. W. D. and G. L. Moltyaner, Twin lake tracer tests: setting, methodology, and hydraulic conductivity distribution, *Water Resour. Res.*, 24, 1585–1612, 1988.

Kirkham, D. and S. B. Affleck, Solute travel times to wells, *Ground Water*, 15, 231–242, 1977.

Klotz, D., K. P. Seiler, H. Moser, and F. Neumaier, Dispersivity and velocity relationship from laboratory and field experiments, *J. Hydrol.*, 45, 169–184, 1980.

Kocher, D. C., *Radioactive Decay Data Tables*, U.S. Department of Energy Technical Information Center, Oak Ridge, Tennessee, 1981.

Konikow, L. F., Modeling solute transport in ground water, in *Environmental Sensing and Assessment: Proceedings of the International Conference*, Article 20-3, Institute for Electrical and Electronic Engineers, Piscataway, New Jersey, 1976.

Konikow, L. F., Modeling Chloride Movement in the Alluvial Aquifer at the Rocky Mountain Arsenal, Colorado, U.S. Geological Survey Water-Supply Paper 2044, 43 pp., 1977.

Konikow, L. F. and J. D. Bredehoeft, Modeling flow and chemical quality changes in an irrigated stream–aquifer system, *Water Resour. Res.*, 10, 546–562, 1974.

Kreft, A. and A. Zuber, On the physical meaning of the dispersion equation and its solution for different initial and boundary conditions, *Chem. Eng. Sci.*, 33, 1471–1480, 1978.

Kreft, A. and A. Zuber, On the use of the dispersion model of fluid flow," *Int. J. Appl. Radiat. Isot.*, 30, 705–708, 1979.

Kreft, A. and A. Zuber, Comments on 'Flux-Averaged and Volume-Averaged Concentrations in Continuum Approaches to Solute Transport', *Water Resour. Res.*, 22, 1157–1158, 1986.

Kreft, A., A. Lenda, B. Turek, A. Zuber, and K. Czauderna, Determination of effective porosities by the two-well pulse method, *Isot. Tech. Ground water Hydrol., Proc. Symp.*, 2, 295–312, 1974.

Kreyszig, E., *Advanced Engineering Mathematics*, John Wiley, 2nd ed., New York, 1967.

Lallemand-Barres, A. and P. Peaudecerf, Recherche des relations entre la valeur de la dispersivité macro-scopique d'un milieu aquifère, ses autres charactéristiques et les conditions de measure, *Bull. Rech. Géol. Minièr.*, 2e Ser., Section III, No. 4, Orleans, France, 1978 (in French).

Lapidus, L. and N. R. Amundson, Mathematics of adsorption in beds: IV. The effect of longitudinal diffusion in ion exchange chromatographic columns," *J. Phys. Chem.*, 56, 984–988, 1952.

Lau, L. K., W. J. Kaufman, and D. K. Todd, Studies of Dispersion in a Radial Flow System, Canal Seepage Research: Dispersion Phenomena in Flow Through Porous Media, Progress Report 3, I.E.R. Ser. 93, Issue 3, Sanit. Eng. Res. Lab., Department of Eng. and School of Public Health, University of California, Berkeley, California, 1957.

Law, J., A statistical approach to the interstitial heterogeneity of sand reservoirs, *Trans. AIME*, 155, 202–222, 1944.

LeBlanc, D. R., Sewage Plume in a Sand and Gravel aquifer, Cape Cod, Massachusetts, U.S. Geological Survey Open File Report, 82–274, 1982, 35pp.

LeBlanc, D. R., S. P. Garabedian, K. M. Hess, L. W. Gelhar, R. D. Quadri, K. G. Stollenwerk, and W. W. Wood, Large-scale natural gradient tracer test in sand and gravel, Cape Cod, Massachusetts, 1. Experimental design and observed tracer movement, *Water Resour. Res.*, 27, 895–910, 1991.

Lee, D. R., J. A. Cherry, and J. F. Pickens, Ground water transport of a salt tracer through sandy lakebed, *Limnol. Oceanogr.*, 25, 46–61, 1980.

Leij, F. J., Skaggs, T. H., and van Genuchten, M. T., Analytical solutions for solute transport in three-dimensional semi-infinite porous media, *Water Resour. Res.*, 27, 2719–2733, 1991.

Leland, D. F. and D. Hillel, Scale effects on measurements of dispersivity in a shallow, unconfined aquifer, Paper presented at *Chapman Conference on Spatial Variability in Hydrologic Modeling*, AGU, Fort Collins, Colorado, July 21–23, 1981.

Lenda, A., and A. Zuber, "Tracer Dispersion in Ground-Water Experiments," in *Isot. Hyrol. Proc. Symp. 1970*, 619–641, 1970.

Leo, A., C. Hansch, and D. Elkins, Partition coefficients and their use, *Chem. Rev.*, 71, 525–621, 1971.

Lindstrom, F. T., R. Haque, V. H. Freed, and L. Boersma, Theory on the movement of some herbicidies in soils: linear diffusion and convection of chemicals in soils, *Environ. Sci. Technol.*, 1, 561–565, 1967.

Loague, K. and G. A. Gander, 1, Spatial variability of infiltration on a small rangeland catchment, *Water Resour. Res.*, 26, 957–971, 1990.

Lohman, S. W., Ground-Water Hydraulics, U.S. geological Survey Professional Paper 708, U.S. Government Printing Office, Washington, DC, 1972, 70pp.

Lumley, J. L. and H. A. Panofsky, *The Structure of Atmospheric Turbulence*, Interscience, New York, 239 pp., 1964.

Lundgren, T. S. and Y. B. Pointin, Turbulent self-diffusion, *Phys. Fluids*, 19, 355–358, 1976.

Luxmore, R. J., B. P. Spaulding, and I. M. Munro, Areal variation and chemical modifications of weathered shale infiltration characteristics, *Soil Sci. Am. J.*, 45, 687–691, 1981.

Lyman, W. J., W. F. Reehl, and D. H. R. Rosenblatt, *Handbook of Chemical Property Estimation Methods*, McGraw-Hill, New York, 1982.

Maasland, M., Theory of fluid flow through anisotropic media, in *(Section II)*, *Drainage of Agricultural Lands*, Eds., Luthin, J. N., American Society of Agronomy, Madison, Wisconsin, 1957, pp. 216–236.

Mackay, D. M., D. L. Freyberg, and P. V. Roberts, A natural gradient experiment on solute transport in a sand aquifer, 1. Approach and overview of plume movement, *Water Resour. Res.*, 22, 2017–2029, 1986.

Marino, M. A., Mass Transport of Solutes in Saturated Porous Media, Water Science and Engineering Papers, No. 2003, Department of Water Science and Engineering, University of California, Davis, California, 36 pp., 1975.

Marino, M. A., Distribution of contaminants in porous media flow, *Water Resour. Res.*, 10, 1013–1018, 1974.

Marle, C., P. Simandoux, J. Pacsirsky, and G. Gaulier, Etude du Desplacement de Fluides Miscibles en Milieu Poreux Stratifie, *Rev. Inst. Francais Petrol.*, 22, 272–294, 1967 (in French).

Matheron, G. and G. de Marsily, Is transport in porous media always diffusive? A counterexample, *Water Resour. Res.*, 16, 901–917, 1980.

Mayer, S. W. and E. R. Tompkins, Theoretical analysis of the column separation process, *J. Am. Chem. Soc.*, 69, 2866–2874, 1947.

McDonald, M. G. and A. W. Harbaugh, A Modular Three-Dimensional Finite-Difference Ground water Flow Model, Open File Report 83–875, U.S. Geological Survey, National Center, Reston, Virginia, 1984.

McMillan, W. D., Theoretical Analysis of Ground Water Basin Operations, *Water Resour. Center Contrib.*, 114, 167 pp., University of California, Berkeley, California, 1966.

Meinzer, O. E., Outline of Ground-Water Hydrology with Definitions, Geological Survey Water Supply Paper 494, U.S. Government Printing Office, Washington, DC, 71 pp., 1923.

Mercado, A., Recharge and Mixing Tests at Yavne 20 Wells Field, Underground Water Storage Study Tech. Rep. 12, Publ. 611, Tahal-Water Plann. For Israel, Tel aviv, 1966.

Mercado, A., The spreading pattern of injected waters in a permeable stratified aquifer, *IAHS AISH Publ.*, 72, 23–36, 1967.

Mercer, J. W. and C. R. Faust, *Ground-Water Modeling*, National Water Well Association, 60 pp., 1981.

Mercer, J. W., D. C. Skipp, and D. Giffin, Basics of Pump and Treat Ground Water Remediation Technology, Robert S. Kerr Environmental Research Laboratory, U.S. Environmental Protection Laboratory, Ada, Oklahoma, 1990.

Mercer, J. W., S. D. Thomas, and B. Ross, Parameters and Variables Appearing in Repository Models, U.S. Nuclear Regulatory Commission, NUREG/CR-3066, Washington, DC, 1982.

Metcalf, R. L., an P. Y. Lu, Partition Coefficients as Measures of Bioaccumulation Potential for Organic Compounds, U.S. EPA Report (in Press), 1978.

Meyer, B. R., C. A. R. Bain, A. S. M. DeJusus, and D. Stephenson, Radiotracer evaluation of ground water dispersion in a multilayered aquifer, *J. Hydrol.*, 50, 259–271, 1981.

Mingelgrin, U. and Z. Gerstl, Reevaluation of partitioning as a mechanism in nonionic adsorption in soils, *J. Environ. Qual.*, 12, 1–11, 1983.

Molinari, J. and P. Peaudecerf, Essais conjoints en laboratoire et sur le terrain en vue d'une approach simplifiée de la prévision des propagations de substances miscibles dans les aquiféres réels, paper presented at Symposium on Hydrodynamics Diffusion and Dispersion in Porous Media, Int. Assoc. For Hydraul. Res., Pavis, Italy, 1977 (in French).

Moltyaner, G. L. and R. W. D. Killey, Twin lake tracer tests: longitudinal dispersion, *Water Resour. Res.*, 24, 1613–1627, 1988a.

Moltyaner, G. L. and R. W. D. Killey, Twin lake tracer tests: transverse dispersion, *Water Resour. Res.*, 24, 1628–1637, 1988b.

Moltyaner, G. L. and C. A. Wills, Local- and plume-scale dispersion in the twin lake 40- and 260-m natural-gradient tracer tests, *Water Resour. Res.*, 27, 2007–2026, 1991.

Molz, F. J., O. Güven, and J. G. Melville, An examination of scale-dependent dispersion coefficients, *Ground Water*, 21, 715–725, 1983.

Molz, F. J., O. Güven, J. G. Melville, R. D. Crocker, and K. T. Matteson, Performance, analysis, and simulation of a two-well tracer test at the Mobile Site, *Water Resour. Res.*, 22, 1031–1037, 1986.

Morris, D. A. and A. I. Johnson, Summary of Hydrologic and Physical Properties of Rock and Soil Materials as Analyzed by the Hydrologic Laboratory of the U.S. Geological Survey, U.S. Geological Survey Water Supply Paper 1839–D, 1967.

Naff, R. L., A Continuum Approach to the Study and Determination of Field Longitudinal Dispersion Coefficients, Ph.D., dissertation, New Mexico Institute of Mining and Technology, Socorro, New Mexico, 176 pp., 1978.

Naff, R. L., T. -C. J. Yeh, and M. W. Kemblowski, A note on the recent natural gradient tracer test at the Borden site, *Water Resour. Res.*, 24, 12, 2099–2103, 1988.

Narasimhan, T. N. and P. A. Witherspoon, Recent developments in modeling ground water systems, presented at the IBM Seminar on Regional Ground water Hydrology and Modeling, Venice, Italy, May 25–26 1976, 35 pp., 1977.

Naymik, T. G. and M. J. Barcelona, Characterization of a contaminant plume in ground water, Meredosia, Illinois, *Ground Water*, 24, 517–526, 1981.

Nelson, J. L., Recent studies at Hanford on soil and mineral reactions in waste disposal, in *Proceedings, Symposium on Ground water Disposal of Radioactive Wastes*, Sanitary Engineering Laboratory, University of California, Berkeley, California, August 25–27, 1959.

Neuman, S. P., Universal scaling of hydraulic conductivities and dispersivities in geologic media, *Water Resour. Res.*, 26, 1749–1758, 1990.

Neuman, S. P., Comment on "A critical review of data on field-scale dispersion in aquifers" by L. W. Gelhar, C. Welty, and K. R. Rehfeldt, *Water Resour. Res.*, 29, 1863–1865, 1993.

Neuman, S. P., Stochastic approach to subsurface flow and transport: A view to the future, in *Subsurface Flow and Transport: A Stochastic Approach*, Eds., Dagan, G. and Neuman, S. P., International Hydrology Series, Cambridge University Press, Cambridge, The United Kingdom, 231–241, 1997.

Neuman, S. P., C. L. Winter, and C. M. Newman, Stochastic theory of field-scale dispersion in anisotropic porous media, *Water Resour. Res.*, 23, 453–466, 1987.

Neuman, S. P. and P. A. Witherspoon, Transient Flow of Ground Water to Wells in Multiple Aquifer Systems, Geotechnical Engineering Report 69–1, University of California, Berkeley, California, January, 182 pp., 1969a.

Neuman, S. P. and P. A. Witherspoon, Theory of flow in a confined two aquifer system, *Water Resour. Res.*, 5, 803–816, 1969b.

Neuman, S. P. and P. A. Witherspoon, Applicability of current theories of flow in leaky aquifers, *Water Resour. Res.*, 5, 817–829, 1969c.

Neuman, S. P. and Y. -K. Zhang, A quasi-linear theory of non-Fickian subsurface dispersion: 1. theoretical analysis with application to isotropic media, *Water Resour. Res.*, 26, 5, 887–902, 1990.

New Zealand Ministry of Works and Development, Water and Soil Division, Movement of Contaminants Into and Through the Heretaunga Plains aquifer, Report, Wellington, New Zealand, 1977.

Nikolaevskij, V. N., Convective diffusion in porous media, *J. Appl. Math. Mech. (P.M.M.)*, 23, 1042–1050, 1959.

Nielsen, D. R., R. D. Jackson, J. W. Cary, and D. D. Evans eds., *Soil Water*, American Society of Agronomy and Soil Science Society of America, Madison, Wisconsin, 1972.

Oakes, D. B. and D. J. Edworthy, Field measurement of dispersion coefficients in the United Kingdom, in *Ground Water Quality, Measurement, Prediction, and Protection*, Water Research Centre, Reading, England, 327–340, 1977.

Ogata, A., Theory of Dispersion in a Granular Medium, U.S. Geological Survey Professional Paper 411–I, 1970.

Ogata, A. and R. B. Banks, A Solution of the Differential Equation of Longitudinal Dispersion in Porous Media, Geological Survey Professional Paper 411–A, United States Government Printing Office, Washington, DC, 7 pp., 1961.

Orlob, G. T. and G. N. Radhakrishna, The effects of entrapped gases on the hydraulic characteristics of porous media, *Trans., Am. Geophys. Union*, 39, 648–659, 1958.

Ou, L. T., J. M. Davidson, and P. S. C. Rao, Rate constants for transformation of pesticides in soil-water systems, in *Retention and Transformation of Pesticides and Phosphorus in Soil-Water Systems: A Review of Data Base*, U.S. EPA, 1980.

Palmer, C. D. and R. L. Johnson, Physical processes controlling the transport of contaminants in the aqueous phase, in *Transport and Fate of Contaminants in the Subsurface*, Seminar Publication, U.S. Environmental Protection Agency, EPA/625/4–89/019, September, chap. 2, 5–22, 1989.

Papadopulos, S. S. and S. P. Larson, Aquifer storage of heated water: II. Numerical simulation of field results, *Ground Water*, 16, 242–248, 1978.

Parker, J. C., Analysis of solute transport in column tracer tests, *Soil Sci. Soc. Am.*, 48, 719–724, 1984.

Parker, J. C. and M.Th. Van Genuchten, Flux-averaged and volume-averaged concentrations in continuum approaches to solute transport, *Water Resour. Res.*, 20, 866–872, 1984.

Parker, J. C. and M.Th. Van Genuchten, Reply, *Water Resour. Res.*, 22, 1159–1160, 1986.

Passioura, J. B., Hydrodynamic dispersion in aggregated media, 1. Theory, *Soil Sci.*, 3, 339–344, 1971.

Passioura, J. B. and D. A. Rose, Hydrodynamic dispersion in aggregated media, 2. Effects of velocity and aggregate size, *Soil Sci.*, 3, 345–351, 1971.

Peaceman, D. W. and Rachford, H. H. Jr. The numerical solution of parabolic and elliptic differential equation, *J. Soc. Appl. Math.*, 3, 28–41, 1955.

Perkins, T. K. and O. C. Johnston, A review of diffusion and dispersion in porous media, *Soc. Petrol. Eng. J.*, 3, 70–84, 1963.

Perlmutter, N. M. and M. Lieber, Dispersal of Plating Wastes and Sewage Contaminants in Ground Water and Surface Water, South Formingdale-Massapequa Area, Nassau, County, New York, U.S. Geological Survey Water Supply Paper 1879–G, 1970.

Philip, J. R., The theory of absorption in aggregated media, *Aust. J. Soil Res.*, 6, 1–19, 1968.

Philip, J. R., Theory of Infiltration, in *Advances in Hydroscience*, Ven Te Chow Ed., Vol. 5, Academic Press, New York, 215–296, 1969.

Pickens, J. F. and G. E. Grisak, Scale-dependent dispersion in a stratified granular aquifer, *Water Resour. Res.*, 17, 1191–1211, 1981a.

Pickens, J. F. and G. E. Grisak, Modeling of scale-dependent dispersion in hydrogeologic systems, *Water Resour. Res.*, 17, 1701–1711, 1981b.

Pinder, G. F., A digital model for aquifer evaluation, *U.S. Geological Survey Techniques Water Resource Investigations*, Book 7, chap. C1, 18 pp., 1970.

Pinder, G. F., A Galerkin—finite element simulation of ground water contamination on long island, *Water Resour. Res.*, 9, 1657–1669, 1973.

Pinder, G. F. and J. D. Bredehoeft, Application of the digital computer for aquifer evaluation, *Water Resour. Res.*, 4, 1069–1093, 1968.

Pinder, G. F. and Cooper, H. H. Jr. A numerical techniques for calculating the transient position of the saltwater front, *Water Resour. Res.*, 6, 875–882, 1970.

Pollock, D. W., Semianalytical computation of path lines for finite-difference models, *Ground Water*, 26, 743–750, 1988.

Pollock, D. W., MODPATH: Documentation of Computer Programs to Compute and Display Pathlines Using Results From the U.S. Geological Survey Modular Three-Dimensional Finite-Difference Ground Water Flow Model (MODFLOW), U.S. Geological Survey OFR 89-381, 81 pp., 1989.

Polubarinova-Kochina, P. Ya., *Theory of Ground Water Movement*, De Wiest, R.J.M., Princeton University Press, Princeton, New Jersey, 1962, 613 pp. (Translated from the Russian by J. M. Roger De Wiest).

Price, H. S., R. S. Varga, and J. E. Warren, Application of oscillation matrices to diffusion-convection equations, *J. Math. Phys.*, 45, 301–311, 1966.

Prickett, T. A., Ground-water computer models - state of the art, *Ground Water*, 17, 167–173, 1979.

Prickett, T. A., and C. G. Lonnquist, Selected digital computer techniques for ground water resource evaluation, *Illinois State Water Surv. Bulletin* 55, 56 pp., 1971.

Prickett, T. A., T. G. Naymik, and C. G. Lonnquist, A 'random walk' solute transport model for selected ground water quality evaluations, *Illinois State Water Surv. Bulletin* 65, 103 pp., 1981.

Priestley, M. B., *Spectral Analysis and Time Series, Vol. 1*, Univariate Series, Academic Press, New York, 653 pp., 1981.

Rabinovitz, D. D. and G. W. Gross, Environmental Tritium as a Hydrometeorologic Tool in the Roswell Basin, New Mexico, Tech. Completion Report, OWRR:A-037-NMEX, N.M. Water Resources Research Institution, Las Cruces, New Mexico, 1972.

Raimondi, P., G. H. F. Gardner, and C. B. Petrick, Effect of pore structure and molecular diffusion on the mixing of miscible liquids, *A.I.Ch.E.Joint Symposium on Fundamental Concepts of Miscible Fluid Displacement*, Part II, Preprint 43, San Francisco, California, Dec. 6–9, 1959.

Rajaram, H. and L. W. Gelhar, Three-dimensional spatial moments analysis of the Borden tracer test, *Water Resour. Res.*, 27, 1239–1251, 1991.

Rao, P. S. C. and J. M. Davidson, Adsorption and movement of selected pesticides at high concentrations in soils, *Water Res.*, 13, 375–380, 1979.

Rao, P. S. C. and J. M. Davidson, Estimation of pesticide retention and transformation parameters required in nonpoint source pollution models, in *Environmental Impact of Nonpoint Source Pollution*, Overcash M. R., and Davidson, J. M., Eds. Ann Arbor Science Publishers Inc., Ann Arbor, Michigan, 23–67, 1980.

Rao, P. S. C., P. Nkedi-Kizza, J. M. Davidson, and L. T. Ou, Retention and transformation of pesticides in relation to nonpoint source pollution from croplands, *Agricultural Management and Water Quality*, Schaller F.W., and Bailey, G.W., Eds., Iowa State University Press, Ames, Iowa, 126–140, 1983.

Reddell, D. L. and D. K. Sunada, Numerical Simulation of Dispersion in Ground water Aquifers, Colorado State University Hydrology Paper 41, 79 pp., 1970.

Reeves, M., D. S. Ward, N. D. Johns and R. M. Cranwell, Theory and Implementation for SWIFT II, The Sandia Waste-Isolation Flow and Transport Model for Fractured Media, Release 4.84, Report NUREG/CR-3328 SAND83–1159, Sandia, Albuquerque, New Mexico, 1986.

Rehfeldt, K. R., Prediction of Macrodispersivity in Heterogeneous Aquifers, Ph.D. dissertation, Department of Civil Engineering, Massachusetts Institute of Technology, Cambridge, Massachusetts, 233 pp., 1988.

Rehfeldt, K. R. and L. W. Gelhar, Stochastic analysis of dispersion in unsteady flow in heterogeneous aquifers, *Water Resour. Res.*, 28, 2085–2099, 1992.

Rehfeldt, K. R., L. W. Gelhar, J. B. Southard, and A. M. Dasinger, "Estimates of Macrodispersivity Based on Analyses of Hydraulic Conductivity Variability at the MADE Site," *Electric Power* Research Institute Report EPRI EN-6405, Project 2485-5, Palo Alto, California, 1989a.

Rehfeldt, K. R., P. Hufschmied, L. W., Gelhar, and M. E. Schaefer, "Measuring Hydraulic Conductivity with the Borehole Flowmeter", *Electric Power Research Institute Report EPRI EN-6511*,Project 2485-5, Palo Alto, California, 1989b.

Remson, I., C. A. Appel, and R. A. Webster, Ground water models solved by digital computer, *Hydraul. Div. Am. Soc. Civil Eng.*, 91, 133–147, 1965.

Reynolds, W. D., Column Studies of Strontium and Cesium Transport Through a Granular Geologic Porous Medium, M. S. Thesis, University of Waterloo, Waterloo, Ontario, 149 pp., 1978.

Roberts, P. V., M. Reinhard, G. D. Hopkins, and R. S. Summers, Advection-Dispersion-Sorption Models for Simulating the Transport of Organic Contaminants, Paper presented at *International Conference on Ground Water Quality Research*, Rice University, Houston, Texas, 1981.

Robertson, J. B., Digital Modeling of Radioactive and Chemical waste Transport in the Snake River Plain Aquifer at the National Reactor Testing Station, Idaho, U.S. Geological Survey Open-File Report IDO-22054, 41 pp., 1974.

Robertson, J. B., and T.B. Barraclough, Radioactive and chemical waste transport in ground water of national reactor testing station: 20-Year Case History and Digital Model, *Underground Waste Manage. Artit. Recharge Prepr. Pap. Int. Symp. 2nd*, 1, 291–322, 1973.

Robin, M. J. L., E. A. Sudicky, R. W. Gillham, and R. G. Kachanaski, Spatial Variability of Strontium Distribution Coefficients and Their Correlation with Hydraulic Conductivity in the Canadian Forces Base Borden Aquifer, *Water Resour. Res.,* 27, 10, 2619–2632, 1991.

Robinson, R. A. and R. H. Stokes, *Electrolyte Solutions*, 2nd ed., Butterworth, London, Great Britain, 1965.

Robson, S. G., Feasibility of Digital Water Quality Modeling Illustrated by Application at Barstow, California, U.S. Geological Survey Water Resources Investigations, 46–73, 1974.

Robson, S. G., Application of Digital Profile Modeling Techniques to Ground-Water Solute Transport at Barstow, California, U. S. Geological Survey Water Supply Paper 2050, 1978.

Rose D. A., Some aspects of the hydrodynamic dispersion of solutes in porous materials, *Journal of Soil Science*, 24, 284–295, 1973.

Rosenberg, J. L., *Theory and Problems of College Chemistry*, Shaum's Outline Series, 6th ed., McGraw-Hill Book Company, New York, 314 pp., 1980.

Rousselot, D., J. P. Sauty, and B. Gaillard, Etude Hydrogéologique de la Zone Industrielle de Blyes-Saint-Vulbas, Rapport Préliminaire No. 5: Characteristiques Hydrodynamiques du Système Aquifère, Report. Jal 77/33, Bur. De Rech. Geol. Et Min., Orleans, France, 1977 (in French).

Rumer, R. R., Longitudinal dispersion in steady and unsteady flow, *J. Hydraul. Div.*, *Proc. Am. Soc. Civil Eng.*, 88, 147–172, July, 1962.

Russo, D., "Geostatistical Approaches to solute transport in heterogeneous fields and its applications to salinity management", *Water Resour. Res.*, 20, 1260–1270, 1984.

Russo D., and E. Bressler, "Soil hydraulic properties as stochastic processes, I, Analysis of field spatial variability", *Soil Sci. Soc. Amer. J.*, 45, 682–687, 1981.

Sagar, B., Dispersion in three dimensions: approximate analytic solutions, *Hydraul. Div. Am. Soc. Civil Eng.*, 108, 47–62, 1982.

Sauty, J. P., Contribution à l'Identification des Paramèters de Dispersion dans les Aquifères Par Interprétation des Expériences de Tracage, Dissertation, Univ. Sci. Et Med. Et Inst. Natl. Polytech. De Grenoble, Grenoble, France, 1977 (in French).

Sauty, J. P., An analysis of hydrodispersive transfer in aquifers, *Water Resour. Res.*, 16, 145–158, 1980.

Sauty, J. P., A. C. Gringarten and P. A. Landel, The effects of thermal dispersion on injection of hot water in aquifers, *Well-Testing Symposium*, Lawrence Berkeley Laboratory, Berkeley, California, 1978.

Scarborough, J. B., *Numerical Mathematical Analysis*, Johns Hopkins Press, Baltimore, Maryland, 1966.

Scheidegger, A. E., General theory of dispersion in porous media, *J. Geophys. Res.*, 66, 3273–3278, 1961.

Scheidegger, A. E., *The Physics of Flow Through Porous Media*, University of Toronto Press, Toronto, Canada, 1963.

Schwarzenbach, R. and J. Westall, Transport of nonpolar organic compounds from surface water to ground water: laboratory sorption studies, *Env. Sci. Tech.*, 15, 1360–1367, 1981.

Shen, H. T., Transient dispersion in uniform porous media flow, *J. Hydraul. Div.*, *Am. Soc. Civil Eng.*, 102, 707–716, 1976.

Segol, G. and G. F. Pinder, Transient simulation of saltwater intrusion in southeastern Florida, *Water Resour. Res.*, 12, 65–70, 1976.

Shamir, U. Y. and D. R. F. Harleman, Numerical and Analytical Solutions of Dispersion Problems in Homogeneous and Layered Aquifers, Report. No. 89, Hydrodynamics Laboratory, Department of Civil Engineering, Massachusetts Institute of Technology, Cambridge, Massachusetts, 1966.

Simmons, C. S., A Stochastic-Convective Ensemble Method for Representing Dispersive Transport in Ground water, Report No. CS–2558, Electric Power Research Institute, Palo Alto, California, 1982a.

Simmons, C. S., A stochastic-convective transport representation of dispersion in one-dimensional porous media systems, *Water Resour. Res.*, 18, 1193–1214, April, 1982b.

Sisson, J. B., and P. J. Wierenga, "Spatial Variability of Steady-State Infiltration Rates as a Stochastic Process", *Soil Sci. Soc. Amer. J.*, 45, pp. 699–704, 1981.

Skibitzke, H. E. and G. M. Robertson, Dispersion in Ground water Flowing Through Heterogeneous Materials, U.S. Geological Survey Professional Paper 386-B, p.5, 1963.

Smith, J.M., *Chemical Engineering Kinetics*, McGraw-Hill, New York, 1970.

Smith, L., "A Stochastic Analysis of Steady State Ground Water Flow in a Bounded Domain", *Ph.D. thesis*, University of British Columbia, Vancouver, Canada, 1978.

Smith, L., Spatial variability of flow parameters in a stratified sand, *Math. Geol.*, 13, 1–21, 1981.

Smith, L. and F. W. Schwartz, Mass transport, 1. A stochastic analysis of macroscopic dispersion, *Water Resour. Res.*, 16, 303–313, 1980.

Spiegel, M. R., *Theory and Problems of Laplace Transforms*, Shaum's Outline Series, McGraw-Hill Book Company, New York, 261 pp., 1965.

Spiegel, M. R., *Mathematical Hanbook of Formulas and Tables*, Shaum's Outline Series, McGraw-Hill Book Company, New York, 271 pp., 1968.

Spiegel, M. R., *Advanced Mathematics for Scientists and Engineers*, Shaum's Outline Series, McGraw-Hill Book Company, New York, 407 pp., 1971.

Spiegel, M. R., *Fourier Analysis With Applications to Boundary Value Problems*, Shaum's Outline Series, McGraw-Hill Book Company, New York, 191 pp., 1974.

Spiegel, M.R., *Probability and Statistics*, Shaum's Outline Series, McGraw-Hill Book Company, New York, 372 pp., 1975.

Sposito, G., and D. A. Barry, On the Dagan model of solute transport in ground water: foundational aspects, *Water Resour. Res.*, 23, 1867–1875, 1987.

Stokes, G. G., On the steady motion of incompressible fluids, *Trans. Cambridge Philos. Soc.*, 7, 439-453, 1842.

Stone, H. K., Iterative solution of implicit approximations of multidimensional partial differential equations, *Soc. Indust. Appl. Math., Jour. Numer. Anal.*, 5, 530–558, 1968.

Sudicky, E. A., A natural gradient experiment of solute transport in a sand aquifer: spatial variability of hydraulic conductivity and its role in the dispersion process, *Water Resour. Res.*, 22, 2069–2082, 1986.

Sudicky, E. A., J. A. Cherry and E. O. Frind, Migration of contaminants in ground water at a landfill: a case study, a natural-gradient dispersion test, *J. Hydrol.*, 63, 81–108, 1983.

Sykes, J. F., S. B. Pahwa, R. B. Lantz and D. S. Ward, Numerical simulation of flow and contaminant migration at an extensively monitored landfill, *Water Resour. Res.*, 18, 1687–1704, 1982.

Sykes, J. F., S. B. Pahwa, D. S. Ward and R. B. Lantz, The validation of SWENT, a geosphere transport model, in *Scientific Computing*, Stepleman, R. et al., Ed., IMACS/North-Holland, Amsterdam, 351–361, 1983.

Tang, D. H., F. W. Schwartz, and L. Smith, Stochastic modeling of mass transport in a random velocity field, *Water Resour. Res.*, 18, 231–244, 1982.

Tanji, K. K., G. R. Dutt, J. L. Paul, and L. D. Doneen, II. A computer method for predicting salt concentrations in soils at variable moisture contents, *Hilgardia 38*, 307–318, 1967.

Taylor, G. I., Diffusion by continuous movement, *Proc. London Math. Soc.*, Vol. Ser. A, 20, 196–211, 1921.

Taylor, G. I., The dispersion of soluble matter in a solvent flowing through a tube, *Proc. Roy. Soc. London*, Ser. A, 219, 186-203, 1953.

Terzaghi, K., *Theoretical Soil Mechanics*, Wiley, New York, 510 pp., 1943.

Theis, C. V., Aquifer and models, in *Proc., National Symp. Ground-water Hydrol.: Am. Water Resour. Assoc. Pro. Ser. 4*, Marino, M. A., Ed., 138–248, 1967.

Thiem, G., *Hydrologische Methoden* (in German), J. M. Gebhardt, Leipzig, Germany, 56 pp., 1906.

Thomas, G. W. and A. R. Swoboda, Anion exclusion effects on chloride movement in soils, *Soil Sci.*, 110, 163–166, 1970.

Trescott, P. C., Documentation of Finite-Difference Method for Simulation of Three-Dimensional Ground-Water Flow, U.S. Geological Survey Open File Report 75–438, 32 pp., 1975.

Trescott, P. C., G. F. Pinder and S. P. Larson, Finite-Difference Model for Aquifer Simulation in Two Dimensions with Results of Numerical Experiments, *U.S. Geological Survey Techniques of Water Resources Investigations, Book 7, chap. C1*, 116 pp., 1976.

Treybal, R. E., *Mass-Transfer Operations*, 3rd ed., McGraw-Hill Book Company, New York, 784 pp., 1980.

Tsang, C. F., The modeling process and validation, *Ground Water*, 29, 825–831, 1991.

U.S. Environmental Protection Agency, Water Quality Assessment: A Screening Procedure for Toxic and Conventional Pollutants in Surface and Ground Water - Part II, EPA/600/6-85/002b, Environmental Research Laboratory, Athens, Georgia, September 1985.

Vaccaro, J. J. and E. L. Bolke, Evaluation of Water Quality Characteristics of Part of the Spokane Aquifer, Washington and Idaho, Using a Solute transport Digital Model, U.S. Geological Survey Open File Report 82–769, 1983.

Valocchi, A. J., P. V. Roberts, G. A. Parks and R. L. Street, Simulation of the transport of ion-exchanging solutes using laboratory-determined chemical parameter values, *Ground Water*, 19, 600–607, 1981.

van der Heijde, P. K. M., Quality Assurance in Computer Simulations of Ground Water Contamination, *GWMI 87-08*, International Ground Water Modeling Center, Holcomb Research Institute, Butler University, Indianapolis, Indiana, *Environmental Software*, Vol. 2, No. 1, 19–25, 1987.

van der Heijde, P. K. M., P. S. Huyakorn, and J. W. Mercer, Testing and Validation of Ground Water Models, *Practical Applications of Ground water Modeling,* NWWA/IGWMC Conference, 19–20, Columbus, Ohio, Aug. 1985.

van Genuchten, M. T., Mass Transfer Studies in Sorbing Porous Media, Ph.D. dissertation, New Mexico State University, Las Cruces, New Mexico, December, 148 pp., 1974.

van Genuchten, M. T., Determining Transport Parameters From Solute Displacement Experiments, Research Report 118, U.S. Salinity Laboratory, Riverside, California, 1980.

van Genuchten, M.T., One-dimensional analytical transport modeling, in *Proceedings of the Symposium on Unsaturated Flow and Transport Modeling*, Arnold, E. M., Gee, G. W., and Nelson, R. W., Eds., Pacific Northwest Laboratory, NUREG/CP-0030, Seattle, Washington, Mar. 22–24, 1982.

van Genuchten, M. T. and W. J. Alves, Analytical Solutions of the One-Dimensional Convective-Dispersive Solute Transport Equation, U.S. Salinity Laboratory, Technical Bulletin Number 1661, Riverside, California, 149 pp., 1982.

van Genuchten, M. T. and J. C. Parker, Boundary conditions for displacement experiments through short laboratory soil columns, *Soil Sci. Soc. Am. J.*, 48, 703–708, 1984.

van Genuchten, M. T., D. H. Tang and R. Guennelon, Some exact solutions for solute transport through soils containing large cylindrical macropores, *Water Resour. Res.*, 20, 335–346, 1984.

van Genuchten, M. T. and P. J. Wierenga, Solute dispersion coefficients and retardation factors, *Methods of Soil Analysis, Part 1, Physical and Mineralogical Methods-Agronomy Monograph No.9*, 2nd ed., American Society of Agronomy, Madison, Wisconsin, 1025–1054, 1986.

Varga, R. S., *Matrix Iterative Analysis*, Prentice-Hall, Englewood Cliffs, New Jersey, 322 pp., 1962.

Vermeulen, T. and N. K. Hiester, Ion exchange chromatography of trace components, *Ind. Eng. Chem.*, 44, 636–651, 1962.

Verruijt, A., *Theory of Ground water Flow*,Gordon and Breach Science Publishers, New York, 190 pp., 1970.

Voice, T. C., C. P. Rice and Weber, W. J. Jr., Effect of solids concentration on the sorptive partitioning of hydrophobic pollutants in aquatic systems, *Environ. Sci. Technol.*, 17, 513–518, 1983.

Viera, S. R., D. R. Nielsen, and J. W. Biggar, "Spatial Variability of Field Measured Infiltration Rate," *Soil Sci. Soc. Amer. J.*, 45, 1040–1048, 1981.

Wachpress, E. L. and G. J. Habetler, An alternating-direction-implicit iteration technique, *J. Soc. Ind. Appl. Math.*, 8, 403–424, 1960.

Walter, G. B., *Convergent Flow Tracer Test at H-6: Waste Isolation Pilot Plant (WIPP)*, Southeast New Mexico (Draft), Hydro Geochem. Inc., Tucson, Arizona, 1983.

Walton, W. C., *Ground water Resource Evaluation*, McGraw-Hill, New York, 644 pp., 1970.

Wang, F. G. and M. P. Anderson, *Introduction to Ground water Modeling: Finite Difference and Finite Element Methods*, W. H. Freeman and Company, San Francisco, California, 237 pp., 1982.

Ward, D. S., Data Input Guide for SWIFT III, The Sandia Waste Isolation Flow and Transport Model for Fractured Media, Release 2.2.5, Modification to Report. NUREG/CR-3162, GeoTrans, Inc., Herndon, Virginia, 1988.

Ward, D. S., D. R. Buss, J. W. Mercer and S. S. Hughes, Evaluation of a ground water corrective action at the chem-dyne hazardous waste site using a telescopic mesh refinement modeling approach, *Water Resour. Res.*, 23, 603-617, 1987.

Ward, D. S., M. Reeves and L. E. Duda, Verification and Field Comparison of the Sandia Waste-Isolation Flow and Transport Model (SWIFT), NUREG/CR-3316, U.S. Nuclear Regulatory Commission, Washington, DC, 1984.

Warren, J. E. and H. S. Price, Flow in heterogeneous porous media, *Soc. Petrol. Eng. J.*, 1, 153–169, 1961.

Warren, J. E. and F. F. Skiba, Macroscopic dispersion, *Trans. Am. Inst. Min. Metall. Pet. Eng.*, 231, 215–230, 1964.

Warrick, A. W., J. W. Biggar and D. R. Nielsen, Simultaneous solute and water transfer for an unsaturated soil, *Water Resour. Res.*, 7, 1216–1225, 1971.

Weaver, W., Jr. *Computer Programs for Structural Analysis*, D. Van Nostrand, Princeton, New Jersey, 1967, 300pp.

Weber, W. J., Jr., P. M. McGinley and L. E. Katz, Sorption phenomena in subsurface systems: concepts, models and effects on contaminant fate and transport, *Water Res.*, 25, 499–528, 1991.

Webster, D. S., J. F. Procter and J. W. Marine, Two-Well Tracer Test in Fractured Crystalline Rock, U.S. Geological Survey Water Supply Paper 1544–I, 1970.

Weeks, O. L., G. L. Stewart and M. E. Weeks, Measurement of non-exchanging pores during miscible displacement in soils, *Soil Sci.*, 122, 139–144, 1976.

Weinstein, H. C., H. L. Stone, and T. V. Kwan, Iterative procedure for solution of systems of parabolic and elliptic equations in three dimensions, *Indus. Engineering Chemistry Fundamentals*, 8, 2, 281–287, 1969.

Welty, C. and L. W. Gelhar, Evaluation of Longitudinal Dispersivity From Tracer Test Data, Report. 320, Ralph M. Parsons Laboratory for Water Resources and Hydrodynamics, Massachusetts Institute of Technology, Cambridge, Massachusetts, 1989.

Werner, A. et al., Nutzung von Grundwasser fur Warmepumpen, Versickerrungstest Aefligen, Versuch 2, 1982/83, *Water and Energy Management. Agency of the State of Bern*, Switzerland, 1983 (in German).

Wiebenga, W. A. et al., Radioisotopes as ground water tracers, *J. Geophy. Res.*, 72, 4081–4091, 1967.

Wilson, L. G., Investigation on the Subsurface Disposal of Waste Effluents at Inland Sites, Res. Develop. Progress Rep. 650, U.S. Dept. of Interior, Washington, DC., 1971.

Wilson, J. L., and P. J. Miller, Two-dimensional plume in uniform ground-water flow, *Journal of Hydraulics Division*, American Society of Civil Engineers, 104, HY4, 503–514, 1978.

Willardson, L. S., and R. L. Hurst, Sample Size Estimates in Permeability Studies", *J. Irrig. Drain. Div. Amer. Soc. Civil Eng.*, 91, IR1, 1–9, 1965.

Winter, C. L., Asymptotic Properties of Mass Transport in Random Porous Media, Ph.D. dissertation, University of Arizona, Tucson, Arizona, 1982.

Wood, W., A geochemical method of determining dispersivity in regional ground water systems, *J. Hydrol.*, 54, 209–224, 1981.

Wood, W. W. and G. G. Ehrlich, Use of Baker's yeast to trace microbial movement in ground water, *Ground Water*, 16, 398–403, 1978.

Woodbury, A.D., and E. A. Sudicky, The Geostatistical characteristics of the Borden Aquifer, *Water Resour. Res.*, 27, 4, 533–546, 1991.

Yeh, G. T., AT123: Analytical Transient One-. Two-, and Three-Dimensional Simulation of Waste Transport in the Aquifer System, Oak Ridge National Laboratory, Report. ORNL-5602, Oak Ridge, Tennessee, 83 pp., 1981.

Zheng, C., MT3D, A Modular Three-Dimensional Transport Model for Simulation of Advection, Dispersion, and Chemical Reactions of Contaminants in Ground water System, Prepared for the U.S. Environmental Protection Agency, 1990.

Zheng, C. and G. D. Bennett, *Applied Contaminant Transport Modeling*, Thomson Publishing Company, New York, 440 pp., 1995.

Zienkiewicz, O. C., *The Finite Element Method*, 3ed., McGraw-Hill, New York, 1977.

Zhang, Y- K. and S. P. Neuman, A quasi-linear theory of non-Fickian and Fickian subsurface dispersion: 2. application to anisotropic media and the borden site, *Water Resour. Res.*, 26, 903–913, 1990.

Zilliox, L. and P. Muntzer, Effects of hydrodynamic processes on the development of ground-water pollution: study on physical models in a saturated porous medium, *Progr. Water Technol.*, 7, 561–568, 1975.

Index

For Product Safety Concerns and Information please contact our EU
representative GPSR@taylorandfrancis.com
Taylor & Francis Verlag GmbH, Kaufingerstraße 24, 80331 München, Germany